UNIVERSAL SPACES AND MAPPINGS

NORTH-HOLLAND MATHEMATICS STUDIES 198

(Continuation of the Notas de Matemática)

Editor: Jan van Mill
Faculteit der Exacte Wetenschappen
Amsterdam, The Netherlands

ELSEVIER

2005

Amsterdam – Boston – Heidelberg – London – New York – Oxford
Paris – San Diego – San Francisco – Singapore – Sydney – Tokyo

UNIVERSAL SPACES AND MAPPINGS

S.D. ILIADIS
Department of Mathematics
Faculty of Science
University of Patras
Patras, Greece

ELSEVIER

2005

Amsterdam – Boston – Heidelberg – London – New York – Oxford
Paris – San Diego – San Francisco – Singapore – Sydney – Tokyo

ELSEVIER B.V.
Radarweg 29
P.O. Box 211, 1000 AE Amsterdam
The Netherlands

ELSEVIER Inc.
525 B Street, Suite 1900
San Diego, CA 92101-4495
USA

ELSEVIER Ltd
The Boulevard, Langford Lane
Kidlington, Oxford OX5 1GB
UK

ELSEVIER Ltd
84 Theobalds Road
London WC1X 8RR
UK

First edition 2005

Library of Congress Cataloging in Publication Data
A catalog record is available from the Library of Congress.

British Library Cataloguing in Publication Data
A catalogue record is available from the British Library.

ISBN: 0-444-51586-0
ISSN: 0304-0208

∞ The paper used in this publication meets the requirements of ANSI/NISO Z39.48-1992 (Permanence of Paper).
Printed in The Netherlands.

Στο γιο μου Δημήτρη

Preface

A space T is said to be *universal* in a class $\mathbb{P}$ of spaces if: (a) $T \in \mathbb{P}$ and (b) for every $X \in \mathbb{P}$ there exists an embedding of X into T. If only the second condition is satisfied, then T is called a *containing* space for $\mathbb{P}$.

The question whether there are universal elements in a given class of spaces is called the *universality problem* for that class. This problem can be posed for any class of spaces that is determined by a topological property.

Problems concerning universal and containing spaces appeared in topology in its early development, when special classes of separable metrizable spaces were considered. At first, the construction of universal elements had a more or less heuristic nature, that is for a given class of spaces one seeks to find a universal element with no general method for constructing such elements, although products of spaces usually play an important role. With the consideration of more general classes of spaces, some methods of construction of universal elements appeared. The methods using factorization theorems seem to be the most important.

We dwell upon the construction of a universal space for the class $\mathbb{P}$ of all normal spaces X with weight $w(X) \leq \tau$ and covering dimension $\dim(X) \leq n$, where τ is an infinite cardinal and n is a non-negative integer. (Instead of dimension dim we can consider some other dimensions). This construction is a typical example of the use of factorization theorems and a general outline can be described as follows. Let $\mathbf{S} \subset \mathbb{P}$ be a set such that each element of $\mathbb{P}$ is homeomorphic to an element of $\mathbf{S}$. For every $X \in \mathbf{S}$ we consider the Čech-Stone compactification $\beta(X)$ of X. It is known that $\dim(\beta(X)) \leq n$. Consider the space Y, which is the free union of spaces $\beta(X)$, $X \in \mathbf{S}$. For this space we have $\dim(Y) \leq n$ and, therefore, for the Čech-Stone compactification $\beta(Y)$ of Y we also have $\dim(\beta(Y)) \leq n$. For every $X \in \mathbf{S}$ we denote by h_X a homeomorphism of X into the Tychonoff cube I^τ and by $h_{\beta(X)}$ the continuous extension of h_X over $\beta(X)$. Let F_Y be the mapping of Y into I^τ, which coincides with $h_{\beta(X)}$ on the subset $\beta(X)$

of Y. Denote by F the continuous extension of F_Y over $\beta(Y)$ and by Z the compact subset $F(\beta(Y))$ of I^τ. Thus, we have a continuous mapping F of the compact space $\beta(Y)$ with $\dim(\beta(Y)) \leq n$ onto the compact space Z with $w(Z) \leq \tau$. By the Mardešić factorization theorem there exists a compact space T, a continuous mapping f of $\beta(Y)$ onto T, and a continuous mapping g of T onto Z such that $w(T) \leq \tau$, $\dim(T) \leq n$, and $F = g \circ f$. Then it is easy to verify that the space T is the required universal element in the class $\mathbb{P}$.

Observe that this construction can be applied to completely regular spaces and also to other dimensions functions, that are preserved by the Čech-Stone compactification of a space. However, the most important fact which is used is that for the considered dimension the corresponding factorization theorem must be true. Though the subspace $F_Y(Y)$ of I^τ is the union of compact subsets of I^τ, we can say nothing about the structure of the space Z because the number of these subsets may be larger than τ. This means that we can also say nothing about the structure of the universal space T, the more so since the main construction of T (contained in the factorization theorem) is carried out in such a manner that T (as a limit space of an inverse system) must satisfy three conditions simultaneously.

In this book we give another approach to the universality problem. Given an indexed collection $\mathbf{S}$ of T_0-spaces of weight less than or equal to a given infinite cardinal τ, we construct by a standard manner some containing T_0-spaces for $\mathbf{S}$ of weight $\leq \tau$. For this purpose for every $X \in \mathbf{S}$ we fix an indexed base $M^X \equiv \{U_\delta^X : \delta \in \tau\}$ and consider the indexed collection $\mathbf{M} \equiv \{M^X : X \in \mathbf{S}\}$, which is called a co-mark of $\mathbf{S}$. Furthermore, we consider an indexed family $\mathrm{R} \equiv \{\sim^s : s \in \mathcal{F}\}$ of equivalence relations on $\mathbf{S}$, where $\mathcal{F}$ is the set of all finite subsets of τ. On the set of all pairs (x, X), where $x \in X \in \mathbf{S}$, we consider an equivalence relation, denoted by $\sim_{\mathrm{R}}^{\mathbf{M}}$, as follows: $(x, X) \sim_{\mathrm{R}}^{\mathbf{M}} (y, Y)$ if and only if $X \sim^s Y$ for every $s \in \mathcal{F}$, and either $x \in U_\delta^X$ and $y \in U_\delta^Y$ or $x \notin U_\delta^X$ and $y \notin U_\delta^Y$ for every $\delta \in \tau$. The set of all equivalence classes of the relation $\sim_{\mathrm{R}}^{\mathbf{M}}$ is denoted by $\mathrm{T}(\mathbf{M}, \mathrm{R})$. For every $s \in \mathcal{F}$, an equivalence class $\mathbf{H}$ of $\sim^s$, and $\delta \in \tau$ we denote by $U_\delta^{\mathrm{T}}(\mathbf{H})$ the set of all elements $\mathbf{a}$ of $\mathrm{T}(\mathbf{M}, \mathrm{R})$ such that there exists a pair (x, X) of $\mathbf{a}$ for which $x \in U_\delta^X$ and $X \in \mathbf{H}$. The set of all such sets is denoted by B^{T}. Under some simple (set-theoretical) conditions on R this set is a base for a topology on the set $\mathrm{T}(\mathbf{M}, \mathrm{R})$ such that the corresponding space is a T_0-space of weight $\leq \tau$. We note that this space is uniquely determined by the co-mark $\mathbf{M}$ and the family R.

For every element X of $\mathbf{S}$ there exists a natural embedding i^X of X into the space $\mathrm{T}(\mathbf{M}, \mathrm{R})$ defined as follows: for every $x \in X$, $i^X(x) = \mathbf{a}$, where $\mathbf{a}$ is the element of $\mathrm{T}(\mathbf{M}, \mathrm{R})$ containing the pair (x, X). Thus, we have constructed Containing Spaces $\mathrm{T}(\mathbf{M}, \mathrm{R})$ for $\mathbf{S}$ of weight $\leq \tau$. (In this book we use capital letters "C" and "S" in the words "Containing Spaces" in order to differentiate between the constructed Containing Spaces and the containing spaces in the general sense defined in the begining of this Preface).

The above construction has some additional properties. Suppose that for every $X \in \mathbf{S}$ we have chosen a subset Q^X of X. Then, we can consider the indexed collection $\mathbf{Q} \equiv \{Q^X : X \in \mathbf{S}\}$ of spaces. For this collection we can also construct the corresponding Containing Spaces. An important property of the given construction is the fact that Containing Spaces constructed for the collection $\mathbf{Q}$, under some (set-theoretical) conditions, can be considered as (specific) subsets of the corresponding Containing Spaces constructed for the collection $\mathbf{S}$. Moreover, we can see that this observation is valid simultaneously for τ many different collections $\mathbf{Q}$. Another important property is the fact that the construction of the specific subsets is commutative with respect to some topological operators on subsets of a space such as the boundary operator, the closure operator, the interior operator, and so on.

The given construction is almost purely set-theoretical in the sense that we do not use any topological property, except for the notion of a base for a topological space. In particular, we do not use separation axioms, products of spaces, partial products of spaces, compactifications, dimensions, factorization theorems and so on. (For simplicity only the considered spaces are assumed to be T_0-spaces).

The constructed Containing Spaces provide us with a new point of view on the universality problem. Suppose that we have a class $\mathbb{P}$ of T_0-spaces of weight $\leq \tau$ defined by a topological property P, that is a space X belongs to $\mathbb{P}$ if and only if some concrete subsets of X satisfy relations included in the definition of the property P. In the universality problem for the class $\mathbb{P}$ of spaces we consider an arbitrary collection $\mathbf{S} \subset \mathbb{P}$ and for some co-mark $\mathbf{M}$ for $\mathbf{S}$ and family R of equivalence relations on $\mathbf{S}$ we seek to find concrete subsets of the space $\mathrm{T}(\mathbf{M}, \mathrm{R})$, which satisfy the relations involved in the definition of the property P. Note: these concrete subsets are usually the above mentioned specific subsets of $\mathrm{T}(\mathbf{M}, \mathrm{R})$. This will mean that the Containing Space $\mathrm{T}(\mathbf{M}, \mathrm{R})$ belongs to $\mathbb{P}$ and, therefore, in $\mathbb{P}$ there exist universal elements. In such a manner we can prove in a unified way the existence of a universal element for many (well-known) classes of spaces.

In the case where the Containing Space $\mathrm{T}(\mathbf{M}, \mathrm{R})$ belongs to $\mathbb{P}$ for "almost all" co-marks $\mathbf{M}$ and families R the class $\mathbb{P}$ is called saturated. One special feature of the book consists in the consideration of just that case. Saturated classes have "something more" than the existence of universal elements. For example, the saturated classes have the *intersection property*, that is the intersection of (not more than τ many) saturated classes of spaces is also a saturated class, while the intersection of classes having universal elements in general may have no universal elements. Thus, with any "new" saturated class we multiply the classes of spaces for which the existence of universal elements is known.

The construction of the specific subsets of Containing Spaces allows to consider classes of spaces with "structure". In the book, as structure we consider a subset of a space, a base for a space, a subset of a space with a so-called positional base, and a representation of a space as a special sum of its subsets. For such classes we define the (new) notion of the universal element considering that the corresponding embeddings of spaces into such an element "preserve" the stucture. In cases where the given structure can be transfered to the corresponding Containing Spaces, we obtain universal elements for the considered classes. Thus, we can define saturated classes of spaces with structure. The consideration of such classes in parallel with the saturated classes of spaces is another special feature of the book. Although we do not investigate the classes of spaces with more general form of structure, the considered cases show that the universality problem for the classes of spaces with structure is a very natural object of research and in this book many results about universal spaces are generalized to universal spaces with structure.

As for the classes of spaces considered, the intersection of saturated classes of spaces with structure is also saturated. This property explains the following peculiarity of the book. Most propositions in this book concerning the property of a class to be saturated have the following form: the class of all T_0-spaces (with structure) satisfying a given property is saturated. This feature of course does not mean that these propositions are true only for the class of T_0-spaces (with structure). By the intersection property of saturated classes they hold for any saturated class of spaces with the same structure. So, for example, new results which concern the class of separable metrizable spaces (which is saturated) are obtained as corollaries of propositions concerning the class of T_0-spaces.

There is another reason to prove statements in the most general class of spaces, that is in the class of all T_0-spaces (with structure). It is natural that the more general a class of spaces is, the "fewer" conditions (as defined by the corresponding property P) are needed to determine this class and, therefore, the fewer conditions are needed to verify the possibility of transferring P to the corresponding Containing Spaces. This point means that the more general a class of spaces is, the "easier" it is to prove that this class is saturated (if really this class is saturated). This also explains why in this book there are many results concerning the universality problem for classes of spaces more general than, for example, the class of completely regular spaces.

Saturated classes are also convenient for constructing other, new, saturated classes. In particular, we use saturated classes to define dimension-like functions of types ind, Ind, and dim satisfying the following *saturation property*: the subclass of the domain of the given function consisting of all elements for which the function is less than or equal to a given integer or a given ordinal is a saturated class. We note that the domains of these functions are not only classes of spaces but classes of spaces with structure as well. Moreover, as the structure on the spaces of a domain becomes more "complicated" the domain admits the definition of more dimension-like functions with the saturation property. In some sense many of the results given in the book can be considered as a step towards obtaining other new results.

The construction of Containing Spaces and the notion of a saturated class can be applied also to universal mappings (preserving the structure), to actions of groups, to isometric universal separable metric spaces, and so on. This approach shows that the "theory of universal mappings" and the "theory of isometrically universal separable metric spaces" are at least as rich as the "theory of universal spaces" and the "theory of universal separable metrizable spaces" respectively.

The important role of the Containing Spaces in the problem of universality appear through the consideration of factorizing τ-spectra. Some simple properties of the limit space of such spectra imply the existence of cofinal subspectra consisting of Containing Spaces. Thus, Containing Spaces appear without any concrete construction.

Considering saturated classes one can see the following (informal) principle: a class is saturated if it is determined by a property in the formulation of which we use (not more than) τ many sets, where τ is the (fixed) weight of the considered spaces. Calling such properties *τ-properties* and the corresponding formulations *τ-formulations*, we illustrate the above with some concrete examples.

The usual definition of the regularity of a space X, that is X is regular if for every $x \in X$ and an open neighbourhood U of x there exists an open subset V of X such that $x \in \mathrm{Cl}_X(V) \subset U$ is not a τ-formulation because the cardinality of X in general is larger than τ. However, the following equivalent definition is a τ-formulation: a space X is regular if there exists a base B of cardinality $\leq \tau$ for open subsets of X such that each element U of B is the union of some elements of B whose closures are contained in U. In this formulation we use (not more than) τ many sets, which means that the regularity of a space is a τ-property.

Similarly, the complete regularity of a space is a τ-property. A space X is completely regular if there exists a base B of cardinality $\leq \tau$ for the open subsets of X such that each element U of B is the union of some elements of B, whose closure are contained in U, and for every such element V there exists a continuous mapping f_V of X into $[0, 1]$ satisfying the following conditions: $f_V(\mathrm{Cl}(V)) \subset \{0\}$ and $f_V(X \setminus U) \subset \{1\}$.

The continuity as well as the openness of a mapping of a space can be defined by a τ-formulation: a mapping f of a space X into a space Y is continuous (respectively, open) if there exists a base B of cardinality $\leq \tau$ for the open subsets of Y (respectively, of X) such that $f^{-1}(U)$ is open in X (respectively, $f(U)$ is open in Y) for every $U \in B$.

Also, the property of a space X to have dimension ind less than or equal to a given integer n has (inductively) a τ-formulation: there exists a base B of cardinality $\leq \tau$ such that for each element U of B we have $\mathrm{ind}(\mathrm{Bd}(U)) \leq n - 1$. Therefore, this proprerty is a τ-property.

We note that both the class of all regular spaces and the class of all completely regular spaces are saturated. The continuity and the openness of the mappings define "saturated classes" of mappings. Also, dimension ind satisfies the saturation property.

On the other hand, the usual definition of compactness, that is a space X is compact if every open cover of X has a finite subcover, is not a τ-formulation because the cardinality of the set of all open covers of X in general is larger than τ. Though the class of all compact spaces is not saturated in this class there exists a universal element. With our point of view the existence of universal elements in the class of all compact spaces can be explained as follows. There exists a saturated class of spaces, namely the class of all completely regular spaces, that contains the class of all compact spaces with the property that each element of this class, in particular, each universal element has a compactification belonging to the same class.

As for compactness, the usual definition of the normality of a space is not a τ-formulation: the number of all pairs of disjoint closed subsets of a space in general is larger than τ. The existence of universal elements in this class has the same explanation as for the class of compact spaces.

We note also that the ordinary definitions of dimensions Ind and dim are not τ-formulations. However, in the class of all normal spaces having dimension Ind or dim less than or equal to a given non-negative integer n there exist universal elements. By our method these results can be obtained as follows. Consider for example dimension Ind. In the saturated class of all completely regular spaces we define a dimension-like function with the saturation property such that: (a) on the subclass of all compact spaces this function coincides with Ind, and (b) each completely regular space has a compactification preserving the value of the function. (This dimension-like function can be considered as a variation of the dimension Ind). Therefore in the class of all completely regular spaces such that the value of the defined function is less than or equal to n there exist universal elements. Consider a compactification of some of these universal elements preserving the value of the function. Then, it is easy to verify that this compactification is a universal element in the class of all normal spaces with the dimension Ind less than or equal to n. (We use here the fact that any normal space has a compactification with the same dimension Ind). The above is also true for dimension dim.

Thus, to construct saturated classes (corresponding to spaces of weight $\leq \tau$) and, therefore, classes in which there exist universal elements, it is important to find properties of spaces which can be formulated using not more than τ many sets. In particular, for classical properties of spaces, which are not τ-properties, it is interesting to find different variations of these properties that have τ-formulations.

The following problem, which arises from the above arguments, is interesting: is it possible to give a formal presentation of the above observations concerning τ-properties and saturated classes?

The book consists of ten chapters. In Chapter 1 we give the main construction of Containing Spaces and we prove some basic properties of this construction.

In Chapter 2 we consider saturated classes of spaces (with structure). As a structure we consider a fixed subset of a space, a fixed base for a space, and a fixed subset of a space with a given positional base.

In Chapter 3 we introduce some dimension-like functions (modulo a class) having the saturation property. We note that the domain of the considered dimension-like functions not only can be a class of spaces but a class of spaces with a given structure as well.

In Chapter 4 we study saturated classes of spaces with structure. We consider as structure some representation of a space as a special sum of its subsets. Some dimension-like functions are also considered.

In Chapter 5 we consider completely regular and compact spaces. In particular, we consider dimensions Ind and dim for these spaces.

In Chapter 6 we consider universal mappings, diagrams, and inverse spectra.

In Chapter 7 we study actions of topological groups.

In Chapter 8 we investigate the connection of Containing Spaces and factorizing τ-spectra.

In Chapter 9 we consider isometric embeddings and isometrically universal spaces for the case of separable metric spaces.

Finally, Chapter 10 is devoted to concluding remarks and to some open problems. Most of the problems are open ended and together with the remarks suggest new aspects of our considerations for a further investigation.

This book is intended for postgraduate students and for researchers with interest in universality problems. Formally, in order to read the book it is not necessary to have special knowledge of the subject. All necessary notions are defined in the text. Some elementary topological and set-theoretical notions, which are not defined, can be found in any textbook on General Topology.

The author thanks Jan van Mill, Maggie Dijkstra, and Jan Dijkstra for their valuable suggestions.

Contents

Preface vii

Introductory remarks 1

Chapter 1. The construction of Containing Spaces 9
- 1.1. Marked spaces and standard equivalence relations 10
- 1.2. The Containing Spaces 14
- 1.3. Specific subspaces of the Containing Spaces 24
- 1.4. Commutative properties of the specific subspaces 29
- 1.5. A generalization of the construction of Containing Spaces 36
- 1.6. Agreement concerning notations of the next chapters 45

Chapter 2. Saturated classes 49
- 2.1. Saturated classes of spaces 51
- 2.2. Saturated classes of subsets 56
- 2.3. Commutative operators 61
- 2.4. Borel type sets 65
- 2.5. Saturated classes of bases 74
- 2.6. Saturated classes of p-bases 78
- 2.7. (ν, μ)-Saturated classes 85

Chapter 3. Dimension-like functions 89
- 3.1. Space dimension-like functions of the type ind 91
- 3.2. Positional dimension-like functions of the type ind 97
- 3.3. Base dimension-like functions of the type ind 106
- 3.4. Base positional dimension-like functions of the type ind 125
- 3.5. Base dimension-like functions of the type Ind 132
- 3.6. Base dimension-like functions of the type dim 144

Chapter 4. Saturated classes of spaces with structure 151
- 4.1. Saturated classes of spaces with structure 152
- 4.2. Dimension-like functions of the type ind 157
- 4.3. D-dimension of spaces 162
- 4.4. Some applications 167

Chapter 5. Completely regular and compact spaces 177
5.1. Completely regular spaces and universality 179
5.2. The base dimension-like function b^n-Ind 193
5.3. Compact spaces and dimension Ind 212
5.4. Compact spaces and dimension dim 225

Chapter 6. Mappings and universality 239
6.1. Mappings with different ranges 240
6.2. Mappings with the same range 256
6.3. Diagrams and inverse spectra 265
6.4. Mappings of compact spaces 278

Chapter 7. Actions of groups 291
7.1. Saturated classes of G-spaces 293
7.2. Spaces with structure and actions of groups 308
7.3. Actions of groups on diagrams 321

Chapter 8. Containing Spaces and factorizing τ-spectra 331
8.1. The spectral theorem 335
8.2. The free union of spaces and factorizing τ-spectra 346
8.3. Saturated classes and factorizing τ-spectra 361
8.4. Structural τ-spectra 380

Chapter 9. Isometries and universality 391
9.1. Isometrically Containing Spaces 393
9.2. Isometrically universal spaces 409
9.3. Completely bounded and compact metric spaces 417
9.4. The metric dimension of metric spaces 423

Chapter 10. Concluding remarks and open problems 431
10.1. Remarks and problems to Chapter 1 433
10.2. Remarks and problems to Chapter 2 439
10.3. Remarks and problems to Chapter 3 452
10.4. Completeness and completion of marked spaces 466
10.5. Containing Spaces and connectedness 474
10.6. Universal sets parametrised by regular zero-dimensional spaces . . 487
10.7. Compactifications and extensions of Containin Spaces 492
10.8. Remarks and problems to Chapters 4-9 497

Bibliography 507

Special symbols 543

Index 551

Introductory remarks

1. Let T be a universal element in a class $\mathbb{P}$ of spaces. Then, the weights and the cardinalities of all elements of $\mathbb{P}$ are restricted by the weight and the cardinality of T, respectively. Thus, for the universality problem we can suppose without loss of generality that the weights of all spaces are less than or equal to a given infinite cardinal. In this case, in order to restrict the cardinalities of spaces it is convenient to consider that all spaces are T_0-spaces.

Agreement. Throughout the book we denote by τ a fixed infinite cardinal. The set of all finite subsets of τ is denoted by $\mathcal{F}$. In particular, the empty set is an element of $\mathcal{F}$. Whenever we consider an arbitrary space, it is assumed that this space is a T_0-space of weight $\leq \tau$.

2. We shall use the symbol "$\equiv$" in order to introduce new notations without mentioning this fact. The following method is used: If we introduce an expression A as a notation of an object (a set, an indexed set, a mapping and so on) and write $A \equiv B$ (or $B \equiv A$), where B is another new expression, then this formula means that B is considered as another notation of the same object. The same is true if we consider a known object with a known expression A as its notation and write $A \equiv B$ (or $B \equiv A$), where B is a new expression.

Below we introduce some of the well-known set-theoretical notions and related notations which we shall use in this book.

3. We shall be concerned with *classes*, *sets*, *collections*, and *families*. A class may not be a set. A set is a class that is an element of another class. A collection and a family are sets. The empty set, denoted by $\emptyset$, is considered to be finite and a finite set is considered to be countable. A finite set consisting of elements $x_0,...,x_n$ is denoted by $\{x_0, ..., x_n\}$. By the *pair* of sets x and y denoted by (x, y) we always mean the *ordered* pair.

Let $X_0,...,X_n$ be a finite number of classes. The *product* of these classes is denoted by $X_0 \times ... \times X_n$ or by $\times\{X_i : i \in \{0, ..., n\}\}$. An element of this product is denoted by $(x_0, ..., x_n)$ and is called an $(n+1)$-*tuple*. It is assumed that $((x_0, ..., x_k), (x_{k+1}, ..., x_n)) = (x_0, ..., x_n)$ for every $k \in \{0, ..., n-1\}$. An $(n+1)$-tuple is called a *triad* or a *tetrad* if $n = 2$ or $n = 3$, respectively. In the case where $X \equiv X_0 = ... = X_n$, the product $X_0 \times ... \times X_n$ is denoted by X^{n+1}. The *union* and the *intersection* of the classes $X_0,...,X_n$ are denoted by $X_0 \cup ... \cup X_n$ and $X_0 \cap ... \cap X_n$ or by $\cup\{X_i : i \in \{0, ..., n\}\}$ and $\cap\{X_i : i \in \{0, ..., n\}\}$, respectively. The *union* and the *intersection* of the elements of a set $\{X_\lambda : \lambda \in \Lambda\}$, where Λ is an infinite set, is denoted by $\cup\{X_\lambda : \lambda \in \Lambda\}$ and $\cap\{X_\lambda : \lambda \in \Lambda\}$, respectively.

4. Let X and Y be two classes. A *mapping* f *of* X *into* Y or a *function* f *from* X *to* Y, denoted by $f : X \to Y$, is a subclass of the product $X \times Y$ satisfying the condition: for every $x \in X$ there exists a unique $y \in Y$ such that $(x, y) \in f$. Instead of $(x, y) \in f$ we write $f(x) = y$. The class X denoted also by D_f is called the *domain* of f and the class Y denoted also by R_f is called the *range* of f. The class of all elements y of Y for which there exists an element x of X such that $(x, y) \in f$ is called the *image-range* of f and is denoted by I_f or by $f(X)$. The mapping f is said to be *onto* if $R_f = I_f$ and *one-to-one* if the condition $f(x_0) = f(x_1)$ implies $x_0 = x_1$. If f is one-to-one, then we denote by f^{-1} the mapping of $f(X)$ onto X such that $f^{-1}(f(x)) = x$ for every $x \in X$. This mapping is called the *inverse* to f mapping. In the case where X and Y are sets we denote by Y^X the set of all mappings of X into Y.

5. Throughout the book we shall widely use indexed sets. Formally, an *indexed set* is a mapping F of a set Λ into a set Y. Hence, F is a subset of $\Lambda \times Y$ and its elements have the form (λ, y_λ), where $\lambda \in \Lambda$ and $y_\lambda = F(\lambda) \in Y$, which means that two elements $(\lambda_0, y_{\lambda_0})$ and $(\lambda_1, y_{\lambda_1})$ of F are distinct if and only if the elements λ_0 and λ_1 of Λ are distinct. However, for simplicity of notations, each element (λ, y_λ) of F is identified with the element y_λ of Y and the indexed set F is denoted also by $\{y_\lambda : \lambda \in \Lambda\}$. In this case the elements of F are considered to be elements of Y (usually supplied with the corresponding elements of Λ as indices). We underline the fact that two such elements of F are considered to be distinct if the corresponding indices are distinct. Thus, in our consideration, the expression $\{y_\lambda : \lambda \in \Lambda\}$

denotes an indexed set as well as a non-indexed set (a subset of Y). This use will cause no confusion because whenever we consider an indexed set we shall emphasize this fact.

Let $F : \Lambda \to Y$ be an indexed set. The indexed set F is said to be an *indication* of Y if $F(\Lambda) = Y$. In the case, where $\Lambda = \omega$, the indexed set $F \equiv \{y_i : i \in \omega\}$ is called a *sequence* of elements of Y and is denoted also by $y_0, y_1, \ldots$. If χ is a one-to-one mapping of a set Γ onto Λ, then we can also consider the indexed set $F \circ \chi : \Gamma \to Y$. In our considerations it is convenient to identify the indexed sets F and $F \circ \chi$.

The set Λ is called the *indexing set* of F. In order to emphasize the use of the set Λ, F is called a Λ-*indexed* set. On the other hand, for simplicity of notations, sometimes we do not indicate one of the sets Λ and Y or both of them. For example, we shall frequently use the following phrase: "Let $\mathbf{S}$ be an indexed collection of spaces". In this case, the indexing set of $\mathbf{S}$ and the range of $\mathbf{S}$ are not indicated explicitly. In the above phrase instead of spaces we shall also consider other objects: elements of a given class, subsets of a space, indexed sets, mappings and so on. We note that an indexed set, in particular, the above indexed collection $\mathbf{S}$, may be the indexing set of another indexed set. (In this case we must remember what elements of the indexing set are considered to be distinct.)

Let $f : X \to Y$ be a mapping of a set X into a set Y, $A \equiv \{U_\lambda : \lambda \in \Lambda\}$ an (indexed) set of subsets of X, and $B \equiv \{V_\lambda : \lambda \in \Lambda\}$ an (indexed) set of subsets of Y. Then, the (indexed) set $\{f(U_\lambda) : \lambda \in \Lambda\}$ is called the *f-image* of the (indexed) set A and the (indexed) set $\{f^{-1}(V_\lambda) : \lambda \in \Lambda\}$ the *f-preimage* of the (indexed) set B.

In the Remarks 10, 11, and 12 given below the most important cases of the using of indexed sets are given.

6. Any *ordinal* α is identified with the set of all ordinals less than α. Thus, for every ordinals β and α the relations $\beta \in \alpha + 1$, $\beta \subset \alpha$, and $\beta \leq \alpha$ are equivalent. An *isolated* ordinal is an ordinal of the form $\alpha + 1$, where α is also an ordinal. A non-isolated ordinal is called *limit*. Therefore, zero is considered a limit ordinal. The class of all ordinals is denoted by $\mathcal{O}$. In parallel with the ordinals we also consider two symbols: -1 and ∞. It is convenient to consider that -1 is less than any ordinal and any ordinal is less than ∞, that is for every $\alpha \in \mathcal{O}$, $-1 < \alpha < \infty$.

Any *cardinal* is identified with the least ordinal of this cardinality. For every cardinal ν we denote by ν^+ the least cardinal greater than ν. The first infinite cardinal is denoted by ω. The elements of ω are also called *non-negative integers* and the elements of ω^+ *countable* ordinals. For every set X we denote by $|X|$ the cardinality of X.

7. An *equivalence relation* on a non-empty class X is a subclass $\sim$ ("tilde") of $X \times X$ satisfying the following conditions: (a) for every $x \in X$, $(x, x) \in \sim$, (b) if $(x, y) \in \sim$ then $(y, x) \in \sim$, and (c) if $(x, y) \in \sim$ and $(y, z) \in \sim$, then $(x, z) \in \sim$. If $(x, y) \in \sim$ then we write also $x \sim y$. A non-empty subclass E of X is said to be an *equivalence class* of $\sim$ or a $\sim$-*equivalence class* if for every $x \in E$ and for every $y \in X$ the condition $x \sim y$ is equivalent to the condition $y \in E$. Any equivalence relation on a class is denoted by the symbol $\sim$ supplied usually with one or more indices. If X is a set, then the set of all equivalence classes of $\sim$ is denoted by $\mathrm{C}(\sim)$.

Equivalence relations are considered also on indexed sets. (In this case it is important to remember what elements of the indexed set are considered distinct.)

8. A *preorder* on a non-empty set Λ is a subset $\prec$ of $\Lambda \times \Lambda$ satisfying the following conditions: (a) for every $\lambda \in \Lambda$, $(\lambda, \lambda) \in \prec$ and (b) if $(\lambda_0, \lambda_1) \in \prec$ and $(\lambda_1, \lambda_2) \in \prec$, then $(\lambda_0, \lambda_2) \in \prec$. If $(\lambda_0, \lambda_1) \in \prec$ then we also write $\lambda_0 \prec \lambda_1$. A preorder $\prec$ is said to be a *partial order* if the conditions $(\lambda_0, \lambda_1) \in \prec$ and $(\lambda_1, \lambda_0) \in \prec$ imply $\lambda_0 = \lambda_1$. A partial order $\prec$ is said to be an *order* if for every $\lambda_0, \lambda_1 \in \Lambda$ either $(\lambda_0, \lambda_1) \in \prec$ or $(\lambda_1, \lambda_0) \in \prec$. The set Λ with a partial order $\prec$ is said to be *directed* if for every $\lambda_0, \lambda_1 \in \Lambda$ there exists $\lambda \in \Lambda$ such that $(\lambda_0, \lambda) \in \prec$ and $(\lambda_1, \lambda) \in \prec$. The set Λ with an order $\prec$ is called *well ordered* if every subset Λ' of Λ has a *minimal* element (that is, Λ' has an element λ'_0 such that $\lambda'_0 \prec \lambda'_1$ for every $\lambda'_1 \in \Lambda'$). In this case, the order $\prec$ is called a *well ordering.*

9. Let X be a set. We denote by $\mathcal{P}(X)$ the set of all subsets of X. A subset $\mathcal{G}$ of $\mathcal{P}(X)$ is said to be a *ring of subsets of* X or briefly a *ring* of X if it is closed under finite unions and finite intersections. If moreover $\mathcal{G}$ is closed under the complements, then $\mathcal{G}$ is called an *algebra* of X. The empty set and the whole set are considered to be elements of any ring. (They can be considered as the intersection and the union, respectively, of the empty number of elements of $\mathcal{G}$.) Obviously, $\mathcal{P}(X)$ is an algebra and the intersection of any number of algebras (or rings) is also an algebra (or a ring). Therefore, the intersection of all algebras (or rings) containing a given subset $\mathcal{G}$ of $\mathcal{P}(X)$ is an algebra (or a ring). This algebra (or ring) is said to be the *minimal* algebra (or ring) containing $\mathcal{G}$ or the algebra (or the ring) *generated* by $\mathcal{G}$. The minimal ring containg $\mathcal{G}$ is denoted by $\mathcal{G}^{\diamondsuit}$. It is easy to see that for every $G \in \mathcal{G}^{\diamondsuit}$ there exists a finite set u consisting of finite subsets of $\mathcal{G}$ such that $G = \cup\{G_v : v \in u\}$, where G_v is the intersection of all elements of the finite subset v of $\mathcal{G}$.

Let Y be also a set. A mapping i of an algebra A of X into $\mathcal{P}(Y)$ is said to be a *homomorphism* if for every elements U and V of A we have $i(X \setminus U) = Y \setminus i(U)$ and $i(U \cup V) = i(U) \cup i(V)$. It is easy to verify that in this case the set $i(A)$ is an algebra of Y. If moreover i is one-to-one, then i is called an *isomorphism.* In this case, it is easy to see that the inverse to i mapping (of $i(A)$ onto A) is also an isomorphism. It is also clear that the composition of two homomorphisms is also a homomorphism and, therefore, the composition of two isomorphisms is an isomorphism.

Below, we give some elementary topological notions and notations.

10. Let X be a space. A family B of open subsets of X is said to be a *base for the open subsets of* X or briefly a *base for* X if any open subset of X is a union of elements of B. Therefore, the empty set is an element of any base for X. (This is convenient for our considerations.) A family C of closed subsets of X is called a *base for the closed subsets of* X if the set consisting of the complements of all elements of C is a base for X. Therefore, $X \in C$. We do not suppose that the cardinality of a base is $\leq \tau$. However, almost all considered bases satisfy this condition. The weight of X is denoted by $w(X)$.

We shall widely use τ*-indexed bases* for the open (closed) subsets of the space X. Any τ-indexed base $M \equiv \{U_\delta : \delta \in \tau\}$ for the open subsets of X is said to be a *mark* of X. (We underline the fact that the indexing set is the set τ even if the weight of X is less than τ.) The space X is said to be *marked* if a mark of X is chosen.

Let B be a base for the open (closed) subsets of X and $M \equiv \{U_\delta : \delta \in \tau\}$ an indication of B. Let also Q be a subset of X. The set

$$B^Q \equiv \{Q \cap U : U \in B\}$$

is called the *trace on* Q *of the base* B. The τ-indexed set

$$M^Q \equiv \{Q \cap U_\delta : \delta \in \tau\}$$

is said to be the *trace on* Q *of the* τ*-indexed set* M. Obviously, M^Q is an indication of B^Q. In particular, if M is a mark of X, then M^Q is a mark of the subspace Q.

Let $M \equiv \{U_\delta : \delta \in \tau\}$ and $N \equiv \{V_\delta : \delta \in \tau\}$ be two indexed sets of subsets of X. It is said that N is an *extension* of M if there exists a mapping θ of τ into itself such that $U_\delta = V_{\theta(\delta)}$, $\delta \in \tau$. In this case, we also say that N is a *θ-extension* M. The mapping θ is called an *indicial mapping* of this extension or an *indicial mapping from M to N*. (We note that in general θ is not unique.)

We denote by $\mathrm{Cl}_X(Q)$, $\mathrm{Bd}_X(Q)$, $\mathrm{Int}_X(Q)$, and $X \setminus Q$ (or $\mathrm{Co}_X(Q)$) the closure, boundary, interior, and the complement of a set Q in X, respectively.

In the book some ordinary notions concerning a space are transferred to an (indexed) collection of spaces. The corresponding new specific notions play an important role throught the book. Below, we give some of them.

11. Let $\mathbf{S}$ be an indexed collection of spaces. (We do not indicate the indexing set of $\mathbf{S}$, which can be an arbitrary set.) An **S**-*indexed set of families* is an indexed set $\mathbf{B} \equiv \{B^X : X \in \mathbf{S}\}$, where B^X is a family of subsets of X. The family B^X is denoted also by $\mathbf{B}(X)$.

An $\mathbf{S}$-indexed set $\mathbf{B} \equiv \{B^X : X \in \mathbf{S}\}$ of families is said to be an *open co-base for* $\mathbf{S}$ or briefly a *co-base for* $\mathbf{S}$) (respectively, a *closed co-base for* $\mathbf{S}$ if B^X is a base for X (respectively, a base for the closed subsets of X) of cardinality less than or equal to τ. (We underline the fact that the cardinality of B^X is $\leq \tau$.)

An **S**-*indexed set of τ-indexed families* is an indexed set

$$\mathbf{M} \equiv \{M^X \equiv \{U_\delta^X : \delta \in \tau\} : X \in \mathbf{S}\},$$

where M^X is a τ-indexed family of subsets of X. The indexed family M^X is denoted also by $\mathbf{M}(X)$.

An $\mathbf{S}$-indexed set $\mathbf{M} \equiv \{M^X : X \in \mathbf{S}\}$ of τ-indexed families is said to be a *co-indication* of an $\mathbf{S}$-indexed set $\mathbf{B} \equiv \{B^X : X \in \mathbf{S}\}$ of families if M^X is an indication of B^X for every $X \in \mathbf{S}$. (We underline the fact that the indexing set of M^X is the set τ.) The above notion is frequently used in the case where $\mathbf{B}$ is a co-base for $\mathbf{S}$. In this case $\mathbf{M}$ is said to be a *co-mark* of $\mathbf{S}$. We note that if $\mathbf{M}$ is a co-mark of $\mathbf{S}$, then for every $X \in \mathbf{S}$, $\mathbf{M}(X)$ is a mark of X.

Let

$$\mathbf{M}^0 \equiv \{\{U_\delta^{0,X} : \delta \in \tau\} : X \in \mathbf{S}\} \text{ and}$$

$$\mathbf{M}^1 \equiv \{\{U_\delta^{1,X} : \delta \in \tau\} : X \in \mathbf{S}\}$$

be two $\mathbf{S}$-indexed sets of τ-indexed families. It is said that $\mathbf{M}^1$ is a *co-extension* of $\mathbf{M}^0$ if there exists a one-to-one mapping θ of τ into itself such that $U_\delta^{0,X} = U_{\theta(\delta)}^{1,X}$ for every $X \in \mathbf{S}$. The mapping θ is called an *indicial mapping* of this co-extension or an *indicial mapping from* $\mathbf{M}^0$ *to* $\mathbf{M}^1$. (We note that in general θ is not unique.)

Suppose that for every $X \in \mathbf{S}$ a subset Q^X of X is given. Then, the $\mathbf{S}$-indexed set $\mathbf{Q} \equiv \{Q^X : X \in \mathbf{S}\}$ is called a *restriction* of $\mathbf{S}$. (Therefore, if X and Y are distinct elements of $\mathbf{S}$, then Q^X and Q^Y are considered to be distinct elements of $\mathbf{Q}$.) The set Q^X is denoted also by $\mathbf{Q}(X)$.

Now, suppose that for every $X \in \mathbf{S}$ another subset K^X of X is given such that $K^X \subset Q^X$. Then, we can also consider the restriction

$$\mathbf{K} \equiv \{K^X : X \in \mathbf{S}\}$$

of $\mathbf{S}$. On the other hand, the restriction $\mathbf{Q}$ can also be considered as an indexed collection of spaces and the indexed set $\mathbf{K}$ can be written as follows:

$$\mathbf{K} = \{K^{Q^X} : Q^X \in \mathbf{Q}\},$$

where $K^{Q^X} = K^X$, which means that $\mathbf{K}$ is a restriction of $\mathbf{Q}$. Thus, $\mathbf{K}$ can be considered simultaneously as a restriction of $\mathbf{S}$ and as a restriction of $\mathbf{Q}$ as well.

Let $\mathbf{Q} \equiv \{Q^X : X \in \mathbf{S}\}$ be a restriction of $\mathbf{S}$ and $\mathbf{M}$ a co-mark of $\mathbf{S}$. The *trace on* $\mathbf{Q}$ *of the co-mark* $\mathbf{M}$ of $\mathbf{S}$ is the co-mark of $\mathbf{Q}$ denoted by $\mathbf{M}|_\mathbf{Q}$ and defined as follows: the mark $(\mathbf{M}|_\mathbf{Q})(Q^X)$ of an element Q^X of $\mathbf{Q}$ is the trace on Q^X of the mark $\mathbf{M}(X)$ of X.

12. Let $\mathbf{S}$ be an indexed collection of sets. (In particular, the elements of $\mathbf{S}$ may be spaces, mappings, and so on.) We shall deal with $\mathcal{F}$-indexed families of equivalence relations on $\mathbf{S}$. For such a family $\mathrm{R} \equiv \{\sim^s : s \in \mathcal{F}\}$ we shall denote by $\mathrm{C}(\mathrm{R})$ the set $\cup\{\mathrm{C}(\sim^s) : s \in \mathcal{F}\}$. We shall frequently refer also to the ring $(\mathrm{C}(\mathrm{R}))^\diamondsuit \equiv \mathrm{C}^\diamondsuit(\mathrm{R})$.

Let

$$\mathrm{R}_0 \equiv \{\sim_0^s : s \in \mathcal{F}\} \text{ and}$$

$$\mathrm{R}_1 \equiv \{\sim_1^s : s \in \mathcal{F}\}$$

be two $\mathcal{F}$-indexed families of equivalence relations on $\mathbf{S}$. It is said that R_1 is a *final refinement* of R_0 if for every $s \in \mathcal{F}$ there exists an element t of $\mathcal{F}$ such that $\sim_1^t \subset \sim_0^s$.

An $\mathcal{F}$-indexed family $\mathrm{R} \equiv \{\sim^s : s \in \mathcal{F}\}$ of equivalence relations on $\mathbf{S}$ is said to be *admissible* if the following conditions are satisfied: (a) if $s \subset t \in \mathcal{F}$, then $\sim^t \subset \sim^s$, (b) for every $s \in \mathcal{F}$ the number of $\sim^s$-equivalence classes is finite, and (c) $\sim^s = \mathbf{S} \times \mathbf{S}$ if $s = \emptyset$. It is easy to verify that for such a family the set $\mathrm{C}^{\diamondsuit}(\mathrm{R})$ is closed also under complements, that is if $\mathbf{H} \in \mathrm{C}^{\diamondsuit}(\mathrm{R})$, then $\mathbf{S} \setminus \mathbf{H} \in \mathrm{C}^{\diamondsuit}(\mathrm{R})$.

Now suppose that $\mathbf{S}$ is an indexed collection of spaces and

$$\mathbf{Q} \equiv \{Q^X : X \in \mathbf{S}\}$$

is a restriction of $\mathbf{S}$. The *trace on* $\mathbf{Q}$ *of an equivalence relation* $\sim$ on $\mathbf{S}$ is the equivalence relation on $\mathbf{Q}$ denoted by $\sim|_{\mathbf{Q}}$ and defined as follows: two elements Q^X and Q^Y of $\mathbf{Q}$ are $\sim|_{\mathbf{Q}}$-equivalent if and only if X and Y are $\sim$-equivalent. Obviously, if $\sim_1$ and $\sim_2$ are two equivalence relations on $\mathbf{S}$ and $\sim_2$ is contained in $\sim_1$, then $\sim_2|_{\mathbf{Q}}$ is contained in $\sim_1|_{\mathbf{Q}}$.

Let $\mathrm{R} \equiv \{\sim^s : s \in \mathcal{F}\}$ be an $\mathcal{F}$-indexed family of equivalence relations on $\mathbf{S}$. The *trace on* $\mathbf{Q}$ *of the family* R is the $\mathcal{F}$-indexed family $\{\sim^s|_{\mathbf{Q}} : s \in \mathcal{F}\}$ of equivalence relations on $\mathbf{Q}$. This family is denoted by $\mathrm{R}|_{\mathbf{Q}}$. Obviously, if R is admissible, then $\mathrm{R}|_{\mathbf{Q}}$ is also admissible. Moreover, if R_0 and R_1 are two $\mathcal{F}$-indexed families of equivalence relations on $\mathbf{S}$ and R_1 is a final refinement of R_0, then $\mathrm{R}_1|_{\mathbf{Q}}$ is a final refinement of $\mathrm{R}_0|_{\mathbf{Q}}$.

The *trace on* $\mathbf{Q}$ *of an element* $\mathbf{H}$ of $\mathrm{C}^{\diamondsuit}(\mathrm{R})$, denoted by $\mathbf{H}|_{\mathbf{Q}}$, is the set of all elements Q^X of $\mathbf{Q}$ for which $X \in \mathbf{H}$. It is easy to see that $\mathbf{H}|_{\mathbf{Q}}$ is an element of $\mathrm{C}^{\diamondsuit}(\mathrm{R}|_{\mathbf{Q}})$. Moreover, if $\mathbf{H} \in \mathrm{C}(\mathrm{R})$, then $\mathbf{H}|_{\mathbf{Q}} \in \mathrm{C}(\mathrm{R}|_{\mathbf{Q}})$.

Chapter 1

The construction of Containing Spaces

In the present chapter we give the main construction of the Containing Spaces and prove some basic properties of this construction.

In Section 1.1 on the class of all marked spaces using algebras of subsets some equivalence relations are defined. These equivalence relations are an important part of the construction of Containing Spaces.

In Section 1.2 we consider an arbitrary indexed collection $\mathbf{S}$ of T_0-spaces (of weight $\leq \tau$) and an arbitrary co-mark $\mathbf{M}$ of $\mathbf{S}$. Using relations considered in Section 1 we define an admissible family $R_{\mathbf{M}}$ of equivalence relations on $\mathbf{S}$ calling $\mathbf{M}$-standard. Furthermore, for an arbitrary admissible family R of equivalence relation on $\mathbf{S}$, which is a final refinement of $R_{\mathbf{M}}$, we construct by a standard manner a Containing Space denoted by $T(\mathbf{M}, R)$. This space is uniquely determined by the collection $\mathbf{S}$, co-mark $\mathbf{M}$, and the family R. For every element X of $\mathbf{S}$ there exists a natural embedding of X into $T(\mathbf{M}, R)$.

In Section 1.3 we consider a restriction $\mathbf{Q}$ of $\mathbf{S}$ and show that if $R|_{\mathbf{Q}}$ is a final refinement of $\mathbf{M}|_{\mathbf{Q}}$-standard family of equivalence relations on $\mathbf{Q}$, then the Containing Space $T(\mathbf{M}|_{\mathbf{Q}}, R|_{\mathbf{Q}})$ can be considered by a natural way as a (specific) subset of the Containing Space $T(\mathbf{M}, R)$.

In Section 1.4 we give the notions of a closed (open) restriction: a restriction $\mathbf{Q}$ is said to be closed (open) if $\mathbf{Q}(X)$ is closed (open) in $X \in \mathbf{S}$. We also give the notion of an $(\mathbf{M}, R)$-complete restriction, which is a generalization of closed and open restrictions. For such restrictions we study commutative properties of specific subsets with respect to some topological operators on spaces. As such operators we consider here the boundary, the closure, the interior, and the complement operator (that is, each subset corresponds to its complement). For example, for the boundary operator the corresponding commutative property can be formulated as follows. Suppose that

$$\mathbf{Q} \equiv \{Q^X : X \in \mathbf{S}\}$$

is a $(\mathbf{M}, \mathrm{R})$-complete restriction. We put

$$\mathbf{Bd}(\mathbf{Q}) \equiv \{\mathrm{Bd}(Q^X) : X \in \mathbf{S}\}.$$

Then, under some simple conditions concerning the corresponding family of equivalence relations on $\mathbf{S}$, we prove the following property:

$$\mathrm{Bd}_{\mathrm{T}}(\mathrm{T}|_{\mathbf{Q}}) = \mathrm{T}|_{\mathbf{Bd}(\mathbf{Q})}.$$

In Section 1.5 we give a generalization of the main construction. Actually, this generalization concerns the families of equivalence relations.

Finally, in Section 1.6 some basic notations which we follow in the next chapters are given.

1.1 Marked spaces and standard equivalence relations

In this section for every $s \in \mathcal{F} \setminus \{\emptyset\}$ on the class of all marked spaces the so-called s-standard equivalence relation is defined and studied. The given results are used in the next section for the construction of Containing Spaces.

The following two lemmas are easily proved. In the sequel, we shall refer to Lemma 1.1.1 in cases where we want to use some relation between elements of an algebra and where it is known that the same relation between the corresponding elements of an isomorphic algebra is true.

1.1.1 Lemma. *Let X and Y be two sets, A an algebra of X and let i be an isomorphism of A into $\mathcal{P}(Y)$. Suppose that U, V, and W are elements of A. Then, the relations $U = X$, $U = \emptyset$, $U \subset V$, and $U \cap V = W$ are equivalent to the relations $i(U) = Y$, $i(U) = \emptyset$, $i(U) \subset i(V)$, and $i(U) \cap i(V) = i(W)$, respectively.* □

1.1.2. Lemma. *Let X and Y be two sets, $\mathcal{G}$ a subset of $\mathcal{P}(X)$, and let A be the minimal algebra of X containing the set $\mathcal{G}$. The following statements are true:*

(1) *If i is a homomorphism of A into $\mathcal{P}(Y)$, then $i(A)$ is the minimal algebra of Y containing the set $i(\mathcal{G})$.*

(2) *If h_1 and h_2 are two homomorphisms of A into $\mathcal{P}(Y)$ such that $h_1(U) = h_2(U)$ for every $U \in \mathcal{G}$, then $h_1 = h_2$.* $\square$

Definition. Let X be a marked space, $\{U_\delta^X : \delta \in \tau\}$ the corresponding mark of X, and $s \in \mathcal{F} \setminus \{\emptyset\}$. By the *$s$-algebra* of X we mean the minimal algebra of X containing the sets U_δ^X, $\delta \in s$. This algebra is denoted by A_s^X. The sets U_δ^X and $X \setminus U_\delta^X$ (which are elements of A_s^X) are denoted also by $X_{(\delta,0)}$ and $X_{(\delta,1)}$, respectively. By definition, if $\emptyset \neq t \subset s$, then $\mathrm{A}_t^X \subset \mathrm{A}_s^X$.

In the sequel, it is convenient to consider an isomorphism of the algebra A_s^X of a marked space X onto an algebra of a set, which is independent on the space X. As such a set we take the set 2^s. (We recall that according to our notations $2 = \{0, 1\}$.)

Notation. Let X be a marked space and $s \in \mathcal{F} \setminus \{\emptyset\}$. For every $f \in 2^s$ we set

$$X_{(s,f)} = \cap\{X_{(\delta,f(\delta))} : \delta \in s\}.$$

As a finite intersection of elements of A_s^X, $X_{(s,f)} \in \mathrm{A}_s^X$. We denote by 2_X^s the subset of 2^s consisting of all elements f of 2^s such that $X_{(s,f)} \neq \emptyset$.

Since the sets $X_{(\delta,0)}$ and $X_{(\delta,1)}$ (for a fixed δ) are disjoint and their union is the space X (true even if $X = \emptyset$) the sets $X_{(s,f)}$ (for a fixed s) are mutually disjoint and their union is also the space X. Therefore, for every $x \in X$ there exists a unique element f of 2^s such that $x \in X_{(s,f)}$. We define a mapping d_s^X of X into 2^s setting

$$d_s^X(x) = f.$$

It is supposed that $d_s^X(X) = \emptyset$ if $X = \emptyset$. Obviously, $2_X^s = d_s^X(X)$.

For every $\delta \in s$ we set

$$u(X, s, \delta) = \{f \in 2_X^s : f(\delta) = 0\}.$$

For every element u of $\mathcal{P}(2_X^s)$ we set

$$X_{(s,u)} \equiv \cup\{X_{(s,f)} : f \in u\}.$$

It is supposed that $X_{(s,u)} = \emptyset$ if $u = \emptyset$. As a finite union of elements of A_s^X, $X_{(s,u)} \in \mathrm{A}_s^X$.

We denote by i_X^s the mapping of the set $\mathcal{P}(2_X^s)$ into A_s^X such that for every $u \in \mathcal{P}(2_X^s)$,

$$i_X^s(u) = X_{(s,u)}.$$

1.1.3. Lemma. *Let X be a marked space and $s \in \mathcal{F} \setminus \{\emptyset\}$. The mapping i_X^s is an isomorphism of the algebra $\mathcal{P}(2_X^s)$ onto the algebra A_s^X such that for every $\delta \in \tau$,*

$$i_X^s(u(X, s, \delta)) = X_{(\delta,0)}. \tag{1.1.1}$$

Proof. Let $v \in \mathcal{P}(2_X^s)$ and $u = 2_X^s \setminus v$. Then, by properties of the sets $X_{(s,f)}$ we have

$$X_{(s,u)} = \cup\{X_{(s,f)} : f \in u\} = \cup\{X_{(s,f)} : f \in 2_X^s \setminus v\} =$$

$$X \setminus (\cup\{X_{(s,f)} : f \in v\}) = X \setminus X_{(s,v)}.$$

This means that $i_X^s(u) = X \setminus i_X^s(v)$.

Let $v, w \in \mathcal{P}(2_X^s)$ and $u = v \cup w$. As the above,

$$X_{(s,u)} = \cup\{X_{(s,f)} : f \in u\} = \cup\{X_{(s,f)} : f \in v \cup w\} =$$

$$\cup\{X_{(s,f)} : f \in v\} \cup (\cup\{X_{(s,f)} : f \in w\}) = X_{(s,v)} \cup X_{(s,w)},$$

that is, $i_X^s(u) = i_X^s(v) \cup i_X^s(w)$. The proving properties show that i_X^s is a homomorphism of $\mathcal{P}(2_X^s)$ into A_s^X.

Since $X_{(s,f)} \neq \emptyset$ for every $f \in 2_X^s$ we have that $X_{(s,u)} \neq X_{(s,v)}$ for distinct elements u and v of $\mathcal{P}(2_X^s)$, that is i_X^s is one-to-one; and, therefore, it is an isomorphism.

Now, we prove the equality (1.1.1). Let $\delta \in s$ and $x \in X_{(\delta,0)}$. For every $\varepsilon \in s$ let $k(\varepsilon)$ denote the unique element of the set $\{0, 1\}$ for which $x \in X_{(\varepsilon,k(\varepsilon))}$. Let f be the element of 2^s such that $f(\varepsilon) = k(\varepsilon)$ for every $\varepsilon \in s$. By the definition of the set $X_{(s,f)}$, $x \in X_{(s,f)}$. On the other hand, by assumption, $k(\delta) = 0$ and therefore $f \in u(X, s, \delta)$. By the definition of the set $X_{(s,u(X,s,\delta))}$, x belongs to this set. Thus, $X_{(\delta,0)} \subset X_{(s,u(X,s,\delta))}$. Conversely, let $x \in X_{(s,u(X,s,\delta))}$. Then, there exists an element f of $u(X,s,\delta)$ such that $x \in X_{(s,f)}$. By the definition of the set $u(X,s,\delta)$, $f(\delta) = 0$. Therefore, $x \in X_{(\delta,0)}$, which means that $X_{(s,u(X,s,\delta))} \subset X_{(\delta,0)}$. Thus, $X_{(s,u(X,s,\delta))} = X_{(\delta,0)}$.

Since A_s^X is the minimal algebra containing the sets $X_{(\delta,0)}$, $\delta \in s$, the equality (1.1.1) implies that $i_X^s(\mathcal{P}(2_X^s)) = \mathrm{A}_s^X$, that is the mapping i_X^s is onto, which completes the proof of the lemma. □

Definition. For every $s \in \mathcal{F} \setminus \{\emptyset\}$ on the class of all marked spaces an equivalence relation, denoted by $\sim_m^s$, is defined as follows: two marked spaces X and Y are $\sim_m^s$-*equivalent* if there exists an isomorphism i of A_s^X onto A_s^Y such that

$$i(X_{(\delta,0)}) = Y_{(\delta,0)}$$

for every $\delta \in s$. This equivalence relation is called *standard*, more precisely, s-*standard*. The isomorphism i, which by Lemma 1.1.2 is uniquely determined, is called *natural*.

Lemma 1.1.2 implies also that $\sim_m^s \subset \sim_m^t$ if $\emptyset \neq t \subset s$. Moreover, the restriction $i|_{\mathrm{A}_t^X}$ of the above natural isomorphism i to A_t^X is the natural isomorphism of A_t^X onto A_t^Y.

1.1.4. Lemma. *Let X and Y be marked spaces and $s \in \mathcal{F} \setminus \{\emptyset\}$. The following condition are equivalent:*

(1) *The marked spaces X and Y are $\sim_m^s$-equivalent.*

(2) $2_X^s = 2_Y^s$.

(3) $u(X,s,\delta) = u(Y,s,\delta)$ *for every $\delta \in s$.*

Proof. (1) ⇒ (2). Let i be the natural isomorphism of A_s^X onto A_s^Y. Then, we have

$$i(X_{(s,f)}) = \cap\{i(X_{(\delta,f(\delta))}) : \delta \in s\} =$$

$$\cap\{Y_{(\delta,f(\delta))}) : \delta \in s\} = Y_{(s,f)}$$

for every $f \in 2_X^s$. Since i is an isomorphism, $Y_{(s,f)} \neq \emptyset$ if $X_{(s,f)} \neq \emptyset$. This fact means that $2_X^s \subset 2_Y^s$. Similarly, $2_Y^s \subset 2_X^s$. Thus, $2_X^s = 2_Y^s$.

(2) ⇒ (3). It follows immediately by the definitions of the sets $u(X,s,\delta)$ and $u(Y,s,\delta)$.

(3) $\Rightarrow$ (2). By Lemmas 1.1.2 and 1.1.3 the set $\mathcal{P}(2^s_X)$ is the minimal algebra of 2^s containing the sets $u(X, s, \delta)$, $\delta \in s$. The same is true for the space Y. This fact means that $\mathcal{P}(2^s_X) = \mathcal{P}(2^s_Y)$; and, therefore, $2^s_X = 2^s_Y$.

(2) $\Rightarrow$ (1). By Lemma 1.1.3 the mapping $i \equiv i^s_Y \circ (i^s_X)^{-1}$ is an isomorphism of A^X_s onto A^Y_s such that $i(X_{(\delta,0)}) = Y_{(\delta,0)}$ for every $\delta \in s$. This fact means that the marked spaces X and Y are $\sim^s_m$-equivalent. □

1.1.5 Lemma. *Let $\{U^X_\delta : \delta \in \tau\}$ be the mark of a marked space X, Q a subspace of X, and let $\{U^Q_\delta : \delta \in \tau\}$ be the trace on Q of the mark $\{U^X_\delta : \delta \in \tau\}$. Then,*

$$d^X_s(x) = d^Q_s(x)$$

for every $x \in Q$ and $s \in \mathcal{F} \setminus \{\emptyset\}$. Therefore, $d^X_s(Q) = d^Q_s(Q)$. (We note that the mapping d^Q_s is constructed with respect to the mark $\{U^Q_\delta : \delta \in \tau\}$ of Q).

Proof. Let $s \in \mathcal{F} \setminus \{\emptyset\}$, $x \in Q$ and $d^X_s(x) = f \in 2^s$. We need to prove that $d^Q_s(x) = f$. By the definition of the mapping d^X_s we have $x \in X_{(s,f)} = \cap\{X_{(\delta,f(\delta))} : \delta \in s\}$. Therefore,

$$x \in Q \cap (\cap\{X_{(\delta,f(\delta))} : \delta \in s\}) = \cap\{Q \cap X_{(\delta,f(\delta))} : \delta \in s\} =$$

$$\cap\{Q_{(\delta,f(\delta))} : \delta \in s\} = Q_{(s,f)}.$$

This means that $d^Q_s(x) = f$. □

1.2 The Containing Spaces

In the present section we give the construction of Containing Spaces. These spaces are used throughout the book.

Definitions. Let $\mathbf{S}$ be an indexed collection of spaces and $\mathbf{M}$ a co-mark of $\mathbf{S}$. For every $s \in \mathcal{F} \setminus \{\emptyset\}$, the s-standard equivalence relation $\sim^s_m$ defines on $\mathbf{S}$ an equivalence relation, denoted by $\sim^s_{\mathbf{M}}$, as follows: two elements X and Y of $\mathbf{S}$ are $\sim^s_{\mathbf{M}}$-*equivalent* if and only if the marked spaces X and Y (with the chosen marks $\mathbf{M}(X)$ and $\mathbf{M}(Y)$, respectively) are $\sim^s_m$-equivalent. We also set $\sim^s_{\mathbf{M}} = \mathbf{S} \times \mathbf{S}$ if $s = \emptyset$. The indexed family

$$\mathrm{R}_{\mathbf{M}} \equiv \{\sim^s_{\mathbf{M}} : s \in \mathcal{F}\}$$

of the above defined equivalence relations is called **M**-*standard*. By Lemma 1.4, $\mathrm{R}_{\mathbf{M}}$ is admissible.

An admissible family R of equivalence relations on **S** is said to be **M**-*admissible* if R is a final refinement of the **M**-standard family $\mathrm{R}_{\mathbf{M}}$.

Agreement. In what follows of this section it is assumed that an arbitrary non-empty indexed collection of spaces (of weight $\leq \tau$) denoted by **S** is given.

It is also assumed that an arbitrary co-mark of **S** is given. This co-mark is denoted by **M**. Moreover, we assume that for every $X \in \mathbf{S}$,

$$\mathbf{M}(X) = \{U_\delta^X : \delta \in \tau\}.$$

Whenever an element X of **S** is considered to be marked, then $\mathbf{M}(X)$ is considered to be the chosen mark of X.

Finally, it is assumed that an **M**-admissible family of equivalence relations on **S** is given. This family is denoted by

$$\mathrm{R} \equiv \{\sim^s : s \in \mathcal{F}\}.$$

Definition. Suppose that **S** contains a non-empty element. Consider the set of all pairs (x, X), where $X \in \mathbf{S}$ and $x \in X$. (Two such pairs (x, X) and (y, Y) are considered to be distinct if either X and Y are distinct elements of the indexed set **S** or X and Y are the same element of the indexed set **S** and $x \neq y$.) On this set, an equivalence relation denoted by $\sim_{\mathbf{M}}^{\mathrm{R}}$ is defined as follows: two pairs (x, X) and (y, Y) are $\sim_{\mathbf{M}}^{\mathrm{R}}$-*equivalent* if

$$X \sim^s Y \text{ and } d_s^X(x) = d_s^Y(y)$$

for every $s \in \mathcal{F} \setminus \{\emptyset\}$. (We note that the condition $d_s^X(x) = d_s^Y(y)$ is true simultaneously for every $s \in \mathcal{F} \setminus \{\emptyset\}$ if and only if for every $\delta \in \tau$ either $x \in U_\delta^X$ and $y \in U_\delta^Y$ or $x \notin U_\delta^X$ and $y \notin U_\delta^Y$.)

Notation. In what follows, the set of all $\sim_{\mathbf{M}}^{\mathrm{R}}$-equivalence classes is denoted by $\mathrm{T}(\mathbf{M}, \mathrm{R}) \equiv \mathrm{T}$ (that is, according to the Introductory notations, $\mathrm{T} = \mathrm{C}(\sim_{\mathbf{M}}^{\mathrm{R}})$). It is assumed that $\mathrm{T}(\mathbf{M}, \mathrm{R}) = \emptyset$ if all elements of **S** are empty.

For every element **H** of $\mathrm{C}^{\diamondsuit}(\mathrm{R})$ (about notation see the Introductory Remarks) the set of all $\mathbf{a} \in \mathrm{T}$, for which there exists an element (x, X) of **a** such that $X \in \mathbf{H}$, is denoted by $\mathrm{T}(\mathbf{M}, \mathrm{R}, \mathbf{H}) \equiv \mathrm{T}(\mathbf{H})$. (Therefore, $\mathrm{T}(\mathbf{H})$

coincides with the set of all $\mathbf{a} \in \mathrm{T}$ such that for every $(x, X) \in \mathbf{a}$, $X \in \mathbf{H}$.) These subsets of T will be called *primary*.

Notation. Let $s \neq \emptyset$ and t be elements of $\mathcal{F}$ such that $\sim^t \subset \sim^s_{\mathbf{M}}$ and let $\mathbf{H}$ be an element of $\mathrm{C}(\sim^t)$. Suppose that an element X of $\mathbf{H}$ is chosen. We denote by $2^s_{\mathbf{H}}$ the set 2^s_X and by $u(\mathbf{H}, s, \delta)$ the set $u(X, s, \delta)$ for every $\delta \in s$. Since $\sim^t \subset \sim^s_{\mathbf{M}}$ and, therefore, $X \sim^s_m Y$ for every $X, Y \in \mathbf{H}$, by Lemma 1.1.4 the sets $2^s_{\mathbf{H}}$ and $u(\mathbf{H}, s, \delta)$ do not depend on a particular choise of a representative from $\mathbf{H}$.

For every $u \in \mathcal{P}(2^s_{\mathbf{H}})$ we denote by $\mathrm{T}_{(s,u)}(\mathbf{H})$ the set of all elements $\mathbf{a}$ of T for which there exists an element (x, X) of $\mathbf{a}$ such that

$$X \in \mathbf{H} \text{ and } x \in X_{(s,u)}.$$

If $u = u(\mathbf{H}, s, \delta)$ for some $\delta \in s$ (and, therefore, $X_{(s,u)} = X_{(\delta,0)}$ for every $X \in \mathbf{H}$), then the set $\mathrm{T}_{(s,u)}(\mathbf{H})$ is also denoted by $\mathrm{T}_{(\delta,0)}(\mathbf{H})$. (We note that since the set $\mathrm{T}_{(\delta,0)}(\mathbf{H})$ coincides with the set of all elements $\mathbf{a}$ of T such that there exists an element (x, X) of $\mathbf{a}$ for which $X \in \mathbf{H}$ and $x \in X_{(\delta,0)}$, this set is independent of the element s of $\mathcal{F}$ for which $\delta \in s$ and $\sim^t \subset \sim^s_{\mathbf{M}}$.)

Finally we set

$$\mathrm{A}^{\mathbf{H}}_s = \{\mathrm{T}_{(s,u)}(\mathbf{H}) : u \in \mathcal{P}(2^s_{\mathbf{H}})\}.$$

Since the set $\mathrm{T}_{(s,u)}(\mathbf{H})$ is a subset of the set $\mathrm{T}(\mathbf{H})$, $\mathrm{A}^{\mathbf{H}}_s$ can be considered as a set of subsets of $\mathrm{T}(\mathbf{H})$.

1.2.1 Lemma. *Let $s, t \in \mathcal{F}$, $s \neq \emptyset$, $\sim^t \subset \sim^s_{\mathbf{M}}$, $\mathbf{H} \in \mathrm{C}(\sim^t)$, and $u \in \mathcal{P}(2^s_{\mathbf{H}})$. Then, the set $\mathrm{T}_{(s,u)}(\mathbf{H})$ coincides with the set of all $\mathbf{a} \in \mathrm{T}$ such that for every $(x, X) \in \mathbf{a}$ we have $X \in \mathbf{H}$ and $x \in X_{(s,u)}$.*

Proof. Consider the set $\mathrm{T}_{(s,u)}(\mathbf{H})$. It suffices to prove that if $\mathbf{a} \in \mathrm{T}_{(s,u)}(\mathbf{H})$ and $(y, Y) \in \mathbf{a}$, then $Y \in \mathbf{H}$ and $y \in Y_{(s,u)}$. Let $\mathbf{a} \in T_{(s,u)}(\mathbf{H})$. There exists an element (x, X) of $\mathbf{a}$ such that $X \in \mathbf{H}$ and $x \in X_{(s,u)}$. Also, there exists an element f of u such that $x \in X_{(s,f)}$, that is, $d^X_s(x) = f$. Now, let $(y, Y) \in \mathbf{a}$. Since $(x, X), (y, Y) \in \mathbf{a}$, by definition, $X \sim^t Y$ and $d^X_s(x) = d^Y_s(y) = f$, which means that $Y \in \mathbf{H}$ and $y \in Y_{(s,f)} \subset Y_{(s,u)}$. $\square$

1.2.2 Lemma. *Let $s, t \in \mathcal{F}$, $s \neq \emptyset$, $\sim^t \subset \sim^s_{\mathbf{M}}$, and $\mathbf{H} \in \mathrm{C}(\sim^t)$. Let also $u, v \in \mathcal{P}(2^s_{\mathbf{H}})$ and $w = 2^s_{\mathbf{H}} \setminus u$. Then, we have*

$$\mathrm{T}_{(s,u)}(\mathbf{H}) \cup \mathrm{T}_{(s,v)}(\mathbf{H}) = \mathrm{T}_{(s,u\cup v)}(\mathbf{H}) \text{ and} \tag{1.2.1}$$

$$\mathrm{T}(\mathbf{H}) \setminus \mathrm{T}_{(s,u)}(\mathbf{H}) = \mathrm{T}_{(s,w)}(\mathbf{H}). \tag{1.2.2}$$

Proof. We prove (1.2.1). Let

$$\mathbf{a} \in \mathrm{T}_{(s,u)}(\mathbf{H}) \cup \mathrm{T}_{(s,v)}(\mathbf{H}).$$

Then, there exists $(x, X) \in \mathbf{a}$ such that $X \in \mathbf{H}$ and either $x \in X_{(s,u)}$ or $x \in X_{(s,v)}$. This means that

$$x \in X_{(s,u)} \cup X_{(s,v)} = X_{(s,u\cup v)}.$$

Hence, $\mathbf{a} \in \mathrm{T}_{(s,u\cup v)}(\mathbf{H})$.

Conversely, let $\mathbf{a} \in \mathrm{T}_{(s,u\cup v)}(\mathbf{H})$. Then, there exists $(x, X) \in \mathbf{a}$ such that $X \in \mathbf{H}$ and $x \in X_{(s,u\cup v)} = X_{(s,u)} \cup X_{(s,v)}$, that is either $x \in X_{(s,u)}$ or $x \in X_{(s,v)}$, which means that $\mathbf{a} \in \mathrm{T}_{(s,u)}(\mathbf{H}) \cup \mathrm{T}_{(s,v)}(\mathbf{H})$. This proves relation (1.2.1). Relation (1.2.2) is proved similarly. □

1.2.3 Corollary. *If $s, t \in \mathcal{F}$, $s \neq \emptyset$, $\sim^t \subset \sim^s_{\mathbf{M}}$, and $\mathbf{H} \in \mathrm{C}(\sim^t)$, then the set $\mathrm{A}^{\mathbf{H}}_s$ is an algebra of the set $\mathrm{T}(\mathbf{H})$.* □

1.2.4 Lemma. *Let $s, t \in \mathcal{F}$, $s \neq \emptyset$, $\sim^t \subset \sim^s_{\mathbf{M}}$ and $\mathbf{H} \in \mathrm{C}(\sim^t)$. Then, the mapping $i^s_{\mathbf{H}}$ of the set $\mathcal{P}(2^s_{\mathbf{H}})$ onto the set $\mathrm{A}^{\mathbf{H}}_s$, for which $i^s_{\mathbf{H}}(u) = \mathrm{T}_{(s,u)}(\mathbf{H})$ for every $u \in 2^s_{\mathbf{H}}$, is an isomorphism.*

Proof. By Lemma 1.2.2, $i^s_{\mathbf{H}}$ is a homomorphism. Therefore, it suffices to prove that $i^s_{\mathbf{H}}$ is one-to-one. Let $u, v \in \mathcal{P}(2^s_{\mathbf{H}})$ and $u \neq v$. Let $X \in \mathbf{H}$. Then, $2^s_{\mathbf{H}} = 2^s_X$ and by Lemma 1.1.3, $X_{(s,u)} \neq X_{(s,v)}$. Hence, there exists an element x of X, which belongs to one of the sets $X_{(s,u)}$ and $X_{(s,v)}$ and does not belong to the other. Let $\mathbf{a}$ be the element of T containing the pair (x, X). Then, by the definition of the sets $\mathrm{T}_{(s,u)}(\mathbf{H})$ and Lemma 1.2.1, $\mathbf{a}$ belongs to one of the sets $\mathrm{T}_{(s,u)}(\mathbf{H})$ and $\mathrm{T}_{(s,v)}(\mathbf{H})$ and does not belong to the other, which means that $\mathrm{T}_{(s,u)}(\mathbf{H}) \neq \mathrm{T}_{(s,v)}(\mathbf{H})$, that is the mapping $i^s_{\mathbf{H}}$ is one-to-one. □

1.2.5 Corollary. *Let $s, t \in \mathcal{F}$, $s \neq \emptyset$, $\sim^t \subset \sim^s_{\mathbf{M}}$, $\mathbf{H} \in \mathrm{C}(\sim^t)$, and $X \in \mathbf{H}$. Then, the mapping $i \equiv i^s_{\mathbf{H}} \circ (i^s_X)^{-1}$ is an isomorphism of A^X_s onto $\mathrm{A}^{\mathbf{H}}_s$ such that $i(X_{(\delta,0)}) = \mathrm{T}_{(\delta,0)}(\mathbf{H})$ for every $\delta \in s$.* □

Notation. For every $\delta \in \tau$ and $\mathbf{H} \in \mathrm{C}^{\diamondsuit}(\mathrm{R})$ (in particular, for every $\mathbf{H} \in \mathrm{C}(\mathrm{R})$) we denote by $U_\delta^{\mathrm{T}}(\mathbf{H})$ the set of all $\mathbf{a} \in \mathrm{T}$ for which there exists an element (x, X) of $\mathbf{a}$ such that $X \in \mathbf{H}$ and $x \in U_\delta^X$. Therefore, if for some element s of $\mathcal{F}$ we have $\delta \in s$ and $\mathbf{H} \in \mathrm{C}(\sim^t)$, where $t \in \mathcal{F}$ and $\sim^t \subset \sim^s_{\mathbf{M}}$, then $U_\delta^{\mathrm{T}}(\mathbf{H}) = \mathrm{T}_{(\delta,0)}(\mathbf{H}) \in \mathrm{A}_s^{\mathbf{H}}$. It is easy to see that such an element s of $\mathcal{F}$ exists if and only if $\mathbf{H} \in \mathrm{C}(\sim^t)$, where $t \in \mathcal{F}$ and $\sim^t \subset \sim^{\{\delta\}}_{\mathbf{M}}$.

Notation. Let κ be a subset of τ. We denote by $\mathrm{B}_*^{\mathrm{T}}$ the set of all sets of the form $U_\delta^{\mathrm{T}}(\mathbf{H})$, where $\delta \in \tau$, $\mathbf{H} \in \mathrm{C}(\sim^t)$, and $\sim^t \subset \sim^{\{\delta\}}_{\mathbf{M}}$ and we put

$$\mathrm{B}_{*,\kappa}^{\mathrm{T}} = \{U_\delta^{\mathrm{T}}(\mathbf{H}) \in \mathrm{B}_*^{\mathrm{T}} : \delta \in \kappa\}.$$

1.2.6 Lemma. *The set $\mathrm{B}_*^{\mathrm{T}}$ is a base for a topology on the set* T. *Moreover, if κ is a subset of τ such that for every $X \in \mathbf{S}$ the set*

$$\{U_\delta^X : \delta \in \kappa\}$$

is a base for X, then the set $\mathrm{B}_{*,\kappa}^{\mathrm{T}}$ is a base for the same topology on T.

Proof. Let κ be a subset of τ such that for every $X \in \mathbf{S}$ the set $\{U_\delta^X : \delta \in \kappa\}$ is a base for X. We shall prove that:

(a) the empty set is an element of the set $\mathrm{B}_{*,\kappa}^{\mathrm{T}}$,

(b) the union of all elements of $\mathrm{B}_{*,\kappa}^{\mathrm{T}}$ is the set T, and

(c) if an element $\mathbf{a}$ of T is contained in the intersection of two elements of $\mathrm{B}_*^{\mathrm{T}}$, then there exists an element of $\mathrm{B}_{*,\kappa}^{\mathrm{T}}$ containing $\mathbf{a}$, which is contained in this intersection.

Let $X \in \mathbf{S}$. Since $\{U_\delta^X : \delta \in \kappa\}$ is a base for X, the empty set is an element of this base; and, therefore, there exists $\delta \in \kappa$ such that $U_\delta^X = \emptyset$. Let s be an element of $\mathcal{F}$ such that $\delta \in s$ and $\sim^s \subset \sim^{\{\delta\}}_{\mathbf{M}}$ and let $\mathbf{H}$ be the $\sim^s$-equivalent class of X. Then, for every $Y \in \mathbf{H}$ the algebra A_s^X is isomorphic (by the natural way) to the algebra A_s^Y, which means that $U_\delta^Y = \emptyset$ and, therefore, $U_\delta^{\mathrm{T}}(\mathbf{H}) = \emptyset$. Since $U_\delta^{\mathrm{T}}(\mathbf{H}) \in \mathrm{B}_{*,\kappa}^{\mathrm{T}}$ we have $\emptyset \in \mathrm{B}_{*,\kappa}^{\mathrm{T}}$.

Now, let $\mathbf{a} \in \mathrm{T}$ and $(x, X) \in \mathbf{a}$. There exists an element δ of κ such that $x \in U_\delta^X$. Let t be an element of $\mathcal{F}$ such that $\sim^t \subset \sim^{\{\delta\}}_{\mathbf{M}}$. Denote by $\mathbf{H}$ the $\sim^t$-equivalence class of X. By definition, $\mathbf{a} \in U_\delta^{\mathrm{T}}(\mathbf{H}) \in \mathrm{B}_{*,\kappa}^{\mathrm{T}} \subset \mathrm{B}_*^{\mathrm{T}}$, which means that the union of all elements of $\mathrm{B}_{*,\kappa}^{\mathrm{T}}$ is the set T.

Finally, let $U_\delta^{\mathrm{T}}(\mathbf{H}), U_\varepsilon^{\mathrm{T}}(\mathbf{L}) \in \mathrm{B}_*^{\mathrm{T}}$ and $\mathbf{a} \in U_\delta^{\mathrm{T}}(\mathbf{H}) \cap U_\varepsilon^{\mathrm{T}}(\mathbf{L})$. Suppose that $q, p \in \mathcal{F}$, $\mathbf{H} \in \mathrm{C}(\sim^q)$, and $\mathbf{L} \in \mathrm{C}(\sim^p)$. Then, $X \in \mathbf{H} \cap \mathbf{L}$ and $x \in U_\delta^X \cap U_\varepsilon^X$. Therefore, there exists an element η of κ such that $x \in U_\eta^X \subset U_\delta^X \cap U_\varepsilon^X$. Let

$s = \{\delta, \varepsilon, \eta\}$ and let t be an element of $\mathcal{F}$ such that $\sim^t \subset \sim^s_{\mathbf{M}}$ and $q \cup p \subset t$. Let also $\mathbf{E}$ be the $\sim^t$-equivalence class of X. Then, $U^{\mathrm{T}}_\eta(\mathbf{E}) \in \mathrm{B}^{\mathrm{T}}_{*,\kappa}$ and $\mathbf{E} \subset \mathbf{H} \cap \mathbf{L}$. By definition, $\mathbf{a} \in U^{\mathrm{T}}_\eta(\mathbf{E})$. Since $U^{\mathrm{T}}_\delta(\mathbf{E}) = \mathrm{T}_{(\delta,0)}(\mathbf{E}) \in \mathrm{A}^{\mathbf{E}}_s$, $U^{\mathrm{T}}_\varepsilon(\mathbf{E}) = \mathrm{T}_{(\varepsilon,0)}(\mathbf{E}) \in \mathrm{A}^{\mathbf{E}}_s$, and $U^{\mathrm{T}}_\eta(\mathbf{E}) = \mathrm{T}_{(\eta,0)}(\mathbf{E}) \in \mathrm{A}^{\mathbf{E}}_s$, by Lemma 1.1.1 and Corollary 1.2.5 we have

$$U^{\mathrm{T}}_\eta(\mathbf{E}) \subset U^{\mathrm{T}}_\delta(\mathbf{E}) \cap U^{\mathrm{T}}_\varepsilon(\mathbf{E}) \subset U^{\mathrm{T}}_\delta(\mathbf{H}) \cap U^{\mathrm{T}}_\varepsilon(\mathbf{L}).$$

The proven properties imply that the set $\mathrm{B}^{\mathrm{T}}_*$ is a base for a topology on T and that the set $\mathrm{B}^{\mathrm{T}}_{*,\kappa}$ is also a base for this topology. □

Definition. The set T equiped with the topology for which the set $\mathrm{B}^{\mathrm{T}}_*$ is a base, will be called the *Containing Space for the indexed collection* **S** *corresponding to the co-mark* **M** *and the family* R. Since $|\mathrm{B}^{\mathrm{T}}_*| \leq \tau$, the weight of T is $\leq \tau$. In what follows, whenever we shall refer to the space T we shall mean just this space.

The subspaces of T of the form $\mathrm{T}(\mathbf{L})$, where $\mathbf{L} \in \mathrm{C}^{\diamondsuit}(\mathrm{R})$, will be called *primary subspaces of* T.

1.2.7 Lemma. *Let $\delta \in \tau$ and $\mathbf{H} \in \mathrm{C}^{\diamondsuit}(\mathrm{R}) \setminus \{\emptyset\}$. The following statements are true:*

(1) *There exists an element t of $\mathcal{F}$ such that* **H** *is a union of $\sim^t$-equivalence classes.*

(2) *The set $U^{\mathrm{T}}_\delta(\mathbf{H})$ is open in* T *and coincides with the set of all* $\mathbf{a} \in \mathrm{T}$ *such that for every $(x, X) \in \mathbf{a}$, $X \in \mathbf{H}$ and $x \in U^X_\delta$.*

(3) *The set $\mathrm{T}(\mathbf{H})$ is simultaneously open and closed in* T.

(4) *The set $\mathrm{Cl}_{\mathrm{T}}(U^{\mathrm{T}}_\delta(\mathbf{H}))$ coincides with the set of all* $\mathbf{a} \in \mathrm{T}$ *such that there exists an element (x, X) of* **a** *for which $X \in \mathbf{H}$ and $x \in \mathrm{Cl}_X(U^X_\delta)$, as well as, with the set of all* $\mathbf{a} \in \mathrm{T}$ *such that for every $(x, X) \in \mathbf{a}$, $X \in \mathbf{H}$ and $x \in \mathrm{Cl}_X(U^X_\delta)$.*

(5) *The set $\mathrm{Bd}_{\mathrm{T}}(U^{\mathrm{T}}_\delta(\mathbf{H}))$ coincides with the set of all* $\mathbf{a} \in \mathrm{T}$ *such that there exists an element (x, X) of* **a** *for which $X \in \mathbf{H}$ and $x \in \mathrm{Bd}_X(U^X_\delta)$ as well as with the set of all* $\mathbf{a} \in \mathrm{T}$ *such that for every $(x, X) \in \mathbf{a}$, $X \in \mathbf{H}$ and $x \in \mathrm{Bd}_X(U^X_\delta)$.*

(We note that in general the sets $U^{\mathrm{T}}_\delta(\mathbf{H})$, $\mathrm{Cl}_{\mathrm{T}}(U^{\mathrm{T}}_\delta(\mathbf{H}))$, and $\mathrm{Bd}_{\mathrm{T}}(U^{\mathrm{T}}_\delta(\mathbf{H}))$ do not belong to any algebra $\mathrm{A}^{\mathbf{H}}_s$.)

Proof. (1). Since $\mathrm{C}^{\diamondsuit}(\mathrm{R})$ is the minimal ring of subsets of **S** containing the set $\mathrm{C}(\mathrm{R})$ there exists (see Introductory Remarks) a finite set u consisting of finite subsets of $\mathrm{C}(\mathrm{R})$ such that

$$\mathbf{H} = \cup\{\mathbf{H}_v : v \in u\},$$

where $\mathbf{H}_v$ is the intersection of all elements of $v \in u$. Suppose that

$$v = \{\mathbf{H}_v^1, ..., \mathbf{H}_v^{k_v}\}.$$

For every $i \in \{1, ..., k_v\}$ we denote by s_v^i an element of the set $\mathcal{F}$ such that $\mathbf{H}_v^i \in \mathrm{C}(\sim^{s_v^i})$ and put

$$s_v = \cup\{s_v^i : i \in \{1, ..., k_v\}\}.$$

Let $t = \cup\{s_v : v \in u\}$ and $\mathbf{E}$ be an element of $\mathrm{C}(\sim^t)$ such that $\mathbf{H} \cap \mathbf{E} \neq \emptyset$. Then, there exists an element v of u such that $\mathbf{E} \cap \mathbf{H}_v \neq \emptyset$ and, therefore, $\mathbf{E} \cap \mathbf{H}_v^i \neq \emptyset$ for every $i \in \{1, ..., k_v\}$. By the choice of t, $\mathbf{E} \subset \mathbf{H}_v^i$ and, therefore, $\mathbf{E} \subset \mathbf{H}_v$. Thus, $\mathbf{E} \subset \mathbf{H}$, which means that $\mathbf{H}$ is a (finite) union of elements of $\mathrm{C}(\sim^t)$.

(2). By (1) there exists an element t of $\mathcal{F}$ such that

$$\mathbf{H} = \mathbf{E}_0 \cup ... \cup \mathbf{E}_n,$$

where $\mathbf{E}_i \in \mathrm{C}(\sim^t)$, $i = 0, ..., n$. Moreover, we can suppose that $\sim^t \subset \sim_{\mathbf{M}}^{\{\delta\}}$. Then,

$$U_\delta^{\mathrm{T}}(\mathbf{H}) = U_\delta^{\mathrm{T}}(\mathbf{E}_0) \cup ... \cup U_\delta^{\mathrm{T}}(\mathbf{E}_n).$$

For every $i = 0, ..., n$ the set $U_\delta^{\mathrm{T}}(\mathbf{E}_i)$ belongs to the base $\mathrm{B}_*^{\mathrm{T}}$ and to the set $\mathrm{A}_{\{\delta\}}^{\mathbf{E}_i}$. Therefore, any such set is open in T and by Lemma 1.2.1 coincides with the set of all $\mathbf{a} \in \mathrm{T}$ such that for every $(x, X) \in \mathbf{a}$, $X \in \mathbf{E}_i$ and $x \in U_\delta^X$. This fact means that the set $U_\delta^{\mathrm{T}}(\mathbf{H})$ is also open in T and coincides with the set of all $\mathbf{a} \in \mathrm{T}$, such that for every $(x, X) \in \mathbf{a}$, $X \in \mathbf{H}$ and $x \in U_\delta^X$.

(3). The openness of the set $\mathrm{T}(\mathbf{H})$ follows by statements (1) and (2) and by relation

$$\mathrm{T}(\mathbf{H}) = \cup\{U_\delta^{\mathrm{T}}(\mathbf{H}) : \delta \in \tau\},$$

which is easily proved. The closeness of this set follows by relation

$$\mathrm{T} \setminus \mathrm{T}(\mathbf{H}) = \cup\{\mathrm{T}(\mathbf{E}) : \mathbf{E} \in \mathrm{C}(\sim^t), \mathbf{E} \not\subset \mathbf{H}\},$$

where t is the element of $\mathcal{F}$ determined in (1). This relation is also easy to prove.

(4). Let $\mathbf{a} \in \mathrm{T}$. First we prove that if for some element (x, X) of $\mathbf{a}$, $X \in \mathbf{H}$ and $x \in \mathrm{Cl}_X(U_\delta^X)$, then $\mathbf{a} \in \mathrm{Cl}_\mathrm{T}(U_\delta^\mathrm{T}(\mathbf{H}))$. Indeed, let $(x, X) \in \mathbf{a}$, $X \in \mathbf{H}$ and $x \in \mathrm{Cl}_X(U_\delta^X)$. Suppose that $U_\varepsilon^\mathrm{T}(\mathbf{L})$ is an arbitrary element of B_*^T containing $\mathbf{a}$. Then, $X \in \mathbf{L}$ and $x \in U_\varepsilon^X$. Therefore, there exists a point y of $U_\delta^X \cap U_\varepsilon^X$. Let $\mathbf{b}$ be the point of T containing the pair (y, X). By the definition of the sets $U_\delta^\mathrm{T}(\mathbf{H})$ and $U_\varepsilon^\mathrm{T}(\mathbf{L})$, $\mathbf{b} \in U_\delta^\mathrm{T}(\mathbf{H}) \cap U_\varepsilon^\mathrm{T}(\mathbf{L})$, which means that $\mathbf{a} \in \mathrm{Cl}_\mathrm{T}(U_\delta^\mathrm{T}(\mathbf{H}))$.

Now, suppose that $\mathbf{a} \in \mathrm{Cl}_\mathrm{T}(U_\delta^\mathrm{T}(\mathbf{H}))$. We prove that for every $(x, X) \in \mathbf{a}$, $X \in \mathbf{H}$ and $x \in \mathrm{Cl}_X(U_\delta^X)$. Let $(x, X) \in \mathbf{a}$. Denote by $\mathbf{L}$ the $\sim^t$-equivalence class of X, where t is the element of $\mathcal{F}$ determined in (1). If $\mathbf{L} \cap \mathbf{H} \neq \emptyset$, then $\mathrm{T}(\mathbf{L})$ is an open neighbourhood (see (2)) of $\mathbf{a}$ such that $\mathrm{T}(\mathbf{L}) \cap U_\delta^\mathrm{T}(\mathbf{H}) = \emptyset$, which is a contradiction. Therefore, $X \in \mathbf{L} \subset \mathbf{H}$. On the other hand, if $x \notin \mathrm{Cl}_X(U_\delta^X)$, then there exists an element ε of τ such that $x \in U_\varepsilon^X$ and $U_\varepsilon^X \cap U_\delta^X = \emptyset$. Let s be an element of $\mathcal{F}$ such that $t \subset s$ and $\sim^s \subset \sim_\mathbf{M}^{\{\delta,\varepsilon\}}$, and let $\mathbf{E}$ be the $\sim^s$-equivalence class of X. Then, $\mathbf{E} \subset \mathbf{L}$ and $\mathbf{a} \in U_\varepsilon^\mathrm{T}(\mathbf{E})$. Since $U_\varepsilon^\mathrm{T}(\mathbf{E}) = \mathrm{T}_{(\varepsilon,0)}(\mathbf{E}) \in \mathrm{A}_s^\mathbf{E}$ and $U_\delta^\mathrm{T}(\mathbf{E}) = \mathrm{T}_{(\delta,0)}(\mathbf{E}) \in \mathrm{A}_s^\mathbf{E}$, by Lemma 1.1.1 and Corollary 1.2.5, $U_\varepsilon^\mathrm{T}(\mathbf{E}) \cap U_\delta^\mathrm{T}(\mathbf{E}) = \emptyset$. It is easy to verify that

$$U_\varepsilon^\mathrm{T}(\mathbf{E}) \cap U_\delta^\mathrm{T}(\mathbf{H}) \subset U_\varepsilon^\mathrm{T}(\mathbf{E}) \cap U_\delta^\mathrm{T}(\mathbf{E}).$$

This fact means that $U_\varepsilon^\mathrm{T}(\mathbf{E}) \cap U_\delta^\mathrm{T}(\mathbf{H}) = \emptyset$, which is a contradiction. The proven properties completes the proof of (4).

(5). Let $\mathbf{a} \in \mathrm{Bd}_\mathrm{T}(U_\delta^\mathrm{T}(\mathbf{H}))$ and $(x, X) \in \mathbf{a}$. Since $\mathbf{a}$ is an element of $\mathrm{Cl}_\mathrm{T}(U_\delta^\mathrm{T}(\mathbf{H})) \setminus U_\delta^\mathrm{T}(\mathbf{H})$ by (2) and (4) it follows that $X \in \mathbf{H}$, $x \notin U_\delta^X$, and $x \in \mathrm{Cl}_X(U_\delta^X)$, that is $x \in \mathrm{Bd}_X(U_\delta^X)$.

Now let $\mathbf{a} \in \mathrm{T}$, $(x, X) \in \mathbf{a}$, $x \in \mathrm{Bd}_X(U_\delta^X)$ and $X \in \mathbf{H}$. Since $x \in \mathrm{Cl}_X(U_\delta^X) \setminus U_\delta^X$ by (4) and the definition of the set $U_\delta^\mathrm{T}(\mathbf{H})$ we have $\mathbf{a} \in \mathrm{Cl}_\mathrm{T}(U_\delta^\mathrm{T}(\mathbf{H})) \setminus U_\delta^\mathrm{T}(\mathbf{H})$, which means that $\mathbf{a} \in \mathrm{Bd}_\mathrm{T}(U_\delta^\mathrm{T}(\mathbf{H}))$. The proven properties completes the proof of (5). □

Remark. We note that if $\mathbf{H} = \emptyset$, then the sets $U_\delta^\mathrm{T}(\mathbf{H})$, $\delta \in \tau$, and the set $\mathrm{T}(\mathbf{H})$ are empty; and therefore, properties (2)–(5) of the above lemma are satisfied.

Notation. Let κ be a subset of τ and $\mathbf{L}$ an element of $\mathrm{C}^\diamondsuit(\mathrm{R})$. We set

$$\mathrm{B}_\kappa^\mathrm{T} = \{U_\delta^\mathrm{T}(\mathbf{H}) : \delta \in \kappa \text{ and } \mathbf{H} \in \mathrm{C}(\mathrm{R})\},$$

$$\mathrm{B}_{\diamondsuit,\kappa}^\mathrm{T} = \{U_\delta^\mathrm{T}(\mathbf{H}) : \delta \in \kappa \text{ and } \mathbf{H} \in \mathrm{C}^\diamondsuit(\mathrm{R})\},$$

$$\mathrm{B}^{\mathbf{L}}_{\kappa} = \{U^{\mathrm{T}}_{\delta}(\mathbf{H}) \in \mathrm{B}^{\mathrm{T}}_{\kappa} : \mathbf{H} \subset \mathbf{L}\},$$

$$\mathrm{B}^{\mathbf{L}}_{\diamondsuit,\kappa} = \{U^{\mathrm{T}}_{\delta}(\mathbf{H}) \in \mathrm{B}^{\mathrm{T}}_{\diamondsuit,\kappa} : \mathbf{H} \subset \mathbf{L}\}.$$

If $\kappa = \tau$, then the families $\mathrm{B}^{\mathrm{T}}_{\kappa}$, $\mathrm{B}^{\mathrm{T}}_{\diamondsuit,\kappa}$, $\mathrm{B}^{\mathbf{L}}_{\kappa}$, and $\mathrm{B}^{\mathbf{L}}_{\diamondsuit,\kappa}$ are denoted by B^{T}, $\mathrm{B}^{\mathrm{T}}_{\diamondsuit}$, $\mathrm{B}^{\mathbf{L}}$, and $\mathrm{B}^{\mathbf{L}}_{\diamondsuit}$, respectively.

Remark. It is easy to verify that for every $\mathbf{L} \in \mathrm{C}^{\diamondsuit}(\mathrm{R})$ we have

$$\mathrm{B}^{\mathbf{L}}_{\diamondsuit,\kappa} = \{U \cap \mathrm{T}(\mathbf{L}) : U \in \mathrm{B}^{\mathrm{T}}_{\diamondsuit,\kappa}\}.$$

On the other hand the relation

$$\mathrm{B}^{\mathbf{L}}_{\kappa} = \{U \cap \mathrm{T}(\mathbf{L}) : U \in \mathrm{B}^{\mathrm{T}}_{\kappa}\}$$

in general is not true.

1.2.8 Corollary. *If for a subset κ of τ and for every $X \in \mathbf{S}$ the family*

$$\{U^{X}_{\delta} : \delta \in \kappa\}$$

is a base for X, then the families $\mathrm{B}^{\mathrm{T}}_{\kappa}$ and $\mathrm{B}^{\mathrm{T}}_{\diamondsuit,\kappa}$ are bases for T. Therefore, for every $\mathbf{L} \in \mathrm{C}^{\diamondsuit}(\mathrm{R})$ the families $\mathrm{B}^{\mathbf{L}}_{\kappa}$ and $\mathrm{B}^{\mathbf{L}}_{\diamondsuit,\kappa}$ are bases for $\mathrm{T}(\mathbf{L})$. $\square$

Definition. Suppose that κ is a subset of τ and $\mathbf{L} \in \mathrm{C}^{\diamondsuit}(\mathrm{R})$. The families $\mathrm{B}^{\mathrm{T}}_{\kappa}$ and $\mathrm{B}^{\mathbf{L}}_{\kappa}$ (respectively, $\mathrm{B}^{\mathrm{T}}_{\diamondsuit,\kappa}$ and $\mathrm{B}^{\mathbf{L}}_{\diamondsuit,\kappa}$) are called *$\kappa$-standard* (respectively, *$(\diamondsuit, \kappa)$-standard*) families.

If moreover for every $X \in \mathbf{S}$ the family $\{U^{X}_{\delta} : \delta \in \kappa\}$ is a base for X, then the above κ-standard and $(\diamondsuit, \kappa)$-standard families are called respectively *κ-standard* and *$(\diamondsuit, \kappa)$-standard bases* for the corresponding spaces T and $\mathrm{T}(\mathbf{L})$.

In the case where $\kappa = \tau$, the considered κ-standard and $(\diamondsuit, \kappa)$-standard families or bases are called *standard* and *$\diamondsuit$-standard* families or bases, respectively.

1.2.9 Proposition. *The space T is a T_0-space (of weight $\leq \tau$).*

Proof. Let $\mathbf{a}$ and $\mathbf{b}$ be two distinct elements of T. We need to find an open subset of T, which contains one of the elements $\mathbf{a}$ and $\mathbf{b}$ and does not contain the other.

Let $(x, X) \in \mathbf{a}$ and $(y, Y) \in \mathbf{b}$. First, we consider the case where for some $s \in \mathcal{F}$, the spaces X and Y are not $\sim^s$-equivalent. In this case, if $\mathbf{H}$ is the $\sim^s$-equivalence class of X, then the open set $\mathrm{T}(\mathbf{H})$ (see Lemma 1.2.7) contains $\mathbf{a}$ and does not contain $\mathbf{b}$.

Now, let $X \sim^s Y$ for every $s \in \mathcal{F}$. Since $\mathbf{a} \neq \mathbf{b}$, $f \equiv d_s^X(x) \neq d_s^Y(y) \equiv g$ for some $s \in \mathcal{F} \setminus \{\emptyset\}$. This fact means that there exists $\delta \in s$ such that $f(\delta) \neq g(\delta)$, that is either $f(\delta) = 0$ and $g(\delta) = 1$ or $f(\delta) = 1$ and $g(\delta) = 0$. Let $\mathbf{H}$ be the $\sim^s$-equivalence class of X and Y. Then, the open set $U_\delta^{\mathrm{T}}(\mathbf{H})$ (see Lemma 1.2.7) contains one of the point $\mathbf{a}$ and $\mathbf{b}$ and does not contain the other. Thus, T is a T_0-space. □

Definition. Let $X \in \mathbf{S}$. Then, for every $x \in X$ there exists a uniquely determined element $\mathbf{a}$ of T which contains the pair (x, X). We define a mapping e_{T}^X of X into T setting

$$e_{\mathrm{T}}^X(x) = \mathbf{a}.$$

The following proposition shows that e_{T}^X is an embedding. This embedding will be called *natural.*

1.2.10 Proposition. *For every $X \in \mathbf{S}$ the mapping e_{T}^X is an embedding of X into* T.

Proof. Let $X \in \mathbf{S}$. If x and y are distinct points of X, then there exists an element δ of τ such that the set U_δ^X contains one of these points and does not contain the other. Therefore, if $s \in \mathcal{F}$ and $\delta \in s$, then $d_s^X(x) \neq d_s^X(y)$. This fact means that the pairs (x, X) and (y, X) belong to distinct elements of the space T, that is the mapping e_{T}^X is one-to-one.

Now, let δ be an arbitrary element of τ and $\mathbf{H}$ be an arbitrary element of $\mathrm{C}^\diamondsuit(\mathrm{R})$ such that $X \in \mathbf{H}$. By the definition of the mapping e_{T}^X, the relation $x \in U_\delta^X$ is equivalent to the relation $e_{\mathrm{T}}^X(x) \equiv \mathbf{a} \in U_\delta^{\mathrm{T}}(\mathbf{H})$. Since the non-indexed set $\{U_\delta^X : \delta \in \tau\}$ is a base for X and B^{T} is a base for T, this fact implies that the mapping e_{T}^X and the inverse to e_{T}^X mapping $(e_{\mathrm{T}}^X)^{-1}$ of $e_{\mathrm{T}}^X(X)$ onto X are continuous, which means that e_{T}^X is an embedding. □

The definitions of the space T and the mappings e_{T}^X imply the following consequence.

1.2.11 Corollary. *The following relation is true*

$$\mathrm{T} = \cup\{e_{\mathrm{T}}^X(X) : X \in \mathbf{S}\}. \ \square$$

1.3 Specific subsets of the Containing Spaces

In this section for a given restriction $\mathbf{Q}$ of $\mathbf{S}$ we define a subset of $\mathrm{T}(\mathbf{M}, \mathrm{R})$ denoted by $\mathrm{T}|_{\mathbf{Q}}$. These specific subsets of $\mathrm{T}(\mathbf{M}, \mathrm{R})$ play an important role throught the book.

Agreement 1. In this section it is supposed that $\mathbf{S}$, $\mathbf{M}$, and R are the same as in Section 1.2. In particular,

$$\mathbf{M} \equiv \{\mathbf{M}(X) \equiv \{U_\delta^X : \delta \in \tau\} : X \in \mathbf{S}\} \text{ and}$$

$$\mathrm{R} = \{\sim^s : s \in \mathcal{F}\}.$$

Moreover, it is supposed that an arbitrary restriction

$$\mathbf{Q} \equiv \{Q^X : X \in \mathbf{S}\}$$

of $\mathbf{S}$ is fixed.

We note that, in general, the trace on $\mathbf{Q}$ of the $\mathbf{M}$-standard family of equivalence relations on $\mathbf{S}$ is not the $\mathbf{M}|_{\mathbf{Q}}$-standard family of equivalence relations on $\mathbf{Q}$. This fact justify the following definition.

Definition. The $\mathbf{M}$-admissible family R of equivalence relations on $\mathbf{S}$ is said to be $(\mathbf{M}, \mathbf{Q})$*-admissible* if $\mathrm{R}|_{\mathbf{Q}}$ is an $\mathbf{M}|_{\mathbf{Q}}$-admissible family of equivalence relations on $\mathbf{Q}$.

1.3.1 Lemma. *The* $\mathbf{M}$*-admissible family* R *of equivalence relations on* $\mathbf{S}$ *is* $(\mathbf{M}, \mathbf{Q})$*-admissible if and only if for every element* s *of* $\mathcal{F} \setminus \{\emptyset\}$ *there exists an element* t *of* $\mathcal{F} \setminus \{\emptyset\}$ *such that for every* $X, Y \in \mathbf{S}$ *the relation* $X \sim^t Y$ *implies relation* $d_s^X(Q^X) = d_s^Y(Q^Y)$.

Proof. Suppose that the condition of the lemma is satisfied. We need to prove that the family R is $(\mathbf{M}, \mathbf{Q})$-admissible. Since R is $\mathbf{M}$-admissible it suffices to prove that the family

$$\mathrm{R}|_{\mathbf{Q}} = \{\sim^s|_{\mathbf{Q}} : s \in \mathcal{F}\}$$

is a final refinement of the $\mathbf{M}|_{\mathbf{Q}}$-standard family of equivalence relations on $\mathbf{Q}$. Denote the last family by

$$\mathrm{R}_{\mathbf{Q}} \equiv \{\sim^s_{\mathbf{Q}} : s \in \mathcal{F}\}.$$

Let $s \in \mathcal{F} \setminus \{\emptyset\}$. Denote by t an element of $\mathcal{F}$ satisfying the condition of the lemma. We prove that the trace on $\mathbf{Q}$ of the equivalence relation $\sim^t$ on $\mathbf{S}$, that is the equivalence relation $\sim^t|_{\mathbf{Q}}$ on $\mathbf{Q}$ is contained in the equivalence relation $\sim^s_{\mathbf{Q}}$ on $\mathbf{Q}$. Indeed, let $Q^X, Q^Y \in \mathbf{Q}$ such that $Q^X \sim^t|_{\mathbf{Q}} Q^Y$. This means that $X \sim^t Y$. By the condition of the lemma, the last relation implies that $d^X_s(Q^X) = d^Y_s(Q^Y)$. By Lemma 1.1.5, $d^{Q^X}_s(Q^X) = d^{Q^Y}_s(Q^Y)$ and by Lemma 1.1.4, $Q^X \sim^s_{\mathbf{Q}} Q^Y$. Thus, the equivalence relation $\sim^t|_{\mathbf{Q}}$ is contained in the equivalence relation $\sim^s_{\mathbf{Q}}$.

Conversely, suppose that the family R is $(\mathbf{M}, \mathbf{Q})$-admissible. Let $s \in \mathcal{F} \setminus \{\emptyset\}$. Since the family $\mathrm{R}|_{\mathbf{Q}}$ is $\mathbf{M}|_{\mathbf{Q}}$-admissible there exists an element t of $\mathcal{F} \setminus \{\emptyset\}$ such that the equivalence relation $\sim^t|_{\mathbf{Q}}$ on $\mathbf{Q}$ is contained in the equivalence relation $\sim^s_{\mathbf{Q}}$ on $\mathbf{Q}$.

Now, let $X, Y \in \mathbf{S}$ such that $X \sim^t Y$. Then, $Q^X \sim^t|_{\mathbf{Q}} Q^Y$ and by the above $Q^X \sim^s_{\mathbf{Q}} Q^Y$. By Lemma 1.1.4 this means that $d^{Q^X}_s(Q^X) = d^{Q^Y}_s(Q^Y)$. This relation and Lemma 1.1.5 imply that $d^X_s(Q^X) = d^Y_s(Q^Y)$. □

Remark. Lemma 1.3.1 implies the existence of $(\mathbf{M}, \mathbf{Q})$-admissible families of equivalence relations on $\mathbf{S}$. For example, such a family is the admissible family $\mathrm{R}_0 \equiv \{\sim^s_0 : s \in \mathcal{F}\}$ for which $X \sim^s_0 Y$ if and only if $X \sim^s_{\mathbf{M}} Y$ and $d^X_s(Q^X) = d^Y_s(Q^Y)$.

Notation. Suppose that R is an $(\mathbf{M}, \mathbf{Q})$-admissible family of equivalence relations on $\mathbf{S}$. Then, besides of the space T we can also consider the Containing Space $\mathrm{T}(\mathbf{M}|_{\mathbf{Q}}, \mathrm{R}|_{\mathbf{Q}})$ for the indexed collection $\mathbf{Q}$ of spaces corresponding to the co-mark

$$\mathbf{M}|_{\mathbf{Q}} = \{\{U^{Q^X}_\delta = Q \cap U^X_\delta : \delta \in \tau\} : Q^X \in \mathbf{Q}\}$$

and the $\mathbf{M}|_{\mathbf{Q}}$-admissible family $\mathrm{R}|_{\mathbf{Q}}$ of equivalence relations on $\mathbf{Q}$. This Containing Space is denoted briefly by $\mathrm{T}|_{\mathbf{Q}}$.

Let $\mathbf{L}$ be an element of $\mathrm{C}^{\diamondsuit}(\mathrm{R})$ and $\mathbf{E} = \mathbf{L}|_{\mathbf{Q}}$. Then, the subspace $\mathrm{T}(\mathbf{M}|_{\mathbf{Q}}, \mathrm{R}|_{\mathbf{Q}}, \mathbf{L}|_{\mathbf{Q}})$ of $\mathrm{T}(\mathbf{M}|_{\mathbf{Q}}, \mathrm{R}|_{\mathbf{Q}})$ is denoted also by $\mathrm{T}(\mathbf{E}) \equiv \mathrm{T}(\mathbf{L}|_{\mathbf{Q}})$.

For the Containing Space $\mathrm{T}|_{\mathbf{Q}}$ the standard, κ-standard, $\diamondsuit$-standard, and $(\diamondsuit, \kappa)$-standard families or bases, where $\kappa \subset \tau$, are denoted by $\mathrm{B}^{\mathrm{T}|_{\mathbf{Q}}}$, $\mathrm{B}^{\mathrm{T}|_{\mathbf{Q}}}_\kappa$, $\mathrm{B}^{\mathrm{T}|_{\mathbf{Q}}}_\diamondsuit$, and $\mathrm{B}^{\mathrm{T}|_{\mathbf{Q}}}_{\diamondsuit,\kappa}$, respectively. If $\mathbf{E}$ is an element of $\mathrm{C}^{\diamondsuit}(\mathrm{R}|_{\mathbf{Q}})$, then by $\mathrm{B}^{\mathbf{E}}$, $\mathrm{B}^{\mathbf{E}}_\kappa$, $\mathrm{B}^{\mathbf{E}}_\diamondsuit$, and $\mathrm{B}^{\mathbf{E}}_{\diamondsuit,\kappa}$ we denote the corresponding families or bases for the subspace $\mathrm{T}(\mathbf{E})$ of $\mathrm{T}|_{\mathbf{Q}}$. The elements of all the above families or bases are denoted by $U^{\mathrm{T}|_{\mathbf{Q}}}_\delta(\mathbf{H})$, where $\delta \in \tau$ and $\mathbf{H} \in \mathrm{C}^{\diamondsuit}(\mathrm{R}|_{\mathbf{Q}})$.

Agreement 2. In the rest of this section it is assumed that R is $(\mathbf{M}, \mathbf{Q})$-admissible.

1.3.2 Lemma. *For every element* $\mathbf{b}$ *of* $\mathrm{T}|_{\mathbf{Q}}$ *there exists a unique element* $\mathbf{a}$ *of* T *such that for every* $x \in Q^X$ *the pair* (x, Q^X) *belongs to* $\mathbf{b}$ *if and only if the pair* (x, X) *belongs to* $\mathbf{a}$.

Proof. Let $\mathbf{b} \in \mathrm{T}|_{\mathbf{Q}}$. Consider an element (y, Q^Y) of $\mathbf{b}$ and denote by $\mathbf{a}$ the element of T containing the pair (y, Y).

Now, let (x, Q^X) be also an element of $\mathbf{b}$. Then,

$$Q^X \sim^s|_{\mathbf{Q}} Q^Y \text{ and } d_s^{Q^X}(x) = d_s^{Q^Y}(y) \tag{1.3.1}$$

for every $s \in \mathcal{F} \setminus \{\emptyset\}$. By Lemma 1.1.5,

$$d_s^X(x) = d_s^{Q^X}(x) \text{ and } d_s^Y(y) = d_s^{Q^Y}(y). \tag{1.3.2}$$

Therefore,

$$X \sim^s Y \text{ and } d_s^X(x) = d_s^Y(y) \tag{1.3.3}$$

for every $s \in \mathcal{F} \setminus \{\emptyset\}$. This fact means that the pairs (x, X) and (y, Y) belong to the same element of the set T. Thus, $(x, X) \in \mathbf{a}$.

Conversely, let $x \in Q^X$ and $(x, X) \in \mathbf{a}$. Then, relation (1.3.3) is true for every $s \in \mathcal{F} \setminus \{\emptyset\}$. Lemma 1.1.5 implies relation (1.3.2). Relations (1.3.2) and (1.3.3) imply relation (1.3.1) for every $s \in \mathcal{F} \setminus \{\emptyset\}$, which means that the pairs (x, Q^X) and (y, Q^Y) belong to the same element of the set $\mathrm{T}|_{\mathbf{Q}}$. Thus, $(x, Q^X) \in \mathbf{b}$.

It is clear that the element $\mathbf{a}$ of T satisfying the condition of the lemma is uniquely determined. □

Notation. Let $\mathbf{b}$ be an arbitrary element of $\mathrm{T}|_{\mathbf{Q}}$ and let $\mathbf{a}$ be the unique element of T satisfying the condition of Lemma 1.3.2. We denote by $e_{\mathrm{T}}^{\mathrm{T}|_{\mathbf{Q}}}$ the mapping of $\mathrm{T}|_{\mathbf{Q}}$ into T such that $e_{\mathrm{T}}^{\mathrm{T}|_{\mathbf{Q}}}(\mathbf{b}) = \mathbf{a}$.

1.3.3 Lemma. *Let* $\delta \in \tau$, $\mathbf{H} \in \mathrm{C}^{\diamondsuit}(\mathrm{R})$, *and* $\mathbf{L} = \mathbf{H}|_{\mathbf{Q}}$. *Then, an element* $\mathbf{b}$ *of* $\mathrm{T}|_{\mathbf{Q}}$ *belongs to* $U_\delta^{\mathrm{T}|_{\mathbf{Q}}}(\mathbf{L})$ *if and only if the element* $\mathbf{a} \equiv e_{\mathrm{T}}^{\mathrm{T}|_{\mathbf{Q}}}(\mathbf{b})$ *belongs to* $U_\delta^{\mathrm{T}}(\mathbf{H})$.

Proof. For every $X \in \mathbf{S}$ the indexed set

$$\{U_\delta^{Q^X} \equiv Q^X \cap U_\delta^X : \delta \in \tau\}$$

is the trace on Q^X of the mark $\mathbf{M}(X)$. Let $\mathbf{b} \in \mathrm{T}|_{\mathbf{Q}}$ and $e_{\mathrm{T}}^{\mathrm{T}|\mathbf{Q}}(\mathbf{b}) = \mathbf{a} \in \mathrm{T}$. Suppose that $\mathbf{b} \in U_\delta^{\mathrm{T}|\mathbf{Q}}(\mathbf{L})$ and let $(x, Q^X) \in \mathbf{b}$. By Lemma 1.3.2, $(x, X) \in \mathbf{a}$. Then, $Q^X \in \mathbf{L}$ and $x \in U_\delta^{Q^X} = Q^X \cap U_\delta^X$ and, therefore, $X \in \mathbf{H}$ and $x \in U_\delta^X$, which means that $\mathbf{a} \in U_\delta^{\mathrm{T}}(\mathbf{H})$.

Similarly we prove that if $\mathbf{a} \in U_\delta^{\mathrm{T}}(\mathbf{H})$, then $\mathbf{b} \in U_\delta^{\mathrm{T}|\mathbf{Q}}(\mathbf{L})$. □

1.3.4. Proposition. *The mapping* $e_{\mathrm{T}}^{\mathrm{T}|\mathbf{Q}}$ *is an embedding of the space* $\mathrm{T}|_{\mathbf{Q}}$ *into the space* T.

Proof. First, we prove that the mapping $e_{\mathrm{T}}^{\mathrm{T}|\mathbf{Q}}$ is one-to-one. Indeed, let $\mathbf{b}_1$ and $\mathbf{b}_2$ be two distinct points of the space $\mathrm{T}|_{\mathbf{Q}}$. Since $\mathrm{T}|_{\mathbf{Q}}$ is a T_0-space (see Proposition 1.2.9) there exist elements δ of τ and $\mathbf{L}$ of $\mathrm{C}^{\diamondsuit}(\mathrm{R}|_{\mathbf{Q}})$ such that one of the points $\mathbf{b}_1$ and $\mathbf{b}_2$ belongs to the open set $U_\delta^{\mathrm{T}|\mathbf{Q}}(\mathbf{L})$ and the other does not belong to this set. Let $\mathbf{H}$ be the element of $\mathrm{C}^{\diamondsuit}(\mathrm{R})$ such that $\mathbf{L}$ is the trace on $\mathbf{Q}$ of $\mathbf{H}$. By Lemma 1.3.3 one of the points $\mathbf{a}_1 \equiv e_{\mathrm{T}}^{\mathrm{T}|\mathbf{Q}}(\mathbf{b}_1)$ and $\mathbf{a}_2 \equiv e_{\mathrm{T}}^{\mathrm{T}|\mathbf{Q}}(\mathbf{b}_2)$ belongs to the set $U_\delta^{\mathrm{T}}(\mathbf{H})$ and the other does not belong to this set. This means that $\mathbf{a}_1 \neq \mathbf{a}_2$, that is $e_{\mathrm{T}}^{\mathrm{T}|\mathbf{Q}}$ is one-to-one.

The continuity of the mappings $e_{\mathrm{T}}^{\mathrm{T}|\mathbf{Q}}$ and $(e_{\mathrm{T}}^{\mathrm{T}|\mathbf{Q}})^{-1}$ follows by Lemma 1.3.3 and by the fact that the elements of the base $\mathrm{B}_{\diamondsuit}^{\mathrm{T}|\mathbf{Q}}$ of the space $\mathrm{T}|_{\mathbf{Q}}$ has the form $U_\delta^{\mathrm{T}|\mathbf{Q}}(\mathbf{L})$ and the elements of the base $\mathrm{B}_{\diamondsuit}^{\mathrm{T}}$ of the space T has the form $U_\delta^{\mathrm{T}}(\mathbf{H})$. Thus, the mapping $e_{\mathrm{T}}^{\mathrm{T}|\mathbf{Q}}$ is an embedding. □

Definition. The embedding $e_{\mathrm{T}}^{\mathrm{T}|\mathbf{Q}}$ is called the *natural embedding* of $\mathrm{T}(\mathbf{M}|_{\mathbf{Q}}, \mathrm{R}|_{\mathbf{Q}})$ into $\mathrm{T}(\mathbf{M}, \mathrm{R})$.

In what follows we identify a point $\mathbf{b}$ of $\mathrm{T}(\mathbf{M}|_{\mathbf{Q}}, \mathrm{R}|_{\mathbf{Q}})$ with the point $e_{\mathrm{T}}^{\mathrm{T}|\mathbf{Q}}(\mathbf{b}) \equiv \mathbf{a}$ of $\mathrm{T}(\mathbf{M}, \mathrm{R})$ and consider the space $\mathrm{T}(\mathbf{M}|_{\mathbf{Q}}, \mathrm{R}|_{\mathbf{Q}})$ as a subspace of the space $\mathrm{T}(\mathbf{M}, \mathrm{R})$ and the embedding $e_{\mathrm{T}}^{\mathrm{T}|\mathbf{Q}}$ as the identical embedding of $\mathrm{T}|_{\mathbf{Q}}$ into T. The subsets of this form will be called *specific* subsets of $\mathrm{T}(\mathbf{M}, \mathrm{R})$.

1.3.5 Lemma. *Let* $X \in \mathbf{S}$ *and let* $e_X^{Q^X}$ *be the identical embedding of* Q^X *into* X. *Then,*

$$e_{\mathrm{T}}^{\mathrm{T}|\mathbf{Q}} \circ e_{\mathrm{T}|\mathbf{Q}}^{Q^X} = e_{\mathrm{T}}^X \circ e_X^{Q^X}. \tag{1.3.4}$$

Proof. Let $x \in Q^X$. Let also

$$(e_{\mathrm{T}}^X \circ e_X^{Q^X})(x) = e_{\mathrm{T}}^X(x) \equiv \mathbf{a}.$$

Then, $\mathbf{a}$ is the point of T containing the pair (x, X). On the other hand, by the definition of the mapping $e_{\mathrm{T}|\mathbf{Q}}^{Q^X}$, $e_{\mathrm{T}|\mathbf{Q}}^{Q^X}(x) \equiv \mathbf{b}$ is the point of $\mathrm{T}|_{\mathbf{Q}}$ containing the pair (x, Q^X). By the construction of the embedding $e_{\mathrm{T}}^{\mathrm{T}|\mathbf{Q}}$ we have $e_{\mathrm{T}}^{\mathrm{T}|\mathbf{Q}}(\mathbf{b}) = \mathbf{a}$. This proves relation (1.3.4). □

The definitions of the mappings e_{T}^X and $e_{\mathrm{T}}^{\mathrm{T}|\mathbf{Q}}$ and the definitions of the subsets $\mathrm{T}(\mathbf{H})$, $U_\delta^{\mathrm{T}}(\mathbf{H})$, and $U_\delta^{\mathrm{T}|\mathbf{Q}}(\mathbf{H}|_{\mathbf{Q}})$ imply the following two corollaries, which are proved similarly to Lemma 1.3.5.

1.3.6. Corollary. *The following relations are true:*

$$\mathrm{T}|_{\mathbf{Q}} = \cup\{e_{\mathrm{T}}^X(Q^X) : X \in \mathbf{S}\} \text{ and}$$

$$\mathrm{T}(\mathbf{H}|_{\mathbf{Q}}) = \cup\{e_{\mathrm{T}}^X(Q^X) : X \in \mathbf{H}\}$$

for every $\mathbf{H} \in \mathrm{C}^{\diamond}(\mathrm{R})$. □

1.3.7 Corollary. *For every* $\delta \in \tau$ *and for every* $\mathbf{H} \in \mathrm{C}^{\diamond}(\mathrm{R})$ *we have*

$$\mathrm{T}(\mathbf{M}|_{\mathbf{Q}}, \mathrm{R}|_{\mathbf{Q}}, \mathbf{H}|_{\mathbf{Q}}) = \mathrm{T}(\mathbf{M}|_{\mathbf{Q}}, \mathrm{R}|_{\mathbf{Q}}) \cap \mathrm{T}(\mathbf{H}) \text{ and}$$

$$\mathrm{T}|_{\mathbf{Q}} \cap U_\delta^{\mathrm{T}}(\mathbf{H}) = U_\delta^{\mathrm{T}|\mathbf{Q}}(\mathbf{H}|_{\mathbf{Q}}). \ \square$$

Definitions. Let Q be a subset of a space X. A family B of open subset of X is said to be a *positional base* or briefly a *p-base* for Q in X if the empty set is an element of B and the set $\{Q \cap V : V \in B\}$ is a base for the open subsets of the subspace Q.

A family B of open subset of X is said to be a *pos-base* for Q in X if the empty set is an element of B and for every $x \in Q$ and an open neighbourhood U of x in X there exists an element V of B such that $x \in V \subset U$.

Any base for the open subsets of X is called also a *ps-base* for Q in X.

The proof of the next lemma is similar to the proof of Lemma 1.2.6.

1.3.8 Lemma. *Let κ be a subset of τ such that the family*

$$\{U_\delta^X : \delta \in \kappa\}$$

is a p-base (respectively, a pos-base) for Q^X in $X \in \mathbf{S}$. Then, the family $\mathrm{B}_\kappa^{\mathrm{T}}$ and, therefore, the family $\mathrm{B}_{\diamondsuit,\kappa}^{\mathrm{T}}$ is a p-base (respectively, a pos-base) for $\mathrm{T}|_{\mathbf{Q}}$ in T. (We note that κ is independent of X) $\square$

1.3.9 Corollary. *Under the conditions of Lemma 1.3.8, for every $\mathbf{L} \in \mathrm{C}^{\diamondsuit}(\mathrm{R})$ the family $\mathrm{B}_\kappa^{\mathbf{L}}$; and, therefore, the family $\mathrm{B}_{\diamondsuit,\kappa}^{\mathbf{L}}$ is a p-base (respectively, a pos-base) for $\mathrm{T}(\mathbf{M}|_{\mathbf{Q}}, \mathrm{R}|_{\mathbf{Q}}, \mathbf{L}|_{\mathbf{Q}})$ in $\mathrm{T}(\mathbf{M}, \mathrm{R}, \mathbf{L})$.* $\square$

1.4 Commutative properties of the specific subspaces

Agreement. It is supposed that $\mathbf{S}$, $\mathbf{M}$, R, and $\mathbf{Q}$ are the same as in Section 1.3 (see Agreements 1 and 2). In particular, it is supposed that the family R is $(\mathbf{M}, \mathbf{Q})$-admissible.

1.4.1 Lemma. *Let $\mathbf{K} \equiv \{K^X : X \in \mathbf{S}\}$ be a restriction of $\mathbf{S}$ such that $K^X \subset Q^X$. (Therefore, $\mathbf{K}$ can be also considered as a restriction of $\mathbf{Q}$). If the family R of equivalence relations on $\mathbf{S}$ is $(\mathbf{M}, \mathbf{Q})$-admissible and $(\mathbf{M}, \mathbf{K})$-admissible, then the family $\mathrm{R}|_{\mathbf{Q}}$ of equivalence relations on $\mathbf{Q}$ is $(\mathbf{M}|_{\mathbf{Q}}, \mathbf{K})$-admissible.*

Proof. Let R be $(\mathbf{M}, \mathbf{Q})$-admissible and $(\mathbf{M}, \mathbf{K})$-admissible. Then, by definition, $\mathrm{R}|_{\mathbf{Q}}$ is $\mathbf{M}|_{\mathbf{Q}}$-admissible. We need to prove that the family $(\mathrm{R}|_{\mathbf{Q}})|_{\mathbf{K}}$ is $(\mathbf{M}|_{\mathbf{Q}})|_{\mathbf{K}}$-admissible. It is easy to verify that $(\mathrm{R}|_{\mathbf{Q}})|_{\mathbf{K}} = \mathrm{R}|_{\mathbf{K}}$ and $(\mathbf{M}|_{\mathbf{Q}})|_{\mathbf{K}} = \mathbf{M}|_{\mathbf{K}}$. Therefore, it suffices to prove that the family $\mathrm{R}|_{\mathbf{K}}$ is $\mathbf{M}|_{\mathbf{K}}$-admissible. But this follows immediately by the condition of the lemma that R is $(\mathbf{M}, \mathbf{K})$-admissible. $\square$

Notation. Let $\mathbf{K} \equiv \{K^X : X \in \mathbf{S}\}$ be a restriction of $\mathbf{S}$ such that $K^X \subset Q^X$. Suppose that the family R is $(\mathbf{M}, \mathbf{Q})$-admissible and $(\mathbf{M}, \mathbf{K})$-admissible. Then, by Lemma 1.4.1, $\mathrm{T}|_{\mathbf{K}}$ can be considered as a subset of $\mathrm{T}|_{\mathbf{Q}}$. The corresponding identical embedding is denoted by $e_{\mathrm{T}|_{\mathbf{Q}}}^{\mathrm{T}|_{\mathbf{K}}}$.

The proof of the next lemma is similar to the proof of Lemma 1.3.5.

1.4.2 Lemma. *Under the conditions of Lemma 1.4.1 the following equality is true:*

$$e_{\mathrm{T}}^{\mathrm{T}|\mathbf{Q}} \circ e_{\mathrm{T}|\mathbf{Q}}^{\mathrm{T}|\mathbf{K}} = e_{\mathrm{T}}^{\mathrm{T}|\mathbf{K}}. \quad \square$$

Definition. A restriction $\mathbf{F} \equiv \{F^X : X \in \mathbf{S}\}$ is said to be an $(\mathbf{M}, \mathrm{R})$-*complete restriction* if the family R is $(\mathbf{M}, \mathbf{F})$-admissible and the subset $\mathrm{T}|_{\mathbf{F}}$ of T satisfies the following condition: for every point $\mathbf{a}$ of $\mathrm{T}|_{\mathbf{F}}$ and for every element (x, X) of $\mathbf{a}$ we have $x \in F^X$.

Definition. Suppose that for every element λ of a set Λ,

$$\mathbf{F}(\lambda) \equiv \{F^X(\lambda) : X \in \mathbf{S}\}$$

is a restriction of $\mathbf{S}$. The *union* (respectively, the *intersection*) *of the restrictions* $\mathbf{F}(\lambda)$ is the restriction $\{F^X : X \in \mathbf{S}\}$ of $\mathbf{S}$ for which

$$F^X = \cup\{F^X(\lambda) : \lambda \in \Lambda\}$$

(respectively,

$$F^X = \cap\{F^X(\lambda) : \lambda \in \Lambda\})$$

for every $X \in \mathbf{S}$. The defined restrictions are denoted respectively by

$$\vee\{\mathbf{F}(\lambda) : \lambda \in \Lambda\} \text{ and } \wedge\{\mathbf{F}(\lambda) : \lambda \in \Lambda\}.$$

1.4.3 Lemma. *Suppose that for every element λ of a set Λ of cardinality $\leq \tau$ an $(\mathbf{M}, \mathrm{R})$-complete restriction $\mathbf{F}(\lambda)$ of $\mathbf{S}$ is given and let $\mathbf{F}$ be either the restriction $\wedge\{\mathbf{F}(\lambda) : \lambda \in \Lambda\}$ or the restriction $\vee\{\mathbf{F}(\lambda) : \lambda \in \Lambda\}$. If the family R is $(\mathbf{M}, \mathbf{F})$-admissible, then $\mathbf{F}$ is an $(\mathbf{M}, \mathrm{R})$-complete restriction.*

Proof. Let

$$\mathbf{F}(\lambda) = \{F^X(\lambda) : X \in \mathbf{S}\},$$

$$\mathbf{F} = \wedge\{\mathbf{F}(\lambda) : \lambda \in \Lambda\} \equiv \{F^X : X \in \mathbf{S}\},$$

$$\mathbf{a} \in \mathrm{T}|_{\mathbf{F}} \equiv \mathrm{T}(\mathbf{M}|_{\mathbf{F}}, \mathrm{R}|_{\mathbf{F}}), \text{ and } (x, X) \in \mathbf{a}.$$

There exists a pair (y, Y) of $\mathbf{a}$ such that $y \in F^Y$. Therefore, $y \in F^Y(\lambda)$ for every $\lambda \in \Lambda$. By Corollary 13.6, $\mathbf{a} \in \mathrm{T}|_{\mathbf{F}(\lambda)}$. Since by assumption $\mathbf{F}(\lambda)$ is an $(\mathbf{M}, \mathrm{R})$-complete restriction, $x \in F^X(\lambda)$ for every $\lambda \in \Lambda$, which means that $x \in F^X$. Thus, $\mathbf{F}$ is also an $(\mathbf{M}, \mathrm{R})$-complete restriction. The case where $\mathbf{F} = \vee\{\mathbf{F}(\lambda) : \lambda \in \Lambda\}$ is proved similarly. □

Notation. We set

$$\mathbf{Cl}(\mathbf{Q}) \equiv \{\mathrm{Cl}_X(Q^X) : X \in \mathbf{S}\},$$

$$\mathbf{Bd}(\mathbf{Q}) \equiv \{\mathrm{Bd}_X(Q^X) : X \in \mathbf{S}\},$$

$$\mathbf{Int}(\mathbf{Q}) \equiv \{\mathrm{Int}_X(Q^X) : X \in \mathbf{S}\},$$

$$\mathbf{Co}(\mathbf{Q}) \equiv \{X \setminus Q^X : X \in \mathbf{S}\}, \text{ and}$$

$$\mathbf{U}_\delta \equiv \{U_\delta^X : X \in \mathbf{S}\},\ \delta \in \tau.$$

Obviously, the indexed sets $\mathbf{Cl}(\mathbf{Q})$, $\mathbf{Bd}(\mathbf{Q})$, $\mathbf{Int}(\mathbf{Q})$, $\mathbf{Co}(\mathbf{Q})$, and $\mathbf{U}_\delta$ are restrictions of $\mathbf{S}$.

Definition. A restriction $\mathbf{F} \equiv \{F^X : X \in \mathbf{S}\}$ of $\mathbf{S}$ is said to be *closed* (respectively, *open*) if for every $X \in \mathbf{S}$, F^X is a closed (respectively, an open) subset of X.

Obviously, the restrictions $\mathbf{Cl}(\mathbf{Q})$, $\mathbf{Bd}(\mathbf{Q})$, and $\mathbf{Co}(\mathbf{U}_\delta)$, for $\delta \in \tau$, are closed and the restrictions $\mathbf{Int}(\mathbf{Q})$ and $\mathbf{U}_\delta$, for $\delta \in \tau$ are open.

1.4.4 Lemma. *For every $\delta \in \tau$ the family* R *is $(\mathbf{M}, \mathbf{U}_\delta)$-admissible and $(\mathbf{M}, \mathbf{Co}(\mathbf{U}_\delta))$-admissible.*

Proof. We prove that R is $(\mathbf{M}, \mathbf{U}_\delta)$-admissible. Let $\delta \in \tau$. Consider an arbitrary element s of $\mathcal{F}$ and set $t = s \cup \{\delta\}$. By Lemma 1.3.1 it suffices to prove that if $X, Y \in \mathbf{S}$ and $X \sim^t Y$, then $d_s^X(U_\delta^X) = d_s^Y(U_\delta^Y)$. Indeed, let $f \in d_s^X(U_\delta^X)$. This means that

$$U_\delta^X \cap (\cap\{X_{(\varepsilon, f(\varepsilon))} : \varepsilon \in s\}) \neq \emptyset.$$

By the choice of t and Lemma 1.1.1,

$$U_\delta^Y \cap (\cap\{Y_{(\varepsilon, f(\varepsilon))} : \varepsilon \in s\}) \neq \emptyset,$$

which means that $f \in d_s^Y(U_\delta^Y)$. Thus, $d_s^X(U_\delta^X) \subset d_s^Y(U_\delta^Y)$. Similarly, $d_s^Y(U_\delta^Y) \subset d_s^X(U_\delta^X)$ and, therefore, $d_s^Y(U_\delta^Y) = d_s^X(U_\delta^X)$. Hence, the family R is $(\mathbf{M}, \mathbf{U}_\delta)$-admissible. Similarly we prove that R is $(\mathbf{M}, \mathbf{Co}(\mathbf{U}_\delta))$-admissible. □

1.4.5 Lemma. *For every $\delta \in \tau$ and $\mathbf{H} \in \mathrm{C}^\diamond(\mathrm{R})$ the following relation is true:*

$$\mathrm{T}|_{\mathbf{U}_\delta} \cap \mathrm{T}(\mathbf{H}) = U_\delta^{\mathrm{T}}(\mathbf{H}).$$

Proof. An element $\mathbf{a}$ of T belongs to $\mathrm{T}|_{\mathbf{U}_\delta} \cap \mathrm{T}(\mathbf{H})$ if and only if there exists an element (x, X) of $\mathbf{a}$ such that $x \in U_\delta^X$ and $X \in \mathbf{H}$, that is if and only if $\mathbf{a} \in U_\delta^{\mathrm{T}}(\mathbf{H})$. □

1.4.6 Lemma. *Suppose that $\mathbf{Q}$ is a closed restriction of $\mathbf{S}$. Then, the following statements are true:*

(1) *The subset $\mathrm{T}|_{\mathbf{Q}}$ of T is closed.*

(2) *If, moreover, R is $(\mathbf{M}, \mathbf{Co}(\mathbf{Q}))$-admissible, then*

$$\mathrm{T}|_{\mathbf{Co}(\mathbf{Q})} = \mathrm{T} \setminus \mathrm{T}|_{\mathbf{Q}}. \tag{1.4.1}$$

(3) *The restriction $\mathbf{Q}$ is an $(\mathbf{M}, \mathrm{R})$-complete restriction.*

Proof. Let $\mathbf{a}$ be a point of T for which there exists an element (x, X) of $\mathbf{a}$ such that $x \notin Q^X$. Since Q^X is closed in X there exists an element δ of τ such that $x \in U_\delta^X$ and $U_\delta^X \cap Q^X = \emptyset$. Let $s = \{\delta\}$.

Since R is $(\mathbf{M}, \mathbf{Q})$-admissible by Lemma 1.3.1 there exists an element t of $\mathcal{F}$ such that $\sim^t \subset \sim_{\mathbf{M}}^s$ and $d_s^X(Q^X) = d_s^Y(Q^Y)$ for every $Y \in \mathbf{S}$ for which $X \sim^t Y$.

Let $\mathbf{H}$ be the $\sim^t$-equivalence class of X. Then, $U_\delta^{\mathrm{T}}(\mathbf{H})$ is an open neighbourhood of $\mathbf{a}$ in T. We prove that

$$U_\delta^{\mathrm{T}}(\mathbf{H}) \cap \mathrm{T}|_{\mathbf{Q}} = \emptyset. \tag{1.4.2}$$

Indeed, in the opposite case there exists a point $\mathbf{b}$ belonging to the set $U_\delta^{\mathrm{T}}(\mathbf{H}) \cap \mathrm{T}|_{\mathbf{Q}}$. Let (y, Q^Y) be an element of $\mathbf{b}$ as a point of $\mathrm{T}|_{\mathbf{Q}}$. By Lemma 1.3.2, $\mathbf{b}$ as a point of the space T contains the pair (y, Y). Since $\mathbf{b} \in U_\delta^T(\mathbf{H})$, we have $Y \in \mathbf{H}$ and $y \in U_\delta^Y$. Therefore, $X \sim^t Y$. By the choice of t, $X \sim_{\mathbf{M}}^s Y$ and $d_s^X(Q^X) = d_s^Y(Q^Y)$. Since $y \in Q^Y$ there exists a point $z \in Q^X$ such that $d_s^X(z) = d_s^Y(y)$. This equality and relation $y \in U_\delta^Y$

imply that $z \in U_\delta^X$. Therefore, $Q^X \cap U_\delta^X \neq \emptyset$, which contradicts the choice of δ. Thus, the relation (1.4.2) is proved.

Now we prove the statements of the lemma.

(1). By Corollary 1.3.6 as the above point $\mathbf{a}$ we can take any point of the set $\mathrm{T} \setminus \mathrm{T}|_{\mathbf{Q}}$. In this case, the relation (1.4.2) implies that the set $\mathrm{T}|_{\mathbf{Q}}$ is closed.

(2). As the point $\mathbf{a}$ we can take any point of the set $\mathrm{T}|_{\mathbf{Co}(\mathbf{Q})}$. In this case, relation (1.4.2) means that $\mathbf{a} \notin \mathrm{T}|_{\mathbf{Q}}$, that is $\mathrm{T}|_{\mathbf{Q}} \cap \mathrm{T}|_{\mathbf{Co}(\mathbf{Q})} = \emptyset$. The Corollary 1.3.6 implies that $\mathrm{T}|_{\mathbf{Q}} \cup \mathrm{T}|_{\mathbf{Co}(\mathbf{Q})} = \mathrm{T}$. The last two relations are equivalent to the relation (1.4.1).

(3). If $\mathbf{Q}$ is not an $(\mathbf{M}, \mathrm{R})$-complete restriction, then there exists a point $\mathbf{a}$ of $\mathrm{T}|_{\mathbf{Q}}$ and an element (x, X) of $\mathbf{a}$ such that $x \notin Q^X$. Then, relation (1.4.2) is true for some open neighbourhood $U_\delta^{\mathrm{T}}(\mathbf{H})$ of $\mathbf{a}$, which is a contradiction. Thus, $\mathbf{Q}$ is an $(\mathbf{M}, \mathrm{R})$-complete restriction. □

1.4.7 Lemma. *Suppose that the restriction* $\mathbf{Q}$ *of* $\mathbf{S}$ *is open and the family* R *is* $(\mathbf{M}, \mathbf{Co}(\mathbf{Q}))$*-admissible. Then:*

(1) *The following relation is true:*

$$\mathrm{T}|_{\mathbf{Co}(\mathbf{Q})} = \mathrm{T} \setminus \mathrm{T}|_{\mathbf{Q}}. \tag{1.4.3}$$

(2) *The subset* $\mathrm{T}|_{\mathbf{Q}}$ *of* T *is open.*

(3) *The restriction* $\mathbf{Q}$ *is an* $(\mathbf{M}, \mathrm{R})$*-complete restriction.*

Proof. (1). The relation (1.4.3) follows immediately by the statement (2) of Lemma 1.4.1 if instead of the restriction $\mathbf{Q}$ we consider the closed restriction $\mathbf{Co}(\mathbf{Q})$. (Note that $\mathbf{Co}(\mathbf{Co}(\mathbf{Q})) = \mathbf{Q}$).

(2). Since the restriction $\mathbf{Co}(\mathbf{Q})$ is closed by Lemma 4.6 the subset $\mathrm{T}|_{\mathbf{Co}(\mathbf{Q})}$ of T is closed. Therefore, by relation (1.4.3) the subset $\mathrm{T}|_{\mathbf{Q}}$ of T is open.

(3). Let $\mathbf{a} \in \mathrm{T}|_{\mathbf{Q}}$ and $(x, X) \in \mathbf{a}$. If $x \notin Q^X$, then $x \in X \setminus Q^X$. By Corollary 1.3.6, $\mathbf{a} \in \mathrm{T}|_{\mathbf{Co}(\mathbf{Q})}$, which contradicts relation (1.4.3). □

1.4.8 Lemma. *Suppose that the family* R *is* $(\mathbf{M}, \mathbf{Cl}(\mathbf{Q}))$*-admissible. Then,*

$$\mathrm{T}|_{\mathbf{Cl}(\mathbf{Q})} = \mathrm{Cl}_{\mathrm{T}}(\mathrm{T}|_{\mathbf{Q}}).$$

Proof. Since the restriction $\mathbf{Cl}(\mathbf{Q})$ is closed by Lemma 1.4.6 the subset $\mathrm{T}|_{\mathbf{Cl}(\mathbf{Q})}$ of T is closed. Therefore, since $\mathrm{T}|_{\mathbf{Q}} \subset \mathrm{T}|_{\mathbf{Cl}(\mathbf{Q})}$, it suffices to prove that $\mathrm{T}|_{\mathbf{Cl}(\mathbf{Q})} \subset \mathrm{Cl}_{\mathrm{T}}(\mathrm{T}|_{\mathbf{Q}})$.

Let $\mathbf{a} \in \mathrm{T}|_{\mathbf{Cl(Q)}}$. Then, by Corollary 1.3.6 there exists an element (x, X) of $\mathbf{a}$ such that $x \in \mathrm{Cl}_X(Q^X)$. Suppose that $\mathbf{a} \notin \mathrm{Cl}_{\mathrm{T}}(\mathrm{T}|_{\mathbf{Q}})$. Then, there exists a neighbourhood $U_\delta^{\mathrm{T}}(\mathbf{H}) \in \mathrm{B}^{\mathrm{T}}$ of $\mathbf{a}$ such that $U_\delta^{\mathrm{T}}(\mathbf{H}) \cap \mathrm{T}|_{\mathbf{Q}} = \emptyset$. Corollary 1.3.6 implies that $U_\delta^Y \cap Q^Y = \emptyset$ for every $Y \in \mathbf{H}$. Since $X \in \mathbf{H}$, $U_\delta^X \cap Q^X = \emptyset$. This fact means that $x \notin \mathrm{Cl}_X(Q^X)$, which is a contradiction. Thus, $\mathbf{a} \in \mathrm{Cl}_{\mathrm{T}}(\mathrm{T}|_{\mathbf{Q}})$ and, therefore, $\mathrm{T}|_{\mathbf{Cl(Q)}} \subset \mathrm{Cl}_{\mathrm{T}}(\mathrm{T}|_{\mathbf{Q}})$. □

1.4.9 Lemma. *Suppose that* $\mathbf{Q}$ *is* $(\mathbf{M}, \mathrm{R})$*-complete and the family* R *is* $(\mathbf{M}, \mathbf{Int}(\mathbf{Q}))$*-admissible and* $(\mathbf{M}, \mathbf{Co}(\mathbf{Int}(\mathbf{Q})))$*-admissible. Then,*

$$\mathrm{T}|_{\mathbf{Int(Q)}} = \mathrm{Int}_{\mathrm{T}}(\mathrm{T}|_{\mathbf{Q}}).$$

Proof. Obviously, the restriction $\mathbf{Int}(\mathbf{Q})$ is open. By assumptions of the lemma and Lemma 1.4.7, the subset $\mathrm{T}|_{\mathbf{Int(Q)}}$ is open in T. Therefore, since $\mathrm{T}|_{\mathbf{Int(Q)}} \subset \mathrm{T}|_{\mathbf{Q}}$, it suffices to prove that $\mathrm{Int}_{\mathrm{T}}(\mathrm{T}|_{\mathbf{Q}}) \subset \mathrm{T}|_{\mathbf{Int(Q)}}$.

Let $\mathbf{a} \in \mathrm{Int}_{\mathrm{T}}(\mathrm{T}|_{\mathbf{Q}})$. There exists an open neighbourhood $U_\delta^{\mathrm{T}}(\mathbf{H}) \in \mathrm{B}^{\mathrm{T}}$ of $\mathbf{a}$ such that $U_\delta^{\mathrm{T}}(\mathbf{H}) \subset \mathrm{Int}_{\mathrm{T}}(\mathrm{T}|_{\mathbf{Q}})$. Let $(x, X) \in \mathbf{a}$. Then, $x \in U_\delta^X$ and $X \in \mathbf{H}$. We prove that $U_\delta^X \subset Q^X$. Indeed in the opposite case there exists a point y belonging to the set $U_\delta^X \setminus Q^X$. Let $\mathbf{b}$ be the point of T containing the pair (y, X). Then, $\mathbf{b} \in U_\delta^{\mathrm{T}}(\mathbf{H})$. Since $\mathbf{Q}$ is an $(\mathbf{M}, \mathrm{R})$-complete restriction, $(y, X) \in \mathbf{b}$, and $y \notin Q^X$ we have $\mathbf{b} \notin \mathrm{T}|_{\mathbf{Q}}$, which contradicts the fact that $U_\delta^{\mathrm{T}}(\mathbf{H}) \subset \mathrm{Int}_{\mathrm{T}}(\mathrm{T}|_{\mathbf{Q}}) \subset \mathrm{T}|_{\mathbf{Q}}$. Thus, $U_\delta^X \subset Q^X$, which means that $x \in \mathrm{Int}_X(Q^X)$ and, therefore, $\mathbf{a} \in \mathrm{T}|_{\mathbf{Int(Q)}}$. □

1.4.10 Lemma. *Suppose that* $\mathbf{Q}$ *is* $(\mathbf{M}, \mathrm{R})$*-complete and the family* R *is* $(\mathbf{M}, \mathbf{Cl}(\mathbf{Q}))$*-admissible,* $(\mathbf{M}, \mathbf{Int}(\mathbf{Q}))$*-admissible,* $(\mathbf{M}, \mathbf{Co}(\mathbf{Int}(\mathbf{Q})))$*-admissible, and* $(\mathbf{M}, \mathbf{Bd}(\mathbf{Q}))$*-admissible. Then,*

$$\mathrm{T}|_{\mathbf{Bd(Q)}} = \mathrm{Bd}_{\mathrm{T}}(\mathrm{T}|_{\mathbf{Q}}). \tag{1.4.4}$$

Proof. Obviously,

$$\mathrm{Bd}_{\mathrm{T}}(\mathrm{T}|_{\mathbf{Q}}) = \mathrm{Cl}_{\mathrm{T}}(\mathrm{T}|_{\mathbf{Q}}) \setminus \mathrm{Int}_{\mathrm{T}}(\mathrm{T}|_{\mathbf{Q}}).$$

Lemmas 1.4.8 and 1.4.9 imply respectively that

$$\mathrm{T}|_{\mathbf{Cl(Q)}} = \mathrm{Cl}_{\mathrm{T}}(\mathrm{T}|_{\mathbf{Q}}) \text{ and } \mathrm{T}|_{\mathbf{Int(Q)}} = \mathrm{Int}_{\mathrm{T}}(\mathrm{T}|_{\mathbf{Q}}).$$

Therefore, it is suffices to prove that

$$\mathrm{T}|_{\mathbf{Bd}(\mathbf{Q})} = \mathrm{T}|_{\mathbf{Cl}(\mathbf{Q})} \setminus \mathrm{T}|_{\mathbf{Int}(\mathbf{Q})}.$$

Let $\mathbf{a} \in \mathrm{T}|_{\mathbf{Bd}(\mathbf{Q})}$. There exists an element (x, X) of $\mathbf{a}$ such that $x \in \mathrm{Bd}_X(Q^X)$. Then, $x \in \mathrm{Cl}_X(Q^X)$ and, therefore, $\mathbf{a} \in \mathrm{T}|_{\mathbf{Cl}(\mathbf{Q})}$. On the other hand, since $\mathbf{Bd}(\mathbf{Q})$ is a closed restriction, by Lemma 1.4.6, $\mathbf{Bd}(\mathbf{Q})$ is an $(\mathbf{M}, \mathrm{R})$-complete restriction. This fact means that for every $(y, Y) \in \mathbf{a}$ we have $y \in \mathrm{Bd}_Y(Q^Y)$, that is $y \notin \mathrm{Int}_Y(Q^Y)$. By Corollary 1.3.6, $\mathbf{a} \notin \mathrm{T}|_{\mathbf{Int}(\mathbf{Q})}$, that is $\mathbf{a} \in \mathrm{T}|_{\mathbf{Cl}(\mathbf{Q})} \setminus \mathrm{T}|_{\mathbf{Int}(\mathbf{Q})}$.

Conversely, let $\mathbf{a} \in \mathrm{T}|_{\mathbf{Cl}(\mathbf{Q})} \setminus \mathrm{T}|_{\mathbf{Int}(\mathbf{Q})}$. Since $\mathbf{Cl}(\mathbf{Q})$ and $\mathbf{Int}(\mathbf{Q})$ are $(\mathbf{M}, \mathrm{R})$-complete restrictions, for every $(x, X) \in \mathbf{a}$ we have

$$x \in \mathrm{Cl}_X(Q^X) \setminus \mathrm{Int}_X(Q^X),$$

that is $x \in \mathrm{Bd}_X(Q^X)$. This fact means that $\mathbf{a} \in \mathrm{T}|_{\mathbf{Bd}(\mathbf{Q})}$. Thus, relation (1.4.4) is proved. □

1.4.11 Corollary. *Suppose that for some $\delta \in \tau$ the family* R *is $(\mathbf{M}, \mathbf{Cl}(\mathbf{U}_\delta))$-admissible and $(\mathbf{M}, \mathbf{Bd}(\mathbf{U}_\delta))$-admissible. Then,*

$$\mathrm{T}|_{\mathbf{Bd}(\mathbf{U}_\delta)} = \mathrm{Bd}_{\mathrm{T}}(\mathrm{T}|_{\mathbf{U}_\delta}).$$

Moreover, for every $\mathbf{L} \in \mathrm{C}^{\diamondsuit}(\mathrm{R})$ we have

$$\mathrm{T}|_{\mathbf{Bd}(\mathbf{U}_\delta)} \cap \mathrm{T}(\mathbf{L}) = \mathrm{Bd}_{\mathrm{T}(\mathbf{L})}(U_\delta^{\mathrm{T}}(\mathbf{L})).$$

Proof. Lemma 1.4.4 implies that the family R is $(\mathbf{M}, \mathbf{U}_\delta)$-admissible, $(\mathbf{M}, \mathbf{Int}(\mathbf{U}_\delta))$-admissible, and $(\mathbf{M}, \mathbf{Co}(\mathbf{Int}(\mathbf{U}_\delta)))$-admissible. By Lemma 1.4.7 the restriction $\mathbf{U}_\delta$ is an $(\mathbf{M}, \mathrm{R})$-complete restriction. (Note that $\mathbf{Int}(\mathbf{U}_\delta) = \mathbf{U}_\delta$.) Therefore, the first part of the corollary follows by Lemma 1.4.10.

The second part of the corollary follows by the first part and Lemma 1.4.5 using the fact that $\mathrm{T}(\mathbf{L})$ is an open and closed subset of T. □

1.4.12 Lemma. *Suppose that $\mathbf{Q}$ is $(\mathbf{M}, \mathrm{R})$-complete and the family* R *is $(\mathbf{M}, \mathbf{Co}(\mathbf{Q}))$-admissible. Then, $\mathbf{Co}(\mathbf{Q})$ is also $(\mathbf{M}, \mathrm{R})$-complete restriction and*

$$\mathrm{T}|_{\mathbf{Co}(\mathbf{Q})} = \mathrm{T} \setminus \mathrm{T}|_{\mathbf{Q}}. \tag{1.4.5}$$

Proof. Let $\mathbf{a} \in \mathrm{T}|_{\mathbf{Co}(\mathbf{Q})}$. There exists an element (x, X) of $\mathbf{a}$ such that $x \in X \setminus Q^X$. Let $(y, Y) \in \mathbf{a}$. If $y \notin Y \setminus Q^Y$, then $\mathbf{a} \in \mathrm{T}|_{\mathbf{Q}}$ and since $\mathbf{Q}$

is a $(\mathbf{M}, \mathrm{R})$-complete restriction we have $x \in Q^X$, which is a contradiction. Therefore, $y \in Y \setminus Q^Y$, which means that $\mathbf{Co}(\mathbf{Q})$ is a $(\mathbf{M}, \mathrm{R})$-complete restriction. This property also implies that $\mathbf{a} \notin \mathrm{T}|_{\mathbf{Q}}$.

By the above $\mathrm{T}|_{\mathbf{Co}(\mathbf{Q})} \subset \mathrm{T} \setminus \mathrm{T}|_{\mathbf{Q}}$. On the other hand $\mathrm{T}|_{\mathbf{Co}(\mathbf{Q})} \cup \mathrm{T}|_{\mathbf{Q}} = \mathrm{T}$. The last two relations imply (1.4.5). □

1.5 A generalization of the construction of Containing Spaces

In this section we generalize the construction of Containing Spaces given in Sections 1.1 and 1.2. This generalization concerns the indexed families R and $\mathrm{R}_{\mathbf{M}}$ of equivalence relations on $\mathbf{S}$. First, instead of the indexing set $\mathcal{F}$ we consider here the set $\mathcal{P}_\nu(\tau)$ consisting of all subsets of τ of cardinality less than a fixed infinite cardinal ν. Furthermore, the elements of the $\mathbf{M}$-standard family is defined here by ν-algebras (see below the definition) instead of algebras. Finally, the property (b) of the definition of an admissible family (see the Introductory Remarks) is replaced by the following: the number of $\sim^s$-equivalence classes is less than a fixed infinite cardinal μ. For Containing Spaces to be T_0-spaces of weight $\leq \tau$ the cardinals ν and μ must satisfy some additional conditions.

The given generalization is carried out in parallel to the construction of Sections 1.1 and 1.2. The definitions, lemmas and propositions considered here have a similar correspondence in mentioned sections. Many of them use the same notations and have the same formulation as the corresponding ones though they are related here to the "infinite case". The proofs of almost all lemmas and propositions are ommited. These proofs are similar to that of the corresponding lemmas and propositions of Sections 1.1 and 1.2.

Agreement 1. In this section we denote by ν and μ two fixed infinite cardinals. It is supposed that the following conditions are satisfied:

(1) The cardinals ν and τ are not π-accessible for any cardinal π less than ν (that is, the sum of less than π many cardinals, which are less than ν or τ is less than ν or τ, respectively).

(2) $|\mathcal{P}_\nu(\tau)| \leq \tau$, (therefore, $\nu \leq \tau$).

(3) $|\mathcal{P}(\mathcal{P}(\pi))| < \mu \leq \tau^+$ for every cardinal π less than ν.

It is easy to see that the above conditions are satisfied if $\nu = \omega$ and μ is an arbitrary cardinal such that $\omega \leq \mu \leq \tau^+$.

Remark. In the case where $\nu = \mu = \omega$, the construction of Containing Spaces given here coincides with that of Sections 1.1 and 1.2. The Containing Spaces corresponding to the case $\nu = \tau = \omega$ and $\mu = \omega^+$ will be used in Chapter 9.

Definitions. By a *ν-algebra of subsets of a set X*, or briefly by a *ν-algebra* of X, we mean a set of subsets of X which is closed under the complements and under the sums of less than ν elements of this set. As for algebras we have: (a) the empty set and the set X are elements of any ν-algebra of X, (b) the set of all subsets of X is a ν-algebra of X, and (c) the intersection of any number of ν-algebras of X is also a ν-algebra of X. Therefore, any given set of subsets of X is contained in a ν-algebra of X. The intersection of all such ν-algebras is called the *minimal ν-algebra* containing the given set of subsets of X.

Let A be a ν-algebra of a set X and i be a mapping of A into the set $\mathcal{P}(Y)$ of all subsets of a set Y. The mapping i is said to be a *homomorphism* if the following conditions are satisfied:

(a) $i(X \setminus U) = Y \setminus i(U)$, $U \in A$, and

(b) $i(\cup\{U_\lambda : \lambda \in \Lambda\}) = \cup\{i(U_\lambda) : \lambda \in \Lambda\}$

for every set Λ of cardinality $< \nu$ and elements U_λ of A. It is easy to verify that if i is an homomorphism, then the set $i(A)$ is a ν-algebra of the set Y.

An one-to-one homomorphism is called an *isomorphism*. It is easy to see that if i is an isomorphism of A onto $i(A) \subset \mathcal{P}(Y)$, then the inverse mapping of $i(A)$ onto $A \subset \mathcal{P}(X)$ is also an isomorphism. It is also clear that the composition of two homomorphisms is a homomorphism; and, therefore, the composition of two isomorphisms is an isomorphism.

The following two lemmas correspond to Lemmas 1.1.1 and 1.1.2 concerning algebras.

1.5.1 Lemma. *Let X and Y be sets, A a ν-algebra of X, and i an isomorphism of A into $\mathcal{P}(Y)$. Suppose that U, V, and U_λ are elements of A, where $\lambda \in \Lambda$ and $|\Lambda| < \nu$. Then, the relations*

$$\cap\{U_\lambda : \lambda \in \Lambda\} = V,\ U = \emptyset,\ U = X, \text{ and } U \subset V$$

are equivalent to the relations

$$\cap\{i(U_\lambda) : \lambda \in \Lambda\} = i(V),\ i(U) = \emptyset,$$

$$i(U) = Y, \text{ and } i(U) \subset i(V),$$

respectively. □

1.5.2 Lemma. *Let X and Y be sets, $\mathcal{G}$ a subset of $\mathcal{P}(X)$, and A the minimal ν-algebra of X containing $\mathcal{G}$. The following statements are true:*

(1) If i is an isomorphism of A into $\mathcal{P}(Y)$, then $i(A)$ is the minimal ν-algebra of Y containing the set $i(\mathcal{G})$.

(2) If h_1 and h_2 are two homomorphisms of A into $\mathcal{P}(Y)$ such that $h_1(U) = h_2(U)$ for every $U \in \mathcal{G}$, then $h_1 = h_2$. □

Definition. Let X be a marked space with a mark $\{U_\delta^X : \delta \in \tau\}$ and let $s \in \mathcal{P}_\nu(\tau) \setminus \{\emptyset\}$. The minimal ν-algebra of X containing the sets U_δ^X and for $\delta \in s$ is called the *s-algebra* of the marked space X and it is denoted by A_s^X.

Notation. Let X be a marked space with a mark $\{U_\delta^X : \delta \in \tau\}$ and let $s \in \mathcal{P}_\nu(\tau) \setminus \{\emptyset\}$. For every $\delta \in \tau$ we set

$$X_{(\delta,0)} = U_\delta^X \text{ and } X_{(\delta,1)} = X \setminus U_\delta^X.$$

For every $f \in 2^s$ we set

$$X_{(s,f)} = \cap\{X_{(\delta,f(\delta))} : \delta \in s\}.$$

We denote by 2_X^s the set of all $f \in 2^s$ such that $X_{(s,f)} \neq \emptyset$.

For every $x \in X$ there exists a unique element f of 2^s such that $x \in X_{(s,f)}$. We define a mapping d_s^X of X into 2^s setting $d_s^X(x) = f$. It is supposed that $d_s^X(X) = \emptyset$ if $X = \emptyset$. Obviously, $d_s^X(X) = 2_X^s$.

For every $\delta \in s$ we set

$$u(X, s, \delta) = \{f \in 2_X^s : f(\delta) = 0\}.$$

For every element u of $\mathcal{P}(2_X^s)$ we set

$$X_{(s,u)} = \cup\{X_{(s,f))} : f \in u\}$$

if $u \neq \emptyset$ and $X_{(s,u)} = \emptyset$ if $u = \emptyset$.

Finally, we denote by $\mathrm{A}(2^s_X)$ the set of all elements u of $\mathcal{P}(2^s_X)$ such that $X_{(s,u)} \in \mathrm{A}^X_s$ and by i^s_X the mapping of $\mathrm{A}(2^s_X)$ into A^X_s such that $i^s_X(u) = X_{(s,u)}$ for every $u \in \mathrm{A}(2^s_X)$. In particular, $\emptyset \in \mathrm{A}(2^s_X)$ and $i^s_X(\emptyset) = \emptyset \in \mathrm{A}^X_s$. Also, for every $f \in 2^s_X$, $\{f\} \in \mathrm{A}(2^s_X)$ and $i^s_X(\{f\}) = X_{(s,f)} \in \mathrm{A}^X_s$.

1.5.3 Lemma. *Let X be a marked space and $s \in \mathcal{P}_\nu(\tau) \setminus \{\emptyset\}$. Then, the set $\mathrm{A}(2^s_X)$ is a ν-algebra of 2^s_X; $u(X, s, \delta) \in \mathrm{A}(2^s_X)$ for every $\delta \in s$; and the mapping i^s_X is an isomorphism of $\mathrm{A}(2^s_X)$ onto A^X_s such that*

$$i^s_X(u(X, s, \delta)) = X_{(\delta,0)}$$

for every $\delta \in s$. □

The above lemma and Lemma 1.5.2 imply the following consequence.

1.5.4 Corollary. *Let X be a marked space and $s \in \mathcal{P}_\nu(\tau) \setminus \{\emptyset\}$. Then, the set $\mathrm{A}(2^s_X)$ is the minimal ν-algebra of 2^s_X containing the sets $u(X, s, \delta)$, $\delta \in s$.* □

Definition. For every $s \in \mathcal{P}_\nu(\tau)$ on the class of all marked spaces an equivalence relation, denoted by $\sim^s_m$, is defined as follows: two marked spaces X and Y is said to be $\sim^s_m$-*equivalent* if either $s = \emptyset$ or $s \neq \emptyset$ and if there exists an isomorphism i of A^X_s onto A^Y_s, called *natural*, such that $i(X_{(\delta,0)}) = Y_{(\delta,0)}$ for every $\delta \in s$. By Lemma 1.5.2 such an isomorphism is uniquely determined.

It is easy to see that if $\emptyset \neq t \subset s \in \mathcal{P}_\nu(\tau)$, X and Y are two $\sim^s_m$-equivalent marked spaces and if i is the corresponding natural isomorphism of A^X_s onto A^Y_s, then $X \sim^t_m Y$ and the restriction to A^X_t of i is the natural isomorphism of A^X_t onto A^Y_t, that is the equivalence relation $\sim^s_m$ is contained in the equivalence relation $\sim^t_m$.

1.5.5 Lemma. *Let X and Y be marked spaces and $s \in \mathcal{P}_\nu(\tau) \setminus \{\emptyset\}$. The following condition are equivalent:*

(1) *X and Y are $\sim^s_m$-equivalent.*
(2) $2^s_X = 2^s_Y$.
(3) *$u(X, s, \delta) = u(Y, s, \delta)$ for every $\delta \in s$.*
(4) $\mathrm{A}(2^s_X) = \mathrm{A}(2^s_Y)$. □

1.5.6 Lemma. *Let s be an element of $\mathcal{P}_\nu(\tau)\setminus\{\emptyset\}$. Then, the cardinality of the set $\mathrm{C}(\sim^s_m)$ of all equivalence classes of the relation $\sim^s_m$ considered* *on a given set of marked spaces is less than or equal to* $|\mathcal{P}(2^s)|$. *Moreover, there exists a set of marked spaces such that* $|\mathrm{C}(\sim^s_m)| = |\mathcal{P}(2^s)|$.

Proof. By Lemma 1.5.5, $|\mathrm{C}(\sim^s_m)| \leq |\mathcal{P}(2^s)|$. We prove that actually $|\mathrm{C}(\sim^s_m)| = |\mathcal{P}(2^s)|$ for some set of marked spaces. For this fact and by the same lemma it suffices to prove that for every subset $A \subset 2^s$ there exists a space X and an indexed base B^X for X such that the subset 2^s_X of 2^s corresponding to the marked space X (with the mark B^X) is equal to A.

Let A be a subset of 2^s. We set $X = A$. For every $\delta \in s$ we set

$$u(A, s, \delta) = \{f \in A : f(\delta) = 0\}.$$

On the set X we consider a topology for which the finite intersections of elements of the set

$$\{u(A, s, \delta) : \delta \in s\} \cup \{X \setminus u(A, s, \delta) : \delta \in s\}$$

is a base for the open sets. Since $|s| \leq \tau$ the weight of the constructed space X is $\leq \tau$.

Let f and g be two distinct elements of X. Then, there exists an element $\delta \in s$ such that $f(\delta) \neq g(\delta)$. Therefore, either $f(\delta) = 0$ and $g(\delta) = 1$ or $f(\delta) = 1$ and $g(\delta) = 0$. Then, either $f \in u(A, s, \delta)$ and $g \notin u(A, s, \delta)$ or $f \notin u(A, s, \delta)$ and $g \in u(A, s, \delta)$. This fact means that X is a T_0-space.

Let $B^X \equiv \{U^X_\delta : \delta \in \tau\}$ be the mark of X such that $U^X_\delta = u(A, s, \delta)$, $\delta \in s$. By definition, an element $f \in 2^s$ belongs to 2^s_X if and only if

$$\cap\{X_{(\delta, f(\delta))} : \delta \in s\} \neq \emptyset.$$

It is easy to verify that the intersection $\cap\{X_{(\delta, f(\delta))} : \delta \in s\}$ is not empty if and only if $f \in A$. Thus, $f \in 2^s_X$ if and only if $f \in A$, which means that $2^s_X = A$. $\square$

Definition. Let $\mathbf{S}$ be an indexed collection of spaces and $\mathbf{M}$ a co-mark of $\mathbf{S}$. For every $s \in \mathcal{P}_\nu(\tau)$ we define on $\mathbf{S}$ an equivalence relation denoted by $\sim^s_{\mathbf{M}}$ as follows: two elements X and Y of $\mathbf{S}$ are $\sim^s_{\mathbf{M}}$-equivalent if and only if either $s = \emptyset$ or $s \neq \emptyset$ and the marked spaces X and Y with the marks $\mathbf{M}(X)$ and $\mathbf{M}(Y)$, respectively, are $\sim^s_m$-equivalent. The family

$$\mathrm{R}^\nu_{\mathbf{M}} \equiv \{\sim^s_{\mathbf{M}} : s \in \mathcal{P}_\nu(\tau)\}$$

of equivalence relations on $\mathbf{S}$ is called $(\mathbf{M},\nu)$-*standard.*

Definitions. Let $\mathbf{S}$ be an indexed collection of spaces. An indexed family

$$\mathrm{R} \equiv \{\sim^s : s \in \mathcal{P}_\nu(\tau)\}$$

of equivalence relations on $\mathbf{S}$ is said to be μ-*admissible* if the following conditions are satisfied: (a) if $s \subset t \in \mathcal{P}_\nu(\tau)$, then the equivalence relation $\sim^t$ is contained in the equivalence relation $\sim^s$, (b) for every $s \in \mathcal{P}_\nu(\tau)$ the number of $\sim^s$-equivalence classes is less than μ, and (c) $\sim^s = \mathbf{S} \times \mathbf{S}$ if $s = \emptyset$.

The set

$$\cup\{\mathrm{C}(\sim^s) : s \in \mathcal{P}_\nu(\tau)\}$$

is denoted by $\mathrm{C}(\mathrm{R})$. The minimal ν-algebra of $\mathbf{S}$ containing the set $\mathrm{C}(\mathrm{R})$ is denoted by $\mathrm{C}_\nu^\diamondsuit(\mathrm{R})$.

By Lemmas 1.5.5 and 1.5.6 and the condition that $|\mathcal{P}(\mathcal{P}(\kappa))| < \mu$ for every cardinal $\kappa < \nu$ (see Agreement 1 of this section) the $(\mathbf{M},\nu)$-standard family of equivalence relations on $\mathbf{S}$ is μ-admissible.

Let $\mathrm{R}_0 \equiv \{\sim_0^s : s \in \mathcal{P}_\nu(\tau)\}$ and $\mathrm{R}_1 \equiv \{\sim_1^s : s \in \mathcal{P}_\nu(\tau)\}$ be two μ-admissible families of equivalence relations on $\mathbf{S}$. It is said that R_1 is a *final refinement* of R_0 if for every $s \in \mathcal{P}_\nu(\tau)$ there exists an element t of $\mathcal{P}_\nu(\tau)$ such that $\sim_1^t \subset \sim_0^s$.

A μ-admissible family R of equivalence relations on $\mathbf{S}$ is said to be $(\mathbf{M},\nu,\mu)$-*admissible* if R is a final refinement of the $(\mathbf{M},\nu)$-standard family $\mathrm{R}_{\mathbf{M}}^\nu$.

Agreement 2. In what follows of this section it is assumed that an arbitrary indexed non-empty collection of spaces (of weight $\leq \tau$) denoted by $\mathbf{S}$ is given.

It is also assumed that an arbitrary co-mark of $\mathbf{S}$ is given. This co-mark is denoted by $\mathbf{M}$. Moreover, we assume that for every $X \in \mathbf{S}$,

$$\mathbf{M}(X) = \{U_\delta^X : \delta \in \tau\}.$$

Finally, it is assumed that an $(\mathbf{M},\nu,\mu)$-admissible family of equivalence relations on $\mathbf{S}$ is given. This family is denoted by

$$\mathrm{R} \equiv \{\sim^s : s \in \mathcal{P}_\nu(\tau)\}.$$

Definition. Suppose that $\mathbf{S}$ contains a non-empty element. On the set of all pairs (x, X), where $X \in \mathbf{S}$ and $x \in X$, we consider an equivalence relation denoted by $\sim_{\mathbf{M}}^{\mathrm{R}}$ and defined as follows: two pairs (x, X) and (y, Y) are $\sim_{\mathbf{M}}^{\mathrm{R}}$-*equivalent* if

$$X \sim^s Y \text{ and } d_s^X(x) = d_s^Y(y)$$

for every $s \in \mathcal{P}_\nu(\tau) \setminus \{\emptyset\}$.

Notation. The set of all equivalence classes of the relation $\sim_{\mathbf{M}}^{\mathrm{R}}$ is denoted by $\mathrm{T}(\mathbf{M}, \mathrm{R}) \equiv \mathrm{T}$. It is supposed that $\mathrm{T}(\mathbf{M}, \mathrm{R}) = \emptyset$ if all elements of $\mathbf{S}$ are empty.

For every element $\mathbf{H}$ of $\mathrm{C}_\nu^\diamondsuit(\mathrm{R})$ the set of all $\mathbf{a} \in \mathrm{T}$ for which there exists an element (x, X) of $\mathbf{a}$ such that $X \in \mathbf{H}$ is denoted by $\mathrm{T}(\mathbf{M}, \mathrm{R}, \mathbf{H}) \equiv \mathrm{T}(\mathbf{H})$. It is easy to see that if $\mathbf{a} \in \mathrm{T}(\mathbf{H})$, then for every $(x, X) \in \mathbf{a}$, $X \in \mathbf{H}$.

Notation. Let $s \neq \emptyset$ and t be elements of $\mathcal{P}_\nu(\tau)$ such that $\sim^t \subset \sim_{\mathbf{M}}^s$ and let $\mathbf{H}$ be an element of $\mathrm{C}(\sim^t)$. Suppose that an element X of $\mathbf{H}$ is chosen. We denote by $2_{\mathbf{H}}^s$ and $\mathrm{A}(2_{\mathbf{H}}^s)$ the sets 2_X^s and $\mathrm{A}(2_X^s)$, respectively, and for every $\delta \in s$ by $u(\mathbf{H}, s, \delta)$ we denote the set $u(X, s, \delta)$. Since $X \sim_{\mathbf{M}}^s Y$ for every $X, Y \in \mathbf{H}$, the sets $2_{\mathbf{H}}^s$, $\mathrm{A}(2_{\mathbf{H}}^s)$, and $u(\mathbf{H}, s, \delta)$ are independent of the element of $\mathbf{H}$ we choose to define them. (See Lemma 1.5.5).

For every $u \in \mathrm{A}(2_{\mathbf{H}}^s)$ we denote by $\mathrm{T}_{(s,u)}(\mathbf{H})$ the set of all elements $\mathbf{a}$ of T for which there exists an element (x, X) of $\mathbf{a}$ such that $X \in \mathbf{H}$ and $x \in X_{(s,u)}$.

Finally we set

$$\mathrm{A}_s^{\mathbf{H}} = \{\mathrm{T}_{(s,u)}(\mathbf{H}) : u \in \mathrm{A}(2_{\mathbf{H}}^s)\}.$$

Since $\mathrm{T}_{(s,u)}(\mathbf{H})$ is a subset of $\mathrm{T}(\mathbf{H})$, $\mathrm{A}_s^{\mathbf{H}}$ can be considered as a set of subsets of $\mathrm{T}(\mathbf{H})$.

1.5.7 Lemma. *Let $s, t \in \mathcal{P}_\nu(\tau)$, $s \neq \emptyset$, $\sim^t \subset \sim_{\mathbf{M}}^s$, $\mathbf{H} \in \mathrm{C}(\sim^t)$, and $u \in \mathrm{A}(2_{\mathbf{H}}^s)$. Then, the set $\mathrm{T}_{(s,u)}(\mathbf{H})$ coincides with the set of all $\mathbf{a} \in \mathrm{T}$ such that for every $(x, X) \in \mathbf{a}$ we have $X \in \mathbf{H}$ and $x \in X_{(s,u)}$.* $\square$

1.5.8 Lemma. *Let $s, t \in \mathcal{P}_\nu(\tau)$, $s \neq \emptyset$, $\sim^t \subset \sim_{\mathbf{M}}^s$, and $\mathbf{H} \in \mathrm{C}(\sim^t)$. Let also Λ be a set of cardinality $< \nu$ and $u, v_\lambda \in \mathrm{A}(2_{\mathbf{H}}^s)$, $\lambda \in \Lambda$. Then, setting*

$$w = 2_{\mathbf{H}}^s \setminus u \text{ and } v = \cup\{v_\lambda : \lambda \in \Lambda\},$$

we have

$$\mathrm{T}(\mathbf{H}) \setminus \mathrm{T}_{(s,u)}(\mathbf{H}) = \mathrm{T}_{(s,w)}(\mathbf{H}) \text{ and}$$

$$\mathrm{T}_{(s,v)}(\mathbf{H}) = \cup\{\mathrm{T}_{(s,v_\lambda)}(\mathbf{H}) : \lambda \in \Lambda\}. \ \Box$$

1.5.9 Corollary. *If $s, t \in \mathcal{P}_\nu(\tau)$, $s \neq \emptyset$, $\sim^t \subset \sim^s_{\mathbf{M}}$, and $\mathbf{H} \in \mathrm{C}(\sim^t)$, then the set $\mathrm{A}^{\mathbf{H}}_s$ is a ν-algebra of the set $\mathrm{T}(\mathbf{H})$.* $\Box$

1.5.10 Lemma. *Let $s, t \in \mathcal{P}_\nu(\tau)$, $s \neq \emptyset$, $\sim^t \subset \sim^s_{\mathbf{M}}$ and $\mathbf{H} \in \mathrm{C}(\sim^t)$. Then, the mapping $i^s_{\mathbf{H}}$ of the set $\mathrm{A}(2^s_{\mathbf{H}})$ onto the set $\mathrm{A}^{\mathbf{H}}_s$, for which $i^s_{\mathbf{H}}(u) = \mathrm{T}_{(s,u)}(\mathbf{H})$ for every $u \in \mathrm{A}(2^s_{\mathbf{H}})$, is an isomorphism.* $\Box$

1.5.11 Corollary. *Let $s, t \in \mathcal{P}_\nu(\tau)$, $s \neq \emptyset$, $\sim^t \subset \sim^s_{\mathbf{M}}$, $\mathbf{H} \in \mathrm{C}(\sim^t)$, and $X \in \mathbf{H}$. Then, the mapping $i \equiv i^s_{\mathbf{H}} \circ (i^s_X)^{-1}$ is an isomorphism of the algebra A^X_s onto the algebra $\mathrm{A}^{\mathbf{H}}_s$ such that for every $\delta \in s$, $i(X_{(\delta,0)}) = \mathrm{T}_{(s,u)}(\mathbf{H})$, where $u = u(\mathbf{H}, s, \delta)$.* $\Box$

Notation. For every $\delta \in \tau$ and $\mathbf{H} \in \mathrm{C}^{\diamondsuit}_\nu(\mathrm{R})$ we denote by $U^{\mathrm{T}}_\delta(\mathbf{H})$ the set of all $\mathbf{a} \in \mathrm{T}$ for which there exists an element (x, X) of $\mathbf{a}$ such that $X \in \mathbf{H}$ and $x \in U^X_\delta$. Therefore, if for some elements s and t of $\mathcal{P}_\nu(\tau)$ we have $\delta \in s$, $\mathbf{H} \in \mathrm{C}(\sim^t)$, and $\sim^t \subset \sim^s_{\mathbf{M}}$, then

$$U^{\mathrm{T}}_\delta(\mathbf{H}) = \mathrm{T}_{(s,u)}(\mathbf{H}) \in \mathrm{A}^{\mathbf{H}}_s,$$

where $u = u(\mathbf{H}, s, \delta)$.

Notation. We denote $\mathrm{B}^{\mathrm{T}}_*$ the set of all subsets of T of the form $U^{\mathrm{T}}_\delta(\mathbf{H})$, where $\delta \in s$ and $\mathbf{H} \in \mathrm{C}(\sim^s)$ for some $t, s \in \mathcal{P}_\nu(\tau)$ such that $\sim^t \subset \sim^s_{\mathbf{M}}\}$. We set

$$\mathrm{B}^{\mathrm{T}} = \{U^{\mathrm{T}}_\delta(\mathbf{H}) : \delta \in \tau \text{ and } \mathbf{H} \in \mathrm{C}(\mathrm{R})\},$$

$$\mathrm{B}^{\mathrm{T}}_{\diamondsuit} = \{U^{\mathrm{T}}_\delta(\mathbf{H}) : \delta \in \tau \text{ and } \mathbf{H} \in \mathrm{C}^{\diamondsuit}_\nu(\mathrm{R})\}.$$

Also for every subset κ of τ we set

$$\mathrm{B}^{\mathrm{T}}_{*,\kappa} = \{U^{\mathrm{T}}_\delta(\mathbf{H}) \in \mathrm{B}^{\mathrm{T}}_* : \delta \in \kappa\},$$

$$\mathrm{B}^{\mathrm{T}}_\kappa = \{U^{\mathrm{T}}_\delta(\mathbf{H}) \in \mathrm{B}^{\mathrm{T}} : \delta \in \kappa\},$$

$$\mathrm{B}^{\mathrm{T}}_{\diamondsuit,\kappa} = \{U^{\mathrm{T}}_{\delta}(\mathbf{H}) \in \mathrm{B}^{\mathrm{T}}_{\diamondsuit} : \delta \in \kappa\}.$$

Since $|\mathcal{P}_{\nu}(\tau)| \leq \tau$ and for every $s \in \mathcal{P}_{\nu}(\tau)$, $|\mathrm{C}(\sim^{s})| < \mu \leq \tau^{+}$ (see Agreement 1 and Lemma 1.5.6) we have $|\mathrm{B}^{\mathrm{T}}_{*}| \leq |\mathrm{B}^{\mathrm{T}}| \leq |\mathrm{B}^{\mathrm{T}}_{\diamondsuit}| \leq \tau$.

1.5.12 Lemma. *The set $\mathrm{B}^{\mathrm{T}}_{*}$ is a base for a topology on the set* T. *Moreover, if κ is a subset of τ such that for every $X \in \mathbf{S}$ the set*

$$\{U^{X}_{\delta} : \delta \in \kappa\}$$

is a base for X, then the set $\mathrm{B}^{\mathrm{T}}_{*,\kappa}$ is a base for the same topology on T. □

Definition. The set T equipped with the topology for which the set $\mathrm{B}^{\mathrm{T}}_{*}$ is a base is called the *Containing Space for the indexed collection* **S** *corresponding to the co-mark* **M** *and the family* R. In what follows whenever we refer to the space T, we mean just this space.

1.5.13 Lemma. *Let $\delta \in \tau$ and $\mathbf{H} \in \mathrm{C}^{\diamondsuit}_{\nu}(\mathrm{R})$. Then, the following statements are true:*

(1) *There exists an element t of $\mathcal{P}_{\nu}(\tau)$ such that* **H** *is a union of $\sim^{t}$-equivalence classes.*

(2) *The set $U^{\mathrm{T}}_{\delta}(\mathbf{H})$ is open in* T *and coincides with the set of all $\mathbf{a} \in \mathrm{T}$ such that for every $(x, X) \in \mathbf{a}$, $X \in \mathbf{H}$ and $x \in U^{X}_{\delta}$.*

(3) *The set* T(**H**) *is simultaneously open and closed in* T.

(4) *The set $\mathrm{Cl}_{\mathrm{T}}(U^{\mathrm{T}}_{\delta}(\mathbf{H}))$ coincides with the set of all $\mathbf{a} \in \mathrm{T}$ such that there exists an element (x, X) of* **a** *for which $X \in \mathbf{H}$ and $x \in \mathrm{Cl}_{X}(U^{X}_{\delta})$ as well as with the set of all $\mathbf{a} \in \mathrm{T}$ such that for every $(x, X) \in \mathbf{a}$, $X \in \mathbf{H}$ and $x \in \mathrm{Cl}_{X}(U^{X}_{\delta})$.*

(5) *The set $\mathrm{Bd}_{\mathrm{T}}(U^{\mathrm{T}}_{\delta}(\mathbf{H}))$ coincides with the set of all $\mathbf{a} \in \mathrm{T}$ such that there exists an element (x, X) of* **a** *for which $X \in \mathbf{H}$ and $x \in \mathrm{Bd}_{X}(U^{X}_{\delta})$ as well as with the set of all $\mathbf{a} \in \mathrm{T}$ such that for every $(x, X) \in \mathbf{a}$, $X \in \mathbf{H}$ and $x \in \mathrm{Bd}_{X}(U^{X}_{\delta})$.* □

1.5.14 Corollary. *The sets B^{T} and $\mathrm{B}^{\mathrm{T}}_{\diamondsuit}$ are bases for the space* T. *Moreover, if for a subset κ of τ and for every $X \in \mathbf{S}$ the set $\{U^{X}_{\delta} : \delta \in \kappa\}$ is a base for X, then the sets $\mathrm{B}^{\mathrm{T}}_{\kappa}$ and $\mathrm{B}^{\mathrm{T}}_{\diamondsuit,\kappa}$ are also bases for* T. □

1.5.15 Proposition. *The Containing Space* T *is a* T_0*-space of the weight $\leq \tau$.* □

Definition. Let $X \in \mathbf{S}$. For every $x \in X$ there exists a uniquely determined element $\mathbf{a}$ of T which contains the pair (x, X). We define a mapping e^X_{T} of X into T setting

$$e^X_{\mathrm{T}}(x) = \mathbf{a}.$$

The next proposition shows that e^X_{T} is an embedding. This embedding is called *natural.*

1.5.16 Proposition. *For every $X \in \mathbf{S}$ the mapping e^X_{T} is an embedding of X into* T. □

1.6 Agreement concerning notations of the next chapters

In the next chapters we shall construct many Containing Spaces corresponding to different indexed collections of spaces, co-marks, and families of equivalence relations. For the corresponding notations we shall follow the notations introduced in the present chapter. Thus, in general outline we shall follow the given below rules.

(1) In the notations of the Containing Spaces, co-marks, families of equivalence relations, standard bases, and their elements we do not indicate the corresponding indexed collections of spaces. These collections will be clear by the context or will be explicitly indicated. The letter "**S**" is used only for the notations of indexed collections of spaces. However, we use other letters for such notations since many restrictions of some given indexed collection of spaces, denoted by different letters, will also be considered as an indexed collection of spaces.

(2) The co-marks and families of equivalence relations on indexed collections of spaces used for the construction of Containing Spaces are always denoted by the letters "**M**" and "R", respectively, and equipped if neccessary with other letters or symbols as indices. Some of the considered indexed co-bases (that is, co-marks) are denoted without use of the letter "**M**". However, the Containing Spaces corresponding to such co-marks are not needed for our considerations. (These co-marks are used in order to define standard bases in Containing Spaces corresponding to extensions of these co-marks).

(3) For the Containing Spaces we shall use two kind of notations. The first kind is the "whole" notation consisting of the letter "T" and in brackets the corresponding co-mark and the family of equivalence relations. For example, if the co-mark is denoted by $\mathbf{M}'_{I\!B}$ and the family of equivalence relations by $\mathrm{R}'_{I\!B}$, then the corresponding Containing Space is denoted by $\mathrm{T}(\mathbf{M}'_{I\!B}, \mathrm{R}'_{I\!B})$. The second kind is the "brief" notation consisting of the letter "T" equipped, if neccessary, with some indices. For example, $\mathrm{T}'_{I\!B}$ can be considered as the brief notation of the space $\mathrm{T}(\mathbf{M}'_{I\!B}, \mathrm{R}'_{I\!B})$. The brief notation of a Containing Space is always defined throughout the book with the symbol "≡" using the whole notation of this space. The next rule is the only exception for such a definition of the brief notation.

(4) There is an "ordinary situation" which is the most frequent case. In these instances, the considered indexed collection of spaces is denoted by the letter "**S**", the co-mark is denoted by the letter "**M**", and the **M**-admissible family of equivalence relations is denoted by the letter "R". All these letters are considered without indices. In this case and only in this case, the brief notation of the corresponding Containing Space is the letter "T" without indices. Because of this fact in the ordinary situation we sometimes use the brief notation without defining it. Moreover, the whole notation of the corresponding Contained Space is not always indicated. We note that in the present chapter we considered exactly this situation.

(5) For the notations of the restrictions of indexed collections of spaces we use different letters equipped if neccessary with some indices. All such letters are capital and bold-faced. For the corresponding specific subsets of Containing Spaces we also use two kinds of the notations. The first kind is the "whole" notation obtained by the whole notation of the Containing Space replacing the corresponding co-mark and the family of equivalence relations by its traces on the restriction. For example, if the Containing Space is denoted by $\mathrm{T}(\mathbf{M}'_{I\!B}, \mathrm{R}'_{I\!B})$ and the restriction is denoted by $\mathbf{U}_\varepsilon$, then the corresponding specific subset is denoted by $\mathrm{T}(\mathbf{M}'_{I\!B}|_{\mathbf{U}_\varepsilon}, \mathrm{R}'_{I\!B}|_{\mathbf{U}_\varepsilon})$.

The second kind is the "brief" notation. This notation is obtained as "the trace" of the brief notation of the Containing Space on the restriction. For example, if the brief notation of the Containing Space $\mathrm{T}(\mathbf{M}'_{I\!B}, \mathrm{R}'_{I\!B})$ is $\mathrm{T}'_{I\!B}$, then the brief notation of the specific subset $\mathrm{T}(\mathbf{M}'_{I\!B}|_{\mathbf{U}_\varepsilon}, \mathrm{R}'_{I\!B}|_{\mathbf{U}_\varepsilon})$ is $\mathrm{T}'_{I\!B}|_{\mathbf{U}_\varepsilon}$.

In a non-ordinary situation the brief notation of a specific subset is always defined through the symbol "≡" using the whole notation of this subset. In the ordinary situation we sometimes use the brief notation of a specific subset without to define it. In this case we do not indicate the whole notation of this subset. This action means, for example, that $\mathrm{T}|_{\mathbf{Q}}$ is the brief notation of the specific subset $\mathrm{T}(\mathbf{M}|_{\mathbf{Q}}, \mathrm{R}|_{\mathbf{Q}})$ of the Containing Space $\mathrm{T}(\mathbf{M}, \mathrm{R})$. We note that the last situation is the most frequent case.

(6) For the primary subspaces we also use the "whole" and "brief" notations. In the ordinary situation the whole notation is $\mathrm{T}(\mathbf{M}, \mathrm{R}, \mathbf{H})$ and the brief is $\mathrm{T}(\mathbf{H})$, where $\mathbf{H}$ is the corresponding element of $\mathrm{C}^{\diamondsuit}(\mathrm{R})$. In the non-ordinary situation the whole and brief notations are obtained by replacing in $\mathrm{T}(\mathbf{M}, \mathrm{R}, \mathbf{H})$ and $\mathrm{T}(\mathbf{H})$ the letters "$\mathbf{M}$", "R"; and "$\mathbf{H}$" by the notations of the corresponding co-mark, the family of equivalence relations, and the corresponding element of the minimal ring containing equivalence classes of all elements of the last family, respectively.

We note that in the brief notations of primary subspaces in the non-ordinary situation the letter "T" is always used without indices, which means that "T" is not the brief notation of the corresponding Containing Space (that is, it is not the same as for the specific subsets). Whenever we use the brief notation of a primary subspace (in a non-ordinary situation), we define this notation by using the symbol "$\equiv$".

For example, if the Containing Space is denoted by $\mathrm{T}(\mathbf{M}'_{I\!B}, \mathrm{R}'_{I\!B})$ and $\mathbf{H}$ is an element of $\mathrm{C}(\mathrm{R}'_{I\!B})$, then the whole notation of the corresponding primary subspace is $\mathrm{T}(\mathbf{M}'_{I\!B}, \mathrm{R}'_{I\!B}, \mathbf{H})$ and the brief is $\mathrm{T}(\mathbf{H})$. In particular, if $\mathbf{U}_\varepsilon$ is a restriction of the considered indexed collection of spaces, then $\mathrm{T}(\mathbf{H}|_{\mathbf{U}_\varepsilon})$ is the brief notation of the primary subspace $\mathrm{T}(\mathbf{M}'_{I\!B}|_{\mathbf{U}_\varepsilon}, \mathrm{R}'_{I\!B}|_{\mathbf{U}_\varepsilon}, \mathbf{H}|_{\mathbf{U}_\varepsilon})$ of the Containing Space $\mathrm{T}(\mathbf{M}'_{I\!B}|_{\mathbf{U}_\varepsilon}, \mathrm{R}'_{I\!B}|_{\mathbf{U}_\varepsilon})$.

(7) In the next chapters in the ordinary situation the κ-standard and $(\diamondsuit, \kappa)$-standard families of (or bases for the) open subsets of the Containing Space T will be denoted (as in Section 1.2 of the present chapter) by $\mathrm{B}^{\mathrm{T}}_{\kappa}$ and $\mathrm{B}^{\mathrm{T}}_{\diamondsuit,\kappa}$, respectively, where κ is a subset of τ. Also, for every element $\mathbf{L}$ of $\mathrm{C}(\mathrm{R})$ or $\mathrm{C}^{\diamondsuit}(\mathrm{R})$ the κ-standard and $(\diamondsuit, \kappa)$-standard families of open subsets of the primary subspace $\mathrm{T}(\mathbf{L})$ of T will be denoted by $B^{\mathbf{L}}_{\kappa}$ and $\mathrm{B}^{\mathbf{L}}_{\diamondsuit,\kappa}$, respectively. Moreover, the elements of all these families will be denoted by $U^{\mathrm{T}}_{\delta}(\mathbf{H})$, where $\delta \in \kappa$ and $\mathbf{H} \in \mathrm{C}(\mathrm{R})$ or $\mathbf{H} \in \mathrm{C}^{\diamondsuit}(\mathrm{R})$, respectively. In the case where $\kappa = \tau$, the letter "κ" is omitted. Of course, instead of the above letters "κ", "$\mathbf{L}$", and "$\mathbf{H}$" we can use other letters or group of symbols. Usually, κ is the image of the set τ under an indicial mapping θ, that is, $\kappa = \theta(\tau)$.

In a non-ordinary situation the notations of all these families and their elements are obtained by the above, replacing the letter "T" by the brief notation of the corresponding Containing Space and the letter "**L**" by the notation of the corresponding element of the minimal ring containing the equivalence classes of all elements of the considered family of equivalence relations.

We note that in the notations of the elements of the above families there are a few cases where instead of the letter "U", we use the letter "V". In these cases the notations will be clearly explained.

Chapter 2

Saturated classes

The basic idea of the book is to find a universal element for a given class $\mathbb{P}$ of spaces between the corresponding Containing Spaces, that is for a suitable indexed collection $\mathbf{S}$ of elements of $\mathbb{P}$ to find a co-mark $\mathbf{M}$ of $\mathbf{S}$ and an $\mathbf{M}$-admissible family R of equivalence relations on $\mathbf{S}$ such that the Containing Space $\mathrm{T}(\mathbf{M}, \mathrm{R})$ is an element of $\mathbb{P}$ (and, therefore, $\mathrm{T}(\mathbf{M}, \mathrm{R})$ will be a univeral element in $\mathbb{P}$). This idea becomes more fruitful if from the class $\mathbb{P}$ of spaces we require something more than the existence of a universal element only. More precisely, we shall consider classes $\mathbb{P}$ of spaces satisfying the following condition: for every indexed collection $\mathbf{S}$ of elements of $\mathbb{P}$ there exists a co-mark $\mathbf{M}^+$ of $\mathbf{S}$ with the property that for every co-mark $\mathbf{M}$, which is a co-extension of $\mathbf{M}^+$, there exists an $\mathbf{M}$-admissible family R^+ of equivalence relations on $\mathbf{S}$ such that for every admissible family R, which is a final refinement of R^+, the Containing Space $\mathrm{T}(\mathbf{M}, \mathrm{R})$ belongs to $\mathbb{P}$. In the case where $\mathbb{P}$ satisfies the above condition, we shall say that $\mathrm{T}(\mathbf{M}, \mathrm{R})$ belongs to $\mathbb{P}$ for "almost all" co-marks $\mathbf{M}$ and families R. Such classes are called saturated. We note that many of the well-known classes of spaces with universal elements are saturated.

In this chapter and in parallel with the classes of spaces, we also study classes of spaces on which a "structure" is given. By a structure on a space X we mean here a given subset Q of X, a given base B for X, or a given subset Q of X with a given p-base B for Q in X. The corresponding classes are called respectively classes of subsets, classes of bases, and classes of p-bases. For these classes we give the new notion of a universal element. Classes of spaces with structure for which the structure can be "transferred" by a standard manner to the Containing Spaces $\mathrm{T}(\mathbf{M}, \mathrm{R})$ for "almost all" co-marks **M** and families R are called saturated.

Saturated classes of spaces (with structure) have, of course, the universality property, that is in any such class there exists a universal element. However, saturated classes have some other very useful properties connected with the fact that these classes are "saturated" by universal elements. For example, saturated classes have the intersection property, that is the intersection of not more than τ many saturated classes is also a saturated class although the intersection of classes having a universal element in general may have no universal element.

Throughout the book we shall use given saturated classes to construct new ones. In particular, in the next chapter we use saturated classes to define dimension-like functions, which in turn define new classes.

In Section 2.1 we define saturated classes of spaces and prove the universality and intersection properties for these classes. We also prove that the class of all regular spaces is a saturated class.

In Section 2.2 we consider saturated classes of subsets. For these classes we introduce the notions of universal and properly universal elements and give the universality and intersection properties. Some concrete classes of subsets are considered: the class of all closed subsets, the class of all open subsets, and the class of all nowhere dense subsets consisting of all pairs (Q, X), where Q is a closed, an open, and a nowhere dense subset of a space X, respectively. We prove that these classes are saturated. We also introduce the notion of a complete saturated class of subsets and prove the existence of properly universal elements for such classes. Any class of subsets defines two classes of spaces, which are called space-component and subset-component of this class. If the class of subsets is saturated, then the space-component and the subset-component are also saturated.

In Section 2.3 for a fixed saturated class $I\!\!R$ of spaces we consider a class $\{\mathrm{O}_X : X \in I\!\!R\}$, where O_X is a topological operator on X. Such a class is called a co-operator on $I\!\!R$. We introduce and study the notion of a commutative co-operator, which are useful for the construction of new saturated classes by the given saturated classes of subsets. In particular, we consider co-operators in cases where O_X is the closure, the boundary, the interior, and the complement topological operators on X proving that these co-operators are commutative.

In Section 2.4 we investigate the Borel type sets. In particular, for $\alpha \in \tau^+$ we prove that the class of all pairs (Q, X), where Q is a subset of a space X of the (additive) multiplicative class α, is a complete saturated class of subsets.

In Section 2.5 we study saturated classes of bases and in Section 2.6 saturated classes of p-bases. For these classes we prove the universality and intersection properties. We also consider concrete saturated classes of bases and p-bases.

Finally, in Section 2.7 using results of Section 1.5 we introduce and study the notion of a (ν, μ)-saturated class. For $\tau = \omega$ the (ω, ω^+)-saturated classes are used in Chapter 9.

2.1 Saturated classes of spaces

Definition. A class $\mathbb{P}$ of spaces is said to be *topological* if for every homeomorphism h of a space X onto a space Y the condition $X \in \mathbb{P}$ implies that $Y \in \mathbb{P}$. In the sequel all considered classes of spaces are assumed to be topological.

Definition. A non-empty class $\mathbb{P}$ of spaces is said to be *saturated* if for every indexed collection $\mathbf{S}$ of elements of $\mathbb{P}$ there exists a co-mark $\mathbf{M}^+$ of $\mathbf{S}$ satisfying the following condition: for every co-extension $\mathbf{M}$ of $\mathbf{M}^+$ there exists an $\mathbf{M}$-admissible family R^+ of equivalence relations on $\mathbf{S}$ such that for every admissible family R of equivalence relations on $\mathbf{S}$, which is a final refinement of R^+, and for every element $\mathbf{L} \in \mathrm{C}^\diamond(\mathrm{R})$ the space $\mathrm{T}(\mathbf{L})$ belongs to $\mathbb{P}$.

The above considered co-mark $\mathbf{M}^+$ is said to be an *initial co-mark* of $\mathbf{S}$ (*corresponding* to the class $\mathbb{P}$) and the family R^+ is said to be an *initial family* of $\mathbf{S}$ (*corresponding* to the co-mark $\mathbf{M}$ and the class $\mathbb{P}$).

Remark. In the above definition any co-extension of an initial co-mark is also an initial co-mark, and any admissible family of equivalence relations on $\mathbf{S}$, which is a final refinement of an initial family, is also an initial family.

The following two lemmas are frequently used in the sequel.

2.1.1 Lemma. *Let $\mathbf{S}$ be an indexed collection of spaces. Suppose that a number of not more than τ many admissible families of equivalence relations on $\mathbf{S}$ is given. Then, there exists an admissible family of equivalence relations on $\mathbf{S}$, which is a final refinement of each given family.*

Proof. Let $\{\mathrm{R}_\delta : \delta \in \tau\}$ be an indication of the given set of families. Suppose that

$$\mathrm{R}_\delta \equiv \{\sim^s_\delta : s \in \mathcal{F}\}.$$

For every $s \in \mathcal{F} \setminus \{\emptyset\}$ we set

$$\sim^s_+ = \cap\{\sim^s_\delta : \delta \in s\}$$

and let

$$\mathrm{R}^+ = \{\sim^s_+ : s \in \mathcal{F}\},$$

where $\sim^s_+ = \mathbf{S} \times \mathbf{S}$ if $s = \emptyset \in \mathcal{F}$. It is easy to verify that R^+ is an admissible family of equivalence relations on $\mathbf{S}$.

We prove that R^+ is a final refinement of R_δ for every $\delta \in \tau$. Let $\delta \in \tau$ and $t \in \mathcal{F}$. Setting $s = t \cup \{\delta\}$, by definition, we have $\sim^s_+ \subset \sim^s_\delta$. Since the family R_δ is admissible, $\sim^s_\delta \subset \sim^t_\delta$. Therefore, $\sim^s_+ \subset \sim^t_\delta$, which proves that R^+ is a final refinement of R_δ. $\square$

2.1.2 Lemma. *Let* $\mathbf{S}$ *be an indexed collection of spaces. Suppose that a number of not more than* τ *many co-marks of* $\mathbf{S}$ *is given. Then, there exists a co-mark of* $\mathbf{S}$*, which is a co-extension of each given co-mark.*

PROOF. Let $\{\mathbf{M}_\delta : \delta \in \tau\}$ be an indication of the set of all given co-marks and let

$$\mathbf{M}_\delta = \{\{V^X_{\delta,\eta} : \eta \in \tau\} : X \in \mathbf{S}\}.$$

Denote by θ_δ, $\delta \in \tau$, a one-to-one mapping of τ into itself such that

$$\theta_{\delta_0}(\tau) \cap \theta_{\delta_1}(\tau) = \emptyset$$

for every distinct elements δ_0 and δ_1 of τ. Let

$$\mathbf{M} \equiv \{\{U^X_\varepsilon : \varepsilon \in \tau\} : X \in \mathbf{S}\}$$

be a co-mark of $\mathbf{S}$ such that if $\varepsilon = \theta_\delta(\eta)$ for some $\varepsilon, \delta, \eta \in \tau$, then $U^X_\varepsilon = V^X_{\delta,\eta}$. It is easy to see that $\mathbf{M}$ is the required co-mark. $\square$

Remark. In the next proposition (as well as in several other places of the book) the notion of the intersection of an infinite number of saturated classes is considered. This notion can be defined using the fact that all considered classes are topological and consist of spaces of weight $\leq \tau$.

2.1.3 Proposition. *The non-empty intersection of not more than τ many saturated classes of spaces is also a saturated class of spaces.*

Proof. Suppose that for every $\delta \in \tau$, $\mathbb{P}_\delta$ is a saturated class of spaces and let $\mathbb{P} = \cap\{\mathbb{P}_\delta : \delta \in \tau\}$. Consider an arbitrary indexed collection $\mathbf{S}$ of elements of $\mathbb{P}$.

Denote by $\mathbf{M}^+_\delta$ an initial co-mark of $\mathbf{S}$ corresponding to the class $\mathbb{P}_\delta$ and let $\mathbf{M}^+$ be a co-extension of $\mathbf{M}^+_\delta$ for every $\delta \in \tau$ (see Lemma 2.1.2). Consider an arbitrary co-mark $\mathbf{M}$ of $\mathbf{S}$, which is a co-extension of $\mathbf{M}^+$. Then, $\mathbf{M}$ is also a co-extension of $\mathbf{M}^+_\delta$. Therefore, there exists an initial family R^+_δ of equivalence relations on $\mathbf{S}$ corresponding to the co-mark $\mathbf{M}$ and the class $\mathbb{P}_\delta$. By Lemma 2.1.1 there exists an admissible family R^+ of equivalence relations on $\mathbf{S}$, which is a final refinement of R^+_δ for every $\delta \in \tau$. Obviously, this family is $\mathbf{M}$-admissible.

We prove that $\mathbf{M}^+$ is an initial co-mark corresponding to the class $\mathbb{P}$ and R^+ is an initial family corresponding to the co-mark $\mathbf{M}$ and the class $\mathbb{P}$. For this purpose we consider an arbitrary admissible family R of equivalence relations on $\mathbf{S}$, which is a final refinement of R^+. Then, R is a final refinement of R^+_δ for every $\delta \in \tau$. We need to prove that for every $\mathbf{L} \in \mathrm{C}^\diamondsuit(\mathrm{R})$ the subspace $\mathrm{T}(\mathbf{L})$ of T belongs to $\mathbb{P}$. Let $\mathbf{L} \in \mathrm{C}^\diamondsuit(\mathrm{R})$. Since $\mathbf{M}^+$ is an initial co-mark corresponding to the class $\mathbb{P}_\delta$ and R^+ is an initial family corresponding to the co-mark $\mathbf{M}$ and the class $\mathbb{P}_\delta$, $\mathrm{T}(\mathbf{L})$ belongs to $\mathbb{P}_\delta$ for every $\delta \in \tau$. By definition, the space $\mathrm{T}(\mathbf{L})$ belongs also to $\mathbb{P}$. Thus, $\mathbb{P}$ is a saturated class of spaces. □

2.1.4. Proposition. *In any saturated class of spaces there exist universal elements.*

PROOF. Let $\mathbb{P}$ be a saturated class of spaces. Since $\mathbb{P}$ is a topological class there exists an indexed collection $\mathbf{S}$ of elements of $\mathbb{P}$ such that every element of $\mathbb{P}$ is homeomorphic to an element of $\mathbf{S}$.

Since $\mathbb{P}$ is a saturated class there exists an initial co-mark $\mathbf{M}^+$ of $\mathbf{S}$ corresponding to the class $\mathbb{P}$. Denote by $\mathbf{M}$ an arbitrary co-extension of $\mathbf{M}^+$ and let R^+ be an initial family of equivalence relation on $\mathbf{S}$ corresponding to the co-mark $\mathbf{M}$ and the class $\mathbb{P}$. Denote also by R an arbitrary admissible family of equivalence relations on $\mathbf{S}$, which is a final refinement of R^+. Then, we can consider the Containing Space T. By construction, this space belongs to $\mathbb{P}$. If Z is an element of $\mathbb{P}$ and f is an homeomorphism of Z onto an element X of $\mathbf{S}$, then $h \equiv e^X_{\mathrm{T}} \circ f$ (where e^X_{T} is the natural embedding of X into T) is an embedding of the space Z into the space T. Thus, T is a universal element in $I\!P$. □

2.1.5 Corollary. *In the class of all spaces there exist universal elements.* □

2.1.6 Proposition. *The class of all regular spaces is saturated.*

Proof. Denote by $I\!P$ the class of all regular spaces and let $\mathbf{S}$ be an indexed collection of elements of $I\!P$ such that each regular space is homeomorphic to an element of $I\!P$. Suppose that

$$\mathbf{N} \equiv \{\{V_\delta^X : \delta \in \tau\} : X \in \mathbf{S}\}$$

is any co-mark of $\mathbf{S}$. Let ϑ_0 and ϑ_1 be two one-to-one mappings of τ into itself such that $\vartheta_0(\tau) \cap \vartheta_1(\tau) = \emptyset$. Let

$$\mathbf{M}^+ \equiv \{\{W_\delta^X : \delta \in \tau\} : X \in \mathbf{S}\}$$

be a co-mark of $\mathbf{S}$ such that $V_\delta^X = W_{\vartheta_0(\delta)}^X$ and $X \setminus \mathrm{Cl}_X(V_\delta^X) = W_{\vartheta_1(\delta)}^X$. Obviously, $\mathbf{M}^+$ is a co-extension of $\mathbf{N}$.

Consider an arbitrary co-mark

$$\mathbf{M} \equiv \{\{U_\delta^X : \delta \in \tau\} : X \in \mathbf{S}\}$$

of $\mathbf{S}$, which is a co-extension of $\mathbf{M}^+$, and let θ be an indicial mapping of this co-extension. Set $\theta_0 = \theta \circ \vartheta_0$ and $\theta_1 = \theta \circ \vartheta_1$. Then, $\mathbf{M}$ is also a co-extension of $\mathbf{N}$ and θ_0 is an indicial mapping from $\mathbf{N}$ to $\mathbf{M}$. Denote by R^+ any $\mathbf{M}$-admissible family of equivalence relations on $\mathbf{S}$. To prove the proposition it suffices to prove that $\mathbf{M}^+$ is an initial co-mark corresponding to the class $I\!P$ and R^+ is an initial family corresponding to the co-mark $\mathbf{M}$ and the class $I\!P$.

Let $\mathrm{R} \equiv \{\sim^s : s \in \mathcal{F}\}$ be an arbitrary admissible family of equivalence relations on $\mathbf{S}$, which is a final refinement of R^+. We need to prove that the space T is regular. (Then, every one of its subspace will be also regular). For this, we consider a point $\mathbf{a}$ of T and an open neighbourhood U of $\mathbf{a}$. It suffices to construct an open subset V of T such that $\mathbf{a} \in V \subset \mathrm{Cl}_{\mathrm{T}}(V) \subset U$. Without loss of generality, we can suppose that $U \equiv U_{\theta_0(\delta)}^{\mathrm{T}}(\mathbf{H}) \in \mathrm{B}_{\theta_0(\tau)}^{\mathrm{T}}$ for some $\delta \in \tau$ and some $\sim^t$-equivalence class $\mathbf{H}$, $t \in \mathcal{F}$. Let $(x, X) \in \mathbf{a}$. Then,

$x \in U^X_{\theta_0(\delta)}$ and $X \in \mathbf{H}$. Also, we have $U^X_{\theta_0(\delta)} = V^X_\delta$. Since X is a regular space there exists an element $\varepsilon \in \tau$ such that

$$x \in V^X_\varepsilon \subset \mathrm{Cl}_X(V^X_\varepsilon) \subset V^X_\delta.$$

Therefore,

$$V^X_\delta \cup (X \setminus \mathrm{Cl}_X(V^X_\varepsilon)) = X$$

or, equivalently,

$$U^X_{\theta_0(\delta)} \cup U^X_{\theta_1(\varepsilon)} = X. \tag{2.1.1}$$

On the other hand, we have

$$U^X_{\theta_0(\varepsilon)} \cap U^X_{\theta_1(\varepsilon)} = \emptyset. \tag{2.1.2}$$

Now, we consider the elements of $\mathbf{S}$ as marked spaces. (The chosen mark for an element Y of $\mathbf{S}$ is considered to be the mark $\mathbf{M}(Y) = \{U^Y_\delta : \delta \in \tau\}$). Let $s = \{\theta_0(\delta), \theta_0(\varepsilon), \theta_1(\varepsilon)\}$. Since R is $\mathbf{M}$-admissible (as a final refinement of R^+) there exists an element q of $\mathcal{F}$ such that $t \subset q$ and $\sim^q \subset \sim^s_{\mathbf{M}}$ (that is, for every $Y, Z \in \mathbf{S}$ if $Y \sim^q Z$, then $Y \sim^s_{\mathbf{M}} Z$). Denote by $\mathbf{E}$ the $\sim^q$-equivalence class of X. It is clear that $\mathbf{E} \subset \mathbf{H}$ and

$$\mathbf{a} \in U^{\mathrm{T}}_{\theta_0(\varepsilon)}(\mathbf{E}). \tag{2.1.3}$$

There exists an isomorphism i of A^X_s onto $\mathrm{A}^{\mathbf{E}}_s$ such that $i(U^X_\eta) = U^{\mathrm{T}}_\eta(\mathbf{E})$, $\eta \in s$. (See Corollary 1.2.5). By Lemma 1.1.1, relation (2.1.1) implies that

$$U^{\mathrm{T}}_{\theta_0(\delta)}(\mathbf{E}) \cup U^{\mathrm{T}}_{\theta_1(\varepsilon)}(\mathbf{E}) = \mathrm{T}(\mathbf{E}) \tag{2.1.4}$$

and relation (2.1.2) implies that $U^{\mathrm{T}}_{\theta_0(\varepsilon)}(\mathbf{E}) \cap U^{\mathrm{T}}_{\theta_1(\varepsilon)}(\mathbf{E}) = \emptyset$, from which it follows that

$$\mathrm{Cl}_{\mathrm{T}}(U^{\mathrm{T}}_{\theta_0(\varepsilon)}(\mathbf{E})) \cap U^{\mathrm{T}}_{\theta_1(\varepsilon)}(\mathbf{E}) = \emptyset. \tag{2.1.5}$$

Relations (2.1.3), (2.1.4), and (2.1.5) imply that the set $V \equiv U^{\mathrm{T}}_{\theta_0(\varepsilon)}(\mathbf{E})$ satisfies the above mentioned conditions, that is

$$\mathbf{a} \in U^{\mathrm{T}}_{\theta_0(\varepsilon)}(\mathbf{E}) \subset \mathrm{Cl}_{\mathrm{T}}(U^{\mathrm{T}}_{\theta_0(\varepsilon)}(\mathbf{E})) \subset U^{\mathrm{T}}_{\theta_0(\delta)}(\mathbf{E}) \subset U^{\mathrm{T}}_{\theta_0(\delta)}(\mathbf{H}),$$

which proves that T is a regular space. Hence, R^+ is an initial family; and, therefore, $\mathbf{M}^+$ is an initial co-mark. Thus, the class of all regular spaces is saturated. □

2.1.7 Corollary. *In the class of all regular spaces there exist universal elements.* □

In particular, for $\tau = \omega$ we have the following consequence.

2.1.8 Corollary. *In the class of all separable metrizable spaces there exist universal elements.* □

2.2 Saturated classes of subsets

Definition. In our considerations by a *class of subsets* we mean a class $I\!P$ consisting of pairs (Q, X), where Q is a subset of a space X. Such a class is said to be *topological* if for every homeomorphism h of a space X onto a space Y the condition $(Q, X) \in I\!P$ implies that $(h(Q), Y) \in I\!P$. In what follows, an arbitrary considered class of subsets is assumed to be topological.

Definition. Let $I\!P$ be a class of subsets. A restriction $\mathbf{Q}$ of an indexed collection $\mathbf{S}$ of spaces is said to be a *$I\!P$-restriction* if $(\mathbf{Q}(X), X) \in I\!P$ for every $X \in \mathbf{S}$. (We recall that $\mathbf{Q} = \{\mathbf{Q}(X) : X \in \mathbf{S}\}$.)

Definition. A non-empty class $I\!P$ of subsets is said to be *saturated* if for every indexed collection $\mathbf{S}$ of spaces and for every $I\!P$-restriction $\mathbf{Q}$ of $\mathbf{S}$ there exists a co-mark $\mathbf{M}^+$ of $\mathbf{S}$ satisfying the following condition: for every co-extension $\mathbf{M}$ of $\mathbf{M}^+$ there exists an $(\mathbf{M}, \mathbf{Q})$-admissible family R^+ of equivalence relations on $\mathbf{S}$ such that for every admissible family R of equivalence relations on $\mathbf{S}$, which is a final refinement of R^+, and elements $\mathbf{L}$ and $\mathbf{H}$ of $C^{\diamondsuit}(R)$, for which $\mathbf{H} \subset \mathbf{L}$, we have $(T(\mathbf{H}|_{\mathbf{Q}}), T(\mathbf{L})) \in I\!P$.

The considered co-mark $\mathbf{M}^+$ is said to be an *initial co-mark* of $\mathbf{S}$ (*corresponding* to the $I\!P$-restriction $\mathbf{Q}$, or *corresponding* to the restriction $\mathbf{Q}$ and the class $I\!P$); and the family R^+ is said to be an *initial family* of $\mathbf{S}$ (*corresponding* to the co-mark $\mathbf{M}$ and the $I\!P$-restriction $\mathbf{Q}$, or *corresponding* to the co-mark $\mathbf{M}$, the restriction $\mathbf{Q}$, and the class $I\!P$).

Definition. Let $\mathbb{P}$ be a class of subsets. An element (Q^T, T) of $\mathbb{P}$ is said to be *universal* in $\mathbb{P}$ if for every element (Q^Z, Z) of $\mathbb{P}$ there exists an embedding h of Z into T such that $Q^Z \subset h^{-1}(Q^T)$.

The following two propositions are proved by a method that is similar to the proofs of Propositions 2.1.3 and 2.1.5, respectively.

2.2.1 Proposition. *In any saturated class of subsets there exist universal elements.* □

2.2.2 Proposition. *The non-empty intersection of not more than τ many saturated classes of subsets is also a saturated class of subsets.* □

Notation. We denote by:

(a) $\mathbb{P}(\mathrm{Cl})$, (b) $\mathbb{P}(\mathrm{Op})$, and (c) $\mathbb{P}(\mathrm{n.dense})$

the classes of subsets consisting of all pairs (Q, X), where Q is a subset of a space X such that: (a) Q is closed, (b) Q is open, and (c) Q is nowhere dense in X, respectively.

2.2.3 Proposition. *The classes* $\mathbb{P}(\mathrm{Cl})$, $\mathbb{P}(\mathrm{Op})$, *and* $\mathbb{P}(\mathrm{n.dense})$ *are saturated classes of subsets.*

Proof. By Lemma 1.4.1 it follows immediately that $\mathbb{P}(\mathrm{Cl})$ is a saturated class. (For every $\mathbb{P}(\mathrm{Cl})$-restriction $\mathbf{Q}$ of an indexed collection $\mathbf{S}$ of spaces we can take any co-mark $\mathbf{M}^+$ of $\mathbf{S}$ as an initial co-mark and for every co-extension $\mathbf{M}$ of $\mathbf{M}^+$ we can take any $(\mathbf{M}, \mathbf{Q})$-admissible family of equivalence relations on $\mathbf{S}$ as an initial family).

Similarly, Lemma 1.4.2 implies that $\mathbb{P}(\mathrm{Op})$ is a saturated class. (In this case and for every $\mathbb{P}(\mathrm{Op})$-restriction $\mathbf{Q}$ of an indexed collection $\mathbf{S}$ of spaces, any co-mark $\mathbf{M}^+$ of $\mathbf{S}$ can be considered as an initial co-mark and for every co-extension $\mathbf{M}$ of $\mathbf{M}^+$ any $(\mathbf{M}, \mathbf{Q})$-admissible and $(\mathbf{M}, \mathbf{Co}(\mathbf{Q}))$-admissible family of equivalence relations on $\mathbf{S}$ can be considered as an initial family.)

Now, we prove that $\mathbb{P}(\mathrm{n.dense})$ is a saturated class. Consider an indexed collection $\mathbf{S}$ of spaces and let

$$\mathbf{Q} \equiv \{Q^X : X \in \mathbf{S}\}$$

be a $\mathbb{P}(\mathrm{n.dense})$-restriction of $\mathbf{S}$. Therefore, Q^X is a nowhere dense subset of $X \in \mathbf{S}$.

Denote by $\mathbf{M}^+$ an arbitrary co-mark of $\mathbf{S}$ and let

$$\mathbf{M} \equiv \{\{U_\delta^X : \delta \in \tau\} : X \in \mathbf{S}\}$$

be an arbitrary co-mark of $\mathbf{S}$, which is a co-extension of $\mathbf{M}^+$. Let R^+ be any $(\mathbf{M}, \mathbf{Q})$-admissible family of equivalence relations on $\mathbf{S}$ and

$$\mathrm{R} \equiv \{\sim^s : s \in \mathcal{F}\}$$

an arbitrary admissible family of equivalence relations on $\mathbf{S}$, which is a final refinement of R^+. Then, R is also $(\mathbf{M}, \mathbf{Q})$-admissible.

To prove the proposition it suffices to prove that $\mathbf{M}^+$ is an initial family corresponding to the $\mathbb{P}$(n.dense)-restriction $\mathbf{Q}$, and R^+ is an initial family corresponding to the co-mark $\mathbf{M}$ and the $\mathbb{P}$(n.dense)-restriction $\mathbf{Q}$. To this end we need to prove that the subset $\mathrm{T}|_{\mathbf{Q}}$ of T is nowhere dense in T (and, therefore, for every $\mathbf{H}, \mathbf{L} \in \mathrm{C}^{\diamondsuit}(\mathrm{R})$ for which $\mathbf{H} \subset \mathbf{L}$, the set $\mathrm{T}|_{\mathbf{Q}} \cap \mathrm{T}(\mathbf{H})$ is nowhere dense in $\mathrm{T}(\mathbf{L})$).

Let U be an open subset of T. In order to prove that the subset $\mathrm{T}|_{\mathbf{Q}}$ of T is nowhere dense it suffices to construct an open subset V of T such that $V \subset U$ and $V \cap \mathrm{T}|_{\mathbf{Q}} = \emptyset$. Without loss of generality, we can suppose that U has the form $U_\delta^{\mathrm{T}}(\mathbf{H})$ for some $\delta \in \tau$ and $\sim^t$-equivalence class $\mathbf{H}$, $t \in \mathcal{F}$. Let $X \in \mathbf{H}$. Since Q^X is nowhere dense in X there exists an element ε of τ such that $U_\varepsilon^X \subset U_\delta^X$ and $U_\varepsilon^X \cap Q^X = \emptyset$. Let $\{\delta, \varepsilon\} = s$. By Lemma 1.3.1 there exists an element q of $\mathcal{F}$ such that $t \subset q$, $\sim^q \subset \sim^s_{\mathbf{M}}$, and $d_s^X(Q^X) = d_s^Y(Q^Y)$ if $X \sim^q Y$. Let $\mathbf{E}$ be the $\sim^q$-equivalence class of X. We show that the set $V \equiv U_\varepsilon^{\mathrm{T}}(\mathbf{E})$ satisfies the above mentioned conditions, that is

$$U_\varepsilon^{\mathrm{T}}(\mathbf{E}) \subset U_\delta^{\mathrm{T}}(\mathbf{H}) \text{ and } U_\varepsilon^{\mathrm{T}}(\mathbf{E}) \cap \mathrm{T}|_{\mathbf{Q}} = \emptyset. \tag{2.2.1}$$

Let Y be an arbitrary element of $\mathbf{E}$. By the choice of q, $X \sim^s_{\mathbf{M}} Y$. By Lemma 1.1.1, $U_\varepsilon^Y \subset U_\delta^Y$. Therefore,

$$U_\varepsilon^{\mathrm{T}}(\mathbf{E}) \subset U_\delta^{\mathrm{T}}(\mathbf{E}) \subset U_\delta^{\mathrm{T}}(\mathbf{H}).$$

Now, we prove that the set $U_\varepsilon^{\mathrm{T}}(\mathbf{E}) \cap \mathrm{T}|_{\mathbf{Q}}$ is empty. Indeed, in the opposite case there exists a point $\mathbf{a}$ of $\mathrm{T}|_{\mathbf{Q}}$ belonging to the set $U_\varepsilon^{\mathrm{T}}(\mathbf{E})$. By Corollary 1.3.6 there exists a pair $(y, Y) \in \mathbf{a}$ such that $y \in Q^Y$. Then, $y \in U_\varepsilon^Y$ and $Y \in \mathbf{E}$. By the choice of q, $X \sim^s_{\mathbf{M}} Y$ and $d_s^X(Q^X) = d_s^Y(Q^Y)$. Therefore, there exists a point $x \in Q^X$ such that $d_s^X(x) = d_s^Y(y)$. This means that $x \in U_\varepsilon^X$, that is $U_\varepsilon^X \cap Q^X \neq \emptyset$, which is a contradiction proving relation (2.2.1). Thus, $\mathrm{T}|_{\mathbf{Q}}$ is nowhere dense in T, which completes the proof of the proposition. □

Definition. A class $I\!\!P$ of subsets is said to be *closed* (respectively, *open*) if for every element (Q, X) of $I\!\!P$ the subset Q of X is closed (respectively, open).

Definitions. A restriction $\mathbf{Q}$ of an indexed collection $\mathbf{S}$ of spaces is said to be *complete* if there exists a co-mark $\mathbf{M}$ of $\mathbf{S}$ and an $(\mathbf{M}, \mathbf{Q})$-admissible family R of equivalence relations on $\mathbf{S}$ such that $\mathbf{Q}$ is an $(\mathbf{M}, \mathrm{R})$-complete restriction.

A class $I\!\!P$ of subsets is said to be *complete* if for every indexed collection $\mathbf{S}$ of spaces any $I\!\!P$-restriction of $\mathbf{S}$ is complete.

Lemmas 1.4.1 and 1.4.2 imply that any closed or any open restriction of any collection of spaces is complete. Therefore, any closed class of subsets or any open class of subsets is complete. In particular, the classes $\mathbb{P}(\mathrm{Cl})$, $\mathbb{P}(\mathrm{Op})$, and $\mathbb{P}(\mathrm{Cl}) \cap \mathbb{P}(\mathrm{n.dense})$ are complete.

2.2.4 Lemma. *Let* $\mathbf{Q}$ *be a restriction of an indexed collection* $\mathbf{S}$ *of spaces. Suppose that* $\mathbf{Q}$ *is an* $(\mathbf{M}_0, \mathrm{R}_0)$*-complete restriction for some co-mark* $\mathbf{M}_0$ *of* $\mathbf{S}$ *and some* $(\mathbf{M}_0, \mathbf{Q})$*-admissible family* R_0 *of equivalence relations on* $\mathbf{S}$*. Then, for every co-extension* $\mathbf{M}$ *of* $\mathbf{M}_0$ *and for every* $(\mathbf{M}, \mathbf{Q})$*-admissible family* R *of equivalence relations on* $\mathbf{S}$*, which is a final refinement of* R_0*,* $\mathbf{Q}$ *is an* $(\mathbf{M}, \mathrm{R})$*-complete restriction.*

Proof. Let $\mathbf{M}$ be a co-extension of the co-mark $\mathbf{M}_0$ and θ an indicial mapping of this co-extension. Let also $\mathrm{R} \equiv \{\sim^s : s \in \mathcal{F}\}$ be an $(\mathbf{M}, \mathbf{Q})$-admissible family of equivalence relations on $\mathbf{S}$, which is a final refinement of the family $\mathrm{R}_0 \equiv \{\sim_0^s : s \in \mathcal{F}\}$. Consider a point $\mathbf{a}$ of $\mathrm{T}|_{\mathbf{Q}}$ and let $(x, X) \in \mathbf{a}$. We need to prove that $x \in \mathbf{Q}(X)$.

There exists an element (y, Y) of $\mathbf{a}$ such that $y \in \mathbf{Q}(Y)$. Let $\mathbf{b}$ be a point of $\mathrm{T}(\mathbf{M}_0, \mathrm{R}_0)$ such that $(y, Y) \in \mathbf{b}$. Since $y \in \mathbf{Q}(Y)$, $\mathbf{b} \in \mathrm{T}(\mathbf{M}_0|_{\mathbf{Q}}, \mathrm{R}_0|_{\mathbf{Q}})$. We prove that $(x, X) \in \mathbf{b}$. Indeed, since R is a final refinement of R_0 and since $X \sim^s Y$ for every $s \in \mathcal{F}$, we have $X \sim_0^t Y$ for every $t \in \mathcal{F}$.

Let t be an arbitrary element of $\mathcal{F}$ and $s = \theta(t)$. For every $Z \in \mathbf{S}$ denote by $\widetilde{\mathrm{A}}_t^Z$ the t-algebra of Z related to the mark $\mathbf{M}_0(Z)$ and by A_s^Z the s-algebra of Z related to the mark $\mathbf{M}(Z)$. Also denote by $\widetilde{d}_t^Z$ the mapping of Z into 2^t constructed for the algebra $\widetilde{\mathrm{A}}_t^Z$ and by d_s^Z the corresponding mapping constructed for the algebra A_s^Z. It is easy to verify that if for some point $z \in Z$, $d_s^Z(z) = f \in 2^s$, then $\widetilde{d}_t^Z(z) = f \circ \theta|_t \in 2^t$, where $\theta|_t$ is the restriction of θ on $t \subset \tau$.

Since (x, X) and (y, Y) are elements of **a** we have $d_s^X(x) = d_s^Y(y)$ for every $s \in \mathcal{F}$. By the above, $\widetilde{d}_t^X(x) = \widetilde{d}_t^Y(y)$ for every $t \in \mathcal{F}$. Therefore, the pairs (x, X) and (y, Y) belong to the same point of $\mathrm{T}(\mathbf{M}_0, \mathrm{R}_0)$. Since $(y, Y) \in \mathbf{b}$ we have $(x, X) \in \mathbf{b}$. Since $\mathbf{Q}$ is an $(\mathbf{M}_0, \mathrm{R}_0)$-complete restriction and $y \in \mathbf{Q}(Y)$ we have $x \in \mathbf{Q}(X)$, completing the proof of the lemma. □

Definition. Let $I\!\!P$ be a class of subsets. An element (Q^T, T) of $I\!\!P$ is said to be *properly universal* in $I\!\!P$ if for every element (Q^Z, Z) of $I\!\!P$ there exists an embedding h of Z into T such that $Q^Z = h^{-1}(Q^T)$.

2.2.5 Proposition. *In any complete saturated class of subsets there exist properly universal elements.*

Proof. Let $I\!\!P$ be a complete saturated class of subsets. Since $I\!\!P$ is a topological class there exists an indexed collection $\mathbf{S}$ of spaces and a $I\!\!P$-restriction $\mathbf{Q} \equiv \{Q^X : X \in \mathbf{S}\}$ of $\mathbf{S}$ such that for every element $(Q^Z, Z) \in I\!\!P$ there exists an element X of $\mathbf{S}$ and a homeomorphism f of Z onto X for which $f(Q^Z) = Q^X$.

Since $I\!\!P$ is complete, $\mathbf{Q}$ is a complete restriction of $\mathbf{S}$. Therefore, there exists a co-mark $\mathbf{M}_0$ of $\mathbf{S}$ and an $(\mathbf{M}_0, \mathbf{Q})$-admissible family R_0 of equivalence relations on $\mathbf{S}$ such that $\mathbf{Q}$ is an $(\mathbf{M}_0, \mathrm{R}_0)$-complete restriction.

On the other hand, since $I\!\!P$ is a saturated class there exists an initial co-mark $\mathbf{M}^+$ of $\mathbf{S}$ corresponding to the $I\!\!P$-restriction $\mathbf{Q}$. Let $\mathbf{M}$ be a co-mark of $\mathbf{S}$, which is a co-extension of $\mathbf{M}^+$ and $\mathbf{M}_0$.

There exists a family R^+ of equivalence relations on $\mathbf{S}$, which is an initial family corresponding to the co-mark $\mathbf{M}$ and the $I\!\!P$-restriction $\mathbf{Q}$. Denote by R an admissible family of equivalence relations on $\mathbf{S}$, which is a final refinement of R^+ and R_0. By Lemma 2.2.4, $\mathbf{Q}$ is an $(\mathbf{M}, \mathrm{R})$-complete restriction.

Now, we consider the Containing Space T and its subset $\mathrm{T}|_{\mathbf{Q}}$. By construction, the pair $(\mathrm{T}|_{\mathbf{Q}}, \mathrm{T})$ is an element of $I\!\!P$. We prove that this element is properly universal in $I\!\!P$.

Indeed, let (Q^Z, Z) be an element of $I\!\!P$. There exists an element X of $\mathbf{S}$ and a homeomorphism f of Z onto X such that $f(Q^Z) = Q^X$. Let e_{T}^X be the natural embedding of X into T and set $h = e_{\mathrm{T}}^X \circ f$. Then, h is a homeomorphism of Z into T. We prove that $Q^Z = h^{-1}(\mathrm{T}|_{\mathbf{Q}})$. It is suffices to prove that

$$(e_{\mathrm{T}}^{X})^{-1}(\mathrm{T}|_{\mathbf{Q}}) \subset Q^{X}. \tag{2.2.2}$$

Let $\mathbf{a} \in \mathrm{T}|_{\mathbf{Q}}$ and $x \in (e_{\mathrm{T}}^{X})^{-1}(\mathbf{a})$. By the definition of the natural embedding e_{T}^{X} we have $(x, X) \in \mathbf{a}$. Since $\mathbf{Q}$ is an $(\mathbf{M}, \mathrm{R})$-complete restriction we have $x \in Q^{X}$ proving relation (2.2.2). Thus, in $I\!\!P$ there exist properly universal elements. □

2.2.6 Corollary. *In the classes* $\mathbb{P}(\mathrm{Cl})$, $\mathbb{P}(\mathrm{Cl}) \cap \mathbb{P}(\text{n.dense})$, *and* $\mathbb{P}(\mathrm{Op})$ *there exist properly universal elements.* □

Definitions. Let $I\!\!P$ be a class of subsets. For every element $(Q, X) \in I\!\!P$ the space X is said to be the *space-coordinate* of this element.

The class of space-coordinates of all elements of $I\!\!P$, is said to be the *space-component* of $I\!\!P$.

The class of all spaces Q such that $(Q, X) \in I\!\!P$ for some space X, is said to be the *subset-component* of $I\!\!P$.

The following propositions are easily proved.

2.2.7 Proposition. *The space-component and subset-component of any saturated class of subsets are also saturated.* □

2.2.8 Proposition. *Let* $I\!\!E$ *be saturated class of spaces and* $I\!\!F$ *a (complete) saturated class of subsets. Then, the classes*

$$\{(Q, X) \in I\!\!F : X \in I\!\!E\} \text{ and } \{(Q, X) \in I\!\!F : Q \in I\!\!E\}$$

are (complete) saturated classes of subsets. □

2.3 Commutative Operators

Agreement. In this section it is supposed that a fixed saturated class of spaces denoted by $I\!\!R$ is given. All considered spaces are assumed to be elements of this class. In particular, the elements of the space-component of any considered class of subsets are elements of $I\!\!R$.

Definition. Suppose that for every space X an operator O_X in X (that is a mapping of the set $\mathcal{P}(X)$ into itself) is given. Then, the class

$$\mathbf{O} \equiv \{\mathrm{O}_X : X \in I\!\!R\} \tag{2.3.1}$$

of all such operators is said to be a *co-operator* on $I\!R$. Such a co-operator is said to be *topological* if for every homeomorphism h of a space X onto a space Y and for every element Q of $\mathcal{P}(X)$ we have

$$h(\mathrm{O}_X(Q)) = \mathrm{O}_Y(h(Q)).$$

In what follows of this section it is assumed that a fixed topological co-operator is given. This co-operator will be denoted by (2.3.1).

Notation. For every indexed collection $\mathbf{S}$ of spaces and a restriction

$$\mathbf{Q} \equiv \{Q^X : X \in \mathbf{S}\}$$

of $\mathbf{S}$ we set

$$\mathbf{O}(\mathbf{Q}) = \{\mathrm{O}_X(Q^X) : X \in \mathbf{S}\}.$$

Obviously, $\mathbf{O}(\mathbf{Q})$ is a restriction of $\mathbf{S}$.

Let $I\!P$ a class of subsets. We denote by $\mathbf{O}(I\!P)$ and $\mathbf{O}^{-1}(I\!P)$ the classes of subsets, which are defined as follows:

$$\mathbf{O}(I\!P) = \{(\mathrm{O}_X(Q), X) : (Q, X) \in I\!P\} \text{ and}$$

$$\mathbf{O}^{-1}(I\!P) = \{(Q, X) : (\mathrm{O}_X(Q), X) \in I\!P\}.$$

It is clear that $\mathbf{O}(I\!P)$ and $\mathbf{O}^{-1}(I\!P)$ are topological classes of subsets.

Definition. The co-operator $\mathbf{O}$ is said to be *commutative* with respect to a restriction $\mathbf{Q}$ of an indexed collection $\mathbf{S}$ of spaces if there exists an initial co-mark $\mathbf{M}^+$ of $\mathbf{S}$ corresponding to the class $I\!R$ satisfying the following condition: for every co-extension $\mathbf{M}$ of $\mathbf{M}^+$ there exists an initial family R^+ of equivalence relations on $\mathbf{S}$ corresponding to the co-mark $\mathbf{M}$ and the class $I\!R$, which is $(\mathbf{M}, \mathbf{Q})$-admissible and $(\mathbf{M}, \mathbf{O}(\mathbf{Q}))$-admissible, such that for every admissible family R of equivalence relations on $\mathbf{S}$, which is a final refinement of R^+, and for every elements $\mathbf{L}$ and $\mathbf{H}$ of $\mathrm{C}^{\diamond}(\mathrm{R})$, for which $\mathbf{H} \subset \mathbf{L}$, we have

$$\mathrm{O}_{\mathrm{T}(\mathbf{L})}(\mathrm{T}(\mathbf{H}|_{\mathbf{Q}})) = \mathrm{T}(\mathbf{H}|_{\mathbf{O}(\mathbf{Q})}).$$

(We note that by construction, the space $\mathrm{T}(\mathbf{L})$ is an element of $I\!R$).

The considered co-mark $\mathbf{M}^+$ is said to be an **O**-*commutative co-mark* (*corresponding* to the restriction **Q**) and the family R^+ is said to be an **O**-*commutative family* (*corresponding* to the co-mark **M** and the restriction **Q**). (We note that any co-extension of $\mathbf{M}^+$ is also **O**-commutative and any admissible family of equivalence relation on **S**, which is a final refinement of R^+, is also an **O**-commutative family).

Definitions. The co-operator **O** is said to be *commutative* with respect to a class $I\!P$ of subsets if it is commutative with respect to any $I\!P$-restriction of any indexed collection of spaces.

The co-operator **O** is said to be *commutative* if it is commutative with respect to any restriction of any indexed collection of spaces.

It is clear that a commutative co-operator is commutative with respect to any class of subsets.

The proof of the following two proposition are similar. We prove only the second.

2.3.1 Proposition. *Let $I\!P$ be a saturated class of subsets. If the co-operator* **O** *is commutative with respect to $I\!P$, then* $\mathbf{O}(I\!P)$ *is a saturated class of subsets.* □

2.3.2 Proposition. *Let $I\!P$ and $I\!F$ be saturated classes of subsets. If the co-operator* **O** *is commutative with respect to $I\!P$, then $I\!P \cap \mathbf{O}^{-1}(I\!F)$ is a saturated class of subsets.*

Proof. Let **S** be an indexed collection of spaces and **Q** a $(I\!P \cap \mathbf{O}^{-1}(I\!F))$-restriction of **S**. Therefore, the restriction **Q** is a $I\!P$-restriction; the restriction $\mathbf{O}(\mathbf{Q})$ of **S** is an $I\!F$-restriction; and the co-operator **O** is commutative with respect to **Q**.

Since $I\!P$ and $I\!F$ are saturated classes and **O** is commutative with respect to $I\!P$ there exists a co-mark $\mathbf{M}^+$ of **S**, which is simultaneously an initial co-mark corresponding to the class $I\!R$; an initial co-mark corresponding to the $I\!P$-restriction **Q**; an initial co-mark corresponding to the $I\!F$-restriction $\mathbf{O}(\mathbf{Q})$; and an **O**-commutative co-mark corresponding to the restriction **Q**. Consider an arbitrary co-extension **M** of the co-mark $\mathbf{M}^+$.

There exists a family R^+ of equivalence relations on **S**, which is simultaneously an initial family corresponding to the co-mark **M** and the class $I\!R$, an initial family corresponding to the co-mark **M** and the $I\!P$-restriction

$\mathbf{Q}$, an initial family corresponding to the co-mark $\mathbf{M}$ and the $\mathbb{F}$-restriction $\mathbf{O}(\mathbf{Q})$, and an $\mathbf{O}$-commutative family corresponding to the co-mark $\mathbf{M}$ and the restriction $\mathbf{Q}$. We show that $\mathbf{M}^+$ is an initial co-mark of $\mathbf{S}$ corresponding to the $(\mathbb{P}\cap\mathbf{O}^{-1}(\mathbb{F}))$-restriction $\mathbf{Q}$ and the family R^+ is an initial family corresponding to the co-mark $\mathbf{M}$ and the $(\mathbb{P}\cap\mathbf{O}^{-1}(\mathbb{F}))$-restriction $\mathbf{Q}$.

Indeed, let R be an admissible family of equivalence relations on $\mathbf{S}$, which is a final refinement of R^+, and let $\mathbf{L}$ and $\mathbf{H}$ be elements of $\mathrm{C}^{\diamond}(\mathrm{R})$ such that $\mathbf{H} \subset \mathbf{L}$. Then, we can consider the space $\mathrm{T}(\mathbf{L})$, which is an element of $\mathbb{R}$, and its subsets $\mathrm{T}(\mathbf{H}|_{\mathbf{O}(\mathbf{Q})})$ and $\mathrm{T}(\mathbf{H}|_{\mathbf{Q}})$.

Since $\mathbb{P}$ and $\mathbb{F}$ are saturated classes we have that $(\mathrm{T}(\mathbf{H}|_{\mathbf{Q}}), \mathrm{T}(\mathbf{L})) \in \mathbb{P}$ and $(\mathrm{T}(\mathbf{H}|_{\mathbf{O}(\mathbf{Q})}), \mathrm{T}(\mathbf{L})) \in \mathbb{F}$. Since $\mathbf{O}$ is commutative with respect to $\mathbb{P}$

$$\mathrm{T}(\mathbf{H}|_{\mathbf{O}(\mathbf{Q})}) = \mathrm{O}_{\mathrm{T}(\mathbf{L})}(\mathrm{T}(\mathbf{H}|_{\mathbf{Q}})).$$

This relation shows that $(\mathrm{T}(\mathbf{H}|_{\mathbf{Q}}), \mathrm{T}(\mathbf{L})) \in \mathbb{P}\cap\mathbf{O}^{-1}(\mathbb{F})$, which means that $\mathbf{M}^+$ is an initial co-mark and R^+ is an initial family. Thus, $\mathbb{P}\cap\mathbf{O}^{-1}(\mathbb{F})$ is a saturated class. □

Notation. We denote by

$$\mathbf{Bd} \equiv \{\mathrm{Bd}_X : X \in \mathbb{R}\},$$

$$\mathbf{Cl} \equiv \{\mathrm{Cl}_X : X \in \mathbb{R}\},$$

$$\mathbf{Int} \equiv \{\mathrm{Int}_X : X \in \mathbb{R}\}, \text{ and}$$

$$\mathbf{Co} \equiv \{\mathrm{Co}_X : X \in \mathbb{R}\},$$

the co-operators on $\mathbb{R}$ such that for every subset Q of a space X, $\mathrm{Bd}_X(Q)$, $\mathrm{Cl}_X(Q)$, $\mathrm{Int}_X(Q)$, and $\mathrm{Co}_X(Q)$ are the boundary, closure, the interior, and the complement of Q in X, respectively.

We note that if the co-operator $\mathbf{O}$ coincides with one of the co-operators: $\mathbf{Cl}$, $\mathbf{Bd}$, $\mathbf{Int}$, and $\mathbf{Co}$, then for every space X, a subset Q of X, and a simultaneously closed and open subset Y of X we have

$$\mathrm{O}_Y(Q \cap Y) = \mathrm{O}_X(Q) \cap Y.$$

Using this fact and Lemmas 1.4.1–1.4.6 we have the following two propositions.

2.3.3 Proposition. *The co-operator* **Cl** *is commutative.* □

2.3.4 Proposition. *The co-operators* **Bd**, **Int**, *and* **Co** *are commutative with respect to any complete class of subsets.* □

2.3.5 Corollary. *If* $\mathbb{F}$ *is a saturated class of subsets, then* $\mathbf{Cl}(\mathbb{F})$ *and* $\mathbf{Cl}^{-1}(\mathbb{F})$ *are also saturated classes of subsets.* □

2.3.6 Corollary. *If* $\mathbb{P}$ *is a complete saturated class of subsets, then the classes* $\mathbf{Bd}(\mathbb{P})$, $\mathbf{Int}(\mathbb{P})$, *and* $\mathbf{Co}(\mathbb{P})$ *are also complete saturated.* □

2.3.7 Corollary. *If* $\mathbb{F}$ *is a saturated class of subsets and* $\mathbb{P}$ *is a complete saturated class of subsets, then* $\mathbb{P} \cap \mathbf{Bd}^{-1}(\mathbb{F})$, $\mathbb{P} \cap \mathbf{Int}^{-1}(\mathbb{F})$, *and* $\mathbb{P} \cap \mathbf{Co}^{-1}(\mathbb{F})$ *are also complete saturated classes of subsets.* □

The proof of the above three corollaries follows by Propositions 2.3.1–2.3.4. We note only the following: in order to prove that $\mathbf{Cl}^{-1}(\mathbb{F})$ (in Corollary 2.3.5) is a saturated class, in the Proposition 2.3.2 we consider that $\mathbb{P}$ is the class $\mathbb{R}$ and use the fact that this class is saturated and that the co-operator **Cl** is commutative with respect to any class. The completeness of $\mathbf{Bd}(\mathbb{P})$ and $\mathbf{Int}(\mathbb{P})$ (in Corollary 2.3.6) follows by the fact that $\mathbf{Bd}(\mathbb{P})$ is a closed class of subsets and $\mathbf{Int}(\mathbb{P})$ is an open class of subsets. The completeness of $\mathbf{Co}(\mathbb{P})$ follows by the fact that if for a collection $\mathbf{S}$ of spaces, a co-mark $\mathbf{M}$ of $\mathbf{S}$, and a family R of equivalence relations on $\mathbf{S}$ a restriction $\mathbf{Q}$ of $\mathbf{S}$ is $(\mathbf{M}, \mathrm{R})$-complete, then the restriction $\mathbf{Co}(\mathbf{Q})$ of $\mathbf{S}$ is also $(\mathbf{M}, \mathrm{R})$-complete. Finally, in Corollary 2.3.7 we use also the fact that any class of subsets, which is contained in a complete class, is also complete.

2.4 Borel type sets

By the family of *Borel sets* of a space X we mean the smallest family of subsets of X containing the closed subsets of X and closed under complements and countable unions (therefore, closed also under countable intersections). We denote this family by $\mathrm{Bor}(X)$. A subset Q of a space X is said to be a *G_δ-set* (respectively, an *F_σ-set*) if Q is the intersection (respectively, the union) of countable many open (respectively, closed) subsets of X.

Let X be a space such that each closed subset of X is a G_δ-set (and therefore each open subset of X is an F_σ-set). It is easy to verify that the family $\mathrm{Bor}(X)$ can be represented as a union of some subfamilies:

$$\mathrm{Bor}(X) = \cup\{F_\gamma : \gamma \in \omega_1\},$$

defined by induction as follows: (a) F_0 is the family of closed subsets of X, (b) if $\gamma \neq 0$ is an odd (respectively, an even) ordinal, then F_γ consists of countable unions (respectively, of countable intersections) of elements of the family $\cup\{F_\beta : \beta \in \gamma\}$. Also,

$$\mathrm{Bor}(X) = \cup\{G_\gamma : \gamma \in \omega_1\},$$

where: (a) G_0 is the family of all open subsets of X, and (b) if $\gamma \neq 0$ is an odd (respectively, an even) ordinal, then G_γ consists of countable intersections (respectively, countable unions) of elements of the family $\cup\{G_\beta : \beta \in \gamma\}$.

The *multiplicative class* γ and the *additive class* γ, where $\gamma \in \omega_1$, are subfamilies of $\mathrm{Bor}(X)$ defined as follows: if γ is an odd ordinal, then the multiplicative class γ coincides with G_γ and the additive class γ with F_γ, and if γ is an even ordinal, then the multiplicative class γ coincides with F_γ and the additive class γ with G_γ.

In the present section using the Borel type process and starting with a given complete saturated class of subsets we construct some new complete saturated classes of subsets.

Agreement. In this section we denote by ν an infinite cardinal such that $\nu \leq \tau$.

Definitions. A mapping Φ of the set ν^+ into the set $\{0, 1\}$ is said to be *ν-directing* if for every ordinal $\gamma \in \nu^+$ we have

$$\Phi(\gamma + 1) = 1 - \Phi(\gamma).$$

Let X be a space, $\mathcal{G}$ a set of subsets of X, and Φ a ν-directing mapping. For every ordinal $\gamma \in \nu^+$ there is a family of subsets of X, denoted by $\mathrm{B}^\gamma_\nu(X, \Phi, \mathcal{G})$, and is defined by induction on γ as follows:

(a) $\mathrm{B}^0_\nu(X, \Phi, \mathcal{G})$ coincides with the family $\mathcal{G}$ and

(b) if $\gamma \neq 0$ and $\Phi(\gamma) = 0$ (respectively, $\Phi(\gamma) = 1$), then a subset Q of X belongs to $\mathrm{B}^\gamma_\nu(X, \Phi, \mathcal{G})$ if Q is a union (respectively, an intersection) of less than or equal to ν many elements of the set $\cup\{\mathrm{B}^\beta_\nu(X, \Phi, \mathcal{G}) : \beta \in \gamma\}$. If $\mathbb{F}$ is a class of subsets and $\mathcal{G}$ coincides with the family of all subsets Q of X such that $(Q, X) \in \mathbb{F}$, then the family $\mathrm{B}^\gamma_\nu(X, \Phi, \mathcal{G})$ is denoted by $\mathrm{B}^\gamma_\nu(X, \Phi, \mathbb{F})$.

Remark. Let X be a space such that each closed subset of X is a G_δ-set. Consider the ω-directing mappings Φ_0 and Φ_1 satisfying conditions: $\Phi_0(\gamma) = 0$ for every limit ordinal $\gamma \in \omega_1$ or for $\gamma = 0$, and $\Phi_1(\gamma) = 1 - \Phi_0(\gamma)$ for every $\gamma \in \omega_1$. Then, it is easy to verify that for an even ordinal γ, $\mathrm{B}^\gamma_\omega(X, \Phi_1, \mathbb{P}(\mathrm{Cl}))$ is the additive class γ; and $\mathrm{B}^\gamma_\omega(X, \Phi_0, \mathbb{P}(\mathrm{Op}))$ is the multiplicative class gamma; for an odd ordinal γ, $\mathrm{B}^\gamma_\omega(X, \Phi_0, \mathbb{P}(\mathrm{Op}))$ is the additive class γ; and $\mathrm{B}^\gamma_\omega(X, \Phi_1, \mathbb{P}(\mathrm{Cl}))$ is the multiplicative class γ.

Definitions. Let Λ be a set considered with the membership relation (a priore given), which is denoted by $\in$. The *level* of an element $\lambda \in \Lambda$, denoted by $l(\lambda)$, is an ordinal defined by induction in such a manner that the following conditions are satisfied:

(a) $l(\lambda) = 0$ if and only if $\lambda \cap \Lambda = \emptyset$, and

(b) $l(\lambda) = \alpha$, where $\alpha > 0$, if and only if α is the minimal ordinal γ, for which $l(\lambda') \in \gamma$ for every $\lambda' \in \lambda \cap \Lambda$.

It is easy to see that the level of every element of Λ is uniquely determined. For every ordinal γ we denote by Λ^γ the set of all elements of Λ, the level of which is γ.

For every element $\lambda \in \Lambda$ we define the notion of the *completion* of λ, which is a subset of Λ denoted by $c(\lambda)$. This notion is defined by induction on the level of λ as follows: if $l(\lambda) = 0$, then $c(\lambda) = \{\lambda\}$ and if $l(\lambda) > 0$, then

$$c(\lambda) = \{\lambda\} \cup (\cup\{c(\lambda') : \lambda' \in \lambda \cap \Lambda\}).$$

The set Λ is said to be *peaked* if there exists an element $\lambda \in \Lambda$ such that $c(\lambda) = \Lambda$. In this case, the element λ is called the *peak* of Λ. It is easy to see that the peak of Λ is uniquely determined. If the level of the peak of Λ is γ, then the set Λ is also called *γ-peaked.*

The set Λ is called a *ν-set* if for every element $\lambda \in \Lambda$ we have $|\lambda \cap \Lambda| \leq \nu$. We note that if Λ is a peaked ν-set then the level of its peak belongs to ν^+ and the cardinality of Λ is $\leq \nu$. These can be proved by induction on the level of the peak of Λ.

Suppose that Λ is a ν-set. We give the notion of a *free element* λ of Λ. This notion is defined by induction on the level of λ. By definition, all elements of the level 0 are *free.* Suppose that we have defined the free elements of the level $< \gamma$, where $\gamma > 0$. Then, an element λ of the level γ is called *free* if the following conditions are satisfied:

(a) $|\lambda \cap \Lambda^\beta| = \nu$ for every $\beta \in \gamma$,

(b) $c(\lambda_1) \cap c(\lambda_2) = \emptyset$ if λ_1 and λ_2 are distinct elements of $\lambda \cap \Lambda$, and
(c) all elements of $\lambda \cap \Lambda$ are free.

A ν-set is said to be *free* if all its elements are free. It is easy to prove that a peaked ν-set is free if and only if its peak is free.

Definition. Suppose that for every $\delta \in \nu$ we have a γ_δ-peaked ν-set Λ_δ such that $\Lambda_{\delta_1} \cap \Lambda_{\delta_2} = \emptyset$ if $\delta_1 \neq \delta_2$, and let λ_δ be the peak of this set. Denote by β the minimal ordinal γ for which $\gamma_\delta \in \gamma$ for every $\delta \in \nu$. It is clear that $\beta \in \nu^+$. Let

$$\Lambda = \{\lambda_p\} \cup (\cup\{\Lambda_\delta : \delta \in \nu\}),$$

where $\lambda_p = \{\lambda_\delta : \delta \in \nu\}$. It is easy to verify that Λ is a β-peaked ν-set and that λ_p is the peak of this set. The set Λ is called the *peaked union* of the sets Λ_δ, $\delta \in \nu$.

2.4.1 Lemma. *For every $\gamma \in \nu^+$ there exists a free γ-peaked ν-set Λ.*

Proof. We prove the lemma by induction on the ordinal γ. For $\gamma = 0$ the lemma is trivial. Let $\gamma > 0$. Suppose that for every $\beta \in \gamma$ we have constructed a free β-peaked ν-set. We prove the existence of such a set for the ordinal γ.

For every $\beta \in \gamma$, let Λ_β be a free β-peaked ν-set and let λ_β be the peak of Λ_β. For every $\delta \in \nu$, let $\Lambda_{(\beta,\delta)}$ be a copy of this set and $\lambda_{(\beta,\delta)}$ be the peak of this copy. We suppose that $\Lambda_{(\beta_1,\delta_1)} \cap \Lambda_{(\beta_2,\delta_2)} = \emptyset$ if either $\beta_1 \neq \beta_2$ or $\delta_1 \neq \delta_2$.

It is easy to verify that the peaked union of the sets $\Lambda_{(\beta,\delta)}$, $\beta \in \gamma$, $\delta \in \nu$, is a free γ-peaked ν-set. □

Definition. Two sets Λ_1 and Λ_2 are called *isomorphic* (with respect to the membership relation) if there exists a one-to-one mapping f of Λ_1 onto Λ_2 such that f and f^{-1} preserve the membership relation.

2.4.2 Lemma. *For every $\gamma \in \nu^+$ any two free γ-peaked ν-sets are isomorphic.*

Proof. We prove the lemma by induction on the ordinal γ. The lemma is trivial if $\gamma = 0$. Suppose that the lemma is true for every $\beta \in \gamma$. We prove it for $\gamma > 0$.

Let Λ_1 and Λ_2 be two free γ-peaked ν-sets and let λ_{1p} and λ_{2p} be the peaks of these sets. Since for every $\beta \in \gamma$, $|\lambda_{1p} \cap \Lambda_1^\beta| = |\lambda_{2p} \cap \Lambda_2^\beta| = \nu$ there exists a one-to-one mapping f_p of the set $\lambda_{1p} \cap \Lambda$ onto the set $\lambda_{2p} \cap \Lambda$ which preserves the level, that is if $\lambda \in \lambda_{1p} \cap \Lambda_1^\beta$, then $f_p(\lambda) \in \lambda_{2p} \cap \Lambda_2^\beta$.

Let $\lambda \in \lambda_{1p} \cap \Lambda_1^\beta$ for some $\beta \in \gamma$. The completions $c(\lambda)$ of λ and $c(f_p(\lambda))$ of $f_p(\lambda)$ are free β-peaked ν-sets with peaks λ and $f_p(\lambda)$, respectively. Since the level of λ is equal to the level of $f_p(\lambda)$ by inductive assumption there exists an isomorphism f_λ of $c(\lambda)$ onto $c(f_p(\lambda))$.

Let f be a mapping of Λ_1 into Λ_2 defined as follows: (a) $f(\lambda_{1p}) = \lambda_{2p}$, and (b) on the set $c(\lambda)$, $\lambda \in \lambda_{1p} \cap \Lambda$, f coincides with the isomorphism f_λ. The mapping f is well defined because the intersection of the completions of two distinct elements of the set λ_{1p} is empty.

It is easy to verify that the mapping f is one-to-one and onto and that the mappings f and f^{-1} preserve the membership relation, that is f is an isomorphism. □

Definition. Let X be a space and Φ a ν-directing mapping. Let also Λ be a ν-set and ψ_X be a mapping of the set Λ into the set $\mathcal{P}(X)$. The mapping ψ_X is said to be Φ-*operational* if for every element λ of Λ of the level $\gamma > 0$ we have

$$\psi_X(\lambda) = \begin{cases} \cup\{\psi_X(\lambda') : \lambda' \in \lambda \cap \Lambda\} & \text{if } \Phi(\gamma) = 0, \\ \cap\{\psi_X(\lambda') : \lambda' \in \lambda \cap \Lambda\} & \text{if } \Phi(\gamma) = 1. \end{cases}$$

By induction on the level of elements of Λ it is easy to prove that every mapping ψ_X^0 of the set Λ^0 into the set $\mathcal{P}(X)$ can be uniquely extended to a Φ-operational mapping ψ_X of Λ into $\mathcal{P}(X)$.

2.4.3 Lemma. *Let X be a space, $\mathcal{G}$ a subset of $\mathcal{P}(X)$, and Φ a ν-directing mapping. A subset Q of X belongs to $\mathrm{B}_\nu^\gamma(X, \Phi, \mathcal{G})$, $\gamma \in \nu^+$, if and only if there exists a (free) γ-peaked ν-set Λ and a Φ-operational mapping $\psi_X : \Lambda \to \mathcal{P}(X)$ such that: (a) $\psi_X(\lambda_p) = Q$, where λ_p is the peak of Λ, and (b) $\psi_X(\Lambda^0) \subset \mathcal{G}$.*

Proof. We prove the lemma by induction on the ordinal γ. For $\gamma = 0$ the lemma is trivial. Suppose that the lemma is true for all ordinals $< \gamma$. We prove the lemma for the ordinal $\gamma > 0$.

Let $Q \in \mathrm{B}_\nu^\gamma(X, \Phi, \mathcal{G})$. Then, by the definition of the family $\mathrm{B}_\nu^\gamma(X, \Phi, \mathcal{G})$, for every $\delta \in \nu$ there exists an element Q_δ of the family

$$\cup\{\mathrm{B}_\nu^\beta(X,\Phi,\mathcal{G}) : \beta\in\gamma\}$$

such that

$$Q=\begin{cases}\cup\{Q_\delta:\delta\in\nu\} & \text{if } \Phi(\gamma)=0,\\ \cap\{Q_\delta:\delta\in\nu\} & \text{if } \Phi(\gamma)=1.\end{cases}$$

By inductive assumption for every element δ of ν there exists a (free) β_δ-peaked ν-set Λ_δ and a Φ-operational mapping ψ_X^δ of the set Λ_δ into the set $\mathcal{P}(X)$ such that: (a) $\beta_\delta\in\gamma$, (b) $\psi_X^\delta(\lambda_\delta)=Q_\delta$, where λ_δ is the peak of Λ_δ, and (c) $\psi_X^\delta(\Lambda_\delta^0)\subset\mathcal{G}$. Without loss of generality, we can suppose that $\Lambda_{\delta_1}\cap\Lambda_{\delta_2}=\emptyset$ if $\delta_1\neq\delta_2$.

For every $(\beta,\delta)\in\gamma\times\nu$ let $\Lambda_{(\beta,\delta)}$ denote: (a) the set Λ_δ if $\beta=\beta_\delta$, and (b) a free β-peaked ν-set if $\beta\neq\beta_\delta$. We suppose that $\Lambda_{(\beta_1,\delta_1)}\cap\Lambda_{(\beta_2,\delta_2)}=\emptyset$ if $(\beta_1,\delta_1)\neq(\beta_2,\delta_2)$.

Let Λ be the peaked union of the sets $\Lambda_{(\beta,\delta)}$, $(\beta,\delta)\in\gamma\times\nu$. By construction, the set Λ is a (free) γ-peaked ν-set. Let λ_p denote the peak of this set. We define the mapping ψ_X of Λ into $\mathcal{P}(X)$ setting

$$\psi_X(\lambda)=\begin{cases}\psi_X^\delta(\lambda) & \text{if } \lambda\in\Lambda_\delta \text{ for some } \delta\in\nu,\\ \emptyset & \text{if } \Phi(\gamma)=0,\ \lambda\neq\lambda_p, \text{ and } \lambda\notin\Lambda_\delta \text{ for any } \delta\in\nu,\\ X & \text{if } \Phi(\gamma)=1,\ \lambda\neq\lambda_p, \text{ and } \lambda\notin\Lambda_\delta \text{ for any } \delta\in\nu,\\ Q & \text{if } \lambda=\lambda_p.\end{cases}$$

It is easy to verify that ψ_X is a Φ-operational mapping satisfying the conditions of the lemma.

Conversely, let Q be a subset of X for which there exists a (free) γ-peaked ν-set Λ and a Φ-operational mapping ψ_X of Λ into $\mathcal{P}(X)$ satisfying the conditions of the lemma. For every $\lambda\in\lambda_p\cap\Lambda$, where λ_p is the peak of Λ, we put $Q_\lambda=\psi_X(\lambda)$. Since the completion $c(\lambda)$ is a β-peaked ν-set, $\beta\in\gamma$, by inductive assumption Q_λ belongs to the family

$$\cup\{\mathrm{B}_\nu^\beta(X,\Phi,\mathcal{G}) : \beta\in\gamma\}.$$

On the other hand, since ψ_X is Φ-operational we have

$$Q=\begin{cases}\cup\{Q_\lambda:\lambda\in\lambda_p\cap\Lambda\} & \text{if } \Phi(\gamma)=0,\\ \cap\{Q_\lambda:\lambda\in\lambda_p\cap\Lambda\} & \text{if } \Phi(\gamma)=1.\end{cases}$$

This means that $Q\in\mathrm{B}_\nu^\gamma(X,\Phi,\mathcal{G})$. □

The following lemma can be easily proved by induction on γ.

2.4.4 Lemma. *Let Y be a subspace of a space X, $\mathcal{G}$ a family of subsets of X, and Φ a ν-directing mapping. If $Q \in \mathrm{B}_\nu^\gamma(X, \Phi, \mathcal{G})$, $\gamma \in \nu^+$, then $Q \cap Y \in \mathrm{B}_\nu^\gamma(Y, \Phi, \mathcal{G}_Y)$, where $\mathcal{G}_Y = \{Q \cap Y : Q \in \mathcal{G}\}$.* $\square$

Notation. Let $I\!F$ be a class of subsets. For every ordinal $\gamma \in \nu^+$ and for every ν-directing mapping Φ we denote by $I\!P(\gamma, \nu, \Phi, I\!F)$ the class of subsets consisting of all pairs (Q, X) such that $Q \in \mathrm{B}_\nu^\gamma(X, \Phi, I\!F)$.

We denote by $I\!P(\mathrm{G}_\delta\text{-set})$ (respectively, by $I\!P(\mathrm{F}_\sigma\text{-set})$) the class of subsets consisting of all pairs (Q, X), where Q is a G_δ-set (respectively, an F_σ-set) of a space X. Obviously, $I\!P(1, \omega, \Phi, I\!P(\mathrm{Op})) = I\!P(\mathrm{G}_\delta\text{-set})$, (respectively, $I\!P(1, \omega, \Phi, I\!P(\mathrm{Cl})) = I\!P(\mathrm{F}_\sigma\text{-set})$), where Φ is a ω^+-directing mapping for which $\Phi(0) = 0$ (respectively, $\Phi(0) = 1$).

2.4.5 Proposition. *Let $I\!F$ be a complete saturated class of subsets and Φ a ν-directing mapping. Then the class $I\!P(\gamma, \nu, \Phi, I\!F)$, $\gamma \in \nu^+$, is also a complete saturated class of subsets.*

Proof. We prove the proposition by induction on the ordinal $\gamma \in \nu$. If $\gamma = 0$, then $I\!P(\gamma, \nu, \Phi, I\!F)$ coincides with the class $I\!F$, and, therefore, the proposition is trivial. Let $\gamma > 0$. Assuming that the proposition is true for every ordinal less than γ we prove it for the ordinal γ.

Let $\mathbf{S}$ be an indexed collection of spaces and $\mathbf{Q} \equiv \{Q^X : X \in \mathbf{S}\}$ a $I\!P(\gamma, \nu, \Phi, I\!F)$-restriction of $\mathbf{S}$. Therefore, for every $X \in \mathbf{S}$ the subset Q^X of X belongs to the family $\mathrm{B}_\nu^\gamma(X, \Phi, I\!F)$.

Let Λ be a free γ-peaked ν-set and let λ_p be the peak of Λ. By Lemma 2.4.3 for every $X \in \mathbf{S}$ there exists a Φ-operational mapping $\psi_X : \Lambda \to \mathcal{P}(X)$ such that: (a) $\psi_X(\lambda_p) = Q^X$ and (b) for every $\lambda \in \Lambda_0$, $(\psi_X(\lambda), X) \in I\!F$. For every $\lambda \in \Lambda \setminus \{\lambda_p\}$ we set $Q^X(\lambda) = \psi_X(\lambda)$ and

$$\mathbf{Q}(\lambda) = \{Q^X(\lambda) : X \in \mathbf{S}\}.$$

Since ψ_X is a Φ-operational mapping we have

$$Q^X = \begin{cases} \cup\{Q^X(\lambda) : \lambda \in \lambda_p \cap \Lambda\} & \text{if } \Phi(\gamma) = 0, \\ \cap\{Q^X(\lambda) : \lambda \in \lambda_p \cap \Lambda\} & \text{if } \Phi(\gamma) = 1. \end{cases} \tag{2.4.1}$$

Therefore,

$$\mathbf{Q} = \begin{cases} \vee\{\mathbf{Q}(\lambda) : \lambda \in \lambda_p \cap \Lambda\} & \text{if } \Phi(\gamma) = 0, \\ \wedge\{\mathbf{Q}(\lambda) : \lambda \in \lambda_p \cap \Lambda\} & \text{if } \Phi(\gamma) = 1. \end{cases} \tag{2.4.2}$$

For a given $\lambda \in \Lambda \setminus \{\lambda_p\}$ we denote by $\beta(\lambda) \equiv \beta$ the level of λ. The completion $c(\lambda)$ of λ is a free β-peaked ν-set the peak of which is the element λ. Using Lemma 2.4.3 it is easy to prove by induction that for every $X \in \mathbf{S}$ the set $\psi_X(\lambda)$ belongs to the family $\mathrm{B}_\nu^\beta(X, \Phi, \mathbb{F})$, that is $(\psi_X(\lambda), X) \in \mathbb{P}(\beta, \nu, \Phi, \mathbb{F})$. Hence, the restriction $\mathbf{Q}(\lambda)$ of $\mathbf{S}$ is a $\mathbb{P}(\beta, \nu, \Phi, \mathbb{F})$-restriction.

By inductive assumption, $\mathbb{P}(\beta, \nu, \Phi, \mathbb{F})$ is a complete saturated class. Therefore, $\mathbf{Q}(\lambda)$ is a complete restriction. Let $\mathbf{M}_\lambda$ be a co-mark of $\mathbf{S}$ and R_λ an $(\mathbf{M}_\lambda, \mathbf{Q}(\lambda))$-admissible family of equivalence relations on $\mathbf{S}$ such that $\mathbf{Q}(\lambda)$ is an $(\mathbf{M}_\lambda, \mathrm{R}_\lambda)$-complete restriction. Also, let $\mathbf{M}_\lambda^+$ be an initial co-mark of $\mathbf{S}$ corresponding to the $\mathbb{P}(\beta, \nu, \Phi, \mathbb{F})$-restriction $\mathbf{Q}(\lambda)$. Since $|\Lambda| \leq \nu \leq \tau$ by Lemma 2.1.2 there exists a co-mark $\mathbf{M}^+$ of $\mathbf{S}$, which is a co-extension of all co-marks $\mathbf{M}_\lambda$ and $\mathbf{M}_\lambda^+$ for every $\lambda \in \Lambda \setminus \{\lambda_p\}$. Consider an arbitrary co-mark $\mathbf{M}$ of $\mathbf{S}$, which is a co-extension of $\mathbf{M}^+$. Then, $\mathbf{M}$ is a co-extension of all co-marks $\mathbf{M}_\lambda$ and $\mathbf{M}_\lambda^+$ for every $\lambda \in \Lambda \setminus \{\lambda_p\}$. Therefore, there exists an $(\mathbf{M}, \mathbf{Q}(\lambda))$-admissible family R_λ^+ of equivalence relation on $\mathbf{S}$, which is an initial family corresponding to the co-mark $\mathbf{M}$ and the $\mathbb{P}(\beta, \nu, \Phi, \mathbb{F})$-restriction $\mathbf{Q}(\lambda)$. By Lemma 2.1.1 there exists an $(\mathbf{M}, \mathbf{Q})$-admissible family R^+ of equivalence relations on $\mathbf{S}$, which is a final refinement of all families R_λ and R_λ^+ for every $\lambda \in \Lambda \setminus \{\lambda_p\}$.

We prove that the co-mark $\mathbf{M}^+$ is an initial co-mark corresponding to the $\mathbb{P}(\beta, \nu, \Phi, \mathbb{F})$-restriction $\mathbf{Q}$ and R^+ is an initial family corresponding to the co-mark $\mathbf{M}$ and the $\mathbb{P}(\beta, \nu, \Phi, \mathbb{F})$-restriction $\mathbf{Q}$. For this purpose we consider an arbitrary admissible family R of equivalence relations on $\mathbf{S}$, which is a final refinement of R^+, and let $\mathbf{L}$ and $\mathbf{H}$ be elements of $\mathrm{C}^\diamond(\mathrm{R})$ such that $\mathbf{H} \subset \mathbf{L}$. Then, R is a final refinement of all families R_λ and R_λ^+ for every $\lambda \in \Lambda \setminus \{\lambda_p\}$. By Lemma 1.2.3, $\mathbf{Q}(\lambda)$ is an $(\mathbf{M}, \mathrm{R})$-complete restriction. By relation (2.4.2) and Lemma 1.4.7, $\mathbf{Q}$ is also an $(\mathbf{M}, \mathrm{R})$-complete restriction. This fact means that $\mathbb{P}(\gamma, \nu, \Phi, \mathbb{F})$ is a complete class.

By inductive assumption $(\mathrm{T}(\mathbf{H}|_{\mathbf{Q}(\lambda)}), \mathrm{T}(\mathbf{L})) \in \mathbb{P}(\beta, \nu, \Phi, \mathbb{F})$, that is $\mathrm{T}(\mathbf{H}|_{\mathbf{Q}(\lambda)})$ belongs to the family $\mathrm{B}_\nu^\beta(\mathrm{T}(\mathbf{L}), \Phi, \mathbb{F})$. We need to prove that $(\mathrm{T}(\mathbf{H}|_{\mathbf{Q}}), \mathrm{T}(\mathbf{L})) \in \mathbb{P}(\gamma, \nu, \Phi, \mathbb{F})$ or, equivalently, that the subset $\mathrm{T}(\mathbf{H}|_{\mathbf{Q}})$ of $\mathrm{T}(\mathbf{L})$ belongs to the family $\mathrm{B}_\nu^\gamma(\mathrm{T}(\mathbf{L}), \Phi, \mathbb{F})$. For this proof, it suffices to show that

$$\mathrm{T}(\mathbf{H}|_{\mathbf{Q}}) = \begin{cases} \cup\{\mathrm{T}(\mathbf{H}|_{\mathbf{Q}(\lambda)}) : \lambda \in \lambda_p \cap \Lambda\} & \text{if } \Phi(\gamma) = 0, \\ \cap\{\mathrm{T}(\mathbf{H}|_{\mathbf{Q}(\lambda)}) : \lambda \in \lambda_p \cap \Lambda\} & \text{if } \Phi(\gamma) = 1. \end{cases}$$

We prove the case $\Phi(\gamma) = 1$. The case $\Phi(\gamma) = 0$ is proved similarly. Let $\mathbf{a} \in \mathrm{T}(\mathbf{H}|_{\mathbf{Q}})$. There exists an element X of $\mathbf{H}$ and a point x of Q^X such that $(x, X) \in \mathbf{a}$. By relation (2.4.1), $x \in Q^X(\lambda)$ for every $\lambda \in \lambda_p \cap \Lambda$. By Corollary 1.3.5, $\mathbf{a} \in \mathrm{T}(\mathbf{H}|_{\mathbf{Q}(\lambda)})$. Thus,

$$\mathrm{T}(\mathbf{H}|_{\mathbf{Q}}) \subset \cap\{\mathrm{T}(\mathbf{H}|_{\mathbf{Q}(\lambda)}) : \lambda \in \lambda_p \cap \Lambda\}.$$

Conversely, let

$$\mathbf{a} \in \cap\{\mathrm{T}(\mathbf{L}|_{\mathbf{Q}(\lambda)}) : \lambda \in \lambda_p \cap \Lambda\}.$$

Then, $\mathbf{a} \in \mathrm{T}(\mathbf{H}|_{\mathbf{Q}(\lambda)})$ for every $\lambda \in \lambda_p \cap \Lambda$. Let $(x, X) \in \mathbf{a}$. Obviously, $X \in \mathbf{H}$. Since $\mathbf{Q}(\lambda)$ is an $(\mathbf{M}, \mathrm{R})$-complete restriction, $x \in Q^X(\lambda)$ and relation (2.4.1) implies that $x \in Q^X$. By Corollary 1.3.5, $\mathbf{a} \in \mathrm{T}|_{\mathbf{Q}}$. Hence,

$$\cap\{\mathrm{T}(\mathbf{H}|_{\mathbf{Q}(\lambda)}) : \lambda \in \lambda_p \cap \Lambda\} \subset \mathrm{T}(\mathbf{H}|_{\mathbf{Q}}),$$

and, therefore,

$$\mathrm{T}(\mathbf{H}|_{\mathbf{Q}}) = \cap\{\mathrm{T}(\mathbf{H}|_{\mathbf{Q}(\lambda)}) : \lambda \in \lambda_p \cap \Lambda\}.$$

Thus, R^+ is an initial family and $\mathbf{M}^+$ is an initial co-mark, which means that $\mathbb{P}(\gamma, \nu, \Phi, \mathbb{F})$ is a saturated class. □

2.4.6 Corollary. *For every ν-directing mapping Φ and for every $\gamma \in \nu^+$ the classes $\mathbb{P}(\gamma, \nu, \Phi, \mathrm{Cl})$ and $\mathbb{P}(\gamma, \nu, \Phi, \mathrm{Op})$ are complete saturated classes of subsets. Therefore, in these classes there exist properly universal elements.* □

In particular, we have the following consequence.

2.4.7 Corollary. *The classes $\mathbb{P}(\mathrm{G}_\delta\text{-set})$ and $\mathbb{P}(\mathrm{F}_\sigma\text{-set})$ are complete saturated classes of subsets. Therefore, in these classes there exist properly universal elements.* □

2.5 Saturated classes of bases

The following lemma is easily proved.

2.5.1 Lemma. *Let* $\mathbf{S}$ *be an indexed collection of spaces,* $\mathbf{M}$ *a co-mark of* $\mathbf{S}$*, and* R *an* $\mathbf{M}$*-admissible family of equivalence relations on* $\mathbf{S}$*. Suppose that* $\mathbf{M}$ *is a co-extension of a co-indexed* $\mathbf{S}$*-family* $\mathbf{N}$*. If* θ_0 *and* θ_1 *are two indicial mappings from* $\mathbf{N}$ *to* $\mathbf{M}$*, then for every* $\mathbf{L} \in \mathrm{C}^{\diamondsuit}(\mathrm{R})$*,* $\mathrm{B}^{\mathbf{L}}_{\diamondsuit,\theta_0(\tau)} = \mathrm{B}^{\mathbf{L}}_{\diamondsuit,\theta_1(\tau)}$. $\square$

Definitions. The class consisting of all pairs (B, X), where B is a base for a space X of cardinality $\leq \tau$, is called the *class of all bases* and is denoted by $\mathbb{P}(\text{base})$. Any subclass of $\mathbb{P}(\text{base})$ is called a *class of bases.* Such a class $\mathbb{P}$ is said to be *topological* if for every homeomorphism h of a space X onto a space Y the condition $(B, X) \in \mathbb{P}$ implies that $(\{h(V) : V \in B\}, Y) \in \mathbb{P}$. In what follows, an arbitrary considered class of bases is assumed to be topological.

Definitions. Let $\mathbb{P}$ be a class of bases. We say that a base B for a space X is a *$\mathbb{P}$-base* if $(B, X) \in \mathbb{P}$.

A co-base $\mathbf{B}$ for a collection $\mathbf{S}$ of spaces is said to be a *$\mathbb{P}$-co-base* if for every $X \in \mathbf{S}$, $\mathbf{B}(X)$ is a $\mathbb{P}$-base for X.

Definition. A non-empty class $\mathbb{P}$ of bases is said to be *saturated* (respectively, *weakly saturated*) if for every indexed collection $\mathbf{S}$ of spaces, for every $\mathbb{P}$-co-base $\mathbf{B}$ for $\mathbf{S}$, and for every (respectively, for some) co-indication $\mathbf{N}$ of $\mathbf{B}$ there exists a co-extension $\mathbf{M}^+$ of $\mathbf{N}$ satisfying the following condition: for every co-extension $\mathbf{M}$ of $\mathbf{M}^+$ there exists an $\mathbf{M}$-admissible family R^+ of equivalence relations on $\mathbf{S}$ such that for every admissible family R of equivalence relations on $\mathbf{S}$, which is a final refinement of R^+, and for every element $\mathbf{L} \in \mathrm{C}^{\diamondsuit}(\mathrm{R})$, we have $(\mathrm{B}^{\mathbf{L}}_{\diamondsuit,\theta(\tau)}, \mathrm{T}(\mathbf{L})) \in \mathbb{P}$, where θ is an indicial mapping from $\mathbf{N}$ to $\mathbf{M}$.

The considered co-mark $\mathbf{M}^+$ is called an *initial co-mark* (*corresponding to the co-indication* $\mathbf{N}$ *of* $\mathbf{B}$ *and the class* $\mathbb{P}$) and the family R^+ is called an *initial family* (*corresponding to the co-mark* $\mathbf{M}$*, the co-indication* $\mathbf{N}$ *of* $\mathbf{B}$*, and the class* $\mathbb{P}$).

Obviously, the class of all bases is saturated and any saturated class of bases is weakly saturated.

2.5.2 Proposition. *The non-empty intersection of not more than τ many saturated classes of bases is also a saturated class of bases.*

Proof. Suppose that for every $\delta \in \tau$, $I\!\!P_\delta$ is a saturated class of bases and let $I\!\!P = \cap\{I\!\!P_\delta : \delta \in \tau\}$. Let $\mathbf{S}$ be an indexed collection of spaces and $\mathbf{B}$ a $I\!\!P$-co-base for $\mathbf{S}$. It is clear that $\mathbf{B}$ is a $I\!\!P_\delta$-co-base for every $\delta \in \tau$. Consider an arbitrary co-indication $\mathbf{N}$ of $\mathbf{B}$.

For every $\delta \in \tau$ denote by $\mathbf{M}_\delta^+$ an initial co-mark of $\mathbf{S}$ corresponding to the class $I\!\!P_\delta$ and let $\mathbf{M}^+$ be a co-extension of all $\mathbf{M}_\delta$. Consider an arbitrary co-extension

$$\mathbf{M} \equiv \{\{U_\delta^X : \delta \in \tau\} : X \in \mathbf{S}\}$$

of $\mathbf{M}^+$. Then, $\mathbf{M}$ is also a co-extension of $\mathbf{N}$. We denote by θ an indicial mapping from $\mathbf{N}$ to $\mathbf{M}$.

Let R_δ^+ be an initial family of equivalence relations on $\mathbf{S}$ corresponding to the co-mark $\mathbf{M}$ and the class $I\!\!P_\delta$. Let also R^+ be a family of equivalence relations on $\mathbf{S}$, which is a final refinement of all R_δ^+. Consider an arbitrary admissible family R of equivalence relations on $\mathbf{S}$, which is a final refinement of R^+, and let $\mathbf{L}$ be an element of $\mathrm{C}^\diamondsuit(\mathrm{R})$.

Since $I\!\!P_\delta$ is saturated for every $\delta \in \tau$, the base $\mathrm{B}^{\mathbf{L}}_{\diamondsuit,\theta(\tau)}$ for the space $\mathrm{T}(\mathbf{L})$ is a $I\!\!P_\delta$-base. By the definition of $I\!\!P$, $\mathrm{B}^{\mathbf{L}}_{\diamondsuit,\theta(\tau)}$ is also a $I\!\!P$-base. Hence, $\mathbf{M}^+$ is an initial co-mark corresponding to the class $I\!\!P$ and R^+ is an initial family corresponding to the co-mark $\mathbf{M}$ and the class $I\!\!P$. Thus, $I\!\!P$ is a saturated class. $\square$

Similarly we can prove the following proposition.

2.5.3. Proposition. *The non-empty intersection of a saturated class of bases and a weakly saturated class of bases is a weakly saturated class of bases.* $\square$

Definition. An element (B^T, T) of a class $I\!\!P$ of bases is said to be *universal in* $I\!\!P$ if for every element (B^X, X) of $I\!\!P$ there exists an embedding h of X into T such that $B^X = \{h^{-1}(V) : V \in B^T\}$.

The following proposition is proved similarly to the Proposition 2.1.4.

2.5.4 Proposition. *In any weakly saturated class of bases there exist universal elements.* $\square$

Definition. Let $I\!\!P$ be a class of bases. Every element (B, X) of $I\!\!P$ is said to be the *space-coordinate* of this element. The class of space-coordinates of all elements of $I\!\!P$ is said to be the *space-component* of $I\!\!P$.

The following two propositions are easily proved.

2.5.5. Proposition. *The space-component of any weakly saturated class of bases is saturated.* □

2.5.6. Proposition. *Let $I\!\!P$ be a (weakly) saturated class of bases and $I\!\!E$ a saturated class of spaces. Then, the class*

$$\{(B, X) \in I\!\!P : X \in I\!\!E\}$$

is a (weakly) saturated class of bases. □

In the next proposition we give a method of construction of saturated classes of bases.

2.5.7 Proposition. *Let $I\!\!K$ be a saturated class of subsets. Then, the class $I\!\!P$ of bases consisting of all pairs $(B, X) \in \mathbb{P}(\text{base})$ such that $(U, X) \in I\!\!K$ for every element U of B, is a saturated class.*

Proof. Let $\mathbf{S}$ be an indexed collection of spaces, $\mathbf{B}$ a $I\!\!P$-co-base for $\mathbf{S}$, and let

$$\mathbf{N} \equiv \{\{V_\eta^X : \eta \in \tau\} : X \in \mathbf{S}\}$$

be an arbitrary co-indication of $\mathbf{B}$.

For every $\eta \in \tau$ we consider the restriction

$$\mathbf{V}_\eta \equiv \{V_\eta^X : X \in \mathbf{S}\}$$

of $\mathbf{S}$. Since $\mathbf{B}$ is a $I\!\!P$-co-base, this restriction is a $I\!\!K$-restriction of $\mathbf{S}$. Since $I\!\!K$ is a saturated class there exists a co-mark $\mathbf{M}^+$ of $\mathbf{S}$ such that for every $\eta \in \tau$, $\mathbf{M}^+$ is an initial co-mark corresponding to the $I\!\!K$-restriction $\mathbf{V}_\eta$. We can also suppose that $\mathbf{M}^+$ is a co-extension of $\mathbf{N}$. Let

$$\mathbf{M} \equiv \{\{U_\delta^X : \delta \in \tau\} : X \in \mathbf{S}\}$$

be an arbitrary co-extension of $\mathbf{M}^+$. Then, $\mathbf{M}$ is also a co-extension of $\mathbf{N}$. Denote by θ an indicial mapping from $\mathbf{N}$ to $\mathbf{M}$.

Let R^+ be a family of equivalence relations on $\mathbf{S}$ such that for every $\eta \in \tau$, R^+ is an initial family corresponding to the co-mark $\mathbf{M}$ and the $I\!\!K$-restriction $\mathbf{V}_\eta$. Let R be an arbitrary admissible family of equivalence relations on $\mathbf{S}$, which is a final refinement of R^+. We prove that $\mathbf{M}^+$ is an initial co-mark of $\mathbf{S}$ corresponding to the co-indication $\mathbf{N}$ of $\mathbf{B}$ and the class $I\!\!P$ and R^+ is an initial family corresponding to the co-mark $\mathbf{M}$, the co-indication $\mathbf{N}$ of $\mathbf{B}$, and the class $I\!\!P$. To this end we denote by $\mathbf{L}$ an arbitrary element of $\mathrm{C}^{\diamondsuit}(\mathrm{R})$ and consider the base $\mathrm{B}^{\mathbf{L}}_{\diamondsuit,\theta(\tau)}$ for $\mathrm{T}(\mathbf{L})$. Any element $U^{\mathrm{T}}_{\delta}(\mathbf{H})$ of this base coincides with the subset $\mathrm{T}(\mathbf{H}|_{\mathbf{V}_\eta})$ of $\mathrm{T}(\mathbf{L})$, where $\eta = \theta^{-1}(\delta)$. By construction, $(\mathrm{T}(\mathbf{H}|_{\mathbf{V}_\eta}), \mathrm{T}(\mathbf{L})) \in I\!\!K$. Therefore, $(\mathrm{B}^{\mathbf{L}}_{\diamondsuit,\theta(\tau)}, \mathrm{T}(\mathbf{L})) \in I\!\!P$.

Hence, $\mathbf{M}^+$ is an initial co-mark and R^+ is an initial family. Thus, $I\!\!P$ is a saturated class of bases. □

Proposition 2.4.5 and Corollaries 2.3.5–2.3.7 imply the following two cconsequences.

2.5.8 Corollary. *Let ν be an infinite cardinal such that $\nu \le \tau$, $\gamma \in \nu^+$, Φ a ν-directing mapping, and let $I\!\!F$ be a complete saturated class of subsets. Then, the non-empty class consisting of all pairs (B, X) of* $\mathbb{P}(\mathrm{base})$ *satisfying one of the following conditions:*

(1) $(U, X) \in \mathbb{P}(\gamma, \nu, \Phi, I\!\!F)$, $U \in B$,
(2) $(U, X) \in \mathbf{Cl}^{-1}(\mathbb{P}(\gamma, \nu, \Phi, I\!\!F))$, $U \in B$,
(3) $(U, X) \in \mathbf{Bd}^{-1}(\mathbb{P}(\gamma, \nu, \Phi, I\!\!F))$, $U \in B$,
(4) $(U, X) \in \mathbf{Int}(\mathbf{Cl}(\mathbb{P}(\mathrm{Op})))$, $U \in B$,

is a saturated class of bases. Therefore, in such a class there exist universal elements. □

2.5.9 Corollary. *The class of all spaces having a base B satisfying one of the given below conditions is saturated and, therefore, in this class there exists a universal space:*

(1) *The elements of B are* F_σ*-sets,*
(2) *The boundaries of elements of B are* G_δ*-sets,*
(3) *The closures of elements of B are* G_δ*-sets,*

(4) *The elements of B are regular open.* (An open subset U of a space X is said to be *regular open* if $U = \mathrm{Int}_X(\mathrm{Cl}(U))$).

Proof. We prove only the second case. The other cases are proved similarly. Let $\mathbb{E}$ be the class of all spaces having a base B satisfying condition (2) of the corollary. Denote by $\mathbb{P}$ the class of subsets consisting of all pairs $(B, X) \in \mathrm{P}(\text{base})$ such that the closure of each element of B is a G_δ-set. By Corollary 2.5.8, $\mathbb{P}$ is a saturated class. Obviously, the space-component of $\mathbb{P}$ coincides with $\mathbb{E}$. By Proposition 2.5.5 the class $\mathbb{E}$ is saturated. □

2.6 Saturated classes of p-bases

Definition. The class consisting of all triads (Q, B, X), where Q is a subset of a space X and B is a p-base for Q in X of cardinality $\leq \tau$, is called the *class of all p-bases* and is denoted by $\mathrm{P}(\text{p-base})$. Any subclass $\mathbb{P}$ of $\mathrm{P}(\text{p-base})$ is called a *class of p-bases.* Such a class $\mathbb{P}$ is said to be *topological* if for every homeomorphism h of a space X onto a space Y the condition $(Q, B, X) \in \mathbb{P}$ implies that $(h(Q), \{h(V) : V \in B\}, Y) \in \mathbb{P}$. In what follows an arbitrary considered class of p-bases is assumed to be topological.

The class of all elements of (Q, B, X) of $\mathrm{P}(\text{p-base})$, where B is a pos-base for Q in X (respectively, a base for X), is called the *class of all pos-bases* (respectively, the *class of all ps-bases*) and is denoted by $\mathrm{P}(\text{pos-base})$ (respectively, by $\mathrm{P}(\text{ps-base})$).

Definition. Let $\mathbf{S}$ be an indexed collection of spaces and $\mathbf{Q}$ a restriction of $\mathbf{S}$. An indexed set

$$\mathbf{B} \equiv \{B^X : X \in \mathbf{S}\}$$

is said to be a *co-p-base* (respectively, a *co-pos-base* or a *co-ps-base*) for $\mathbf{Q}$ in $\mathbf{S}$ if for every $X \in \mathbf{S}$, B^X is a p-base (respecively, a pos-base or a ps-base) for $\mathbf{Q}(X)$ in X.

Definition. Let $\mathbb{P}$ be a class of p-bases. We say that a p-base B for a subset Q in a space X is a *$\mathbb{P}$-p-base* if $(Q, B, X) \in \mathbb{P}$.

A co-p-base $\mathbf{B}$ for a restriction $\mathbf{Q}$ in an indexed collection $\mathbf{S}$ of spaces is called a *$\mathbb{P}$-co-p-base* if for every $X \in \mathbf{S}$, $\mathbf{B}(X)$ is a $\mathbb{P}$-p-base for $\mathbf{Q}(X)$ in X.

Definition. A non-empty class $I\!\!P$ of p-bases is said to be *saturated* (respectively, *weakly saturated*) if for every indexed collection $\mathbf{S}$ of spaces, a restriction $\mathbf{Q}$ of $\mathbf{S}$, a $I\!\!P$-co-p-base $\mathbf{B}$ for $\mathbf{Q}$ in $\mathbf{S}$, and for every (respectively, for some) co-indication $\mathbf{N}$ of $\mathbf{B}$ there exists a co-mark $\mathbf{M}^+$, which is a co-extension of $\mathbf{N}$, satisfying the following condition: for every co-extension $\mathbf{M}$ of $\mathbf{M}^+$ there exists an $(\mathbf{M}, \mathbf{Q})$-admissible family R^+ of equivalence relations on $\mathbf{S}$ such that for every admissible family R of equivalence relations on $\mathbf{S}$, which is a final refinement of R^+, and for every elements $\mathbf{L}$, $\mathbf{H}$, and $\mathbf{E}$ of $\mathrm{C}^{\diamondsuit}(\mathrm{R})$ for which $\mathbf{E} \subset \mathbf{H} \subset \mathbf{L}$, we have $(\mathrm{T}(\mathbf{E}|_{\mathbf{Q}}), \mathrm{B}^{\mathbf{H}}_{\diamondsuit,\theta(\tau)}, \mathrm{T}(\mathbf{L})) \in I\!\!P$, where θ is an indicial mapping from $\mathbf{N}$ to $\mathbf{M}$.

The considered co-mark $\mathbf{M}^+$ is called an *initial co-mark* (*corresponding to the restriction* $\mathbf{Q}$, *the co-indication* $\mathbf{N}$ *of* $\mathbf{B}$, *and the class* $I\!\!P$) and the family R^+ is called an *initial family* (*corresponding to the co-mark* $\mathbf{M}$, *the restriction* $\mathbf{Q}$, *the co-indication* $\mathbf{N}$ *of* $\mathbf{B}$, *and the class* $I\!\!P$).

It is easy to see that the class $\mathbb{P}$(p-base) of all bases as well as the classes $\mathbb{P}$(pos-base) and $\mathbb{P}$(ps-base) are saturated. Also, any saturated class of p-bases is also weakly saturated.

The following proposition is proved similarly to Proposition 2.5.2.

2.6.1 Proposition. *The following statements are true:*

(1) *The non-empty intersection of not more than τ many saturated classes of p-bases is also a saturated class of p-bases.*

(2) *The non-empty intersection of a saturated class of p-bases and a weakly saturated class of p-bases is weakly saturated.* □

Definition. Let $I\!\!P$ be a class of p-bases. For every element (Q, B, X) of $I\!\!P$ the space X is said to be the *space-coordinate* and the set Q the *subset-coordinate* of this element. The class consisting of space-coordinates of all element of $I\!\!P$ is said to be the *space-component* of $I\!\!P$ and the class consisting of subset-coordinates of all elements of $I\!\!P$ is said to be the *subset-component* of $I\!\!P$.

The class of subsets consisting of all pairs (Q, X) such that $(Q, B, X) \in I\!\!P$ for some p-base B for Q in X is called the *adjacent to* $I\!\!P$ *class of subsets.*

Definition. A class $I\!\!P$ of p-bases is said to be *complete* (respectively, *closed* or *open*) if the adjacent to $I\!\!P$ class of subsets is complete (respectively, closed or open).

Definition. An element (Q^T, B^T, T) of a class $I\!P$ of p-bases is said to be *universal* in $I\!P$ (respectively, *properly universal* in $I\!P$) if for every element $(Q^Z, B^Z, Z) \in I\!P$ there exists an embedding h of Z into T such that $Q^Z \subset h^{-1}(Q^T)$ (respectively, $Q^Z = h^{-1}(Q^T)$) and

$$B^Z = \{h^{-1}(V) : V \in B^T\}.$$

2.6.2 Proposition. *In any (complete) weakly saturated class of p-bases there exist (properly) universal elements.*

Proof. Let $I\!P$ be a complete weakly saturated class of p-bases. Since $I\!P$ is a topological class there exists a set $I\!P_0$ of elements of $I\!P$ such that for every element $(Q^Y, B^Y, Y) \in I\!P$ there exists an element $(Q^X, B^X, X) \in I\!P_0$ and a homeomorphism h of Y onto X such that $h(Q^Y) = Q^X$ and $B^Y = \{h^{-1}(V) : V \in B^X\}$.

Denote by $\mathbf{S}$ the indexed collection of spaces which is defined as follows. The indexing set of $\mathbf{S}$ is the set $I\!P_0$ and the element of $\mathbf{S}$ having as index an element (Q^X, B^X, X) of $I\!P_0$ is the space X. Therefore, for every $X \in \mathbf{S}$ there exist a unique determined subset Q^X of X and a unique determined p-base B^X for Q^X in X such that $(Q^X, B^X, X) \in I\!P_0$.

Consider the restriction

$$\mathbf{Q} \equiv \{Q^X : X \in \mathbf{S}\} \text{ of } \mathbf{S}$$

and the $I\!P$-co-p-base

$$\mathbf{B} \equiv \{B^X : X \in \mathbf{S}\}$$

for $\mathbf{Q}$ in $\mathbf{S}$. Since $I\!P$ is a weakly saturated class there exists a co-indication $\mathbf{N}$ of $\mathbf{B}$ and an initial co-mark $\mathbf{M}^+$ of $\mathbf{S}$ corresponding to the restriction $\mathbf{Q}$, the co-indication $\mathbf{N}$ of $\mathbf{B}$, and the class $I\!P$. (Therefore, $\mathbf{M}^+$ is a co-extension of $\mathbf{N}$). Let $\mathbf{M}$ be an arbitrary co-extension of $\mathbf{M}^+$. Then, $\mathbf{M}$ is also a co-extension of $\mathbf{N}$. Denote by θ an indicial mapping from $\mathbf{N}$ to $\mathbf{M}$.

Let R^+ be an initial family of equivalence relations on $\mathbf{S}$ corresponding to the co-mark $\mathbf{M}$, the restriction $\mathbf{Q}$, the co-indication $\mathbf{N}$ of $\mathbf{B}$, and the class $I\!P$. Let also R be an arbitrary family of equivalence relations on $\mathbf{S}$, which is a final refinement of R^+.

Since $I\!P$ is a complete class by Lemma 2.2.3 we can suppose that $\mathbf{M}$ and R are chosen in such a manner that $\mathbf{Q}$ is an $(\mathbf{M}, \mathrm{R})$-complete restriction.

By construction, $(\mathrm{T}|_{\mathbf{Q}}, \mathrm{B}^{\mathrm{T}}_{\diamondsuit,\theta(\tau)}, \mathrm{T}) \in I\!\!P$. We prove that this element is properly universal in $I\!\!P$. Indeed, let $(Q^Y, B^Y, Y) \in I\!\!P$. There exists an element (Q^X, B^X, X) of $I\!\!P_0$ and a homeomorphism h of Y onto X such that $h(Q^Y) = Q^X$ and $B^Y = \{h^{-1}(V) : V \in B^X\}$. Denote by e^X_{T} the natural embedding of X into T and set $f = e^X_{\mathrm{T}} \circ h$.

We prove that $f^{-1}(Q^{\mathrm{T}}) = Q^Y$ and

$$B^Y = \{f^{-1}(V) : V \in \mathrm{B}^{\mathrm{T}}_{\diamondsuit,\theta(\tau)}\}.$$

To this end it suffices to prove that $(e^X_{\mathrm{T}})^{-1}(Q^{\mathrm{T}}) = Q^X$ and

$$B^X = \{(e^X_{\mathrm{T}})^{-1}(V) : V \in \mathrm{B}^{\mathrm{T}}_{\diamondsuit,\theta(\tau)}\}.$$

However, the first relation follows immediately by the fact that $\mathbf{Q}$ is an $(\mathbf{M}, \mathrm{R})$-complete restriction and the second by the fact that $I\!\!P$ is a weakly saturated class. Thus, in $I\!\!P$ there exist properly universal elements. The case, where $I\!\!P$ is a weakly saturated class of p-bases, is proved similarly. □

The next proposition is proved similarly to Proposition 2.5.7.

2.6.3 Proposition. *Let $I\!\!P$ be a saturated class of p-bases and $I\!\!K$ a saturated class of subsets. Then, the class of p-bases consisting of all elements $(Q, B, X) \in \mathrm{I\!P}(\text{p-base})$ such that $(U, X) \in I\!\!K$ for every $U \in B$ is a saturated class.* □

The following three propositions are easily proved.

2.6.4 Proposition. *Let $I\!\!P$ be a weakly saturated class of p-bases. Then, the space-component of $I\!\!P$, the subset-component of $I\!\!P$, and the adjacent to $I\!\!P$ class of subsets are saturated classes.* □

2.6.5 Proposition. *Let $I\!\!P$ be a (complete) (weakly) saturated class of p-bases, $I\!\!F$ a saturated class of subsets, and $I\!\!E$ a saturated class of spaces. Then, the classes*

$$\{(Q, B, X) \in I\!\!P : (Q, X) \in I\!\!F\},$$

$$\{(Q, B, X) \in I\!\!P : Q \in I\!\!E\}, \text{ and}$$

$$\{(Q, B, X) \in I\!\!P : X \in I\!\!E\}$$

are also (complete) (weakly) saturated classes of p-bases. □

2.6.6 Proposition. *Let $I\!P$ be a (complete)(weakly) saturated class of p-bases. Then, the class of all elements (Q, B, X) of $I\!P$ such that B is a pos-base for* **Q** *in X (respectively, B is a base for X) are also (complete) (weakly) saturated classes of p-bases.* □

Now, we define different "kinds of regularity" of p-bases.

Definitions and notations. Let B be a p-base for a subset Q in a space X. Then, B is said to be: (a) a T_3-p-*base*, (b) a T_3^s-p-*base*, (c) a T_3^c-p-*base*, and (d) a T_3^h-p-*base* if for every point $x \in Q$ and for every neighbourhood $U \in B$ of x there exists an element $V \in B$ such that $x \in V \subset U$ and

(a) $\mathrm{Cl}_X(V) \subset U$,
(b) $\mathrm{Cl}_X(V) \cap Q \subset U$,
(c) $\mathrm{Cl}_X(V \cap Q) \subset U$, and
(d) $\mathrm{Cl}_Q(V \cap Q) \subset U$,

respectively. We denote by: (a) $\mathbb{P}(T_3$-p-base), (b) $\mathbb{P}(T_3^s$-p-base), (c) $\mathbb{P}(T_3^c$-p-base), and (d) $\mathbb{P}(T_3^h$-p-base) the class of p-bases consisting of all triads (Q, B, X) such that Q is a subset of a space X and B is: (a) a T_3-p-base, (b) a T_3^s-p-base, (c) a T_3^c-p-base, and (d) a T_3^h-p-base), respectively.

2.6.6 Proposition. *The classes*

$$\mathbb{P}(\mathrm{T}_3\text{-p-base}),\ \mathbb{P}(\mathrm{T}_3^s\text{-p-base}),$$

$$\mathbb{P}(\mathrm{T}_3^c\text{-p-base}),\ \textit{and}\ \mathbb{P}(\mathrm{T}_3^h\text{-p-base})$$

are saturated classes of p-bases.

Proof. We prove the proposition for the class $\mathbb{P} \equiv \mathbb{P}(T_3^s$-p-base). The proofs for the other classes are similar.

Let **S** be an indexed collection of spaces, $\mathbf{Q} \equiv \{Q^X : X \in \mathbf{S}\}$ a restriction of **S**, and **B** a $\mathbb{P}$-co-p-base for **Q** in **S**. Denote by

$$\mathbf{N} \equiv \{\{V_\delta^X : \delta \in \tau\} : X \in \mathbf{S}\}$$

an arbitrary co-indication of $\mathbf{B}$. Let ϑ_0 and ϑ_1 be two one-to-one mappings of τ into itself such that $\vartheta_0(\tau) \cap \vartheta_1(\tau) = \emptyset$. Let

$$\mathbf{M}^+ \equiv \{\{W_\delta^X : \delta \in \tau\} : X \in \mathbf{S}\}$$

be a co-mark of $\mathbf{S}$ such that $V_\delta^X = W_{\vartheta_0(\delta)}^X$ and $X \setminus \mathrm{Cl}_X(V_\delta^X) = W_{\vartheta_1(\delta)}^X$. Obviously, $\mathbf{M}^+$ is a co-extension of $\mathbf{N}$. Let also

$$\mathbf{M} \equiv \{\{U_\delta^X : \delta \in \tau\} : X \in \mathbf{S}\}$$

be an arbitrary co-extension of $\mathbf{M}^+$ and θ an inditial mapping from $\mathbf{M}^+$ to $\mathbf{M}$. Set $\theta_0 = \theta \circ \vartheta_0$ and $\theta_1 = \theta \circ \vartheta_1$. Then, $\mathbf{M}$ is also a co-extension of $\mathbf{N}$ and θ_0 is an indicial mapping from $\mathbf{N}$ to $\mathbf{M}$.

Now, we consider the elements of $\mathbf{S}$ as marked spaces. The chosen mark of an element $X \in \mathbf{S}$ is considered to be the mark $\mathbf{M}(X)$. For every $s \in \mathcal{F} \setminus \{\emptyset\}$ and $X \in \mathbf{S}$ the minimal algebra of X containing the set

$$\mathrm{A}_s^X \cup \{Q^X \cap A : A \in \mathrm{A}_s^X\},$$

where A_s^X is the s-algebra of the marked space X, is denoted by $\widetilde{\mathrm{A}}_s^X$

For every $s \in \mathcal{F}$ we define an equivalence relation $\sim_+^s$ on $\mathbf{S}$ considering that two elements X and Y of $\mathbf{S}$ are $\sim_+^s$-equivalent if either $s = \emptyset$ or $s \neq \emptyset$ and: (a) $X \sim_{\mathbf{M}}^s Y$ and $d_s^X(Q^X) = d_s^Y(Q^Y)$. (Therefore, we can consider the natural isomorphism i of A_s^X onto A_s^Y;) and (b) there exists an isomorphism $\widetilde{i}$ of $\widetilde{\mathrm{A}}_s^X$ onto $\widetilde{\mathrm{A}}_s^Y$ such that $\widetilde{i}(A) = i(A)$ and $\widetilde{i}(Q^X \cap A) = Q^Y \cap i(A)$ for every $A \in \mathrm{A}_s^X$. It is easy to verify that $\sim_+^s$ is really an equivalence relation on $\mathbf{S}$. Moreover, the family $\mathrm{R}^+ \equiv \{\sim_+^s : s \in \mathcal{F}\}$ is an $(\mathbf{M}, \mathbf{Q})$-admissible family of equivalence relations on $\mathbf{S}$.

We prove that $\mathbf{M}^+$ is an initial co-mark corresponding to the restriction $\mathbf{Q}$, the co-indication $\mathbf{N}$ of $\mathbf{B}$, and the class $\mathbb{P}$ and R^+ is an initial family corresponding to the co-mark $\mathbf{M}$, the restriction $\mathbf{Q}$, the co-indication $\mathbf{N}$ of $\mathbf{B}$, and the class $\mathbb{P}$. Indeed, let $\mathrm{R} \equiv \{\sim^s : s \in \mathcal{F}\}$ be an arbitrary admissible family of equivalence relations on $\mathbf{S}$, which is a final refinement of R^+. Consider three elements $\mathbf{L}$, $\mathbf{H}$, and $\mathbf{E}$ of $\mathrm{C}^\diamondsuit(\mathrm{R})$ such that $\mathbf{E} \subset \mathbf{H} \subset \mathbf{L}$. We need to prove that the p-base $\mathrm{B}_{\diamondsuit,\theta_0(\tau)}^{\mathbf{H}}$ for $\mathrm{T}(\mathbf{E}|_{\mathbf{Q}})$ in $\mathrm{T}(\mathbf{L})$ is a T_3^s-p-base. To this end it suffices to prove that the p-base $\mathrm{B}_{\diamondsuit,\theta_0(\tau)}^{\mathrm{T}}$ for $\mathrm{T}|_{\mathbf{Q}}$ in T is a T_3^s-p-base.

Let $\mathbf{a}$ be a point of $\mathrm{T}|_{\mathbf{Q}}$ and U an open neighbourhood of $\mathbf{a}$. We need to find an element V of $\mathrm{B}_{\diamondsuit,\theta_0(\tau)}^{\mathrm{T}}$ such that

$$\mathbf{a} \in V \subset \mathrm{Cl}_{\mathrm{T}}(V) \cap \mathrm{T}|_{\mathbf{Q}} \subset U.$$

Without loss of generality we can suppose that $V = U^{\mathrm{T}}_{\theta_0(\delta)}(\mathbf{H}_0) \in \mathrm{B}^{\mathrm{T}}_{\diamondsuit,\theta_0(\tau)}$, where $\mathbf{H}_0 \in \mathrm{C}(\sim^t)$ for some $t \in \mathcal{F}$. Hence, $U^X_{\theta_0(\delta)} = V^X_\delta \in \mathbf{N}(X)$ for every $X \in \mathbf{H}_0$. There exists an element (x, X) of $\mathbf{a}$ such that $x \in Q^X \cap V^X_\delta$ and $X \in \mathbf{H}_0$ (see Lemma 1.2.7). Since $\mathbf{B}(X)$ is a T^s_3-base for Q^X in X there exists an element ε of τ such that $x \in V^X_\varepsilon$ and

$$\mathrm{Cl}_X(V^X_\varepsilon) \cap Q^X \subset V^X_\delta$$

or, equivalently, $x \in U^X_{\theta_0(\varepsilon)}$ and

$$\mathrm{Cl}_X(U^X_{\theta_0(\varepsilon)}) \cap Q^X \subset U^X_{\theta_0(\delta)}. \tag{2.6.1}$$

Let $s = \{\theta_0(\delta), \theta_0(\varepsilon), \theta_1(\varepsilon)\}$. Since the family R is a final refinement of R^+ there exists an element q of $\mathcal{F}$ such that $\sim^q \subset \sim^s_+$ and $t \subset q$. Let $\mathbf{H}_1$ be the $\sim^q$-equivalence class of X. Then, $\mathbf{H}_1 \subset \mathbf{H}_0$. We prove that the set $V \equiv U^{\mathrm{T}}_{\theta_0(\varepsilon)}(\mathbf{H}_1)$ satisfies the above mentioned conditions, that is $\mathbf{a} \in U^{\mathrm{T}}_{\theta_0(\varepsilon)}(\mathbf{H}_1)$ and

$$\mathrm{Cl}_{\mathrm{T}}(U^{\mathrm{T}}_{\theta_0(\varepsilon)}(\mathbf{H}_1)) \cap \mathrm{T}|_{\mathbf{Q}} \subset U^{\mathrm{T}}_{\theta_0(\delta)}(\mathbf{H}_1) \subset U^{\mathrm{T}}_{\theta_0(\delta)}(\mathbf{H}_0). \tag{2.6.2}$$

It is suffices to prove the first inclusion. Let $\mathbf{b} \in \mathrm{Cl}_{\mathrm{T}}(U^{\mathrm{T}}_{\theta_0(\varepsilon)}(\mathbf{H}_1)) \cap \mathrm{T}|_{\mathbf{Q}}$. There exists an element (y, Y) of $\mathbf{b}$ such that $y \in Q^Y$ and $Y \in \mathbf{H}_1$. By Lemma 1.2.7, $y \in \mathrm{Cl}_Y(U^Y_{\theta_0(\varepsilon)})$ and, therefore, $y \in \mathrm{Cl}_Y(U^Y_{\theta_0(\varepsilon)}) \cap Q^Y$. By construction, the sets $U^Y_{\theta_0(\delta)}$ and $\mathrm{Cl}_Y(U^Y_{\theta_0(\varepsilon)}) = Y \setminus U^Y_{\theta_1(\varepsilon)}$ are elements of A^Y_s. Moreover, we have $i(U^X_{\theta_0(\delta)}) = U^Y_{\theta_0(\delta)}$ and

$$i(\mathrm{Cl}_X(U^X_{\theta_0(\varepsilon)})) = i(X \setminus U^X_{\theta_1(\varepsilon)}) = Y \setminus U^Y_{\theta_1(\varepsilon)} = \mathrm{Cl}_Y(U^Y_{\theta_0(\varepsilon)}),$$

where i is the natural isomorphism of A^X_s onto A^Y_s. Therefore,

$$\widetilde{i}(\mathrm{Cl}_X(U^X_{\theta_0(\varepsilon)}) \cap Q^X) = \mathrm{Cl}_Y(U^Y_{\theta_0(\varepsilon)}) \cap Q^Y.$$

Since $\widetilde{i}$ is an isomorphism by Lemma 1.1.1 relation (2.6.1) implies that

$$\mathrm{Cl}_Y(U^Y_{\theta_0(\varepsilon)}) \cap Q^Y \subset U^Y_{\theta_0(\delta)},$$

which means that $y \in U^{Y}_{\theta_0(\delta)}$. Therefore, $\mathbf{b} \in U^{\mathrm{T}}_{\theta_0(\delta)}(\mathbf{H}_1)$ proving relation (2.6.2).

The above show that $\mathrm{B}^{\mathrm{T}}_{\diamondsuit,\theta_0(\tau)}$ is a T_3^s-p-base. Hence, R^+ is an initial family, and, therefore, $\mathbf{M}^+$ is an initial co-mark. Thus, $\mathbb{P}$ is a saturation class of p-bases. □

2.7 (ν, μ)-Saturated classes

In the present section we shall use the notations, notions, and results of Section 1.5.

Agreement. In this section it is assumed that ν and μ are two cardinals satisfying conditions (1)–(3) of the Agreement 1 of Section 1.5.

Definition. A class $\mathbb{P}$ of spaces is said to be (ν, μ)-*saturated* if for every indexed collection $\mathbf{S}$ of elements of $\mathbb{P}$ there exists a co-mark $\mathbf{M}^+$ of $\mathbf{S}$ satisfying the following condition: for every co-extension $\mathbf{M}$ of $\mathbf{M}^+$ there exists an $(\mathbf{M}, \nu, \mu)$-admissible family R^+ of equivalence relations on $\mathbf{S}$ such that for every μ-admissible family R of equivalence relations on $\mathbf{S}$, which is a final refinement of R^+, and an element $\mathbf{L}$ of $\mathrm{C}^{\diamondsuit}_{\nu}(\mathrm{R})$, the space $\mathrm{T}(\mathbf{M}, \mathrm{R}, \mathbf{L})$ belongs to $\mathbb{P}$.

The co-mark $\mathbf{M}^+$ is said to be an *initial* co-mark (*corresponding to the class* $\mathbb{P}$) and the family R^+ is said to be an *initial* family (*corresponding to the co-mark* $\mathbf{M}$ *and the class* $\mathbb{P}$).

Remark. We note that the notion of (ω, ω)-saturated class coincides with the notion of a saturated class given in Section 2.1.

The following three propositions are proved similarly to the corresponding propositions of Section 2.1.

2.7.1 Proposition. *In any non-empty (ν, μ)-saturated class of spaces there exists a universal element.* □

2.7.2 Proposition. *The intersection of not more than τ many (ν, μ)-saturated classes is a (ν, μ)-saturated class.* □

2.7.3 Proposition. *The class of regular spaces is (ν, μ)-saturated.* □

Definition. Let σ be a cardinal. A space X is said to be a P_σ-*space* if any intersection of less than σ many open subsets of X is open. It is easy to see that a space X is a P_σ-space if and only if for some (every) base B for open subsets of X any intersection of less than σ many elements of B is open.

2.7.4 Lemma. *There exists a one-to-one mapping χ of $\mathcal{P}_\nu(\tau)\setminus\{\emptyset\}$ into τ having the following property: for every space X any base B^X for X of cardinality $\leq \tau$ has an indication*

$$N^X \equiv \{V_\varepsilon^X : \varepsilon \in \tau\}$$

such that

$$V_{\chi(s)}^X = \cap\{V_\varepsilon^X : \varepsilon \in s\},\ s \in \mathcal{P}_\nu(\tau) \setminus \{\emptyset\}.$$

Proof. For every ordinal δ we denote by $\mathcal{P}_\nu(\delta\times\tau)$ the set of all subsets of the set $\delta \times \tau$ of cardinality $< \nu$ and by $\mathcal{P}_\nu^\delta(\delta \times \tau)$ the set consisting of all elements s of $\mathcal{P}_\nu(\delta \times \tau)$ such that for every $\delta_0 \in \delta$ there exists an element $(\delta_1, \eta) \in s$ for which $\delta_0 \leq \delta_1$. Since τ is not ν-accessible (see the Agreement) for every element s of $\mathcal{P}_\nu(\tau \times \tau)$ there exists a unique element δ of τ such that $s \in \mathcal{P}_\nu^\delta(\delta \times \tau)$. Thus, we have

$$\mathcal{P}_\nu(\tau \times \tau) = \cup\{\mathcal{P}_\nu^\delta(\delta \times \tau) : \delta \in \tau\} \text{ and}$$

$$\mathcal{P}_\nu^{\delta_0}(\delta_0 \times \tau) \cap \mathcal{P}_\nu^{\delta_1}(\delta_1 \times \tau) = \emptyset$$

for every $\delta_0, \delta_1 \in \tau$, $\delta_0 \neq \delta_1$. (The second relation follows by the definition of the including sets).

By the assumptions of the Agreement it follows that $|\mathcal{P}_\nu^\delta(\delta \times \tau)| = \tau$, $\delta \neq 0$. Therefore, for every $\delta \in \tau \setminus \{0\}$ there exists a one-to-one mapping χ_δ of $\mathcal{P}_\nu^\delta(\delta \times \tau)$ onto the set $\{\delta\} \times \tau$.

For the definition of χ we consider a one-to-one mapping ψ of $\tau \times \tau$ onto τ. Let s be a non-empty element of $\mathcal{P}_\nu(\tau)$. There exists an element $\delta \neq 0$ of τ such that $\psi^{-1}(s) \in \mathcal{P}_\nu^\delta(\delta \times \tau)$. We set $\chi(s) = \psi(\chi_\delta(\psi^{-1}(s)))$. It is easy to verify that χ is one-to-one.

Now, let B^X be a base of cardinality $\leq \tau$ for the open subsets of a space X. Let

$$\{N_{(0,\eta)}^X : \eta \in \tau\}$$

be an indication of B^X. For every $\delta \in \tau \setminus \{0\}$ we define by induction on δ an indexed family $\{V_{(\delta,\eta)}^X : \eta \in \tau\}$ of elements of B^X setting

$$V^X_{(\delta,\eta)} = \cap\{V^X_{(\delta',\eta')} : (\delta', \eta') \in \chi_\delta^{-1}(\delta, \eta)\}.$$

The indication N^X of B^X is defined setting $V^X_\varepsilon = V^X_{\psi^{-1}(\varepsilon)}$. It is easy to verify that this indication and the mapping χ satisfy the condition of the lemma. □

2.7.5 Proposition. *The class of all* P_ν*-spaces is* (ν, μ)*-saturated.*

Proof. Let $\mathbf{S}$ be an indexed collection of P_ν-spaces. For every $X \in \mathbf{S}$ we consider the indexed base $N^X \equiv \{V^X_\varepsilon : \varepsilon \in s\}$ constructed in Lemma 2.7.4 and set

$$\mathbf{M}^+ = \{N^X : X \in \mathbf{S}\}.$$

Denote by

$$\mathbf{M} \equiv \{\{U^X_\delta : \delta \in \tau\} : X \in \mathbf{S}\}$$

an arbitrary co-extension of $\mathbf{M}^+$ and by θ an indicial mapping of this co-extension.

Let R^+ be an $(\mathbf{M}, \nu, \mu)$-admissible family of equivalence relations on $\mathbf{S}$. Denote by

$$\mathrm{R} \equiv \{\sim^s : s \in \mathcal{P}_\nu(\tau)\}$$

an arbitrary $(\mathbf{M}, \nu)$-admissible family of equivalence relations on $\mathbf{S}$, which is a final refinement of R^+.

To prove the proposition it suffices to prove that the Containing Space $\mathrm{T} \equiv \mathrm{T}(\mathbf{M}, \mathrm{R})$ is a P_ν-space. For this, we consider the base $\mathrm{B}^{\mathrm{T}}_{\theta(\tau)}$ of the Containing Space T and prove that any intersection of less than ν many elements of this base is open. Indeed, let $\{U^{\mathrm{T}}_{\delta_\gamma}(\mathbf{H}_\gamma) : \gamma \in \Gamma\}$ be a set of elements of $\mathrm{B}^{\mathrm{T}}_{\theta(\tau)}$, where $|\Gamma| < \nu$. Set $s = \{\delta_\gamma : \gamma \in \Gamma\}$ and $\delta = \theta(\chi(\theta^{-1}(s)))$, where χ is the mapping of Lemma 2.7.4. For every $\gamma \in \Gamma$ denote by t_γ an element of $\mathcal{P}_\nu(\tau)$ such that $\mathbf{H}_\gamma$ is a $\sim^{t_\gamma}$-equivalence class and consider the element $t \equiv \cup\{t_\gamma : \gamma \in \Gamma\}$ of $\mathcal{P}_\nu(\tau)$.

It is easy to prove that the set $\mathbf{L} \equiv \cap\{\mathbf{H}_\gamma : \gamma \in \Gamma\}$ is the union of $\sim^t$-equivalence classes. Then, we have

$$\cap\{U^{\mathrm{T}}_{\delta_\gamma}(\mathbf{H}_\gamma) : \gamma \in \Gamma\} =$$

$$\cup\{U_\delta^{\mathrm{T}}(\mathbf{H}) : \mathbf{H} \in \mathrm{C}(\sim^t) \text{ and } \mathbf{H} \subset \mathbf{L}\}. \tag{2.7.1}$$

Indeed, if $\mathbf{a}$ is an element of the left side of (2.7.1) and $(x, X) \in \mathbf{a}$, then $x \in U_{\delta_\gamma}^X$ for every $\gamma \in \Gamma$ and $X \in \mathbf{L}$. Therefore,

$$x \in \cap\{U_{\delta_\gamma}^X : \gamma \in \Gamma\} = U_{\chi(\delta)}^X,$$

which means that $\mathbf{a} \in U_\delta^{\mathrm{T}}(\mathbf{H})$, where $\mathbf{H}$ is the $\sim^t$-equivalence class containing X, that is, $\mathbf{a}$ belongs to the right side of (2.7.1). The converse inclusion is proved similarly. Thus, any intersection of less than ν many elements of $\mathrm{B}_{\theta(\tau)}^{\mathrm{T}}$ is open. This, completes the proof of the proposition. □

2.7.6 Corollary. *The class of all regular* P_ν*-spaces is a* (ν, μ)*-saturated class. Therefore, in this class there exist universal elements.* □

Chapter 3

Dimension-like functions

Dimension theory gives us the most interesting classes of spaces in which there exist universal elements. In this chapter we use saturated classes to define *dimension-like functions* of types ind, Ind, and dim or briefly *dimensions*, which in turn define new saturated classes. Besides the functions with as domain the class of all spaces which we call *space* dimension-like functions (Section 3.1), we consider also the so-called *positional* dimension-like functions with as domain the class of all subsets (Section 3.2), *base* dimension-like functions with as domain the class of all bases (Sections 3.3, 3.5, and 3.6), and *base positional* dimension-like functions with as domain the class of all p-bases (Section 3.4). The range of all these functions is considered to be the class $\mathcal{O} \cup \{-1, \infty\}$.

The words "of types ind, Ind, and dim" are used in order to underline the fact that the given definitions of the considered dimension-like functions correspond to those of the ordinary dimensions ind, Ind, and dim.

Concerning the definitions of dimension-like functions that are considered we note the following. For the definition of any such function, we use some (saturated) classes called *defining* classes. One of them (used in the notations as a lower index) is a subclass of the domain consisting exactly of all elements on which the value of the function is equal to -1. In this sense we can say that any considered function is a dimension modulo a given class. Also, the same class satisfies some kind of "heredity" with respect to other classes. This hereditary property is needed so that the given definition is well defined. (However, usually the fact that a given definition is well defined will not be mentioned.) Finally, we note that whenever a dimension-like function is considered, then it is supposed that the defining classes satisfy the assumptions of the corresponding definition.

Although we use the word "dimension" for the considered functions we do not investigate the standard problems concerning dimensions. We also do not investigate the relations between different introduced dimensions or between introduced and ordinary dimensions. We study these functions only from the universality point of view. More exactly, we use the given dimensions for the construction of new (weakly) saturated classes.

The main results of this chapter will generally have the following form: for a given dimension-like function, whose defining classes are saturated, and for a given ordinal $\alpha \in \tau^+$ (or integer $n \in \omega$) the subclass of the domain consisting of all elements on which the value of the function is less than or equal to α (or to n) is a (weakly) saturated class. Using Containing Spaces unifies the proofs of the corresponding propositions. Because of this fact we do not give many proofs proving only some representative propositions.

Notation. For every dimension-like function df and for every $\alpha \in \{-1\} \cup \mathcal{O}$, we denote by $\mathbb{P}(df \leq \alpha)$ the class of all elements E of the domain such that $df(E) \leq \alpha$.

Below, for the notations of concrete dimension-like functions instead of "df" we use more complicated expressions. In particular, in these expressions we use the symbols "ind", "Ind", and "dim" in order to underline the type of the function. Also, we use the letter "p" for the positional dimension-like functions, the letter "b" for base dimension-like functions, and the letters "b-p" for the base positional dimension-like functions.

Definition. Let df be a dimension-like function. It is said that df satisfies the (*weak*) *saturation property for an element* α of the set $\{-1\} \cup \mathcal{O}$ if the class $\mathbb{P}(df \leq \alpha)$ is a (weakly) saturated class.

Notation. For every $\alpha \in \tau^+$ we denote by $\alpha[*]$ an arbitrary fixed mapping of the set τ onto the set $\{-1\} \cup \alpha$ such that

$$|(\alpha[*])^{-1}(\beta)| = \tau$$

for every $\beta \in \{-1\} \cup \alpha$. By $\alpha[*]$, where $\alpha = -1$, we denote the unique mapping of the set τ onto the set $\{-1\}$. For every $\alpha \in \{-1\} \cup \tau^+$ and for every $\delta \in \tau$ the element $\alpha[*](\delta)$ is denoted also by $\alpha[\delta]$. We note that $\alpha[\delta] < \alpha$ for every $\alpha \in \tau^+$ and $\delta \in \tau$.

Remark. We recall that by $\mathbb{P}$(subset) (respectively, by $\mathbb{P}$(base) and $\mathbb{P}$(p-base)) we denote the class of all subsets (respectively, the class of all bases and the class of all p-bases), that is the class of all pairs (Q, X), where Q is a subset of a space X (respectively, the class of all pairs (B, X), where B is a base for a space X of cardinality $\leq \tau$, and the class of all triads (Q, B, X), where Q is a subset of a space X and B is a p-base for Q in X of cardinality $\leq \tau$).

Also, by $\mathbb{P}$(pos-base) (respectively, by $\mathbb{P}$(ps-base)) we denote the class of all pos-bases (respectively, the class of all ps-bases), that is, the class of all triads (Q, B, X), where Q is a subset of a space X and B is a pos-base for Q in X (respectively, B is a base for X).

We recall also that any base for a space X as well as any p-base for a subset Q of a space X contains by definition the empty set. In particular, in all the above cases the set B contains the empty set.

Notation. We denote by $\mathbb{P}(\emptyset)$ the (non-empty) class of spaces consisting of the empty space only; by $\mathbb{P}(\emptyset, \emptyset)$ the (non-empty) class of subsets consisting of the pair $(\emptyset, \emptyset)$ only; by $\mathbb{P}(\emptyset, \text{space})$ the class of subsets consisting exactly of all pairs $(\emptyset, X)$, where X is a space; by $\mathbb{P}(\{\emptyset\}, \emptyset)$ the (non-empty) class of bases consisting of the pair $(\{\emptyset\}, \emptyset)$ only; by $\mathbb{P}(\emptyset, \{\emptyset\}, \emptyset)$ the (non-empty) class of p-bases consisting of the triad $(\emptyset, \{\emptyset\}, \emptyset)$ only; and by $\mathbb{P}(\emptyset, \{\emptyset\}, \text{space})$ the class of p-bases consisting of all triads $(\emptyset, \{\emptyset\}, X)$, where X is a space.

We note that $\mathbb{P}(\emptyset)$, $\mathbb{P}(\emptyset, \emptyset)$, $\mathbb{P}(\emptyset, \text{space})$, $\mathbb{P}(\{\emptyset\}, \emptyset)$, $\mathbb{P}(\emptyset, \{\emptyset\}, \emptyset)$, and $\mathbb{P}(\emptyset, \{\emptyset\}, \text{space})$ are saturated classes.

3.1 Space dimension-like functions of the type ind

Definition. The *small inductive dimension* denoted by ind is the unique function such that the domain is the class of all spaces and the range is the class $\mathcal{O} \cup \{-1, \infty\}$ and that satisfies the following conditions: (a) $\text{ind}(X) = -1$ if and only if $X = \emptyset$ and (b) $\text{ind}(X) \leq \alpha$, where α is an ordinal, if and only if there exists a base B for X such that $\text{ind}(\text{Bd}_X(U)) < \alpha$ for every element $U \in B$ (and, therefore, $\text{ind}(X) = \infty$ if and only if the inequality $\text{ind}(X) \leq \alpha$ is not true for every ordinal α).

Remark. Usually the dimension ind is considered for regular spaces or for some more restrictive class of spaces. In the above given definition of ind we do not use the regularity of spaces, means that we can formally apply it to any space.

Definition. A class $\mathbb{E}$ of spaces is said to be $\mathbb{B}$-*rim-hereditary*, where $\mathbb{B}$ is a class of bases if for every space X belonging to $\mathbb{E}$ there exists a $\mathbb{B}$-base B for X such that the subspace $\mathrm{Bd}_X(U)$ of X belongs to $\mathbb{E}$ for every $U \in B$.

We note that if $\mathbb{E}$ is a $\mathbb{B}$-rim-hereditary, then since $\emptyset \in B$ we have that $\emptyset \in \mathbb{E}$.

Definition. Let $\mathbb{B}$ be a class of bases and $\mathbb{E}$ a $\mathbb{B}$-rim-hereditary class of spaces. We denote by $\mathrm{ind}_{\mathbb{E}}^{\mathbb{B}}$ the dimension-like function with as domain the class of all spaces and as range the class $\mathcal{O} \cup \{-1, \infty\}$ satisfying the following conditions:

$$\text{(a) } \mathrm{ind}_{\mathbb{E}}^{\mathbb{B}}(X) = -1$$

if and only if the space X belongs to $\mathbb{E}$, and

$$\text{(b) } \mathrm{ind}_{\mathbb{E}}^{\mathbb{B}}(X) \le \alpha,$$

where α is an ordinal, if and only if there exists a $\mathbb{B}$-base B for X such that $\mathrm{ind}_{\mathbb{E}}^{\mathbb{B}}(\mathrm{Bd}_X(U)) < \alpha$ for every $U \in B$. Therefore, $\mathrm{ind}_{\mathbb{E}}^{\mathbb{B}}(X) = \infty$ if and only if for every ordinal α the inequality $\mathrm{ind}_{\mathbb{E}}^{\mathbb{B}}(X) \le \alpha$ is not true. We note that $\mathrm{ind}_{\mathbb{E}}^{\mathbb{B}}(\emptyset) = -1$.

It is easy to verify that if $\mathbb{B} = \mathbb{P}(\text{base})$ and $\mathbb{E} = \mathbb{P}(\emptyset)$, then the dimension $\mathrm{ind}_{\mathbb{E}}^{\mathbb{B}}$ coincides with the small inductive dimension ind. (Obviously, the class $\mathbb{P}(\emptyset)$ of spaces is $\mathbb{P}(\text{base})$-rim-hereditary.)

Remark. In order to prove that the given definition of the dimension $\mathrm{ind}_{\mathbb{E}}^{\mathbb{B}}$ is well defined we need to prove that if for a space X we have $\mathrm{ind}_{\mathbb{E}}^{\mathbb{B}}(X) = -1$, then we also have $\mathrm{ind}_{\mathbb{E}}^{\mathbb{B}}(X) \le 0$. However, this follows immediately by the fact that the class $\mathbb{E}$ is $\mathbb{B}$-rim-hereditary.

The following Lemmas 3.1.1 and 3.1.3 will be used not only for the space dimension-like functions but for the other type of functions as well.

3.1.1 Lemma. *Let df be a mapping of a class $\mathbb{P}$ into the class $\{-1, \infty\} \cup \mathcal{O}$ satisfying the following condition: for every element E of $\mathbb{P}$ for which $df(E) \in \mathcal{O}$ there exists a subset B of $\mathbb{P}$ of cardinality $\le \tau$*

such that $df(E)$ is the minimal ordinal β for which $df(F) \in \{-1\} \cup \beta$ for every $F \in B$. Then, $df(\mathbb{P}) \subset \{-1, \infty\} \cup \tau^+$.

Proof. Suppose that the lemma is not true. Denote by E an element of $\mathbb{P}$ such that $df(E)$ is the minimal element of the class $df(\mathbb{P}) \cap (\mathcal{O} \setminus \tau^+)$. By the condition of the lemma there exists a subset B of $\mathbb{P}$ of cardinality $\leq \tau$ having the property: $df(E)$ is the minimal ordinal β for which $df(F) \in \beta$ for every $F \in B$. The condition $df(F) \in \mathcal{O} \setminus \tau^+$ for some $F \in B$ contradicts the choice of E. Therefore, $df(F) \in \tau^+$ for all $F \in B$. Since the cardinality of B is $\leq \tau$ there exists an ordinal $\alpha \in \tau^+$ such that $df(F) \in \alpha$ for every $F \in B$, which contradicts the property of the ordinal $df(E)$. This contradiction proves the lemma. □

3.1.2 Corollary. *If X is a space such that*

$$\mathrm{ind}_{\mathbb{E}}^{\mathbb{B}}(X) \equiv \alpha \neq \infty,$$

then $\alpha \in \{-1\} \cup \tau^+$. □

3.1.3 Lemma. *Let B be a set of cardinality $\leq \tau$ and df a mapping of B into the set $\{-1\} \cup \alpha$, where $\alpha \in \tau^+$, such that for some element $F_\emptyset$ of B, $df(F_\emptyset) = -1$. Then, there exists an indication $N \equiv \{F_\delta : \delta \in \tau\}$ of B such that $df(F_\delta) \leq \alpha[\delta]$ for every $\delta \in \tau$.*

Proof. Let $\{G_\delta : \delta \in \tau\}$ be an indication of B. Denote by θ a mapping of the set τ into itself such that for every $\delta \in \tau$,

$$\theta(\delta) \in \tau \setminus \{\theta(\varepsilon) : \varepsilon \in \delta\} \text{ and } df(G_\delta) = \alpha[\theta(\delta)].$$

Such a mapping can be constructed by induction on $\delta \in \tau$ using the fact that the cardinality of the set $(\alpha[*])^{-1}(df(G_\delta))$ is equal to τ for every $\delta \in \tau$. Then, the indication

$$N \equiv \{F_\delta : \delta \in \tau\}$$

of B, where $F_\delta = G_{\theta^{-1}(\delta)}$ if $\delta \in \theta(\tau)$ and $F_\delta = F_\emptyset$ if $\delta \notin \theta(\tau)$, is the required indication of the set B. □

The above lemma implies the following consequence. (As the element $F_\emptyset$ of this lemma we consider the empty set).

3.1.4 Corollary. *Let X be a space and $\alpha \in \{-1\} \cup \tau^+$ such that $\mathrm{ind}^{\mathbb{B}}_{\mathbb{E}}(X) \leq \alpha$. Then, there exists a $\mathbb{B}$-base B for X and an indication*

$$N \equiv \{W^X_\delta : \delta \in \tau\}$$

of B such that for every $\delta \in \tau$,

$$\mathrm{ind}^{\mathbb{B}}_{\mathbb{E}}(\mathrm{Bd}_X(W_\delta)) \leq \alpha[\delta]. \ \square$$

3.1.5 Proposition. *Suppose that $\mathbb{B}$ and $\mathbb{E}$ are saturated classes. Then for every $\alpha \in \{-1\} \cup \tau^+$ the class $\mathbb{P}(\mathrm{ind}^{\mathbb{B}}_{\mathbb{E}} \leq \alpha)$ is also saturated.*

Proof. Let $\mathbb{B}$ and $\mathbb{E}$ be saturated classes. We prove the proposition by induction on α. Let $\alpha = -1$. A space X belongs to $\mathbb{P}(\mathrm{ind}^{\mathbb{B}}_{\mathbb{E}} \leq -1)$, or equivalently, $\mathrm{ind}^{\mathbb{B}}_{\mathbb{E}}(X) = -1$ if and only if X belongs to $\mathbb{E}$, which means that $\mathbb{E} = \mathbb{P}(\mathrm{ind}^{\mathbb{B}}_{\mathbb{E}} \leq -1)$. Hence, $\mathbb{P}(\mathrm{ind}^{\mathbb{B}}_{\mathbb{E}} \leq -1)$ is a saturated class of spaces.

Now, let $\alpha \in \tau^+$. Suppose that for every $\beta \in \{-1\} \cup \alpha$ the class $\mathbb{P}(\mathrm{ind}^{\mathbb{B}}_{\mathbb{E}} \leq \beta)$ is saturated. We prove that $\mathbb{P} \equiv \mathbb{P}(\mathrm{ind}^{\mathbb{B}}_{\mathbb{E}} \leq \alpha)$ is also such a class.

Indeed, let $\mathbf{S}$ be an indexed collection of elements of $\mathbb{P}$. By Corollary 3.1.4 for every $X \in \mathbf{S}$ there exists a $\mathbb{B}$-base B^X for X and an indication

$$N^X \equiv \{W^X_\eta : \eta \in \tau\}$$

of B^X such that

$$\mathrm{ind}^{\mathbb{B}}_{\mathbb{E}}(\mathrm{Bd}_X(W^X_\eta)) \leq \alpha[\eta]. \tag{3.1.1}$$

We set

$$\mathbf{B} = \{B^X : X \in \mathbf{S}\} \text{ and}$$

$$\mathbf{N} = \{\{W^X_\eta : \eta \in \tau\} : X \in \mathbf{S}\}.$$

Obviously, $\mathbf{B}$ is a $\mathbb{B}$-co-base for $\mathbf{S}$ and $\mathbf{N}$ is a co-indication of $\mathbf{B}$.

Since $\mathbb{B}$ is a saturated class of bases there exists an initial co-mark $\mathbf{M}^+_{\mathbb{B}}$ of $\mathbf{S}$ corresponding to the co-indication $\mathbf{N}$ of $\mathbf{B}$ and the class $\mathbb{B}$. In particular, $\mathbf{M}^+_{\mathbb{B}}$ is a co-extension of $\mathbf{N}$.

For every $\eta \in \tau$ we set

$$\mathbf{W}_\eta = \{W_\eta^X : X \in \mathbf{S}\}$$

and consider the restriction $\mathbf{S}_\eta \equiv \mathbf{Bd}(\mathbf{W}_\eta)$ of $\mathbf{S}$. By relation (3.1.1), $\mathbf{S}_\eta$ is an indexed collection of elements of the class $\mathbb{P}_\eta \equiv \mathbb{P}(\mathrm{ind}_{\mathbb{E}}^{\mathbb{B}} \leq \alpha[\eta])$. By inductive assumption, this class is saturated. Therefore, there exists an initial co-mark $\mathbf{M}_{\mathbf{S}_\eta}^+$ of $\mathbf{S}_\eta$ corresponding to the class $\mathbb{P}_\eta$. Denote by $\mathbf{M}_\eta$ a co-mark of $\mathbf{S}$ such that its trace on $\mathbf{S}_\eta$ is a co-extension of the co-mark $\mathbf{M}_{\mathbf{S}_\eta}^+$. The existence of such a co-mark is easily proved.

There exists a co-mark $\mathbf{M}^+$ of $\mathbf{S}$, which a co-extension of the co-marks $\mathbf{M}_{\mathbb{B}}^+$ and $\mathbf{M}_\eta$ for every $\eta \in \tau$. In particular, $\mathbf{M}^+$ is a co-extension of $\mathbf{N}$. Consider an arbitrary co-extension

$$\mathbf{M} \equiv \{\{U_\delta^X : \delta \in \tau\} : X \in \mathbf{S}\}$$

of $\mathbf{M}^+$. Then, $\mathbf{M}$ is also a co-extension of the co-marks $\mathbf{M}_{\mathbb{B}}^+$ and $\mathbf{M}_\eta$. In particular, $\mathbf{M}$ is a co-extension of $\mathbf{N}$. Denote by θ an indicial mapping from $\mathbf{N}$ to $\mathbf{M}$. It is easy to verify that the co-mark $\mathbf{M}|_{\mathbf{S}_\eta}$ (that is, the trace on $\mathbf{S}_\eta$ of $\mathbf{M}$) is a co-extension of the co-mark $\mathbf{M}_{\mathbf{S}_\eta}^+$ of $\mathbf{S}_\eta$.

Let $\mathrm{R}_{\mathbb{B}}^+$ be an initial family of equivalence relations on $\mathbf{S}$ corresponding to the co-mark $\mathbf{M}$, the co-indication $\mathbf{N}$ of $\mathbf{B}$, and the class $\mathbb{B}$. Let also $\mathrm{R}_{\mathbf{S}_\eta}^+$ be an initial family of equivalence relations on $\mathbf{S}_\eta$ corresponding to the co-mark $\mathbf{M}|_{\mathbf{S}_\eta}$ and the class $\mathbb{P}_\eta$. Denote by R_η the family of equivalence relations on $\mathbf{S}$ such that the trace on $\mathbf{S}_\eta$ of R_η is the family $\mathrm{R}_{\mathbf{S}_\eta}^+$.

There exists an admissible family R^+ of equivalence relations on $\mathbf{S}$, which is simultaneously a final refinement of the families $\mathrm{R}_{\mathbb{B}}^+$ and R_η for every $\eta \in \tau$. Moreover, setting

$$\mathbf{U}_\delta = \{U_\delta^X : X \in \mathbf{S}\}$$

we can suppose that for every $\delta \in \theta(\tau)$, R^+ is $(\mathbf{M}, \mathbf{Cl}(\mathbf{U}_\delta))$-admissible and $(\mathbf{M}, \mathbf{Bd}(\mathbf{U}_\delta))$-admissible. We note that by Lemma 1.4.4 the family R^+ is also $(\mathbf{M}, \mathbf{U}_\delta)$-admissible $(\mathbf{M}, \mathbf{Co}(\mathbf{U}_\delta))$-admissible.

We prove that $\mathbf{M}^+$ is an initial co-mark corresponding to the class $\mathbb{P}$ and R^+ is an initial family corresponding to the co-mark $\mathbf{M}$ of $\mathbf{S}$ and the class $\mathbb{P}$. Indeed, let R be an arbitrary admissible family of equivalence relations on $\mathbf{S}$, which is a final refinement of R^+. Then, we can consider the Containing Space T and its subsets $\mathrm{T}|_{\mathbf{U}_\delta}$ and $\mathrm{T}|_{\mathbf{Bd}(\mathbf{U}_\delta)}$, $\delta \in \theta(\tau)$.

We need to prove that for every $\mathbf{L} \in \mathrm{C}^\diamondsuit(\mathrm{R})$ the subspace $\mathrm{T}(\mathbf{L})$ of T belongs to $\mathbb{P}$. Let $\mathbf{L} \in \mathrm{C}^\diamondsuit(\mathrm{R})$. Since $\mathbb{B}$ is a saturated class, the base

$\mathrm{B}^{\mathbf{L}}_{\diamondsuit,\theta(\tau)}$ for $\mathrm{T}(\mathbf{L})$ is a $I\!B$-base. Consider an element $U^{\mathrm{T}}_{\delta}(\mathbf{H})$ of $\mathrm{B}^{\mathbf{L}}_{\diamondsuit,\theta(\tau)}$ and set $\eta = \theta^{-1}(\delta)$. Then, $\mathbf{U}_{\delta} = \mathbf{W}_{\eta}$ and $\mathbf{Bd}(\mathbf{U})_{\delta} = \mathbf{S}_{\eta}$. It suffices to prove that

$$\mathrm{ind}^{I\!B}_{I\!E}(\mathrm{Bd}_{\mathrm{T}(\mathbf{L})}(U^{\mathrm{T}}_{\delta}(\mathbf{H}))) \leq \alpha[\eta] < \alpha. \tag{3.1.2}$$

Obviously, the conditions of Lemmas 1.4.7 and 1.4.10 are satisfied if as the restriction $\mathbf{Q}$ of these lemmas we consider the restriction $\mathbf{U}_{\delta}$. Therefore,

$$\mathrm{T}|_{\mathbf{Bd}(\mathbf{U}_{\delta})} = \mathrm{Bd}_{\mathrm{T}}(\mathrm{T}|_{\mathbf{U}_{\delta}}). \tag{3.1.3}$$

Then, we have the following relation:

$$\mathrm{Bd}_{\mathrm{T}}(U^{\mathrm{T}}_{\delta}(\mathbf{H})) = \mathrm{Bd}_{\mathrm{T}}(\mathrm{T}|_{\mathbf{U}_{\delta}} \cap \mathrm{T}(\mathbf{H})) =$$

$$\mathrm{Bd}_{\mathrm{T}}(\mathrm{T}|_{\mathbf{U}_{\delta}}) \cap \mathrm{T}(\mathbf{H}) = \mathrm{T}|_{\mathbf{Bd}(\mathbf{U}_{\delta})} \cap \mathrm{T}(\mathbf{H}) =$$

$$\mathrm{T}(\mathbf{M}|_{\mathbf{S}_{\eta}}, \mathrm{R}|_{\mathbf{S}_{\eta}}, \mathbf{H}|_{\mathbf{S}_{\eta}}).$$

The first equality of the above relation follows by Lemma 1.4.5, the second equality by the fact that $\mathrm{T}(\mathbf{H})$ is simultaneously closed and open in T, the third by relation (3.1.3), and the last by the definition of the subset $\mathrm{T}|_{\mathbf{Bd}(\mathbf{U}_{\delta})}$ of T (see the definition of the subset $\mathrm{T}|_{\mathbf{Q}}$ of T in Chapter I).

Since $I\!P_{\eta}$ is a saturated class, by construction of the co-mark $\mathbf{M}$ and the family R, the subspace $\mathrm{T}(\mathbf{M}|_{\mathbf{S}_{\eta}}, \mathrm{R}|_{\mathbf{S}_{\eta}}, \mathbf{H}|_{\mathbf{S}_{\eta}})$ of $\mathrm{T}(\mathbf{M}|_{\mathbf{S}_{\eta}}, \mathrm{R}|_{\mathbf{S}_{\eta}})$ belongs to $I\!P_{\eta}$, that is,

$$\mathrm{Bd}_{\mathrm{T}}(U^{\mathrm{T}}_{\delta}(\mathbf{H})) = \mathrm{Bd}_{\mathrm{T}(\mathbf{L})}(U^{\mathrm{T}}_{\delta}(\mathbf{H})) \in I\!P_{\eta},$$

which means that relation (3.1.2) are satisfied. Thus,

$$\mathrm{ind}^{I\!B}_{I\!E}(\mathrm{T}(\mathbf{L})) \leq \alpha,$$

that is, the space $\mathrm{T}(\mathbf{L})$ belongs to $I\!P$. Hence, $\mathbf{M}^{+}$ is an initial co-mark of $\mathbf{S}$, R^{+} is an initial family, and therefore $I\!P$ is a saturated class of spaces. □

3.1.6 Corollary. *For every ordinal $\alpha \in \tau^{+}$ in the class of all spaces of dimension* ind *less than or equal to α there exists a universal element.* □

Propositions 2.1.3 and 2.1.6 imply the following consequence.

3.1.7 Corollary. *For every ordinal $\alpha \in \tau^+$ in the class of all regular spaces of dimension* ind *less than or equal to α there exists a universal element.* □

In particular, for $\tau = \omega$ we have the following consequence.

3.1.8 Corollary. *For every countable ordinal α in the class of all separable metrizable spaces of dimension* ind *less than or equal to α there exist universal elements.* □

The following corollary follows by the fact that the class of all completely regular spaces are saturated (see Chapter 5).

3.1.9 Corollary. *For every ordinal $\alpha \in \tau^+$ in the class of all completely regular spaces of dimension* ind *less than or equal to α there exists a universal element.* □

3.2 Positional dimension-like functions of the type ind

Dimension-like functions of the type ind with as domain the class of all subsets can be defined in different ways. Below we give some of them.

Definition. Let $\mathbb{D}$ be a class of p-bases. A class $\mathbb{F}$ of subsets is said to be *$\mathbb{D}$-p_0-rim-hereditary* (respectively, *$\mathbb{D}$-p_1-rim-hereditary*) if for every element (Q, X) of $\mathbb{F}$ there exists a $\mathbb{D}$-p-base B for Q in X such that for every $U \in B$ we have

$$(Q \cap \mathrm{Bd}_X(U), \mathrm{Bd}_X(U)) \in \mathbb{F}$$

(respectively,

$$(Q \cap \mathrm{Bd}_X(U), X) \in \mathbb{F}).$$

We note that if $\mathbb{F}$ is a non-empty $\mathbb{D}$-p_0-rim-hereditary class, then the pair $(\emptyset, \emptyset)$ is an element of $\mathbb{F}$. This follows by the fact that any p-base for a subset Q of a space X contains by definition the empty set. Also, if $\mathbb{F}$ is a non-empty $\mathbb{D}$-p_1-rim-hereditary class, then $(\emptyset, \emptyset) \in \mathbb{F}$.

Definition. Let $i \in \{0,1\}$. For every class $\mathbb{D}$ of p-bases and for every $\mathbb{D}$-p_i-rim-hereditary class $\mathbb{F}$ of subsets we denote by $\mathrm{p}_i\text{-ind}^{\mathbb{D}}_{\mathbb{F}}$ the dimension-like function that has as domain the class of all subsets and as range the class $\mathcal{O} \cup \{-1, \infty\}$ satisfying the following conditions:

$$\text{(a) } \mathrm{p}_i\text{-ind}^{\mathbb{D}}_{\mathbb{F}}(Q, X) = -1$$

if and only if $(Q, X) \in \mathbb{F}$ and

$$\text{(b) } \mathrm{p}_i\text{-ind}^{\mathbb{D}}_{\mathbb{F}}(Q, X) \leq \alpha,$$

where α is an ordinal if and only if there exists a $\mathbb{D}$-p-base B for Q in X such that for every $U \in B$ we have

$$\mathrm{p}_i\text{-ind}^{\mathbb{D}}_{\mathbb{F}}(Q \cap \mathrm{Bd}_X(U), \mathrm{Bd}_X(U)) < \alpha \text{ for } i = 0, \text{ and}$$

$$\mathrm{p}_i\text{-ind}^{\mathbb{D}}_{\mathbb{F}}(Q \cap \mathrm{Bd}_X(U), X) < \alpha \text{ for } i = 1.$$

In the case, where $\mathbb{D} = \mathbb{P}(\text{p-base})$ and $\mathbb{F} = \mathbb{P}(\emptyset, \emptyset)$ (respectively, $\mathbb{F} = \mathbb{P}(\emptyset, \text{space})$) the dimension-like function $\mathrm{p}_0\text{-ind}^{\mathbb{D}}_{\mathbb{F}}$ (respectively, $\mathrm{p}_1\text{-ind}^{\mathbb{D}}_{\mathbb{F}}$) is denoted by p_0-ind (respectively, by p_1-ind).

If $\mathbb{D}$ is considered to be a class of pos-bases (respectively, a class of ps-bases), then the dimension-like function $\mathrm{p}_i\text{-ind}^{\mathbb{D}}_{\mathbb{F}}$ is denoted by $\mathrm{pos}_i\text{-ind}^{\mathbb{D}}_{\mathbb{F}}$ (respectively, by $\mathrm{ps}_i\text{-ind}^{\mathbb{D}}_{\mathbb{F}}$). Moreover, in the case where $\mathbb{D} = \mathbb{P}(\text{pos-base})$ and $\mathbb{F} = \mathbb{P}(\emptyset, \emptyset)$ (respectively, $\mathbb{F} = \mathbb{P}(\emptyset, \text{space})$) the dimension-like function $\mathrm{p}_0\text{-ind}^{\mathbb{D}}_{\mathbb{F}}$ (respectively, $\mathrm{p}_1\text{-ind}^{\mathbb{D}}_{\mathbb{F}}$) is denoted by pos_0-ind (respectively, by pos_1-ind). If in the above we replace the class $\mathbb{P}(\text{pos-base})$ by the class $\mathbb{P}(\text{ps-base})$, then we obtain the dimension-like functions ps_0-ind and ps_1-ind.

We observe that the class $\mathbb{P}(\emptyset, \emptyset)$ is $\mathbb{P}(\text{p-base})$-$\mathrm{p}_0$-rim-hereditary as well as $\mathbb{P}(\text{pos-base})$-$\mathrm{p}_0$-rim-hereditary and $\mathbb{P}(\text{ps-base})$-$\mathrm{p}_0$-rim-hereditary, and the class $\mathbb{P}(\emptyset, \text{space})$ is $\mathbb{P}(\text{p-base})$-$\mathrm{p}_1$-rim-hereditary, $\mathbb{P}(\text{pos-base})$-$\mathrm{p}_1$-rim-hereditary, and $\mathbb{P}(\text{ps-base})$-$\mathrm{p}_1$-rim-hereditary.

Lemma 3.1.1 implies the following consequence

3.2.1 Corollary. *If Q is a subset of a space X such that*

$$\mathrm{p}_i\text{-ind}^{\mathbb{D}}_{\mathbb{F}}(Q, X) \equiv \alpha_i \neq \infty$$

then $\alpha \in \{-1\} \cup \tau^+$, $i = 0, 1$. □

3.2.2 Lemma. *If Q is a subset of a space X such that*

$$\mathrm{p}_1\text{-}\mathrm{ind}_{\mathbb{F}}^{\mathbb{D}}(Q, X) \equiv \alpha \neq \infty \tag{3.2.1}$$

then $\mathrm{p}_1\text{-}\mathrm{ind}_{\mathbb{F}}^{\mathbb{D}}(\emptyset, X) = -1$, *that is* $(\emptyset, X) \in \mathbb{F}$.

Proof. Let Q be a subset of a space X satisfying relation (3.2.1). This means that there exists a $\mathbb{D}$-p-base B for Q in X such that

$$\mathrm{p}_1\text{-}\mathrm{ind}_{\mathbb{F}}^{\mathbb{D}}(Q \cap \mathrm{Bd}_X(U), X) < \alpha$$

for every element U of B. In particular, for $U = \emptyset$ we have

$$\mathrm{p}_1\text{-}\mathrm{ind}_{\mathbb{F}}^{\mathbb{D}}(\emptyset, X) \equiv \alpha_\emptyset < \alpha.$$

If $\alpha_\emptyset \neq -1$, then there exists a $\mathbb{D}$-p-base $B_\emptyset$ for the empty set in X such that

$$\mathrm{p}_1\text{-}\mathrm{ind}_{\mathbb{F}}^{\mathbb{D}}(\emptyset \cap \mathrm{Bd}_X(V), X) < \alpha_\emptyset$$

for every $V \in B_\emptyset$. Therefore, $\mathrm{p}_1\text{-}\mathrm{ind}_{\mathbb{F}}^{\mathbb{D}}(\emptyset, X) < \alpha_\emptyset$, which is a contradiction. Thus, $\mathrm{p}_1\text{-}\mathrm{ind}_{\mathbb{F}}^{\mathbb{D}}(\emptyset, X) = -1$. □

Lemma 3.1.3 implies the following consequence.

3.2.3 Corollary. *Let $i \in \{0, 1\}$. Suppose that Q is a subset of a space X such that*

$$\mathrm{p}_i\text{-}\mathrm{ind}_{\mathbb{F}}^{\mathbb{D}}(Q, X) \equiv \alpha \in \{-1\} \cup \tau^+.$$

Then, there exists a $\mathbb{D}$-p-base B for Q in X and an indication

$$N \equiv \{V_\varepsilon : \varepsilon \in \tau\}$$

of B such that for every $\varepsilon \in \tau$,

$$\mathrm{p}_i\text{-}\mathrm{ind}_{\mathbb{F}}^{\mathbb{D}}(Q \cap \mathrm{Bd}_X(V_\varepsilon), \mathrm{Bd}_X(V_\varepsilon)) < \alpha[\varepsilon] \textit{ for } i = 0, \textit{ and}$$

$$\mathrm{p}_i\text{-}\mathrm{ind}_{\mathbb{F}}^{\mathbb{D}}(Q \cap \mathrm{Bd}_X(V_\varepsilon), X) < \alpha[\varepsilon] \textit{ for } i = 1. \ \square$$

We note that in the proof of the above corollary as element $F_\emptyset$ of Lemma 3.1.3 we can consider the pair $(\emptyset, \emptyset)$ for $i = 0$ and the pair $(\emptyset, X)$ for $i = 1$ (see Lemma 3.2.2).

3.2.4 Proposition. *If $\mathbb{D}$ and $\mathbb{F}$ are saturated classes, then*

$$\mathbb{P}(\text{p}_i\text{-ind}_{\mathbb{F}}^{\mathbb{D}} \leq \alpha),\ \alpha \in \{-1\} \cup \tau^+,$$

is also a saturated class. In particular, the classes

$$\mathbb{P}(\text{p}_i\text{-ind} \leq \alpha),\ \mathbb{P}(\text{pos}_i\text{-ind} \leq \alpha),\ \textit{and}$$

$$\mathbb{P}(\text{ps}_i\text{-ind} \leq \alpha),\ i \in \{0, 1\},$$

are saturated.

Proof. We prove the proposition for the case $i = 0$. The proof for the case $i = 1$ is similar.

Let $\mathbb{D}$ and $\mathbb{F}$ be saturated classes. It is easy to see that

$$\mathbb{F} = \mathbb{P}(\text{p}_0\text{-ind}_{\mathbb{F}}^{\mathbb{D}} \leq -1)$$

and, therefore, $\mathbb{P}(\text{p}_0\text{-ind}_{\mathbb{F}}^{\mathbb{D}} \leq -1)$ is a saturated class of subsets.

Now, let $\alpha \in \tau^+$. Supposing that for every $\beta \in \{-1\} \cup \alpha$ the class $\mathbb{P}(\text{p}_0\text{-ind}_{\mathbb{F}}^{\mathbb{D}} \leq \beta)$ is saturated we prove that

$$\mathbb{P} \equiv \mathbb{P}(\text{p}_0\text{-ind}_{\mathbb{F}}^{\mathbb{D}} \leq \alpha)$$

is also such a class.

Let $\mathbf{S}$ be an indexed collection of spaces and

$$\mathbf{Q} \equiv \{Q^X : X \in \mathbf{S}\}$$

a $\mathbb{P}$-restriction of $\mathbf{S}$. By Corollary 3.2.3 for every $X \in \mathbf{S}$ there exists a $\mathbb{D}$-p-base B^X for Q^X in X and an indication

$$N^X \equiv \{V_\varepsilon^X : \varepsilon \in \tau\}$$

of B^X such that

$$\text{p}_0\text{-ind}_{\mathbb{F}}^{\mathbb{D}}(Q^X \cap \text{Bd}_X(V_\varepsilon^X), \text{Bd}_X(V_\varepsilon)) \leq \alpha[\varepsilon] \tag{3.2.2}$$

for every $\varepsilon \in \tau$. Let

$$\mathbf{B} = \{B^X : X \in \mathbf{S}\} \text{ and } \mathbf{N} = \{N^X : X \in \mathbf{S}\}.$$

Oviously, $\mathbf{B}$ is a $\mathbb{D}$-co-p-base for $\mathbf{Q}$ in $\mathbf{S}$ and $\mathbf{N}$ is a co-indication of $\mathbf{B}$.

Since $\mathbb{D}$ is a saturated class there exists an initial co-mark $\mathbf{M}^+_{\mathbb{D}}$ of $\mathbf{S}$ corresponding to the restriction $\mathbf{Q}$, the co-indication $\mathbf{N}$ of $\mathbf{B}$, and the class $\mathbb{D}$. In particular, $\mathbf{M}^+_{\mathbb{D}}$ is a co-extension of $\mathbf{N}$.

For every $\varepsilon \in \tau$ we consider the following restrictions of $\mathbf{S}$:

$$\mathbf{V}_\varepsilon \equiv \{V^X_\varepsilon : X \in \mathbf{S}\},$$

$$\mathbf{S}_\varepsilon \equiv \mathbf{Bd}(\mathbf{V}_\varepsilon), \text{ and}$$

$$\mathbf{K}_\varepsilon \equiv \{Q^X \cap \mathrm{Bd}_X(V^X_\varepsilon) : X \in \mathbf{S}\}.$$

By relation (3.2.2), $\mathbf{K}_\varepsilon$ is also a $\mathbb{P}_\varepsilon$-restriction of $\mathbf{S}_\varepsilon$, where

$$\mathbb{P}_\varepsilon = \mathbb{P}(\mathrm{p_O\text{-}ind}^{\mathbb{D}}_{\mathbb{L}} \leq \alpha[\varepsilon]).$$

By inductive assumption $\mathbb{P}_\varepsilon$ is a saturated class of subsets. Therefore, there exists an initial co-mark $\mathbf{M}^+_{\mathbf{S}_\varepsilon}$ of $\mathbf{S}_\varepsilon$ corresponding to the $\mathbb{P}_\varepsilon$-restriction $\mathbf{K}_\varepsilon$. Denote by $\mathbf{M}^+_\varepsilon$ a co-mark of $\mathbf{S}$ such that its trace on $\mathbf{S}_\varepsilon$ is a co-extension of $\mathbf{M}^+_{\mathbf{S}_\varepsilon}$. The existence of such a co-mark is easily proved.

Let $\mathbf{M}^+$ be a co-mark of $\mathbf{S}$, which is a co-extension of the co-mark $\mathbf{M}^+_{\mathbb{D}}$ and any co-mark $\mathbf{M}^+_\varepsilon$, $\varepsilon \in \tau$. Consider an arbitrary co-mark

$$\mathbf{M} \equiv \{\{U^X_\delta : \delta \in \tau\} : X \in \mathbf{S}\}$$

of $\mathbf{S}$, which is a co-extension of $\mathbf{M}^+$. In particular, $\mathbf{M}$ is a co-extension of $\mathbf{N}$. Denote by θ an indicial mapping from $\mathbf{N}$ to $\mathbf{M}$. Obviously, the trace on $\mathbf{S}_\varepsilon$ of the co-mark $\mathbf{M}$, that is the co-mark $\mathbf{M}|_{\mathbf{S}_\varepsilon}$ is a co-extension of $\mathbf{M}^+_{\mathbf{S}_\varepsilon}$.

There exists an initial family $\mathrm{R}^+_{\mathbb{D}}$ of equivalence relations on $\mathbf{S}$ corresponding to the co-mark $\mathbf{M}$, the restriction $\mathbf{Q}$, the co-indication $\mathbf{N}$ of $\mathbf{B}$, and the class $\mathbb{D}$. Also, there exists an initial family $\mathrm{R}^+_{\mathbf{S}_\varepsilon}$ of equivalence relations on $\mathbf{S}_\varepsilon$ corresponding to the co-mark $\mathbf{M}|_{\mathbf{S}_\varepsilon}$ and the $\mathbb{P}_\varepsilon$-restriction $\mathbf{K}_\varepsilon$. Denote by R^+_ε the admissible family of equivalence relations on $\mathbf{S}$ such that its trace on $\mathbf{S}_\varepsilon$ is the family $\mathrm{R}^+_{\mathbf{S}_\varepsilon}$, that is, $\mathrm{R}^+_\varepsilon|_{\mathbf{S}_\varepsilon} = \mathrm{R}^+_{\mathbf{S}_\varepsilon}$.

Let R^+ be an admissible family of equivalence relations on $\mathbf{S}$, which is a final refinement of $\mathrm{R}^+_{\mathbb{D}}$ and any family R^+_ε, $\varepsilon \in \tau$. Therefore, R^+ is $(\mathbf{M}, \mathbf{Q})$-admissible. Setting

$$\mathbf{U}_\delta \equiv \{U_\delta^X : X \in \mathbf{S}\} \text{ and}$$

$$\mathbf{G}_\delta \equiv \{Q^X \cap \mathrm{Bd}(U_\delta^X) : X \in \mathbf{S}\},\ \delta \in \theta(\tau),$$

without loss of generality we can suppose that the family R^+ is $(\mathbf{M}, \mathbf{G}_\delta)$-admissible, $(\mathbf{M}, \mathbf{Cl}(\mathbf{U}_\delta))$-admissible and $(\mathbf{M}, \mathbf{Bd}(\mathbf{U}_\delta))$-admissible. We note that the family R^+ is also $(\mathbf{M}, \mathbf{U}_\delta)$-admissible and $(\mathbf{M}, \mathbf{Co}(\mathbf{U}_\delta))$-admissible (see Lemma 1.4.4).

We prove that $\mathbf{M}^+$ is an initial co-mark of $\mathbf{S}$ corresponding to the $\mathbb{P}$-restriction $\mathbf{Q}$ and R^+ is an initial family corresponding to the co-mark $\mathbf{M}$ and the $\mathbb{P}$-restriction $\mathbf{Q}$. Indeed, let R be an arbitrary admissible family of equivalence relations on $\mathbf{S}$, which is a final refinement of R^+. Then, we can consider the Containing Space T and its subsets $\mathrm{T}|_{\mathbf{Q}}$, $\mathrm{T}|_{\mathbf{U}_\delta}$, $\mathrm{T}|_{\mathbf{G}_\delta}$, and $\mathrm{T}|_{\mathbf{Bd}(\mathbf{U}_\delta)}$, $\delta \in \theta(\tau)$. We can also consider the Containing Space

$$\mathrm{T}_\varepsilon \equiv \mathrm{T}(\mathbf{M}|_{\mathbf{S}_\varepsilon}, \mathrm{R}|_{\mathbf{S}_\varepsilon})$$

and its subset

$$\mathrm{T}((\mathbf{M}|_{\mathbf{S}_\varepsilon})|_{\mathbf{K}_\varepsilon}, (\mathrm{R}|_{\mathbf{S}_\varepsilon})|_{\mathbf{K}_\varepsilon}) = \mathrm{T}(\mathbf{M}|_{\mathbf{K}_\varepsilon}, \mathrm{R}|_{\mathbf{K}_\varepsilon}) = \mathrm{T}|_{\mathbf{K}_\varepsilon}.$$

We note that the first equality of the above relation follows by Corolary 1.4.2. Since $\mathbb{P}_\varepsilon$ is a saturated class, for every $\mathbf{H} \in \mathrm{C}^\diamondsuit(\mathrm{R})$ we have

$$(\mathrm{T}(\mathbf{M}|_{\mathbf{K}_\varepsilon}, \mathrm{R}|_{\mathbf{K}_\varepsilon}, \mathbf{H}|_{\mathbf{K}_\varepsilon}), \mathrm{T}(\mathbf{M}|_{\mathbf{S}_\varepsilon}, \mathrm{R}|_{\mathbf{S}_\varepsilon}, \mathbf{H}|_{\mathbf{S}_\varepsilon})) \in \mathbb{P}_\varepsilon. \tag{3.2.3}$$

By Corollary 1.3.7,

$$\mathrm{T}(\mathbf{M}|_{\mathbf{K}_\varepsilon}, \mathrm{R}|_{\mathbf{K}_\varepsilon}, \mathbf{H}|_{\mathbf{K}_\varepsilon}) = \mathrm{T}|_{\mathbf{K}_\varepsilon} \cap \mathrm{T}(\mathbf{H}) \text{ and}$$

$$\mathrm{T}(\mathbf{M}|_{\mathbf{S}_\varepsilon}, \mathrm{R}|_{\mathbf{S}_\varepsilon}, \mathbf{H}|_{\mathbf{S}_\varepsilon}) = \mathrm{T}|_{\mathbf{S}_\varepsilon} \cap \mathrm{T}(\mathbf{H}) =$$

$$\mathrm{T}|_{\mathbf{Bd}(\mathbf{V}_\varepsilon)} \cap \mathrm{T}(\mathbf{H}). \tag{3.2.4}$$

Let $\mathbf{L}$ and $\mathbf{E}$ be elements of $\mathrm{C}^\diamondsuit(\mathrm{R})$, $\mathbf{E} \subset \mathbf{L}$. We need to prove that

$$(\mathrm{T}(\mathbf{E}|_{\mathbf{Q}}), \mathrm{T}(\mathbf{L})) \in \mathbb{P}.$$

By construction, the p-base $\mathrm{B}^{\mathbf{E}}_{\diamondsuit,\theta(\tau)}$ for $\mathrm{T}(\mathbf{E}|_{\mathbf{Q}})$ in $\mathrm{T}(\mathbf{L})$ is a $\mathbb{D}$-p-base. Therefore, to prove that $(\mathrm{T}(\mathbf{E}|_{\mathbf{Q}}), \mathrm{T}(\mathbf{L})) \in \mathbb{P}$ it suffices to prove that

$$\text{p}_O\text{-ind}^{\mathbb{D}}_{\mathbb{F}}(\text{T}', \text{Bd}_{\text{T}(\mathbf{L})}(U^{\text{T}}_{\delta}(\mathbf{H}))) < \alpha,$$

where $\text{T}' = \text{T}(\mathbf{E}|_{\mathbf{Q}}) \cap \text{Bd}_{\text{T}(\mathbf{L})}(U^{\text{T}}_{\delta}(\mathbf{H}))$, for every $U^{\text{T}}_{\delta}(\mathbf{H}) \in \text{B}^{\mathbf{E}}_{\theta(\tau)}$.

Consider an element $U^{\text{T}}_{\delta}(\mathbf{H})$ of $\text{B}^{\mathbf{E}}_{\diamondsuit,\theta(\tau)}$ and set $\varepsilon = \theta^{-1}(\delta)$. We note that the restrictions $\mathbf{U}_{\delta}$, $\mathbf{Bd}(\mathbf{U}_{\delta})$, and $\mathbf{G}_{\delta}$ coincide with the restrictions $\mathbf{V}_{\varepsilon}$, $\mathbf{Bd}(\mathbf{V}_{\varepsilon})$, and $\mathbf{K}_{\varepsilon}$ of $\mathbf{S}$, respectively.

All conditions of Lemmas 1.4.7 and 1.4.10 are satisfied if as the restriction $\mathbf{Q}$ of these lemmas we consider the restriction $\mathbf{U}_{\delta}$. Therefore,

$$\text{T}|_{\mathbf{Bd}(\mathbf{U}_{\delta})} = \text{Bd}_{\text{T}}(\text{T}|_{\mathbf{U}_{\delta}}). \tag{3.2.5}$$

We prove that

$$\text{T}|_{\mathbf{G}_{\delta}} = \text{T}|_{\mathbf{Q}} \cap \text{T}|_{\mathbf{Bd}(\mathbf{U}_{\delta})}. \tag{3.2.6}$$

By Corollary 1.3.6, $\text{T}|_{\mathbf{G}_{\delta}} \subset \text{T}|_{\mathbf{Q}} \cap \text{T}|_{\mathbf{Bd}(\mathbf{U}_{\delta})}$. Conversely, suppose that

$$\mathbf{a} \in \text{T}|_{\mathbf{Q}} \cap \text{T}|_{\mathbf{Bd}(\mathbf{U}_{\delta})}.$$

Then, there exists an element (x, X) of $\mathbf{a}$ such that $x \in Q^X$. By Lemma 1.4.6, $\mathbf{Bd}(\mathbf{U}_{\delta})$ as a closed restriction of $\mathbf{S}$ is an $(\mathbf{M}, \text{R})$-complete restriction of $\mathbf{S}$. Therefore, $x \in \text{Bd}_X(U^X_{\delta})$. Thus, $x \in Q^X \cap \text{Bd}_X(U^X_{\delta})$, and by Corollary 1.3.6, $\mathbf{a} \in \text{T}|_{\mathbf{G}_{\delta}}$, which proves relation (3.2.6).

Since $U^{\text{T}}_{\delta}(\mathbf{H}) = \text{T}|_{\mathbf{U}_{\delta}} \cap \text{T}(\mathbf{H})$ (see Lemma 1.4.5) relation (3.2.5) implies

$$\begin{aligned} &\text{Bd}_{\text{T}}(U^{\text{T}}_{\delta}(\mathbf{H})) = \text{Bd}_{\text{T}}(\text{T}|_{\mathbf{U}_{\delta}} \cap \text{T}(\mathbf{H})) = \\ &\text{Bd}_{\text{T}}(\text{T}|_{\mathbf{U}_{\delta}}) \cap \text{T}(\mathbf{H}) = \text{T}|_{\mathbf{Bd}(\mathbf{U}_{\delta})} \cap \text{T}(\mathbf{H}) = \\ &\text{T}|_{\mathbf{Bd}(\mathbf{V}_{\varepsilon})} \cap \text{T}(\mathbf{H}). \end{aligned} \tag{3.2.7}$$

Since $\text{T}(\mathbf{E}|_{\mathbf{Q}}) = \text{T}|_{\mathbf{Q}} \cap \text{T}(\mathbf{E})$ (see Corollary 1.3.7) and $\mathbf{H} \subset \mathbf{E}$ the above relation and relation (3.2.6) imply

$$\begin{aligned} &\text{T}(\mathbf{E}|_{\mathbf{Q}}) \cap \text{Bd}_{\text{T}}(U^{\text{T}}_{\delta}(\mathbf{H})) = \\ &(\text{T}|_{\mathbf{Q}} \cap \text{T}(\mathbf{E})) \cap (\text{T}|_{\mathbf{Bd}(\mathbf{U}_{\delta})} \cap \text{T}(\mathbf{H})) = \\ &\text{T}|_{\mathbf{G}_{\delta}} \cap \text{T}(\mathbf{H}) = \text{T}|_{\mathbf{K}_{\varepsilon}} \cap \text{T}(\mathbf{H}). \end{aligned} \tag{3.2.8}$$

Relations (3.2.3), (3.2.4), (3.2.7), and (3.2.8) imply that

$$\text{p}_0\text{-ind}_{\mathbb{F}}^{\mathbb{D}}(\text{T}', \text{Bd}_{\text{T}(\mathbf{L})}(U_\delta^{\text{T}}(\mathbf{H}))) \leq \alpha[\varepsilon] < \alpha.$$

Hence, $\mathbf{M}^+$ is an initial co-mark of $\mathbf{S}$ and R^+ is an initial family. Thus, $\mathbb{P}(\text{p}_0\text{-ind}_{\mathbb{L}}^{\mathbb{D}} \leq \alpha)$ is a saturated class of subsets. □

Formally, in the definition of the dimension-like functions $\text{p}_i\text{-ind}_{\mathbb{F}}^{\mathbb{D}}$ the class $\mathbb{D}$ of p-bases can be replaced by a class of bases. Below, we give the corresponding definitions. However, the obtained dimensions are particular cases of the above defined.

Definition. Let $\mathbb{B}$ be a class of bases. A class $\mathbb{F}$ of subsets is said to be $\mathbb{B}$-p_0*-rim-hereditary* (respectively, $\mathbb{B}$-p_1*-rim-hereditary*) if for every element $(Q, X) \in \mathbb{F}$ there exists a $\mathbb{B}$-base B for X such that

$$(Q \cap \text{Bd}_X(U), \text{Bd}_X(U)) \in \mathbb{F}$$

for every $U \in B$ (respectively,

$$(Q \cap \text{Bd}_X(U), X) \in \mathbb{F}).$$

Definition. Let $\mathbb{B}$ be a class of bases and $\mathbb{F}$ a $\mathbb{B}$-p_i-rim-hereditary class of subsets. We denote by $\text{p}_i\text{-ind}_{\mathbb{F}}^{\mathbb{B}}$, $i \in \{0, 1\}$, the unigue dimension-like function that has as domain the class of all subsets and as range the class $\mathbb{O} \cup \{-1, \infty\}$ satisfying the following conditions:

(a) $\text{p}_i\text{-ind}_{\mathbb{F}}^{\mathbb{B}}(Q, X) = -1$

if and only if $(Q, X) \in \mathbb{F}$ and

(b) $\text{p}_i\text{-ind}_{\mathbb{F}}^{\mathbb{B}}(Q, X) \leq \alpha$,

where α is an ordinal if and only if there exists a $\mathbb{B}$-base B for X such that for every $U \in B$ we have

$\text{p}_i\text{-ind}_{\mathbb{F}}^{\mathbb{B}}(Q \cap \text{Bd}_X(U), \text{Bd}_X(U)) < \alpha$ for $i = 0$, and

$\text{p}_i\text{-ind}_{\mathbb{F}}^{\mathbb{B}}(Q \cap \text{Bd}_X(U), X) < \alpha$ for $i = 1$.

Remarks. 1. Let $\mathbb{E}$ be a class of spaces and $\mathbb{B}$ a class of bases. Set

$$\mathbb{F} = \{(\emptyset, X) : X \in \mathbb{E}\}.$$

It is easy to see that: (a) the class $\mathbb{E}$ is saturated if and only if the class $\mathbb{F}$ of subsets is saturated and (b) $\mathbb{E}$ is $\mathbb{B}$-rim-hereditary if and only if the class $\mathbb{F}$ of subsets is $\mathbb{B}$-p_0-rim-hereditary. Also, it is easy to verify that for every space X,

$$\mathrm{ind}_{\mathbb{E}}^{\mathbb{B}}(X) = \mathrm{p}_0\text{-}\mathrm{ind}_{\mathbb{F}}^{\mathbb{B}}(\emptyset, X).$$

Thus, the dimension p_0-$\mathrm{ind}_{\mathbb{F}}^{\mathbb{B}}$ can be considered as a generalization of the dimension $\mathrm{ind}_{\mathbb{E}}^{\mathbb{B}}$.

2. Let $\mathbb{B}$ be a class of bases and $\mathbb{F}$ a class of subsets. Denote by $\mathbb{D}$ the class of ps-bases consisting of all triads $(Q, B, X) \in \mathbb{P}(\text{p-base})$ such that (B, X) in $\mathbb{B}$. It is easy to see that: (a) $\mathbb{B}$ is saturated if and only if $\mathbb{D}$ is saturated and (b) $\mathbb{F}$ is $\mathbb{B}$-p_i-rim-hereditary, where $i = 0, 1$, if and only if $\mathbb{F}$ is $\mathbb{D}$-p_i-rim-hereditary. It is easy to verify that for every $(Q, X) \in \mathbb{P}(\text{subset})$,

$$\mathrm{p}_i\text{-}\mathrm{ind}_{\mathbb{F}}^{\mathbb{B}}(Q, X) = \mathrm{ps}_i\text{-}\mathrm{ind}_{\mathbb{F}}^{\mathbb{D}}(Q, X).$$

Therefore, the dimension p_i-$\mathrm{ind}_{\mathbb{F}}^{\mathbb{B}}$ can be considered as a particular case of the dimension ps_i-$\mathrm{ind}_{\mathbb{F}}^{\mathbb{D}}$.

Any positional dimension-like function defines some space dimension-like functions as follows.

Definition. Let $\mathbb{K}$ be a class of subsets and *df* a positional dimension-like function. We denote by $\mathbb{K}$-*df* the dimension-like function that has as domain the class of all spaces and as range the class $\mathcal{O} \cup \{-1, \infty\}$, which is defined as follows: for every space X, $\mathbb{K}$-$df(X)$ is the minimal element α of $\mathcal{O} \cup \{-1, \infty\}$ for which there exists a subset Q of X such that $(Q, X) \in \mathbb{K}$ and $df(Q, X) \leq \alpha$.

If $\mathbb{K}$ is the class of all subsets and *df* one of the functions p_i-ind, pos_i-ind, and ps_i-ind, where $i = 0, 1$, then the function $\mathbb{K}$-*df* is denoted by s-p_i-ind, s-pos_i-ind, and s-ps_i-ind, respectively.

3.2.5 Proposition. *If $\mathbb{K}$ is a saturation class of subsets and df is a positional dimension-like function satisfying the saturation property for an ordinal α of $\{-1\} \cup \tau^+$, then $\mathbb{K}$-df also satisfies the saturation property for the ordinal α.*

Proof. Let $\mathbf{S}$ be an indexed collection of elements of the class

$$\mathbb{P} \equiv \mathbb{P}(\mathbb{K}\text{-}df \leq \alpha).$$

Therefore, for every $X \in \mathbb{P}$ there exists a subset Q^X of X such that $(Q^X, X) \in \mathbb{K}$ and $df(Q^X, X) \leq \alpha$. This fact means that

$$\mathbf{Q} \equiv \{Q^X : X \in \mathbf{S}\}$$

is simultaneously a $\mathbb{K}$-restriction and a $\mathbb{P}_\alpha$-restriction of $\mathbf{S}$, where $\mathbb{P}_\alpha = \mathbb{P}(df \leq \alpha)$. Since $\mathbb{K}$ and $\mathbb{P}_\alpha$ are saturated classes, there exists a co-mark $\mathbf{M}^+$ of $\mathbf{S}$, which an initial co-mark of $\mathbf{S}$ corresponding to $\mathbb{K}$-restriction $\mathbf{Q}$ and an initial co-mark corresponding to $\mathbb{P}_\alpha$-restriction $\mathbf{Q}$. Denote by $\mathbf{M}$ an arbitrary co-extension of $\mathbf{M}^+$ and let R^+ be simultaneously an initial family of equivalence relations on $\mathbf{S}$ corresponding to the co-mark $\mathbf{M}$ and the $\mathbb{K}$-restriction $\mathbf{Q}$ and an initial family corresponding to the co-mark $\mathbf{M}$ and the $\mathbb{P}_\alpha$-restriction $\mathbf{Q}$. Then, for an arbitrary admissible family R of equivalence relations on $\mathbf{S}$, which is a final refinement of R^+, we have $df(\mathrm{T}|_\mathbf{Q}, \mathrm{T}) \leq \alpha$ and $(\mathrm{T}|_\mathbf{Q}, \mathrm{T}) \in \mathbb{K}$, or, equivalently, $\mathbb{K}\text{-}df(\mathrm{T}) \leq \alpha$, that is $\mathrm{T} \in \mathbb{P}$. This means that $\mathbb{P}$ is a saturated class. □

3.2.6 Corollary. *For every $\alpha \in \{-1\} \cup \tau^+$ and $i \in \{0, 1\}$ the classes*

$$\mathbb{P}(\text{s-p}_i\text{-ind} \leq \alpha),\ \mathbb{P}(\text{s-pos}_i\text{-ind} \leq \alpha),\ \textit{and}$$

$$\mathbb{P}(\text{s-ps}_i\text{-ind} \leq \alpha)$$

are saturated classes of spaces. □

3.3 Base dimension-like functions of the type ind

In this section base dimension-like functions of the type ind are considered. Such functions can be defined in many different ways. The most natural it seems to be the following.

Definition. Let $\mathbb{B}$ be a class of bases. A class $\mathbb{L}$ of bases is said to be $\mathbb{B}$-b_0-*rim-hereditary* if for every element (A, X) of $\mathbb{L}$ there exists a $\mathbb{B}$-base B for X such that for every element V of B,

$$(\{\mathrm{Bd}_X(V) \cap W : W \in A\}, \mathrm{Bd}_X(V)) \in \mathbb{L}.$$

We note that since the empty set is an element of B the above relation implies that $(\{\emptyset\}, \emptyset) \in \mathbb{L}$.

Definition. Let $\mathbb{B}$ be a class of bases and $\mathbb{L}$ a $\mathbb{B}$-b_0-rim-hereditary class of bases. We denote by $\mathrm{b}_0\text{-ind}_{\mathbb{L}}^{\mathbb{B}}$ the unique dimension-like function with as domain the class of all bases and as range the class $\mathcal{O} \cup \{-1, \infty\}$ such that the following conditions are satisfied:

$$\text{(a) } \mathrm{b}_0\text{-ind}_{\mathbb{L}}^{\mathbb{B}}(A, X) = -1$$

if and only if $(A, X) \in \mathbb{L}$, and

$$\text{(b) } \mathrm{b}_0\text{-ind}_{\mathbb{L}}^{\mathbb{B}}(A, X) \leq \alpha,$$

where α is an ordinal if and only if there exists a $\mathbb{B}$-base B for X such that for every $V \in B$,

$$\mathrm{b}_0\text{-ind}_{\mathbb{L}}^{\mathbb{B}}(\{\mathrm{Bd}_X(V) \cap W : W \in A\}, \mathrm{Bd}_X(V)) < \alpha.$$

If $\mathbb{B} = \mathbb{P}(\text{base})$ and $\mathbb{L} = \mathbb{P}(\{\emptyset\}, \emptyset)$, then the dimension-like function $\mathrm{b}_0\text{-ind}_{\mathbb{L}}^{\mathbb{B}}$ is denoted by b_0-ind. (Obviously, $\mathbb{L} = \mathbb{P}(\{\emptyset\}, \emptyset)$ is $\mathbb{P}(\text{base})$-b_0-rim-hereditary.)

Lemma 3.1.1 implies the following concequence.

3.3.1 Corollary. *Let A be a base for a space X such that*

$$\mathrm{b}_0\text{-ind}_{\mathbb{L}}^{\mathbb{B}}(A, X) \equiv \alpha \in \{-1\} \cup \mathcal{O}.$$

Then, $\alpha \in \{-1\} \cup \tau^+$. □

3.3.2 Proposition. *The following statements are true:*
(1) *If $\mathbb{B}$ and $\mathbb{L}$ are saturated classes, then the class*

$$\mathbb{P}(\mathrm{b}_0\text{-ind}_{\mathbb{L}}^{\mathbb{B}} \leq \alpha),\ \alpha \in \{-1\} \cup \tau^+,$$

is also saturated. In particular, $\mathbb{P}(\mathrm{b}_0\text{-ind} \leq \alpha)$ is a saturated class of bases.
(2) *If $\mathbb{B}$ is weakly saturated and $\mathbb{L}$ is a saturated class, then the class*

$$\mathbb{P}(\mathrm{b}_0\text{-ind}_{\mathbb{L}}^{\mathbb{B}} \leq n),\ n \in \{-1\} \cup \omega,$$

is saturated.

Proof. We prove only statement (1). Statement (2) is proved similarly. The proof is carried out by induction on α. Let $\mathbb{B}$ and $\mathbb{L}$ be saturated classes. For $\alpha = -1$ the class

$$\mathbb{P}_\alpha \equiv \mathbb{P}(\text{b}_0\text{-ind}_{\mathbb{L}}^{\mathbb{B}} \leq \alpha)$$

coincides with $\mathbb{L}$ and, therefore, it is saturated. Consider an ordinal $\alpha \in \tau^+$ and suppose that for every $\beta \in \{-1\} \cup \alpha$ the statement is true.

Let $\mathbf{S}$ be an indexed collection of spaces,

$$\mathbf{A} \equiv \{A^X : X \in \mathbf{S}\}$$

a $\mathbb{P}_\alpha$-co-base for $\mathbf{S}$, and

$$\mathbf{N} \equiv \{N^X \equiv \{W_\varepsilon^X : \varepsilon \in \tau\} : X \in \mathbf{S}\}$$

an arbitrary co-indication of $\mathbf{A}$. Thus, for every $X \in \mathbf{S}$ we have

$$\text{b}_0\text{-ind}_{\mathbb{L}}^{\mathbb{B}}(A^X, X) \leq \alpha.$$

Denote by B^X a $\mathbb{B}$-base for X such that for every $V \in B^X$,

$$\text{b}_0\text{-ind}_{\mathbb{L}}^{\mathbb{B}}(A_V^X, \text{Bd}_X(V)) < \alpha,$$

where $A_V^X = \{\text{Bd}_X(V) \cap W : W \in A^X\}$. There exists an indication

$$K^X \equiv \{V_\eta^X : \eta \in \tau\}$$

of B^X (see Lemma 3.1.3) such that

$$\text{b}_0\text{-ind}_{\mathbb{L}}^{\mathbb{B}}(A_\eta^X, \text{Bd}_X(V_\eta^X)) \leq \alpha[\eta], \tag{3.3.1}$$

where $A_\eta^X = \{\text{Bd}_X(V_\eta^X) \cap W : W \in A^X\}$. (As an element $F_\emptyset$ of Lemma 3.1.3 we use here the pair $(\{\emptyset\}, \emptyset)$.)

Consider the $\mathbb{B}$-co-base

$$\mathbf{B} \equiv \{B^X : X \in \mathbf{S}\}$$

for $\mathbf{S}$ and its co-indication

$$\mathbf{K} \equiv \{K^X \equiv \{V_\eta^X : \eta \in \tau\} : X \in \mathbf{S}\}.$$

For every $\eta \in \tau$ we set:

$$\mathbf{V}_\eta = \{V_\eta^X : X \in \mathbf{S}\} \text{ and } \mathbf{S}_\eta = \mathbf{Bd}(\mathbf{V}_\eta).$$

Consider the traces on $\mathbf{S}_\eta$ of $\mathbf{A}$ and $\mathbf{N}$:

$$\mathbf{A}|_{\mathbf{S}_\eta} = \{\{\mathrm{Bd}_X(V_\eta^X) \cap W : W \in A^X\} : \mathrm{Bd}_X(V_\eta^X) \in \mathbf{S}_\eta\},$$

$$\mathbf{N}|_{\mathbf{S}_\eta} = \{\{\mathrm{Bd}_X(V_\eta^X) \cap W_\varepsilon^X : \varepsilon \in \tau\} : \mathrm{Bd}_X(V_\eta^X) \in \mathbf{S}_\eta\}.$$

It is clear that $\mathbf{A}|_{\mathbf{S}_\eta}$ is a $\mathbb{P}_{\alpha[\eta]}$-co-base for $\mathbf{S}_\eta$ and $\mathbf{N}|_{\mathbf{S}_\eta}$ is a co-indication of $\mathbf{A}|_{\mathbf{S}_\eta}$.

Let $\mathbf{M}^+_{\mathbf{S}_\eta}$ be a co-mark of $\mathbf{S}_\eta$, which is an initial co-mark corresponding to the co-indication $\mathbf{N}|_{\mathbf{S}_\eta}$ of $\mathbf{A}|_{\mathbf{S}_\eta}$ and the class $\mathbb{P}_{\alpha[\eta]}$. Without loss of generality we can consider a mapping $\theta^{\mathbf{N}}$, which is independent of η, as an indicial mapping from $\mathbf{N}|_{\mathbf{S}_\eta}$ to $\mathbf{M}^+_{\mathbf{S}_\eta}$.

Let $\mathbf{M}^+$ be a co-mark of $\mathbf{S}$ satisfying the following conditions: (a) $\mathbf{M}^+$ is a co-extension of $\mathbf{N}$, (b) for every $\eta \in \tau$ the co-mark $\mathbf{M}^+|_{\mathbf{S}_\eta}$ of $\mathbf{S}_\eta$ is a co-extension of $\mathbf{M}^+_{\mathbf{S}_\eta}$, and (c) $\mathbf{M}^+$ is an initial co-mark corresponding to the co-indication $\mathbf{K}$ of $\mathbf{B}$ and the class $\mathbb{B}$. Without loss of generality we can suppose that $\theta^{\mathbf{N}}$ is an indicial mapping from $\mathbf{N}$ to $\mathbf{M}^+$. By $\theta^{\mathbf{K}}$ we denote an indicial mapping from $\mathbf{K}$ to $\mathbf{M}^+$.

Denote by

$$\mathbf{M} \equiv \{\{U_\delta^X : \delta \in \tau\} : X \in \mathbf{S}\}$$

an arbitrary co-extension of $\mathbf{M}^+$ and let θ^+ be an indicial mapping from $\mathbf{M}^+$ to $\mathbf{M}$. Setting

$$\theta = \theta^+ \circ \theta^{\mathbf{N}} \text{ and } \vartheta = \theta^+ \circ \theta^{\mathbf{K}}$$

θ is an indicial mapping from $\mathbf{N}$ to $\mathbf{M}$ and ϑ is an indicial mapping from $\mathbf{K}$ to $\mathbf{M}$. Moreover, θ is an indicial mapping from $\mathbf{N}|_{\mathbf{S}_\eta}$ to $\mathbf{M}|_{\mathbf{S}_\eta}$.

Let R^+ be an $\mathbf{M}$-admissible family of equivalence relations on $\mathbf{S}$ such that:

(a) for every $\eta \in \tau$ the family $\mathrm{R}^+|_{\mathbf{S}_\eta}$ of equivalence relation on $\mathbf{S}_\eta$ is an initial family corresponding to the co-mark $\mathbf{M}|_{\mathbf{S}_\eta}$, the co-indication $\mathbf{N}|_{\mathbf{S}_\eta}$ of $\mathbf{A}|_{\mathbf{S}_\eta}$, and the class $\mathbb{P}_{\alpha[\eta]}$

(b) the family R^+ is an initial family corresponding to the co-mark $\mathbf{M}$, the co-indication $\mathbf{K}$ of $\mathbf{B}$, and

(c) R^+ is $(\mathbf{M}, \mathbf{Cl}(\mathbf{U}_\delta))$-admissible and $(\mathbf{M}, \mathbf{Bd}(\mathbf{U}_\delta))$-admissible for every $\delta \in \tau$, where

$$\mathbf{U}_\delta \equiv \{U_\delta^X : X \in \mathbf{S}\}.$$

We note that by Lemma 1.4.4, the family R^+ is also $(\mathbf{M}, \mathbf{U}_\delta)$-admissible and $(\mathbf{M}, \mathbf{Co}(\mathbf{U}_\delta))$-admissible. Denote by

$$\mathrm{R} \equiv \{\sim^s : s \in \mathcal{F}\}$$

an arbitrary admissible family of equivalence relations on $\mathbf{S}$, which is a final refinement of R^+.

For every $\mathbf{E} \in \mathrm{C}^\diamondsuit(\mathrm{R})$ and $\eta \in \tau$ we denote by $\mathrm{B}_{\diamondsuit,\theta(\tau)}^{\mathbf{E}|\mathbf{s}_\eta}$ the $(\diamondsuit, \theta(\tau))$-standard base for the subspace

$$\mathrm{T}(\mathbf{M}|\mathbf{s}_\eta, \mathrm{R}|\mathbf{s}_\eta, \mathbf{E}|\mathbf{s}_\eta) \equiv \mathrm{T}(\mathbf{E}|\mathbf{s}_\eta)$$

of the Containing Space

$$\mathrm{T}(\mathbf{M}|\mathbf{s}_\eta, \mathrm{R}|\mathbf{s}_\eta) \equiv \mathrm{T}|\mathbf{s}_\eta.$$

An element $U_\delta^{\mathrm{T}|\mathbf{s}_\eta}(\mathbf{F}|\mathbf{s}_\eta)$ of the $\diamondsuit$-standard base for the space $\mathrm{T}|\mathbf{s}_\eta$ belongs to $\mathrm{B}_{\diamondsuit,\theta(\tau)}^{\mathbf{E}|\mathbf{s}_\eta}$ if and only if $\delta \in \theta(\tau)$ and $\mathbf{F}$ is an element of $\mathrm{C}^\diamondsuit(\mathrm{R})$ such that $\mathbf{F} \subset \mathbf{E}$.

By inductive assumption the class $\mathbb{P}_{\alpha[\eta]}$ is saturated. Therefore, by the choice of the co-mark $\mathbf{M}^+$, the family R^+, and the indicial mapping θ relation (3.3.1) implies that

$$(\mathrm{B}_{\diamondsuit,\theta(\tau)}^{\mathbf{E}|\mathbf{s}_\eta}, \mathrm{T}(\mathbf{E}|\mathbf{s}_\eta)) \in \mathbb{P}_{\alpha[\eta]}. \tag{3.3.2}$$

Now, for an arbitrary fixed element $\mathbf{L}$ of $\mathrm{C}^\diamondsuit(\mathrm{R})$ we consider the subspace $\mathrm{T}(\mathbf{L})$ of T and its $(\diamondsuit, \theta(\tau))$-standard base $\mathrm{B}_{\diamondsuit,\theta(\tau)}^{\mathbf{L}} \equiv A^{\mathbf{L}}$. We need to prove that

$$(A^{\mathbf{L}}, \mathrm{T}(\mathbf{L})) \in \mathbb{P}_\alpha.$$

For this, we consider the $(\diamondsuit, \vartheta(\tau))$-standard base $\mathrm{B}_{\diamondsuit,\vartheta(\tau)}^{\mathbf{L}}$ for $\mathrm{T}(\mathbf{L})$. By construction,

$$(\mathrm{B}^{\mathbf{L}}_{\diamondsuit,\vartheta(\tau)}, \mathrm{T}(\mathbf{L})) \in \mathbb{B}.$$

Therefore, it suffices to prove that for every element V of $\mathrm{B}^{\mathbf{L}}_{\diamondsuit,\vartheta(\tau)}$ we have

$$(\{\mathrm{Bd}_{\mathrm{T}(\mathbf{L})}(V) \cap W : W \in A^{\mathbf{L}}\}, \mathrm{Bd}_{\mathrm{T}(\mathbf{L})}(V)) \in \mathbb{P}_\beta \tag{3.3.3}$$

for some $\beta \in \{-1\} \cup \alpha$.

Let $V \equiv U^{\mathrm{T}}_{\delta}(\mathbf{E})$ be an arbitrary element of $\mathrm{B}^{\mathbf{L}}_{\diamondsuit,\vartheta(\tau)}$, which means that $\delta \in \vartheta(\tau)$ and $\mathbf{E} \subset \mathbf{L}$. Let $\eta = \vartheta^{-1}(\delta)$. Then,

$$\mathbf{U}_\delta = \mathbf{V}_\eta \text{ and } \mathbf{Bd}(\mathbf{U}_\delta) = \mathbf{S}_\eta.$$

Since the family R^+ (and, therefore, the family R) satisfies all conditions of Corollary 1.4.11 we have

$$\mathrm{Bd}_{\mathrm{T}(\mathbf{L})}(U^{\mathrm{T}}_{\delta}(\mathbf{E})) = \mathrm{T}_{\mathbf{Bd}(\mathbf{U}_\delta)} \cap \mathrm{T}(\mathbf{E}) = \mathrm{T}(\mathbf{E}|_{\mathbf{S}_\eta}).$$

Let $U^{\mathrm{T}}_{\varepsilon}(\mathbf{H})$ be an element of $A^{\mathbf{L}}$. The above relation and Corollary 1.3.7 imply that

$$\mathrm{Bd}_{\mathrm{T}(\mathbf{L})}(U^{\mathrm{T}}_{\delta}(\mathbf{E})) \cap U^{\mathrm{T}}_{\varepsilon}(\mathbf{H}) = (\mathrm{T}|_{\mathbf{Bd}(\mathbf{U}_\delta)} \cap \mathrm{T}(\mathbf{E})) \cap U^{\mathrm{T}}_{\varepsilon}(\mathbf{H}) =$$

$$\mathrm{T}|_{\mathbf{Bd}(\mathbf{U}_\delta)} \cap U^{\mathrm{T}}_{\varepsilon}(\mathbf{E} \cap \mathbf{H}) = U^{\mathrm{T}|_{\mathbf{Bd}(\mathbf{U}_\delta)}}_{\varepsilon}((\mathbf{E} \cap \mathbf{H})|_{\mathbf{Bd}(\mathbf{U}_\delta)}) =$$

$$U^{\mathrm{T}|_{\mathbf{S}_\eta}}_{\varepsilon}(\mathbf{F}|_{\mathbf{S}_\eta}),$$

where $\mathbf{E} \cap \mathbf{H} = \mathbf{F} \in \mathrm{C}^{\diamondsuit}(\mathrm{R})$. By the choice of the indicial mappings θ it follows that $U^{\mathrm{T}|_{\mathbf{S}_\eta}}_{\varepsilon}(\mathbf{F}|_{\mathbf{S}_\eta})$ is an element of the $(\diamondsuit, \theta(\tau))$-standard base $\mathrm{B}^{\mathbf{E}|_{\mathbf{S}_\eta}}_{\diamondsuit,\theta(\tau)}$ for $\mathrm{T}(\mathbf{E}|_{\mathbf{S}_\eta})$. It is easy to verify that any element of $\mathrm{B}^{\mathbf{L}|_{\mathbf{S}_\eta}}_{\diamondsuit,\theta(\tau)}$ is the intersection of $\mathrm{Bd}_{\mathrm{T}(\mathbf{L})}(U^{\mathrm{T}}_{\delta}(\mathbf{E}))$ with some element of $A^{\mathbf{L}}$. Hence, we have

$$\{\mathrm{Bd}_{\mathrm{T}(\mathbf{L})}(U^{\mathrm{T}}_{\delta}(\mathbf{E})) \cap W : W \in A^{\mathbf{L}}\} = \mathrm{B}^{\mathbf{E}|_{\mathbf{S}_\eta}}_{\diamondsuit,\theta(\tau)}.$$

Thus, relation (3.3.2) implies relation (3.3.3). This fact completes the proof of the proposition. □

Below, we give a modification of the above given dimension-like function.

Definition. Let $\mathbb{B}$ be a class of bases. A class $\mathbb{L}$ of bases is said to be $\mathbb{B}$-b_1-*rim-hereditary* if for every element (A, X) of $\mathbb{L}$ there exists a $\mathbb{B}$-base B for X with the property that for every element V of B there exists a positional base A_U for $\mathrm{Bd}_X(V)$ in X such that $A_U \subset A$ and

$$(\{\mathrm{Bd}_X(V) \cap W : W \in A_U\}, \mathrm{Bd}_X(V)) \in \mathbb{L}.$$

Definition. Let $\mathbb{B}$ be a class of bases and $\mathbb{L}$ a $\mathbb{B}$-b_1-rim-hereditary class of bases. We denote by $\mathrm{b}_1\text{-ind}_{\mathbb{L}}^{\mathbb{B}}$ the unique dimension-like function that has as domain the class of all bases and as range the class $\mathcal{O} \cup \{-1, \infty\}$ such that the following conditions are satisfied:

$$\text{(a) } \mathrm{b}_1\text{-ind}_{\mathbb{L}}^{\mathbb{B}}(A, X) = -1$$

if and only if $(A, X) \in \mathbb{L}$, and

$$\text{(b) } \mathrm{b}_1\text{-ind}_{\mathbb{L}}^{\mathbb{B}}(A, X) \leq \alpha,$$

where α is an ordinal if and only if there exists a $\mathbb{B}$-base B for X with the property that for every $V \in B$ there exists a positional base A_U for $\mathrm{Bd}_X(V)$ in X such that $A_U \subset A$ and

$$\mathrm{b}_1\text{-ind}_{\mathbb{L}}^{\mathbb{B}}(\{\mathrm{Bd}_X(V) \cap W : W \in A_U\}, \mathrm{Bd}_X(V)) < \alpha.$$

If $\mathbb{B} = \mathbb{P}(\text{base})$ and $\mathbb{L} = \mathbb{P}(\{\emptyset\}, \emptyset)$, then the dimension-like function $\mathrm{b}_1\text{-ind}_{\mathbb{L}}^{\mathbb{B}}$ is denoted by b_1-ind. (Obviously, $\mathbb{L} = \mathbb{P}(\{\emptyset\}, \emptyset)$ is $\mathbb{P}(\text{base})$-b_1-rim-hereditary).

Lemma 3.1.1 implies the following concequence.

3.3.3 Corollary. *Let A be a base for a space X such that*

$$\mathrm{b}_1\text{-ind}_{\mathbb{L}}^{\mathbb{B}}(A, X) \equiv \alpha \in \{-1\} \cup \mathcal{O}.$$

Then, $\alpha \in \{-1\} \cup \tau^+$. □

3.3.4 Proposition. *The following statement are true:*
(1) *If $\mathbb{B}$ is saturated and $\mathbb{L}$ is a weakly saturated class, then*

$$\mathbb{P}(\mathrm{b}_1\text{-ind}_{\mathbb{L}}^{\mathbb{B}} \leq \alpha),\ \alpha \in \{-1\} \cup \tau^+,$$

is a weakly saturated class. In particular, $\mathbb{P}(\mathrm{b}_1\text{-ind} \le \alpha)$ *is a weakly saturated class of bases.*

(2) *If* $\mathbb{B}$ *and* $\mathbb{L}$ *are weakly saturated classes, then*

$$\mathbb{P}(\mathrm{b}_1\text{-ind}_{\mathbb{L}}^{\mathbb{B}} \le n),\ n \in \{-1\} \cup \omega,$$

is also weakly saturated. □

Two other base dimention-like functions are given below. These functions are obtained from the preceding ones by "changing the roles" of the bases A and B including in its definitions.

Definition. Let $\mathbb{B}$ be a class of bases. A class $\mathbb{L}$ of bases is said to be $\mathbb{B}$-b^0-*rim-hereditary* if for every $(A, X) \in \mathbb{L}$ there exists a $\mathbb{B}$-base B for X such that for every $U \in A$,

$$(\{\mathrm{Bd}_X(U) \cap V : V \in B\}, \mathrm{Bd}_X(U)) \in \mathbb{L}.$$

Definition. Let $\mathbb{B}$ be a class of bases and $\mathbb{L}$ a $\mathbb{B}$-b^0-rim-hereditary class of bases. We denote by $\mathrm{b}^0\text{-ind}_{\mathbb{L}}^{\mathbb{B}}$ the unique dimension-like function that has as domain the class of all bases and as range the class $\mathcal{O} \cup \{-1, \infty\}$ such that the following conditions are satisfied:

$$\text{(a) } \mathrm{b}^0\text{-ind}_{\mathbb{L}}^{\mathbb{B}}(A, X) = -1$$

if and only if $(A, X) \in \mathbb{L}$, and

$$\text{(b) } \mathrm{b}^0\text{-ind}_{\mathbb{L}}^{\mathbb{B}}(A, X) \le \alpha,$$

where α is an ordinal if and only if there exists a $\mathbb{B}$-base B for X such that for every $U \in A$,

$$\mathrm{b}^0\text{-ind}_{\mathbb{L}}^{\mathbb{B}}(\{\mathrm{Bd}_X(U) \cap V : V \in B\}, \mathrm{Bd}_X(U)) < \alpha.$$

If $\mathbb{B} = \mathbb{P}(\text{base})$ and $\mathbb{L} = \mathbb{P}(\{\emptyset\}, \emptyset)$, then the dimension $\mathrm{b}^0\text{-ind}_{\mathbb{L}}^{\mathbb{B}}$ is denoted by b^0-ind. (Obviously, $\mathbb{L} = \mathbb{P}(\{\emptyset\}, \emptyset)$ is $\mathbb{P}(\text{base})$-b^0-rim-hereditary.)

Lemma 3.1.1 implies the following consequence.

3.3.5 Corollary. *Let* A *be a base for a space* X *such that*

$$\mathrm{b}^0\text{-ind}_{\mathbb{L}}^{\mathbb{B}}(A, X) \equiv \alpha \in \{-1\} \cup \mathcal{O}.$$

Then $\alpha \in \tau^+$. □

3.3.6 Proposition. *If* $\mathbb{B}$ *is weakly saturated and* $\mathbb{L}$ *is a saturated class, then*

$$\mathbb{P}(\mathrm{b}^0\text{-ind}_{\mathbb{L}}^{\mathbb{B}} \leq n),\ n \in \{-1\} \cup \omega,$$

is a saturated class. In particular, the class $\mathbb{P}(\mathrm{b}^0\text{-ind} \leq n)$ *is a saturated class of bases.* □

Definition. Let $\mathbb{B}$ be a class of bases. A class $\mathbb{L}$ of bases is said to be $\mathbb{B}$-b^1*-rim-hereditary* if for every $(A, X) \in \mathbb{L}$ there exists a $\mathbb{B}$-base B for X with the property that for every $U \in A$ there exists a positional base B_U for $\mathrm{Bd}_X(U)$ in X such that $B_U \subset B$ and

$$(\{\mathrm{Bd}_X(U) \cap V : V \in B_U\}, \mathrm{Bd}_X(U)) \in \mathbb{L}.$$

Definition. Let $\mathbb{B}$ be a class of bases and $\mathbb{L}$ a $\mathbb{B}$-b^1-rim-hereditary class of bases. We denote by $\mathrm{b}^1\text{-ind}_{\mathbb{L}}^{\mathbb{B}}$ the unique dimension-like function that has as domain the class of all bases and as range the class $\mathcal{O} \cup \{-1, \infty\}$ such that the following conditions are satisfied:

$$\text{(a) } \mathrm{b}^1\text{-ind}_{\mathbb{L}}^{\mathbb{B}}(A, X) = -1$$

if and only if $(A, X) \in \mathbb{L}$, and

$$\text{(b) } \mathrm{b}^1\text{-ind}_{\mathbb{L}}^{\mathbb{B}}(A, X) \leq \alpha,$$

where α is an ordinal if and only if there exists a $\mathbb{B}$-base B for X such that for every $U \in A$ there exists a positional base B_U for $\mathrm{Bd}_X(U)$ in X for which $B_U \subset B$ and

$$\mathrm{b}^1\text{-ind}_{\mathbb{L}}^{\mathbb{B}}(\{\mathrm{Bd}_X(U) \cap V : V \in B_U\}, \mathrm{Bd}_X(U)) < \alpha.$$

If $\mathbb{B} = \mathbb{P}(\mathrm{base})$ and $\mathbb{L} = \mathbb{P}(\{\emptyset\}, \emptyset)$, then the dimension-like function $\mathrm{b}^1\text{-ind}_{\mathbb{L}}^{\mathbb{B}}$ is denoted by b^1-ind. (Obviously, $\mathbb{L} = \mathbb{P}(\{\emptyset\}, \emptyset)$ is $\mathbb{P}(\mathrm{base})$-$\mathrm{b}^1$-rim-hereditary.)

Lemma 3.1.1 implies the following consequence.

3.3.7 Corollary. *Let A be a base for a space X such that*

$$\text{b}^1\text{-ind}_{\mathbb{L}}^{\mathbb{B}}(A, X) \equiv \alpha \in \{-1\} \cup \mathcal{O}.$$

Then, $\alpha \in \tau^+$. □

3.3.8 Proposition. *The following statements are true:*
(1) *If $\mathbb{B}$ is saturated and $\mathbb{L}$ is a weakly saturated class, then*

$$\mathbb{P}(\text{b}^1\text{-ind}_{\mathbb{L}}^{\mathbb{B}} \leq \alpha),\ \alpha \in \{-1\} \cup \tau^+,$$

is a weakly saturated class. In particular, $\mathbb{P}(\text{b}^1\text{-ind} \leq n)$ *is a weakly saturated class of bases.*
(2) *If $\mathbb{B}$ and $\mathbb{L}$ are saturated classes, then*

$$\mathbb{P}(\text{b}^1\text{-ind}_{\mathbb{L}}^{\mathbb{B}} \leq n),\ n \in \{-1\} \cup \omega,$$

is also a saturated class. □

The next base dimension-like function is obtained from the above function $\text{b}^0\text{-ind}_{\mathbb{L}}^{\mathbb{B}}$ replacing the class $\mathbb{B}$ of bases by a class of p-bases.

Definition. Let $\mathbb{D}$ be a class of p-bases. A class $\mathbb{L}$ of bases is said to be $\mathbb{D}$-b^{p}-*rim-hereditary* if for every $(A, X) \in \mathbb{L}$ and for every $U \in A$ there exists a $\mathbb{D}$-p-base B for $\text{Bd}_X(U)$ in X such that

$$(\{\text{Bd}_X(U) \cap V : V \in B\}, \text{Bd}_X(U)) \in \mathbb{L}.$$

Definition. Let $\mathbb{D}$ be a class of p-bases and $\mathbb{L}$ a $\mathbb{D}$-b^{p}-rim-hereditary class of bases. We denote by $\text{b}^{\text{p}}\text{-ind}_{\mathbb{L}}^{\mathbb{D}}$ the unique dimension-like function that has as domain the class of all bases and as range the class $\mathcal{O} \cup \{-1, \infty\}$ such that the following conditions are satisfied:

$$\text{(a) } \text{b}^{\text{p}}\text{-ind}_{\mathbb{L}}^{\mathbb{D}}(A, X) = -1$$

if and only if $(A, X) \in \mathbb{L}$, and

$$\text{(b) } \text{b}^{\text{p}}\text{-ind}_{\mathbb{L}}^{\mathbb{D}}(A, X) \leq \alpha,$$

where α is an ordinal if and only if for every $U \in A$ there exists a $\mathbb{D}$-p-base B_U for $\text{Bd}_X(U)$ in X such that

$$\mathrm{b^p\text{-}ind}_{\mathbb{L}}^{\mathbb{D}}(\{\mathrm{Bd}_X(U) \cap V : V \in B_U\}, \mathrm{Bd}_X(U)) < \alpha.$$

If $\mathbb{D}$ is considered to be a class of pos-bases or a class of ps-bases, then the dimension-like function $\mathrm{b^p\text{-}ind}_{\mathbb{L}}^{\mathbb{D}}$ is denoted by $\mathrm{b^{pos}\text{-}ind}_{\mathbb{L}}^{\mathbb{D}}$ and $\mathrm{b^{ps}\text{-}ind}_{\mathbb{L}}^{\mathbb{D}}$, respectively.

If $\mathbb{D} = \mathbb{P}(\text{p-base})$ (respectively, $\mathbb{D} = \mathbb{P}(\text{pos-base})$ or $\mathbb{D} = \mathbb{P}(\text{ps-base})$) and $\mathbb{L} = \mathbb{P}(\{\emptyset\}, \emptyset)$, then the dimension $\mathrm{b^p\text{-}ind}_{\mathbb{L}}^{\mathbb{D}}$ is denoted by $\mathrm{b^p\text{-}ind}$ (respectively, by $\mathrm{b^{pos}\text{-}ind}$ or by $\mathrm{b^{ps}\text{-}ind}$). (Obviously, $\mathbb{P}(\{\emptyset\}, \emptyset)$ is $\mathbb{P}(\text{p-base})$-$\mathrm{b^p}$-rim-hereditary, $\mathbb{P}(\text{pos-base})$-$\mathrm{b^p}$-rim-hereditary, and $\mathbb{P}(\text{ps-base})$-$\mathrm{b^p}$-rim-hereditary.)

Lemma 3.1.1 implies the following consequence.

3.3.9 Corollary. *Let A be a base for a space X such that*

$$\mathrm{b^p\text{-}ind}_{\mathbb{L}}^{\mathbb{D}}(A, X) \equiv \alpha \in \{-1\} \cup \mathcal{O}.$$

Then $\alpha \in \tau^+$. □

3.3.10 Proposition. *The following statements are true:*

(1) *If $\mathbb{D}$ is saturated and $\mathbb{L}$ is a weakly saturated class, then*

$$\mathbb{P}(\mathrm{b^p\text{-}ind}_{\mathbb{L}}^{\mathbb{D}} \leq \alpha),\ \alpha \in \{-1\} \cup \tau^+,$$

is a weakly saturated class. In particular, the classes

$$\mathbb{P}(\mathrm{b^p\text{-}ind} \leq \alpha),\ \mathbb{P}(\mathrm{b^{pos}\text{-}ind} \leq \alpha),\ \text{and}\ \mathbb{P}(\mathrm{b^{ps}\text{-}ind} \leq \alpha)$$

are weakly saturated.

(2) *If $\mathbb{B}$ and $\mathbb{L}$ are saturated classes, then*

$$\mathbb{P}(\mathrm{b^p\text{-}ind}_{\mathbb{L}}^{\mathbb{B}} \leq n),\ n \in \{-1\} \cup \omega,$$

is also a saturated class. In particular, the classes

$$\mathbb{P}(\mathrm{b^p\text{-}ind} \leq n),\ \mathbb{P}(\mathrm{b^{pos}\text{-}ind} \leq n),\ \text{and}\ \mathbb{P}(\mathrm{b^{ps}\text{-}ind} \leq n)$$

are saturated. □

Below, we define two "more special" base dimension-like functions.

Definition. A class $\mathbb{L}$ of bases is said to be b(0)-*rim-hereditary* if for every $(A, X) \in \mathbb{L}$ and $U \in A$ we have

$$(\{\mathrm{Bd}_X(U) \cap V : V \in A\}, \mathrm{Bd}_X(U)) \in \mathbb{L}.$$

Definition. For every b(0)-rim-hereditary class $\mathbb{L}$ of bases we denote by b(0)-ind$_{\mathbb{L}}$ the unique dimension-like function that has as domain the class of all bases and as range the class $\mathcal{O} \cup \{-1, \infty\}$ satisfying the following conditions:

$$\text{(a) b(0)-ind}_{\mathbb{L}}(A, X) = -1$$

if and only if the pair (A, X) belongs to $\mathbb{L}$, and

$$\text{(b) b(0)-ind}_{\mathbb{L}}(A, X) \leq \alpha,$$

where α is an ordinal if and only if for every element U of A,

$$\text{b(0)-ind}_{\mathbb{L}}(\{\mathrm{Bd}_X(U) \cap V : V \in A\}, \mathrm{Bd}_X(U)) < \alpha.$$

If $\mathbb{L} = \mathbb{P}(\{\emptyset\}, \emptyset)$, then the dimension b(0)-ind$_{\mathbb{L}}$ is denoted by b(0)-ind. (Obviously, the class $\mathbb{L} = \mathbb{P}(\{\emptyset\}, \emptyset)$ is b(0)-rim-hereditary.)

Lemma 3.1.1 implies the following consequence.

3.3.11 Corollary. *Let A be a base for a space X such that*

$$\text{b(0)-ind}_{\mathbb{L}}(A, X) \equiv \alpha \in \{-1\} \cup \mathcal{O}.$$

Then, $\alpha \in \tau^+$. □

3.3.12 Proposition. *If $\mathbb{L}$ is a saturated class, then the class*

$$\mathbb{P}(\text{b(0)-ind}_{\mathbb{L}} \leq n),\ n \in \{-1\} \cup \omega,$$

is also a saturated class. In particular, the class $\mathbb{P}(\text{b(0)-ind} \leq n)$ is a saturated class of bases. □

Definition. A class $\mathbb{L}$ of bases is said to be $\mathbb{D}$-b(1)-*rim-hereditary,* where $\mathbb{D}$ is a class of p-bases if for every $(A, X) \in \mathbb{L}$ and for every $U \in A$ there exists a $\mathbb{D}$-p-base A_U for $\mathrm{Bd}_X(U)$ in X such that $A_U \subset A$ and

$$(\{\mathrm{Bd}_X(U) \cap V : V \in A_U\}, \mathrm{Bd}_X(U)) \in \mathbb{L}.$$

Definition. For every class $\mathbb{D}$ of p-bases and $\mathbb{D}$-b(1)-rim-hereditary class $\mathbb{L}$ of bases we denote by b(1)-$\mathrm{ind}_{\mathbb{L}}^{\mathbb{D}}$ the unique dimension-like function that has as domain the class of all bases and as range the class $\mathcal{O} \cup \{-1, \infty\}$ satisfying the following conditions:

$$\text{(a) b(1)-}\mathrm{ind}_{\mathbb{L}}^{\mathbb{D}}(A, X) = -1$$

if and only if $(A, X) \in \mathbb{L}$, and

$$\text{(b) b(1)-}\mathrm{ind}_{\mathbb{L}}^{\mathbb{D}}(A, X) \leq \alpha,$$

where α is an ordinal if and only if for every $U \in A$ there exists a $\mathbb{D}$-p-base A_U for $\mathrm{Bd}_X(U)$ in X such that $A_U \subset A$ and

$$\text{b(1)-}\mathrm{ind}_{\mathbb{L}}^{\mathbb{D}}(\{\mathrm{Bd}_X(U) \cap V : V \in A_U\}, \mathrm{Bd}_X(U)) < \alpha.$$

If $\mathbb{D} = \mathbb{P}(\text{p-base})$ and $\mathbb{L} = \mathbb{P}(\{\emptyset\}, \emptyset)$, then the dimension b(1)-$\mathrm{ind}_{\mathbb{L}}^{\mathbb{D}}$ is denoted by b(1)-ind. (Obviously, the class $\mathbb{L} = \mathbb{P}(\{\emptyset\}, \emptyset)$ is $\mathbb{P}(\text{p-base})$-b(1)-rim-hereditary.)

Lemma 3.1.1 imlies the following consequence.

3.3.13 Corollary. *Let A be a base for a space X such that*

$$\text{b(1)-}\mathrm{ind}_{\mathbb{L}}(A, X) \equiv \alpha \in \{-1\} \cup \mathcal{O}.$$

Then $\alpha \in \{-1\} \cup \tau^+$. □

3.3.14 Proposition. *If $\mathbb{L}$ is a weakly saturated class, then the class*

$$\mathbb{P}(\text{b(1)-}\mathrm{ind}_{\mathbb{L}} \leq n),\ n \in \{-1\} \cup \omega,$$

is also a weakly saturated class. In particular, $\mathbb{P}(\text{b(1)-ind} \leq n)$ is a weakly saturated class of bases.

Proof. Let $\mathbb{L}$ be a weakly saturated class. We prove by induction on n that the class $\mathbb{P}_n \equiv \mathbb{P}(\text{b(1)-}\mathrm{ind}_{\mathbb{L}} \leq n)$ is weakly saturated. It is clear that $\mathbb{L} = \mathbb{P}_{-1}$; and, therefore, $\mathbb{P}_{-1}$ is a weakly saturated class. Let $n \in \omega$.

Suppose that for every $m \in \{-1\} \cup n$, $\mathbb{P}_m$ is a weakly saturated class. We prove that $\mathbb{P}_n$ is also a weakly saturated class.

First, we set

$$\widetilde{\tau} = \cup\{\tau^k : k \in \omega \setminus \{0\}\}.$$

We note that if $\eta \equiv (\eta_0, ..., \eta_{k-1})$ is an element of $\tau^k \subset \widetilde{\tau}$, then (η, ε) is an element of $\tau^{k+1} \subset \widetilde{\tau}$ and, conversely, each element of τ^{k+1} has the form (η, ε), where $\eta \in \tau^k$ and $\varepsilon \in \tau$.

Now, let $\mathbf{S}$ be a collection of spaces and

$$\mathbf{A} \equiv \{A^X : X \in \mathbf{S}\}$$

a $\mathbb{P}_n$-co-base for $\mathbf{S}$. For every $X \in \mathbf{S}$ we shall construct an auxiliary indication

$$N^X_{\widetilde{\tau}} \equiv \{V^X_\eta : \eta \in \widetilde{\tau}\}$$

of A^X. (We note that the indexing set of this indication is the set $\widetilde{\tau}$.) The elements V^X_η of $N^X_{\widetilde{\tau}}$ will be constructed by induction on $k \in \omega \setminus \{0\}$, for which $\eta \in \tau^k$. For $k = 1$ the elements V^X_η, $\eta \in \tau^1 \subset \widetilde{\tau}$, are defined in such a manner that the indexed set

$$\{V^X_\eta : \eta \in \tau^1\}.$$

is an indication of A^X.

Suppose that the elements V^X_η of $N^X_{\widetilde{\tau}}$ have been constructed for all indices $\eta \in \tau^k$, $k \in \tau \setminus \{0\}$. We construct the elements V^X_η, $\eta \in \tau^{k+1}$. For this, for every $\eta \in \tau^k$ we set

$$\mathbf{V}_\eta = \{V^X_\eta : X \in \mathbf{S}\} \text{ and}$$

$$\mathbf{S}_\eta = \mathbf{Bd}(\mathbf{V}_\eta) = \{\mathrm{Bd}_X(V^X_\eta) : X \in \mathbf{S}\}.$$

Furthermore, for every $\eta \in \tau^k$ we denote by B^X_η a subset of A^X, which is a p-base for $\mathrm{Bd}_X(V^X_\eta)$ in X, such that

$$\text{b(1)-ind}_{\mathbb{L}}(A^{\mathrm{Bd}_X(V^X_\eta)}, \mathrm{Bd}_X(V^X_\eta)) \leq n - 1,$$

where

$$A^{\mathrm{Bd}_X(V_\eta^X)} = \{\mathrm{Bd}_X(V_\eta^X) \cap V : V \in B_\eta^X\},$$

that is

$$\mathbf{A}_{\mathbf{S}_\eta} \equiv \{A^{\mathrm{Bd}_X(V_\eta^X)} : \mathrm{Bd}_X(V_\eta^X) \in \mathbf{S}_\eta\}$$

is a $\mathbb{P}_{n-1}$-co-base for $\mathbf{S}_\eta$. Since by assumption $\mathbb{P}_{n-1}$ is a weakly saturated class there exists:

(a) a co-indication

$$\mathbf{N}_{\mathbf{S}_\eta} \equiv \{\{V_\varepsilon^{X,\eta} : \varepsilon \in \tau\} : X \in \mathbf{S}\}$$

of $\mathbf{A}_{\mathbf{S}_\eta}$, and

(b) an initial co-mark $\mathbf{M}_{\mathbf{S}_\eta}$ of $\mathbf{S}_\eta$ corresponding to the co-indication $\mathbf{N}_{\mathbf{S}_\eta}$ of $\mathbf{A}_{\mathbf{S}_\eta}$ and the class $\mathbb{P}_{n-1}$.

Now, for every element (η, ε) of $\tau^{k+1} \subset \widetilde{\tau}$, where $\eta \in \tau^k$ and $\varepsilon \in \tau$, we denote by $V_{(\eta,\varepsilon)}^X$ an element of B_η^X such that

$$V_{(\eta,\varepsilon)}^X \cap \mathrm{Bd}_X(V_\eta^X) = V_\varepsilon^{X,\eta}.$$

(We note that in general the element $V_{(\eta,\varepsilon)}^X$ of B_η^X is not uniquely determined by the above relation.) Thus, we have define the elements V_η^X of $N_{\widetilde{\tau}}^X$ for all $\eta \in \tau^{k+1}$. It is easy to see that the above construction determines the indication $N_{\widetilde{\tau}}^X$.

By construction it follows that for every $\eta \in \widetilde{\tau}$ the (non-indexed) set

$$A_\eta^X \equiv \{V_{(\eta,\varepsilon)}^X : \varepsilon \in \tau\}$$

is a p-base for $\mathrm{Bd}_X(V_\eta^X)$ in X such that $A_\eta^X \subset A^X$ and

$$A^{\mathrm{Bd}_X(V_\eta^X)} = \{\mathrm{Bd}_X(V_\eta^X) \cap V : V \in A_\eta^X\}.$$

(We note that the set A_η^X may not coincide with the set B_η^X.) Therefore,

$$\mathbf{A}_\eta \equiv \{A_\eta^X : X \in \mathbf{S}\}$$

is a co-p-base for $\mathbf{S}_\eta$ in $\mathbf{S}$.

Let ϑ be an one-to-one mapping of τ onto the set $\widetilde{\tau}$. For every $X \in \mathbf{S}$ and $\delta \in \tau$ we set

$$W_\delta^X = V_{\vartheta(\delta)}^X.$$

Thus, we have the following co-indication of $\mathbf{A}$:

$$\mathbf{N} \equiv \{W_\delta^X : \delta \in \tau\}.$$

The above constructed co-mark $\mathbf{M}_{\mathbf{S}_\eta}$ of $\mathbf{S}_\eta$ (as an initial co-mark) is a co-extension of $\mathbf{N}_{\mathbf{S}_\eta}$. We denote by $\theta_{\mathbf{S}_\eta}$ an indicial mapping of this co-extension.

Now, let $\mathbf{M}^+$ be a co-mark of $\mathbf{S}$ such that $\mathbf{M}^+$ is a co-extension of $\mathbf{N}$, and $\mathbf{M}^+|_{\mathbf{S}_\eta}$ is a co-extension of the co-mark $\mathbf{M}_{\mathbf{S}_\eta}$ for every $\eta \in \widetilde{\tau}$. Without loss of generality we can consider that an indicial mapping from $\mathbf{N}$ to $\mathbf{M}^+$ denoted by $\theta_{\mathbf{N}}^+$ and an indicial mapping from $\mathbf{M}_{\mathbf{S}_\eta}$ to $\mathbf{M}^+|_{\mathbf{S}_\eta}$ denoted by $\theta_{\mathbf{S}_\eta}^+$ satisfy the following condition:

$$\theta_{\mathbf{N}}^+(\vartheta^{-1}(\eta, \varepsilon)) = \theta_{\mathbf{S}_\eta}^+(\theta_{\mathbf{S}_\eta}(\varepsilon)) \tag{3.3.4}$$

for every $\eta \in \widetilde{\tau}$ and for every $\varepsilon \in \tau$.

Denote by

$$\mathbf{M} \equiv \{\{U_\delta^X : \delta \in \tau\} : X \in \mathbf{S}\}$$

an arbitrary co-extension of $\mathbf{M}^+$ and by θ^+ an indicial mapping from $\mathbf{M}^+$ to $\mathbf{M}$.

For every $\eta \in \widetilde{\tau}$ let $\mathrm{R}_{\mathbf{S}_\eta}^+$ be a family of equivalence relations on $\mathbf{S}_\eta$, which is an initial family corresponding to the co-indication $\mathbf{N}_{\mathbf{S}_\eta}$ of $\mathbf{B}_{\mathbf{S}_\eta}$, the co-mark $\mathbf{M}|_{\mathbf{S}_\eta}$ of $\mathbf{S}_\eta$, and the class $I\!\!P_{n-1}$. We denote by R_η^+ the unique family of equivalence relations on $\mathbf{S}$ such that

$$\mathrm{R}_\eta^+|_{\mathbf{S}_\eta} = \mathrm{R}_{\mathbf{S}_\eta}^+.$$

Let R^+ be an $\mathbf{M}$-admissible family of equivalence relations on $\mathbf{S}$, which is a final refinement of any family R_η^+, $\eta \in \widetilde{\tau}$, and let R be an arbitrary admissible family of equivalence relations on $\mathbf{S}$, which is a final refinement of R^+.

Consider the $(\diamondsuit, \theta(\tau))$-standard base $\mathrm{B}_{\diamondsuit,\theta(\tau)}^{\mathrm{T}} \equiv A$ of the space $\mathrm{T}(\mathbf{M}, \mathrm{R})$, where $\theta = \theta^+ \circ \theta_{\mathbf{N}}^+$. We need to prove that $(A, \mathrm{T}) \in \mathrm{I\!P}_n$. For this fact it suffices to prove that for every element $U \equiv U_\delta^{\mathrm{T}}(\mathbf{H})$ of A there exists a p-base A_U for $\mathrm{Bd}_{\mathrm{T}}(U_\delta^{\mathrm{T}}(\mathbf{H}))$ in T such that $A_U \subset A$ and

$$\text{b(1)-ind}_{\mathbb{L}}(B, \text{Bd}_{\text{T}}(U_\delta^{\text{T}}(\mathbf{H}))) \leq n - 1. \tag{3.3.5}$$

where

$$B = \{\text{Bd}_{\text{T}}(U_\delta^{\text{T}}(\mathbf{H})) \cap V : V \in A_U\}.$$

Let $U \equiv U_\delta^{\text{T}}(\mathbf{H})$ be an arbitrary fixed element of A and

$$\eta = \vartheta^{-1}((\theta_{\mathbf{N}}^{+})^{-1}((\theta^{+})^{-1}(\delta))).$$

Then, for every $X \in \mathbf{S}$ we have $U_\delta^X = V_\eta^X$. Without loss of generality we can suppose that the family R^{+} (and, therefore, the family R) satisfies all conditions of Corollary 1.4.1 if instead of $\mathbf{Q}$ of this corollary we consider the restriction

$$\mathbf{U}_\delta \equiv \{U_\delta^X : X \in \mathbf{S}\} = \{V_\eta^X : X \in \mathbf{S}\} = \mathbf{V}_\eta$$

of $\mathbf{S}$. Therefore,

$$\text{Bd}_{\text{T}}(U_\delta^{\text{T}}(\mathbf{H})) = \text{T}|_{\mathbf{Bd}(\mathbf{U}_\delta)} \cap \text{T}(\mathbf{H}) = \text{T}|_{\mathbf{S}_\eta} \cap \text{T}(\mathbf{H}|_{\mathbf{S}_\eta}). \tag{3.3.6}$$

We set

$$\tau_\eta = \{(\eta, \varepsilon) \in \widetilde{\tau} : \varepsilon \in \tau\} \text{ and}$$

$$\kappa = \theta^{+}(\theta_{\mathbf{N}}^{+}(\vartheta(\tau_\eta))).$$

By relation (3.3.4) we have

$$\kappa = \theta^{+}(\theta_{\mathbf{S}_\eta}^{+}(\theta_{\mathbf{S}_\eta}(\tau))).$$

Denote by A_U the set of all elements $U_\xi^{\text{T}}(\mathbf{L})$ of A such that $\mathbf{L} \subset \mathbf{H}$ and $\xi \in \kappa$ and let $U_\xi^{\text{T}}(\mathbf{L})$ be an arbitrary fixed element of A_U. Then, relation (3.3.6) and Corollary 1.3.7 imply that

$$\text{Bd}_{\text{T}}(U_\delta^{\text{T}}(\mathbf{H})) \cap U_\xi^{\text{T}}(\mathbf{L}) =$$

$$(\text{T}|_{\mathbf{Bd}(\mathbf{U}_\delta)} \cap \text{T}(\mathbf{H})) \cap U_\xi^{\text{T}}(\mathbf{L}) = U_\xi^{\text{T}|_{\mathbf{S}_\eta}}(\mathbf{E}|_{\mathbf{S}_\eta}),$$

where $\mathbf{E} = \mathbf{L} \cap \mathbf{H}$. This relation means that

$$\{\mathrm{Bd}_{\mathrm{T}}(U_{\delta}^{\mathrm{T}}(\mathbf{H})) \cap U : U \in A_U\} \subset \mathrm{B}_{\diamondsuit,\kappa}^{\mathrm{T}|\mathrm{s}_\eta},$$

where $\mathrm{B}_{\diamondsuit,\kappa}^{\mathrm{T}|\mathrm{s}_\eta}$ is the $(\diamondsuit,\kappa)$-standard base for the open subsets of the Containing Space $\mathrm{T}(\mathbf{M}|_{\mathbf{S}_\eta}, \mathrm{R}|_{\mathbf{S}_\eta})$. It is easy to see that the converse inclusion is also true. Therefore,

$$\{\mathrm{Bd}_{\mathrm{T}}(U_{\delta}^{\mathrm{T}}(\mathbf{H})) \cap U : U \in A_U\} = \mathrm{B}_{\diamondsuit,\kappa}^{\mathrm{T}|\mathrm{s}_\eta}. \tag{3.3.7}$$

By construction,

$$(\mathrm{B}_{\diamondsuit,\kappa}^{\mathrm{T}|\mathrm{s}_\eta}, \mathrm{Bd}_{\mathrm{T}}(U_{\delta}^{\mathrm{T}}(\mathbf{H}))) \in \mathbb{P}_{n-1}. \tag{3.3.8}$$

Relations (3.3.7) and (3.3.8) imply (3.3.5). This fact completes the proof of the proposition. □

Any base dimension-like function defines some space dimension-like functions. This fact can be done as follows.

Definition. Let $\mathbb{M}$ be a class of bases and *df* a base dimension-like function. We denote by $\mathbb{M}$-*df* the dimension-like function with as domain the class of all spaces and as range the class $\mathcal{O} \cup \{-1, \infty\}$ such that for every space X, $\mathbb{M}$-*df*(X) is the minimal element α of $\mathcal{O} \cup \{-1, \infty\}$ for which there exists an element (B, X) of $\mathbb{M}$ such that *df*$(B, X) \leq \alpha$ (if there is no such element (B, X) of $\mathbb{M}$, then it is supposed that $\mathbb{M}$-*df*$(X) = \infty$).

If $\mathbb{M}$ coincides with $\mathbb{P}$(base), then the function $\mathbb{M}$-*df* is denoted by s-*df*. In the cases, where *df* coincides with one of the functions b_0-ind, b_1-ind, b^0-ind, b^1-ind, b^{p}-ind, $\mathrm{b}^{\mathrm{pos}}$-ind, b^{ps}-ind, b(0)-ind, and b(1)-ind, we have respectively the following space dimension-like functions: s-b_0-ind, s-b_1-ind, s-b^0-ind, s-b^1-ind, s-b^{p}-ind, s-$\mathrm{b}^{\mathrm{pos}}$-ind, s-$\mathrm{b}^{\mathrm{ps}}$-ind, s-b(0)-ind, and s-b(1)-ind.

3.3.15 Proposition. *If $\mathbb{M}$ is a weakly saturated (respectively, saturated) class of bases and df is a base dimension-like function satisfying the saturation (respectively, the weak saturation) property for an element α of the set $\{-1\} \cup \tau^+$, then the space dimension-like function $\mathbb{M}$-df satisfies the saturation property for α.*

Proof. Suppose that $\mathbb{M}$ is a weakly saturated class and

$$\mathbb{P}_\alpha \equiv \mathbb{P}(df \leq \alpha)$$

is saturated. Let **S** be an indexed collection of elements of

$$\mathbb{P} \equiv \mathbb{P}(\mathbb{M}\text{-}df \leq \alpha).$$

Thus, for every $X \in \mathbb{P}$ there exists a base B^X for X such that $(B^X, X) \in \mathbb{M}$ and $df(B^X, X) \leq \alpha$.

Since $\mathbb{M}$ is weakly saturated there exists a co-mark $\mathbf{M}^+_{\mathbb{M}}$, which is an initial co-mark of **S** corresponding to a co-indication $\mathbf{N} \equiv \{N^X : X \in \mathbf{S}\}$ of $\mathbf{B} \equiv \{B^X : X \in \mathbf{S}\}$ and to the class $\mathbb{M}$. Since $\mathbb{P}$ is a saturated class there exists a co-mark $\mathbf{M}^+_{\mathbb{P}}$ of **S**, which is an initial co-mark corresponding to the co-indication **N** of **B** and the class $\mathbb{P}$. Denote by $\mathbf{M}^+$ a co-mark of **S**, which is a co-extension of $\mathbf{M}^+_{\mathbb{M}}$ and $\mathbf{M}^+_{\mathbb{P}}$. Let **M** be an arbitrary co-extension of $\mathbf{M}^+$ and θ an indicial mapping from **N** to **M**.

Denote by R^+ an admissible family of equivalence relations on **S**, which is an initial family corresponding to the co-mark **M**, the co-indication **N** of **B**, and the class $\mathbb{M}$ as well as an initial family corresponding to the co-mark **M**, the co-indication **N** of **B**, and the class $\mathbb{P}$. Let R be an arbitrary admissible family of equivalence relations on **S**, which is a final refinement of R^+.

Then, the $(\diamondsuit, \theta(\tau))$-standard base $\mathrm{B}^{\mathrm{T}}_{\diamondsuit,\theta(\tau)}$ for T is an $\mathbb{M}$-base and

$$df(\mathrm{B}^{\mathrm{T}}_{\diamondsuit,\theta(\tau)}, \mathrm{T}) \leq \alpha.$$

This fact means that $\mathrm{T} \in \mathbb{P}$ and, therefore, $\mathbb{P}$ is a saturated class. The proof of the proposition is similar if $\mathbb{M}$ is saturated and $\mathbb{P}(df \leq \alpha)$ is a weakly saturated class. □

3.3.16 Corollary. *The following statements are true:*
(1) *For every* $\alpha \in \{-1\} \cup \tau^+$ *the classes*

$$\mathbb{P}(\text{s-b}_0\text{-ind} \leq \alpha),\ \mathbb{P}(\text{s-b}_1\text{-ind} \leq \alpha),\ \mathbb{P}(\text{s-b}^{\text{p}}\text{-ind} \leq \alpha),$$

$$\mathbb{P}(\text{s-b}^{\text{pos}}\text{-ind} \leq \alpha),\ \textit{and}\ \mathbb{P}(\text{s-b}^{\text{ps}}\text{-ind} \leq \alpha)$$

are saturated classes.
(2) *For every* $n \in \{-1\} \cup \omega$ *the classes*

$$\mathbb{P}(\text{s-b}^0\text{-ind} \leq n),\ \mathbb{P}(\text{s-b}^1\text{-ind} \leq n),$$

$$\mathbb{P}(\text{s-b}(0)\text{-ind} \leq n), \text{ and } \mathbb{P}(\text{s-b}(1)\text{-ind} \leq n)$$

are saturated classes. □

3.4 Base positional dimension-like functions of the type ind

Combining the definitions of some positional dimension-like functions and some base dimension-like functions we can define many dimension-like functions that has as domain the class of all p-bases. In the present section we define some of these. The proofs of the corresponding statements are ommited. (They can be done combining the proofs of the corresponding propositions given in Sections 3.2 and 3.3).

Definition. Let $\mathbb{B}$ be a class of bases. A class $\mathbb{N}$ of p-bases is said to be $\mathbb{B}$-b_0-p_0-*rim-hereditary* (respectively, $\mathbb{B}$-b_0-p_1-*rim-hereditary*) if for every element (Q, A, X) of $\mathbb{N}$ there exists a $\mathbb{B}$-base B for X such that for every element U of B we have

$$(Q \cap \mathrm{Bd}_X(U), A_U, \mathrm{Bd}_X(U)) \in \mathbb{N},$$

where $A_U = \{\mathrm{Bd}_X(U) \cap V : V \in A\}$ (respectively,

$$(Q \cap \mathrm{Bd}_X(U), A, X) \in \mathbb{N}).$$

Definition. Let $i \in \{0, 1\}$. For every class $\mathbb{B}$ of bases and for every $\mathbb{B}$-b_0-p_i-rim-hereditary class $\mathbb{N}$ of p-bases we denote by $\mathrm{b}_0\text{-}\mathrm{p}_i\text{-ind}_{\mathbb{N}}^{\mathbb{B}}$ the unique dimension-like function that has as domain the class of all p-bases and as range the class $\mathcal{O} \cup \{-1, \infty\}$, which satisfies the following two conditions:

$$\text{(a) } \mathrm{b}_0\text{-}\mathrm{p}_i\text{-ind}_{\mathbb{N}}^{\mathbb{B}}(Q, A, X) = -1$$

if and only if the triad (Q, A, X) belongs to $\mathbb{N}$, and

$$\text{(b) } \mathrm{b}_0\text{-}\mathrm{p}_i\text{-ind}_{\mathbb{N}}^{\mathbb{B}}(Q, A, X) \leq \alpha,\ \alpha \in \mathcal{O}$$

if and only if there exists a $\mathbb{B}$-base B for X such that for every $U \in B$ we have

$$\mathrm{b}_0\text{-}\mathrm{p}_i\text{-ind}_{\mathbb{N}}^{\mathbb{B}}(Q \cap \mathrm{Bd}_X(U), A_U, \mathrm{Bd}_X(U)) < \alpha \text{ for } i = 0,$$

where $A_U = \{\mathrm{Bd}_X(U) \cap V : V \in A\}$, and

$$\mathrm{b}_0\text{-}\mathrm{p}_i\text{-}\mathrm{ind}_{\mathbb{N}}^{\mathbb{B}}(Q \cap \mathrm{Bd}_X(U), A, X) < \alpha \text{ for } i = 1.$$

If $\mathbb{B} = \mathbb{P}(\text{base})$ and $\mathbb{N} = \{(\emptyset, \{\emptyset\}, \emptyset)\}$ or $\mathbb{N} = \{(\emptyset, \{\emptyset\}, \text{space})\}$, then the dimensions $\mathrm{b}_0\text{-}\mathrm{p}_0\text{-}\mathrm{ind}_{\mathbb{L}}^{\mathbb{B}}$ and $\mathrm{b}_0\text{-}\mathrm{p}_1\text{-}\mathrm{ind}_{\mathbb{L}}^{\mathbb{B}}$ are denoted by b_0-p_0-ind and b_0-p_1-ind, respectively. Obviously the class $\mathbb{N} \equiv \{(\emptyset, \{\emptyset\}, \emptyset)\}$ (respectively, $\mathbb{N} \equiv \{(\emptyset, \{\emptyset\}, \text{space})\}$) is $\mathbb{B}$-b_0-p_0-rim-hereditary (respectively, $\mathbb{B}$-b_0-p_1-rim-hereditary).

Lemma 3.1.1 implies the following consequence.

3.4.1 Corollary. *Let $i \in \{0, 1\}$. If A is a p-base for a subset Q in a space X such that*

$$\mathrm{b}_0\text{-}\mathrm{p}_i\text{-}\mathrm{ind}_{\mathbb{N}}^{\mathbb{B}}(Q, A, X) \equiv \alpha \in \{-1\} \cup \mathcal{O},$$

then $\alpha \in \{-1\} \cup \tau^+$. □

3.4.2 Proposition. *The following statements are true:*

(1) *If $\mathbb{B}$ and $\mathbb{N}$ are saturated classes, then the class*

$$\mathbb{P}(\mathrm{b}_0\text{-}\mathrm{p}_i\text{ -}\mathrm{ind}_{\mathbb{N}}^{\mathbb{B}} \leq \alpha),\ \alpha \in \{-1\} \cup \tau^+,\ i \in \{0, 1\},$$

is also saturated. In particular, the class $\mathbb{P}(\mathrm{b}_0\text{-}\mathrm{p}_i\text{-ind} \leq \alpha)$ is a saturated class of p-bases.

(2) *If $\mathbb{B}$ is a weakly saturated class and $\mathbb{N}$ is a saturated class, then*

$$\mathbb{P}(\mathrm{b}_0\text{-}\mathrm{p}_i\text{-}\mathrm{ind}\ _{\mathbb{N}}^{\mathbb{B}} \leq n),\ n \in \{-1\} \cup \omega,\ i \in \{0, 1\}$$

is a saturated class of p-bases. □

Definition. Let $\mathbb{B}$ be a class of bases. A class $\mathbb{N}$ of p-bases is said to be $\mathbb{B}$-b_1-p_0-*rim-hereditary* (respectively, $\mathbb{B}$-b_1-p_1-*rim-hereditary*) if for every element (Q, A, X) of $\mathbb{N}$ there exists a $\mathbb{B}$-base B for X such that for every element U of B there exists a p-base A_U for $Q \cap \mathrm{Bd}_X(U)$ in X for which $A_U \subset A$ and

$$(Q \cap \mathrm{Bd}_X(U), A'_U, \mathrm{Bd}_X(U)) \in \mathbb{N},$$

where $A'_U = \{\mathrm{Bd}_X(U) \cap V : V \in A_U\}$ (respectively,

$(Q \cap \mathrm{Bd}_X(U), A_U, X) \in \mathbb{N})$.

Definition. Let $\mathbb{B}$ be a class of bases and $\mathbb{N}$ a $\mathbb{B}$-b_1-p_i-rim-hereditary class of p-bases. We denote by b_1-p_i-$\mathrm{ind}^{\mathbb{B}}_{\mathbb{N}}$, where $i \in \{0, 1\}$, the unique dimension-like function that has as domain the class of all p-bases and as range the class $\mathcal{O} \cup \{-1, \infty\}$ such that:

$$\text{(a) } \mathrm{b}_1\text{-}\mathrm{p}_i\text{-}\mathrm{ind}^{\mathbb{B}}_{\mathbb{N}}(Q, A, X) = -1$$

if and only if $(Q, A, X) \in \mathbb{N}$, and

$$\text{(b) } \mathrm{b}_1\text{-}\mathrm{p}_i\text{-}\mathrm{ind}^{\mathbb{B}}_{\mathbb{N}}(Q, A, X) \leq \alpha,\ \alpha \in \mathcal{O}$$

if and only if there exists a $\mathbb{B}$-base B for X such that for every $U \in B$ there exists a p-base A_U for $Q \cap \mathrm{Bd}_X(U)$ in X for which $A_U \subset A$ and

$$\mathrm{b}_1\text{-}\mathrm{p}_i\text{-}\mathrm{ind}^{\mathbb{B}}_{\mathbb{N}}(Q \cap \mathrm{Bd}_X(U), A'_U, \mathrm{Bd}_X(U)) < \alpha \text{ for } i = 0,$$

where $A'_U = \{\mathrm{Bd}_X(U) \cap V : V \in A_U\}$, and

$$\mathrm{b}_1\text{-}\mathrm{p}_i\text{-}\mathrm{ind}^{\mathbb{B}}_{\mathbb{N}}(Q \cap \mathrm{Bd}_X(U), A_U, X) < \alpha \text{ for } i = 1.$$

In the case where $\mathbb{B} = \mathbb{P}(\text{base})$ and $\mathbb{N} = \{(\emptyset, \{\emptyset\}, \emptyset)\}$ (respectively, $\mathbb{N} = \{(\emptyset, \{\emptyset\}, \text{space})\}$), then the dimension b_1-p_0-$\mathrm{ind}^{\mathbb{B}}_{\mathbb{L}}$ (respectively, b_1-p_1-$\mathrm{ind}^{\mathbb{B}}_{\mathbb{L}}$) is denoted by b_1-p_0-ind (respectively, by b_1-p_1-ind). Obviously the class $\mathbb{N} \equiv \{(\emptyset, \{\emptyset\}, \emptyset)\}$ (respectively, $\mathbb{N} \equiv \{(\emptyset, \{\emptyset\}, \text{space})\}$) is $\mathbb{B}$-b_1-p_0-rim-hereditary (respectively, $\mathbb{B}$-b_1-p_1-rim-hereditary).

Lemma 3.1.1 implies the following consequence.

3.4.3 Corollary. *Let $i \in \{0, 1\}$. If A is a p-base for a subset Q of a space X such that*

$$\mathrm{b}_1\text{-}\mathrm{p}_i\text{-}\mathrm{ind}^{\mathbb{B}}_{\mathbb{N}}(Q, A, X) \equiv \alpha \in \{-1\} \cup \mathcal{O}$$

then $\alpha \in \{-1\} \cup \tau^+$. □

3.4.4 Proposition. *The following statements are true:*
(1) *If $\mathbb{B}$ is saturated and $\mathbb{N}$ is a weakly saturated class, then*

$$\mathbb{P}(\mathrm{b}_1\text{-}\mathrm{p}_i\text{ -}\mathrm{ind}^{\mathbb{B}}_{\mathbb{N}} \leq \alpha),\ \alpha \in \{-1\} \cup \tau^+, \text{ where } i \in \{0, 1\}$$

is a weakly saturated class. In particular, $\mathbb{P}(\mathrm{b}_1\text{-p}_i\text{-ind} \leq \alpha)$ *is a weakly saturated class of p-bases.*

(2) *If* $\mathbb{B}$ *and* $\mathbb{N}$ *are weakly saturated classes, then*

$$\mathbb{P}(\mathrm{b}_1\text{-p}_i\text{-ind}^{\mathbb{B}}_{\mathbb{N}} \leq n),\ n \in \{-1\} \cup \omega,\ \text{where } i \in \{0, 1\},$$

is a weakly saturated class. □

Now, in the above definitions we change the role of the bases under notation "A" with the bases under notation "B" obtaining in a such manner some new base positional dimension-like functions.

Definition. Let $\mathbb{B}$ be a class of bases. A class $\mathbb{N}$ of p-bases is said to be $\mathbb{B}$-b^0-p_0-*rim-hereditary* (respectively, $\mathbb{B}$-b^0-p_1-*rim-hereditary*) if for every element (Q, A, X) of $\mathbb{N}$ there exists a $\mathbb{B}$-base B for X such that for every element U of A we have

$$(Q \cap \mathrm{Bd}_X(U), B_U, \mathrm{Bd}_X(U)) \in \mathbb{N},$$

where $B_U = \{\mathrm{Bd}_X(U) \cap V : V \in B\}$ (respectively,

$$(Q \cap \mathrm{Bd}_X(U), B, X) \in \mathbb{N}).$$

Definition. Let $\mathbb{B}$ be a class of bases and $\mathbb{N}$ a $\mathbb{B}$-b^0-p_i-rim-hereditary class of p-bases. We denote by b^0-p_i-$\mathrm{ind}^{\mathbb{B}}_{\mathbb{N}}$, $i \in \{0, 1\}$, the unique dimension-like function that has as domain the class of all p-bases and as range the class $\mathcal{O} \cup \{-1, \infty\}$ such that:

$$\text{(a) } \mathrm{b}^0\text{-p}_i\text{-ind}^{\mathbb{B}}_{\mathbb{N}}(Q, A, X) = -1$$

if and only if $(Q, A, X) \in \mathbb{N}$, and

$$\text{(b) } \mathrm{b}^0\text{-p}_i\text{-ind}^{\mathbb{B}}_{\mathbb{N}}(Q, A, X) \leq \alpha,\ \alpha \in \mathcal{O}$$

if and only if there exists a $\mathbb{B}$-base B for X such that for every element U of A we have

$$\mathrm{b}^0\text{-p}_i\text{-ind}^{\mathbb{B}}_{\mathbb{N}}(Q \cap \mathrm{Bd}_X(U), B_U, \mathrm{Bd}_X(U)) < \alpha \text{ for } i = 0,$$

where $B_U = \{\mathrm{Bd}_X(U) \cap V : V \in B\}$, and

$$\mathrm{b}^0\text{-}\mathrm{p}_i\text{-}\mathrm{ind}_{I\!N}^{I\!B}(Q \cap \mathrm{Bd}_X(U), B, X) < \alpha \text{ for } i = 1.$$

In the case, where $I\!B = \mathbb{P}(\mathrm{base})$ and $I\!N = \{(\emptyset, \{\emptyset\}, \emptyset)\}$ (respectively, $I\!N = \{(\emptyset, \{\emptyset\}, \mathrm{space})\}$), the dimension b^0-p_0-$\mathrm{ind}_{I\!L}^{I\!B}$ (respectively, b^0-p_1-$\mathrm{ind}_{I\!L}^{I\!B}$) is denoted by b^0-p_0-ind (respectively, by b^0-p_1-ind). Obviously the class $I\!N \equiv \{(\emptyset, \{\emptyset\}, \emptyset)\}$ (respectively, $I\!N \equiv \{(\emptyset, \{\emptyset\}, \mathrm{space})\}$) is $I\!B$-b^0-p_0-rim-hereditary (respectively, $I\!B$-b^0-p_1-rim-hereditary).

Lemma 3.1.1 implies the following consequence.

3.4.5 Corollary. *Let $i \in \{0, 1\}$. If A is a p-base for a subset Q of space X such that*

$$\mathrm{b}^0\text{-}\mathrm{p}_i\text{-}\mathrm{ind}_{I\!N}^{I\!B}(Q, A, X) \equiv \alpha \in \{-1\} \cup \mathcal{O}$$

then $\alpha \in \tau^+$. □

3.4.6. Proposition. *Let $i \in \{0, 1\}$. If $I\!B$ is weakly saturated and $I\!N$ is a saturated class, then*

$$\mathbb{P}(\mathrm{b}^0\text{-}\mathrm{p}_i\text{-}\mathrm{ind}_{I\!N}^{I\!B} \leq n),\ n \in \{-1\} \cup \omega$$

is a saturated class. In particular, the class $\mathbb{P}(\mathrm{b}^0\text{-}\mathrm{p}_i\text{-}\mathrm{ind} \leq n)$ is a saturated class of p-bases. □

Definition. Let $I\!B$ be a class of bases. A class $I\!N$ of p-bases is said to be $I\!B$-b^1-p_0-*rim-hereditary* (respectively, $I\!B$-b^1-p_1-*rim-hereditary*) if for every element (Q, A, X) of $I\!N$ there exists a $I\!B$-base B for X such that for every element U of A there exists a p-base B_U for $Q \cap \mathrm{Bd}_X(U)$ in X for which $B_U \subset B$ and

$$(Q \cap \mathrm{Bd}_X(U), B'_U, \mathrm{Bd}_X(U)) \in I\!N,$$

where $B'_U = \{\mathrm{Bd}_X(U) \cap V : V \in B_U\}$, (respectively,

$$(Q \cap \mathrm{Bd}_X(U), B_U, X) \in I\!N).$$

Definition. Let $I\!B$ be a class of bases and $I\!N$ a $I\!B$-b^1-p_i-rim-hereditary class of p-bases. We denote by b^1-p_i-$\mathrm{ind}_{I\!N}^{I\!B}$, $i \in \{0, 1\}$ the unique dimension-like function that has as domain the class of all p-bases and as range the class $\mathcal{O} \cup \{-1, \infty\}$ such that:

(a) $\text{b}^1\text{-p}_i\text{-ind}^{\mathbb{B}}_{\mathbb{N}}(Q, A, X) = -1$

if and only if $(Q, A, X) \in \mathbb{N}$, and

(b) $\text{b}^1\text{-p}_i\text{-ind}^{\mathbb{B}}_{\mathbb{N}}(Q, A, X) \leq \alpha$, $\alpha \in \mathcal{O}$

if and only if there exists a $\mathbb{B}$-base B for X such that for every $U \in A$ there exists a p-base B_U for $Q \cap \text{Bd}_X(U)$ in X for which $B_U \subset B$ and

$$\text{b}^1\text{-p}_i\text{-ind}^{\mathbb{B}}_{\mathbb{N}}(Q \cap \text{Bd}_X(U), B'_U, \text{Bd}_X(U)) < \alpha \text{ for } i = 0,$$

where $B'_U = \{\text{Bd}_X(U) \cap V : V \in B_U\}$, and

$$\text{b}^1\text{-p}_i\text{-ind}^{\mathbb{B}}_{\mathbb{N}}(Q \cap \text{Bd}_X(U), B_U, X) < \alpha \text{ for } i = 1.$$

In the case where $\mathbb{B} = \mathbb{P}(\text{base})$ and $\mathbb{N} = \{(\emptyset, \{\emptyset\}, \emptyset)\}$ (respectively, $\mathbb{N} = \{(\emptyset, \{\emptyset\}, \text{space})\}$), then the dimension $\text{b}^1\text{-p}_0\text{-ind}^{\mathbb{B}}_{\mathbb{L}}$ (respectively, $\text{b}^1\text{-p}_1\text{-ind}^{\mathbb{B}}_{\mathbb{L}}$) is denoted by $\text{b}^1\text{-p}_0\text{-ind}$ (respectively, by $\text{b}^1\text{-p}_1\text{-ind}$). Obviously the class $\mathbb{N} \equiv \{(\emptyset, \{\emptyset\}, \emptyset)\}$ (respectively, $\mathbb{N} \equiv \{(\emptyset, \{\emptyset\}, \text{space})\}$) is $\mathbb{B}$-b^1-p_0-rim-hereditary (respectively, $\mathbb{B}$-b^1-p_1-rim-hereditary).

Lemma 3.1.1 implies the following consequence.

3.4.7 Corollary. *Let $i \in \{0, 1\}$. If A is a base for a subset Q of a space X such that*

$$\text{b}^1\text{-p}_i\text{-ind}^{\mathbb{B}}_{\mathbb{N}}(Q, A, X) \equiv \alpha \in \{-1\} \cup \mathcal{O}$$

then $\alpha \in \tau^+$. □

3.4.8 Proposition. *Let $i \in \{0, 1\}$. The following statements are true:*
(1) If $\mathbb{B}$ is saturated and $\mathbb{N}$ is a weakly saturated class, then

$$\mathbb{P}(\text{b}^1\text{-p}_i\text{ -ind}^{\mathbb{B}}_{\mathbb{N}} \leq \alpha), \ \alpha \in \{-1\} \cup \tau^+,$$

is a weakly saturated class. In particular, $\mathbb{P}(\text{b}^1\text{-p}_i\text{-ind} \leq n)$ is a weakly saturated class of p-bases.
(2) If $\mathbb{B}$ and $\mathbb{N}$ are saturated classes, then

$$\mathbb{P}(\text{b}^1\text{-p}_i\text{ -ind}^{\mathbb{B}}_{\mathbb{N}} \leq n), \ n \in \{-1\} \cup \omega$$

is also a saturated class. □

Many distinct base positional dimension-like functions can be obtained if in the definitions of the given functions we replace the class $\mathbb{B}$ of bases by a class $\mathbb{D}$ of p-bases and modify these definitions by a suitable manner.

Any base positional dimension-like function determines some space and positional dimension-like functions. This fact can be done as follows.

Definition. Let *df* be a base positional dimension-like function and $\mathbb{M}$ a class of p-bases. We denote by $\mathbb{M}^s$-*df* (respectively, by $\mathbb{M}^p$-*df*) the space (respectively, positional) dimension-like function, which is defined as follows: for every space X (respectively, for every pair (Q, X) of $\mathbb{P}$(subset)), $\mathbb{M}^s$-*df*(X) (respectively, $\mathbb{M}^p$-*df*(Q, X)) is the minimal element α of $\{0, \infty\} \cup \mathcal{O}$ for which there exist a subset Q of X and its p-base in X (respectively, a p-base B for Q in X) such that $(Q, B, X) \in \mathbb{M}$ and *df*$(Q, B, X) \leq \alpha$.

Suppose that $\mathbb{M}$ is the class of all p-bases and *df* is one of the function: b_0-p_0-ind, b_0-p_1-ind, b_1-p_0-ind, b_1-p_1-ind, b^0-p_0-ind, b^0-p_1-ind, b^1-p_0-ind, and b^1-p_1-ind. In this case the function $\mathbb{M}^s$-*df* is denoted respectively by s-b_0-p_0-ind, s-b_0-p_1-ind, s-b_1-p_0-ind, s-b_1-p_1-ind, s-b^0-p_0-ind, s-b^0-p_1-ind, s-b^1-p_0-ind, and s-b^1-p_1-ind. Similarly, the function $\mathbb{M}^p$-*df* is denoted respectively by p-b_0-p_0-ind, p-b_0-p_1-ind, p-b_1-p_0-ind, p-b_1-p_1-ind, p-b^0-p_0-ind, p-b^0-p_1-ind, p-b^1-p_0-ind, and p-b^1-p_1-ind.

3.4.9 Proposition. *If $\mathbb{M}$ is a weakly saturated (respectively, saturated) class of p-bases and df is a base positional dimension-like function satisfying the saturation (respectively, the weak saturation) property for an element α of the set $\{-1\} \cup \tau^+$, then the space dimension-like function $\mathbb{M}^s$-df and the positional dimension-like function $\mathbb{M}^p$-df satisfy the saturation property for α.* □

3.4.10 Corollary. *The following statements are true:*
(1) *For every $\alpha \in \{-1\} \cup \tau^+$ the classes*

$$\mathbb{P}(\text{s-b}_0\text{-p}_0\text{-ind} \leq \alpha),\ \mathbb{P}(\text{p-b}_0\text{-p}_0\text{-ind} \leq \alpha),$$

$$\mathbb{P}(\text{s-b}_0\text{-p}_1\text{-ind} \leq \alpha),\ \mathbb{P}(\text{p-b}_0\text{-p}_1\text{-ind} \leq \alpha),$$

$$\mathbb{P}(\text{s-b}_1\text{-p}_0\text{-ind} \leq \alpha),\ \mathbb{P}(\text{p-b}_1\text{-p}_0\text{-ind} \leq \alpha),$$

$$\mathbb{P}(\text{s-b}_1\text{-p}_1\text{-ind} \leq \alpha) \text{ and } \mathbb{P}(\text{p-b}_1\text{-p}_1\text{-ind} \leq \alpha)$$

are saturated classes.

(2) *For every* $n \in \{-1\} \cup \omega$ *the classes*

$$\mathbb{P}(\text{s-b}^0\text{-p}_0\text{-ind} \leq n), \mathbb{P}(\text{p-b}^0\text{-p}_0\text{-ind} \leq n),$$

$$\mathbb{P}(\text{s-b}^0\text{-p}_1\text{-ind} \leq n), \mathbb{P}(\text{p-b}^0\text{-p}_1\text{-ind} \leq n),$$

$$\mathbb{P}(\text{s-b}^1\text{-p}_0\text{-ind} \leq n), \mathbb{P}(\text{p-b}^1\text{-p}_0\text{-ind} \leq n),$$

$$\mathbb{P}(\text{s-b}^0\text{-p}_1\text{-ind} \leq n) \text{ and } \mathbb{P}(\text{p-b}^0\text{-p}_1\text{-ind} \leq n)$$

are saturated classes. □

3.5 Base dimension-like functions of the type ind

Definition. The *large inductive dimension* denoted by Ind is the unique function with as domain the class of all spaces and as range the class $\mathcal{O} \cup \{-1, \infty\}$ satisfying the following conditions: (a) $\text{Ind}(X) = -1$ if and only if $X = \emptyset$ and (b) $\text{Ind}(X) \leq \alpha$, where $\alpha \in \mathcal{O}$ if and only if for every pair (K, V) of subsets of X, where K is closed, V is open, and $K \subset V$, there exists an open subset W of X such that $K \subset W \subset V$ and $\text{Ind}(\text{Bd}_X(W)) < \alpha$ (and, therefore, $\text{Ind}(X) = \infty$ if and only if for every ordinal α the inequality $\text{Ind}(X) \leq \alpha$ is not true).

Remarks. 1. The definition of the dimension Ind usually is given for normal space. However, the given definition of Ind does not use the normality of a space and, therefore, formally can be applied to all spaces.

2. We note that the cardinality of the set of all pairs (K, V) considered in the definition of Ind may be large than τ, which means that the given formulation of dimension Ind is not a τ-formulation. (See the Preface about this informal notion.)

In the present section we consider dimension-like functions of type Ind with the domain the class of all bases. The peculiarity of these dimensions is the fact that in their definitions we restrict ourselves by considering only special pairs (K, V) of subsets of a space X, the number of which is less than or equal to τ. These pairs are generated by a given base of the space.

Definition. Let B be a base for a space X. A pair (K, V) of subsets of X is said to be $B^\vee$-*proper* if K is the closure of the union of finite elements of B, V is the union of finite elements of B, and $K \subset V$. Obviously, the pair $(\emptyset, \emptyset)$ of subsets of X is $B^\vee$-proper.

Definition. Let $N \equiv \{V_\eta^X : \eta \in \tau\}$ be an indication of a base B for a space X. For every element v of $\mathcal{F}$ we set

$$K(N, v) = \cup\{\mathrm{Cl}(V_\eta^X) : \eta \in v\}$$

and

$$V(N, v) = \cup\{V_\eta^X : \eta \in v\}.$$

It is considered that $K(N, v) = V(N, v) = \emptyset$ if $v = \emptyset$. A pair (k, v) of elements of $\mathcal{F}$ is said to be $N^\vee$-*proper* if the pair $(K(N, k), V(N, v))$ of subsets of X is $B^\vee$-proper.

Definition. Let $\mathbb{F}$ be a class of subsets and $\mathbb{B}$ a class of bases. It is said that $\mathbb{B}$ is $\mathbb{F}$-$\vee$-*hereditary* if for every element $(B, X) \in \mathbb{B}$ and for every $B^\vee$-proper pair (K, V) of subsets of X there exists an element (W, X) of $\mathbb{F}$ such that $K \subset W \subset V$ and

$$(\{V \cap \mathrm{Bd}_X(W) : V \in B\}, \mathrm{Bd}_X(W)) \in \mathbb{B}.$$

Obviously, the class $\mathbb{P}(\text{base})$ of all bases is $(\mathbb{P}(\text{Op}))^\vee$-hereditary. (For a $B^\vee$-proper pair (K, V) we set $(W, X) = (V, X)$).

Definition. For every class $\mathbb{F}$ of open subsets and for every $\mathbb{F}$-$\vee$-hereditary class $\mathbb{B}$ of bases we denote by $\mathrm{b}^\vee\text{-}\mathrm{Ind}_{\mathbb{B}}^{\mathbb{F}}$ the unique dimension-like function that has as domain the class of all bases and as range the class $\mathcal{O} \cup \{-1, \infty\}$ satisfying the following condition:

$$\text{(a) } \mathrm{b}^\vee\text{-}\mathrm{Ind}_{\mathbb{B}}^{\mathbb{F}}(B, X) = -1$$

if and only if $(B, X) \in \mathbb{B}$, and

$$\text{(b) } \mathrm{b}^\vee\text{-}\mathrm{Ind}_{\mathbb{B}}^{\mathbb{F}}(B, X) \leq \alpha,\ \alpha \in \mathcal{O}$$

if and only if for every $B^\vee$-proper pair (K, V) of subsets of X there exists an element (W, X) of $\mathbb{F}$ such that $K \subset W \subset V$ and

$$\text{b}^{\vee}\text{-Ind}^{\mathbb{F}}_{\mathbb{B}}(\{V \cap \text{Bd}_X(W) : V \in B\}, \text{Bd}_X(W)) < \alpha$$

(and, therefore, $\text{b}^{\vee}\text{-Ind}^{\mathbb{F}}_{\mathbb{B}}(B, X) = \infty$ if and only if for every ordinal α the inequality $\text{b}^{\vee}\text{-Ind}^{\mathbb{F}}_{\mathbb{B}}(B, X) \leq \alpha$ is not true).

If $\mathbb{F} = \mathbb{P}(\text{Op})$ and $\mathbb{B} = \mathbb{P}(\{\emptyset\}, \emptyset)$, then the function $\text{b}^{\vee}\text{-Ind}^{\mathbb{F}}_{\mathbb{B}}$ is denoted by $\text{b}^{\vee}$-Ind. (It is easy to see that $\mathbb{P}(\{\emptyset\}, \emptyset)$ is $\mathbb{P}(\text{Op})$-$\vee$-hereditary.)

3.5.1 Lemma. *Let* $(B, X) \in \mathbb{P}(\text{base})$ *and* $\alpha \in \{-1\} \cup \mathcal{O}$. *If*

$$\text{b}^{\vee}\text{-Ind}^{\mathbb{F}}_{\mathbb{B}}(B, X) \leq \alpha,$$

then $(\emptyset, X) \in \mathbb{F}$ *and* $(\{\emptyset\}, \emptyset) \in \mathbb{B}$.

Proof. Let

$$\text{b}^{\vee}\text{-Ind}^{\mathbb{F}}_{\mathbb{B}}(B, X) \leq \alpha \neq \infty.$$

The pair $(K, V) \equiv (\emptyset, \emptyset)$ of subsets of X is $B^{\vee}$-proper. The subset $W \equiv \emptyset$ of X is the only subset of X for which $K \subset W \subset V$. Since $\text{b}^{\vee}\text{-Ind}^{\mathbb{F}}_{\mathbb{B}}(B, X) \leq \alpha$ we have $(\emptyset, X) \in \mathbb{F}$ and $\text{b}^{\vee}\text{-Ind}^{\mathbb{F}}_{\mathbb{B}}(\{\emptyset\}, \emptyset) < \alpha$. The last condition implies (by induction on α) that $(\{\emptyset\}, \emptyset) \in \mathbb{B}$. □

Lemma 3.1.1 implies the following consequence.

3.5.2 Corollary. *Let* $(B, X) \in \mathbb{P}(\text{base})$ *and* $\alpha \in \{-1\} \cup \mathcal{O}$. *If*

$$\text{b}^{\vee}\text{-Ind}^{\mathbb{F}}_{\mathbb{B}}(B, X) = \alpha,$$

then $\alpha \in \{-1\} \cup \tau^{+}$. □

Definition. It is said that a class $\mathbb{F}$ of subsets satisfies the *Finite Free Union Condition* if for every space X, a finite number of mutually disjoint open subspaces X_i, $i \in j \in \omega$, whose union is the space X, and a subset U_i of X_i the conditions $(U_i, X_i) \in \mathbb{F}$, $i \in j$, imply that the pair (U, X) belongs to $\mathbb{F}$, where $U = \cup\{U_i : i \in j\}$.

Obviously, the class $\mathbb{P}(\text{Op})$ satisfies the Finite Free Union Condition.

Definition. It is a said that a class $\mathbb{B}$ of bases satisfies the *Finite Free Union Condition* if for every space X, a base B for X, and a finite number mutually disjoint open subspaces X_i, $i \in j \in \omega$, whose union is the space X, the conditions $(B_i, X_i) \in \mathbb{B}$, where $B_i = \{U \cap X_i : U \in B\}$, $i \in j$, imply that $(B, X) \in \mathbb{B}$. (We note that B is not uniquely determined by the sets B_i.)

Obviously, the classes $\mathbb{P}(\text{base})$ and $\mathbb{B} \equiv \{(\{\emptyset\}, \emptyset)\}$ satisfy the Finite Free Union Condition.

3.5.3 Lemma. *The class*

$$\mathbb{P}(\text{b}^{\vee}\text{-Ind}^{\mathbb{F}}_{\mathbb{B}} \leq \alpha),\ \alpha \in \{-1\} \cup \tau^{+},$$

satisfies the Finite Free Union Condition provided that the classes $\mathbb{F}$ and $\mathbb{B}$ satisfy the Finite Free Union Condition.

Proof. Let $\mathbb{F}$ and $\mathbb{B}$ be classes satisfying the Finite Free Union Condition. We prove the lemma by induction on α. For $\alpha = -1$ the class $\mathbb{P}(\text{b}^{\vee}\text{-Ind}^{\mathbb{F}}_{\mathbb{B}} \leq \alpha)$ coincides with the class $\mathbb{B}$ and, therefore, satisfies the Finite Free Union Condition. Supposing that for an ordinal α of τ^{+} the class $\mathbb{P}(\text{b}^{\vee}\text{-Ind}^{\mathbb{F}}_{\mathbb{B}} \leq \beta)$ satisfies the Finite Free Union Condition for every $\beta \in \{-1\} \cup \alpha$ we prove that the class $\mathbb{P}(\text{b}^{\vee}\text{-Ind}^{\mathbb{F}}_{\mathbb{B}} \leq \alpha)$ satisfies also this condition.

Let X be a space, B a base for X, and X_i, $i \in j \in \omega$, mutually disjoint open subspaces of X, whose union is X. Let also

$$(B_i, X_i) \in \mathbb{P}(\text{b}^{\vee}\text{-Ind}^{\mathbb{F}}_{\mathbb{B}} \leq \alpha),$$

where $B_i = \{X_i \cap U : U \in B\}$. We need to prove that

$$(B, X) \in \mathbb{P}(\text{b}^{\vee}\text{-Ind}^{\mathbb{F}}_{\mathbb{B}} \leq \alpha).$$

Let (K, V) be a $B^{\vee}$-proper pair of subsets of X. It is easy to see that the pair (K_i, V_i), where $K_i = K \cap X_i$ and $V_i = V \cap X_i$ is a $B^{\vee}_i$-proper pair of subsets of X_i. Therefore, there exists an element (U_i, X_i) of $\mathbb{F}$ such that $K_i \subset U_i \subset V_i$ and

$$(B'_i, \text{Bd}_{X_i}(U_i)) \in \mathbb{P}(\text{b}^{\vee}\text{-Ind}^{\mathbb{F}}_{\mathbb{B}} \leq \beta_i)$$

for some $\beta_i \in \{-1\} \cup \alpha$, where

$$B'_i = \{U \cap \text{Bd}_{X_i}(U_i) : U \in B_i\}.$$

Let $\beta = \max\{\beta_i : i \in j\}$. Then, for every $i \in j$,

$$(B'_i, \mathrm{Bd}_{X_i}(U_i)) \in \mathbb{P}(\mathrm{b}^{\vee}\text{-}\mathrm{Ind}_{\mathbb{B}}^{\mathbb{F}} \le \beta). \tag{3.5.1}$$

Since $\mathbb{F}$ satisfies the Finite Free Union Condition we have that $(W, X) \in \mathbb{F}$, where $W = \cup\{U_i : i \in j\}$. Obviously, $K \subset W \subset V$.

It is clear that the set $\mathrm{Bd}_X(W)$ is the union of mutually disjoint subsets $\mathrm{Bd}_{X_i}(U_i)$, $i \in j$. Setting

$$B' = \{U \cap \mathrm{Bd}_X(W) : U \in B\}$$

we have

$$B'_i = \{U \cap \mathrm{Bd}_{X_i}(U_i) : U \in B'\}$$

for every $i \in j$. Since the class $\mathbb{P}(\mathrm{b}^{\vee}\text{-}\mathrm{Ind}_{\mathbb{B}}^{\mathbb{F}} \le \beta)$ satisfies the Finite Free Union Condition, relation (3.5.1) implies that

$$(B', \mathrm{Bd}_X(W)) \in \mathbb{P}(\mathrm{b}^{\vee}\text{-}\mathrm{Ind}_{\mathbb{B}}^{\mathbb{F}} \le \beta).$$

This fact means that $(B, X) \in \mathbb{P}(\mathrm{b}^{\vee}\text{-}\mathrm{Ind}_{\mathbb{B}}^{\mathbb{F}} \le \alpha)$, which completes the proof of the lemma. □

3.5.4 Proposition. *The class*

$$\mathbb{P}(\mathrm{b}^{\vee}\text{-}\mathrm{Ind}_{\mathbb{B}}^{\mathbb{F}} \le n),\ n \in \{-1\} \cup \omega,$$

is a saturated class satisfying the Finite Free Union Condition provided that $\mathbb{F}$ and $\mathbb{B}$ are saturated classes satisfying the Finite Free Union Condition.

Proof. Let $\mathbb{F}$ and $\mathbb{B}$ be saturated classes satisfying the Finite Free Union Condition. We prove that for every $n \in \{-1\} \cup \omega$ the class $\mathbb{P}_n \equiv \mathbb{P}(\mathrm{b}^{\vee}\text{-}\mathrm{Ind}_{\mathbb{B}}^{\mathbb{F}} \le n)$ is also such a class. By Lemma 5.3 this class satisfies the Finite Free Union Condition. Therefore, it suffices to prove that the class $\mathbb{P}(\mathrm{b}^{\vee}\text{-}\mathrm{Ind}_{\mathbb{B}}^{\mathbb{F}} \le n)$ is saturated. We prove this fact by induction on n.

For $n = -1$ the class $\mathbb{P}(\mathrm{b}^{\vee}\text{-}\mathrm{Ind}_{\mathbb{B}}^{\mathbb{F}} \le n)$ coincides with the class $\mathbb{B}$ and, therefore, it is saturated. Supposing that for every $m \in \{-1\} \cup n$ the class $\mathbb{P}(\mathrm{b}^{\vee}\text{-}\mathrm{Ind}_{\mathbb{B}}^{\mathbb{F}} \le m)$ is saturated we prove that the class $\mathbb{P}(\mathrm{b}^{\vee}\text{-}\mathrm{Ind}_{\mathbb{B}}^{\mathbb{F}} \le n)$ is also saturated.

Let $\mathbf{S}$ be an indexed collection of spaces,

$$\mathbf{B} \equiv \{B^X : X \in \mathbf{S}\}$$

a $\mathbb{P}_n$-co-base for $\mathbf{S}$, and

$$\mathbf{N} \equiv \{N^X \equiv \{V_\eta^X : \eta \in \tau\} : X \in \mathbf{S}\}$$

a co-indication of $\mathbf{B}$.

Let $\{(k_\varepsilon, v_\varepsilon) : \varepsilon \in \tau\}$ be an indication of the set $\mathcal{F} \times \mathcal{F}$. For every $X \in \mathbf{S}$ and for every $\varepsilon \in \tau$ denote by W_ε^X an open subset of X such that:

(a) if the pair $(k_\varepsilon, v_\varepsilon)$ is $(N^X)^\vee$-proper, then

$$(W_\varepsilon^X, X) \in \mathbb{F},\ K(N^X, k_\varepsilon) \subset W_\varepsilon^X \subset V(N^X, v_\varepsilon), \text{ and}$$

$$(\{U \cap \mathrm{Bd}_X(W_\varepsilon^X) : U \in B^X\}, \mathrm{Bd}_X(W_\varepsilon^X)) \in \mathbb{P}_{n-1} \text{ and} \quad (3.5.2)$$

(b) if the pair $(k_\varepsilon, v_\varepsilon)$ is not $(N^X)^\vee$-proper, then $W_\varepsilon^X = \emptyset$.

We note that in the last case by Lemma 3.5.1 we have $(W_\varepsilon^X, X) \in \mathbb{F}$ and

$$(B, \mathrm{Bd}_X(W_\varepsilon^X)) = (\{\emptyset\}, \emptyset) \in \mathbb{B} \subset \mathbb{P}_{n-1},$$

where $B = \{U \cap \mathrm{Bd}_X(W_\varepsilon^X) : U \in B^X\}$.

For every $\varepsilon \in \tau$ we consider the restrictions

$$\mathbf{K}(\varepsilon) \equiv \{K(N^X, k_\varepsilon) : X \in \mathbf{S}\},$$

$$\mathbf{V}(\varepsilon) \equiv \{V(N^X, v_\varepsilon) : X \in \mathbf{S}\}, \text{ and}$$

$$\mathbf{W}(\varepsilon) \equiv \{W^X(\varepsilon) : X \in \mathbf{S}\}$$

of $\mathbf{S}$. Obviously, the restriction $\mathbf{K}(\varepsilon)$ is closed, $\mathbf{V}(\varepsilon)$ is open, and $\mathbf{W}(\varepsilon)$ is an open $\mathbb{F}$-restriction. Denote by $\mathbf{S}_\varepsilon$ the indexed collection

$$\mathbf{Bd}(\mathbf{W}(\varepsilon)) = \{\mathrm{Bd}_X(W^X(\varepsilon)) : X \in \mathbf{S}\}$$

of spaces. For every element $Y \equiv \mathrm{Bd}_X(W^X(\varepsilon))$ of $\mathbf{S}_\varepsilon$ we set

$$B^Y = \{V \cap Y : V \in B^X\}$$

and consider the indication

$$N^Y \equiv \{V_\eta^Y \equiv V_\eta^X \cap Y : \eta \in \tau\}$$

of B^Y. Then, by relation (3.5.2),

$$\mathbf{B}_\varepsilon \equiv \{B^Y : Y \in \mathbf{S}_\varepsilon\}$$

is a $\mathbb{P}_{n-1}$-co-base for $\mathbf{S}_\varepsilon$ and

$$\mathbf{N}_\varepsilon \equiv \{\{V_\eta^Y : \eta \in \tau\} : Y \in \mathbf{S}_\varepsilon\}$$

is a co-indication of $\mathbf{B}_\varepsilon$.

By inductive assumption $\mathbb{P}_{n-1}$ is a saturated class. Therefore, there exists an initial co-mark $\mathbf{M}_\varepsilon^+$ of $\mathbf{S}_\varepsilon$ corresponding to the co-indication $\mathbf{N}_\varepsilon$ of $\mathbf{B}_\varepsilon$ and the class $\mathbb{P}_{n-1}$. Without loss of generality we can suppose that a one-to-one mapping $\theta^{\mathbf{N}}$ of τ into itself with the condition $|\tau \setminus \theta^{\mathbf{N}}(\tau)| = \tau$ is an indicial mapping from $\mathbf{N}_\varepsilon$ to $\mathbf{M}_\varepsilon^+$ for every $\varepsilon \in \tau$.

For every $\varepsilon \in \tau$ it is easy to construct a co-mark $\mathbf{M}_\varepsilon^{\mathbf{S}}$ of the collection $\mathbf{S}$ such that:

(a) the restriction to $\mathbf{S}_\varepsilon$ of $\mathbf{M}_\varepsilon^{\mathbf{S}}$ is the co-mark $\mathbf{M}_\varepsilon^+$ of $\mathbf{S}_\varepsilon$,
(b) $\mathbf{M}_\varepsilon^{\mathbf{S}}$ is an extension of $\mathbf{N}$, and
(c) $\theta^{\mathbf{N}}$ is an indicial mapping from $\mathbf{N}$ to $\mathbf{M}_\varepsilon^{\mathbf{S}}$.

We denote by $\theta_\varepsilon^{\mathbf{S}}$ an one-to-one mapping of τ into itself such that:

$$(1)\ \theta_\varepsilon^{\mathbf{S}}(\delta) = \delta \text{ for every } \delta \in \theta^{\mathbf{N}}(\tau), \text{ that is, } \theta_\varepsilon^{\mathbf{S}} \circ \theta^{\mathbf{N}} = \theta^{\mathbf{N}},$$

$$(2)\ \theta_{\varepsilon_0}^{\mathbf{S}}(\tau \setminus \theta^{\mathbf{N}}(\tau)) \cap \theta_{\varepsilon_1}^{\mathbf{S}}(\tau \setminus \theta^{\mathbf{N}}(\tau)) = \emptyset, \text{ if } \varepsilon_0 \neq \varepsilon_1, \text{ and}$$

$$(3)\ |\tau \setminus (\cup\{\theta_\varepsilon^{\mathbf{N}}(\tau) : \varepsilon \in \tau\})| = \tau.$$

Let $\mathbf{M}^+$ be a co-mark of $\mathbf{S}$ such that:

(a) $\mathbf{M}^+$ is an initial co-mark corresponding to $\mathbb{F}$-restriction $\mathbf{W}_\varepsilon$,
(b) $\mathbf{M}^+$ is a co-extension of $\mathbf{M}_\varepsilon^{\mathbf{S}}$, and
(c) $\theta_\varepsilon^{\mathbf{S}}$ is an indicial mapping from $\mathbf{M}_\varepsilon^{\mathbf{S}}$ to $\mathbf{M}^+$.

Then, $\mathbf{M}^+|_{\mathbf{S}_\varepsilon}$ is a co-extension of $\mathbf{M}_\varepsilon^+$ and $\theta_\varepsilon^{\mathbf{S}}$ is an indicial mapping of this co-extension. Let also

$$\mathbf{M} \equiv \{\{U_\delta^X : \delta \in \tau\} : X \in \mathbf{S}\}$$

be a co-extension of $\mathbf{M}^+$ and θ^+ an indicial mapping from $\mathbf{M}^+$ to $\mathbf{M}$. Then, for every $\varepsilon \in \tau$ the restriction to $\mathbf{S}_\varepsilon$ of the co-mark $\mathbf{M}$, that is the co-mark $\mathbf{M}|_{\mathbf{S}_\varepsilon} \equiv \mathbf{M}_\varepsilon$ is a co-extension of the co-mark $\mathbf{M}^+|_{\mathbf{S}_\varepsilon}$ and θ^+ is an indicial mapping of this co-extension. Setting $\theta = \theta^+ \circ \theta_\varepsilon^{\mathbf{S}} \circ \theta^{\mathbf{N}}$ we have that θ is an indicial mapping from $\mathbf{N}$ to $\mathbf{M}$ as well as an indicial mapping from $\mathbf{N}_\varepsilon$ to $\mathbf{M}_\varepsilon$ for every $\varepsilon \in \tau$.

Furthermore, R_ε^+ be an initial family of equivalence relation on $\mathbf{S}_\varepsilon$ corresponding to the co-mark $\mathbf{M}_\varepsilon$, the co-indication $\mathbf{N}_\varepsilon$ of $\mathbf{B}_\varepsilon$, and the class $\mathbb{P}_{n-1}$.

We denote by R^+ a family of equivalence relations on $\mathbf{S}$ such that for every $\varepsilon \in \tau$ we have:

(a) R^+ is
$(\mathbf{M}, \mathbf{K}(\varepsilon))$-admissible,
$(\mathbf{M}, \mathbf{W}(\varepsilon))$-admissible,
$(\mathbf{M}, \mathbf{Co}(\mathbf{W}(\varepsilon)))$-admissible,
$(\mathbf{M}, \mathbf{Cl}(\mathbf{W}(\varepsilon)))$-admissible, and
$(\mathbf{M}, \mathbf{Bd}(\mathbf{W}(\varepsilon)))$-admissible,

(b) the restriction to $\mathbf{S}_\varepsilon$ of the family R^+ is a final refinement of the family R_ε^+, and

(c) R^+ is an initial family corresponding to the co-mark $\mathbf{M}$ and the $\mathbb{F}$-restriction $\mathbf{W}(\varepsilon)$.

Note that R^+ is also $(\mathbf{M}, \mathbf{V}(\varepsilon))$-admissible and $(\mathbf{M}, \mathbf{Co}(\mathbf{V}(\varepsilon)))$-admissible.

Let $\mathrm{R} \equiv \{\sim^s : s \in \mathcal{F}\}$ be an $\mathbf{M}$-admissible family of equivalence relations on $\mathbf{S}$, which is a final refinement of R^+, and denote by R_ε, $\varepsilon \in \tau$, the restriction to $\mathbf{S}_\varepsilon$ of this family. Then, R_ε is a final refinement of R_ε^+.

To prove the proposition we need to prove that if $\mathbf{L} \in \mathrm{C}^\diamond(\mathrm{R})$, then

$$(\mathrm{B}^{\mathbf{L}}_{\diamond,\kappa}, \mathbf{L}) \in \mathbb{P}_n,$$

where $\kappa = \theta(\tau)$. Let $\mathbf{L} \in \mathrm{C}^\diamond(\mathrm{R})$ and let (K, V) be an arbitrary $(\mathrm{B}^{\mathbf{L}}_{\diamond,\kappa})^\vee$-proper pair of subsets of $\mathrm{T}(\mathbf{L})$. It suffices to find an open subset W of $\mathrm{T}(\mathbf{L})$ such that $(W, \mathrm{T}(\mathbf{L})) \in \mathbb{F}$, $K \subset W \subset V$, and

$$(\{U \cap \mathrm{Bd}_{\mathrm{T}(\mathbf{L})}(W) : U \in \mathrm{B}^{\mathbf{L}}_{\diamond,\kappa}\}, \mathrm{Bd}_{\mathrm{T}(\mathbf{L})}(W)) \in \mathbb{P}_{n-1}.$$

Let $N = \{U_\zeta : \zeta \in \tau\}$ be an indication of the base $\mathrm{B}^{\mathbf{L}}_{\diamond,\kappa}$. For every $\zeta \in \tau$ we denote by $\delta(\zeta)$ and $\mathbf{H}(\zeta)$ an element of κ and the element of $\mathrm{C}^\diamond(\mathrm{R})$, respectively, such that $U_\zeta = U^{\mathrm{T}}_{\delta(\zeta)}(\mathbf{H}(\zeta))$. Obviously, $\mathbf{H}(\zeta) \subset \mathbf{L}$.

There exist elements k and v of $\mathcal{F}$ such that $K = K(N, k)$ and $V = V(N, v)$. For every $X \in \mathbf{L}$, we denote by k^X (respectively, by v^X) the set of all elements η of τ such that there exists an element ζ of k (respectively, an element ζ of v) for which $X \in \mathbf{H}(\zeta)$ and $\eta = \theta^{-1}(\delta(\zeta))$. First we prove the following relations:

$$K = \cup\{e^X_{\mathrm{T}}(K(N^X, k^X)) : X \in \mathbf{L}\} \text{ and}$$

$$V = \cup\{e_{\mathrm{T}}^{X}(V(N^X, v^X)) : X \in \mathbf{L}\}. \tag{3.5.3}$$

Consider the first relation. Let $\mathbf{a} \in K$. By the definition of K there exists an element ζ of k such that

$$\mathbf{a} \in \mathrm{Cl}_{\mathrm{T}(\mathbf{L})}(U_\zeta) = \mathrm{Cl}_{\mathrm{T}}(U^{\mathrm{T}}_{\delta(\zeta)}(\mathbf{H}(\zeta))).$$

By Lemma 1.2.7(4) there exists an element (x, X) of $\mathbf{a}$ such that $X \in \mathbf{H}(\zeta)$ and

$$x \in \mathrm{Cl}_X(U^X_{\delta(\zeta)}) = \mathrm{Cl}_X(V^X_\eta),$$

where $\eta = \theta^{-1}(\delta(\zeta))$. This means that $\eta \in k^X$ and $x \in K(N^X, k^X)$, which implies that

$$\mathbf{a} \in e_{\mathrm{T}}^{X}(K(N^X, k^X)).$$

Thus,

$$\mathbf{a} \in \cup\{e_{\mathrm{T}}^{X}(K(N^X, k^X)) : X \in \mathbf{L}\}.$$

Conversely, let $\mathbf{a}$ be an element of the right side of the first equation of (3.5.3). There exists an element X of $\mathbf{L}$ such that

$$\mathbf{a} \in e_{\mathrm{T}}^{X}(K(N^X, k^X)) = \cup\{e_{\mathrm{T}}^{X}(\mathrm{Cl}(V^X_\eta)) : \eta \in k^X\}.$$

There exists $\eta \in k^X$ such that $\mathbf{a} \in e_{\mathrm{T}}^{X}(\mathrm{Cl}_X(V^X_\eta))$. This means that there exists $(x, X) \in \mathbf{a}$ such that

$$x \in \mathrm{Cl}_X(V^X_\eta) = \mathrm{Cl}_X(U^X_{\theta(\eta)}).$$

Since $\eta \in k^X$, there exists an element ζ of k such that $X \in \mathbf{H}(\zeta)$ and $\eta = \theta^{-1}(\delta(\zeta))$. By Lemma 1.2.7(4),

$$\mathbf{a} \in \mathrm{Cl}_{\mathrm{T}}(U^{\mathrm{T}}_{\delta(\zeta)}(\mathbf{H}(\zeta))) =$$

$$\mathrm{Cl}_{\mathrm{T}(\mathbf{L})}(U^{\mathrm{T}}_{\delta(\zeta)}(\mathbf{H}(\zeta))) = \mathrm{Cl}_{\mathrm{T}(\mathbf{L})}(U_\zeta).$$

Therefore, $\mathbf{a} \in K(N, k) = K$. The second relation of (3.5.3) is proved similarly.

Thus, for every $\mathbf{H} \in \mathrm{C}^{\diamond}(\mathrm{R})$ such that $\mathbf{H} \subset \mathbf{L}$ we have

$$K \cap \mathrm{T}(\mathbf{H}) = \cup\{e^X_{\mathrm{T}}(K(N^X, k^X)) : X \in \mathbf{H}\} \text{ and}$$

$$V \cap \mathrm{T}(\mathbf{H}) = \cup\{e^X_{\mathrm{T}}(V(N^X, v^X)) : X \in \mathbf{H}\}. \tag{3.5.4}$$

Denote by t an element of $\mathcal{F}$ such that

$$\{\delta(\zeta) : \zeta \in k\} \subset t, \ \{\delta(\zeta) : \zeta \in v\} \subset t,$$

and for every t-equivalence class $\mathbf{H}$ the condition $\mathbf{H} \cap \mathbf{H}(\zeta) \neq \emptyset$ for some $\zeta \in k \cup v$ implies that $\mathbf{H} \subset \mathbf{H}(\zeta)$. For every $X, Y \in \mathbf{L}$ by the definition of the sets k^X, v^X, k^Y, and v^Y it follows that if $X \sim^t Y$, then $k^X = k^Y$ and $v^X = v^Y$.

Let $\mathbf{H}$ be an arbitrary $\sim^t$-equivalence class such that $\mathbf{H} \subset \mathbf{L}$. We set $k^{\mathbf{H}} = k^X$ and $v^{\mathbf{H}} = v^X$, where X is an element of $\mathbf{H}$. By the above, $k^{\mathbf{H}}$ and $v^{\mathbf{H}}$ are independent of the element X of $\mathbf{H}$. There exists an element $\varepsilon(\mathbf{H}) \in \tau$ such that

$$(k_{\varepsilon(\mathbf{H})}, v_{\varepsilon(\mathbf{H})}) = (k^{\mathbf{H}}, v^{\mathbf{H}}).$$

Consider the subsets $\mathrm{T}|_{\mathbf{W}(\varepsilon)}$ and $\mathrm{T}|_{\mathbf{Bd}(\mathbf{W}(\varepsilon))}$ of T. Since R^+ is an initial family corresponding to the co-mark $\mathbf{M}$ and the $I\!\!F$-restriction $\mathbf{W}(\varepsilon(\mathbf{H}))$ (and since R is a final refinement of R^+) we have

$$(\mathrm{T}|_{\mathbf{W}(\varepsilon(\mathbf{H}))} \cap \mathrm{T}(\mathbf{H}), \mathrm{T}(\mathbf{H})) \in I\!\!F.$$

Since $I\!\!F$ satisfies the Finite Free Union Condition we have $(W, \mathrm{T}(\mathbf{L})) \in I\!\!F$, where

$$W = \cup\{\mathrm{T}|_{\mathbf{W}(\varepsilon(\mathbf{H}))} \cap \mathrm{T}(\mathbf{H}) : \mathbf{H} \in \mathrm{C}(\sim^t) \text{ and } \mathbf{H} \subset \mathbf{L}\}.$$

By Corollary 1.3.6,

$$\mathrm{T}|_{\mathbf{W}(\varepsilon(\mathbf{H}))} \cap \mathrm{T}(\mathbf{H}) = \cup\{e^X_{\mathrm{T}}(W^X(\varepsilon(\mathbf{H}))) : X \in \mathbf{H}\}.$$

This relation and relations (3.5.4) imply that

$$K \cap \mathrm{T}(\mathbf{H}) \subset \mathrm{T}|_{\mathbf{W}(\varepsilon(\mathbf{H}))} \cap \mathrm{T}(\mathbf{H}) \subset V \cap \mathrm{T}(\mathbf{H})$$

and, therefore, $K \subset W \subset V$.

Thus, it remains to be proved that

$$(\{U \cap \mathrm{Bd}_{\mathrm{T}(\mathbf{L})}(W) : U \in \mathrm{B}^{\mathbf{L}}_{\diamondsuit,\kappa}\}, \mathrm{Bd}_{\mathrm{T}(\mathbf{L})}(W)) \in \mathbb{P}_{n-1}.$$

Since $\mathrm{T}(\mathbf{H})$ is open and closed in $\mathrm{T}(\mathbf{L})$, the set $\mathrm{Bd}_{\mathrm{T}(\mathbf{L})}(W)$ is the union of all sets $\mathrm{Bd}_{\mathrm{T}(\mathbf{H})}(\mathrm{T}|_{\mathbf{W}(\varepsilon(\mathbf{H}))} \cap \mathrm{T}(\mathbf{H}))$, where $\mathbf{H} \in \mathrm{C}(\sim^t)$ and $\mathbf{H} \subset \mathbf{L}$.

All conditions of Lemma 1.4.10 are satisfied if instead of the restriction $\mathbf{Q}$ we consider the restriction $\mathbf{W}(\varepsilon)$. Therefore,

$$\mathrm{Bd}_{\mathrm{T}}(\mathrm{T}|_{\mathbf{W}(\varepsilon(\mathbf{H}))}) = \mathrm{T}|_{\mathbf{Bd}(\mathbf{W}(\varepsilon(\mathbf{H})))} = \mathrm{T}(\mathbf{M}_{\varepsilon(\mathbf{H})}, \mathrm{R}_{\varepsilon(\mathbf{H})}).$$

Thus,

$$\mathrm{Bd}_{\mathrm{T}(\mathbf{H})}(\mathrm{T}|_{\mathbf{W}(\varepsilon(\mathbf{H}))} \cap \mathrm{T}(\mathbf{H})) = \mathrm{Bd}_{\mathrm{T}}(\mathrm{T}|_{\mathbf{W}(\varepsilon(\mathbf{H}))} \cap \mathrm{T}(\mathbf{H})) =$$

$$\mathrm{Bd}_{\mathrm{T}}(\mathrm{T}|_{\mathbf{W}(\varepsilon(\mathbf{H}))}) \cap \mathrm{T}(\mathbf{H}) = \mathrm{T}|_{\mathbf{Bd}(\mathbf{W}(\varepsilon(\mathbf{H})))} \cap \mathrm{T}(\mathbf{H}) =$$

$$\mathrm{T}(\mathbf{M}_{\varepsilon(\mathbf{H})}, \mathrm{R}_{\varepsilon(\mathbf{H})}, \mathbf{L}_{\varepsilon(\mathbf{H})}),$$

where $\mathbf{L}_{\varepsilon(\mathbf{H})}$ is the trace on $\mathbf{W}(\varepsilon)$ of $\mathbf{H}$. Since the class $\mathbb{P}_{n-1}$ is saturated we have

$$(\mathrm{B}^{\mathbf{L}_{\varepsilon(\mathbf{H})}}_{\diamondsuit,\kappa}, \mathrm{T}(\mathbf{M}_{\varepsilon(\mathbf{H})}, \mathrm{R}_{\varepsilon(\mathbf{H})}, \mathbf{L}_{\varepsilon(\mathbf{H})})) \in \mathbb{P}_{n-1},$$

where $\mathrm{B}^{\mathbf{L}_{\varepsilon(\mathbf{H})}}_{\diamondsuit,\kappa}$ is the $(\diamondsuit, \kappa)$-standard base for $\mathrm{T}(\mathbf{M}_{\varepsilon(\mathbf{H})}, \mathrm{R}_{\varepsilon(\mathbf{H})}, \mathbf{L}_{\varepsilon(\mathbf{H})})$. On the other hand, by Lemma 1.3.3,

$$\{U \cap \mathrm{T}(\mathbf{M}_{\varepsilon(\mathbf{H})}, \mathrm{R}_{\varepsilon(\mathbf{H})}, \mathbf{L}_{\varepsilon(\mathbf{H})}) : U \in \mathrm{B}^{\mathbf{L}}_{\diamondsuit,\kappa}\} = \mathrm{B}^{\mathbf{L}_{\varepsilon(\mathbf{H})}}_{\diamondsuit,\kappa}.$$

Since $\mathbb{P}_{n-1}$ satisfies the Finite Free Union Condition we have

$$(\{U \cap \mathrm{Bd}_{\mathrm{T}(\mathbf{L})}(W) : U \in \mathrm{B}^{\mathbf{L}}_{\diamondsuit,\kappa}\}, \mathrm{Bd}_{\mathrm{T}(\mathbf{L})}(W)) \in \mathbb{P}_{n-1},$$

which completes the proof of the proposition. □

We observe that the dimension $\mathrm{b}^{\vee}\text{-}\mathrm{Ind}^{\mathbb{F}}_{\mathbb{B}}$ depends of the types of pairs (K, V) considered in the definition of the dimension. Below, we modify the definition of the $B^{\vee}$-proper pair and define in such a manner some new dimension-like functions.

Definition. Let B be a base for a space X. A pair (K, V) of subsets of X is said to be B-*proper* if K is the closure of an element of B and V is an element of B such that $K \subset V$.

If in the above definition instead of B we consider the base $B^{\diamondsuit}$ for X, then we obtain the notion of $B^{\diamondsuit}$-*proper* pair of subsets of X. (We recall that $B^{\diamondsuit}$ is the minimal set of subsets of X containing all elements of B and closed under finite unions and finite intersections.)

Definition. Let $\mathbb{F}$ be a class of subsets and $\mathbb{B}$ a class of bases. If in the definition of $\mathbb{F}$-$\vee$-hereditarity instead of the $B^{\vee}$-proper pairs we consider the B-proper pairs (respectively, the $B^{\diamondsuit}$-proper pairs), then we obtain the notion of $\mathbb{F}$-*hereditarity* (respectively, the notion of $\mathbb{F}$-$\diamondsuit$-*hereditarity*).

Definition. If in the definition of the dimension-like function $\text{b}^{\vee}\text{-Ind}^{\mathbb{F}}_{\mathbb{B}}$ instead of the $B^{\vee}$-proper pairs we consider the B-proper pairs (respectively, the $B^{\diamondsuit}$-proper pairs), then we obtain a new dimension-like function denoted by $\text{b-Ind}^{\mathbb{F}}_{\mathbb{B}}$ (respectively, by $\text{b}^{\diamondsuit}\text{-Ind}^{\mathbb{F}}_{\mathbb{B}}$).

The proof of the following proposition is similar to the proof of Proposition 3.5.4.

3.5.5 Proposition. *For every* $n \in \{-1\} \cup \omega$ *the classes*

$$\mathbb{P}(\text{b-Ind}^{\mathbb{F}}_{\mathbb{B}} \leq n) \text{ and } \mathbb{P}(\text{b}^{\diamondsuit}\text{-Ind}^{\mathbb{F}}_{\mathbb{B}} \leq n)$$

are saturated classes of bases satisfying the Finite Free Union Condition provided that $\mathbb{F}$ *and* $\mathbb{B}$ *are saturated classes satisfying the Finite Free Union Condition.* □

Notation. Consider the following space dimension-like functions: s-$\text{b-Ind}^{\mathbb{F}}_{\mathbb{B}}$, s-$\text{b}^{\vee}\text{-Ind}^{\mathbb{F}}_{\mathbb{B}}$, and s-$\text{b}^{\diamondsuit}\text{-Ind}^{\mathbb{F}}_{\mathbb{B}}$. (See Section 3.3, where to any base dimension-like function *df* we correspond a space dimension-like function s-*df*.) If $\mathbb{F} = \mathbb{P}(\text{Op})$ and $\mathbb{B} = \mathbb{P}(\text{base})$, then these functions are denoted by s-b-Ind, s-$\text{b}^{\vee}$-Ind, and s-$\text{b}^{\diamondsuit}$-Ind, respectively. (Obviously, $\mathbb{P}(\text{base})$ is $\mathbb{P}(\text{Op})$-hereditary, $\mathbb{P}(\text{Op})$-$\vee$-hereditary, and $\mathbb{P}(\text{Op})$-$\diamondsuit$-hereditary.)

3.5.6 Corollary. *For every* $n \in \{-1\} \cup \omega$ *the classes*

$$\mathbb{P}(\text{s-b-Ind} \leq n),\ \mathbb{P}(\text{s-b}^{\vee}\text{-Ind} \leq n), \text{ and}$$

$$\mathbb{P}(\text{s-b}^{\diamondsuit}\text{-Ind} \leq n)$$

are saturated classes of spaces. □

3.6 Base dimension-like functions of the type dim

Definition. The *dimension* dim is the unique function that has as domain the class of all spaces and as range the set $\{-1, \infty\} \cup \omega$ satisfying the following condition: $\dim(X) \leq n$, $n \in \{-1\} \cup \omega$, if and only if for every finite open cover c of X there exists a finite open cover r of X such that each element of r is contained in an element of c and the intersection of any $n+2$ distinct element of r is empty (and, therefore, $\dim(X) = \infty$ if and only if for every $n \in \omega$ the relation $\dim(X) \leq n$ is not true).

Remarks. 1. The above given definition of the dimension dim is applied usually to the class of regular spaces or to the more restricted class of spaces. However, in this definition we do not use the regularity of a space and, therefore, we can apply it to the class of all spaces.

2. We note that the cardinality of the set of all open covers c considered in the above definition in general is large than τ, which means that the given formulation is not τ-formulation.

In the present section we consider dimension-like functions of the type dim with the domain the class of all bases. In the definition of these functions finite open covers c of a space are also considered. However, the elements of these covers are considered to be elements of the given base for the space. This is the reason why the domain of the introducing functions is the class of all bases.

Definitions. Let X be a space. A *cover* of X is a non-empty set of subsets of X, whose union is X.

A cover r of X is said to be a *refinement* of a cover c of X if each element of r is contained in an element of c.

The *order* of a cover r of X is:

(a) the integer -1 if r consists of the empty set only, (and, therefore, $X = \emptyset$);

(b) an integer $n \in \omega$ if the intersection of any $n+2$ distinct elements of r is empty and there exist $n+1$ distinct elements of r, whose intersection is not empty; and

(c) the symbol ∞ if for every $n \in \omega$ there exist n distinct elements of r, whose intersection is not empty.

Let B be a base for a space X. A cover c of X is said to be *B-cover* if all elements of c are elements of B.

Let $I\!F$ be a class of subsets. A cover r of a space X is said to be $I\!F$-*cover* if $(V, X) \in I\!F$ for every $V \in r$. An $I\!F$-cover of a space X, which is a refinement of a cover c of X, is called $I\!F$-*refinement* of c.

Definition. For every class $I\!F$ of subsets we denote by b-dim$^{I\!F}$ the (unique) function that has as domain the class of all bases and as range the set $\{-1, \infty\} \cup \omega$ satisfying the following condition:

$$\text{b-dim}^{I\!F}(B, X) \leq n,\ n \in \{-1\} \cup \omega$$

if and only if for every finite B-cover c of the space X there exists a finite $I\!F$-refinement of c of the order $\leq n$.

Definition. It is said that a class $I\!F$ of subsets satisfies the *Finite Union Condition* if the conditions $(F_i, X) \in I\!F$, $i \in j \in \omega$, imply that

$$(\cup\{F_i : i \in j\}, X) \in I\!F.$$

It is said that a class $I\!F$ of subsets satisfies the *Empty Subset Condition* if the condition $(Q, X) \in I\!F$ implies that $(\emptyset, X) \in I\!F$.

3.6.1 Proposition. *The class*

$$I\!P(\text{b-dim}^{I\!F} \leq n),\ n \in \{-1\} \cup \omega,$$

is a saturated class provided that $I\!F$ is a complete saturated class of subsets satisfying the Finite Union Condition and the Empty Subset Condition.

Proof. Let $I\!F$ be a complete saturated class of subsets satisfying the Finite Union Condition and the Empty Subset Condition, and let n be an element of $\{-1\} \cup \omega$. We prove that the class

$$I\!P_n \equiv I\!P(\text{b-dim}^{I\!F} \leq n)$$

is saturated. Let $\mathbf{S}$ be an indexed collection of spaces and

$$\mathbf{B} \equiv \{B^X : X \in \mathbf{S}\}$$

a $I\!P_n$-co-base for $\mathbf{S}$. Consider an arbitrary co-indication

$$\mathbf{N} \equiv \{N^X \equiv \{V_\eta^X : \eta \in \tau\} : X \in \mathbf{S}\}$$

of $\mathbf{B}$. Let $q \equiv \{\eta_0, ..., \eta_k\}$ be an element of $\mathcal{F} \setminus \{\emptyset\}$. For every $X \in \mathbf{S}$ we consider the indexed set

$$V^X(q) \equiv \{V^X_\eta : \eta \in q\}.$$

First, we suppose that the (non-indexed) set $V^X(q)$ is a cover (and, therefore, a B^X-cover) of X. In this case we construct an indexed set

$$W^X(q) \equiv \{W^X(q, \eta) : \eta \in q\}$$

such that: (a) $W^X(q, \eta) \subset V^X_\eta$ for every $\eta \in q$ and (b) the non-indexed set $W^X(q)$ is an $\mathbb{F}$-refinement of $V^X(q)$ of the order $\leq n$. This construction is done as follows. Since $V^X(q)$ is a B-cover of X and $(B^X, X) \in \mathbb{P}_n$, there exists a finite $\mathbb{F}$-refinement r^X_q of $V^X(q)$ of the order $\leq n$. Then, the set $W^X(q, \eta_0)$ is considered to be the union of all elements of r^X_q, which are contained in $V^X_{\eta_0}$. If there is no such elements, then it is supposed that $W^X(q, \eta_0) = \emptyset$. The set $W^X(q, \eta_i)$, $i \in \{1, ..., k\}$, is considered to be the union of all elements of r^X_q, which are contained in $V^X_{\eta_i}$ and are not contained in any set V^X_η, $\eta \in \{\eta_0, ..., \eta_{i-1}\}$. If there is no such elements, then it is supposed that $W^X(q, \eta_i) = \emptyset$. Since r^X_q is an $\mathbb{F}$-cover of the order $\leq n$ and since $\mathbb{F}$ satisfies the Finite Union Condition and the Empty Subset Condition, the set $W^X(q)$ is also $\mathbb{F}$-cover of X of the order $\leq n$. Moreover, by construction, $W^X(q, \eta) \subset V^X_\eta$, $\eta \in q$. We note that the cover $W^X(q)$ has also the following property: if $\eta_0, ..., \eta_{n+1}$ are distinct elements of q such that $W^X(q, \eta_i) \neq \emptyset$, $i \in \{0, ..., n+1\}$, then $W^X(q, \eta_0), ..., W^X(q, \eta_{n+1})$ are distinct elements of $W^X(q)$ and, therefore,

$$W^X(q, \eta_0) \cap ... \cap W^X(q, \eta_{n+1}) = \emptyset.$$

Now, if $V^X(q)$ is not a cover of X, then we set $W^X(q, \eta) = \emptyset$, $\eta \in q$. Thus, for every $q \in \mathcal{F} \setminus \{\emptyset\}$ and for every $\eta \in q$ it is constructed an $\mathbb{F}$-restriction of $\mathbf{S}$:

$$\mathbf{W}(q, \eta) \equiv \{W^X(q, \eta) : X \in \mathbf{S}\}.$$

Let $\mathbf{M}^+$ be a co-mark of $\mathbf{S}$, which is a co-extension of $\mathbf{N}$, and let $\theta^{\mathbf{N}}$ be an indicial mapping of this co-extension. Without loss of generality we can suppose that for every $q \in \mathcal{F} \setminus \{\emptyset\}$ and for every $\eta \in q$, $\mathbf{M}^+$ is an initial co-mark corresponding to the restriction $\mathbf{W}(q, \eta)$ of $\mathbf{S}$ and the class

$\mathbb{F}$. Moreover, since $\mathbb{F}$ is a complete class we can suppose that there exists a family R_0^+ of equivalence relations on $\mathbf{S}$ such that for every $q \in \mathcal{F} \setminus \{\emptyset\}$ and for every $\eta \in q$ we have: (a) the family R_0^+ is $(\mathbf{M}^+, \mathbf{W}(q,\eta))$-admissible and (b) the restriction $\mathbf{W}(q,\eta)$ of $\mathbf{S}$ is $(\mathbf{M}^+, \mathrm{R}_0^+)$-complete.

We prove that $\mathbf{M}^+$ is an initial co-mark corresponding to the indication $\mathbf{N}$ of $\mathbf{B}$ and the class $\mathbb{P}_n$. For this we consider a co-mark

$$\mathbf{M} \equiv \{\{U_\delta^X : \delta \in \tau\} : X \in \mathbf{S}\}$$

of $\mathbf{S}$, which is a co-extension of $\mathbf{M}^+$, and denote by θ^+ an indicial mapping of this co-extension. Then. $\mathbf{M}$ is a co-extension of $\mathbf{N}$ and $\theta \equiv \theta^+ \circ \theta^{\mathbf{N}}$ is an indicial mapping from $\mathbf{N}$ to $\mathbf{M}$.

For every $q \in \mathcal{F} \setminus \{\emptyset\}$ and for every $\eta \in q$ we consider a family $\mathrm{R}_{q,\eta}^+$ of equivalence relations on $\mathbf{S}$, which is an initial family corresponding to the co-mark $\mathbf{M}$, the restriction $\mathbf{W}(q,\eta)$ of $\mathbf{S}$, and the class $\mathbb{F}$. Denote by R^+ a family of equivalence relations on $\mathbf{S}$, which is a final refinement of all families $\mathrm{R}_{q,\eta}^+$. Moreover, we suppose that R^+ is a final refinement of the family R_0^+.

To prove that $\mathbf{M}^+$ is initial it suffices to prove that R^+ is an initial family corresponding to the co-mark $\mathbf{M}$, the indication $\mathbf{N}$ of $\mathbf{B}$, and the class $\mathbb{P}_n$. For this we consider a family

$$\mathrm{R} \equiv \{\sim^s : s \in \mathcal{F}\}$$

of equivalence relation on $\mathbf{S}$, which is a final refinement of R^+. We need to prove that for every $\mathbf{H} \in \mathrm{C}^\diamondsuit(\mathrm{R})$,

$$(\mathrm{B}^{\mathbf{H}}_{\diamondsuit,\theta(\tau)}, \mathrm{T}(\mathbf{H})) \in \mathbb{P}_n,$$

where $\mathrm{B}^{\mathbf{H}}_{\diamondsuit,\theta(\tau)}$ is the $(\diamondsuit, \theta(\tau))$-standard base for the subspace $\mathrm{T}(\mathbf{H})$ of $\mathbf{T}$.

Let $\mathbf{H} \in \mathrm{C}^\diamondsuit(\mathrm{R})$ and

$$c \equiv \{U_{\delta_0}^{\mathrm{T}}(\mathbf{H}_0), ..., U_{\delta_k}^{\mathrm{T}}(\mathbf{H}_k)\}$$

be a finite $\mathrm{B}^{\mathbf{H}}_{\diamondsuit,\theta(\tau)}$-cover of $\mathrm{T}(\mathbf{H})$, where $\delta_0, ..., \delta_k \in \theta(\tau)$ and $\mathbf{H}_0, ..., \mathbf{H}_k \in \mathrm{C}^\diamondsuit(\mathrm{R})$. (Therefore, $\mathbf{H}_0 \subset \mathbf{H}$,...,$\mathbf{H}_k \subset \mathbf{H}$.) We set $s = \{\delta_0, ..., \delta_k\}$ (it is possible that for distinct i and j, $\delta_i = \delta_j$) and $q = \theta^{-1}(s)$.

For every $X \in \mathbf{H}$ denote by q^X the set of all elements η of q such that for some $i \in \{0, ..., k\}$, $\theta(\eta) = \delta_i$ and $X \in \mathbf{H}_i$. It is easy to see that $q^X \neq \emptyset$ and that the set

$$V^X(q^X) \equiv \{V_\eta^X : \eta \in q^X\}$$

is a B^X-cover of X. Therefore,

$$W^X(q^X) \equiv \{W^X(q^X, \eta) : \eta \in q^X\}$$

is an $\mathbb{F}$-cover of X.

Let t be an element of $\mathcal{F}$ such that if $\mathbf{L}$ is $\sim^t$-equivalence class and $\mathbf{L} \cap \mathbf{H}_i \neq \emptyset$ for some $i \in \{0, ..., k\}$, then $\mathbf{L} \subset \mathbf{H}_i$. Obviously, if $X \sim^t Y$, then $q^X = q^Y$.

Let $\mathbf{L}$ be an element of $\mathrm{C}(\sim^t)$ such that $\mathbf{L} \subset \mathbf{H}$. We set $q^{\mathbf{L}} = q^X$, where $X \in \mathbf{L}$. By the above, $q^{\mathbf{L}}$ is independent of the element X of $\mathbf{L}$. Consider the cover

$$c(\mathbf{L}) \equiv \{U_{\delta_i}^{\mathrm{T}}(\mathbf{H}_i) \cap \mathrm{T}(\mathbf{L}) = U_{\delta_i}^{\mathrm{T}}(\mathbf{L}) : \delta_i \in \theta(q^{\mathbf{L}}), \mathbf{L} \subset \mathbf{H}_i\}$$

of $\mathrm{T}(\mathbf{L})$. We prove that

$$r(\mathbf{L}) \equiv \{\mathrm{T}|_{\mathbf{W}(q^{\mathbf{L}},\eta)} \cap \mathrm{T}(\mathbf{L}) : \eta \in q^{\mathbf{L}}\}$$

is an $\mathbb{F}$-refinement of $c(\mathbf{L})$ of the order $\leq n$.

First we note that since $\mathbb{F}$ is saturated by the choice of the family R^+ for every $\eta \in q^{\mathbf{L}}$ we have

$$(\mathrm{T}|_{\mathbf{W}(q^{\mathbf{L}},\eta)} \cap \mathrm{T}(\mathbf{L}), \mathrm{T}) \in \mathbb{F},$$

that is $r(\mathbf{L})$ is an $\mathbb{F}$-cover. For every $X \in \mathbf{L}$ and for every $\eta \in q^{\mathbf{L}}$ we have

$$W^X(q^{\mathbf{L}}, \eta) \subset V_\eta^X$$

and, therefore,

$$e_{\mathrm{T}}^X(W^X(q^{\mathbf{L}}, \eta)) \subset e_{\mathrm{T}}^X(V_\eta^X),$$

where e_{T}^X is the natural embeding of X into T. By Corollary 1.3.6 we have

$$\mathrm{T}|_{\mathbf{W}(q^{\mathbf{L}},\eta)} \cap \mathrm{T}(\mathbf{L}) \subset U_{\delta_i}^{\mathrm{T}}(\mathbf{L}),$$

where i is an element of the set $\{0, ..., k\}$ such that $\delta_i = \theta(\eta)$ and $\mathbf{L} \subset \mathbf{H}_i$, which means that $r(\mathbf{L})$ is an $\mathbb{F}$-refinement of $c(\mathbf{L})$.

Now, we prove that the order of $r(\mathbf{L})$ is $\leq n$. Indeed, in the opposite case, there exists $n+2$ distinct elements $\eta_0, ..., \eta_{n+1}$ of $q^{\mathbf{L}}$ such that

$$(\mathrm{T}|_{\mathbf{W}(q^{\mathbf{L}},\eta_0)} \cap \mathrm{T}(\mathbf{L})) \cap ... \cap (\mathrm{T}|_{\mathbf{W}(q^{\mathbf{L}},\eta_{n+1})} \cap \mathrm{T}(\mathbf{L})) \neq \emptyset.$$

Let $\mathbf{a}$ be an element of the above intersection and let $(x, X) \in \mathbf{a}$. Then

$$\mathbf{a} \in \mathrm{T}|_{\mathbf{W}(q^{\mathbf{L}},\eta_0)} \cap ... \cap \mathrm{T}|_{\mathbf{W}(q^{\mathbf{L}},\eta_{n+1})}.$$

Since the restrictions $\mathbf{W}(q^{\mathbf{L}}, \eta_i)$, $i \in \{0, ..., n+1\}$, are $(\mathbf{M}^+, \mathrm{R}_0^+)$-complete, by Lemma 2.2.3 these restrictions are also $(\mathbf{M}, \mathrm{R})$-complete. Therefore,

$$x \in W^X(q^X, \eta_0) \cap ... \cap W^X(q^X, \eta_{n+1}).$$

Since $W^X(q^X, \eta_0), ..., W^X(q^X, \eta_{n+1})$ are distinct elements of $W^X(q^X)$ the above contradicts the fact that the cover $W^X(q)$ of X has order $\leq n$. Thus, $r(\mathbf{L})$ has order $\leq n$.

Now, it is easy to verify that the set

$$r \equiv \cup\{r(\mathbf{L}) : \mathbf{L} \in \mathrm{C}(\sim^t) \text{ and } \mathbf{L} \subset \mathbf{H}\}$$

is an $\mathbb{F}$-refinement of c of the order $\leq n$, which completes the proof of the proposition. □

Notation. If $\mathbb{F}$ is the class $\mathbb{P}(\mathrm{Op})$ (respectively, the class $\mathbb{P}(\mathrm{Cl})$), then the function b-dim$^{\mathbb{F}}$ is denoted by b-dim^{Op} (respectively, by b-dim^{Cl}).

Denote by $\mathbb{P}(\mathrm{rCl})$ the class of subsets consisting of all pairs (Q, X), where Q is a regular closed subset of a space X. (A subset Q of a space X is called *regular closed* if $Q = \mathrm{Cl}(\mathrm{Int}(Q))$.) It is easy to verify that $\mathbb{P}(\mathrm{rCl})$ is a complete saturated class of subsets satisfying the Finite Union Condition and the Empty Subset Condition.

If $\mathbb{F} = \mathbb{P}(\mathrm{rCl})$, then the function b-dim$^{\mathbb{F}}$ is denoted by b-dim^{rCl}.

3.6.2 Corollary. *For every $n \in \{-1\} \cup \omega$ in the classes*

$$\mathbb{P}(\text{b-dim}^{\mathrm{Op}} \leq n),$$

$$\mathbb{P}(\text{b-dim}^{\mathrm{Cl}} \leq n), \text{ and}$$

$$\mathbb{P}(\text{b-dim}^{\mathrm{rCl}} \leq n)$$

there exist universal elements. □

Notation. For every class $\mathbb{M}$ of bases we can consider the space dimension-like function $\mathbb{M}$-b-dim$^{\mathbb{F}}$ (see Section 3.3). If $\mathbb{M}$ is the class of all bases and $\mathbb{F}$ is one of the classes $\mathbb{P}(\mathrm{Op})$, $\mathbb{P}(\mathrm{Cl})$, and $\mathbb{P}(\mathrm{rCl})$, then the function $\mathbb{M}$-b-dim$^{\mathbb{F}}$ is denoted by s-b-dim^{Op}, s-b-dim^{Cl}, and s-b-dim^{rCl}, respectively.

3.6.3 Proposition. *For every $n \in \{-1\} \cup \omega$ the classes*

$\mathbb{P}(\text{s-b-dim}^{\mathrm{Op}} \leq n)$,

$\mathbb{P}(\text{s-b-dim}^{\mathrm{Cl}} \leq n)$, and

$\mathbb{P}(\text{s-b-dim}^{\mathrm{rCl}} \leq n)$

are saturated classes of spaces. Therefore, in these classes there exist universal elements. □

Chapter 4

Saturated classes of spaces with structure

In the present chapter classes of spaces with "structure", different from that of Chapter 2, are considered. By a structure on a space we mean here a fixed representation of the space as a union of its subsets having some additional properties.

In Section 4.1 we introduce the notion of a (properly) universal element and the notion of a (complete) saturated class of spaces with structure. Saturated classes have the intersection and the universality properties. Some conditions on the structure are given in order that the class of all spaces with such structure is to be (complete) saturated.

In Section 4.2 we define dimension-like functions of the type ind. The domain of these functions is the class of all spaces with structure. We prove that the defined functions satisfy the saturation property.

In Section 4.3 we give the notion of the so-called D-dimension of a space and prove that this dimension satisfies the universality property. The definition of the D-dimension is based on the dimension ind. However, instead of ind we can take many other dimensions considered in Chapter 3.

Section 4.4 is devoted to some applications of the results given in Section 4.1. We give a series of corollaries concerning the existence of universal elements (preserving the structure) in some concrete classes. These classes are related to the notions of countable-dimensionality, strongly countable-dimensionality, locally finite-dimensionality, and their generalizations. The formulations of the corollaries have "more usual form" than that of the corresponding propositions. In the formulations of the corollaries we use the class of all spaces and the dimension ind. However, instead of the class of all spaces we can take any saturated class of spaces and (independently) instead of the dimension ind we can take many other dimensions considered in Chapter 3.

4.1 Saturated classes of spaces with structure

Remark. Below, we shall use the notions of the ν-set, ν-directing mapping, and Φ-operational mapping as well as the notations Λ^α, $\alpha \in \mathcal{O}$, which were introduced in Section 2.4. In that section the cardinal ν was infinite while here it can be finite. For a finite cardinal ν we shall use exactly the same definitions of the ν-set, ν-directing mapping, and Φ-operational mapping as for an infinte cardinal given in Section 2.4.

Agreement. In the present section it is supposed that: (a) a fixed cardinal ν such that $1 \leq \nu \leq \tau$, (b) a fixed ν-directing mapping Φ, (c) a fixed ν-set Λ of cardinality $\leq \tau$, (d) a fixed preorder $\prec$ on Λ, and (e) a fixed mapping $\mathbf{p}$ of Λ into the class of all topological classes of subsets are given. For every $\lambda \in \Lambda$ the class $\mathbf{p}(\lambda)$ is also denoted by $\mathbb{P}_\lambda$.

Definition. A mapping ψ_X of the set Λ into the set $\mathcal{P}(X)$ of all subsets of a space X is said to be a $(\Phi, \Lambda, \prec, \mathbf{p})$-*structure* or briefly a *structure* on X if the following conditions are satisfied:

(1) ψ_X is a Φ-operational mapping,

(2) $\cup\{\psi_X(\lambda) : \lambda \in \Lambda\} = X$,

(3) $\psi_X(\lambda_0) \subset \psi_X(\lambda_1)$ if $\lambda_0 \prec \lambda_1$, and

(4) $(\psi_X(\lambda), X) \in \mathbb{P}_\lambda$ for every $\lambda \in \Lambda$.

By a *space with structure* we mean a pair (ψ_X, X), where ψ_X is a structure on a space X.

In the case, where $\Lambda = \Lambda^0$ the notion of $(\Phi, \Lambda, \prec, \mathbf{p})$-structure is independent of Φ. In this case instead of $(\Phi, \Lambda, \prec, \mathbf{p})$-structure we write $(\Lambda, \prec, \mathbf{p})$-structure.

Definition. A class $\mathbb{P}$ of spaces with structure is said to be *topological* if for every homeomorphism h of a space X onto a space Y the condition $(\psi_X, X) \in \mathbb{P}$ implies that $(h_{\mathcal{P}} \circ \psi_X, Y) \in \mathbb{P}$, where $h_{\mathcal{P}}$ is the mapping of $\mathcal{P}(X)$ into $\mathcal{P}(Y)$ such that $h_{\mathcal{P}}(Q) = h(Q)$, $Q \in \mathcal{P}(X)$. If $(\psi_X, X) \in \mathbb{P}$, then we say that ψ_X is a $\mathbb{P}$-*structure* on the space X.

In the sequel, all considered classes of spaces with structure are assumed to be topological. We shall denote by $\mathbb{P}(\Phi, \Lambda, \prec, \mathbf{p})$ the class of all spaces with structure. Obviously, $\mathbb{P}(\Phi, \Lambda, \prec, \mathbf{p})$ is a topological class. In the case, where $\Lambda = \Lambda^0$ the class $\mathbb{P}(\Phi, \Lambda, \prec, \mathbf{p})$ is denoted by $\mathbb{P}(\Lambda, \prec, \mathbf{p})$.

Definition. Let $\mathbf{S}$ be an indexed collection of spaces. Any indexed collection

$$\boldsymbol{\Psi} \equiv \{\psi_X : X \in \mathbf{S}\},$$

where ψ_X is a structure on X, is called a $(\Phi, \Lambda, \prec, \mathbf{p})$-*co-structure* on $\mathbf{S}$ or briefly a *co-structure* on $\mathbf{S}$. Such a collection $\boldsymbol{\Psi}$ is called a $\mathbb{P}$-*co-structure* on $\mathbf{S}$, where $\mathbb{P}$ is a class of spaces with structure, if $(\psi_X, X) \in \mathbb{P}$ for every $X \in \mathbf{S}$.

Notation. Let $\boldsymbol{\Psi} \equiv \{\psi_X : X \in \mathbf{S}\}$ be a co-structure on an indexed collection $\mathbf{S}$ of spaces. For every $\lambda \in \Lambda$ we denote by $\boldsymbol{\Psi}(\lambda)$ the restriction $\{\psi_X(\lambda) : X \in \mathbf{S}\}$ of $\mathbf{S}$.

Definition. Let $\boldsymbol{\Psi}$ be a co-structure on an indexed collection $\mathbf{S}$ of spaces and let $\mathbf{M}$ be a co-mark of $\mathbf{S}$. An admissible family R of equivalence relations on $\mathbf{S}$ is said to be $(\mathbf{M}, \boldsymbol{\Psi})$-*admissible* if R is $(\mathbf{M}, \boldsymbol{\Psi}(\lambda))$-admissible for every $\lambda \in \Lambda$.

Definition. A non-empty class $\mathbb{P}$ of spaces with structure is said to be *saturated* if for every indexed collection $\mathbf{S}$ of spaces and for every $\mathbb{P}$-co-structure $\boldsymbol{\Psi}$ on $\mathbf{S}$ there exists a co-mark $\mathbf{M}^+$ of $\mathbf{S}$ satisfying the following condition: for every co-extension $\mathbf{M}$ of $\mathbf{M}^+$ there exists an $(\mathbf{M}, \boldsymbol{\Psi})$-admissible family R^+ of equivalence relations on $\mathbf{S}$ such that for every admissible family R of equivalence relations on $\mathbf{S}$, which is a final refinement of R^+, the pair $(\psi_{\mathrm{T}(\mathbf{L})}, \mathrm{T}(\mathbf{L}))$ belongs to $\mathbb{P}$, where $\mathbf{L}$ is an element of $\mathrm{C}^\diamond(\mathrm{R})$ and $\psi_{\mathrm{T}(\mathbf{L})}$ is the mapping of Λ into $\mathcal{P}(\mathrm{T}(\mathbf{L}))$ such that

$$\psi_{\mathrm{T}(\mathbf{L})}(\lambda) = \mathrm{T}(\mathbf{M}|_{\boldsymbol{\Psi}(\lambda)}, \mathrm{R}|_{\boldsymbol{\Psi}(\lambda)}, \mathbf{L}|_{\boldsymbol{\Psi}(\lambda)})$$

for every $\lambda \in \Lambda$.

The co-mark $\mathbf{M}^+$ is said to be an *initial co-mark of* $\mathbf{S}$ (*corresponding* to the $\mathbb{P}$-co-structure $\boldsymbol{\Psi}$ of $\mathbf{S}$), and the family R^+ is said to be an *initial family* (*corresponding* to the co-mark $\mathbf{M}$ and the $\mathbb{P}$-co-structure $\boldsymbol{\Psi}$). The mapping $\psi_{\mathrm{T}(\mathbf{L})}$ is said to be the *containing structure* on $\mathrm{T}(\mathbf{L})$ (*corresponding* to the family R, the co-mark $\mathbf{M}$, the $\mathbb{P}$-co-structure $\mathbf{\Psi}$, and the subspace $\mathrm{T}(\mathbf{L})$ of T).

The proofs of the following two propositions are similar to the proofs of the corresponding propositions of Chapter 2.

4.1.1 Proposition. *The non-empty intersection of not more than τ many saturated classes of spaces with structure is also a saturated class of spaces with structure.* □

Definition. Let $\mathbb{P}$ be a class of spaces with structure. An element (ψ_T, T) of $\mathbb{P}$ is said to be *universal* (respectively, *properly universal*) if for every element (ψ_X, X) of $\mathbb{P}$ there exists an embedding h of X into T such that $\psi_X(\lambda) \subset h^{-1}(\psi_T(\lambda))$ (respectively, $\psi_X(\lambda) = h^{-1}(\psi_T(\lambda))$) for every $\lambda \in \Lambda$.

Definition. A class $\mathbb{P}$ of spaces with structure is said to be *complete* if for every indexed collection $\mathbf{S}$ of spaces and for every $\mathbb{P}$-co-structure $\mathbf{\Psi}$ on $\mathbf{S}$ the restriction $\mathbf{\Psi}(\lambda)$ of $\mathbf{S}$ is complete for every $\lambda \in \Lambda$.

4.1.2 Proposition. *In each (complete) saturated class of spaces with structure there exist (properly) universal elements.* □

Definition. Let $\mathbb{P}$ be a saturated class of spaces with structure and $E \equiv (\psi_X, X)$ an element of $\mathbb{P}$. The space X is said to be the *space-coordinate* of the element E. The class consisting of space-coordinates of all elements of $\mathbb{P}$ is said to be the *space-component* of $\mathbb{P}$.

The next two propositions are easy to prove.

4.1.3 Proposition. *The space-component of any saturated class of spaces with structure is a saturated class of spaces.* □

4.1.4 Proposition. *Suppose that $\mathbb{P}$ is a saturated class of spaces with structure and $\mathbb{E}$ is a saturated class of spaces. Then, the subclass of $\mathbb{P}$ consisting of all elements (ψ_X, X) of $\mathbb{P}$ such that $X \in \mathbb{E}$ is also a saturated class of spaces with structure.* □

4.1.5 Proposition. *Suppose that P_λ, $\lambda \in \Lambda$ is a saturated class. If $\Lambda = \Lambda^0$, then $\mathrm{P}(\Lambda, \prec, \mathbf{p})$ is also a saturated class.*

Proof. Suppose that $\Lambda = \Lambda^0$. Let $\mathbf{S}$ be an indexed collection of spaces and $\boldsymbol{\Psi}$ a $\mathbb{P}$-co-structure on $\mathbf{S}$, where $\mathbb{P} = \mathbb{P}(\Lambda, \prec, \mathbf{p})$.

For every $\lambda \in \Lambda$ we denote by $\mathbf{M}^+_\lambda$ an initial co-mark of $\mathbf{S}$ corresponding to the restriction $\boldsymbol{\Psi}(\lambda)$ and to the class $\mathbb{P}_\lambda$. Let $\mathbf{M}^+$ be a co-mark of $\mathbf{S}$, which is a co-extension of all co-marks $\mathbf{M}^+_\lambda$, $\lambda \in \Lambda$, and let $\mathbf{M}$ be an arbitrary co-extension of $\mathbf{M}^+$.

For every $\lambda \in \Lambda$ let R^+_λ be an initial family of equivalence relations on $\mathbf{S}$ corresponding to the co-mark $\mathbf{M}$, to the restriction $\boldsymbol{\Psi}(\lambda)$, and to the class $\mathbb{P}_\lambda$. Let R^+ be a family of equivalence relation on $\mathbf{S}$, which a final refinement of all families R^+_λ, where $\lambda \in \Lambda$. It is easy to see that R^+ is $(\mathbf{M}, \boldsymbol{\Psi})$-admissible. Denote by R an arbitrary family of equivalence relations on $\mathbf{S}$, which is a final refinement of R^+.

To prove the proposition it suffices to prove that $(\psi_{\mathrm{T}(\mathbf{H})}, \mathrm{T}(\mathbf{H})) \in \mathbb{P}$ for every $\mathbf{H} \in \mathrm{C}^\diamondsuit(\mathrm{R})$, where $\psi_{\mathrm{T}(\mathbf{H})}$ is the containing structure corresponding to the family R, the co-mark $\mathbf{M}$, the $\mathbb{P}$-co-structure $\boldsymbol{\Psi}$, and the subspace $\mathrm{T}(\mathbf{H})$ of T.

Let $\mathbf{H} \in \mathrm{C}^\diamondsuit(\mathrm{R})$, $\mathbf{a} \in \mathrm{T}(\mathbf{H})$ and $(x, X) \in \mathbf{a}$. Therefore, $X \in \mathbf{H}$. Since ψ_X is a structure on X there exists $\lambda \in \Lambda$ such that $x \in \psi_X(\lambda)$. Then,

$$\mathbf{a} \in \mathrm{T}(\mathbf{M}|_{\boldsymbol{\Psi}(\lambda)}, \mathrm{R}|_{\boldsymbol{\Psi}(\lambda)}, \mathbf{H}|_{\boldsymbol{\Psi}(\lambda)}) = \psi_{\mathrm{T}(\mathbf{H})}(\lambda),$$

which means that condition (2) of the definition of the structure is satisfied.

Let $\lambda_0 \prec \lambda_1$. Since for every $X \in \mathbf{S}$, $\psi_X(\lambda_0) \subset \psi_X(\lambda_1)$ we have

$$\mathrm{T}(\mathbf{M}|_{\boldsymbol{\Psi}(\lambda_0)}, \mathrm{R}|_{\boldsymbol{\Psi}(\lambda_0)}, \mathbf{H}|_{\boldsymbol{\Psi}(\lambda_0)}) \subset$$

$$\mathrm{T}(\mathbf{M}|_{\boldsymbol{\Psi}(\lambda_1)}, \mathrm{R}|_{\boldsymbol{\Psi}(\lambda_1)}, \mathbf{H}|_{\boldsymbol{\Psi}(\lambda_1)}),$$

that is $\psi_{\mathrm{T}(\mathbf{H})}(\lambda_0) \subset \psi_{\mathrm{T}(\mathbf{H})}(\lambda_1)$, which means that condition (3) of the definition of the structure is also satisfied.

Condition (4) follows by the choice of the co-mark $\mathbf{M}^+$ and the family R^+. Thus, $\psi_{\mathrm{T}(\mathbf{H})}$ is a structure completing the proof of the proposition. □

4.1.6 Proposition. *Suppose that $\mathbb{P}_\lambda$, $\lambda \in \Lambda$, is a saturated class. If, moreover, $\mathbb{P}_\lambda$ is complete for every $\lambda \in \Lambda^0$, then $\mathbb{P}(\Phi, \Lambda, \prec, \mathbf{p})$ is a complete saturated class.*

Proof. Suppose that $\mathbb{P}_\lambda$ is complete for every $\lambda \in \Lambda^0$. Let $\mathbf{S}$ be an indexed collection of spaces and $\boldsymbol{\Psi}$ a $\mathbb{P}$-co-structure on $\mathbf{S}$, where $\mathbb{P}$ denotes the class $\mathbb{P}(\Phi, \Lambda, \prec, \mathbf{p})$.

Since $\mathbb{P}_\lambda$, $\lambda \in \Lambda^0$, is complete the restriction $\mathbf{\Psi}(\lambda)$ of $\mathbf{S}$ is also complete. Using Lemmas 1.4.7 and 2.2.3 we can prove (by induction on the ordinal α for which $\lambda \in \Lambda^\alpha$) that $\mathbf{\Psi}(\lambda)$ is a complete restriction of $\mathbf{S}$ for every $\lambda \in \Lambda$. Therefore, $\mathbb{P}$ is a complete class. Thus, for every $\lambda \in \Lambda$ there exists a co-mark $\mathbf{M}_\lambda$ of $\mathbf{S}$ and a family R_λ of equivalence relations on $\mathbf{S}$ such that the restriction $\mathbf{\Psi}(\lambda)$ is $(\mathbf{M}_\lambda, \mathrm{R}_\lambda)$-complete.

For every $\lambda \in \Lambda$ we denote by $\mathbf{M}_\lambda^+$ an initial co-mark of $\mathbf{S}$ corresponding to the restriction $\mathbf{\Psi}(\lambda)$ and the class $\mathbb{P}(\lambda)$. Moreover, we suppose that $\mathbf{M}_\lambda^+$ is a co-extension of $\mathbf{M}_\lambda$. Let $\mathbf{M}^+$ be a co-mark of $\mathbf{S}$, which is a co-extension of all co-marks $\mathbf{M}_\lambda^+$, $\lambda \in \Lambda$, and let $\mathbf{M}$ be an arbitrary co-extension of $\mathbf{M}^+$.

For every $\lambda \in \Lambda$ we denote by R_λ^+ an initial family of equivalence relations on $\mathbf{S}$ corresponding to the co-mark $\mathbf{M}$, the restriction $\mathbf{\Psi}(\lambda)$, and the class $\mathbb{P}_\lambda$. Moreover, we suppose that R_λ^+ is a final refinement of R_λ, $\lambda \in \Lambda$. Let R^+ be a family of equivalence relations on $\mathbf{S}$, which is a final refinement of all families R_λ^+, and let R be an arbitrary family of equivalence relations on $\mathbf{S}$, which is a final refinement of R^+.

To prove the proposition it suffices to prove that $(\psi_{\mathrm{T}(\mathbf{H})}, \mathrm{T}(\mathbf{H})) \in \mathbb{P}$ for every $\mathbf{H} \in \mathrm{C}^\diamondsuit(\mathrm{R})$, where $\psi_{\mathrm{T}(\mathbf{H})}$ is the containing structure corresponding to the family R, the co-mark $\mathbf{M}$, the $\mathbb{P}$-co-structure $\mathbf{\Psi}$, and the subspace $\mathrm{T}(\mathbf{H})$ of T.

First, we note that for every $\lambda \in \Lambda$ the restriction $\mathbf{\Psi}(\lambda)$ is $(\mathbf{M}, \mathrm{R})$-complete. This follows by the choice of $\mathbf{M}$ and R.

Now, let $\mathbf{H} \in \mathrm{C}^\diamondsuit(\mathrm{R})$. We must prove that $\psi_{\mathrm{T}(\mathbf{H})}$ is a structure on $\mathrm{T}(\mathbf{H})$. Conditions (2), (3), and (4) of the definition of the structure are proved as in Proposition 4.1.5. It remains to prove that the mapping $\psi_{\mathrm{T}(\mathbf{H})}$ is Φ-operational, that is for every $\lambda \in \Lambda \setminus \Lambda^0$ the following relation are satisfied:

$$\psi_{\mathrm{T}(\mathbf{H})}(\lambda) = \begin{cases} \cup\{\psi_{\mathrm{T}(\mathbf{H})}(\lambda') : \lambda' \in \lambda\} & \text{if } \Phi(\lambda) = 0, \\ \cap\{\psi_{\mathrm{T}(\mathbf{H})}(\lambda') : \lambda' \in \lambda\} & \text{if } \Phi(\lambda) = 1. \end{cases} \tag{4.1.1}$$

Let $\lambda \in \Lambda \setminus \Lambda^0$. Suppose that $\Phi(\lambda) = 1$. Let $\mathbf{a} \in \psi_{\mathrm{T}(\mathbf{H})}(\lambda)$ and $(x, X) \in \mathbf{a}$. Since the restriction $\mathbf{\Psi}(\lambda)$ is $(\mathbf{M}, \mathrm{R})$-complete, $x \in \psi_X(\lambda)$. Since ψ_X is Φ-operational and $\Phi(\lambda) = 1$ we have $x \in \psi_X(\lambda')$ for every $\lambda' \in \lambda$. This means that $\mathbf{a} \in \psi_{\mathrm{T}(\mathbf{H})}(\lambda')$ for every $\lambda' \in \lambda$. Therefore,

$$\mathbf{a} \in \cap\{\psi_{\mathrm{T}(\mathbf{H})}(\lambda') : \lambda' \in \lambda\}. \tag{4.1.2}$$

Conversely, suppose that relation (4.1.2) is true. Let $(x, X) \in \mathbf{a}$. Since for every $\lambda' \in \lambda$ the restriction $\mathbf{\Psi}(\lambda')$ is $(\mathbf{M}, \mathrm{R})$-complete we have

$$x \in \cap\{\psi_X(\lambda') : \lambda' \in \lambda\}.$$

Since ψ_X is Φ-operational and $\Phi(\lambda) = 1$ we have $x \in \psi_X(\lambda)$, which means that $x \in \psi_{\mathrm{T}(\mathbf{H})}(\lambda)$. Thus, relation (4.1.1) for the case $\Phi(\lambda) = 1$ is proved. The case where $\Phi(\lambda) = 0$ is proved similarly. The proof of the proposition is completed. □

4.2 Dimension-like functions of the type ind

Agreement. In this section, as in the preceding section, we suppose that: (a) a fixed cardinal ν such that $1 \leq \nu \leq \tau$, (b) a fixed ν-directing mapping Φ, (c) a fixed ν-set Λ of cardinality $\leq \tau$, (d) a fixed preorder $\prec$ on Λ, and (e) a fixed mapping $\mathbf{p}$ of Λ into the class of all topological classes of subsets, are given. For every $\lambda \in \Lambda$ the class $\mathbf{p}(\lambda)$ is also denote by $I\!\!P_\lambda$.

By a *structure* on a space we everywhere mean a $(\Phi, \Lambda, \prec, \mathbf{p})$-structure.

Definition. Let ψ_X be a structure on a space X and K a subset of X. The mapping ψ_K of Λ into the set $\mathcal{P}(K) \subset \mathcal{P}(X)$ such that

$$\psi_K(\lambda) = \psi_X(\lambda) \cap K,\ \lambda \in \Lambda$$

is called the *range-restriction* of ψ_X to K.

Definition. Let $I\!\!R$ be a class of spaces with structure and $I\!\!B$ a class of bases. The class $I\!\!R$ is said to be *$I\!\!B$-str-rim-hereditary* if for every $I\!\!R$-structure ψ_X on a space X there exists a $I\!\!B$-base B for X such that for every $U \in B$ the range-restriction $\psi_{\mathrm{Bd}_X(U)}$ of ψ_X to $\mathrm{Bd}_X(U)$ is also an $I\!\!R$-structure on $\mathrm{Bd}_X(U)$.

Definition. For every class $I\!\!B$ of bases and a $I\!\!B$-str-rim-hereditary class $I\!\!R$ of spaces with structure we denote by $(\Phi, \Lambda, \prec, \mathbf{p})\text{-ind}_{I\!\!R}^{I\!\!B}$ or briefly by $\text{str-ind}_{I\!\!R}^{I\!\!B}$ the dimension-like function with the domain the class of all spaces with structure and the range the class $\{-1, \infty\} \cup \mathcal{O}$ satisfying the following conditions:

$$\text{(a) str-ind}_{I\!\!R}^{I\!\!B}(\psi_X, X) = -1$$

if and only if $(\psi_X, X) \in I\!\!R$, and

(b) $\text{str-ind}_{\mathbb{R}}^{\mathbb{B}}(\psi_X, X) \leq \alpha$,

where $\alpha \in \mathcal{O}$, if and only if there exists a $\mathbb{B}$-base B for X such that for every $U \in B$ the range-restriction $\psi_{\text{Bd}_X(U)}$ of ψ_X to $\text{Bd}_X(U)$ is a structure on $\text{Bd}_X(U)$ and

$$\text{str-ind}_{\mathbb{R}}^{\mathbb{B}}(\psi_{\text{Bd}_X(U)}, \text{Bd}_X(U)) < \alpha.$$

In what follows whenever we use the notation $\text{str-ind}_{\mathbb{R}}^{\mathbb{B}}$, then it is supposed that $\mathbb{B}$ is a class of bases and $\mathbb{R}$ is a $\mathbb{B}$-str-rim-hereditary class of spaces with structure.

Remark. As in Chapter 3, the $\mathbb{B}$-str-rim-hereditarity of the class $\mathbb{R}$ considered in the above definition is needed in order that the defined dimension-like function be well-defined.

Lemma 3.1.1 implies the following consequence.

4.2.1 Corollary. *Suppose that*

$$\text{str-ind}_{\mathbb{R}}^{\mathbb{B}}(\psi_X, X) = \alpha \in \{-1\} \cup \mathcal{O}.$$

Then $\alpha \in \{-1\} \cup \tau^+$. □

4.2.2 Proposition. *The class*

$$\mathbb{P}(\text{str-ind}_{\mathbb{R}}^{\mathbb{B}} \leq \alpha),\ \alpha \in \{-1\} \cup \tau^+,$$

is a saturated class provided that $\mathbb{B}$ *and* $\mathbb{R}$ *are saturated classes.*

Proof. Let $\mathbb{B}$ and $\mathbb{R}$ be saturated classes. Setting

$$\mathbb{P}_\alpha \equiv \mathbb{P}(\text{str-ind}_{\mathbb{R}}^{\mathbb{B}} \leq \alpha)$$

we prove the proposition by induction on α. For $\alpha = -1$ the class $\mathbb{P}_\alpha$ coincides with the class $\mathbb{R}$ and, therefore, it is saturated. We prove the proposition for an $\alpha \in \tau^+$ supposing that it is true for all $\beta \in \{-1\} \cup \alpha$.

Consider an $\alpha \in \tau^+$. Let $\mathbf{S}$ be an indexed collection of spaces. For every $X \in \mathbf{S}$ let ψ_X be a $\mathbb{P}_\alpha$-structure on X and B^X a $\mathbb{B}$-base for X such that for every $U \in B$ the range-restriction $\psi_{\text{Bd}_X(U)}$ of ψ_X to $\text{Bd}_X(U)$ is a structure on $\text{Bd}_X(U)$ and

$$\text{str-ind}^{I\!B}_{I\!R}(\psi_{\text{Bd}_X(U)}, \text{Bd}_X(U)) < \alpha.$$

For every $X \in \mathbf{S}$ we consider an indication

$$N^X \equiv \{V^X_\varepsilon : \varepsilon \in \tau\}$$

of B^X such that

$$\text{str-ind}^{I\!B}_{I\!R}(\psi_{\text{Bd}_X(V^X_\varepsilon)}, \text{Bd}_X(V^X_\varepsilon)) \leq \alpha[\varepsilon].$$

(About notation $\alpha[\varepsilon]$ see Section 3.1). The existence of such an indication follows by Lemma 3.1.3.

Consider the $I\!B$-co-base

$$\mathbf{B} \equiv \{B^X : X \in \mathbf{S}\}$$

for $\mathbf{S}$, its co-indication

$$\mathbf{N} \equiv \{N^X : X \in \mathbf{S}\},$$

and the $\mathbb{P}_\alpha$-co-structure

$$\boldsymbol{\Psi} \equiv \{\psi_X : X \in \mathbf{S}\}.$$

For every $\varepsilon \in \tau$ we set

$$\mathbf{V}_\varepsilon = \{V^X_\varepsilon : X \in \mathbf{S}\},$$

$$\mathbf{S}_\varepsilon \equiv \mathbf{Bd}(\mathbf{V}_\varepsilon) = \{\text{Bd}_X(V^X_\varepsilon) : X \in \mathbf{S}\}, \text{ and}$$

$$\boldsymbol{\Psi}_\varepsilon = \{\psi_{\text{Bd}_X(V^X_\varepsilon)} : \text{Bd}(V^X_\varepsilon) \in \mathbf{S}_\varepsilon\}.$$

Obviously, $\mathbf{S}_\varepsilon$ is a restriction of $\mathbf{S}$ and $\boldsymbol{\Psi}_\varepsilon$ is a $\mathbb{P}_{\alpha[\varepsilon]}$-co-structure on $\mathbf{S}_\varepsilon$.

We denote by $\mathbf{M}^+$ a co-mark of $\mathbf{S}$ such that: (a) $\mathbf{M}^+$ is a co-extension of $\mathbf{N}$ and $\theta^{\mathbf{N}}$ is an indicial mapping of this co-extension and (b) for every $\varepsilon \in \tau$ the restriction $\mathbf{M}^+|_{\mathbf{S}_\varepsilon}$ of $\mathbf{M}^+$ to $\mathbf{S}_\varepsilon$ is an initial co-mark corresponding to the $\mathbb{P}_{\alpha[\varepsilon]}$- co-structure $\boldsymbol{\Psi}_\varepsilon$. Let

$$\mathbf{M} \equiv \{\{U^X_\delta : \delta \in \tau\} : X \in \mathbf{S}\}$$

be an arbitrary co-extension of $\mathbf{M}^+$ and θ^+ an indicial mapping of this co-extension.

Let R^+ be a family of equivalence relations on $\mathbf{S}$ satisfying the following properties: (a) R^+ is an $(\mathbf{M}, \mathbf{\Psi})$-admissible family and (b) for every $\varepsilon \in \tau$ the restriction $\mathrm{R}^+|_{\mathbf{S}_\varepsilon}$ of R^+ to $\mathbf{S}_\varepsilon$ is an initial family corresponding to the co-mark $\mathbf{M}|_{\mathbf{S}_\varepsilon}$ of $\mathbf{S}_\varepsilon$ and the $\mathbb{P}_{\alpha[\varepsilon]}$-co-structure $\mathbf{\Psi}_\varepsilon$ on $\mathbf{S}_\varepsilon$. Denote by R an arbitrary admissible family of equivalence relations on $\mathbf{S}$, which is a final refinement of R^+.

Let $\mathbf{H} \in \mathrm{C}^\diamondsuit(\mathrm{R})$. Consider the subspace $\mathrm{T}(\mathbf{H})$ of the Containing Space T and let $\psi_{\mathrm{T}(\mathbf{H})}$ be the containing structure corresponding to the family R, the co-mark $\mathbf{M}$, the $\mathbb{P}$-co-structure $\mathbf{\Psi}$, and the subspace $\mathrm{T}(\mathbf{H})$ of T. To prove the proposition we need to prove that $(\psi_{\mathrm{T}(\mathbf{H})}, \mathrm{T}(\mathbf{H})) \in \mathbb{P}_\alpha$. For this it suffices to prove that for every element U of the $(\diamondsuit, \kappa)$-standard base $\mathrm{B}^{\mathbf{H}}_{\diamondsuit,\kappa}$ for $\mathrm{T}(\mathbf{H})$, where $\kappa = (\theta^+ \circ \theta^{\mathrm{N}})(\tau)$, we have

$$\text{str-ind}^{\mathbb{B}}_{\mathbb{R}}(\psi_{\mathrm{Bd}_{\mathrm{T}(\mathbf{H})}(U)}, \mathrm{Bd}_{\mathrm{T}(\mathbf{H})}(U)) < \alpha, \tag{4.2.1}$$

where $\psi_{\mathrm{Bd}_{\mathrm{T}(\mathbf{H})}(U)}$ is the range-restriction of $\psi_{\mathrm{T}(\mathbf{H})}$ to $\mathrm{Bd}_{\mathrm{T}(\mathbf{H})}(U)$.

Let $U \equiv U^{\mathrm{T}}_\delta(\mathbf{H}) \in \mathrm{B}^{\mathbf{H}}_{\diamondsuit,\kappa}$, where $\delta \in \kappa$ and $\mathbf{H} \in \mathrm{C}^\diamondsuit(\mathrm{R})$, and let $\delta = (\theta^+ \circ \theta^{\mathrm{N}})(\varepsilon)$. Denote by ψ' the containing structure corresponding to the family $\mathrm{R}|_{\mathbf{S}_\varepsilon}$, the co-mark $\mathbf{M}|_{\mathbf{S}_\varepsilon}$ of $\mathbf{S}_\varepsilon$, the $\mathbb{P}_{\alpha[\varepsilon]}$-co-structure $\mathbf{\Psi}_\varepsilon$, and the subspace $\mathrm{T}' \equiv \mathrm{T}(\mathbf{M}|_{\mathbf{S}_\varepsilon}, \mathrm{R}|_{\mathbf{S}_\varepsilon}, \mathbf{H}|_{\mathbf{S}_\varepsilon})$ of the Containing Space $\mathrm{T}(\mathbf{M}|_{\mathbf{S}_\varepsilon}, \mathrm{R}|_{\mathbf{S}_\varepsilon})$. By inductive assumption the class $\mathbb{P}_{\alpha[\varepsilon]}$ is saturated. Therefore, $(\psi', \mathrm{T}') \in \mathbb{P}_{\alpha[\varepsilon]}$, that is

$$\text{str-ind}^{\mathbb{B}}_{\mathbb{R}}(\psi', \mathrm{T}') < \alpha.$$

To prove relation (4.2.1) it suffices to prove that: (a) the set $\mathrm{Bd}_{\mathrm{T}(\mathbf{H})}(U)$ coincides with the set T' and (b) the mapping $\psi_{\mathrm{Bd}_{\mathrm{T}(\mathbf{H})}(U)}$ coincides with the mapping ψ'. By Lemma 1.2.7 the set $\mathrm{Bd}_{\mathrm{T}(\mathbf{H})}(U) = \mathrm{Bd}_{\mathrm{T}}(U)$ coincides with the set of all point $\mathbf{a}$ of T such that there exists a pair (x, X) of $\mathbf{a}$ for which $X \in \mathbf{H}$ and $x \in \mathrm{Bd}_X(U^X_\delta) = \mathrm{Bd}_X(V^X_\varepsilon)$. On the other hand, by definition of the natural embeddings $e^{\mathrm{T}|\mathbf{Q}}_{\mathrm{T}}$ (see Section 1.3) the set T' coincides with the set of all points $\mathbf{a}$ of T such that there exists a pair (x, X) of $\mathbf{a}$ for which $\mathrm{Bd}_X(V^X_\varepsilon) \in \mathbf{H}|_{\mathbf{S}_\varepsilon} \subset \mathbf{S}_\varepsilon$ or equivalently $X \in \mathbf{H} \subset \mathbf{S}$, and $x \in \mathrm{Bd}_X(V^X_\varepsilon)$. Therefore, the set $\mathrm{Bd}_{\mathrm{T}(\mathbf{H})}(U)$ coincides with the set T'.

Now we prove that the mapping $\psi_{\mathrm{Bd}_{\mathrm{T}(\mathbf{H})}(U)}$ coincides with the mapping ψ'. Let $\lambda \in \Lambda$. Then,

$$\psi_{\mathrm{Bd}_{\mathrm{T}(\mathbf{H})}(U)}(\lambda) = \psi_{\mathrm{T}}(\lambda) \cap \mathrm{Bd}_{\mathrm{T}(\mathbf{H})}(U).$$

By the definition of the containing structures $\psi_{\mathrm{T}(\mathbf{H})}$ and ψ' we have

$$\psi_{\mathrm{T}(\mathbf{H})}(\lambda) = \mathrm{T}(\mathbf{M}|_{\mathbf{\Psi}(\lambda)}, \mathrm{R}|_{\mathbf{\Psi}(\lambda)}, \mathbf{H}|_{\mathbf{\Psi}(\lambda)}) \text{ and}$$

$$\psi'(\lambda) = \mathrm{T}((\mathbf{M}|_{\mathbf{S}_\varepsilon})|_{\mathbf{\Psi}_\varepsilon(\lambda)}, (\mathrm{R}|_{\mathbf{S}_\varepsilon})|_{\mathbf{\Psi}_\varepsilon(\lambda)}, (\mathbf{H}|_{\mathbf{S}_\varepsilon})|_{\mathbf{\Psi}_\varepsilon(\lambda)}).$$

Since $\psi_{\mathrm{Bd}_X(V_\varepsilon^X)}(\lambda) \subset \mathrm{Bd}_X(V_\varepsilon^X)$, $X \in \mathbf{S}$, we have

$$\psi'(\lambda) = \mathrm{T}(\mathbf{M}|_{\mathbf{\Psi}_\varepsilon(\lambda)}, \mathrm{R}|_{\mathbf{\Psi}_\varepsilon(\lambda)}, \mathbf{H}|_{\mathbf{\Psi}_\varepsilon(\lambda)}).$$

Also, since $\mathrm{Bd}_X(V_\varepsilon^X) \cap \psi_X(\lambda) = \psi'_\varepsilon(\lambda)$, $X \in \mathbf{S}$, we have

$$\psi'(\lambda) = \mathrm{T}(\mathbf{M}|_{\mathbf{S}_\varepsilon \wedge \mathbf{\Psi}(\lambda)}, \mathrm{R}|_{\mathbf{S}_\varepsilon \wedge \mathbf{\Psi}(\lambda)}, \mathbf{H}|_{\mathbf{S}_\varepsilon \wedge \mathbf{\Psi}(\lambda)}),$$

where

$$\mathbf{S}_\varepsilon \wedge \mathbf{\Psi}(\lambda) = \{\mathrm{Bd}_X(V_\varepsilon^X) \cap \psi_X(\lambda) : X \in \mathbf{S}\}.$$

On the other hand, we have

$$\mathrm{T}(\mathbf{M}|_{\mathbf{S}_\varepsilon \wedge \mathbf{\Psi}(\lambda)}, \mathrm{R}|_{\mathbf{S}_\varepsilon \wedge \mathbf{\Psi}(\lambda)}, \mathbf{H}|_{\mathbf{S}_\varepsilon \wedge \mathbf{\Psi}(\lambda)}) =$$

$$\mathrm{T}(\mathbf{M}|_{\mathbf{S}_\varepsilon}, \mathrm{R}|_{\mathbf{S}_\varepsilon}, \mathbf{H}|_{\mathbf{S}_\varepsilon}) \cap \mathrm{T}(\mathbf{M}|_{\mathbf{\Psi}(\lambda)}, \mathrm{R}|_{\mathbf{\Psi}(\lambda)}, \mathbf{H}|_{\mathbf{\Psi}(\lambda)}). \tag{4.2.2}$$

Indeed, let $\mathbf{a}$ belongs to the left side of the equation (4.2.2). Then, there exists $(x, X) \in \mathbf{a}$ such that $X \in \mathbf{H}$ and $x \in \mathrm{Bd}_X(V_\varepsilon^X) \cap \psi_X(\lambda)$. By Corollary 1.3.5, $\mathbf{a}$ belongs to the right side of (4.2.2). Conversely, suppose that $\mathbf{a}$ belongs to the right side of (4.2.2). Then, there exists $(x, X) \in \mathbf{a}$ such that $X \in \mathbf{H}$ and $x \in \psi_X(\lambda)$. The restriction $\mathbf{S}_\varepsilon$ of $\mathbf{S}$, as a closed restriction, is complete. Without loss of generality, we can suppose that $\mathbf{S}_\varepsilon$ is $(\mathbf{M}, \mathrm{R})$-complete for every $\varepsilon \in \tau$. In this case, $x \in \mathrm{Bd}_X(V_\varepsilon^X)$. Thus,

$$x \in \mathrm{Bd}_X(V_\varepsilon^X) \cap \psi_X(\lambda),$$

which means that $\mathbf{a}$ belongs to the left side of (4.2.2). Relation (4.2.2) immediately implies that

$$\psi(\lambda) \cap \mathrm{Bd}_{\mathrm{T}}(U) = \psi'(\lambda),$$

which completes the proof of the proposition. □

4.3 D-Dimension of spaces

Notation. Each ordinal α can be uniquely represented as the union $\beta+n$, where β is a limit ordinal (in particular, β may be zero) and n is a non-negative integer, that is an element of ω. In this section the limit ordinal β will be denoted by $\beta(\alpha)$ and the integer n by $n(\alpha)$.

By $\mathcal{O}^*$ we denote the class consisting of all ordinals and all formal expressions $\beta+(-1) \equiv \beta-1$, where β is a limit ordinal. If $\beta = 0$, then the expression $\beta+(-1)$ is identified with the symbol (integer) -1. Therefore, any element of $\mathcal{O}^*$ can be uniquely represented as the union $\beta+n$, where β is a limit ordinal and $n \in \{-1\} \cup \omega$.

In the class $\mathcal{O}^* \cup \{\infty\}$ we define an order denoted by $\leq$ as follows. If α is an element of $\mathcal{O}^* \cup \{\infty\}$, then we set $\alpha \leq \infty$. If $\beta+n$ and β_0+n_0 are two elements of $\mathcal{O}^*$, then we set $\beta+n \leq \beta_0+n_0$ if either $\beta < \beta_0$ or $\beta = \beta_0$ and $n \leq n_0$. (We recall that $-1 \leq \alpha$ for every $\alpha \in \mathcal{O} \cup \{\infty\}$.)

Definition. Let β be a limit ordinal and $n \in \{-1\} \cup \omega$. A family

$$A \equiv \{A_\alpha : \alpha \in \beta+1\} \tag{4.3.1}$$

of subsets of X is said to be a $(\beta+n)$-*representation* of X if the following conditions are satisfied:

(1) $X = \cup\{A_\alpha : \alpha \in \beta+1\}$,
(2) Each subset A_α is closed,
(3) $\mathrm{ind}(A_\alpha) \leq n(\alpha)$ for every $\alpha \in \beta$ and $\mathrm{ind}(A_\beta) \leq n$,
(4) For every $\delta \in \beta+1$ the set $\cup\{A_\alpha : \alpha \in (\beta+1)\setminus\delta\}$ is closed in X,
(5) For every limit ordinal $\delta \in \beta+1$ the following equality is true:

$$\cup\{A_\alpha : \alpha \in (\beta+1)\setminus\delta\} =$$
$$\cap\{\cup\{A_\alpha : \alpha \in (\beta+1)\setminus\varepsilon\} : \varepsilon \in \delta\}.$$

(Observe that for $n \geq 0$, $\beta+n$ is an ordinal while for $n=-1$, $\beta+(-1) = \beta-1$ is a formal expression).

Remark. We note that if a space X has a $(\beta+n)$-representation and $\beta+n \leq \beta_0+n_0$, where β_0 is a limit ordinal and $n_0 \in \{-1\} \cup \omega$, then X has also a (β_0+n_0)-representation. Indeed, let (4.3.1) be a $(\beta+n)$-representation of X. If $\beta = \beta_0$, then (4.3.1) is also (β_0+n_0)-representation of X. If $\beta < \beta_0$, then it is easy to verify that the family $\{A_\alpha : \alpha \in \beta_1+1\}$, where $A_\alpha = A_\beta$ if $\alpha = \beta + n$ and $A_\alpha = \emptyset$ for all $\alpha \in (\beta_0 + 1) \setminus (\beta \cup \{\beta + n\})$ is a $(\beta_0 + n_0)$-representation of X.

4.3.1 Lemma. *If a space X has a $(\beta_0 + n_0)$-representation for some element $\beta_0 + n_0$ of calO*, then X has also a $(\beta + n)$-representation, where $\beta \in \tau^+$.*

Proof. Let

$$\{A_\alpha : \alpha \in \beta_0 + 1\}$$

be a $(\beta_0 + n_0)$-representation of a space X. Denote by β_0' the subset of $\beta_0 + 1$ consisting of the element β_0 and all elements α of β_0 such that

$$A_\alpha \not\subset \cup\{A_\gamma : \gamma \in (\beta_0 + 1) \setminus (\alpha + 1)\}.$$

For every $\alpha \in \beta_0 + 1$ we set

$$A_\alpha^+ = \cup\{A_\delta : \delta \in (\beta_0 + 1) \setminus \alpha\}$$

and for every $\alpha \in (\beta_0 + 1) \cap \beta_0'$ we set

$$A_\alpha^* = \cup\{A_\delta : \delta \in \beta_0' \setminus \alpha\}.$$

We prove that for every $\alpha \in (\beta_0 + 1) \cap \beta_0'$, $A_\alpha^+ = A_\alpha^*$. It is clear that $A_\alpha^* \subset A_\alpha^+$. Let $x \in A_\alpha^+$. If $x \in A_{\beta_0}$, then $x \in A_\alpha^*$. Let $x \notin A_{\beta_0}$. Using condition (5) of the above definition we can prove that there exists an element δ of $\beta_0 + 1$ such that $x \in A_\delta$ and

$$x \notin \cup\{A_\varepsilon : \varepsilon \in (\beta_0 + 1) \setminus (\alpha + 1)\}.$$

This fact means that $\delta \in \beta_0'$ and, therefore, $x \in A_\alpha^*$. Thus, $A_\alpha^+ = A_\alpha^*$. This equation shows that the set $A_{\alpha+1}^*$ is a closed subset of X and $A_\alpha \not\subset A_{\alpha+1}^*$. We suppose that $A_{\alpha+1}^*$ is the empty set if $\alpha = \beta_0$.

Now we prove that $|\beta_0'| \leq \tau$. Indeed, for every $\alpha \in \beta_0'$ we take a point x_α of A_α, which does not belong to the set $A_{\alpha+1}^*$. Let B be a base for X of cardinality $\leq \tau$. Then, for every $\alpha \in \beta_0'$ we can find an element U_α such that $x_\alpha \in U_\alpha$ and $x_\alpha \notin U_\delta$ for every $\delta \in \beta_0' \setminus (\alpha + 1)$. Obviously, if α_0 and α_1 are distinct elements of β_0', then U_{α_0} and U_{α_1} are distinct elements of B. This means that $|\beta_0'| \leq \tau$.

The well ordering of $\beta_0 + 1$ defines a well ordering on β_0'. Now, we define a one-to-one mapping θ of β_0' into τ^+ preserving the order. Let γ be the first element of β_0'. Then, we set $\theta(\gamma) = n(\gamma)$. Suppose that we have defined the ordinal $\theta(\gamma)$ for all $\gamma \in \alpha \cap \beta_0'$. Then, the ordinal $\theta(\alpha)$ is defined to be the minimal ordinal δ such that $n(\alpha) \leq n(\delta)$ and $\theta(\gamma) < \delta$ for all elements γ of $\beta_0' \cap \alpha$.

Let β be the first limit ordinal such that $\theta(\beta_0) \leq \beta$ and let n be an element of ω such that $n(\beta_0) \leq n$. We prove that X has a $(\beta + n)$-representation. Indeed, let $\delta \in \beta + 1$. If $\delta = \theta(\alpha)$, then we set $B_\delta = A_\alpha$. If $\delta \neq \theta(\alpha)$ for every $\alpha \in \beta_0'$, then we set $B_\delta = \emptyset$. It is not difficult to verify that the family $\{B_\delta : \delta \in \beta + 1\}$ is a $(\beta + 1)$-representation of X. $\square$

Notation. For every limit ordinal β and an element n of $\{-1\} \cup \omega$ we denote by $\mathbb{P}((\beta + n)\text{-rep.})$ the class consisting of all pairs (A^X, X), where X is a space and A^X is a $(\beta + n)$-representation of X.

Definition. Let (A^X, X) and (A^Y, Y) be elements of $\mathbb{P}((\beta + n)\text{-rep.})$, where

$$A^X = \{A_\alpha^X : \alpha \in \beta + 1\} \text{ and } A^Y = \{A_\alpha^Y : \alpha \in \beta + 1\}.$$

A homeomorphism h of X into Y is said to be an *embedding* of the pair (A^X, X) into the pair (A^Y, Y) if for every $\alpha \in \beta + 1$ we have $h^{-1}(A_\alpha^Y) = A_\alpha^X$.

Definition. An element (A^T, T) of the class $\mathbb{P}((\beta + n)\text{-rep.})$ is said to be *universal* in this class if for every element (A^X, X) of $\mathbb{P}((\beta + n)\text{-rep.})$ there exists an embedding of (A^X, X) into (A^T, T).

4.3.2 Proposition. *For every limit ordinal $\beta \in \tau^+$ and an element n of $\{-1\} \cup \omega$ in the class $\mathbb{P}((\beta + n)\text{-rep.})$ there exists a universal element.*

Proof. Let β be a limit ordinal and $n \in \{-1\} \cup \omega$. Consider a set Θ of cardinality $|\beta + 1|$ such that $\Theta = \Theta^0$ and let φ be a one-to-one mapping of $\beta + 1$ onto Θ. For every $\delta \in \beta + 1$ we set $\lambda_\delta^1 = \varphi((\beta + 1) \setminus \delta)$ and let

$$\Theta_1 = \{\lambda_\delta^1 : \delta \in \beta + 1\}.$$

For every limit ordinal $\delta \in \beta + 1$ we set $\lambda_\delta^2 = \{\lambda_\varepsilon^1 : \varepsilon \in \delta\}$ and let

$$\Theta_2 = \{\lambda_\delta^2 : \delta \text{ is a limit ordinal of } \beta + 1\}.$$

Set $\Lambda = \Theta \cup \Theta_1 \cup \Theta_2$. It is easy to see that Λ is a ν-set of cardinality $\leq \tau$ such that $\Lambda^0 = \Theta$, $\Lambda^1 = \Theta_1$, and $\Lambda^2 = \Theta_2$, where $\nu = |\beta + 1|$.

On the set Λ we define a preorder as follows: $\lambda_1 \prec \lambda_2$ if and only if there exists a limit ordinal $\delta \in \beta + 1$ such that $\lambda_1 = \lambda_\delta^2$ and $\lambda_2 = \lambda_\delta^1$. It is easy to verify that $\prec$ is really a preorder.

We define a mapping $\mathbf{p}$ of the set Λ into the class of all classes of subsets as follows. If $\lambda \in \Lambda^0 \setminus \{\varphi(\beta)\}$, then $\mathbf{p}(\lambda)$ is the class of subsets consisting of all pairs (Q, X) such that Q is closed in X and $\text{ind}(Q) \leq n(\varphi^{-1}(\lambda))$. If $\lambda = \varphi(\beta) \in \Lambda^0$, then $\mathbf{p}(\lambda)$ is the class of subsets consisting of all pairs (Q, X) such that Q is closed in X and $\text{ind}(Q) \leq n$. If $\lambda = \lambda_\delta^1 \in \Lambda^1$, where $\delta = 0$, then $\mathbf{p}(\lambda)$ is the class of subsets consisting of all pairs (Q, X) such that $Q = X$ (and, therefore, Q is closed in X). If $\lambda \in \Lambda^1 \setminus \{\lambda_0^1\}$, then $\mathbf{p}(\lambda) = \mathbb{P}(\text{Cl})$. Finally, if $\lambda \in \Lambda^2$, then $\mathbf{p}(\lambda) = \mathbb{P}(\text{subset})$. It is easy to see that for every $\lambda \in \Lambda$ the class $\mathbf{p}(\lambda)$ is a complete saturated class.

Let Φ be a ν-directing mapping such that $\Phi(0) = 1$. By Proposition 4.1.6 the class $\mathbb{P}(\Phi, \Lambda, \prec, \mathbf{p})$ is a complete saturated class and, therefore, in this class there exists a properly universal element. Let (ψ_T, T) be such an element.

For every $\alpha \in \beta + 1$ we set $A_\alpha^T = \psi_T(\varphi(\alpha))$ and consider the family

$$A^T \equiv \{A_\alpha^T : \alpha \in \beta + 1\}.$$

We prove that this family is a $(\beta + n)$-representation of T.

Indeed, since $\Phi(0) = 1$ and $\lambda_0^1 \in \Lambda^1$ we have

$$\psi_T(\lambda_0^1) = \cup\{\psi_T(\lambda) : \lambda \in \lambda_0^1\} = \cup\{A_\alpha^T : \alpha \in \beta + 1\}.$$

By the definition of the class $\mathbf{p}(\lambda_0^1)$ we have

$$\psi_T(\lambda_0^1) = T = \cup\{A_\alpha^T : \alpha \in \beta + 1\}.$$

Therefore, property (1) of the definition of the $(\beta + n)$-representation is satisfied. Since $\mathbf{p}(\lambda) \subset \mathbb{P}(\text{Cl})$ for every $\lambda \in \Lambda^0$, property (2) is also satisfied. Property (3) follows immediately by the definition of the classes $\mathbf{p}(\lambda)$ for $\lambda \in \Lambda^0$.

Let $\delta \in \beta + 1$. Then $\lambda_\delta^1 \in \Lambda^1$ and, therefore,

$$\psi_T(\lambda_\delta^1) = \cup\{\psi_T(\lambda) : \lambda \in \lambda_\delta^1)\} = \cup\{A_\alpha^T : \alpha \in (\beta + 1) \setminus \delta\}$$

is closed in T, that is property (4) of the above mentioned definition is satisfied.

Let $\delta \in \beta+1$ be a limit ordinal. Since $\lambda_\delta^2 \in \Lambda^2$ and $\Phi(2)=1$ we have

$$\psi_T(\lambda_\delta^2) = \cap\{\psi_T(\lambda_\varepsilon^1) : \varepsilon \in \delta\} =$$

$$\cap\{\cup\{A_\alpha^T : \alpha \in (\beta+1)\setminus \varepsilon\} : \varepsilon \in \delta\}.$$

On the other hand,

$$\psi_T(\lambda_\delta^1) = \cup\{A_\alpha^T : \alpha \in (\beta+1)\setminus\delta\}.$$

Obviously, $\psi_T(\lambda_\delta^1) \subset \psi_T(\lambda_\delta^2)$. Since $\lambda_\delta^2 \prec \lambda_\delta^1$ we have $\psi_T(\lambda_\delta^2) \subset \psi_T(\lambda_\delta^1)$, that is $\psi_T(\lambda_\delta^2) = \psi_T(\lambda_\delta^1)$, which means that property (5) of the definition of the $(\beta+n)$-representation is satisfied.

Thus, the family A^T is a $(\beta+n)$-representation of T, that is $(A^T, T) \in \mathbb{P}((\beta+n)\text{-rep.})$. We prove that the pair (A^T, T) is universal in the class $\mathbb{P}((\beta+n)\text{-rep.})$

Indeed, let (A^X, X) be an element of $\mathbb{P}((\beta+n)\text{-rep.})$ and let

$$A^X \equiv \{A_\alpha^X : \alpha \in \beta+1\}.$$

We define an operational mapping ψ_X of Λ into $\mathcal{P}(X)$ setting $\psi_X(\lambda) = A^X_{\varphi^{-1}(\lambda)}$ for every $\lambda \in \Lambda^0$. It is not dificult to prove that ψ_X defines a $(\Phi, \Lambda, \prec, \mathbf{p})$-structure on X. Therefore,

$$(\psi_X, X) \in \mathbb{P}(\Phi, \Lambda, \prec, \mathbf{p}).$$

Since (ψ_T, T) is a properly universal element, there exists an embedding h of X into T such that $h^{-1}(A_\alpha^T) = A_\alpha^X$ for every $\alpha \in \beta+1$, which completes the proof of the proposition. □

Definition. We denote by D the unique dimension-like function with the domain the class of all spaces and the range the class $\mathcal{O}^* \cup \{\infty\}$ defined by the following condition: $\mathrm{D}(X) \leq \beta+n \in \mathcal{O}^*$, where β is a limit ordinal and n is an element of $\{-1\}\cup\omega$, if and only if the space X has a $(\beta+n)$-representation. We denote by $\mathbb{P}(\mathrm{D} \leq \beta+n)$ the class of all spaces X such that $\mathrm{D}(X) \leq \beta+n$.

4.3.3 Corollary. *For every $\beta + n \in \mathcal{O}^*$, where β is a limit ordinal and n is an element of $\{-1\} \cup \omega$, in the class $\mathbb{P}(\mathrm{D} \leq \beta + n)$ there exists a universal element.* □

4.3.4 Proposition. *Proposition 4.3.2 and Corollary 4.3.3 hold if we replace independently the class of all spaces by an arbitrary saturated class of spaces and the dimension ind (considered in the definition of the $(\beta+n)$-representation) by any of the following dimensions:*

s-p_0-ind, s-p_1-ind, s-pos_0-ind, s-pos_1-ind,

s-ps_0-ind, s-ps_1-ind, s-b_0-ind, s-b_1-ind,

s-b^0-ind, s-b^1-ind, s-b^{p}-ind,

s-$\mathrm{b}^{\mathrm{pos}}$-ind, s-$\mathrm{b}^{\mathrm{ps}}$-ind, s-b(0)-ind, s-b(1)-ind,

s-b_0-p_0-ind, s-b_0-p_1-ind, s-b_1-p_0-ind,

s-b_1-p_1-ind, s-b^0-p_0-ind, s-b^0-p_1-ind,

s-b^1-p_0-ind, *and* s-b^1-p_1-ind.

In particular, the mentioned proposition and corollary hold if instead of the class of all spaces we consider the class of all regular spaces or the class of all completely regular spaces (see Corollary 5.1.6) or the class of all separable metrizable spaces if $\tau = \omega$. □

4.4 Some applications

Agreement. Unless we explicity specify otherwise, everywhere in this section we mean by dimension the small inductive dimension. So, if we say that a space X is zero-dimensional or finite-dimensional we mean that $\mathrm{ind}(X) = 0$ or $\mathrm{ind}(X)$ is finite, respectively.

We denote by ν a fixed cardinal such that $1 \leq \nu \leq \tau$.

Definition. An indexed set $\{X_\delta : \delta \in \{-1\} \cup \nu\}$, where $X_{-1} \equiv X$ is a space and X_δ, $\delta \in \nu$, is a subset of X is called a *ν-sequence*. Below, such an indexed set is denoted by $(X, X_0, ..., X_\delta, ...)_{\delta \in \nu}$.

Definition. Let $\overline{\mathrm{S}}$ be a class of ν-sequences. An element

$$(T, T_0, ..., T_\delta, ...)_{\delta \in \nu}$$

of $\overline{\mathrm{S}}$ is called *universal* (respectively, *properly universal*) in $\overline{\mathbf{S}}$ if for every

$$(X, X_0, ..., X_\delta, ...)_{\delta \in \nu} \in \overline{\mathrm{S}}$$

there exists an embedding h of X into T such that $X_\delta \subset h^{-1}(T_\delta)$ (respectively, $X_\delta = h^{-1}(T_\delta)$) for every $\delta \in \nu$.

Definition. Let ξ be a mapping of the set ν into the set $\{-1\} \cup \tau^+$. We define the classes:

$$\text{(a) } \overline{\mathrm{S}}(\nu, \xi), \text{ (b) } \overline{\mathrm{S}}(\nu, \xi, \mathrm{Cl}), \text{ (c) } \overline{\mathrm{S}}(\nu, \xi, \mathrm{Op}),$$

$$\text{(d) } \overline{\mathrm{S}}(\nu, \leq, \xi), \text{ (e) } \overline{\mathrm{S}}(\nu, \leq, \xi, \mathrm{Cl}), \text{ and (f) } \overline{\mathrm{S}}(\nu, \leq, \xi, \mathrm{Op}), \quad (4.4.1)$$

consisting of ν-sequences as follows.

A ν-sequence $(X, X_0, ..., X_\delta, ...)_{\delta \in \nu}$ belongs to the class $\overline{\mathrm{S}}(\nu, \xi)$ if

$$X = \cup\{X_\delta : \delta \in \nu\} \text{ and } \mathrm{ind}(X_\delta) \leq \xi(\delta)$$

for every $\delta \in \nu$.

An element $(X, X_0, ..., X_\delta, ...)_{\delta \in \nu}$ of $\overline{\mathrm{S}}(\nu, \xi)$ belongs to the class $\overline{\mathrm{S}}(\nu, \xi, \mathrm{Cl})$ (respectively, to the class $\overline{\mathrm{S}}(\nu, \xi, \mathrm{Op})$) if X_δ is a closed (respectively, an open) subset of X.

An element $(X, X_0, ..., X_\delta, ...)_{\delta \in \nu}$ of $\overline{\mathrm{S}}(\nu, \xi)$ (respectively, of $\overline{\mathrm{S}}(\nu, \xi, \mathrm{Cl})$ or of $\overline{\mathrm{S}}(\nu, \xi, \mathrm{Op})$) belongs to $\overline{\mathrm{S}}(\nu, \leq, \xi)$ (respectively, to $\overline{\mathrm{S}}(\nu, \leq, \xi, \mathrm{Cl})$ or to $\overline{\mathrm{S}}(\nu, \leq, \xi, \mathrm{Op})$) if $X_\delta \subset X_\varepsilon$ for every $\delta \in \varepsilon \in \nu$.

Obviously, the ν-sequence $(X, X_0, ..., X_\delta, ...)_{\delta \in \nu}$, where $X = X_\delta = \emptyset$, $\delta \in \nu$, belongs to all the above classes, which means that these classes are not empty. We note that the class $\overline{\mathrm{S}}(\nu, \xi)$ contains all other classes.

Now we show that the introduced classes can be considered as classes of the form $\mathbb{P}(\Lambda, \prec, \mathbf{p})$ for some concrete Λ, $\prec$, and $\mathbf{p}$. (The classes $\mathbb{P}(\Lambda, \prec, \mathbf{p})$ are independent of any ν-directing mapping.) For this, we denote by Λ a set of cardinality ν such that $\Lambda = \Lambda^0$ and by ξ_Λ an one-to-one mapping of Λ onto ν.

On the set Λ we consider two preorders. The first, denoted by $\prec_t$, is the trivial partial order. (A partial order $\prec$ on a set is *trivial* if the condition

$\lambda_0 \prec \lambda_1$ implies $\lambda_0 = \lambda_1$ for every elements λ_0 and λ_1 of this set.) The second, denoted by $\prec_o$, is an order on Λ, which is defined as follows: for two elements λ_0 and λ_1 of Λ we set $\lambda_0 \prec_o \lambda_1$ if and only if $\xi_\Lambda(\lambda_0) \leq \xi_\Lambda(\lambda_1)$.

Furthermore, we define three mappings $\mathbf{p}_0$, $\mathbf{p}_{\mathrm{Cl}}$, and $\mathbf{p}_{\mathrm{Op}}$ of Λ into the class of all classes of subsets as follows. Let $\lambda \in \Lambda$. Then, $\mathbf{p}_0(\lambda)$ is the class of subsets consisting of all pairs (Q, X) such that $\mathrm{ind}(Q) \leq \xi(\xi_\Lambda(\lambda))$. The class $\mathbf{p}_{\mathrm{Cl}}(\lambda)$ (respectively, $\mathbf{p}_{\mathrm{Op}}(\lambda)$) consists of all pairs (Q, X) such that Q is closed (respectively, open) in X and $\mathrm{ind}(Q) \leq \xi(\xi_\Lambda(\lambda))$. It is easy to see that $\mathbf{p}_0(\lambda)$ is a saturated class and $\mathbf{p}_{\mathrm{Cl}}(\lambda)$ and $\mathbf{p}_{\mathrm{Op}}(\lambda)$ are complete saturated classes of subsets.

Thus, we can consider the following classes of spaces with structure:

$$\text{(a) } \mathbb{P}(\Lambda, \prec_t, \mathbf{p}_0),\ \text{(b) } \mathbb{P}(\Lambda, \prec_t, \mathbf{p}_{\mathrm{Cl}}),$$

$$\text{(c) } \mathbb{P}(\Lambda, \prec_t, \mathbf{p}_{\mathrm{Op}}),\ \text{(d) } \mathbb{P}(\Lambda, \prec_o, \mathbf{p}_0),$$

$$\text{(e) } \mathbb{P}(\Lambda, \prec_o, \mathbf{p}_{\mathrm{Cl}}),\ \text{and (f) } \mathbb{P}(\Lambda, \prec_o, \mathbf{p}_{\mathrm{Op}}). \tag{4.4.2}$$

We note that the class $\mathbb{P}(\Lambda, \prec_t, \mathbf{p}_0)$ contains all other classes.

The following lemma is easy to verify.

4.4.1 Lemma. *Let $(X, X_0, ..., X_\delta, ...)_{\delta \in \nu}$ be a ν-sequence and ψ_X the mapping of Λ into $\mathcal{P}(X)$ such that $\psi_X(\lambda) = X_{\xi_\Lambda(\lambda)}$, $\lambda \in \Lambda$. Then, the correspondence*

$$(X, X_0, ..., X_\delta, ...)_{\delta \in \nu} \to (\psi_X, X)$$

has the following properties:

(1) A ν-sequence $(X, X_0, ..., X_\delta, ...)_{\delta \in \nu}$ belongs to a class of (4.4.1) if and only if the pair (ψ_X, X) belongs to the corresponding class of (4.4.2).

(2) A ν-sequence $(X, X_0, ..., X_\delta, ...)_{\delta \in \nu}$ is (properly) universal in a class of (4.4.1) if the pair (ψ_X, X) is (properly) universal in the corresponding class of (4.4.2). □

In what follows we shall identify each element $(X, X_0, ..., X_\delta, ...)_{\delta \in \nu}$ of $\overline{\mathrm{S}}(\nu, \xi)$ with the corresponding element (ψ_X, X) of $\mathbb{P}(\Lambda, \prec_t, \mathbf{p}_0)$. In this case, each class of (4.4.1) will coincide with the corresponding class of (4.4.2). Thus, Propositions 4.1.5 and 4.1.6 can be formulated as follows.

4.4.2 Proposition. *The classes* $\overline{\mathrm{S}}(\nu,\xi)$ *and* $\overline{\mathrm{S}}(\nu,\leq,\xi)$ *are saturated classes and, therefore, in these classes there exist universal elements.* □

4.4.3 Proposition. *The classes*

$$\overline{\mathrm{S}}(\nu,\xi,\mathrm{Cl}),\ \overline{\mathrm{S}}(\nu,\xi,\mathrm{Op}),\ \overline{\mathrm{S}}(\nu,\leq,\xi,\mathrm{Cl}),\ \textit{and}\ \overline{\mathrm{S}}(\nu,\leq,\xi,\mathrm{Op})$$

are complete saturated classes and, therefore, in these classes there exist properly universal elements. □

Definition. Let $\overline{\mathrm{S}}$ be a class consisting of ν-sequences and

$$E \equiv (X, X_0, ..., X_\delta, ...)_{\delta\in\nu}$$

an element of $\overline{\mathrm{S}}$. The space X is said to be the *space-coordinate* of E. The class consisting of space -coordinates of all elements of $\overline{\mathrm{S}}$ is said to be the *space-component* of $\overline{\mathrm{S}}$. The space-components of the classes (4.4.1) are denoted by

$$\text{(a) } \mathrm{S}(\nu,\xi),\ \text{(b) } \mathrm{S}(\nu,\xi,\mathrm{Cl}),\ \text{(c) } \mathrm{S}(\nu,\xi,\mathrm{Op}),$$

$$\text{(d) } \mathrm{S}(\nu,\leq,\xi),\ \text{(e) } \mathrm{S}(\nu,\leq,\xi,\mathrm{Cl}),\ \text{and (f) } \mathrm{S}(\nu,\leq,\xi,\mathrm{Op}), \quad (4.4.3)$$

respectively.

Proposition 4.1.3 implies the following proposition.

4.4.4 Proposition. *The classes* (4.4.3) *are saturated and, therefore, in these classes there exist universal spaces.* □

Below, we give a series of corollaries of the above propositions concerning the existence of universal elements in some concrete classes. Each of these classes coincide with some of the above considered classes of ν-sequences or with some of the space-components of them. However, we do not investigate if some of the considered classes coincide.

Definitions. A space X is said to be *countable-dimensional* if X can be represented as a countable union of finite-dimensional subspaces.

A space X is said to be *strongly countable-dimensional* if X can be represented as a countable union of closed finite-dimensional subspaces.

A space X is said to be *locally finite-dimensional* if each point $x \in X$ has an open finite-dimensional neighbourhood.

It is easy to verify that a space X is locally finite-dimensional if and only if X can be represented as a union of not more than τ many open finite-dimensional subspaces.

The following lemma can be proved by induction.

4.4.5 Lemma. *Let X be a space such that ind$(X) < \infty$. Then, X can be represented as a union of τ many zero-dimensional subspaces.* □

4.4.6 Corollary. *In the class of all spaces, which can be represented as a union of τ many subspaces, whose dimension ind is less than ∞, there exists a universal space.*

Proof. Let S be the family of spaces considered in the corollary. By Lemma 4.4.5 a space X belongs to this family if and only if X can be represented as a union of τ many zero-dimensional subspaces.

Let ξ be the mapping of the set τ into $\{-1\} \cup \tau^+$ such that $\xi(\delta) = 0$ for every $\delta \in \tau$. Consider the class $\mathrm{S}(\tau, \xi)$ (it is supposed that $\nu = \tau$). By definition, a space X belongs to this class if and only if there exists a τ-sequence $(X, X_0, ..., X_\delta, ...)_{\delta \in \tau}$ such that

$$X = \cup\{X_\delta : \delta \in \tau\} \text{ and } \mathrm{ind}(X_\delta) \leq 0,$$

that is if and only if X belongs to the class S. Thus, $\mathrm{S} = \mathrm{S}(\tau, \xi)$. By Proposition 4.4.4 in the class $\mathrm{S}(\tau, \xi) = \mathrm{S}$ there exists a universal space. □

4.4.7 Corollary. *Let $\beta \in \tau^+$. In the class of all spaces which can be represented as the union of ν many (closed) subspaces of dimension $\leq \beta$ there exists a universal space.* □

In particular,

4.4.8 Corollary. *In the class of all spaces which can be represented as the union of ν many (closed) zero-dimensional subspaces there exists a universal space.* □

Definition. An ordinal β is called *ν-accessible* if there exists a mapping ξ of the set ν into the set β such that $\sup\{\xi(\delta) : \delta \in \nu\} = \beta$.

4.4.9 Corollary. *Let $\beta \in \tau^+$ be a ν-accessible limit ordinal. In the class of all spaces which can be represented as the union of ν many subspaces of dimension $< \beta$ there exists a universal space.*

Proof. Since β is ν-accessible there exists a mapping ξ of ν into β such that $\sup\{\xi(\delta) : \delta \in \nu\} = \beta$. Moreover, without loss of generality, we can suppose that $|\xi^{-1}(\xi(\delta))| = \nu$ for every $\delta \in \nu$.

Let S be the class of spaces considered in the corollary. By Proposition 4.4.4. it suffices to prove that $S = \mathrm{S}(\nu, \xi)$. Obviously $\mathrm{S}(\nu, \xi) \subset S$.

For the proof of the converse inclusion consider an element X of S. Then for every $\delta \in \nu$ there exists a subspace Q_δ of X such that $\mathrm{ind}(Q_\delta) < \beta$ and $X = \cup\{Q_\delta : \delta \in \nu\}$. We construct a one-to-one mapping φ of the set ν into itself such that $\mathrm{ind}(Q_\delta) \leq \xi(\varphi(\delta))$ for every $\delta \in \nu$. For this we define by induction on δ the ordinal $\varphi(\delta)$. Let $\varphi(0)$ be the least ordinal of the set $\xi^{-1}(\alpha)$, where α is the least ordinal of the set $\xi(\nu)$ for which $\mathrm{ind}(Q_0) \leq \alpha$. Suppose that we have defined the ordinals $\varphi(\delta)$ for all $\delta \in \varepsilon \in \nu$. Then $\varphi(\varepsilon)$ is defined as the least ordinal of the set $\xi^{-1}(\alpha) \setminus \{\varphi(\delta) : \delta \in \varepsilon\}$, where α is the least ordinal of the set $\xi(\nu)$ for which $\mathrm{ind}(Q_\varepsilon) \leq \alpha$. Note that by the choice of the mapping ξ the set $\xi^{-1}(\alpha) \setminus \{\varphi(\delta) : \delta \in \varepsilon\}$ is not empty. It is easy to verify that the constructed mapping φ has the required properties.

Set $X_\delta = Q_{\varphi^{-1}(\delta)}$ if $\delta \in \varphi(\nu)$ and $X_\delta = \emptyset$ if $\delta \notin \varphi(\nu)$. Then we have $X = \cup\{X_\delta : \delta \in \nu\}$ and $\mathrm{ind}(X_\delta) \leq \xi(\delta)$, that is $X \in \mathrm{S}(\nu, \xi)$. Thus, $S = \mathrm{S}(\nu, \xi)$. □

In particular (the case $\nu = \tau$ and $\beta = \omega$), we have the following consequence.

4.4.10. Corollary. *In the class of all countable-dimensional spaces there exists a universal space.* □

The proof of the following corollary is similar to that of Corollary 4.9.

4.4.11 Corollary. *Let $\beta \in \tau^+$ be a ν-accessible limit ordinal. In the class of all spaces which can be represented as the union of ν many (open) closed subspaces of dimension $< \beta$ there exists a universal space.* □

In particular (the cases: (a) $\nu = \beta = \omega$ and (b) $\nu = \tau$ and $\beta = \omega$), we have the following two consequences.

4.4.12 Corollary. *In the class of all strongly countable-dimensional spaces there exists a universal space.* □

4.4.13 Corollary. *In the class of all locally finite-dimensional spaces there exists a universal space.* □

Since every limit ordinal β of the set τ^+ is τ-accessible we have the following two corollaries.

4.4.14 Corollary. *Let $\beta \in \tau^+$ be a limit ordinal. In the class of all spaces which can be represented as the union of τ many subspaces of dimension $< \beta$ there exists a universal space.* □

4.4.15 Corollary. *Let $\beta \in \tau^+$ be a limit ordinal. In the class of all spaces which can be represented as the union of τ many (open) closed subspaces of dimension $< \beta$ there exists a universal space.* □

In the next corollaries we consider spaces with some fixed subsets, which actually means that we consider spaces with "structure".

4.4.16. Corollary. *There exists a space T and a fixed sequence $T_0, T_1, ...$ of subspaces of T for which*

$$T = T_0 \cup T_1 \cup ... \text{ and } \operatorname{ind}(T_0) \leq 0,\ \operatorname{ind}(T_1) \leq 1, ...$$

such that for every space X and for every sequence $X_0, X_1, ...$ of subspaces of X for which

$$X = X_0 \cup X_1 \cup ... \text{ and } \operatorname{ind}(X_0) \leq 0,\ \operatorname{ind}(X_1) \leq 1, ...$$

there exists an embedding h of X into T for which

$$X_0 \subset h^{-1}(T_0),\ X_1 \subset h^{-1}(T_1), ... \ \square$$

4.4.17 Corollary. *There exists a space T and a fixed sequence $T_0, T_1, ...$ of (open) closed subspaces of T for which*

$$T = T_0 \cup T_1 \cup ... \text{ and } \operatorname{ind}(T_0) \leq 0,\ \operatorname{ind}(T_1) \leq 1, ...$$

such that for every space X and for every sequence $X_0, X_1, ...$ of subspaces of X for which

$$X = X_0 \cup X_1 \cup ... \text{ and } \operatorname{ind}(X_0) \leq 0,\ \operatorname{ind}(X_1) \leq 1, ...$$

there exists an embedding h of X into T for which

$$X_0 = h^{-1}(T_0),\ X_1 = h{-}1(T_1), \ldots. \ \square$$

4.4.18 Corollary. *There exists a space T and a fixed sequence $T_0, T_1, \ldots$ of zero-dimensional subspaces (respectively, of zero-dimensional closed subspaces) of T for which $T = T_0 \cup T_1 \cup \ldots$ such that for every space X and for every sequence $X_0, X_1, \ldots$ of zero-dimensional subspaces (respectively, of zero-dimensional closed subspaces) of X for which $X = X_0 \cup X_1 \cup \ldots$ there exists an embedding h of X into T for which*

$$X_0 \subset h^{-1}(T_0),\ X_1 \subset h^{-1}(T_1), \ldots$$

(respectively,

$$X_0 = h^{-1}(T_0),\ X_1 = h^{-1}(T_1), \ldots). \ \square$$

4.4.19 Corollary. *There exists a space T and a fixed increasing sequence $T_0 \subset T_1 \subset \ldots$ of subspaces of T for which*

$$T = T_0 \cup T_1 \cup \ldots \text{ and } \mathrm{ind}(T_0) \leq 0,\ \mathrm{ind}(T_1) \leq 1, \ldots$$

such that for every space X and for every increasing sequence $X_0 \subset X_1 \subset \ldots$ of subspaces of X for which

$$X = X_0 \cup X_1 \cup \ldots \text{ and } \mathrm{ind}(X_0) \leq 0,\ \mathrm{ind}(X_1) \leq 1, \ldots$$

there exists an embedding h of X into T for which

$$X_0 \subset h^{-1}(T_0),\ X_1 \subset h^{-1}(T_1), \ldots. \ \square$$

4.4.20 Corollary. *There exists a space T and a fixed increasing sequence $T_0 \subset T_1 \subset \ldots$ of (open) closed subspaces of T for which*

$$T = T_0 \cup T_1 \cup \ldots \text{ and } \mathrm{ind}(T_0) \leq 0,\ \mathrm{ind}(T_1) \leq 1, \ldots$$

such that for every space X and for every increasing sequence $X_0 \subset X_1 \subset \ldots$ of subspaces of X for which

$$X = X_0 \cup X_1 \cup \ldots \text{ and } \mathrm{ind}(X_0) \leq 0,\ \mathrm{ind}(X_1) \leq 1, \ldots$$

there exists an embedding h of X into T for which

$$X_0 = h^{-1}(T_0),\ X_1 = h^{-1}(T_1), \dots . \ \square$$

4.4.21 Corollary. *There exists a space T and a fixed increasing sequence $T_0 \subset T_1 \subset \dots$ of zero-dimensional subspaces (respectively, of zero-dimensional closed subspaces) of T for which $T = T_0 \cup T_1 \cup \dots$ such that for every space X and for every increasing sequence $X_0 \subset X_1 \subset \dots$ of zero-dimensional subspaces (respectively, of zero-dimensional closed subspaces) of X for which $X = X_0 \cup X_1 \cup \dots$ there exists an embedding h of X into T for which*

$$X_0 \subset h^{-1}(T_0),\ X_1 \subset h^{-1}(T_1), \dots$$

(respectively,

$$X_0 = h^{-1}(T_0),\ X_1 = h^{-1}(T_1), \dots). \ \square$$

4.4.22 Corollary. *In the class of all spaces which can be represented as the union of an increasing sequence $X_0 \subset X_1 \subset \dots$ of zero-dimensional (closed) subspaces there exists a universal space.* $\square$

4.4.23 Corollary. *Let $\omega_1 \leq \tau$. There exists a space T and a fixed increasing transfinite sequence*

$$T_0 \subset T_1 \subset \dots T_\delta \subset \dots,\ \delta \in \omega_1,$$

of zero-dimensional subspaces of T such that for every space X and for every increasing transfinite sequence

$$X_0 \subset X_1 \subset \dots X_\delta \subset \dots,\ \delta \in \omega_1,$$

of zero-dimensional subspaces of X there exists an embedding h of X into T for which

$$X_0 \subset h^{-1}(T_0),\ X_1 \subset h^{-1}(T_1), \dots,\ h(X_\delta) \subset T_\delta, \dots,\ \delta \in \omega_1.$$

Proof. The proof of the corollary follows from the fact that in the class $\overline{\mathrm{S}}(\omega_1, \leq, \xi)$, where ξ is the mapping of ω_1 into τ^+ for which $\xi(\delta) = 0$ for every $\delta \in \omega_1$, there exists a universal element. $\square$

4.4.24 Corollary. *Let $\omega_1 \le \tau$. In the class of all spaces which can be represented as the union of an increasing transfinite sequence*

$$X_0 \subset X_1 \subset ...X_\delta \subset ..., \ \delta \in \omega_1,$$

of zero-dimensional subspaces there exists a universal space. □

4.4.25 Corollary. *Let n be a natural number. There exists a space T and a fixed zero-dimensional subspaces $T_0, ..., T_n$ of T for which $T = T_0 \cup ... \cup T_n$ such that for every space X and for every zero-dimensional subspaces $X_0, ..., X_n$ of X for which $X = X_0 \cup ... \cup X_n$ there exists an embedding h of X into T satisfying conditions*

$$h(X_0) \subset T_0, ..., h(X_n) \subset T_n. \square$$

4.4.26 Proposition. *Propositions 4.4.2, 4.4.3 and 4.4.4 and Corollaries 4.4.6–4.4.25 hold if we replace independently the class of all spaces by an arbitrary saturated class of spaces and the dimension ind by one of the following dimensions:*

s-$\mathrm{p_0}$-ind, s-$\mathrm{p_1}$-ind, s-$\mathrm{pos_0}$-ind, s-$\mathrm{pos_1}$-ind,

s-$\mathrm{ps_0}$-ind, s-$\mathrm{ps_1}$-ind, s-$\mathrm{b_0}$-ind, s-$\mathrm{b_1}$-ind,

s-$\mathrm{b^p}$-ind, s-$\mathrm{b^{pos}}$-ind, s-$\mathrm{b^{ps}}$-ind, s-$\mathrm{b_0}$-$\mathrm{p_0}$-ind,

s-$\mathrm{b_0}$-$\mathrm{p_1}$-ind, s-$\mathrm{b_1}$-$\mathrm{p_0}$-ind, *and* s-$\mathrm{b_1}$-$\mathrm{p_1}$-ind.

Moreover, Corollaries 4.4.8, 4.4.10, and 4.4.12–4.4.25 hold if we replace dimension ind by one of the following dimensions:

s-$\mathrm{b^0}$-ind, s-$\mathrm{b^1}$-ind, s-b(0)-ind, s-b(1)-ind,

s-$\mathrm{b^0}$-$\mathrm{p_0}$-ind, s-$\mathrm{b^0}$-$\mathrm{p_1}$-ind,

s-$\mathrm{b^1}$-$\mathrm{p_0}$-ind, *and* s-$\mathrm{b^1}$-$\mathrm{p_1}$-ind.

In particular, the above mentioned propositions and corollaries hold if instead of the class of all spaces we consider the class of all regular spaces or the class of all completely regular spaces (see Corollary 5.1.6) or the class of all separable metrizable spaces if $\tau = \omega$. □

Chapter 5

Completely regular and compact spaces

In this chapter we consider completely regular spaces and compact spaces. For the investigation of these spaces we use the notion of a normal base for the closed (open) subsets of a space. The technical part of our investigation is the construction of some special indications of normal bases.

In Section 5.1 we first prove that the existence of a normal base (of cardinality $\leq \tau$) for the closed subsets of a space characterize the completely regular spaces. Using some special indications of the normal bases we prove that the class of bases consisting of all pairs (A, X), where A is a base for the open subsets of a space X which is contained in a normal base of cardinality $\leq \tau$, is saturated. Therefore, the class of all completely regular spaces being the space-component of this class is also saturated.

In Section 5.2 we give the notion of a (b^n-Ind $\leq \alpha$)-dimensional normal base for the closed (open) subsets of a space, where $\alpha \in \{-1\} \cup \mathcal{O}$. We characterize such bases using the subsequent specification of the special indications (considering in Section 1) of the normal bases. The notion of a (b^n-Ind $\leq \alpha$)-dimensional base defines a base dimensional-like function denoted by b^n-Ind and a space dimension-like function denoted by s-b^n-Ind. These functions are defined as follows: for every $(A, X) \in \mathbb{P}(\text{base})$,

$$\mathrm{b}^n\text{-Ind}(A, X) \leq \alpha$$

if and only if A is contained in a (b^n-Ind $\leq \alpha$)-dimensional normal base for X, and for every space X,

$$\text{s-b}^n\text{-Ind}(X) \leq \alpha$$

if and only if in X there exists a $(\text{b}^n\text{-Ind} \leq \alpha)$-dimensional normal base. Using the given characterization of the $(\text{b}^n\text{-Ind} \leq \alpha)$-dimensional normal bases we prove that the class $\mathbb{P}(\text{b}^n\text{-Ind} \leq m)$, where $m \in \{-1\} \cup \omega$, is a saturated class of bases. Therefore, the class consisting of all spaces X such that $\text{s-b}^n\text{-Ind}(X) \leq m$ being the space-component of this class is also saturated. We give the notion of $(\text{s-b}^n\text{-Ind} \leq \alpha)$-uniform class of spaces (consisting of spaces X with $\text{s-b}^n\text{-Ind}\ (X) \leq \alpha$) and prove that for such a class there exists a Containing Space T with $\text{s-b}^n\text{-Ind}\ (T) \leq \alpha$. Finally, for a normal space X we prove that

$$\text{s-b}^n\text{-Ind}(X) \leq \text{Ind}(X).$$

In Section 5.3 we consider compact spaces. To a given normal base C for the closed subsets of a space X, we let a compact extension of X denoted by $c(X, C)$ and a normal base for the closed subsets of $c(X, C)$ denoted by $c(C)$. Furthermore, we prove that if C is a $(\text{b}^n\text{-Ind} \leq \alpha)$-dimensional base for X, then $c(C)$ is also $(\text{b}^n\text{-Ind} \leq \alpha)$-dimensional. On the other hand, for a compact space X we prove that

$$\text{Ind}(X) \leq \text{s-b}^n\text{-Ind}(X).$$

This relation and the inverse relation of Section 5.2 concerning normal spaces imply that for a compact space X,

$$\text{Ind}(X) = \text{s-b}^n\text{-Ind}(X).$$

Thus, if C is a $(\text{b}^n\text{-Ind} \leq \alpha)$-dimensional normal base for a space X, then $\text{Ind}(c(X, C)) \leq \alpha$. Using this result and the fact that the above defined class $\mathbb{P}(\text{b}^n\text{-Ind} \leq m)$, where $m \in \{-1\} \cup \omega$, is saturated we prove that in the class of all compact spaces X with $\text{Ind}(X) \leq m \in \{-1\} \cup \omega$ there exists a universal element. (We note that a Containing Space for an indexed collection of compact spaces may be not compact.) Actually, we prove the following more general result. For every cardinal $\nu \leq \tau$ in the class of all completely regular spaces X, which can be represented as a union:

$$X = \cup\{X_\delta : \delta \in \nu\},$$

where X_δ is a compact subspace of X with

$$\operatorname{Ind}(X_\delta) \leq m_\delta \in \{-1\} \cup \omega,$$

there exists a universal element T. Moreover, the space T can be represented as a union:

$$T = \cup\{T_\delta : \delta \in \nu\},$$

where T_δ is a compact subspace of T with $\operatorname{Ind}(T_\delta) \leq m_\delta$, such that for every space X of this class (with a given as the above representation) there exists an embedding h of X into T for which $h^{-1}(T_\delta) = X_\delta$, $\delta \in \nu$.

In Section 5.4 we consider compact spaces and dimension dim. In parallel to the case of the dimension Ind we give the corresponding notions and results related to dimension dim. (The elements of the corresponding covers are considered to be elements of a given normal base for open subsets.) In particular, we prove that in the class of all compact spaces or in the class of all normal spaces X with $\dim(X) \leq m \in \{-1\} \cup \omega$ there exists a universal element.

5.1 Completely regular spaces and universality

We start giving the notion of a normal base for a space and proving that the existence of such bases characterize completely regular spaces.

Definition. Let B and C be two families of subsets of a space X. The family B is said to be *complementary* to C if B consists of the complements of all elements of C. (In this case, C is of course complementary to B.)

Suppose that the cardinalities of B and C are less than or equal to τ. An indication $\{B_\delta : \delta \in \tau\}$ of B is said to be *complementary* to an indication $\{C_\delta : \delta \in \tau\}$ of C if $B_\delta = X \setminus C_\delta$ for every $\delta \in \tau$. Obviously, in this case the indication $\{C_\delta : \delta \in \tau\}$ of C is complementary to the indication $\{B_\delta : \delta \in \tau\}$ of B and the family C is complementary to the family B.

Definition. Let (G, K) be a pair of disjoint closed subsets of a space X. A pair (L, H) of closed subsets of X is called a *screening* of (G, K) (in the space X) if the following conditions are satisfied:

$$G \cap H = \emptyset,\ L \cap K = \emptyset \text{ and } L \cup H = X.$$

Definition. A base C for the closed subsets of a space X is said to be *normal* if the following conditions are satisfied:

(1) C is a ring: C is closed under finite unions and finite intersections, that is, $C = C^{\diamond}$.

(2) C is *disjunctive*: For every Q in C and a point x of X not in Q there is an F in C such that $x \in F$ and $F \cap Q = \emptyset$.

(3) C is *base-normal*: For every pair (G, K) of disjoint elements of C there exists a pair (L, H) of elements of C, which is a screening of (G, K).

A base B for the open subsets of a space X is said to be *normal* if the complementary to B base for the closed subsets of X is normal.

Remarks. The following observations are easy to verify:

1. The empty set and the whole space belong to any of its normal bases for the closed (open) subsets.

2. In the condition (2) of the above definition as the set Q we can consider an arbitrary closed subset of X.

3. If a base C for the closed subsets of a space X contains a disjunctive base, then C is also disjunctive.

4. If Q is an element of a normal base C for the closed subsets of a space X, then the set $\{Q \cap F : F \in C\}$ is a normal base for the closed subsets of the subspace Q of X.

Definition. A closed subset G of a space X is said to be a *zero-set* if there exists a real function g on X such that $g^{-1}(0) = G$.

5.1.1 Proposition. *A space X is completely regular if and only if X has a normal base of cardinality $\leq \tau$ for the closed subsets.*

Proof. Let X be a completely regular space. It is easy to verify that the set of all zero-sets of X is a normal base for the closed subsets of X. However, the cardinality of this base in general may be larger than τ.

Below, we prove that any normal base for the closed subsets of X contains such a base of cardinality $\leq \tau$. Indeed, let C be a normal base for the closed subsets of X. By C^{-1} we denote any base of cardinality $\leq \tau$ for the closed subsets ox X, which is contained in C.

During the proof of the lemma a pair (G, K) of elements of C^{-1} is called *separated* if there exists an element F of C such that $G \subset X \setminus F \subset K$. In this case, F is called a *separator* for (G, K).

For each separated pair of elements of C^{-1} we choose a fixed separator and denote by C^0 the union of the set of all such separators and the set C^{-1}. Obviously, C^0 is a base of cardinality $\leq \tau$ for the closed subsets of X. We prove that C^0 is disjunctive.

Indeed, let Q be an element of C^0 and x a point of $X \setminus Q$. Since C^{-1} is a base for the closed subsets of X there exists an element Q' of C^{-1} such that $Q \subset Q'$ and $x \notin Q'$. Since C is disjunctive there exists an element F' of C such that $x \in F'$ and $F' \cap Q' = \emptyset$. Since C is base-normal there exist elements L and H of C such that the pair (L, H) is a screening of the pair (F', Q'). Let H' be an element of C^{-1} such that $x \notin H'$ and $H \subset H'$. Therefore, we have $Q' \subset X \setminus L \subset H'$, which means that the pair (Q', H') is separated. Let F be the choosing separator of (Q', H'). Then, $x \in F$ and $Q \cap F = \emptyset$, which means that C^0 is disjunctive.

By induction on $n \in \omega$ we can construct a sequence C^0, C^1,...,C^n,... of subsets of C of cardinality $\leq \tau$ such that for every $n > 0$ we have:

(a) the set C^n contains the set $(C^{n-1})^\diamondsuit$ and

(b) for every pair (G, K) of disjoint elements of C^{n-1} there exists a screening (L, H) of (G, K), where L and H are elements of C^n.

It is easy to verify that the set $\cup\{C^n : n \in \omega\}$ is the required normal base of cardinality $\leq \tau$ for the closed subsets of X.

Conversely, let X be a space having a normal base C for the closed subsets of X. We need to prove that X is completely regular. Let a be a point of X and K a closed subset of X such that $a \notin K$. Without loss of generality we can suppose that $K \in C$. Since C is disjunctive there exists an element $G \in C$ such that $a \in G$ and $G \cap K = \emptyset$.

Let $\mathcal{Q}$ be the set of rational numbers of the open interval $(0, 1)$ of real numbers. Using the fact that C is base-normal, for every rational number $r \in \mathcal{Q}$ we can construct a screening (L_r, H_r) of (G, K) such that if $r_i < r_j$, then $L_{r_i} \cap H_{r_j} = \emptyset$. Let $L = \cup\{L_r : r \in \mathcal{Q}\}$.

Setting

$$g(x) = \begin{cases} 1 & \text{if } x \in X \setminus L, \\ \inf\{r \in \mathcal{Q} : x \in L_r\} & \text{if } x \in L \end{cases}$$

we define a continuous mapping $g : X \to [0, 1]$ such that $g(a) = 0$ and $g(x) = 1$ for every $x \in K$, which proves that X is completely regular. □

The auxiliary notion given in the next definition is convenient to use for the construction of some special indications of normal bases. These indications are used in the construction of Containing Spaces.

Definition. Any triad $(\vartheta_\vee, \vartheta_\wedge, \vartheta)$, where $\vartheta_\vee$ and $\vartheta_\wedge$ are mappings of the set $\mathcal{F} \setminus \{\emptyset\}$ into the set τ and ϑ is a mapping of τ^2 into τ^2, is said to be an *admissible* triad.

Definition. Let $(\vartheta_\vee, \vartheta_\wedge, \vartheta)$ be an admissible triad. Let also X be a completely regular space, C a normal base of cardinality $\leq \tau$ for the closed subsets of X, and B the complementary to C base for the open subsets of X.

An indication $\{C_\delta : \delta \in \tau\}$ of C is said to be $(\vartheta_\vee, \vartheta_\wedge, \vartheta)$-*normal* if the following conditions are satisfied:

(a) for every $\{\delta_0, ..., \delta_{n-1}\} \in \mathcal{F} \setminus \{\emptyset\}$ we have

$$C_{\delta_0} \cup ... \cup C_{\delta_{n-1}} = C_\eta \text{ and } C_{\delta_0} \cap ... \cap C_{\delta_{n-1}} = C_\varepsilon,$$

where

$$\eta = \vartheta_\vee(\{\delta_0, ..., \delta_{n-1}\}) \text{ and } \varepsilon = \vartheta_\wedge(\{\delta_0, ..., \delta_{n-1}\}),$$

(b) for every $(\delta_0, \delta_1) \in \tau^2$ with $(\varepsilon_0, \varepsilon_1) = \vartheta(\delta_0, \delta_1)$ the pair $(C_{\varepsilon_0}, C_{\varepsilon_1})$ is a screening of the pair $(C_{\delta_0}, C_{\delta_1})$ if $C_{\delta_0} \cap C_{\delta_1} = \emptyset$ and $(C_{\varepsilon_0}, C_{\varepsilon_1}) = (\emptyset, \emptyset)$ if $C_{\delta_0} \cap C_{\delta_1} \neq \emptyset$, and

(c) $C_0 = X$ and $C_1 = \emptyset$.

An indication of B is said to be $(\vartheta_\vee, \vartheta_\wedge, \vartheta)$-*normal* if the corresponding complementary indication of C is $(\vartheta_\vee, \vartheta_\wedge, \vartheta)$-normal.

The following auxiliary notations are used for the construction of some special admissible triads considered in Lemma 5.1.3. (These notations will also be used in the next sections.)

Notation. For every $k \in \omega$ we consider the set

$$(k+1) \times \tau = \{(i, \delta) : i \in k+1, \delta \in \tau\}$$

and denote:

(a) by $\mathcal{F}_k$ the set of all non-empty finite subsets of $(k+1) \times \tau$, which intersect the subset $\{k\} \times \tau$ of $(k+1) \times \tau$ and

(b) by P_k the set of all pairs $((i_0, \delta_0), (i_1, \delta_1))$ of elements of $(k+1) \times \tau$ such that $k = \max\{i_0, i_1\}$.

Remark. We note that the sets $\mathcal{F}_0$,$\mathcal{F}_1$,...,$\mathcal{F}_k$,... are mutually disjoint and the set $\cup\{\mathcal{F}_k : k \in \omega\}$ coincides with the set of all non-empty finite subsets of the set $\omega \times \tau$.

Similarly, the sets P_0,P_1,...,P_k,... are mutually disjoint and the set $\cup\{P_k : k \in \omega\}$ coincides with the set of all pairs of elements of $\omega \times \tau$.

The following lemma, which is easily proved, are used in the proof of Lemma 5.1.3.

5.1.2 Lemma. *For every $k \in \omega$ there exist:*

(a) *two one-to-one mappings $\vartheta_{\vee,k}$ and $\vartheta_{\wedge,k}$ of $\mathcal{F}_k$ into the set $\{k+1\} \times \tau$ and*

(b) *a mapping ϑ_k of P_k into the set $(\{k+1\} \times \tau)^2$ such that:*

(1) *The mappings $\pi_1^k \circ \vartheta_k$ and $\pi_2^k \circ \vartheta_k$ are one-to-one, where π_1^k and π_2^k are respectively the first and the second projections of the set $(\{k+1\} \times \tau)^2$ onto $\{k+1\} \times \tau$ (that is,*

$$\pi_1^k((k+1, \delta), (k+1, \varepsilon)) = (k+1, \delta) \text{ and}$$

$$\pi_2^k((k+1, \delta), (k+1, \varepsilon)) = (k+1, \varepsilon)).$$

(2) *The sets $\mathrm{I}(\pi_1^k \circ \vartheta_k)$, $\mathrm{I}(\pi_2^k \circ \vartheta_k)$, $\mathrm{I}(\vartheta_{\vee,k})$, and $\mathrm{I}(\vartheta_{\wedge,k})$ are mutually disjoint.* (See the Introductory Remarks about the last notations). □

The following lemma is used in the proof of Proposition 5.1.5. (This lemma will also be used in the next sections).

5.1.3 Lemma. *There exist an admissible triad $(\vartheta_\vee, \vartheta_\wedge, \vartheta)$ and a mapping $\overline{\theta}$ of τ into itself such that for every base D^X of cardinality $\leq \tau$ for the closed subsets of a space X, its indication*

$$K^X \equiv \{K_\delta^X : \delta \in \tau\},$$

and a normal base C^X of cardinality $\leq \tau$ for the closed subsets of X containing D^X there exists a $(\vartheta_\vee, \vartheta_\wedge, \vartheta)$-normal indication

$$F^X \equiv \{F_\delta^X : \delta \in \tau\}$$

of C^X with the property that $K_\delta^X = F_{\overline{\theta}(\delta)}^X$ for every $\delta \in \tau$.

Proof. The proof of the lemma consists of two parts. In the first part we construct the corresponding admissible triad $(\vartheta_\vee, \vartheta_\wedge, \vartheta)$ and the mapping $\overline{\theta}$. This construction is independent of the considered space. In the second part for a given space X and for the corresponding D^X, K^X, and C^X we construct a $(\vartheta_\vee, \vartheta_\wedge, \vartheta)$-normal indication F^X of C^X satisfying the condition of the lemma.

Let ϖ be a one-to-one mapping of the set $\omega \times \tau$ onto τ such that $\varpi(0,0) = 0$ and $\varpi(0,1) = 1$. For every $k \in \omega$ we consider the mappings $\vartheta_{\vee,k}$, $\vartheta_{\wedge,k}$, and ϑ_k mentioned in Lemma 5.1.2. The required admissible triad $(\vartheta_\vee, \vartheta_\wedge, \vartheta)$ and the mapping $\overline{\theta}$ are constructed as follows.

First we define the mappings $\vartheta_\vee$ and $\vartheta_\wedge$. Let $\{\delta_0, ..., \delta_{n-1}\}$ be an element of the set $\mathcal{F} \setminus \{\emptyset\}$ and $\varpi^{-1}(\delta_j) = (i_j, \varepsilon_j)$, $j = 0, ..., n-1$. Setting

$$k = \max\{i_0, ..., i_{n-1}\}$$

we have $\{(i_0, \varepsilon_0), ..., (i_{n-1}, \varepsilon_{n-1})\} \in \mathcal{F}_k$. If

$$\vartheta_{\vee,k}(\{(i_0, \varepsilon_0), ..., (i_{n-1}, \varepsilon_{n-1})\}) = (k+1, \delta),$$

$$\vartheta_{\wedge,k}(\{(i_0, \varepsilon_0), ..., (i_{n-1}, \varepsilon_{n-1})\}) = (k+1, \delta'),$$

$$\varpi(k+1, \delta) = \eta, \text{ and } \varpi(k+1, \delta') = \eta',$$

then we set $\vartheta_\vee(\{\delta_0, ..., \delta_{n-1}\}) = \eta$ and $\vartheta_\wedge(\{\delta_0, ..., \delta_{n-1}\}) = \eta'$.

Futhermore, we define the mapping ϑ. Let $(\delta_0, \delta_1) \in \tau^2$, $\varpi^{-1}(\delta_0) = (i_0, \eta_0)$, and $\varpi^{-1}(\delta_1) = (i_1, \eta_1)$. If $k = \max\{i_0, i_1\}$, then $((i_0, \eta_0), (i_1, \eta_1)) \in P_k$. Suppose that

$$\vartheta_k((i_0, \eta_0), (i_1, \eta_1)) = ((k+1, \eta^0), (k+1, \eta^1)),$$

$$\varpi(k+1, \eta^0) = \varepsilon_0 \text{ and } \varpi(k+1, \eta^1) = \varepsilon_1.$$

Then, we set $\vartheta(\delta_0, \delta_1) = (\varepsilon_0, \varepsilon_1)$.

Finally, we define the mapping $\overline{\theta}$. Let θ' be a mapping of τ into the set $\omega \times \tau$ such that

$$\theta'(\tau) \subset (\{0\} \times \tau) \setminus \{(0,0), (0,1)\} \subset \omega \times \tau \text{ and}$$

$$|(\{0\} \times \tau) \setminus \theta'(\tau)| = \tau.$$

(We note that the last condition is not used in the proof of the present lemma. It will be used in the next sections.) Then, we set $\overline{\theta} = \varpi \circ \theta'$.

Now, let X, D^X, $K^X \equiv \{K^X_\delta : \delta \in \tau\}$, and C^X be the same as in the formulation of the lemma. Let $\{C^0_\delta : \delta \in \tau\}$ be an indication of C^X such that $C_0^0 = X$, $C_1^0 = \emptyset$ and $C_\varepsilon^0 = K_\delta^X$ if $(0,\varepsilon) = \theta'(\delta)$. By induction, for every $k \in \omega$ it is easy to construct an indexed set $\{C_\delta^k : \delta \in \tau\}$ consisting of elements of C^X such that the following conditions are satisfied:

(a) if $\{(i_0, \delta_0), ..., (i_m, \delta_m)\}$ is an element of $\mathcal{F}_k$ and

$$(k+1, \varepsilon) = \vartheta_{\vee,k}(\{(i_0, \delta_0), ..., (i_m, \delta_m)\}),$$

then $C_\varepsilon^{k+1} = C_{\delta_0}^{i_0} \cup ... \cup C_{\delta_m}^{i_m}$;

(b) if $\{(i_0, \delta_0), ..., (i_m, \delta_m)\}$ is an element of $\mathcal{F}_k$ and

$$(k+1, \varepsilon) = \vartheta_{\wedge,k}(\{(i_0, \delta_0), ..., (i_m, \delta_m)\}),$$

then $C_\varepsilon^{k+1} = C_{\delta_0}^{i_0} \cap ... \cap C_{\delta_m}^{i_m}$; and

(c) for every element $((i_0, \delta_0), (i_1, \delta_1))$ of P_k for which

$$((k+1, \varepsilon_0), (k+1, \varepsilon_1)) = \vartheta_k((i_0, \delta_0), (i_1, \delta_1))$$

the pair $(C_{\varepsilon_0}^{k+1}, C_{\varepsilon_1}^{k+1})$ is a screening of the pair $(C_{\delta_0}^{i_0}, C_{\delta_1}^{i_1})$ if $C_{\delta_0}^{i_0} \cap C_{\delta_1}^{i_1} = \emptyset$, and $(C_{\varepsilon_0}^{k+1}, C_{\varepsilon_1}^{k+1}) = (\emptyset, \emptyset)$ if $C_{\delta_0}^{i_0} \cap C_{\delta_1}^{i_1} \neq \emptyset$.

For every $\delta \in \tau$ we set $F_\delta^X = C_\varepsilon^k$, where $(k, \varepsilon) = \varpi^{-1}(\delta)$. It is easy to verify that $F^X \equiv \{F_\delta^X : \delta \in \tau\}$ is a $(\vartheta_\vee, \vartheta_\wedge, \vartheta)$-normal indication of C^X such that $K_\delta^X = F_{\bar{\theta}(\delta)}^X$. $\square$

Now we extend some of the above given notions concerning a space to an indexed collection of spaces.

Definition. Let $\mathbf{S}$ be an indexed collection of spaces, $\mathbf{B}$ a co-base for $\mathbf{S}$, and $\mathbf{C}$ a closed co-base for $\mathbf{S}$. The co-base $\mathbf{B}$ for $\mathbf{S}$ is said to be *complementary* to $\mathbf{C}$ if $\mathbf{B}(X)$ is complementary to $\mathbf{C}(X)$ for every $X \in \mathbf{S}$.

A co-indication $\mathbf{N}$ of $\mathbf{B}$ is said to be *complementary* to a co-indication $\mathbf{F}$ of $\mathbf{C}$ if $\mathbf{N}(X)$ is complementary to $\mathbf{F}(X)$ for every $X \in \mathbf{S}$.

Definitions. Let $\mathbf{S}$ be an indexed collection of spaces, $\mathbf{C}$ a closed co-base for $\mathbf{S}$, and $\mathbf{B}$ the complementary to $\mathbf{C}$ open co-base for $\mathbf{S}$. Let also $(\vartheta_\vee, \vartheta_\wedge, \vartheta)$ be an admissible triad.

The closed co-base $\mathbf{C}$ is said to be *normal* if for every $X \in \mathbf{S}$, $\mathbf{C}(X)$ is a normal base for the closed subsets of X. The open co-base $\mathbf{B}$ is said to be *normal* if the closed co-base $\mathbf{C}$ is normal. (We note that by the definition of a co-base for $\mathbf{S}$ the cardinality of $\mathbf{C}(X)$ is less than or equal to τ.)

A co-indication $\mathbf{F}$ of $\mathbf{C}$ is said to be $(\vartheta_\vee, \vartheta_\wedge, \vartheta)$-*normal,* if for every $X \in \mathbf{S}$, $\mathbf{F}(X)$ is a $(\vartheta_\vee, \vartheta_\wedge, \vartheta)$-normal indication of $\mathbf{C}(X)$. A co-indication $\mathbf{N}$ of $\mathbf{B}$ is said to be $(\vartheta_\vee, \vartheta_\wedge, \vartheta)$-*normal* if the comlementary to $\mathbf{N}$ co-indication of $\mathbf{C}$ is $(\vartheta_\vee, \vartheta_\wedge, \vartheta)$-normal.

Definition and notation. Let $\mathbf{S}$ be an indexed collection of spaces,

$$\mathbf{M} \equiv \{\{U_\delta^X : \delta \in \tau\} : X \in \mathbf{S}\}$$

a co-mark of $\mathbf{S}$, and R an $\mathbf{M}$-admissible family of equivalence relations on $\mathbf{S}$. Suppose that for some subset κ of τ and for every $X \in \mathbf{S}$ the family $\{U_\delta^X : \delta \in \kappa\}$ is base for X. Then, the κ-standard family $\mathrm{B}_\kappa^\mathrm{T}$ of T is a base for T. We note that the $(\diamondsuit, \kappa)$-standard base $\mathrm{B}_{\diamondsuit,\kappa}^\mathrm{T}$ for T is contained in the base $(\mathrm{B}_\kappa^\mathrm{T})^\diamondsuit \equiv \mathrm{B}_\kappa^{\mathrm{T},\diamondsuit}$ for T. However, in general, $\mathrm{B}_{\diamondsuit,\kappa}^\mathrm{T}$ does not coincide with $\mathrm{B}_\kappa^{\mathrm{T},\diamondsuit}$.

We denote by $\mathrm{C}_\kappa^{\mathrm{T},\diamondsuit}$ the base for the closed subsets of T, which is complementary to $\mathrm{B}_\kappa^{\mathrm{T},\diamondsuit}$. It is easy to see that $\mathrm{C}_\kappa^{\mathrm{T},\diamondsuit} = (\mathrm{C}_\kappa^\mathrm{T})^\diamondsuit$, where $\mathrm{C}_\kappa^\mathrm{T}$ is the base for the closed subsets of T, which is complementary to $\mathrm{B}_\kappa^\mathrm{T}$. The bases $\mathrm{B}_\kappa^{\mathrm{T},\diamondsuit}$ and $\mathrm{C}_\kappa^{\mathrm{T},\diamondsuit}$ are called $(\kappa, \diamondsuit)$-*standard* .

Similarly, for every $\mathbf{L} \in \mathrm{C}^\diamondsuit(\mathrm{R})$ we consider the base $\mathrm{B}_\kappa^{\mathbf{L},\diamondsuit} \equiv (\mathrm{B}_\kappa^\mathbf{L})^\diamondsuit$ for the open subsets of the space $\mathrm{T}(\mathbf{L})$ and denote by $\mathrm{C}_\kappa^{\mathbf{L},\diamondsuit}$ the complementary to $\mathrm{B}_\kappa^{\mathbf{L},\diamondsuit}$ base for the closed subsets of $\mathrm{T}(\mathbf{L})$. It is also easy to verify that $\mathrm{C}_\kappa^{\mathbf{L},\diamondsuit} = (\mathrm{C}_\kappa^\mathbf{L})^\diamondsuit$, where $\mathrm{C}_\kappa^\mathbf{L}$ is the base for the closed subsets of $\mathrm{T}(\mathbf{L})$ complementary to $\mathrm{B}_\kappa^\mathbf{L}$. The bases $\mathrm{B}_\kappa^{\mathbf{L},\diamondsuit}$ and $\mathrm{C}_\kappa^{\mathbf{L},\diamondsuit}$ are also called $(\kappa, \diamondsuit)$-*standard.*

It is easy to prove that for every $\mathbf{L} \in \mathrm{C}^\diamondsuit(\mathrm{R})$ we have

$$\mathrm{B}_\kappa^{\mathbf{L},\diamondsuit} = \{\mathrm{T}(\mathbf{L}) \cap U : U \in \mathrm{B}_\kappa^{\mathrm{T},\diamondsuit}\}$$

and, therefore,

$$\mathrm{C}_\kappa^{\mathbf{L},\diamondsuit} = \{\mathrm{T}(\mathbf{L}) \cap U : U \in \mathrm{C}_\kappa^{\mathrm{T},\diamondsuit}\}.$$

Finally, we note that the introduced notations: $\mathrm{B}_\kappa^{\mathrm{T},\diamondsuit}$, $\mathrm{B}_\kappa^{\mathbf{L},\diamondsuit}$, $\mathrm{C}_\kappa^{\mathrm{T},\diamondsuit}$, $\mathrm{C}_\kappa^{\mathbf{L},\diamondsuit}$ will also be used in the case where instead of the letters "T", "**L**", and "κ" we shall use other letters or symbols for the corresponding objects.

The following lemma is the most important part of the proof of Proposition 5.1.5. (This lemma will also be used in the next section.)

5.1.4 Lemma. *Let* $\mathbf{S}$ *be an indexed collection of spaces,* $\mathbf{B}$ *a normal open co-base for* $\mathbf{S}$*, and* $\mathbf{N}$ *a* $(\vartheta_\vee, \vartheta_\wedge, \vartheta)$*-normal co-indication of* $\mathbf{B}$*, where* $(\vartheta_\vee, \vartheta_\wedge, \vartheta)$ *is an admissible triad. Then, for every co-extension* $\mathbf{M}$ *of* $\mathbf{N}$ *and an* $\mathbf{M}$*-admissible family* R *of equivalence relations on* $\mathbf{S}$ *the set* $\mathrm{C}^{\mathrm{T},\diamondsuit}_{\theta(\tau)}$ *is a normal base for the closed subsets of* T*, where* θ *is an indicial mapping from* $\mathbf{N}$ *to* $\mathbf{M}$*. Therefore, for every* $\mathbf{L} \in \mathrm{C}^{\diamondsuit}(\mathrm{R})$ *the set* $\mathrm{B}^{\mathbf{L},\diamondsuit}_{\theta(\tau)}$ *is a normal base for the open subsets of the space* $\mathrm{T}(\mathbf{L})$.

Proof. Let

$$\mathbf{B} = \{B^X : X \in \mathbf{S}\} \text{ and}$$

$$\mathbf{N} = \{\{V^X_\varepsilon : \varepsilon \in \tau\} : X \in \mathbf{S}\}.$$

Denote by

$$\mathbf{C} \equiv \{C^X : X \in \mathbf{S}\}$$

the complementary to $\mathbf{B}$ closed co-base for $\mathbf{S}$ and by

$$\{\{C^X_\varepsilon : \varepsilon \in \tau\} : X \in \mathbf{S}\}$$

the complementary to $\mathbf{N}$ co-indication of $\mathbf{C}$. Therefore, C^X is the complementary to B^X base for the closed subsets of X and $C^X_\varepsilon = X \setminus V^X_\varepsilon$ for every $X \in \mathbf{S}$ and $\varepsilon \in \tau$. Let

$$\mathbf{M} \equiv \{\{U^X_\delta : \delta \in \tau\} : X \in \mathbf{S}\}$$

be a co-extension of $\mathbf{N}$ and θ an indicial mapping from $\mathbf{N}$ to $\mathbf{M}$. Let also

$$\mathrm{R} \equiv \{\sim^s : s \in \mathcal{F}\}$$

be an $\mathbf{M}$-admissible family of equivalence relations on $\mathbf{S}$. We need to prove that $\mathrm{C}^{\mathrm{T},\diamondsuit}_{\theta(\tau)}$ is a normal base for the open subsets of T.

By construction, $\mathrm{B}^{\mathrm{T},\diamondsuit}_{\theta(\tau)}$ is a base for the open subsets of T and, therefore, $\mathrm{C}^{\mathrm{T},\diamondsuit}_{\theta(\tau)}$ is a base for the closed subset of T. Moreover, $\mathrm{C}^{\mathrm{T},\diamondsuit}_{\theta(\tau)}$ is a ring. To prove that $\mathrm{C}^{\mathrm{T},\diamondsuit}_{\theta(\tau)}$ is disjunctive and base-normal we first prove that every element U of $\mathrm{B}^{\mathrm{T},\diamondsuit}_{\theta(\tau)}$ has the form:

$$U = U^{\mathrm{T}}_{\delta_0}(\mathbf{L}_0) \cup ... \cup U^{\mathrm{T}}_{\delta_n}(\mathbf{L}_n), \tag{5.1.1}$$

where $n \in \omega$, $\delta_0, ..., \delta_n \in \theta(\tau)$, and $\mathbf{L}_0, ..., \mathbf{L}_n$ are distinct elements of $\mathrm{C}(\sim^s)$ for some $s \in \mathcal{F}$. Indeed, there exists a finite set u consisting of finite subsets of $\mathrm{B}^{\mathrm{T}}_{\theta(\tau)}$ such that

$$U = \cup\{U_v : v \in u\},$$

where U_v is the intersection of all elements of $v \in u$. Suppose that for every $v \in u$ we have

$$v = \{U^{\mathrm{T}}_{\eta^i_v}(\mathbf{L}^i_v) : i \in k_v \in \omega\},$$

where $\eta^i_v \in \theta(\tau)$ and $\mathbf{L}^i_v \in \mathrm{C}(\mathrm{R})$. Consider the (non-indexed) set

$$\{\theta^{-1}(\eta^i_v) : i \in k_v\}$$

and put

$$\delta_v = \theta(\vartheta_\wedge(\{\theta^{-1}(\eta^i_v) : i \in k_v\})).$$

Then, we have

$$\cap\{U^{\mathrm{T}}_{\eta^i_v}(\mathbf{L}^i_v) : i \in k_v\} = U^{\mathrm{T}}_{\delta_v}(\mathbf{L}_v) = U_v,$$

where

$$\mathbf{L}_v = \cap\{\mathbf{L}^i_v : i \in k_v\} \in \mathrm{C}^{\diamond}(\mathrm{R}).$$

Therefore,

$$U = \cup\{U^{\mathrm{T}}_{\delta_v}(\mathbf{L}_v) : v \in u\}.$$

For every $v \in u$ we represent $\mathbf{L}_v$ as a union of elements of $\mathrm{C}(\sim^{s_v})$ for some $s_v \in \mathcal{F}$ (see Lemma 1.2.7) and set $s = \cup\{s_v : v \in u\}$. Let $\mathbf{L}_0, ..., \mathbf{L}_n$ be the elements of $\mathrm{C}(\sim^s)$, which intersect the set $\cup\{\mathbf{L}_v : v \in u\}$. Consider the element $\mathbf{L}_0$ and let u_0 be the subset of u consisting of all elements v of u such that $\mathbf{L}_v \cap \mathbf{L}_0 \neq \emptyset$. Supposing that $u_0 = \{v_0, ..., v_m\}$ we set

$$\delta_0 = \theta(\vartheta_\vee(\{\theta^{-1}(\delta_{v_0}), ..., \theta^{-1}(\delta_{v_m})\})).$$

Similarly, we can define the elements $\delta_1, ..., \delta_m$ of $\theta(\tau)$. Then, it is easy to verify that for the defined $n \in \omega$, $\delta_0, ..., \delta_n \in \theta(\tau)$, $s \in \mathcal{F}$, and $\mathbf{L}_0, ...\mathbf{L}_n \in \mathrm{C}(\sim^s)$ the relation (5.1.1) is true.

It is easy to see that for every $\mathbf{L} \in \mathrm{C}^{\diamondsuit}(\mathrm{R})$ we have $\mathrm{T}(\mathbf{L}) \in \mathrm{B}^{\mathrm{T},\diamondsuit}_{\theta(\tau)}$ and, therefore, $\mathrm{T}(\mathbf{L}) \in \mathrm{C}^{\mathrm{T},\diamondsuit}_{\theta(\tau)}$. Also, for every $U^{\mathrm{T}}_{\delta}(\mathbf{L}) \in \mathrm{B}^{\mathrm{T},\diamondsuit}_{\theta(\tau)}$ we have

$$\mathrm{T}(\mathbf{L}) \setminus U^{\mathrm{T}}_{\delta}(\mathbf{L}) = \mathrm{T}(\mathbf{L}) \cap (\mathrm{T} \setminus U^{\mathrm{T}}_{\delta}(\mathbf{L})) \in \mathrm{C}^{\mathrm{T},\diamondsuit}_{\theta(\tau)}.$$

We note that by Lemma 1.2.7, $\mathbf{L} = \mathbf{L}_0 \cup ... \cup \mathbf{L}_k$, where $\mathbf{L}_0$,...,$\mathbf{L}_k$ are $\sim^t$-equivalence classes for some $t \in \mathcal{F}$, and, therefore,

$$\mathrm{T}(\mathbf{L}) = U^{\mathrm{T}}_{\delta}(\mathbf{L}_0) \cup ... \cup U^{\mathrm{T}}_{\delta}(\mathbf{L}_k),$$

where $\delta = \theta(1)$. On the other hand,

$$\mathrm{T}(\mathbf{L}) = \mathrm{T} \setminus \mathrm{T}(\mathbf{H}), \text{ where } \mathbf{H} = \mathbf{S} \setminus \mathbf{L} \in \mathrm{C}^{\diamondsuit}(\mathrm{R}).$$

Now, we prove that $\mathrm{C}^{\mathrm{T},\diamondsuit}_{\theta(\tau)}$ is disjunctive. Let $\mathbf{a} \in \mathrm{T}$, $G \in \mathrm{C}^{\mathrm{T},\diamondsuit}_{\theta(\tau)}$, and $\mathbf{a} \notin G$. Then, $\mathrm{T} \setminus G$ has the form (5.1.1):

$$\mathrm{T} \setminus G = U^{\mathrm{T}}_{\delta_0}(\mathbf{L}_0) \cup ... \cup U^{\mathrm{T}}_{\delta_n}(\mathbf{L}_n),$$

where $n \in \omega$, $\delta_0, ..., \delta_n \in \theta(\tau)$, and $\mathbf{L}_0, ..., \mathbf{L}_n$ are distinct elements of $\mathrm{C}(\sim^s)$ for some $s \in \mathcal{F}$. Let $(x, X) \in \mathbf{a}$. There exists $i \in \{0, ..., n\}$ such that $X \in \mathbf{L}_i$ and $x \in U^X_{\delta_i}$. Since C^X is disjunctive there exists $\varepsilon \in \tau$ such that $x \in C^X_{\varepsilon} \in C^X$ and $C^X_{\varepsilon} \cap (X \setminus U^X_{\delta_i}) = \emptyset$. Let $\delta = \theta(\varepsilon)$ (and, therefore, $U^X_{\delta} = X \setminus C^X_{\varepsilon}$) and set $q = s \cup \{\delta\}$. There exists $t \in \mathcal{F}$ such that $\sim^t \subset \sim^q_{\mathbf{M}}$. Denote by $\mathbf{L}$ the $\sim^t$-equivalence class of X.

Since U^X_{δ} and $U^X_{\delta_i}$ are elements of the q-algebra A^X_q, which is isomorphic (by a natural way) to the algebra $\mathrm{A}^{\mathbf{L}}_q$, relation $U^X_{\delta} \cup U^X_{\delta_i} = X$ implies $U^{\mathrm{T}}_{\delta}(\mathbf{L}) \cup U^{\mathrm{T}}_{\delta_i}(\mathbf{L}) = \mathrm{T}(\mathbf{L})$. Therefore,

$$(\mathrm{T}(\mathbf{L}) \setminus U^{\mathrm{T}}_{\delta}(\mathbf{L})) \cap G =$$

$$(\mathrm{T}(\mathbf{L}) \setminus U^{\mathrm{T}}_{\delta}(\mathbf{L})) \cap (\mathrm{T}(\mathbf{L}) \setminus U^{\mathrm{T}}_{\delta_i}(\mathbf{L})) = \emptyset.$$

Since $\mathbf{a} \in \mathrm{T}(\mathbf{L}) \setminus U^{\mathrm{T}}_{\delta}(\mathbf{L}) \in \mathrm{C}^{\mathrm{T},\diamondsuit}_{\theta(\tau)}$, the above relation shows that $\mathrm{C}^{\mathrm{T},\diamondsuit}_{\theta(\tau)}$ is disjunctive.

Finally, we prove that $\mathrm{C}^{\mathrm{T},\diamondsuit}_{\theta(\tau)}$ is base-normal. Let G and K be disjoint elements of $\mathrm{C}^{\mathrm{T},\diamondsuit}_{\theta(\tau)}$. The elements $\mathrm{T}\setminus G$ and $\mathrm{T}\setminus K$ of $\mathrm{B}^{\mathrm{T},\diamondsuit}_{\theta(\tau)}$ has the form (5.1.1):

$$\mathrm{T}\setminus G = U^{\mathrm{T}}_{\delta_0}(\mathbf{L}_0)\cup ... \cup U^{\mathrm{T}}_{\delta_n}(\mathbf{L}_n) \text{ and}$$

$$\mathrm{T}\setminus K = U^{\mathrm{T}}_{\delta'_0}(\mathbf{L}'_0)\cup ... \cup U^{\mathrm{T}}_{\delta'_m}(\mathbf{L}'_m),$$

where $n, m \in \omega$, $\delta_0, ..., \delta_n, \delta'_0, ..., \delta'_m \in \theta(\tau)$, $\mathbf{L}_0, ..., \mathbf{L}_m$ are distinct elements of $\mathrm{C}(\sim^{s_0})$ for some $s_0 \in \mathcal{F}$, and $\mathbf{L}'_0, ..., \mathbf{L}'_m$ are distinct elements of $\mathrm{C}(\sim^{s_1})$ for some $s_1 \in \mathcal{F}$. Without loss of generality we can suppose that $s = s_0 = s_1$ and $\sim^s \subset \sim^q_{\mathbf{M}}$, where q is an element of $\mathcal{F}$ containing the elements $\theta(0), \theta(1), \delta_0, ..., \delta_n, \delta'_0, ..., \delta'_m$ of τ and the images, under the mapping θ, of elements of all pairs $\vartheta(\theta^{-1}(\delta), \theta^{-1}(\delta'))$, where $\delta \in \{\theta(0), \theta(1), \delta_0, ..., \delta_n\}$ and $\delta' \in \{\theta(0), \theta(1), \delta'_0, ..., \delta'_m\}$.

Let $\mathbf{L} \in \mathrm{C}(\sim^s)$. There exist

$$\delta^{\mathbf{L}}_0, \delta^{\mathbf{L}}_1 \in \{\theta(0), \theta(1), \delta_0, ..., \delta_n, \delta'_0, ..., \delta'_m\}$$

such that

$$\mathrm{T}(\mathbf{L}) \cap (\mathrm{T}\setminus G) = U^{\mathrm{T}}_{\delta^{\mathbf{L}}_0}(\mathbf{L}) \text{ and}$$

$$\mathrm{T}(\mathbf{L}) \cap (\mathrm{T}\setminus K) = U^{\mathrm{T}}_{\delta^{\mathbf{L}}_1}(\mathbf{L}).$$

(We note that if $\mathrm{T}(\mathbf{L}) \cap (\mathrm{T}\setminus G) = \emptyset$, then we can set $\delta^{\mathbf{L}}_0 = \theta(0)$ and if $\mathrm{T}(\mathbf{L}) \cap (\mathrm{T}\setminus G) = \mathrm{T}(\mathbf{L})$, then we can set $\delta^{\mathbf{L}}_0 = \theta(1)$. The same is true if G is replaced by K.)

Set $\eta^{\mathbf{L}}_0 = \theta^{-1}(\delta^{\mathbf{L}}_0)$, $\eta^{\mathbf{L}}_1 = \theta^{-1}(\delta^{\mathbf{L}}_1)$, and $\vartheta(\eta^{\mathbf{L}}_0, \eta^{\mathbf{L}}_1) = (\varepsilon^{\mathbf{L}}_0, \varepsilon^{\mathbf{L}}_1)$. Let $X \in \mathbf{L}$. Then, we have

$$X \setminus U^X_{\delta^{\mathbf{L}}_0} = X \setminus V^X_{\eta^{\mathbf{L}}_0} = C^X_{\eta^{\mathbf{L}}_0},$$

$$X \setminus U^X_{\delta^{\mathbf{L}}_1} = X \setminus V^X_{\eta^{\mathbf{L}}_1} = C^X_{\eta^{\mathbf{L}}_1}, \text{ and}$$

$$C^X_{\eta^{\mathbf{L}}_0} \cap C^X_{\eta^{\mathbf{L}}_1} = \emptyset.$$

By properties of the mapping ϑ the pair

$$(C^X_{\varepsilon_0^{\mathbf{L}}} \equiv X \setminus U^X_{\theta(\varepsilon_0^{\mathbf{L}})}, C^X_{\varepsilon_1^{\mathbf{L}}} \equiv X \setminus U^X_{\theta(\varepsilon_1^{\mathbf{L}})})$$

is a screening of the pair $(C^X_{\eta_0^{\mathbf{L}}}, C^X_{\eta_1^{\mathbf{L}}})$, that is

$$(X \setminus U^X_{\theta(\varepsilon_1^{\mathbf{L}})}) \cap (X \setminus U^X_{\delta_0^{\mathbf{L}}}) = \emptyset,$$

$$(X \setminus U^X_{\theta(\varepsilon_0^{\mathbf{L}})}) \cap (X \setminus U^X_{\delta_1^{\mathbf{L}}}) = \emptyset, \text{ and}$$

$$(X \setminus U^X_{\theta(\varepsilon_0^{\mathbf{L}})}) \cup (X \setminus U^X_{\theta(\varepsilon_1^{\mathbf{L}})}) = X.$$

Since $U^X_{\theta(\varepsilon_0^{\mathbf{L}})}$, $U^X_{\theta(\varepsilon_1^{\mathbf{L}})}$, $U^X_{\delta_0^{\mathbf{L}}}$, $U^X_{\delta_1^{\mathbf{L}}}$, and its complements to X are elements of the algebra A_q^X and since this algebra is isomorphic (by a natural way) to the algebra $\mathrm{A}_q^{\mathbf{L}}$ we have

$$(\mathrm{T}(\mathbf{L}) \setminus U^{\mathrm{T}}_{\theta(\varepsilon_0^{\mathbf{L}})}(\mathbf{L})) \cap (\mathrm{T}(\mathbf{L}) \setminus U^{\mathrm{T}}_{\delta_1^{\mathbf{L}}}(\mathbf{L})) = \emptyset,$$

$$(\mathrm{T}(\mathbf{L}) \setminus U^{\mathrm{T}}_{\theta(\varepsilon_1^{\mathbf{L}})}(\mathbf{L})) \cap (\mathrm{T}(\mathbf{L}) \setminus U^{\mathrm{T}}_{\delta_0^{\mathbf{L}}}(\mathbf{L})) = \emptyset, \text{ and}$$

$$(\mathrm{T}(\mathbf{L}) \setminus U^{\mathrm{T}}_{\theta(\varepsilon_0^{\mathbf{L}})}(\mathbf{L})) \cup (\mathrm{T}(\mathbf{L}) \setminus U^{\mathrm{T}}_{\theta(\varepsilon_1^{\mathbf{L}})}(\mathbf{L})) = \mathrm{T}(\mathbf{L}). \tag{5.1.2}$$

Let

$$L = \cup\{\mathrm{T}(\mathbf{L}) \setminus U^{\mathrm{T}}_{\theta(\varepsilon_0^{\mathbf{L}})}(\mathbf{L}) : \mathbf{L} \in \mathrm{C}(\sim^s)\} \text{ and}$$

$$H = \cup\{\mathrm{T}(\mathbf{L}) \setminus U^{\mathrm{T}}_{\theta(\varepsilon_1^{\mathbf{L}})}(\mathbf{L}) : \mathbf{L} \in \mathrm{C}(\sim^s)\}.$$

Obviously L and H are elements of $\mathrm{C}^{\mathrm{T},\diamondsuit}_{\theta(\tau)}$. It is easy to verify that relations (5.1.2) imply

$$L \cap K = \emptyset,\ H \cap G = \emptyset, \text{ and } L \cup H = \mathrm{T},$$

which means that (L, H) is a screening of (G, K). Thus, $\mathrm{C}^{\mathrm{T},\diamondsuit}_{\theta(\tau)}$ is a normal base for the closed subsets of T. □

Notation. We denote by $\mathbb{P}(\subset$normal base$)$ the class of bases consisting of all pairs $(A, X) \in \mathbb{P}(\text{base})$ such that there exists a normal base B of cardinality $\leq \tau$ for the open subsets of X containing the base A.

5.1.5 Proposition. *The class* $\mathbb{P}(\subset\text{normal base})$ *is saturated.*

Proof. Let $\mathbf{S}$ be an indexed collection of spaces and

$$\mathbf{A} \equiv \{A^X : X \in \mathbf{S}\}$$

an open $\mathbb{P}$-co-base for $\mathbf{S}$, where $\mathbb{P} = \mathbb{P}(\subset\text{normal base})$. Let also

$$\mathbf{F} \equiv \{\{W_\delta^X : \delta \in \tau\} : X \in \mathbf{S}\}$$

be an arbitrary co-indication of $\mathbf{A}$.

For every $X \in \mathbf{S}$ consider a normal base B^X for the open subsets of X containing the base A^X. By Lemma 5.1.3 there exists an admissible triad $(\vartheta_\vee, \vartheta_\wedge, \vartheta)$, a mapping $\overline{\theta}$ of τ into itself, and a $(\vartheta_\vee, \vartheta_\wedge, \vartheta)$-normal indication

$$N^X \equiv \{V_\delta^X : \delta \in \tau\}$$

of B^X such that $V_{\overline{\theta}(\delta)}^X = W_\delta^X$ for every $\delta \in \tau$. We set

$$\mathbf{B} = \{B^X : X \in \mathbf{S}\} \text{ and}$$

$$\mathbf{N} = \{N^X : X \in \mathbf{S}\}.$$

Then, $\mathbf{N}$ is a $(\vartheta_\vee, \vartheta_\wedge, \vartheta)$-normal co-indication of $\mathbf{B}$, which is a co-extension of $\mathbf{F}$, and $\overline{\theta}$ is an indicial mapping of this co-extension.

Now, let $\mathbf{M}^+$ be a co-mark of $\mathbf{S}$, which is a co-extension of $\mathbf{N}$, $\mathbf{M}$ an arbitrary co-extension of $\mathbf{M}^+$, and θ an indicial mapping from $\mathbf{N}$ to $\mathbf{M}$. Then, $\theta_{\mathbf{A}} \equiv \theta \circ \overline{\theta}$ is an indicial mapping from $\mathbf{F}$ to $\mathbf{M}$.

Let also R^+ be an $\mathbf{M}$-admissible family of equivalence relation on $\mathbf{S}$ and R an admissible family of equivalence relations on $\mathbf{S}$, which is a final refinement of R^+.

To prove the proposition it suffices to show that for every $\mathbf{L} \in \mathrm{C}^\diamondsuit(\mathrm{R})$,

$$(\mathrm{B}^{\mathbf{L}}_{\diamondsuit,\theta_{\mathbf{A}}(\tau)}, \mathrm{T}(\mathbf{L})) \in \mathbb{P},$$

that is the base $\mathrm{B}^{\mathbf{L}}_{\diamondsuit,\theta_{\mathbf{A}}(\tau)}$ is contained in a normal base B for the open subsets of $\mathrm{T}(\mathbf{L})$. But this follows immediately by the fact that

$$\mathrm{B}^{\mathbf{L}}_{\diamondsuit,\theta_{\mathbf{A}}(\tau)} \subset \mathrm{B}^{\mathbf{L}}_{\diamondsuit,\theta(\tau)} \subset \mathrm{B}^{\mathbf{L},\diamondsuit}_{\theta(\tau)}$$

and Lemma 5.1.4. As the base B we can take the base $B^{T,\diamondsuit}_{\theta(\tau)}$. □

Propositions 5.1.1 and 3.5.5 imply the following consequence.

5.1.6 Corollary. *The class consisting of all completely regular spaces is saturated.* □

5.1.7 Corollary. *In the class of all completely regular spaces there exist universal elements.* □

5.2 The base dimension-like function b^n-Ind

Definition. Let C be a normal base of cardinality $\leq \tau$ for the closed subsets of a space X. The base C is said to be (b^n-Ind ≤ -1)-*dimensional* if $X = \emptyset$. By induction, the base C is said to be (b^n-Ind $\leq \alpha$)-*dimensional*, where $\alpha \in \mathcal{O}$, if for every pair (G, K) of disjoint elements of C there exists a screening (L, H) of (G, K) by elements of C such that the normal base

$$\{(L \cap H) \cap F : F \in C\}$$

for the closed subsets of the space $L \cap H$ is (b^n-Ind $\leq \beta$)-dimensional, where $\beta \in \{-1\} \cup \alpha$. (We note that the ordinal β depends on the screening (L, H).)

Let B be a normal base of cardinality $\leq \tau$ for the open subsets of a space X. The base B is said to be (b^n-Ind $\leq \alpha$)-*dimensional* if the base for the closed subsets of X, which is complementary to B, is (b^n-Ind $\leq \alpha$)-dimensional.

Below, we give some auxiliary notions and notations needed for some special indications of the (b^n-Ind $\leq \alpha$)-dimensional normal bases.

Notation. We denote by $\overline{\tau}$ the set

$$\{\emptyset\} \cup (\cup\{(\tau^2)^k : k \in \omega \setminus \{0\}\}).$$

For an element $\psi \equiv ((\varepsilon_1, \eta_1), ..., (\varepsilon_n, \eta_n))$ of $\overline{\tau}$ it is assumed that $\psi = \emptyset$ if $n = 0$. Let

$$\psi \equiv ((\varepsilon_1, \eta_1), ..., (\varepsilon_n, \eta_n)) \text{ and } \chi \equiv ((\varepsilon'_1, \eta'_1), ..., (\varepsilon'_k, \eta'_k))$$

be two elements of $\overline{\tau}$. We denote by $\chi\psi$ the element

$$((\varepsilon_1', \eta_1'), ..., (\varepsilon_k', \eta_k'), (\varepsilon_1, \eta_1), ..., (\varepsilon_n, \eta_n))$$

of $\overline{\tau}$. It is supposed that $\chi\psi = \psi$ if $\chi = \emptyset$ and $\chi\psi = \chi$ if $\psi = \emptyset$.

Definition. An indexed set

$$\Theta \equiv \{\vartheta_\psi : \psi \in \Psi\}$$

of mappings of τ^2 into itself, where Ψ is a subset of $\overline{\tau}$, is said to be a *branching family of mappings* if:

(a) the empty set is an element of Ψ and

(b) an element

$$((\varepsilon_1, \eta_1), ..., (\varepsilon_k, \eta_k), (\varepsilon, \eta)) \in (\tau^2)^{k+1}, \text{ where } k \in \omega,$$

belongs to Ψ if and only if $\psi \equiv ((\varepsilon_1, \eta_1), ..., (\varepsilon_k, \eta_k))$ is an element of Ψ and (ε, η) belongs to the set $\vartheta_\psi(\tau^2)$.

Definition. Let $\Theta \equiv \{\vartheta_\psi : \psi \in \Psi\}$ be a branching family of mappings. A mapping

$$\Delta : \Psi \to \{-1\} \cup \mathcal{O}$$

is said to be an *expansion* of Θ, if for every element

$$((\varepsilon_1, \eta_1), ..., (\varepsilon_k, \eta_k), (\varepsilon, \eta)) \in \Psi, \ k \in \omega,$$

the condition

$$\Delta((\varepsilon_1, \eta_1), ..., (\varepsilon_k, \eta_k)) = \beta$$

implies that

$$\Delta((\varepsilon_1, \eta_1), ..., (\varepsilon_k, \eta_k), (\varepsilon, \eta)) \in \{-1\} \cup \beta.$$

(By the definition of a branching family of mappings, $((\varepsilon_1, \eta_1), ..., (\varepsilon_k, \eta_k))$ is an element of Ψ.)

An expansion Δ of Θ is said to be an *α-expansion*, where $\alpha \in \{-1\} \cup \mathcal{O}$, if $\Delta(\emptyset) = \alpha$.

Notation. Let $\Theta \equiv \{\vartheta_\psi : \psi \in \Psi\}$ be a branching family of mappings, Δ an expansion of Θ, and χ an element of Ψ.

We denote by Ψ^χ the subset of $\overline{\mathcal{T}}$ consisting of all elements ψ such that $\chi\psi \in \Psi$ and put

$$\Theta^\chi \equiv \{\vartheta_{\chi\psi} : \psi \in \Psi^\chi\}.$$

Also we denote by Δ^χ the mapping of Ψ^χ into $\{-1\} \cup \mathcal{O}$ defined by the relation

$$\Delta^\chi(\psi) = \Delta(\chi\psi).$$

It is easy to verify the following lemma.

5.2.1 Lemma. *Let*

$$\Theta \equiv \{\vartheta_\psi : \psi \in \Psi\}$$

be a branching family of mappings. Then, for every $\chi \in \Psi$ the indexed set

$$\Theta^\chi \equiv \{\vartheta_{\chi\psi} : \psi \in \Psi^\chi\}$$

is also a branching family of mappings. Moreover, if Δ is an expansion of Θ, then the mapping Δ^χ is an expansion of Θ^χ. $\square$

Notation. Let $N \equiv \{C_\delta : \delta \in \tau\}$ be an indexed set of subsets of a space X. For every element $\psi \equiv ((\varepsilon_1, \eta_1), ..., (\varepsilon_k, \eta_k))$ of $\overline{\mathcal{T}}$ we denote by X_ψ^N the space X if $k = 0$ (that is, if $\psi = \emptyset$) and the subset

$$\cap\{C_{\varepsilon_i} \cap C_{\eta_i} : i \in \{1, ..., k\}\}$$

of X if $k \in \omega \setminus \{0\}$.

Definition. Any triad $(\vartheta_\vee, \vartheta_\wedge, \Theta)$, where $\vartheta_\vee$ and $\vartheta_\wedge$ are mappings of $\mathcal{F} \setminus \{\emptyset\}$ into τ and Θ is a branching family of mappings, is called an *admissible branching triad.*

Definitions. Let $(\vartheta_\vee, \vartheta_\wedge, \Theta)$ be an admissible branching triad, and Δ an expansion of Θ. Suppose that C is a normal base of cardinality $\leq \tau$ for the closed subsets of a space X and $N \equiv \{C_\delta : \delta \in \tau\}$ is an indication of this base.

The indication N of C is said to be $(\vartheta_\vee, \vartheta_\wedge, \Theta)$-*normal* if for every element ψ of Ψ the indication

$$\{X_\psi^N \cap C_\delta : \delta \in \tau\}$$

of the base

$$\{X_\psi^N \cap F : F \in C\}$$

for the closed subsets of the space X_ψ^N is $(\vartheta_\vee, \vartheta_\wedge, \vartheta_\psi)$-normal. (About the notion of a $(\vartheta_\vee, \vartheta_\wedge, \vartheta_\psi)$-normal indication see Section 5.1.)

The indication N of the base C is said to be $(\vartheta_\vee, \vartheta_\wedge, \Theta, \Delta)$-*normal* if N is $(\vartheta_\vee, \vartheta_\wedge, \Theta)$-normal and for every element ψ of Ψ the condition $\Delta(\psi) = 0$ implies $X_\psi^N = \emptyset$.

It is easy to verify the following lemma.

5.2.2 Lemma. *Let $(\vartheta_\vee, \vartheta_\wedge, \Theta)$ be an admissible branching triad, where*

$$\Theta = \{\vartheta_\psi : \psi \in \Psi\},$$

and Δ an expansion of Θ. If

$$N \equiv \{C_\delta : \delta \in \tau\}$$

is a $(\vartheta_\vee, \vartheta_\wedge, \Theta, \Delta)$-normal indication of a base C for the closed subsets of a space X, then for every $\chi \in \Psi$ the indication

$$\{X_\chi^N \cap C_\delta : \delta \in \tau\}$$

of $\{X_\chi^N \cap F : F \in C\}$ is $(\vartheta_\vee, \vartheta_\wedge, \Theta^\chi, \Delta^\chi)$-normal. $\square$

The following lemma, which is an analog to Lemma 5.1.2, can easily be proved by induction on k. (The sets $\mathcal{F}_k$ and P_k using in this lemma were introduced in Section 1.)

5.2.3 Lemma. *For every $k \in \omega$ there exist:*
(a) *two one-to-one mappings $\vartheta_{\vee,k}$ and $\vartheta_{\wedge,k}$ of $\mathcal{F}_k$ into the set $\{k+1\} \times \tau$,*
(b) *a set Ψ_k consisting of m-tuples of elements of the set*

$$((k+1) \times \tau)^2, 0 \le m \le k,$$

(It is supposed that a 0-tuple is the empty set and, therefore, $\Psi_0 = \{\emptyset\}$.*),*

(c) *a mapping* $\vartheta_{\psi,k}$, $\psi \in \Psi_k$, *that has as domain the set* $((k+1) \times \tau)^2$ *and as range the set* $(\{k+1\} \times \tau)^2$, *and*

(d) *a mapping*

$$\vartheta_{\psi,k},\ \psi \in \cup\{\Psi_j : j \in k\},$$

that has as domain the set P_k *and as range the set* $(\{k+1\} \times \tau)^2$ *such that the following conditions are satisfied:*

(1) *For every* $\psi \in \cup\{\Psi_j : j \in k+1\}$ *the mappings* $\pi_1^k \circ \vartheta_{\psi,k}$ *and* $\pi_2^k \circ \vartheta_{\psi,k}$ *are one-to-one, where* π_1^k *and* π_2^k *are respectively the first and the second projection of the set* $(\{k+1\} \times \tau)^2$ *onto the set* $\{k+1\} \times \tau$.

(2) *The sets* $\mathrm{I}(\vartheta_{\vee,k})$, $\mathrm{I}(\vartheta_{\wedge,k})$, $\mathrm{I}(\pi_1^k \circ \vartheta_{\psi,k})$ *and* $\mathrm{I}(\pi_2^k \circ \vartheta_{\psi,k})$ *(considered simultaneously for all* ψ*) are mutually disjoint, where* $\psi \in \cup\{\Psi_j : j \in k+1\}$.

(3) *An* m*-tuple*

$$(((i_1, \delta_1), (i'_1, \delta'_1)), \ldots, ((i_m, \delta_m), (i'_m, \delta'_m)))$$

of elements of

$$((k+1) \times \tau)^2,\ k \in \omega \setminus \{0\},\ 1 \le m \le k,$$

belongs to Ψ_k *if and only if*

$$\chi \equiv (((i_1, \delta_1), (i'_1, \delta'_1)), \ldots, ((i_{m-1}, \delta_{m-1}), (i'_{m-1}, \delta'_{m-1})))$$

belongs to $\cup\{\Psi_j : j \in k\}$ *and the pair* $((i_m, \delta_m), (i'_m, \delta'_m))$ *belongs to the set* $\mathrm{I}(\vartheta_{\chi,k-1})$. *(It is supposed that* $\chi = \emptyset$ *if* $m = 1$.*)* □

Remarks. 1. In the above lemma, the domains of all mappings $\vartheta_{\vee,0}$, $\vartheta_{\vee,1}$,... are mutually disjoint and their union is the set of all non-empty finite subsets of $\omega \times \tau$. The same is true for the mappings $\vartheta_{\wedge,0}$, $\vartheta_{\wedge,1}$,....

2. For every $\psi \in \Psi_k$, $k \in \omega$, the domains of all mappings $\vartheta_{\psi,k}$, $\vartheta_{\psi,k+1}$, ... are mutually disjoint and their union is the set $(\omega \times \tau)^2$.

3. By condition (3) of Lemma 5.2.3 it follows that if

$$(((i_1, \delta_1), (i'_1, \delta'_1)), \ldots, ((i_m, \delta_m), (i'_m, \delta'_m))) \in \Psi_k,\ k \in \omega \setminus \{0\},$$

then $i_1 = i'_1 < \ldots < i_m = i'_m = k$.

5.2.4 Lemma. *Let* D^X *be a base for the closed subsets of a space* X,

$$K^X \equiv \{K^X_\delta : \delta \in \tau\}$$

an indication of D^X, C^X *a normal base for the closed subsets of* X *containing the base* D^X, *and* $\alpha \in \{-1\} \cup \mathcal{O}$. *Then,* C^X *is* (b^n-Ind $\leq \alpha$)-*dimensional if and only if for some admissible branching triad* $(\vartheta_\vee, \vartheta_\wedge, \Theta)$, *an* α-*expansion* Δ *of* Θ, *and a mapping* $\overline{\theta}$ *of the set* τ *into itself there exists a* $(\vartheta_\vee, \vartheta_\wedge, \Theta, \Delta)$-*normal indication*

$$F^X \equiv \{F^X_\delta : \delta \in \tau\}$$

of C^X *such that* $K^X_\delta = F^X_{\overline{\theta}(\delta)}$ *for every* $\delta \in \tau$. *Moreover, if* $\alpha \in \{-1\} \cup \omega$, *then the triad* $(\vartheta_\vee, \vartheta_\wedge, \Theta)$, *the* α-*expansion* Δ, *and the mapping* $\overline{\theta}$ *can be considered to be independent of the space* X.

Proof. Suppose that for some admissible branching triad $(\vartheta_\vee, \vartheta_\wedge, \Theta)$, some α-expansion Δ of Θ, and some mapping $\overline{\theta}$ of τ into itself,

$$F^X \equiv \{F^X_\delta : \delta \in \tau\}$$

is a $(\vartheta_\vee, \vartheta_\wedge, \Theta, \Delta)$-normal indication of C^X such that $K^X_\delta = F^X_{\overline{\theta}(\delta)}$, $\delta \in \tau$. We prove by induction on α that C^X is (b^n-Ind $\leq \alpha$)-dimensional.

Let $\alpha = -1$. Then, $\Delta(\emptyset) = -1$. Since F^X is $(\vartheta_\vee, \vartheta_\wedge, \Theta, \Delta)$-normal, $X_\emptyset = X = \emptyset$. This fact means that C^X is (b^n-Ind ≤ -1)-dimensional.

Supposing that our suggestion is true for all $\beta \in \{-1\} \cup \alpha$ we prove it for the ordinal α. Let (G, K) be a pair of disjoint elements of C^X. There exists an element $(\varepsilon', \eta') \in \tau^2$ such that $G = F^X_{\varepsilon'}$ and $K = F^X_{\eta'}$. Set $\psi \equiv (\varepsilon, \eta) = \vartheta_\emptyset(\varepsilon', \eta')$. Consider the set $X^{F^X}_\psi = F^X_\varepsilon \cap F^X_\eta$. By Lemma 5.2.2, the indication

$$\{X^{F^X}_\psi \cap F^X_\delta : \delta \in \tau\}$$

of the base

$$\{X^{F^X}_\psi \cap F : F \in C^X\}$$

for $X^{F^X}_\psi$ is $(\vartheta_\vee, \vartheta_\wedge, \Theta^\psi, \Delta^\psi)$-normal. Since Δ is an expansion of Θ we have

$$\Delta^\psi(\emptyset) = \Delta(\psi) \equiv \beta \in \{-1\} \cup \alpha.$$

By inductive assumption, the indication

$$\{X_{\psi}^{F^X} \cap F_{\delta}^X : \delta \in \tau\}$$

of the set

$$\{X_{\psi}^{F^X} \cap F : F \in C^X\}$$

is (b^n-Ind $\leq \beta$)-dimensional, which proves that the indication F^X of C^X is (b^n-Ind $\leq \alpha$)-dimensional.

Conversely, suppose that C^X is (b^n-Ind $\leq \alpha$)-dimensional. We prove that for some admissible branching triad $(\vartheta_\vee, \vartheta_\wedge, \Theta)$, some α-expansion Δ of Θ, and some mapping $\overline{\theta}$ of τ into itself there exists a $(\vartheta_\vee, \vartheta_\wedge, \Theta, \Delta)$-normal indication

$$F^X \equiv \{F_{\delta}^X : \delta \in \tau\}$$

of C^X such that $K_{\delta}^X = F_{\overline{\theta}(\delta)}^X$, $\delta \in \tau$.

First, we shall construct the corresponding admissible branching triad $(\vartheta_\vee, \vartheta_\wedge, \Theta)$. Denote by ϖ an one-to-one mapping of $\omega \times \tau$ onto τ such that $\varpi(0,0) = 0$ and $\varpi(0,1) = 1$ and consider the sets and mappings constructed in Lemma 5.2.3. The mappings $\vartheta_\vee$ and $\vartheta_\wedge$ are defined as in Lemma 5.1.3.

The indexing set Ψ of the branching family Θ is defined as follows: an element $((\varepsilon_1, \eta_1), ..., (\varepsilon_m, \eta_m))$ of $\overline{\tau}$ belongs to Ψ if and only if either $m = 0$ or $m \neq 0$ and the element

$$((\varpi^{-1}(\varepsilon_1), \varpi^{-1}(\eta_1)), ..., (\varpi^{-1}(\varepsilon_m), \varpi^{-1}(\eta_m))$$

of τ belongs to $\cup\{\Psi_j : j \in \omega\}$.

For every $\psi \in \Psi$ the corresponding element ϑ_ψ of Θ is defined as follows. Let

$$\psi = ((\varepsilon_1, \eta_1), ..., (\varepsilon_m, \eta_m)) \text{ and}$$

$$\chi = (((\varpi^{-1}(\varepsilon_1), \varpi^{-1}(\eta_1)), ..., (\varpi^{-1}(\varepsilon_m), \varpi^{-1}(\eta_m))).$$

There exists $k \in \omega$ such that $\chi \in \Psi_k$. Consider an element (δ_0, δ_1) of τ^2. Suppose that $\varpi^{-1}(\delta_0) = (i_0, \eta_0)$ and $\varpi^{-1}(\delta_1) = (i_1, \eta_1)$ and let $j = \max\{i_0, i_1\}$. If $j \leq k$, then we set

$$\vartheta_{\psi,k}((i_0,\eta_0),(i_1,\eta_1)) = ((k+1,\eta^0),(k+1,\eta^1)),$$

$$\varpi(k+1,\eta^0) = \varepsilon_0, \text{ and } \varpi(k+1,\eta^1) = \varepsilon_1.$$

If $k < j$, then we set

$$\vartheta_{\psi,j}((i_0,\eta_0),(i_1,\eta_1)) = ((j+1,\eta^0),(j+1,\eta^1)),$$

$$\varpi(j+1,\eta^0) = \varepsilon_0, \text{ and } \varpi(j+1,\eta^1) = \varepsilon_1.$$

In both cases we set $\vartheta_\psi(\delta_0,\delta_1) = (\varepsilon_0,\varepsilon_1)$. It is easy to verify that the constructed family $\Theta \equiv \{\vartheta_\psi : \psi \in \Psi\}$ is a branching family of mappings.

The mapping $\overline{\theta}$ is defined as follows. Let θ' be an arbitrary mapping of τ into the set $\omega \times \tau$ such that

$$\theta'(\tau) \subset (\{0\} \times \tau) \setminus \{(0,0),(0,1)\} \text{ and}$$

$$|\tau \setminus \theta'(\tau)| \leq \tau.$$

Then, we set $\overline{\theta} = \varpi \circ \theta'$.

In order to construct the required indication of C^X we first construct indexed sets $\{C^k_\delta : \delta \in \tau\}$, $k \in \omega$, consisting of elements of C^X and satisfying some conditions. For the formulation of these conditions we introduce the following notation. For every

$$\chi \equiv (((i_1,\delta_1),(i_1,\delta'_1)), ..., ((i_m,\delta_m),(i_m,\delta'_m))) \in \Psi_k$$

we set $X_\chi = X$ if $k = 0$ (therefore, $\chi = \emptyset$) and

$$X_\chi = \cap\{C^{i_j}_{\delta_j} \cap C^{i_j}_{\delta'_j} : j \in \{1,...,m\}\} \text{ if } k \neq 0.$$

Now, we give the corresponding conditions.

(1) The indexed set $\{C^0_\delta : \delta \in \tau\}$ is an indication of C^X such that $C^0_0 = X$, $C^0_1 = \emptyset$ and $K^X_\delta = C^0_{\theta'(\delta)}$, if $\theta'(\delta) = (0,\delta)$, $\delta \in \tau$.

(2) If $\{(i_0,\delta_0),...,(i_m,\delta_m)\}$ is an element of $\mathcal{F}_k$ and

$$(k+1,\varepsilon) = \vartheta_{\vee,k}(\{(i_0,\delta_0),...,(i_m,\delta_m)\}),$$

then $C^{k+1}_\varepsilon = C^{i_0}_{\delta_0} \cup ... \cup C^{i_m}_{\delta_m}$.

(3) If $\{(i_0,\delta_0),...,(i_m,\delta_m)\}$ is an element of $\mathcal{F}_k$ and

$$(k+1,\varepsilon) = \vartheta_{\wedge,k}(\{(i_0,\delta_0),...,(i_m,\delta_m)\}),$$

then $C_\varepsilon^{k+1} = C_{\delta_0}^{i_0} \cap ... \cap C_{\delta_m}^{i_m}$.

(4) If $((i,\delta),(i',\delta'))$ is an element of the domain of a mapping $\vartheta_{\chi,k}$, where

$$\chi \in \cup\{\Psi_j : j \in k+1\}, \text{ and}$$

$$((k+1,\varepsilon),(k+1,\varepsilon')) = \vartheta_{\chi,k}((i,\delta),(i',\delta')),$$

then:

(a) the pair $(C_\varepsilon^{k+1}, C_{\varepsilon'}^{k+1})$ is a screening of the pair $(X_\chi \cap C_\delta^i, X_\chi \cap C_{\delta'}^{i'})$ in the space X_χ if $(X_\chi \cap C_\delta^i) \cap (X_\chi \cap C_{\delta'}^{i'}) = \emptyset$ and

(b) $(C_\varepsilon^{k+1}, C_{\varepsilon'}^{k+1}) = (\emptyset,\emptyset)$ if $(X_\chi \cap C_\delta^i) \cap (X_\chi \cap C_{\delta'}^{i'}) \neq \emptyset$.

(5) For every element

$$\psi \equiv (((i_1,\delta_1),(i_1,\delta_1')),...,((i_m,\delta_m),(i_m,\delta_m')))$$

of the set

$$\cup\{\Psi_j : j \in k+1\},$$

where $0 \leq m \leq k$, there exists an ordinal α_ψ such that:

(a) if $m = 0$, then $\alpha_\psi = \alpha$,

(b) if $1 \leq m$, then $\alpha_\psi \in \alpha_\chi$, where

$$\chi = (((i_1,\delta_1),(i_1,\delta_1')),...,((i_{m-1},\delta_{m-1}),(i_{m-1},\delta_{m-1}')))$$

(therefore, $\chi = \emptyset$ if $m = 1$), and

(c) the set

$$\{X_\psi \cap F : F \in C^X\}$$

is a (b^n-Ind $\leq \alpha_\psi$)-dimensional base for the closed subsets of X_ψ.

The construction of the indexed sets $\{C_\delta^k : \delta \in \tau\}$, $k \in \omega$, can be done by induction on k as follows: For $k = 0$ the indexed set $\{C_\delta^0 : \delta \in \tau\}$ is determined by condition (1). Suppose that the indexed sets $\{C_\delta^j : \delta \in \tau\}$ are constructed for all $j \in k+1$ such that condition (5) is satisfied and conditions (2), (3), and (4) concerning the mappings $\vartheta_{\vee,k}$, $\vartheta_{\wedge,k}$, and $\vartheta_{\psi,k}$, respectively, are satisfied if $k > 0$ is replaced by $k-1$. We construct the indexed set $\{C_\delta^{k+1} : \delta \in \tau\}$ such that conditions (2), (3), (4) are satisfied and condition (5) is satisfied if k is replaced by $k+1$.

For an element ε of τ, for which the pair $(k+1, \varepsilon)$ belongs to the set $\mathrm{I}(\vartheta_{\vee,k})$ or to the set $\mathrm{I}(\vartheta_{\wedge,k})$, the set C_ε^{k+1} is determined by condition (2) or by condition (3), respectively.

Suppose that ε and ε' are two elements of τ such that

$$((k+1,\varepsilon),(k+1,\varepsilon')) = \vartheta_{\chi,k}((i,\delta),(i',\delta'))$$

for some $\chi \in \cup\{\Psi_j : j \in k+1\}$. Let

$$\chi = (((i_1,\delta_1),(i_1,\delta_1')),\dots,((i_m,\delta_m),(i_m,\delta_m'))),\ 0 \le m \le k.$$

By assumption, the set $\{X_\chi \cap F : F \in C^X\}$ is an $(\mathrm{b}^n\text{-Ind} \le \alpha_\chi)$-dimensional base for the closed subsets of X_χ. Therefore, if the sets $X_\chi \cap C_\delta^i$ and $X_\chi \cap C_{\delta'}^{i'}$ are disjoint, then there exists two elements G and K of C^X such that the pair $(X_\chi \cap G, X_\chi \cap K)$ is a screening of the pair $(X_\chi \cap C_\delta^i, X_\chi \cap C_{\delta'}^{i'})$ in the space X_χ and the set

$$\{(X_\chi \cap (G \cap K)) \cap F : F \in C^X\}$$

is a $(\mathrm{b}^n\text{-Ind} \le \beta)$-dimensional base for the closed subsets of the space $X_\chi \cap (G \cap K)$, where $\beta \in \alpha_\chi$. We set $C_\varepsilon^{k+1} = G$, $C_{\varepsilon'}^{k+1} = K$, and $\alpha_\psi = \beta$, where

$$\psi = (((i_1,\delta_1),(i_1,\delta_1')),\dots,((i_m,\delta_m),(i_m,\delta_m')),((k+1,\varepsilon),(k+1,\varepsilon'))).$$

If $(X_\chi \cap C_\delta^i) \cap (X_\chi \cap C_{\delta'}^{i'}) \neq \emptyset$, then we set $C_\varepsilon^{k+1} = \emptyset$, $C_{\varepsilon'}^{k+1} = \emptyset$, and $\alpha_\psi = -1$, where ψ is the above.

For an element ε of τ, which does not belong to the sets $\mathrm{I}(\vartheta_{\vee,k})$, $\mathrm{I}(\vartheta_{\wedge,k})$, $\mathrm{I}(\pi_1 \circ \vartheta_{\psi,k})$, and $\mathrm{I}(\pi_2 \circ \vartheta_{\psi,k})$, where $\psi \in \cup\{\Psi_j : j \in k+1\}$, the set C_ε^{k+1} is an arbitrary element of C^X.

The required indication $\{F_\delta^X : \delta \in \tau\}$ of C^X is defined as follows. For every $\delta \in \tau$ we set $F_\delta^X = C_\varepsilon^k$, where $(k,\varepsilon) = \varpi^{-1}(\delta)$.

Finally, we define the corresponding α-expansion Δ of Θ. Let

$$\psi \equiv ((\varepsilon_1,\eta_1),\dots,(\varepsilon_m,\eta_m))$$

be an element of Ψ. If $m = 0$, that is, if $\psi = \emptyset$, then we set $\Delta(\psi) = \alpha$. If $0 < m$ and

$$\chi \equiv ((\varpi^{-1}(\varepsilon_1), \varpi^{-1}(\eta_1)), ..., (\varpi^{-1}(\varepsilon_m), \varpi^{-1}(\eta_m)) \in \cup\{\Psi_j : j \in \omega\},$$

then we set $\Delta(\psi) = \alpha_\chi$.

It is not dificult to verify that the constructed indication

$$\{F_\delta^X : \delta \in \tau\}$$

of C^X is $(\vartheta_\vee, \vartheta_\wedge, \Theta, \Delta)$-normal and that $K_\delta^X = F_{\overline{\theta}(\delta)}^X$, $\delta \in \tau$.

Now, suppose that $\alpha \in \{-1\} \cup \omega$. First, we observe that the above construction of the admissible branching triad $(\vartheta_\vee, \vartheta_\wedge, \Theta)$ is independent of the considered space X while the construction of the α-expansion Δ depends on this space. However, for the considered case, the α-expansion Δ of Θ can be defined as follows: for an element $\psi \equiv ((\varepsilon_1, \eta_1), ..., (\varepsilon_m, \eta_m))$ of Ψ we set $\Delta(\psi) = \alpha$ if $m = 0$ and $\Delta(\psi) = \max\{-1, \alpha - m\}$ if $0 < m$. This construction of Δ is independent of the considered space. The proof of the lemma is completed. □

The following lemma can easily be proved by induction.

5.2.5 Lemma. *Let X be a space and C a normal base for the closed subsets of X. Let also $G, G_1, ..., G_n$ be elements of C, $n \in \omega \setminus \{0\}$, such that the sets $G_1, ..., G_n$ are mutually disjoint and*

$$G = \cup\{G_i : i \in n + 1\}.$$

Then, the normal base $\{G \cap F : F \in C\}$ for the closed subsets of the space G is (b^n-Ind $\leq \alpha$)-dimensional, where $\alpha \in \{-1\} \cup \mathcal{O}$, if and only if for every $i \in n+1$ the normal base $\{G_i \cap F : F \in C\}$ for the closed subsets of the space G_i is (b^n-Ind $\leq \alpha$)-dimensional. □

Below we extend some of the preceding definitions concerning a space to an indexed collection of spaces.

Definition. Let $\mathbf{S}$ be an indexed collection of spaces. A normal closed co-base $\mathbf{C} \equiv \{C^X : X \in \mathbf{S}\}$ for $\mathbf{S}$ is said to be *(b^n-Ind $\leq \alpha$)-dimensional,* where $\alpha \in \{-1\} \cup \mathcal{O}$, if for every $X \in \mathbf{S}$, C^X is (b^n-Ind $\leq \alpha$)-dimensional.

A normal open co-base $\mathbf{B}$ for $\mathbf{S}$ is said to be *(b^n-Ind $\leq \alpha$)-dimensional,* where $\alpha \in \{-1\} \cup \mathcal{O}$, if the comlementary to $\mathbf{B}$ closed co-base for $\mathbf{S}$ is (b^n-Ind $\leq \alpha$)-dimensional.

Definition. Let $\mathbf{S}$ be an indexed collection of spaces, $\mathbf{C}$ a normal co-base for $\mathbf{S}$, and $\mathbf{B}$ the complementary to $\mathbf{C}$ open co-base for $\mathbf{S}$. Let also $(\vartheta_\vee, \vartheta_\wedge, \Theta)$ be an admissible branching triad and Δ an expansion of Θ.

A co-indication $\mathbf{F}$ of $\mathbf{C}$ is said to be $(\vartheta_\vee, \vartheta_\wedge, \Theta, \Delta)$-*normal* if for every $X \in \mathbf{S}$, $\mathbf{F}(X)$ is a $(\vartheta_\vee, \vartheta_\wedge, \Theta, \Delta)$-normal indication of $\mathbf{C}(X)$.

A co-indication $\mathbf{N}$ of $\mathbf{B}$ is said to be $(\vartheta_\vee, \vartheta_\wedge, \Theta, \Delta)$-*normal* if the complementary to $\mathbf{N}$ co-indication of $\mathbf{C}$ is $(\vartheta_\vee, \vartheta_\wedge, \Theta, \Delta)$-normal.

5.2.6 Lemma. *Let $\mathbf{S}$ be an indexed collection of spaces, $\mathbf{B}$ a normal open co-base for $\mathbf{S}$, and $\mathbf{N}$ a $(\vartheta_\vee, \vartheta_\wedge, \Theta, \Delta)$-normal co-indication of $\mathbf{B}$, where $(\vartheta_\vee, \vartheta_\wedge, \Theta)$ is an admissible branching triad and Δ is an α-expansion of Θ, $\alpha \in \{-1\} \cup \mathcal{O}$. Then, there exists a co-mark $\mathbf{M}^+$ of $\mathbf{S}$, which is a co-extension of $\mathbf{N}$, satisfying the following condition: for every co-extension $\mathbf{M}$ of $\mathbf{M}^+$ there exists an $\mathbf{M}$-admissible family R^+ of equivalence relations on $\mathbf{S}$ such that for every admissible family R of equivalence relations on $\mathbf{S}$, which is a final refinement of R^+, the set $\mathrm{C}^{\mathrm{T},\diamondsuit}_{\theta(\tau)}$ is a $(\mathrm{b}^n\text{-Ind} \le \alpha)$-dimensional normal base for the closed subsets of T, where θ is an indicial mapping from $\mathbf{N}$ to $\mathbf{M}$.*

Proof. We prove the lemma by induction on α. The Lemma is trivial if $\alpha = -1$. Supposing that the lemma is true for every element of $\{-1\} \cup \mathcal{O}$, which is less than $\alpha \in \mathcal{O}$, we prove it for α. Let

$$\mathbf{B} = \{B^X : X \in \mathbf{S}\} \text{ and}$$

$$\mathbf{N} = \{\{V^X_\varepsilon : \varepsilon \in \tau\} : X \in \mathbf{S}\}.$$

Denote by

$$\mathbf{C} \equiv \{C^X : X \in \mathbf{S}\}$$

the complementary to $\mathbf{B}$ closed co-base for $\mathbf{S}$ and by

$$\mathbf{F} \equiv \{\{C^X_\varepsilon : \varepsilon \in \tau\} : X \in \mathbf{S}\}$$

the complementary to $\mathbf{N}$ co-indication of $\mathbf{C}$. Therefore, C^X is the complementary to B^X base for the closed subsets of X and $C^X_\varepsilon = X \setminus V^X_\varepsilon$ for every $X \in \mathbf{S}$ and $\varepsilon \in \tau$.

Suppose that $\Theta = \{\vartheta_\psi : \psi \in \Psi\}$. For every $\psi \equiv (\varepsilon_0, \varepsilon_1) \in \vartheta_\emptyset(\tau^2)$ we set

$$X_\psi = C^X_{\varepsilon_0} \cap C^X_{\varepsilon_1},$$

$$B^X_\psi = \{X_\psi \cap V : V \in B^X\},$$

$$N^X_\psi = \{X_\psi \cap V^X_\varepsilon : \varepsilon \in \tau\},$$

$$C^X_\psi = \{X_\psi \cap C : C \in C^X\},$$

$$F^X_\psi = \{X_\psi \cap C^X_\varepsilon : \varepsilon \in \tau\},$$

$$\mathbf{S}_\psi = \{X_\psi : X \in \mathbf{S}\},$$

$$\mathbf{B}_\psi = \{B^X_\psi : X \in \mathbf{S}\},$$

$$\mathbf{N}_\psi = \{N^X_\psi : X \in \mathbf{S}\},$$

$$\mathbf{C}_\psi = \{C^X_\psi : X \in \mathbf{S}\}, \text{ and}$$

$$\mathbf{F}_\psi = \{F^X_\psi : X \in \mathbf{S}\}.$$

We note that:

(a) $\mathbf{S}_\psi$ is a closed restriction of $\mathbf{S}$;

(b) $\mathbf{B}_\psi$ is a (b^n-Ind $\leq \alpha_\psi$)-dimensional open co-base for $\mathbf{S}_\psi$, where $\alpha_\psi = \Delta(\psi) \in \alpha$;

(c) $\mathbf{N}_\psi$ is a $(\vartheta_\vee, \vartheta_\wedge, \Theta^\psi, \Delta^\psi)$-normal co-indication of $\mathbf{B}_\psi$ (see Lemma 5.2.2);

(d) $\mathbf{C}_\psi$ is the complementary to $\mathbf{B}_\psi$ closed co-base for $\mathbf{S}_\psi$ and

(e) $\mathbf{F}_\psi$ is the complementary to $\mathbf{N}_\psi$ co-indication of $\mathbf{C}_\psi$.

By inductive assumption, there exists a co-mark $\mathbf{M}^+_\psi$ of $\mathbf{S}_\psi$, which is a co-extension of $\mathbf{N}_\psi$, satisfying the following condition: for every co-extension $\mathbf{M}_\psi$ of $\mathbf{M}^+_\psi$ there exists an $\mathbf{M}_\psi$-admissible family R^+_ψ of equivalence relations on $\mathbf{S}_\psi$ such that for every admissible family R_ψ of equivalence relations on $\mathbf{S}_\psi$, which is a final refinement of the family R^+_ψ, the $(\theta_\psi(\tau), \diamondsuit)$-standard base $\mathrm{C}^{\mathrm{T}_\psi, \diamondsuit}_{\theta_\psi(\tau)}$ for the closed subsets of the Containing Space

$$\mathrm{T}_\psi \equiv \mathrm{T}(\mathbf{M}_\psi, \mathrm{R}_\psi)$$

is a (b^n-Ind $\leq \alpha_\psi$)-dimensional normal base for T_ψ, where θ_ψ is an indicial mapping from $\mathbf{N}_\psi$ to $\mathbf{M}_\psi$. Moreover, by Lemma 5.2.5 for every $\mathbf{L}_\psi \in \mathrm{C}^\diamondsuit(\mathrm{R}_\psi)$ the set

$$\mathrm{C}^{\mathbf{L}_\psi,\diamondsuit}_{\theta_\psi(\tau)} = \{\mathrm{T}(\mathbf{L}_\psi) \cap C : C \in \mathrm{C}^{\mathrm{T}_\psi,\diamondsuit}_{\theta_\psi(\tau)}\}$$

is a (b^n-Ind $\leq \alpha_\psi$)-dimensional normal base for the closed subsets of the space $\mathrm{T}(\mathbf{L}_\psi)$.

Denote by $\mathbf{M}^+$ a co-mark of $\mathbf{S}$ such that: (a) $\mathbf{M}^+$ is a co-extension of $\mathbf{N}$, and (b) for every $\psi \in \Psi$ the restriction to $\mathbf{S}_\psi$ of $\mathbf{M}^+$ is a co-extension of $\mathbf{M}^+_\psi$. We prove that $\mathbf{M}^+$ satisfies the required condition.

Indeed, let

$$\mathbf{M} \equiv \{\{U^X_\delta : \delta \in \tau\} : X \in \mathbf{S}\}$$

be a co-extension of $\mathbf{M}^+$. Then, $\mathbf{M}|_{\mathbf{S}_\psi}$ (the restriction to $\mathbf{S}_\psi$ of $\mathbf{M}$) is a co-extension of $\mathbf{M}^+_\psi$. Let R^+_ψ be the above considered $\mathbf{M}|_{\mathbf{S}_\psi}$-admissible family of equivalence relations on $\mathbf{S}_\psi$. (As co-mark $\mathbf{M}_\psi$ we consider the co-mark $\mathbf{M}|_{\mathbf{S}_\psi}$). Let $\mathrm{R}^{+,\psi}$ be the admissible family of equivalence relations on $\mathbf{S}$ such that its restriction to $\mathbf{S}_\psi$ is the family R^+_ψ. Denote by R^+ an $\mathbf{M}$-admissible family of equivalence relations on $\mathbf{S}$, which is a final refinement of all families $\mathrm{R}^{+,\psi}$, $\psi \in \Psi$. (Therefore, R^+ is $(\mathbf{M}, \mathbf{S}_\psi)$-admissible for every $\psi \in \Psi$.) Finally, let R be an admissible family of equivalence relations on $\mathbf{S}$, which is a final refinement of R^+.

We must prove that the set $\mathrm{C}^{\mathrm{T},\diamondsuit}_{\theta(\tau)}$ is a (b^n-Ind $\leq \alpha$)-dimensional normal base for the closed subsets of the Containing Space T, where θ is an indicial mapping from $\mathbf{N}$ to $\mathbf{M}$. By Lemma 5.1.4, $\mathrm{C}^{\mathrm{T},\diamondsuit}_{\theta(\tau)}$ is a normal base for the closed subsets of T. Thus, it is suffices to prove that $\mathrm{C}^{\mathrm{T},\diamondsuit}_{\theta(\tau)}$ is (b^n-Ind $\leq \alpha$)-dimensional.

Let (G, K) be a pair of disjoint elements of $\mathrm{C}^{\mathrm{T},\diamondsuit}_{\theta(\tau)}$. As in Lemma 5.1.4, the elements $\mathrm{T} \setminus G$ and $\mathrm{T} \setminus K$ of $\mathrm{B}^{\mathrm{T},\diamondsuit}_{\theta(\tau)}$ has the form:

$$\mathrm{T} \setminus G = U^{\mathrm{T}}_{\delta_0}(\mathbf{L}_0) \cup ... \cup U^{\mathrm{T}}_{\delta_n}(\mathbf{L}_n) \text{ and}$$

$$\mathrm{T} \setminus K = U^{\mathrm{T}}_{\delta'_0}(\mathbf{L}'_0) \cup ... \cup U^{\mathrm{T}}_{\delta'_m}(\mathbf{L}'_m),$$

where $n, m \in \omega$, $\delta_0, ..., \delta_n, \delta'_0, ..., \delta'_m \in \theta(\tau)$, $\mathbf{L}_0, ..., \mathbf{L}_n$ are distinct elements of $\mathrm{C}(\sim^s)$ for some $s \in \mathcal{F}$, and $\mathbf{L}'_0, ..., \mathbf{L}'_m$ are also distinct elements of

$\mathrm{C}(\sim^s)$. Also we can suppose that $\sim^s \subset \sim^q_{\mathbf{M}}$, where q is an element of $\mathcal{F}$ containing the elements $\theta(0), \theta(1), \delta_0, ..., \delta_n, \delta'_0, ..., \delta'_m$ of τ and the images, under the mapping θ, of elements of all pairs $\vartheta_\emptyset(\theta^{-1}(\delta), \theta^{-1}(\delta'))$, where $\delta \in \{\theta(0), \theta(1), \delta_0, ..., \delta_n\}$ and $\delta' \in \{\theta(0), \theta(1), \delta'_0, ..., \delta'_m\}$.

Let $\mathbf{L} \in \mathrm{C}(\sim^s)$. There exist elements $\delta_0^{\mathbf{L}}, \delta_1^{\mathbf{L}}$ of

$$\{\theta(0), \theta(1), \delta_0, ..., \delta_n, \delta'_0, ..., \delta'_m\}$$

such that

$$\mathrm{T}(\mathbf{L}) \cap (\mathrm{T} \setminus G) = U^{\mathrm{T}}_{\delta_0^{\mathbf{L}}}(\mathbf{L}) \text{ and}$$

$$\mathrm{T}(\mathbf{L}) \cap (\mathrm{T} \setminus K) = U^{\mathrm{T}}_{\delta_1^{\mathbf{L}}}(\mathbf{L}).$$

Set $\eta_0^{\mathbf{L}} = \theta^{-1}(\delta_0^{\mathbf{L}})$, $\eta_1^{\mathbf{L}} = \theta^{-1}(\delta_1^{\mathbf{L}})$, and $\vartheta_\emptyset(\eta_0^{\mathbf{L}}, \eta_1^{\mathbf{L}}) = (\varepsilon_0^{\mathbf{L}}, \varepsilon_1^{\mathbf{L}}) \equiv \psi^{\mathbf{L}}$. Then, for every $X \in \mathbf{S}$ we have

$$C^X_{\eta_0^{\mathbf{L}}} = X \setminus V^X_{\eta_0^{\mathbf{L}}} = X \setminus U^X_{\delta_0^{\mathbf{L}}},$$

$$C^X_{\eta_1^{\mathbf{L}}} = X \setminus V^X_{\eta_1^{\mathbf{L}}} = X \setminus U^X_{\delta_1^{\mathbf{L}}},$$

$$C^X_{\varepsilon_0^{\mathbf{L}}} = X \setminus U^X_{\theta(\varepsilon_0^{\mathbf{L}})},$$

$$C^X_{\varepsilon_1^{\mathbf{L}}} = X \setminus U^X_{\theta(\varepsilon_1^{\mathbf{L}})}, \text{ and}$$

$$C^X_{\eta_0^{\mathbf{L}}} \cap C^X_{\eta_1^{\mathbf{L}}} = \emptyset.$$

By properties of the mapping $\vartheta_\emptyset$ the pair $(C^X_{\varepsilon_0^{\mathbf{L}}}, C^X_{\varepsilon_1^{\mathbf{L}}})$ is a screening of the pair $(C^X_{\eta_0^{\mathbf{L}}}, C^X_{\eta_1^{\mathbf{L}}})$. Moreover, the set

$$\{X_{\psi^{\mathbf{L}}} \cap C : C \in C^X\}$$

is a (b^n-Ind $\leq \alpha_{\psi^{\mathbf{L}}}$)-dimensional normal base for the closed subsets of the space $X_{\psi^{\mathbf{L}}}$.

Setting for simplicity of the notations $\psi^{\mathbf{L}} = \psi$ we consider the restriction

$$\mathbf{S}_\psi = \{X_\psi : X \in \mathbf{S}\}$$

of $\mathbf{S}$, the co-mark $\mathbf{M}|_{\mathbf{S}_\psi}$ of $\mathbf{S}_\psi$, and the $\mathbf{M}|_{\mathbf{S}_\psi}$-admissible family $\mathrm{R}|_{\mathbf{S}_\psi}$ of equivalence relations on $\mathbf{S}_\psi$. By the construction of $\mathbf{M}$ and R the Containing Space

$$\mathrm{T}|_{\mathbf{S}_\psi} \equiv \mathrm{T}(\mathbf{M}|_{\mathbf{S}_\psi}, \mathrm{R}|_{\mathbf{S}_\psi})$$

is a subset of T coinciding with the set

$$(\mathrm{T} \setminus U^{\mathrm{T}}_{\theta(\varepsilon_0^{\mathbf{L}})}) \cap (\mathrm{T} \setminus U^{\mathrm{T}}_{\theta(\varepsilon_1^{\mathbf{L}})}).$$

Therefore, if $\mathbf{L}|_{\mathbf{S}_\psi}$ is the restriction to $\mathbf{S}_\psi$ of $\mathbf{L}$, then

$$\mathrm{T}|_{\mathbf{S}_\psi} \cap \mathrm{T}(\mathbf{L}) = \mathrm{T}(\mathbf{L}|_{\mathbf{S}_\psi}) =$$

$$(\mathrm{T}(\mathbf{L}) \setminus U^{\mathrm{T}}_{\theta(\varepsilon_0^{\mathbf{L}})}(\mathbf{L})) \cap (\mathrm{T}(\mathbf{L}) \setminus U^{\mathrm{T}}_{\theta(\varepsilon_1^{\mathbf{L}})}(\mathbf{L})).$$

By inductive assumption, the $(\theta_\psi(\tau), \diamondsuit)$-standard base $\mathrm{C}^{\mathrm{T}|_{\mathbf{S}_\psi}, \diamondsuit}_{\theta_\psi(\tau)}$ of the Containing Space $\mathrm{T}|_{\mathbf{S}_\psi}$ is a $(\mathrm{b}^n\text{-Ind} \leq \alpha_\psi)$-dimensional normal base for the closed subsets of $\mathrm{T}|_{\mathbf{S}_\psi}$. It is easy to verify that

$$\mathrm{C}^{\mathrm{T}|_{\mathbf{S}_\psi}, \diamondsuit}_{\theta_\psi(\tau)} = \{\mathrm{T}|_{\mathbf{S}_\psi} \cap C : C \in \mathrm{C}^{\mathrm{T}, \diamondsuit}_{\theta(\tau)}\}.$$

By Lemma 5.2.5 the set

$$\mathrm{C}^{\mathbf{L}||_{\mathbf{S}_\psi}, \diamondsuit} = \{\mathrm{T}(\mathbf{L}|_{\mathbf{S}_\psi}) \cap C : C \in \mathrm{C}^{\mathrm{T}, \diamondsuit}_{\theta(\tau)}\}$$

is also a $(\mathrm{b}^n\text{-Ind} \leq \alpha_\psi)$-dimensional normal base for the closed subsets of the space $\mathrm{T}(\mathbf{L}|_{\mathbf{S}_\psi})$.

Now, let

$$L = \cup\{\mathrm{T}(\mathbf{L}) \setminus U^{\mathrm{T}}_{\theta(\varepsilon_0^{\mathbf{L}})}(\mathbf{L}) : \mathbf{L} \in \mathrm{C}(\sim^s)\} \text{ and}$$

$$H = \cup\{\mathrm{T}(\mathbf{L}) \setminus U^{\mathrm{T}}_{\theta(\varepsilon_1^{\mathbf{L}})}(\mathbf{L}) : \mathbf{L} \in \mathrm{C}(\sim^s)\}.$$

Obviously L and H are elements of $\mathrm{C}^{\mathrm{T}, \diamondsuit}_{\theta(\tau)}$ and

$$L \cap H = \cup\{\mathrm{T}(\mathbf{L}|_{\mathbf{S}_\psi}) =$$

$$(\mathrm{T}(\mathbf{L}) \setminus U^{\mathrm{T}}_{\theta(\varepsilon^{\mathbf{L}}_0)}(\mathbf{L})) \cap (\mathrm{T}(\mathbf{L}) \setminus U^{\mathrm{T}}_{\theta(\varepsilon^{\mathbf{L}}_1)}(\mathbf{L})) : L \in \mathrm{C}(\sim^s)\}.$$

As in Lemma 5.1.4 we prove that (L, H) is a screening of (G, K). Setting

$$\alpha^{L \cap H} = \max\{\alpha^{\mathbf{L}} : \mathbf{L} \in \mathrm{C}(\sim^s)\},$$

we have that $\alpha^{L\cap H} \in \alpha$ and that the set

$$\{\mathrm{T}(\mathbf{L}|_{\mathbf{s}_\psi}) \cap C : C \in \mathrm{C}^{\mathrm{T},\diamondsuit}_{\theta(\tau)}\}$$

is a $(\mathrm{b}^n\text{-Ind} \leq \alpha^{L\cap H})$-dimensional normal base for the closed subsets of the space $\mathrm{T}(\mathbf{L}|_{\mathbf{s}_\psi})$. By Lemma 5.2.5 the set

$$\{(L \cap H) \cap C : C \in \mathrm{C}^{\mathrm{T},\diamondsuit}_{\theta(\tau)}\}$$

is also a $(\mathrm{b}^n\text{-Ind} \leq \alpha^{L\cap H})$-dimensional normal base for the closed subsets of the space $L \cap H$. Thus, $\mathrm{C}^{\mathrm{T},\diamondsuit}_{\theta(\tau)}$ is a $(\mathrm{b}^n\text{-Ind} \leq \alpha)$-dimensional normal base. This complete the proof of the lemma. □

Definition. We denote by b^n-Ind the base dimensional-like function that has as domain the class of all bases and as range the class $\{-1\} \cup \mathcal{O}$ satisfying the following condition:

$$\mathrm{b}^n\text{-Ind}(A, X) \leq \alpha, \ \alpha \in \{-1\} \cup \mathcal{O},$$

if and only if there exists a $(\mathrm{b}^n\text{-Ind} \leq \alpha)$-dimensional normal base for the open subsets of X containing the base A.

As in Chapter 3 we have the following lemma.

5.2.7 Corollary. *If for a pair* $(A, X) \in \mathbb{P}(\text{base})$ *we have*

$$\mathrm{b}^n\text{-Ind}(A, X) \equiv \alpha \in \{-1\} \cup \mathcal{O},$$

then $\alpha \in \{-1\} \cup \tau^+$. □

5.2.8 Proposition. *The class*

$$\mathbb{P}(\mathrm{b}^n\text{-Ind} \leq m), \ m \in \{-1\} \cup \omega,$$

is saturated.

Proof. Let $\mathbf{S}$ be an indexed collection of spaces. Suppose that $\mathbf{A}$ is a $\mathbb{P}$-co-base for $\mathbf{S}$, where $\mathbb{P} \equiv \mathbb{P}(\mathrm{b}^n\text{-Ind} \leq m)$, and $\mathbf{F}$ is a co-indication of $\mathbf{A}$. By the definition of the class $\mathbb{P}$ there exists a $(\mathrm{b}^n\text{-Ind} \leq m)$-dimensional co-base $\mathbf{B}$ for $\mathbf{S}$ such that $\mathbf{A}(X) \subset \mathbf{B}(X)$ for every $X \in \mathbf{S}$.

By Lemma 5.2.3 there exists an admissible branching triad $(\theta_\vee, \theta_\wedge, \Theta)$ and an m-expansion Δ of Θ, which are independent of the elements of $\mathbf{S}$ as well as a $(\theta_\vee, \theta_\wedge, \Theta, \Delta)$-normal co-indication $\mathbf{N}$ of the co-base $\mathbf{B}$, which is a co-extension of $\mathbf{F}$.

Let $\mathbf{M}^+$, $\mathbf{M}$, R^+, R, and θ are the same as in Lemma 5.2.6. Denote also by $\theta^{\mathbf{F}}$ an indicial mapping from $\mathbf{F}$ to $\mathbf{N}$ and set $\overline{\theta} = \theta \circ \theta^{\mathbf{F}}$. We need to prove that $(\mathrm{B}^{\mathbf{L}}_{\diamondsuit, \overline{\theta}(\tau)}, \mathrm{T}(\mathbf{L})) \in \mathbb{P}$, $\mathbf{L} \in \mathrm{C}^{\diamondsuit}(\mathrm{R})$, that is the base $\mathrm{B}^{\mathbf{L}}_{\diamondsuit, \overline{\theta}(\tau)}$ is contained in a base B for $\mathrm{T}(\mathbf{L})$, which is $(\mathrm{b}^n\text{-Ind} \leq m)$-dimensional. In the case, where $\mathrm{T} = \mathbf{L}$ the fact that $(\mathrm{B}^{\mathrm{T}}_{\diamondsuit, \overline{\theta}(\tau)}, \mathrm{T})$ belongs to $\mathbb{P}$ follows immediately by Lemma 5.2.6. As base B we can take the base $\mathrm{B}^{\mathrm{T}, \diamondsuit}_{\theta(\tau)}$. For the other cases it follows by Lemma 5.2.5. The proof of the proposition is completed. □

Definition. We denote by s-b^n-Ind the dimension-like function with the domain the class of all spaces and the range the class $\{-1, \infty\} \cup \mathcal{O}$ satisfying the following condition:

$$\text{s-b}^n\text{-Ind}(X) \leq \alpha,\ \alpha \in \{-1\} \cup \mathcal{O},$$

if and only if there exists an $(\mathrm{b}^n\text{-Ind} \leq \alpha)$-dimensional normal base of cardinality $\leq \tau$ for the closed subsets of X.

Since the space-component of $\mathbb{P}(\mathrm{b}^n\text{-Ind} \leq m)$ is $\mathbb{P}(\text{s-b}^n\text{-Ind} \leq m)$, Proposition 5.2.8 implies the following consequence.

5.2.9 Corollary. *The class*

$$\mathbb{P}(\text{s-b}^n\text{-Ind} \leq m),\ m \in \{-1\} \cup \omega,$$

is saturated. □

Definition. A class $I\!\!P$ of spaces is said to be (s-b^n-Ind $\leq \alpha$)*-uniform*, where $\alpha \in \{-1\} \cup \tau^+$, if for every indexed collection $\mathbf{S}$ of elements of $I\!\!P$ there exists a co-base for $\mathbf{S}$, which has a $(\vartheta_\vee, \vartheta_\wedge, \Theta, \Delta)$-normal co-indication for some admissible branching triad $(\vartheta_\vee, \vartheta_\wedge, \Theta)$ and an α-expansion Δ of the branching family Θ of mappings.

Lemma 5.2.4 implies the following consequence.

5.2.10 Corollary. *For every* $m \in \{-1\} \cup \omega$ *the class of all spaces* X *for which* s-b^n-Ind $(X) \leq m$ *is* (s-b^n-Ind $\leq m$)*-uniform.* $\square$

The proof of the following proposition is similar to the proof of Proposition 5.2.8.

5.2.11 Proposition. *Let* $\alpha \in \{-1\} \cup \tau^+$. *For any* (s-$\mathrm{b}^n$-Ind $\leq \alpha$)*-uniform class of spaces there exists a Containing Space with*

$$\text{s-b}^n\text{-Ind}(X) \leq \alpha. \ \square$$

5.2.12 Proposition. *For every normal space* X *we have*

$$\text{s-b}^n\text{-Ind}(X) \leq \text{Ind}(X).$$

Proof. Let X be a normal space. The proposition is trivial if Ind $(X) = \infty$. Suppose that $\text{Ind}(X) \leq \alpha$, where $\alpha \in \{-1\} \cup \mathcal{O}$. We need to construct a (b^n-Ind $\leq \alpha$)-dimensional normal base for the closed subsets of the space X.

By Lemma 5.2.4 it suffices to construct an admissible branching triad $(\vartheta_\vee, \vartheta_\wedge, \Theta)$, an α-expansion Δ of Θ, and a $(\vartheta_\vee, \vartheta_\wedge, \Theta, \Delta)$-normal indication $F^X \equiv \{F^X_\delta : \delta \in \tau\}$ of some base C^X for the closed subsets of X.

The construction of the normal triad $(\vartheta_\vee, \vartheta_\wedge, \Theta)$ is similar to the construction of the corresponding triad of Lemma 5.2.4.

For the construction of C^X and F^X we first construct indexed sets

$$\{C^k_\delta : \delta \in \tau\},\ k \in \omega,$$

consisting of closed subsets of X. The construction of these indexed sets is similar to the construction of the corresponding indexed sets of the proof of Lemma 5.2.4. However, we must make the following changes.

As the indexed set $\{C^0_\delta : \delta \in \tau\}$ in condition (1) of the proof of Lemma 5.2.4 we consider any indication of an arbitrary disjunctive base of cardinality $\leq \tau$ for the closed subsets of X satisfying conditions $C^0_0 = X$ and $C^0_1 = \emptyset$.

The case (c) in condition (5) is replaced by the following:

(c) $\mathrm{Ind}(X_\psi) \leq \alpha_\psi$.

Then, in the corresponding part of the proof of Lemma 5.2.4 (concerning this condition) as elements of the screening (G, K) of $(C^i_\delta, C^{i'}_{\delta'})$, we can take any closed subsets of X such that

$$\mathrm{Ind}(X_\chi \cap (G \cap K)) \leq \beta \in \alpha_\psi.$$

Now, as in Lemma 5.2.4 using the indexed sets $\{C^k_\delta : \delta \in \tau\}$, $k \in \omega$, we define the indexed set $\{F^X_\delta : \delta \in \tau\}$ and the α-expansion Δ of Θ. Finally, as the base C^X we consider the (non-indexed) set $\{F^X_\delta : \delta \in \tau\}$. Then, $\{F^X_\delta : \delta \in \tau\}$ is a $(\vartheta_\vee, \vartheta_\wedge, \Theta, \Delta)$-indication of C^X. By Lemma 5.2.4 the base C^X is (b^n-Ind $\leq \alpha$)-dimensional. □

5.3 Compact spaces and dimension Ind

We start giving results concerning the compactification of a completely regular space corresponding to a given normal base for the closed subsets of the space.

Definitions. Let X be a space and C a normal base for the closed subsets of X. A non-empty subset φ of C consisting of non-empty elements is said to be a *filter* of C if the following conditions are satisfied: (a) if $G \in \varphi$, $K \in C$, and $G \subset K$, then $K \in \varphi$ and (b) if $G, K \in \varphi$, then $G \cap K \in \varphi$.

A filter φ of C is said to be *maximal* if the condition $\varphi \subset \varphi'$, where φ' is a filter of C, implies that $\varphi = \varphi'$.

It is clear that any filter of C contains the set X and does not contain the empty set.

Notation. Let X be a space and C a normal base for the closed subsets of X. For every $x \in X$ we denote by φ^C_x the subset of C consisting of all elements G such that $x \in G$.

For every $G \in C$ we denote by $c(G, C)$ the set of all maximal filter φ of C for which $G \in \varphi$. In particular, $c(X, C)$ denotes the set of all maximal filters of C and $c(\emptyset, C)$ denotes the empty set.

Also, we set $c(C) = \{c(G, C) : G \in C\}$.

The following lemma is easy to prove.

5.3.1 Lemma. *Let X be a space and C a normal base for the closed subsets of X. The following properties are true:*

(1) *Each filter of C is contained in a maximal filter.*

(2) *For every $x \in X$ the set φ_x^C is a maximal filter.*

(3) *If $G \in C$ and φ is a maximal filter such that $G \notin \varphi$, then there exists an element $K \in \varphi$ such that $G \cap K = \emptyset$.*

(4) *If $G, K \in C$ and $G \cup K \in \varphi$, where φ is a maximal filter, then either $G \in \varphi$ or $K \in \varphi$.*

(5) *If φ_0 and φ_1 are two distinct maximal filters, then there exists an element $G \in \varphi_0$ and an element $K \in \varphi_1$ such that $G \cap K = \emptyset$ (and, therefore, $G \notin \varphi_1$ and $K \notin \varphi_0$).* □

5.3.2 Lemma. *Let X be a space and C a normal base for the closed subsets of X. Then, the set $c(C)$ is a normal base for the closed subsets of a topology on the set $c(X, C)$.*

Proof. First, we prove that $c(C)$ is a base for the closed subsets of a topology on the set $c(X, C)$. Denote by $c(\cap, C)$ the set of all subsets of $c(X, C)$, which can be represented as an intersection of elements of $c(C)$. We prove that this set satisfies the following properties:

(1) $c(X, C) \in c(\cap, C)$ and $\emptyset \in c(\cap, C)$.

(2) The intersection of any elements of $c(\cap, C)$ is also an element of $c(\cap, C)$.

(3) The finite union of elements of $c(\cap, C)$ is also an element of $c(\cap, C)$.

Properties (1) and (2) are clear. Property (3) follows by the relation

$$(\cap\{c(G_\lambda, C) : \lambda \in \Lambda\}) \cup (\cap\{c(G_{\lambda'}, C) : \lambda' \in \Lambda'\}) =$$

$$\cap\{c(G_\lambda \cup G_{\lambda'}, C) : \lambda \in \Lambda, \lambda' \in \Lambda'\}. \tag{5.3.1}$$

We prove this relation. Let φ be an element of the left part of this relation. Then, either for every $\lambda \in \Lambda$, we have $\varphi \in c(G_\lambda, C)$ and, therefore, $G_\lambda \in \varphi$, or for every $\lambda' \in \Lambda'$, we have $f \in c(G_{\lambda'}, C)$ and, therefore, $G_{\lambda'} \in \varphi$. In both cases, $G_\lambda \cup G_{\lambda'} \in \varphi$ for every $\lambda \in \Lambda$ and for every $\lambda' \in \Lambda'$, which means that φ belongs to the right part of the relation (5.3.1). Conversely, let φ be an element of the right part of the relation (5.3.1). Then, for every $\lambda \in \Lambda$ and for every $\lambda' \in \Lambda'$, we have $G_\lambda \cup G_{\lambda'} \in \varphi$. If f does not belong to the left part of (5.3.1), then there exists an element $\lambda \in \Lambda$ and an element $\lambda' \in \Lambda'$ such that $\varphi \notin c(G_\lambda, C)$ and $\varphi \notin c(G_{\lambda_\prime}, C)$. Therefore, $G_\lambda \notin \varphi$ and $G_{\lambda'} \notin f$. By Lemma 5.3.1, $G_\lambda \cup G_{\lambda'} \notin \varphi$, which is a contradiction.

Properties (1)–(3) show that $c(\cap, C)$ is the set of the closed subsets of a topology on $c(X, C)$ and, therefore, $c(C)$ is the base for the closed subsets of this topology. We prove that the defined space $c(X, C)$ is a T_0-space. Indeed, let φ_0 and φ_1 be two distinct elements of $c(X, C)$. Then (see Lemma 5.3.1), there exists an element $G \in \varphi_0$ such that $G \notin \varphi_1$. This means that $\varphi_0 \in c(G, C)$ and $\varphi_1 \notin c(G, C)$, which shows that $c(X, C)$ is a T_0-space.

Now, we prove that $c(C)$ is a normal base for the closed subsets of the space $c(X, C)$. For this purpose it suffices to prove the following properties:

(4) $c(G, C) \cap c(K, C) = c(G \cap K, C)$.

(5) $c(G, C) \cup c(K, C) = c(G \cup K, C)$.

(6) If $\varphi \in c(X, C)$, $G \in C$, and $\varphi \notin c(G, C)$, then there exists $K \in C$ such that $\varphi \in c(K, C)$ and $c(G, C) \cap c(K, C) = \emptyset$.

(7) Let G, K, L and H are elements of C. Then, $G \cap K = \emptyset$ and the pair (L, H) is a screening of the pair (G, K) (in the space X) if and only if $c(G, C) \cap c(K, C) = \emptyset$ and the pair $(c(L, C), c(H, C))$ is a screening of the pair $(c(G, C), c(K, C))$ (in the space $c(X, C)$).

We prove property (4). Let $\varphi \in c(G, C) \cap c(K, C)$. Then, $G, K \in \varphi$. Therefore, $G \cap K \in \varphi$, which means that $\varphi \in c(G \cap K, C)$. Conversely, if $\varphi \in c(G \cap K, C)$, then $G \cap K \in \varphi$ and, therefore, $G \in \varphi$ and $K \in \varphi$, which means that $\varphi \in c(G, C) \cap c(K, C)$. The property (5) is similarly proved.

We prove property (6). Let $\varphi \in c(X, C)$, $c(G, C) \in c(C)$, and $\varphi \notin c(G, C)$. Then, $G \notin \varphi$ and, therefore (see Lemma 5.3.1), there exists an element $K \in \varphi$ such that $G \cap K = \emptyset$, which means that $\varphi \in c(K, C)$ and $c(G, C) \cap c(K, C) = \emptyset$.

Finally, we prove property (7). By property (4) the condition $G \cap K = \emptyset$ is equivalent to the condition $c(G, C) \cap c(K, C) = \emptyset$. If $G \cap K = \emptyset$ and the pair (L, H) is a screening of the pair (G, K), then we have $G \cap H = \emptyset$, $K \cap L = \emptyset$, and $L \cup H = X$. By properties (4) and (5) these relations are equivalent to relations $c(G, C) \cap c(H, C) = \emptyset$, $c(K, C) \cap c(L, C) = \emptyset$, and $c(L, C) \cup c(H, C) = c(X, C)$, respectively. This proves property (7). □

5.3.3 Lemma. *Let C be a normal base for the closed subsets of a space X. Then, the space $c(X, C)$ is compact.*

Proof. To prove the lemma it suffices to prove that if

$$\{c(G_\lambda, C) : \lambda \in \Lambda\}$$

is a family of elements of $c(C)$ with the empty intersection, then there exists a finite subfamily with the empty intersection. Let $\{c(G_\lambda, C) : \lambda \in \Lambda\}$ be such a family. Suppose that every finite subfamily of this family has the non-empty intersection. Consider the family $\{G_\lambda : \lambda \in \Lambda\}$ of elements of C. Then, every finite subfamily of this family also has the non-empty intersection (see property (4) of Lemma 5.3.2). This fact means that the set φ' consisting of all elements of C containing the intersection of a finite number of elements of the family $\{G_\lambda : \lambda \in \Lambda\}$, is a filter of C. Let φ be a maximal filter of C containing φ'. Then, we have $\varphi \in c(G_\lambda, C)$ for every $\lambda \in \Lambda$, which is a contradiction . □

Notation. Let C be a normal base for the closed subsets of a space X. We denote by i_C^X the mapping of X into $c(X, C)$ defined by relation $i_C^X(x) = \varphi_x^C$, $x \in X$.

The following lemma follows immediately by the fact that a point $x \in X$ belongs to an element G of C if and only if φ_x^C belongs to $c(G, C)$.

5.3.4 Lemma. *Let C be a normal base for the closed subsets of a space X. Then, the mapping i_C^X is an embedding of X into $c(X, C)$ and the set $i_C^X(X)$ is a dense subset of $c(X, C)$. Therefore, if X is compact, then $i_c^X(X) = c(X, C)$.* □

In what follows we identify the point $x \in X$ with the point φ_x^C of $c(X, C)$ and consider the space X as a dense subset of $c(X, C)$, that is $c(X, C)$ is considered to be a compactification of X (corresponding to the normal base C) and i_C^X is considered to be the identical mapping of X into $c(X, C)$.

Let C be a normal base for the closed subsets of a space X. Let also Q be a fixed element of C. Then, the set

$$C^Q \equiv \{Q \cap G : G \in C\}$$

is a normal base for the closed subsets of the space Q. Therefore, we can consider the space $c(Q, C^Q)$. Since $C^Q \subset C$ each element φ' of $c(Q, C^Q)$ is a filter of C. We have the following lemma.

5.3.5 Lemma. *Let C be a normal base for the closed subsets of a space X and Q an element of C. The following statements are true:*

(1) *For every element* φ' *of* $c(Q, C^Q)$ *there exists a unique element* φ *of* $c(X, C)$ *such that* $\varphi' \subset \varphi$.

(2) *The mapping* i *of* $c(Q, C^Q)$ *into* $c(X, C)$, *for which* $i(\varphi') = \varphi$, *where* φ' *is an arbitrary element of* $c(Q, C^Q)$ *and* φ *is the corresponding element of* $c(X, C)$ *containing* φ', *is an embedding (calling natural).*

Proof. (1). Let φ' be an element of $c(Q, C^Q)$. Suppose that there exist two distinct maximal filters φ_0 and φ_1 of C such that $\varphi' \subset \varphi_0 \cap \varphi_1$. There exists an element $G \in \varphi_0$ such that $G \notin \varphi_1$ (see Lemma 5.3.1). Then, $Q \cap G \notin \varphi'$ because in the opposite case $Q \cap G \in \varphi_1$ and therefore $G \in \varphi_1$. This means that there exists an element $K \in \varphi' \subset \varphi_0$ such that $K \cap G = \emptyset$, which contradicts the fact that φ_0 is a filter.

(2). It is clear that if φ' and φ'' are two distinct elements of $c(Q, C^Q)$, then $i(\varphi')$ and $i(\varphi'')$ are distinct elements of $c(X, C)$. Since $c(Q, C^Q)$ is compact (see Lemma 5.3.3) it suffices to prove that i is continuous, that is it suffices to prove that $c(Q \cap G, C^Q) = i^{-1}(c(G, C))$ or equivalently $i(c(Q \cap G, C^Q)) = c(G, C)$ for every $G \in C$. Let $G \in C$, $\varphi' \in c(Q \cap G, C^Q)$, and $i(\varphi') = \varphi$. Then, $Q \cap G \in \varphi'$. Hence $Q \cap G \in \varphi$ and, therefore, $G \in \varphi$, which means that $\varphi \in c(G, C)$ and $i(c(Q \cap G, C^Q)) \subset c(G, C)$. Conversely, let $\varphi \in c(G, C)$ and $i(\varphi') = \varphi$. Then, $G \in \varphi$. If $Q \cap G \notin \varphi'$, then there exists an element K of φ' such that $K \cap (Q \cap G) = \emptyset$ (see Lemma 5.3.1). Since $K \subset Q$ we have $K \cap G = \emptyset$, which contradicts the fact that K and G are elements of φ. □

In what follows for every $Q \in C$ we identify a point φ' of $c(Q, C^Q)$ with the point φ of $c(X, C)$ for which $\varphi' \subset \varphi$ and consider $c(Q, C^Q)$ as a subspace of $c(X, C)$. Therefore, the natural embedding i of $c(Q, C^Q)$ into $c(X, C)$ is considered to be the identical mapping of $c(Q, C^Q)$ into the space $c(X, C)$.

5.3.6 Lemma. *Let* C *be a* (b^n-Ind $\leq \alpha$)*-dimensional normal base for the closed subsets of a space* X, *where* $\alpha \in \{-1\} \cup \mathcal{O}$. *Then, the base* $c(C)$ *for the closed subsets of the space* $c(X, C)$ *is also* (b^n-Ind $\leq \alpha$)*-dimensional.*

Proof. We prove the lemma by induction on α. Obviously, the lemma is true if $\alpha = -1$. Let $\alpha \in \mathcal{O}$ and suppose that the lemma is true for all ordinals less than α. We prove it for α.

Let $(c(G, C), c(K, C))$ be a pair of disjoint elements of $c(C)$. Then, (G, K) is a pair of disjoint elements of C. Therefore, there exists a screening (L, H) of (G, K) by elements of C (in the space X) such that the set

$$C^{L\cap H} \equiv \{(L\cap H)\cap F : F\in C\}$$

is an (b^n-Ind $\leq \beta$)-dimensional normal base for the closed subsets of the space $L\cap H$, where $\beta \in \{-1\}\cup\alpha$. By inductive assumption, the base

$$c(C^{L\cap H}) = \{c((L\cap H)\cap F, C^{L\cap H}) : F\in C\}$$

for the closed subsets of $c(L\cap H, C^{L\cap H})$ is (b^n-Ind $\leq \beta$)-dimensional.

Now, we observe that the pair $(c(L,C), c(H,C))$ is a screening of the pair $(c(G,C), c(K,C))$ in the space $c(X,C)$ (see properties (4) and (5) of the proof of Lemma 5.3.2). We have

$$c(L,C)\cap c(H,C) = c(L\cap H, C) \text{ and}$$

$$c(C^{L\cap H}) = \{c(L\cap H, C)\cap F : F\in c(X,C)\},$$

which means that the base $c(C)$ is (b^n-Ind $\leq \alpha$)-dimensional. □

5.3.7 Lemma. *Let X be a space, C a normal base for the closed subsets of X, and (G_0, K_0) a pair of disjoint closed subsets of X such that G_0 is compact. Then, there exist elements G and K of C such that $G_0\subset G$, $K_0\subset K$, and $G\cap K = \emptyset$.*

Proof. Since C is a base for the closed subsets of X, for every $x\in G_0$ there exists an element $K_x\in C$ such that $x\notin K_x$ and $K_0\subset K_x$. Since C is disjunctive for every $x\in G_0$ there exists an element $G_x\in C$ such that $G_x\cap K_x = \emptyset$. Therefore, there exists a screening (L_x, H_x) of (G_x, K_x) by elements of C. Then, $x\notin H_x$ and the set

$$\{(X\setminus H_x)\cap G_0 : x\in G_0\}$$

is a cover of G_0 (by open subsets of X). Let $x_0, ..., x_n$, $n\in\omega$, be points of G_0 such that $G_0\subset G_{x_0}\cup ...\cup G_{x_n}$. It is easy to verify that the sets $G\equiv G_{x_0}\cup ...\cup G_{x_n}$ and $K\equiv H_{x_0}\cap ...\cap H_{x_n}$ satisfy the conditions of the lemma. □

5.3.8 Proposition. *For every compact Hausdorff space X we have*

$$\mathrm{Ind}(X) = \text{s-b}^n\text{-Ind}(X).$$

Proof. Let X be a compact Hausdorff space. By Proposition 5.2.9 it suffices to prove the following statement: if X has a $(\mathrm{b}^n\text{-Ind} \leq \alpha)$-dimensional normal base for the closed subsets of X, where $\alpha \in \{-1\} \cup \mathcal{O}$, then $\mathrm{Ind}(X) \leq \alpha$.

We prove this statement by induction on α. For $\alpha = -1$ it is trivial. Suppose that the our statement is true for all elements of the set $\{-1\} \cup \alpha$, where $\alpha \in \mathcal{O}$. We prove it for α.

Let C be a $(\mathrm{b}^n\text{-Ind} \leq \alpha)$-dimensional normal base for the closed subsets of the space X. Let G_0 and K_0 be two closed subsets of X such that $G_0 \cap K_0 = \emptyset$. By Lemma 5.3.7 there exist elements G and K of C such that $G_0 \subset G$, $K_0 \subset K$, and $G \cap K = \emptyset$. Since C is $(\mathrm{b}^n\text{-Ind} \leq \alpha)$-dimensional there exists a screening (L, H) of (G, K) by elements of C such that the set

$$\{(L \cap H) \cap F : F \in C\}$$

is an $(\mathrm{b}^n\text{-Ind} \leq \beta)$-dimensional normal base for the closed subsets of $L \cap H$, where $\beta \in \{-1\} \cup \alpha$. By inductive assumption, $\mathrm{Ind}(L \cap H) \leq \beta$, which proves that $\mathrm{Ind}(X) \leq \alpha$. □

5.3.9 Proposition. *Let $m \in \{-1\} \cup \omega$. In the class of all compact Hausdorff spaces X with* $\mathrm{Ind}(X) \leq m$ *there exists a universal element.*

Proof. Let $\mathbb{P}$ be the considered class. By Proposition 5.3.8 the class $\mathbb{P}$ is a subclass of the class $\mathbb{P}(\text{s-b}^n\text{-Ind} \leq m)$. By Proposition 5.2.11 in the last class there exists universal elements. Let T be such an element. Then, we have $\text{s-b}^n\text{-Ind}(T) \leq m$, that is, the space T has a $(\mathrm{b}^n\text{-Ind} \leq m)$-dimensional base C for the closed subsets. Consider the compact space $c(T, C)$. By Proposition 5.3.6 the base $c(C)$ for this space is $(\mathrm{b}^n\text{-Ind} \leq m)$-dimensional. By Proposition 5.3.8, $\mathrm{Ind}(c(T, C)) \leq m$. Therefore, $c(T, C) \in \mathbb{P}$, which means that $c(T, C)$ is a universal element in $\mathbb{P}$. □

5.3.10 Lemma. *Let C^0 be a normal base of cardinality $\leq \tau$ for the closed subsets of a space X and C^{-1} a set of cardinality $\leq \tau$ consisting of compact subsets of X. Then, the minimal ring C of X containing the set $C^{-1} \cup C^0$ is also a normal base of cardinality $\leq \tau$ for the closed subsets of X. Moreover, if C^0 is* $(\mathrm{b}^n\text{-Ind} \leq \alpha)$*-dimensional, where $\alpha \in \{-1\} \cup \mathcal{O}$, then C is also* $(\mathrm{b}^n\text{-Ind} \leq \alpha)$*-dimensional.*

Proof. First we note that the ring $C = (C^0 \cup C^{-1})^\diamond$ is a disjunctive base for the closed subsets of X. Therefore, in order to prove that C is a normal base it suffices to prove that C is base-normal.

Without loss of generality we can assume that $C^{-1} = (C^{-1})^{\diamondsuit}$ and, therefore, $\emptyset \in C^{-1}$. In this case it is easy to verify that any element Q of C has the form:

$$\cup\{Q_i^0 \cap Q_i^{-1} : i \in n \in \omega\},$$

where $Q_i^0 \in C^0$ and $Q_i^{-1} \in C^{-1}$.

Let (G, K) be a pair of disjoint elements of C. Then,

$$G = \cup\{G_i^0 \cap G_i^{-1} : i \in n \in \omega\} \text{ and}$$

$$K = \cup\{K_j^0 \cap K_j^{-1} : j \in m \in \omega\},$$

where $G_i^0, K_j^0 \in C^0$ and $G_i^{-1}, K_j^{-1} \in C^{-1}$. For every $i \in n$ and for every $j \in m$ there exist two elements L_{ij} and H_{ij} of C^0 such that $G_i^0 \cap G_i^{-1} \subset L_{ij}$, $K_j^0 \cap K_j^{-1} \subset H_{ij}$, and $L_{ij} \cap H_{ij} = \emptyset$.

Indeed, let $i \in n$ and $j \in m$. If $G_i^{-1} = \emptyset$ and $K_j^{-1} = \emptyset$, then $G_i^0 \cap K_j^0 = \emptyset$. In this case we set $L_{ij} = G_i^0$ and $H_{ij} = K_j^0$. If $G_i^{-1} \neq \emptyset$, then in Lemma 5.3.7 we set $C = C^0$, $G_0 = G_i^0 \cap G_i^{-1}$ and $K_0 = K_j^0 \cap K_j^{-1}$. In this case L_{ij} and H_{ij} are considered to be respectively the sets G and K of Lemma 5.3.7. Similarly, we define the sets L_{ij} and H_{ij} if $K_j^{-1} \neq \emptyset$.

Let

$$L_i = \cap\{L_{ij} : j \in m\} \text{ and } H_j = \cap\{H_{ij} : i \in n\}.$$

Let also

$$L = \cup\{L_i : i \in n\} \text{ and } H = \cup\{H_j : j \in m\}.$$

It is easy to verify that $G \subset L$, $K \subset H$, and $L \cap H = \emptyset$. Since L and H are elements of C^0 there exists a screening (L', H') of the pair (L, H) such that $L', H' \in C^0$. Then, (L', H') is a sreening of (G, K) by elements of C. Thus, C is base-normal.

Now, suppose that C^0 is (b^n-Ind $\leq \alpha$)-dimensional, where $\alpha \in \{-1\} \cup \mathcal{O}$. We prove by induction on α that C is also (b^n-Ind $\leq \alpha$)-dimensional. This is trivial if $\alpha = -1$. Let $\alpha \in \mathcal{O}$ and suppose that for all elements of α our suggestion is true.

Let (G, K) be a disjoint pair of elements of C. By the preceding there exists a pair (L, H) of disjoint elements of C^0 such that $G \subset L$ and $K \subset H$.

Since C^0 is (b^n-Ind$\leq \alpha$)-dimensional there exists a screening (L', H') of (L, H) such that $L', H' \in C^0$ and the set $C^0_P \equiv \{P \cap F : F \in C^0\}$, where $P = L' \cap H'$, is a (b^n-Ind $\leq \beta$)-dimensional base for the closed subsets of P, where $\beta \in \{-1\} \cup \alpha$.

It is easy to verify that the set

$$C_P \equiv \{P \cap F : F \in C\}$$

is the minimal ring of P containing the set $C^0_P \cup C^{-1}_P$, where

$$C^{-1}_P = \{P \cap F : F \in C^{-1}\}.$$

By inductive assumption, C_P is (b^n-Ind $\leq \beta$)-dimensional, which means that C is (b^n-Ind $\leq \alpha$)-dimensional. □

5.3.11 Lemma. *Let X be a completely regular space such that*

$$X = \cup\{X_\lambda : \lambda \in \Lambda\},$$

where $|\Lambda| \leq \tau$, X_λ is a compact subspace of X, and

$$\mathrm{Ind}(X_\lambda) \equiv \alpha_\lambda \in \{-1\} \cup \mathcal{O}.$$

Then, there exists a normal base C^X of cardinality $\leq \tau$ for the closed subsets of X such that for every $\lambda \in \Lambda$ the set

$$\{X_\lambda \cap F : F \in C^X\}$$

is a (b^n-Ind $\leq \alpha_\lambda$)*-dimensional normal base for the closed subsets of X_λ.*

Proof. Let C^0 be an arbitrary normal base of cardinality $\leq \tau$ for the closed subsets of X. By Proposition 5.3.8 for every $\lambda \in \Lambda$ there exists a (b^n-Ind $\leq \alpha_\lambda$)-dimensional normal base $C^{0,\lambda}$ of cardinality $\leq \tau$ for the closed subsets of the space X_λ. Denote by C^X the base of cardinality $\leq \tau$ for the closed subsets of X, which is the minimal ring of X containing the set

$$C^0 \cup (\cup\{C^{0,\lambda} : \lambda \in \Lambda\}).$$

By Lemma 5.3.10, C^X is normal.

It is easy to verify that for every $\lambda \in \Lambda$ the set

$$C^{X,\lambda} \equiv \{X_\lambda \cap F : F \in C^X\}$$

coincides with the minimal ring of X_λ containing the set

$$\{X_\lambda \cap F : F \in C^0 \cup (\cup\{C^{0,\lambda'} : \lambda' \in \Lambda\})\}.$$

Since the last set contains $C^{0,\lambda}$, $C^{X,\lambda}$ is a $(\mathrm{b}^n\text{-Ind} \le \alpha_\lambda)$-dimensional normal base of cardinality $\le \tau$ (see Lemma 5.3.10). □

The proof of the following lemma is omitted. This proof can be done using the method of the proof of Lemma 5.2.4.

5.3.12 Lemma. *Let X be a completely regular space such that*

$$X = \cup\{X_\lambda : \lambda \in \Lambda\},$$

where $|\Lambda| \le \tau$, X_λ is a compact subspace of X, and

$$\mathrm{Ind}(X_\lambda) \equiv \alpha_\lambda \in \{-1\} \cup \mathcal{O}.$$

Let also C^X be a base of cardinality $\le \tau$ for the closed subsets of X such that for every $\lambda \in \Lambda$ the set

$$C^{X,\lambda} \equiv \{X_\lambda \cap F : F \in C^X\}$$

is a $(\mathrm{b}^n\text{-Ind} \le \alpha_\lambda)$-dimensional normal base for the closed subsets of X_λ. Then, there exist:

(a) *an admissible branching triad $(\vartheta_\vee, \vartheta_\wedge, \Theta)$,*
(b) *a branching family Θ_λ and an α_λ-expansion of Δ_λ, $\lambda \in \Lambda$,*
(c) *a mapping θ_Λ of Λ into τ, and*
(d) *a $(\vartheta_\vee, \vartheta_\wedge, \Theta)$-normal indication*

$$F^X \equiv \{F^X_\delta : \delta \in \tau\}$$

of the base C^X
such that for every $\lambda \in \Lambda$ we have $F^X_{\theta_\Lambda(\lambda)} = X_\lambda$ and

$$F^{X,\lambda} \equiv \{F^X_\delta \cap X_\lambda : \delta \in \tau\}$$

is a $(\vartheta_\vee, \vartheta_\wedge, \Theta_\lambda, \Delta_\lambda)$- normal indication of $C^{X,\lambda}$. Moreover, if $\alpha \in \{-1\} \cup \omega$, then we can suppose that the admissible branching triad $(\vartheta_\vee, \vartheta_\wedge, \Theta)$, the *branching family Θ_λ, the α-expansion Δ_λ of Θ_λ, and the mapping θ_Λ are independent of the space X.* $\Box$

Notation. Let ν be a cardinal such that $1 \leq \nu \leq \tau$, ξ a mapping of ν into the set $\{-1\} \cup \omega$, and $\prec$ a partial order on the set ν. We denote by $\mathbb{P}(\mathrm{Ind}, \nu, \xi, \prec)$ the class consisting of all ν-sequences $(X, X_0, ..., X_\eta, ...)_{\eta\in\nu}$ such that:

(a) X is a completely regular space,
(b) X_η is a compact subset of X,
(c) $X = \cup\{X_\eta : \eta \in \nu\}$,
(d) $X_{\eta_0} \subset X_{\eta_1}$ if $\eta_0 \prec \eta_1$, and
(e) $\mathrm{Ind}(X_\eta) \leq \xi(\eta)$.

(The above notion of a ν-sequence and the notion of a properly universal element for a class of ν-sequences, using in the next proposition, were introduced in Section 4.4.3.)

5.3.13 Proposition. *In the class $\mathbb{P}(\mathrm{Ind}, \nu, \xi, \prec)$ there exists a properly universal element.*

Proof. Let $\mathbf{S}$ be a collection of elements of $\mathbb{P} \equiv \mathbb{P}(\mathrm{Ind}, \nu, \xi, \prec)$ such that for every element $(Y, Y_0, Y_1, ..., Y_\eta, ...)_{\eta\in\nu}$ of $\mathbb{P}$ there exists an element $(X, X_0, X_1, ..., X_\eta, ...)_{\eta\in\nu}$ of $\mathbf{S}$ and a homeomorphism h of Y onto X such that $h^{-1}(X_\eta) = Y_\eta$.

In Lemma 5.3.12 we set $\alpha = m$ and $\Lambda = \nu$ and consider the admissible branching triad $(\vartheta_\vee, \vartheta_\wedge, \Theta)$, the branching family Θ_η, the m-expansion Δ_η of Θ_η, and the mapping θ_ν (which are independent of the space X).

By Lemma 5.3.11 for every element $(X, X_0, X_1, ..., X_\eta, ...)_{\eta\in\nu}$ of $\mathbf{S}$ there exists a normal base C^X of cardinality $\leq \tau$ for the closed subsets of X such that for every $\eta \in \nu$ the set

$$C^{X,\eta} \equiv \{X_\eta \cap F : F \in C^X\}$$

is a (b^n-Ind $\leq \xi(\eta)$)-dimensional normal base for the closed subsets of X_η. Denote by $F^X \equiv \{F^X_\delta : \delta \in \tau\}$ the $(\vartheta_\vee, \vartheta_\wedge, \Theta)$-normal indication of C^X such that for every $\eta \in \nu$ we have that $F^X_{\theta_\nu(\eta)} = X_\eta$ and that the indexed set

$$F^{X,\eta} \equiv \{F^X_\delta \cap X_\eta : \delta \in \tau\}$$

is a $(\vartheta_\vee, \vartheta_\wedge, \Theta_\eta, \Delta_\eta)$- normal indication of X_η. (See Lemma 5.3.12).

Furthermore, we set

$$\mathbf{S}_\eta = \{X_\eta : X \in \mathbf{S}\},$$

$$\mathbf{C} = \{C^X : X \in \mathbf{S}\}, \text{ and}$$

$$\mathbf{F} = \{F^X : X \in \mathbf{S}\}.$$

Also for every $\eta \in \nu$ we set

$$\mathbf{C}^\eta = \{C^{X,\eta} : X \in \mathbf{S}\} \text{ and}$$

$$\mathbf{F}^\eta = \{F^{X,\eta} : X \in \mathbf{S}\}.$$

We denote by $\mathbf{B}$ the complementary to $\mathbf{C}$ open co-base for $\mathbf{S}$ and by $\mathbf{N}$ the complementary to $\mathbf{F}$ co-indication of $\mathbf{B}$. Also, we denote by $\mathbf{B}_\eta$ the complementary (in the indexed collection $\mathbf{S}_\eta$) to $\mathbf{C}^\eta$ open co-base for $\mathbf{S}_\eta$ and by $\mathbf{N}_\eta$ the complementary to $\mathbf{F}^\eta$ co-indication of $\mathbf{B}^\eta$.

Now, we denote by $\mathbf{M}$ a co-mark of $\mathbf{S}$ satisfying the following conditions:

(a) $\mathbf{M}$ is an initial co-mark corresponding to the co-indication $\mathbf{N}$ of $\mathbf{B}$ and the class $\mathbb{P}(\subset\text{n.base})$ of bases;

(b) the trace $\mathbf{M}|_{\mathbf{S}_\eta}$ of $\mathbf{M}$ on $\mathbf{S}_\eta$, $\eta \in \nu$, is an initial co-mark for $\mathbf{S}_\eta$ corresponding to the co-indication $\mathbf{N}_\eta$ of $\mathbf{B}_\eta$ and the class $\mathbb{P}(\text{b}^n\text{-Ind} \leq \xi(\eta))$; and

(c) an indicial mapping from $\mathbf{N}$ to $\mathbf{M}$ denoted by θ is also an indicial mapping from $\mathbf{N}_\eta$ to $\mathbf{M}_\eta$ for every $\eta \in \nu$.

Also, we denote by R an $\mathbf{M}$-admissible family of equivalence relations on $\mathbf{S}$ satisfying the following conditions:

(a) the family R is an initial family corresponding to the co-mark $\mathbf{M}$, the co-indication $\mathbf{N}$ of $\mathbf{B}$, and the class $\mathbb{P}(\subset\text{normal base})$ of bases and

(b) for every $\eta \in \nu$ the trace $\mathrm{R}|_{\mathbf{S}_\eta}$ of R on $\mathbf{S}_\eta$ is an initial family of equivalence relations on $\mathbf{S}_\eta$ corresponding to the co-mark $\mathbf{M}_\eta$, the co-indication $\mathbf{N}_\eta$ of $\mathbf{B}_\eta$, and $\mathbb{P}(\text{b}^n\text{-Ind} \leq \xi(\eta))$.

In the Containing Space T we consider the $\theta(\tau)$-standard base $\mathrm{B}^{\mathrm{T}}_{\theta(\tau)}$ and in the Containing Space $\mathrm{T}_\eta \equiv \mathrm{T}|_{\mathbf{S}_\eta}$ we consider the $\theta(\tau)$-standard base $\mathrm{B}^{\mathrm{T}\eta}_{\theta(\tau)}$. By the choice of $\mathbf{M}$ and R the base $\mathrm{B}^{\mathrm{T}}_{\theta(\tau)}$ is contained in a normal base B for the open subsets of T. Similarly, for every $\eta \in \nu$ the base $\mathrm{B}^{\mathrm{T}\eta}_{\theta(\tau)}$ is contained in a $(\text{b}^n\text{-Ind} \leq \xi(\eta))$-dimensional base B_η for the open subsets of the space T_η. Without loss of generality we can suppose that

$$B = \mathrm{B}^{\mathrm{T},\diamondsuit}_{\theta(\tau)} \text{ and } B_\eta = (\mathrm{B}^{\mathrm{T}_\eta}_{\theta(\tau)})^\diamondsuit \equiv \mathrm{B}^{\mathrm{T}_\eta,\diamondsuit}_{\theta(\tau)}.$$

Since for every $\eta \in \nu$,

$$\mathrm{B}^{\mathrm{T}_\eta}_{\theta(\tau)} = \{\mathrm{T}_\eta \cap U : U \in \mathrm{B}^{\mathrm{T}}_{\theta(\tau)}\}$$

we have

$$\mathrm{B}^{\mathrm{T}_\eta,\diamondsuit}_{\theta(\tau)} = \{\mathrm{T}_\eta \cap U : U \in \mathrm{B}^{\mathrm{T},\diamondsuit}_{\theta(\tau)}\}.$$

We recall that by $\mathrm{C}^{\mathrm{T},\diamondsuit}_{\theta(\tau)}$ is denoted the complementary to $\mathrm{B}^{\mathrm{T},\diamondsuit}_{\theta(\tau)}$ normal base for the closed subsets of T and for every $\eta \in \nu$ by $\mathrm{C}^{\mathrm{T}_\eta,\diamondsuit}_{\theta(\tau)}$ is denoted the complementary to $\mathrm{B}^{\mathrm{T}_\eta,\diamondsuit}_{\theta(\tau)}$ normal base for the closed subsets of T_η. Obviously, we have

$$\mathrm{C}^{\mathrm{T}_\eta,\diamondsuit}_{\theta(\tau)} = \{\mathrm{T}_\eta \cap U : U \in \mathrm{C}^{\mathrm{T},\diamondsuit}_{\theta(\tau)}\}.$$

Now, we consider the compact space $c(\mathrm{T}, \mathrm{C}^{\mathrm{T},\diamondsuit}_{\theta(\tau)})$ (see Lemma 5.3.3) and its closed subsets $c(\mathrm{T}_\eta, \mathrm{C}^{\mathrm{T}_\eta,\diamondsuit}_{\theta(\tau)})$, $\eta \in \nu$ (see Lemma 5.3.5). We note that for every $\eta \in \nu$ the set T_η belongs to $\mathrm{C}^{\mathrm{T},\diamondsuit}_{\theta(\tau)}$. Moreover, by Lemma 5.3.6 we have

$$\mathrm{Ind}(c(\mathrm{T}_\eta, \mathrm{C}^{\mathrm{T}_\eta,\diamondsuit}_{\theta(\tau)})) \leq \xi(\eta).$$

We set

$$\mathrm{Y} = \mathrm{T} \cup (\cup\{c(\mathrm{T}_\eta, \mathrm{C}^{\mathrm{T}_\eta,\diamondsuit}_{\theta(\tau)}) : \eta \in \nu\})$$

and for every $\eta \in \nu$ we set

$$\mathrm{Y}_\eta = c(\mathrm{T}_\eta, \mathrm{C}^{\mathrm{T}_\eta,\diamondsuit}_{\theta(\tau)}).$$

It is easy to verify that the ν-sequence

$$(\mathrm{Y}, \mathrm{Y}_0, ..., \mathrm{Y}_\eta, ...)_{\eta \in \nu}$$

is the required properly universal element in the class $\mathbb{P}$. $\square$

5.4 Completely regular spaces and dimension dim

Definition. Let X be a completely regular space and $m \in \{-1, \infty\} \cup \omega$. A normal base B for the open subsets of X is called $(\mathrm{b}^n\text{-dim} \leq m)$-*dimensional* if for every finite open cover c of X by elements of B there exists a finite open cover of X by elements of B, which is a refinement of c and has order $\leq m$.

A normal base C for the closed subsets of X is called $(\mathrm{b}^n\text{-dim} \leq m)$-*dimensional* if the complementary to C normal base for the open subsets of X is $(\mathrm{b}^n\text{-dim} \leq m)$-dimensional.

Definition. Any tetrad $(\vartheta_\vee, \vartheta_\wedge, \vartheta, \vartheta_{\mathcal{F}})$, where $(\vartheta_\vee, \vartheta_\wedge, \vartheta)$ is an admissible triad and $\vartheta_{\mathcal{F}}$ is a mapping of $\mathcal{F} \setminus \{\emptyset\}$ into itself such that $|\vartheta_{\mathcal{F}}(s)| = |s|$, is called an *admissible* tetrad.

Definition. Let X be a space, B^X a $(\mathrm{b}^n\text{-dim} \leq m)$-dimensional base for the open subsets of X, C^X the complementary to B^X base for the closed subsets of X, and let $(\vartheta_\vee, \vartheta_\wedge, \vartheta, \vartheta_{\mathcal{F}})$ be an admissible tetrad. An indication $\{U_\delta^X : \delta \in \tau\}$ of B^X is said to be $(\vartheta_\vee, \vartheta_\wedge, \vartheta, \vartheta_{\mathcal{F}}, m)$-*normal* if:

(a) the indication $\{U_\delta^X : \delta \in \tau\}$ of B^X is $(\vartheta_\vee, \vartheta_\wedge, \vartheta)$-normal and

(b) if $\{\delta_0, ..., \delta_k\} \in \mathcal{F} \setminus \{\emptyset\}$, $\{U_{\delta_0}^X, ..., U_{\delta_k}^X\}$ is a cover of X, and

$$\vartheta_{\mathcal{F}}(\{\delta_0, ..., \delta_k\}) = \{\varepsilon_0, ..., \varepsilon_k\},$$

where $\delta_i < \delta_j$ and $\varepsilon_i < \varepsilon_j$ for $i < j$, then $\{U_{\varepsilon_0}^X, ..., U_{\varepsilon_k}^X\}$ is a cover of X of the order $\leq m$ such that $U_{\varepsilon_i}^X \subset U_{\delta_i}^X$, $i \in \{0, ..., k\}$.

An indication F^X of C^X is said to be $(\vartheta_\vee, \vartheta_\wedge, \vartheta, \vartheta_{\mathcal{F}}, m)$-*normal* if the complementary to F^X indication of B^X is $(\vartheta_\vee, \vartheta_\wedge, \vartheta, \vartheta_{\mathcal{F}}, m)$-normal.

The following auxiliary lemma is clear. (The sets $\mathcal{F}_k$ and P_k of this lemma were defined in Section 5.1.)

5.4.1 Lemma. *For every $k \in \omega$ there exist:*

(a) *two one-to-one mappings $\vartheta_{\vee,k}$ and $\vartheta_{\wedge,k}$ of $\mathcal{F}_k$ into the set $\{k+1\} \times \tau$,*

(b) *a mapping ϑ_k of P_k into the set $(\{k+1\} \times \tau)^2$, and*

(c) *a mapping $\vartheta_{\mathcal{F},k}$ of $\mathcal{F}_k$ into the set of all non-empty finite subsets of the set $\{k+1\} \times \tau$*

such that:

(1) *The mappings* $\pi_1^k \circ \vartheta_k$ *and* $\pi_2^k \circ \vartheta_k$ *are one-to-one, where* π_1^k *and* π_2^k *are respectively the first and the second projections of the set* $(\{k+1\} \times \tau)^2$ *onto* $\{k+1\} \times \tau$.

(2) *For every element* s *of* $\mathcal{F}_k$, $|\vartheta_{\mathcal{F},k}(s)| = |s|$.

(3) *The sets* $\mathrm{I}(\pi_1^k \circ \vartheta_k)$, $\mathrm{I}(\pi_2^k \circ \vartheta_k)$, $\mathrm{I}(\vartheta_{\vee,k})$, $\mathrm{I}(\vartheta_{\wedge,k})$, *and* $\mathrm{I}^*(\vartheta_{\mathcal{F},k})$ *are mutually disjoint, where* $\mathrm{I}^*(\vartheta_{\mathcal{F},k})$ *is the union of images of all elements of* $\mathcal{F}_k$. □

5.4.2 Lemma. *Let* D^X *be a base for the closed subsets of a space* X,

$$K^X \equiv \{K_\delta^X : \delta \in \tau\}$$

an indication of D^X, C^X *a normal base for the closed subsets of* X *containing the base* D^X, *and* $m \in \{-1\} \cup \omega$. *Then,* C^X *is* (b^n-dim $\leq m$)-*dimensional if and only if for some admissible tetrad* $(\vartheta_\vee, \vartheta_\wedge, \vartheta, \vartheta_{\mathcal{F}})$ *and some mapping* $\overline{\theta}$ *of* τ *into itself there exists a* $(\vartheta_\vee, \vartheta_\wedge, \vartheta, \vartheta_{\mathcal{F}}, m)$-*normal indication*

$$F^X \equiv \{F_\delta^X : \delta \in \tau\}$$

of C^X *such that* $K_\delta^X = F_{\overline{\theta}(\delta)}^X$ *for every* $\delta \in \tau$. *Moreover, the admissible tetrad* $(\vartheta_\vee, \vartheta_\wedge, \vartheta, \vartheta_{\mathcal{F}})$ *and the mapping* $\overline{\theta}$ *can be considered to be independent of the space* X.

Proof. In general outline the proof of the lemma is similar to the proof of Lemma 5.1.3. Instead of Lemma 5.1.2 using in the proof of Lemma 5.1.3 we use here the Lemma 5.4.1.

It is easy to verify that if for some admissible tetrad $(\vartheta_\vee, \vartheta_\wedge, \vartheta, \vartheta_{\mathcal{F}})$ there exists a $(\vartheta_\vee, \vartheta_\wedge, \vartheta, \vartheta_{\mathcal{F}}, m)$-normal indication of the base C^X, then C^X is (b^n-dim $\leq m$)-dimensional.

To prove the converse assertion we first construct the corresponding admissible tetrad $(\vartheta_\vee, \vartheta_\wedge, \vartheta, \vartheta_{\mathcal{F}}, m)$ and the mapping $\overline{\theta}$.

Let ϖ be an one-to-one mapping of the set $\omega \times \tau$ onto τ such that $\varpi(0,0) = 0$ and $\varpi(0,1) = 1$. The mappings $\vartheta_\vee$, $\vartheta_\wedge$, ϑ, and $\overline{\theta}$ are constructed in the same manner as in Lemma 5.1.3.

We construct the mapping $\vartheta_{\mathcal{F}}$. Let $\{\delta_1, ..., \delta_n\}$ be an element of the set $\mathcal{F} \setminus \{\emptyset\}$, $|\{\delta_1, ..., \delta_n\}| = n$. Suppose that

$$(\varpi)^{-1}(\delta_1) = (k_1, \varepsilon_1), ..., (\varpi)^{-1}(\delta_n) = (k_n, \varepsilon_n), \text{ and}$$

$$k = \max\{k_1, ..., k_n\}.$$

Then, $\{(k_1, \varepsilon_1), ..., (k_n, \varepsilon_n)\}$ is an element of the set $\mathcal{F}_k$. Let

$$\vartheta_{\mathcal{F},k}(\{(k_1, \varepsilon_1), ..., (k_n, \varepsilon_n)\}) = \{(k+1, \varepsilon^1), ..., (k+1, \varepsilon^n)\}$$

and $\varpi(k+1, \varepsilon^1) = \delta^1$,..., $\varpi(k+1, \varepsilon^n) = \delta^n$. Then, $|\{\delta^1, ..., \delta^n\}| = n$ and we set

$$\vartheta_{\mathcal{F}}(\{\delta_1, ..., \delta_n\}) = \{\delta^1, ..., \delta^n\}.$$

Thus, the constructed admissible tetrad $(\vartheta_\vee, \vartheta_\wedge, \vartheta, \vartheta_{\mathcal{F}})$ and the mapping $\overline{\theta}$ are independent of any space.

It is easy to verify that if there exists a $(\vartheta_\vee, \vartheta_\wedge, \vartheta, \vartheta_{\mathcal{F}}, m)$-normal indication of C^X, then C^X is (b^n-dim $\leq m$)-dimensional.

Now, suppose that C^X is (b^n-dim $\leq m$)-dimensional. We prove that there exist a $(\vartheta_\vee, \vartheta_\wedge, \vartheta, \vartheta_{\mathcal{F}}, m)$-normal indication $F^X \equiv \{F^X_\delta : \delta \in \tau\}$ such that $K^X_\delta = F^X_{\overline{\theta}(\delta)}$. In order to construct this indication we first construct indexed sets $\{C^k_\delta : \delta \in \tau\}$, $k \in \omega$. These indexed sets are constructed by induction on k in the same manner as in Lemma 5.1.3. However, here we require one more condition. This condition can be formulated as follows. Suppose that

$$\vartheta_{\mathcal{F},k}(\{(k_1, \varepsilon_1), ..., (k_n, \varepsilon_n)\}) = \{(k+1, \varepsilon^1), ..., (k+1, \varepsilon^n)\}$$

where $\{(k_1, \varepsilon_1), ..., (k_n, \varepsilon_n)\}$ is an element of $\mathcal{F}_k$ such that

$$|\{(k_1, \varepsilon_1), ..., (k_n, \varepsilon_n)\}| = n.$$

Therefore,

$$|\{(k+1, \varepsilon^1), ..., (k+1, \varepsilon^n)\}| = n.$$

Suppose also that $\varpi(k_1, \varepsilon_1) = \delta_1$,..., $\varpi(k_n, \varepsilon_n) = \delta_n$, $\varpi(k+1, \varepsilon^1) = \delta^1$,..., $\varpi(k+1, \varepsilon^1) = \delta^1$. Therefore,

$$\vartheta_{\mathcal{F}}(\{\delta_1, ..., \delta_n\}) = \{\delta^1, ..., \delta^n\}.$$

Without loss of generality we can suppose that $\delta_1 < ... < \delta_n$. Let

$$\{\delta^1, ..., \delta^n\} = \{\eta^1, ..., \eta^n\},$$

where $\eta^1 < ... < \eta^n$. If the set

$$\{X \setminus C_{\varepsilon_1}^{k_1}, ..., X \setminus C_{\varepsilon_n}^{k_n}\} \tag{5.4.1}$$

is not a cover of X, then we set $C_{\varepsilon^1}^{k+1} = ... = C_{\varepsilon^n}^{k+1} = \emptyset$. Suppose that the set (5.4.1) is a cover of X. Then, there exists an open cover

$$\{V_1, ..., V_p\} \tag{5.4.2}$$

of X, which is a refinement of (5.4.1) and has order $\leq m$. We denote by U_1 the union of all elements of the cover (5.4.2), which are contained in $X \setminus C_{\varepsilon_1}^{k_1}$. If there is no such elements, then we set $U_1 = \emptyset$. By induction, we denote by U_i, $i \in \{2, ..., n\}$ the union of all elements of the cover (5.4.2), which are contained in the set $X \setminus C_{\varepsilon_i}^{k_i}$ and are not contained in any set $X \setminus C_{\varepsilon_1}^{k_1}, ..., X \setminus C_{\varepsilon_{i-1}}^{k_{i-1}}$. Thus, the set $\{U_1, ..., U_n\}$ is a cover of X, which is a refinement of the cover (5.4.1) and has order $\leq m$. Moreover, $U_i \subset X \setminus C_{\varepsilon_i}^{k_i}$. In this case we set

$$C_{\eta^i}^{k+1} = X \setminus U_i, \, i \in \{1, ..., n\}.$$

Then, the required $(\vartheta_\vee, \vartheta_\wedge, \vartheta, \vartheta_{\mathcal{F}}, m)$-normal indication $\{F_\delta^X : \delta \in \tau\}$ of C^X can be constructed in an analogous manner as in Lemma 5.1.3. □

The following lemma can be easily proved.

5.4.3 Lemma. *Let X be a space and C a normal base for the closed subsets of X. Let $G, G_1, ..., G_n$ be elements of C, $n \in \omega \setminus \{0\}$, such that the sets $G_1, ..., G_n$ are mutually disjoint and*

$$G = \cup\{G_i : i \in n+1\}.$$

Then, the normal base $\{G \cap F : F \in C\}$ for the closed subsets of the space G is (b^n-dim $\leq m$)*-dimensional, where $m \in \{-1\} \cup \omega$, if and only if for every $i \in n+1$ the normal base $\{G_i \cap F : F \in C\}$ for the closed subsets of the space G_i is* (b^n-dim $\leq m$)*-dimensional.* □

Definitions. Let **S** be a collection of spaces, **B** a normal open co-base for **S**, **N** a co-indication of **B**, **C** the complementary to **B** normal closed co-base for **S**, and **F** the complementary to **N** co-indication of **C**.

The open co-base $\mathbf{B}$ is said to be (b^n-dim $\leq m$)-*dimensional* if for every $X \in \mathbf{S}$, the base $\mathbf{B}(X)$ of X is (b^n-dim $\leq m$)-dimensional. In this case the closed co-base $\mathbf{C}$ is also called (b^n-dim $\leq m$)-*dimensional.*

Suppose that $\mathbf{B}$ is (b^n-dim $\leq m$)-dimensional and let $(\vartheta_\vee, \vartheta_\wedge, \vartheta, \vartheta_\mathcal{F})$ be an admissible tetrad. The co-indication $\mathbf{N}$ of the co-base $\mathbf{B}$ is said to be a $(\vartheta_\vee, \vartheta_\wedge, \vartheta, \vartheta_\mathcal{F}, m)$-*normal* if for every $X \in \mathbf{S}$, $\mathbf{N}(X)$ is $(\vartheta_\vee, \vartheta_\wedge, \vartheta, \vartheta_\mathcal{F}, m)$-normal indication of $\mathbf{B}(X)$ In this case the co-indication $\mathbf{F}$ of $\mathbf{C}$ is also called $(\vartheta_\vee, \vartheta_\wedge, \vartheta, \vartheta_\mathcal{F}, m)$-*normal.*

It is easy to see that if the co-indication $\mathbf{N}$ of $\mathbf{B}$ is $(\vartheta_\vee, \vartheta_\wedge, \vartheta, \vartheta_\mathcal{F}, m)$-normal, then it also $(\vartheta_\vee, \vartheta_\wedge, \vartheta)$-normal.

5.4.4 Lemma. *Suppose that* $\mathbf{S}$ *is an indexed collection of spaces;* $\mathbf{B}$ *is a* (b^n-dim $\leq m$)*-dimensional normal open co-base for* $\mathbf{S}$*, where* $m \in \{-1\} \cup \omega$*; and* $\mathbf{N}$ *is a* $(\vartheta_\vee, \vartheta_\wedge, \vartheta, \vartheta_\mathcal{F}, m)$*-normal co-indication of* $\mathbf{B}$*, where* $(\vartheta_\vee, \vartheta_\wedge, \vartheta, \vartheta_\mathcal{F})$ *is an admissible tetrad. Then, for every co-mark* $\mathbf{M}$ *of* $\mathbf{S}$*, which is a co-extension of* $\mathbf{N}$*, and for every* $\mathbf{M}$*-admissible family* R *of equivalence relations on* $\mathbf{S}$ *the set* $\mathrm{C}_{\theta(\tau)}^{\mathrm{T},\diamondsuit}$ *is a* (b^n-dim $\leq m$)*-dimensional normal base for the closed subsets of* T*, where* θ *is an indicial mapping from* $\mathbf{N}$ *to* $\mathbf{M}$*.*

Proof. Suppose that

$$\mathbf{B} = \{B^X : X \in \mathbf{S}\} \text{ and}$$

$$\mathbf{N} = \{\{V_\varepsilon^X : \varepsilon \in \tau\} : X \in \mathbf{S}\}.$$

Let

$$\mathbf{M} \equiv \{\{U_\delta^X : \delta \in \tau\} : X \in \mathbf{S}\}$$

be a co-extension of $\mathbf{N}$ and θ an indicial mapping from $\mathbf{N}$ to $\mathbf{M}$. Let also

$$\mathrm{R} \equiv \{\sim^s : s \in \mathcal{F}\}$$

be an $\mathbf{M}$-admissible family of equivalence relations on $\mathbf{S}$. We need to prove that $\mathrm{B}_{\theta(\tau)}^{\mathrm{T},\diamondsuit}$ is a (b^n-dim $\leq m$)-dimensional normal base for the open subsets of T.

Since $\mathbf{N}$ is also a $(\vartheta_\vee, \vartheta_\wedge, \vartheta)$-normal indication of $\mathbf{B}$ by Lemma 5.1.4 $\mathrm{B}_{\theta(\tau)}^{\mathrm{T},\diamondsuit}$ is a normal base for the open subsets of T. Therefore, it is sufficient to prove that the base $\mathrm{B}_{\theta(\tau)}^{\mathrm{T},\diamondsuit}$ is (b^n-dim $\leq m$)-dimensional, that is for every cover c of T by elements of $\mathrm{B}^{\mathrm{T},\diamondsuit}_{\theta(\tau)}$ there exists an open cover r of T by elements of $\mathrm{B}^{\mathrm{T},\diamondsuit}_{\theta(\tau)}$, which is a refinement of c and has order $\leq m$.

Let

$$c \equiv \{U_0, ..., U_n\}$$

be a cover of T by elements of $\mathrm{B}^{\mathrm{T},\diamondsuit}_{\theta(\tau)}$. Since each element $U \neq \mathrm{T}$ of $\mathrm{B}^{\mathrm{T},\diamondsuit}_{\theta(\tau)}$ is a finite union of elements of $\mathrm{B}^{\mathrm{T}}_{\theta(\tau)}$ (see the proof of Lemma 5.1.4) we can suppose that all elements of c are elements of $\mathrm{B}^{\mathrm{T}}_{\theta(\tau)}$.

Let $U_i = U^{\mathrm{T}}_{\delta_i}(\mathbf{H}_i)$, $i \in \{0, ..., n\}$, where $\delta_i \in \theta(\tau)$ and $\mathbf{H}_i \in \mathrm{C(R)}$. Let $\varepsilon_i = \theta^{-1}(\delta_i)$, $i \in \{0, ..., n\}$. (It is possible that $|\{\delta_0, ..., \delta_n\}| < n$ and, therefore, $|\{\varepsilon_0, ..., \varepsilon_n\}| < n$.) Without loss of generality we can suppose that there exists an element s of $\mathcal{F}$ such that $\mathbf{H}_i$ is a $\sim^s$-equivalence class for every $i \in \{0, ..., n\}$, $\{\delta_0, ..., \delta_n\} \subset s$, and $\theta(\vartheta_{\mathcal{F}}(t)) \subset s$, for every subset t of the set $\{\varepsilon_0, ..., \varepsilon_n\}$.

Let $\mathbf{H}$ be a $\sim^s$-equivalence class. Denote by $i(\mathbf{H})$ the set of all integers i of $\{0, ..., n\}$ such that $\mathbf{H}_i = \mathbf{H}$. Then, the set

$$c^{\mathbf{H}} \equiv \{U^{\mathrm{T}}_{\delta_i}(\mathbf{H}) : i \in i(\mathbf{H})\}$$

is a cover of $\mathrm{T}(\mathbf{H})$ by elements of $\mathrm{B}^{\mathrm{T}}_{\theta(\tau)}$. We note that $\delta_i \neq \delta_{i'}$ and, therefore, $\varepsilon_i \neq \varepsilon_{i'}$ if i and i' are distinct elements of $i(\mathbf{H})$.

Let $X \in \mathbf{H}$. Then, the set

$$c^X \equiv \{U^X_{\delta_i} : i \in i(\mathbf{H})\} = \{V^X_{\varepsilon_i} : i \in i(\mathbf{H})\}$$

is an open cover of X by elements of B^X. Without loss of generality we can suppose that $\varepsilon_i < \varepsilon_{i'}$ if $i < i'$. Let

$$\vartheta_{\mathcal{F}}(\{\varepsilon_i : i \in i(\mathbf{H})\} = \{\eta_i : i \in i(\mathbf{H})\},$$

where $\eta_i < \eta_{i'}$ if $i < i'$.

Since $\mathbf{N}(X)$ is a $(\vartheta_\vee, \vartheta_\wedge, \vartheta, \vartheta_{\mathcal{F}}, m)$-normal indication of B^X the set

$$r^X \equiv \{V^X_{\eta_i} : i \in i(\mathbf{H})\}$$

is an open cover of X by elements of B^X having order $\leq m$ and satisfying the condition $V^X_{\eta_i} \subset V^X_{\varepsilon_i}$.

Let Y be another element of $\mathbf{H}$. Since $X \sim^s Y$ by the choice of s the set

$$c^Y \equiv \{U^Y_{\delta_i} : i \in i(\mathbf{H})\} = \{V^Y_{\varepsilon_i} : i \in i(\mathbf{H})\}$$

is a cover of Y by elements of B^Y and the set

$$r^Y \equiv \{V^Y_{\varepsilon_i} : i \in i(\mathbf{H})\}$$

is a cover of Y by elements of B^Y having order $\leq m$ and satisfying the condition $V^Y_{\eta_i} \subset V^Y_{\varepsilon_i}$.

Setting $\theta(\eta_i) = \delta^i$ we have that the set

$$r^{\mathbf{H}} \equiv \{U^{\mathrm{T}}_{\delta^i}(\mathbf{H}) : i \in i(\mathbf{H})\}$$

is a cover of $\mathrm{T}(\mathbf{H})$ by elements of $\mathrm{B}^{\mathrm{T}}_{\theta(\tau)}$ having order $\leq m$ and satisfying the condition $U^{\mathrm{T}}_{\delta^i}(\mathbf{H}) \subset U^{\mathrm{T}}_{\delta_i}(\mathbf{H})\}$. Then, the set

$$r \equiv \cup\{r^{\mathbf{H}} : \mathbf{H} \in \mathrm{C}(\sim^s)\}$$

is a cover of T by elements of $\mathrm{B}^{\mathrm{T}}_{\theta(\tau)}$ having order $\leq m$, which is a refinement of the cover c.

Thus the base $\mathrm{B}^{\mathrm{T},\diamond}_{\theta(\tau)}$ is (b^n-dim $\leq m$)-dimensional, which completes the proof of the lemma. □

Definition. We denote by b^n-dim the base dimension-like function that has as domain the class of all bases and as range the class $\{-1, \infty\} \cup \omega$ satisfying the following condition:

$$\mathrm{b}^n\text{-dim}(A, X) \leq m, \ m \in \{-1\} \cup \omega,$$

if and only if A is contained in a (b^n-dim $\leq m$)-dimensional normal base for the open subsets of the space X.

The proof of the next proposition is similar to the proof of Proposition 5.1.5. Instead of Lemma 5.1.4 using in the proof of Proposition 5.1.5 we use here the Lemma 5.4.4.

5.4.5 Proposition. *The class*

$$\mathbb{P}(\mathrm{b}^n\text{-dim} \leq m), \ m \in \{-1\} \cup \omega,$$

is a saturated class of bases. □

Definition. We denote by s-b^n-dim the space dimension-like function that has as domain the class of all spaces and as range the class $\{-1, \infty\} \cup \omega$ satisfying the following condition: s-b^n-dim$(X) \leq m$, $m \in \{-1\} \cup \omega$, if and only if the space X has a (b^n-dim $\leq m$)-dimensional normal base for the open subsets of X.

The following corollary follows by the fact the class $\mathbb{P}$(s-b^n-dim $\leq m$) is the space-component of the class $\mathbb{P}$(b^n-dim $\leq m$) of bases.

5.4.6 Corollary. *The class*

$$\mathbb{P}(\text{s-b}^n\text{-dim} \leq m),\ m \in \{-1\} \cup \omega,$$

is a saturated class of spaces. □

5.4.7 Proposition. *If a space X is normal, then*

$$\text{s-b}^n\text{-dim}(X) \leq \dim(X).$$

Proof. Let X be a normal space. The proposition is true if $\dim(X) = \infty$. Suppose that $\dim(X) = m$, where $m \in \{-1\} \cup \omega$. To prove the proposition we need to prove the existence of a (b^n-dim $\leq m$)-dimensional normal base of cardinality $\leq \tau$ for the open subsets of X.

Let B^0 be a normal base of cardinality $\leq \tau$ for the open subsets of X. By induction on $n \in \omega$ we can construct a sequence $B^0, B^1, ..., B^n, ...$ of sets of cardinality $\leq \tau$ consisting of open subsets of X such that for every $n > 0$ we have:

(a) the set B^n contains the set $(B^{n-1})^\diamond$,

(b) for every two elements U_0 and U_1 of B^{n-1}, whose union is X, there exist two elements V_0 and V_1 of B^n such that the pair $(X \setminus V_0, X \setminus V_1)$ is a screening of the pair $(X \setminus U_0, X \setminus U_1)$, and

(c) for every cover $c \equiv \{U_0, ..., U_k\}$ of X by elements of B^{n-1} there exists a cover $r \equiv \{V_0, ..., V_p\}$ of X by elements of B^n, which is a refinement of c and has order $\leq m$.

In order to satisfy conditions (b) and (c) we use respectively the fact that X is normal and the fact that $\dim(X) \leq m$. It is easy to verify that the set $B \equiv \cup\{B^n : n \in \omega\}$ is the required normal base of cardinality $\leq \tau$ for the open subsets of X. □

5.4.8 Proposition. *If X is a compact Hausdorff space, then*

$$\dim(X) \le \text{s-b}^n\text{-dim}(X).$$

Proof. Let X be a compact Hausdorff space such that

$$\text{s-b}^n\text{-dim}(X) \le m.$$

Therefore, there exists a $(\text{b}^n\text{-dim} \le m)$-dimensional normal base B of cardinality $\le \tau$ for open subsets of X.

We need to prove that $\dim(X) \le m$. Let c be a finite open cover of X. Since X is compact there exists a finite cover c^B of X by elements of B, which is a refinement of c. Since B is $(\text{b}^n\text{-dim} \le m)$-dimensional there exists a finite cover r^B of X by elements of B, which is a refinement of c^B and, therefore, a refinement of c and has order $\le m$. This fact proves the proposition. □

Propositions 5.4.7 and 5.4.8 imply the following consequence.

5.4.9 Corollary. *For a compact Hausdorff space X we have*

$$\text{s-b}^n\text{-dim}(X) = \dim(X). \ \square$$

5.4.10 Proposition. *Let C be an $(\text{b}^n\text{-dim} \le m)$-dimensional normal base for the closed subsets of a space X, where $m \in \{-1\} \cup \omega$. Then, $c(C)$ is also $(\text{b}^n\text{-dim} \le m)$-dimensional normal base for the closed subsets of the space $c(X, C)$.*

Proof. Denote by B the complementary to C base for the open subsets of X and by B^c the complementary to $c(C)$ base for the open subsets of $c(X, C)$. It suffices to prove that for every finite cover $c \equiv \{U_0, ..., U_k\}$ of $c(X, C)$ by elements of B^c there exists a finite cover r of $c(X, C)$ by elements of B^c, which is a refinement of c and has order $\le m$.

For every $i \in \{0, ..., k\}$ let $c(X, C) \setminus U_i = c(F_i)$, where $F_i \in C$. Then, the set

$$c^X \equiv \{X \setminus F_0, ..., X \setminus F_k\}$$

is an open cover of X. Indeed, otherwise there exists a point x of X such that $x \in F_0 \cap ... \cap F_k$. Therefore, the point φ_x^C of $c(X, C)$ (that is, the maximal filter consisting of all elements of C containing the point x of X) belongs to all sets $c(F_i)$, $i \in \{0, ..., k\}$, which means that the set $\{U_0, ..., U_k\}$ is not a cover of $c(X, C)$.

Since B is (b^n-dim $\leq m$)-dimensional there exists a cover

$$r^X \equiv \{V_0, ..., V_n\}$$

of X by elements of B, which is a refinement of c^X and has order $\leq m$. We set $G_i = X \setminus V_i$, $i \in \{0, ..., n\}$ and prove that the set

$$r \equiv \{c(X, C) \setminus c(G_0), ..., c(X, C) \setminus c(G_n)\}$$

is the required cover of $c(X, C)$. Indeed, let $\{V_{i_1}, ..., V_{i_j}\}$ be a subset of r^X. Then,

$$V_{i_1} \cap ... \cap V_{i_j} \neq \emptyset$$

if and only if the set $\{G_{i_1}, ..., G_{i_j}\}$ is not a cover of X, that is if and only if (see Lemma 5.3.1) the set $\{c(G_{i_1}), ..., c(G_{i_j})\}$ is not a cover of $c(X, C)$ or, equivalently, if and only if

$$(c(X, C) \setminus c(G_{i_1})) \cap ... \cap (c(X, C) \setminus c(G_{i_j})) \neq \emptyset.$$

This means that the order of r is $\leq m$.

On the other hand, $c(X, C) \setminus c(G_{j_0}) \subset U_{j_1}$, where $j_0 \in \{1, ..., n\}$ and $j_1 \in \{1, ..., k\}$, if and only if $c(F_{j_1}) = c(X, C) \setminus U_{j_1} \subset c(G_{j_0})$, that is if and only if $F_{j_1} \subset G_{j_0}$ or, equivalently, $V_{j_0} \subset X \setminus F_{j_1}$. This fact means that r is a refinement of c. Thus, r is the required cover of $c(X, C)$. □

5.4.11 Proposition. *Let $m \in \{-1\} \cup \omega$. In the class of all compact Hausdorff spaces with dimension* dim $\leq m$ *there exist universal elements.*

Proof. Let $I\!P$ be the considered class. By Corollary 5.4.9 the class $I\!P$ is a subclass of the class $\mathbb{P}$(s-b^n-dim $\leq m$). By Corollary 5.4.6 in the last class there exist universal elements. Let T be such an element. Then, we have s-b^n-dim$(T) \leq m$, that is T has a (b^n-dim $\leq m$)-dimensional normal base C for the closed subsets of T. Consider the compact Hausdorff space $c(T, C)$. By Proposition 5.4.10, $\dim(T) \leq m$. Therefore, $c(T, C) \in I\!P$, which means that $c(T, C)$ is a universal element in $I\!P$. □

5.4.12 Lemma. *Let* $m \in \{-1\} \cup \omega$. *Let also* C^0 *be a* (b^n-dim $\leq m$)-*dimensional normal base of cardinality* $\leq \tau$ *for the closed subsets of a space* X *and* C^{-1} *a set of cardinality* $\leq \tau$ *consisting of compact subsets of* X. *Then, the minimal ring* C *containing the set* $C^0 \cup C^{-1}$ *is a* (b^n-dim $\leq m$)-*dimensional normal base for the closed subsets of* X.

Proof. Without loss of generality we can suppose that C^{-1} is closed under finite union and finite intersection. Denote by B^0 the complementary to C^0 base for the open subsets of X and by B^{-1} the set consisting of the complements of all elements of C^{-1}.

By Lemma 5.3.10 the base C is normal. Therefore, it suffices to prove only that C is (b^n-dim $\leq m$)-dimensional. For this fact we denote by B the complementary to C normal base for the open subsets of X and consider a finite cover c of X by elements of B. We need to construct a cover r of X by elements of B, which has order $\leq m$.

Since B is the minimal ring containing the sets B^0 and B^{-1} each element of B and, therefore, each element of c is a finite union of the sets of the form $U^0 \cap U^{-1}$, where $U^0 \in B^0$ and $U^{-1} \in B^{-1}$. Replacing if it is nessesary the cover c by another cover, which is a refinement of c, we can suppose, without loss of generality, that all elements of c have such a form.

Let $c = \{U_0, ..., U_k\}$ and let $U_i = U_i^0 \cap U_i^{-1}$, where $U_i^0 \in B^0$ and $U_i^{-1} \in B^{-1}$. Therefore, $U_i = U_i^0 \setminus K_i^{-1}$, where $K_i^{-1} = X \setminus U_i^{-1} \in C^{-1}$.

Now, we construct successively covers $c_i = \{U_{i,0}, ..., U_{i,k}\}$ of X, $i \in \{0, ..., k\}$, having the following properties:

(a) $U_{0,j} = U_j$ if $j \in \{1, ..., k\}$, and $U_{0,0}$ is an element of B^0, which is contained in U_0 and

(b) $U_{i,j} = U_{i-1,j}$ if $i \in \{1, ..., k\}$ and $j \in \{0, ...k\} \setminus \{i\}$, and $U_{i,i}$ is an element of B^0, which is contained in $U_{i-1,i} = U_i$.

First, we construct the cover c_0, that is the set $U_{0,0}$. Consider the element U_0 of c. Since $U_0 = U_0^0 \setminus K_0^{-1}$ we have

$$K_0^{-1} \subset \cup\{U_j : j \in \{1, ..., k\}\} \equiv W.$$

By the compactness of the set K_0^{-1} and since the base B^0 is a ring there exists an element V of B^0 such that $K_0^{-1} \subset V \subset W$. Since C^0 is disjunctive for every $x \in K_0^{-1}$ there exist an element F_x of C^0 such that $x \in F_x$ and $F_x \cap (X \setminus V) = \emptyset$. Since C^0 is base-normal there exists a screening (H_x, G_x) of the pair $(F_x, X \setminus V)$ by elements of C^0. Therefore, the set

$\{X \setminus G_x : x \in K_0^{-1}\}$ is a cover of K_0^{-1} by open subsets of X, which means that there exists a finite subset $\{x_0, ..., x_p\}$ of K_0^{-1} such that

$$K_0^{-1} \subset \cup\{X \setminus G_x : x \in \{x_0, ..., x_p\}\}.$$

We set

$$U_{0,0} = \cap\{X \setminus H_x : x \in \{x_0, ..., x_p\}\}.$$

Then, $U_{0,0} \in B$. For every $x \in K_0^{-1}$ we have $(X \setminus G_x) \cap (X \setminus F_x) = \emptyset$, which means that $K_0^{-1} \cap U_{0,0} = \emptyset$, that is $U_{0,0} \subset U_0$. On the other hand, since $X \setminus V \subset X \setminus H_x$ for every $x \in K_0^{-1}$, we have $X \setminus V \subset U_{0,0}$, which means that the set $c_0 \equiv \{U_{0,0}, U_1, ..., U_k\}$ is a cover of X. Moreover, c_0 is a refinement of c.

Furthermore, the construction of the cover c_i is similar to the construction of c_0. It is nessesary only to replace in the construction the cover c by the cover c_{i-1}.

Finally, to prove the lemma we note that the cover c_k consists of elements of B^0 and it is a refinement of c. Since B^0 is (b^n-dim $\leq m$)-dimensional there exists a finite cover r by elements of B^0, which is a refinement of c and has order $\leq m$. Thus, r is the requered cover. □

The next lemma is proved similarly to the proof of Lemma 5.3.11. Instead of Lemma 5.3.10 using in the proof of Lemma 5.3.11 we use here Lemma 5.4.12.

5.4.13 Lemma. *Let X be a completely regular space and*

$$X = \cup\{X_\lambda : \lambda \in \Lambda\}$$

such that $|\Lambda| \leq \tau$, X_λ is a compact subspace of X, and

$$\dim(X_\lambda) \equiv m_\lambda \in \{-1\} \cup \omega.$$

Then, there exists a normal base C^X of cardinality $\leq \tau$ for the closed subsets of X such that for every $\lambda \in \Lambda$ the set

$$\{X_\lambda \cap F : F \in C^X\}$$

is a (b^n-dim $\leq m_\lambda$)*-dimensional normal base for the closed subsets of the space X_λ.* □

The proof of the following lemma is omitted. It can be done using the method of the proof of Lemma 5.2.4.

5.4.14 Lemma. *Let X be a completely regular space and*

$$X = \cup\{X_\lambda : \lambda \in \Lambda\}$$

such that $|\Lambda| \leq \tau$, X_λ is a compact subspace of X, and

$$\dim(X_\lambda) \equiv m_\lambda \in \{-1\} \cup \omega.$$

Let also C^X be a base of cardinality $\leq \tau$ for the closed subsets of X_λ such that for every $\lambda \in \Lambda$ the set

$$C^{X,\lambda} \equiv \{X_\lambda \cap F : F \in C^X\}$$

is a (b^n-dim $\leq m_\lambda$)-dimensional normal base for the closed subsets of X_λ. Then, there exist:

(a) *an admissible triad $(\vartheta_\vee, \vartheta_\wedge, \vartheta)$,*
(b) *a mapping $\vartheta_{\mathcal{F},\lambda}$ of $\mathcal{F} \setminus \{\emptyset\}$ into itself, $\lambda \in \Lambda$,*
(c) *a mapping θ_Λ of Λ into τ, and*
(d) *a $(\vartheta_\vee, \vartheta_\wedge, \vartheta)$-normal indication*

$$F^X \equiv \{F^X_\delta : \delta \in \tau\} \text{ of } C^X$$

such that for every $\lambda \in \Lambda$ we have $F^X_{\theta_\Lambda(\lambda)} = X_\lambda$ and

$$F^{X,\lambda} \equiv \{F^X_\delta \cap X_\lambda : \delta \in \tau\}$$

is a $(\vartheta_\vee, \vartheta_\wedge, \vartheta, \vartheta_{\mathcal{F},\lambda}, m_\lambda)$-normal indication of $C^{X,\lambda}$. Moreover, the admissible triad $(\vartheta_\vee, \vartheta_\wedge, \vartheta)$ and the mappings $\vartheta_{\mathcal{F},\lambda}$, $\lambda \in \Lambda$, are independent of the space X. □

Notation. Let ν be a cardinal such that $1 \leq \nu \leq \tau$, ξ a mapping of ν into the set $\{-1\} \cup \omega$, and $\prec$ a partial order on the set ν. We denote by $\mathbb{P}(\dim, \nu, \xi, \prec)$ the class consisting of all ν-sequences $(X, X_0, ..., X_\eta, ...)_{\eta \in \nu}$ such that

(a) X is a completely regular space,
(b) X_η is a compact subset of X,
(c) $X = \cup\{X_\eta : \eta \in \nu\}$,

(d) $X_{\eta_0} \subset X_{\eta_1}$ if $\eta_0 \prec \eta_1$, and
(e) $\dim(X_\eta) \leq \xi(\eta)$.

The proof of the next proposition is omitted. This proof is similar to the proof of Proposition 5.3.13.

5.4.15 Proposition. *In the class* $\mathbb{P}(\dim, \nu, \xi, \prec)$ *there exists a properly universal element.* □

Chapter 6

Mappings and universality

In this chapter we consider mappings of spaces. Whenever we consider an arbitrary mapping it is assumed that this mapping is continuous. We study the universality problem for classes consisting of mappings. For this purpose we apply the main construction of Containing Spaces separately to the class consisting of the domains and to the class consisting of the ranges of all mappings of the considered class.

In Section 6.1 we consider classes consisting of mappings with different ranges. We give the notion of a saturated class and prove the intersection and the universality properties of such classes. We also prove the following results: (a) the class of the domains and the class of the ranges of all elements of a saturated class of mappings are saturated classes of spaces, and (b) the class of all (open) mappings whose domains and ranges belong to a given saturated classes of spaces is saturated. (We note that the property of a mapping to be continuous or to be open is a τ-property.) We give some variations of the closeness of a mapping (which have τ-formulations) and prove the last mentioned result for such mappings. Finally, we prove that for every complete saturated class $\mathbb{P}$ of subsets and a saturated class $\mathbb{F}$ of mappings the class

$$\{(f^{-1}(P^f), D_f) : f \in \mathbb{F}, (P^f, R_f) \in \mathbb{P}\}$$

is a complete saturated class of subsets. (We recall that by D_f and R_f we denote the domain and the range of the mapping f).

In Section 6.2 we consider classes consisting of mappings that have as range a fixed space denoted by Y. Such mappings are called Y-mappings. For classes of Y-mappings we also give the notion of a saturated class and prove the intersection and the universality properties. Furthermore, we prove the following results: (a) the class of the domains of all elements of a saturated class of Y-mappings is saturated, (b) the class of all (open) Y-mappings whose domains belong to a given saturated class of spaces is saturated, and (c) for every saturated class $I\!F$ of mappings and for some subset P of Y (satisfying a given condition) the class $\{(f^{-1}(P), D_f) : f \in I\!F\}$ is a complete saturated class of subsets. In particular, the class consisting of all pairs (Q, X) where Q is a zero set of a space X is a (complete) saturated class of subsets. Finally, we give some notions concerning the separation of the points and the closed subsets of a space by Y-mappings. These notions define classes consisting of spaces whose points and closed subsets can be separated in the corresponding sense. We prove that these classes are saturated. In particular, we give another proof that the class of all completely regular spaces is saturated.

In Section 6.3 we present the notion of a commutative diagram of mappings on a (fixed) set Λ of cardinality $\leq \tau$ with a (fixed) order $\prec$ which we will simply call a diagram. In the case where the order $\prec$ is directed, diagrams are called inverse spectra. On the other hand, if Λ consists of two elements λ_0 and λ_1 and $(\lambda_0, \lambda_1) \in \prec$, diagrams can be identified with mappings. We generalize results of Section 6.1 to diagrams. We also prove that in the class of the limit spaces of all elements of a saturated class of inverse spectra there exist universal elements. In particular, in the class of the limit spaces of all inverse spectra whose spaces belong to a given saturated class of spaces and mappings to a given saturated class of mappings there exist universal elements.

In Section 6.4 we consider the following classes: the class $I\!P$(compact) consisting of all compact spaces; the class $I\!P_c(\mathrm{Ind} \leq n)$, where $n \in \omega$, consisting of all compact elements of the class $I\!P(\mathrm{Ind} \leq n)$; and the class $I\!P_c(\dim \leq m)$, where $m \in \omega$, consisting of all compact elements of the class $I\!P(\dim \leq n)$. (We note that these classes are not saturated.) For these classes we prove some of the results of Sections 6.1–6.3.

6.1 Saturated classes of mappings with different ranges

Definitions. Let f and g be two mappings of spaces. A pair (i, j), where i is an embedding of D_g into D_f and j is an embedding of R_g into R_f such that

$$f \circ i = j \circ g,$$

is said to be an *embedding of g into f*.

An element f of a class $\mathbb{F}$ of mappings is said to be *universal* in $\mathbb{F}$ if for every $g \in \mathbb{F}$ there exists an embedding of g into f.

Definitions. Let $\mathbb{F}$ be a class of mappings. The class $\mathbb{F}$ is said to be *topological* if for every mappings f and g, for every homeomorphism i of D_g onto D_f, and for every homeomorphism j of R_g onto R_f such that $f \circ i = j \circ g$ the condition $f \in \mathbb{F}$ implies $g \in \mathbb{F}$. In the sequel all considered classes of mappings are assumed to be topological.

The classes of spaces

$$\{D_f : f \in \mathbb{F}\} \text{ and } \{R_f : f \in \mathbb{F}\}$$

are called the *domain-class* and the *range-class* of $\mathbb{F}$, respectively.

Definition. Let $\mathbf{F}$ be an indexed collection of mappings. The indexed collections

$$\mathbf{S}_d \equiv \{D_f : f \in \mathbf{F}\} \text{ and } \mathbf{S}_r \equiv \{R_f : f \in \mathbf{F}\}$$

are called the $\mathbf{F}$-*domain* and the $\mathbf{F}$-*range*, respectively.

Definitions. Let $\mathbf{F}$ be an indexed collection of mappings, $\mathbf{S}_d$ the $\mathbf{F}$-domain, and $\mathbf{S}_r$ the $\mathbf{F}$-range. Suppose that for every $f \in \mathbf{F}$ we have a τ-indexed family $\{U^f_\delta : \delta \in \tau\}$ of subsets of $\mathbf{S}_d$ and a τ-indexed family $\{V^f_\delta : \delta \in \tau\}$ of subsets of $\mathbf{S}_r$ and consider the $\mathbf{F}$-indexed sets

$$\mathbf{M}^d \equiv \{\{U^f_\delta : \delta \in \tau\} : f \in \mathbf{F}\} \text{ and}$$

$$\mathbf{M}^r \equiv \{\{V^f_\delta : \delta \in \tau\} : f \in \mathbf{F}\}.$$

The $\mathbf{F}$-indexed set

$$\{\{f^{-1}(V^f_\delta) : \delta \in \tau\} : f \in \mathbf{F}\}$$

is called the $\mathbf{F}$-*preimage* of $\mathbf{M}^r$ and the $\mathbf{F}$-indexed set

$$\{\{f(U^f_\delta) : \delta \in \tau\} : f \in \mathbf{F}\}$$

is called the **F**-*image* of $\mathbf{M}^d$.

The pair $(\mathbf{M}^d, \mathbf{M}^r)$ is said to be an **F**-*co-mark* if $\mathbf{M}^d$ is a co-mark of $\mathbf{S}_d$ and $\mathbf{M}^r$ is a co-mark of $\mathbf{S}_r$. If, moreover, $\mathbf{M}^d$ is a co-extension of the **F**-preimage of $\mathbf{M}^r$, then the **F**-co-mark $(\mathbf{M}^d, \mathbf{M}^r)$ is said to be *proper*.

An **F**-co-mark $(\mathbf{M}^d, \mathbf{M}^r)$ is said to be a *co-extension* of an **F**-co-mark $(\mathbf{M}^{d,+}, \mathbf{M}^{r,+})$ if the co-mark $\mathbf{M}^d$ is a co-extension of $\mathbf{M}^{d,+}$ and the co-mark $\mathbf{M}^r$ is a co-extension of $\mathbf{M}^{r,+}$.

Definitions. Let **F** be an indexed collection of mappings, $\mathbf{S}_d$ the **F**-domain, and $\mathbf{S}_r$ the **F**-range. Let also

$$\mathrm{R} = \{\sim^s : s \in \mathcal{F}\}$$

be an admissible family of equivalence relations on **F**. We denote by

$$\mathrm{R}^d = \{\sim^s_d : s \in \mathcal{F}\} \text{ and } \mathrm{R}^r = \{\sim^s_r : s \in \mathcal{F}\}$$

the admissible families of equivalence relations on $\mathbf{S}_d$ and $\mathbf{S}_r$, respectively, which are defined as follows:

$$D_f \sim^s_d D_g \text{ and } R_f \sim^s_r R_g$$

if and only if $f \sim^s g$ for every $f, g \in \mathbf{F}$ and $s \in \mathcal{F}$. The family R^d (respectively, R^r) is called the family of equivalence relations on the **F**-domain (respectively, on the **F**-range) *induced* by the family R.

Let $\mathbf{H} \in \mathrm{C}(\mathrm{R})$. We denote by $\mathbf{H}^d$ (respectively, by $\mathbf{H}^r$) the element of $\mathrm{C}(\mathrm{R}^d)$ (respectively, of $\mathrm{C}(\mathrm{R}^r)$), which is defined as follows: for an element f of **F** we have $D_f \in \mathbf{H}^d$ (respectively, $R_f \in \mathbf{H}^r$) if and only if $f \in \mathbf{H}$. The equivalence class $\mathbf{H}^d$ (respectively, $\mathbf{H}^r$) is called the equivalence class of $\mathrm{C}(\mathrm{R}^d)$ (respectively, of $\mathrm{C}(\mathrm{R}^r)$) *induced* by **H**.

Let $(\mathbf{M}^d, \mathbf{M}^r)$ be an **F**-co-mark. The family R is said to be $(\mathbf{M}^d, \mathbf{M}^r)$-*admissible* if R^d is $\mathbf{M}^d$-admissible and R^r is $\mathbf{M}^r$-admissible.

6.1.1 Lemma. *Let* **F** *be an indexed collection of mappings. Suppose that* $(\mathbf{M}^d, \mathbf{M}^r)$ *is a proper* **F***-co-mark and* R *an* $(\mathbf{M}^d, \mathbf{M}^r)$*-admissible family of equivalence relations on* **F**. *Then, there exists a unique mapping* f^{T} *of the containing space* $\mathrm{T}(\mathbf{M}^d, \mathrm{R}^d)$ *into Containing Space* $\mathrm{T}(\mathbf{M}^r, \mathrm{R}^r)$ *such that for every* $f \in \mathbf{F}$ *the pair* (i_f, j_f) *(where* i_f *is the natural embedding of* D_f *into* $\mathrm{T}(\mathbf{M}^d, \mathrm{R}^d)$ *and* j_f *is the natural embedding of* R_f *into* $\mathrm{T}(\mathbf{M}^r, \mathrm{R}^r)$*) is an embedding of* f *into* f^{T}. *Therefore, the mapping* f^{T} *is defined by the* *condition: if* $(x, D_f) \in \mathbf{a} \in \mathrm{T}(\mathbf{M}^d, \mathrm{R}^d)$ *and* $(f(x), R_f) \in \mathbf{b} \in \mathrm{T}(\mathbf{M}^r, \mathrm{R}^r)$, *then* $f^{\mathrm{T}}(\mathbf{a}) = \mathbf{b}$. *Moreover, if all mappings of* $\mathbf{F}$ *are onto, then the mapping* f^{T} *is also onto.*

Proof. Let

$$\mathbf{M}^d = \{\{U_\delta^f : \delta \in \tau\} : f \in \mathbf{F}\},$$

$$\mathbf{M}^r = \{\{V_\delta^f : \delta \in \tau\} : f \in \mathbf{F}\}, \text{ and}$$

$$\mathrm{R} = \{\sim^s : s \in \mathcal{F}\}.$$

Therefore,

$$\mathrm{R}^d = \{\sim_d^s : s \in \mathcal{F}\} \text{ and } \mathrm{R}^r = \{\sim_r^s : s \in \mathcal{F}\}.$$

Let also

$$\mathbf{M}^{-r} \equiv \{\{W_\delta^f : \delta \in \tau\} : f \in \mathbf{F}\}$$

be the $\mathbf{F}$-preimage of $\mathbf{M}^r$.

Let $\mathbf{a}$ be an arbitrary point of the Containing Space $\mathrm{T}(d) \equiv \mathrm{T}(\mathbf{M}^d, \mathrm{R}^d)$. Suppose that (x, D_f) and (y, D_g) are elements of $\mathbf{a}$. We prove that the pairs $(f(x), R_f)$ and $(g(y), R_g)$ belong to the same point of $\mathrm{T}(r) \equiv \mathrm{T}(\mathbf{M}^r, \mathrm{R}^r)$.

Indeed, we have $D_f \sim_d^s D_g$ for every $s \in \mathcal{F}$. Therefore, $R_f \sim_r^s R_g$ for every $s \in \mathcal{F}$. On the other hand, for every $\delta \in \tau$ either $x \in U_\delta^f$ and $y \in U_\delta^g$ or $x \notin U_\delta^f$ and $y \notin U_\delta^g$. In particular, for every $\delta \in \tau$ either $x \in W_\delta^f$ and $y \in W_\delta^g$ or $x \notin W_\delta^f$ and $y \notin W_\delta^g$, which means that for every $\delta \in \tau$ either $f(x) \in V_\delta^f$ and $g(y) \in V_\delta^g$ or $f(x) \notin V_\delta^f$ and $g(y) \notin V_\delta^g$.

Thus, the pairs $(f(x), R_f)$ and $(g(y), R_g)$ belong to the same point of $\mathrm{T}(r)$. Let $\mathbf{b}$ be this point. Setting $f^{\mathrm{T}}(\mathbf{a}) = \mathbf{b}$ we define a mapping f^{T} of the set $\mathrm{T}(d)$ into the set $\mathrm{T}(r)$. We prove that f^{T} is a continuous mapping. To this end it suffices to prove that the set $(f^{\mathrm{T}})^{-1}(V_\delta^{\mathrm{T}(r)}(\mathbf{H}^r))$, where $\mathbf{H} \in \mathrm{C}(\mathrm{R})$ and $V_\delta^{\mathrm{T}(r)}(\mathbf{H}^r)$ is an element of the standard base $B^{\mathrm{T}(r)}$ for $\mathrm{T}(r)$, is open in $\mathrm{T}(d)$.

Let $\mathbf{H} \in \mathrm{C}(\mathrm{R})$ and $V_\delta^{\mathrm{T}(r)}(\mathbf{H}^r) \in \mathrm{B}^{\mathrm{T}(r)}$. Let θ be an indicial mapping from $\mathbf{M}^{-r}$ to $\mathbf{M}^d$ and $\theta(\delta) = \varepsilon$. We prove that

$$(f^{\mathrm{T}})^{-1}(V_\delta^{\mathrm{T}(r)}(\mathbf{H}^r)) = U_\varepsilon^{\mathrm{T}(d)}(\mathbf{H}^d). \tag{6.1.1}$$

We note that $U_\varepsilon^{\mathrm{T}(d)}(\mathbf{H}^d)$ is the corresponding element of the standard base $\mathrm{B}^{\mathrm{T}(d)}$ for $\mathrm{T}(d)$. Indeed, let

$$\mathbf{a} \in (f^{\mathrm{T}})^{-1}(V_\delta^{\mathrm{T}(r)}(\mathbf{H}^r))$$

and $(x, D_f) \in \mathbf{a}$. Then,

$$(f(x), R_f) \in f^{\mathrm{T}}(\mathbf{a}) \in V_\delta^{\mathrm{T}(r)}(\mathbf{H}^r),$$

which means that $f(x) \in V_\delta^f$ and $R_f \in \mathbf{H}^r$. Then, $x \in W_\delta^f = U_\varepsilon^f$ and $D_f \in \mathbf{H}^d$, that is $\mathbf{a} \in U_\varepsilon^{\mathrm{T}(d)}(\mathbf{H}^d)$.

Conversely, let $\mathbf{a} \in U_\varepsilon^{\mathrm{T}(d)}(\mathbf{H}^d)$ and $(x, D_f) \in \mathbf{a}$. Then, $x \in U_\varepsilon^f = W_\delta^f$ and $D_f \in \mathbf{H}^d$. Therefore, $f(x) \in V_\delta^f$ and $R_f \in \mathbf{H}^r$, which means that $f^{\mathrm{T}}(\mathbf{a}) \in V_\delta^{\mathrm{T}(r)}(\mathbf{H}^r)$, that is $\mathbf{a} \in (f^{\mathrm{T}})^{-1}(V_\delta^{\mathrm{T}(r)}(\mathbf{H}^r))$, which completes the proof of relation (1∗). Thus, $(f^{\mathrm{T}})^{-1}(V_\delta^{\mathrm{T}(r)}(\mathbf{H}))$ is an open subset of $\mathrm{T}(d)$.

Now, let $f \in \mathbf{F}$. Let i_f be the natural embedding of D_f into $\mathrm{T}(d)$ and j_f the natural embedding of R_f into $\mathrm{T}(r)$. We prove that

$$f^{\mathrm{T}} \circ i_f = j_f \circ f. \tag{6.1.2}$$

Indeed, let $x \in D_f$, $(x, D_f) \in \mathbf{a} \in \mathrm{T}(d)$, and $(f(x), R_f) \in \mathbf{b} \in \mathrm{T}(r)$. Then, $i_f(x) = \mathbf{a}$ and $f^{\mathrm{T}}(\mathbf{a}) = \mathbf{b}$. On the other hand, $j_f(f(x)) = \mathbf{b}$, which proves relation (6.1.2). Thus, the pair (i_f, j_f) is an embedding of f into f^{T}. We note that relation (6.1.2) uniquely determines the mapping f^{T}. Finally, it is easy to see that if all mappings f of $\mathbf{F}$ are onto, then f^{T} is also onto. □

Definition. The unique mapping

$$f^{\mathrm{T}} : \mathrm{T}(\mathbf{M}^d, \mathrm{R}^d) \to \mathrm{T}(\mathbf{M}^r, \mathrm{R}^r)$$

of the above lemma is called the *containing mapping* for $\mathbf{F}$ (*corresponding to the pair* $(\mathbf{M}^d, \mathbf{M}^r)$ and the family R). It is easy to see that if $\mathbf{H} \in \mathrm{C}^\diamond(\mathrm{R})$, then

$$f^{\mathrm{T}}(\mathrm{T}(\mathbf{M}^d, \mathrm{R}^d, \mathbf{H}^d) \subset \mathrm{T}(\mathbf{M}^r, \mathrm{R}^r, \mathbf{H}^r).$$

We denote by $f_{\mathbf{H}}^{\mathrm{T}}$ the mapping of $\mathrm{T}(\mathbf{M}^d, \mathrm{R}^d, \mathbf{H}^d)$ into $\mathrm{T}(\mathbf{M}^r, \mathrm{R}^r, \mathbf{H}^r)$ such that

$$f^{\mathrm{T}}_{\mathbf{H}}(\mathbf{a}) = f^{\mathrm{T}}(\mathbf{a}),\ \mathbf{a} \in \mathrm{T}(\mathbf{M}^d, \mathrm{R}^d, \mathbf{H}^d).$$

The mapping $f^{\mathrm{T}}_{\mathbf{H}}$ is called the *containing mapping* for $\mathbf{F}$ (*corresponding* to the pair $(\mathbf{M}^d, \mathbf{M}^r)$, the family R, and the equivalence class $\mathbf{H}$).

Definition. A non-empty class $\mathbb{F}$ of mappings is said to be *saturated* if for every indexed collection $\mathbf{F}$ of elements of $\mathbb{F}$ there exists a proper $\mathbf{F}$-co-mark $(\mathbf{M}^{d,+}, \mathbf{M}^{r,+})$ satisfying the following condition: for every proper $\mathbf{F}$-co-mark $(\mathbf{M}^d, \mathbf{M}^r)$, which is a co-extension of $(\mathbf{M}^{d,+}, \mathbf{M}^{r,+})$, there exists an $(\mathbf{M}^d, \mathbf{M}^r)$-admissible family R^+ of equivalence relations on $\mathbf{F}$ such that for every family R, which is a final refinement of R^+, and an element $\mathbf{H}$ of $\mathrm{C}^{\diamond}(\mathrm{R})$ the containing mapping

$$f^{\mathrm{T}}_{\mathbf{H}} : \mathrm{T}(\mathbf{M}^d, \mathrm{R}^d, \mathbf{H}^d) \to \mathrm{T}(\mathbf{M}^r, \mathrm{R}^r, \mathbf{H}^r)$$

corresponding to the pair $(\mathbf{M}^d, \mathbf{M}^r)$, the family R and the equivalence class $\mathbf{H}$ belongs to $\mathbb{F}$.

The proper $\mathbf{F}$-co-mark $(\mathbf{M}^{d,+}, \mathbf{M}^{r,+})$ is said to be an *initial* $\mathbf{F}$-co-mark (*corresponding* to the class $\mathbb{F}$), and the family R^+ is an *initial* family *corresponding* to the $\mathbf{F}$-co-mark $(\mathbf{M}^{d,+}, \mathbf{M}^{r,+})$ and the class $\mathbb{F}$.

6.1.2 Lemma. *Let* $\mathbf{F}$ *be an indexed collection of mappings. Suppose that* R^+_d *and* R^+_r *are admissible families of equivalence relations on* $\mathbf{S}_d$ *and* $\mathbf{S}_r$, *respectively. Then, there exists an admissible family* R *of equivalence relations on* $\mathbf{F}$ *such that* R^d *is a final refinement of* R^+_d *and* R^r *is a final refinement of* R^+_r.

Proof. Let

$$\mathrm{R}^+_d = \{\sim^s_{+,d}\colon s \in \mathcal{F}\} \text{ and}$$

$$\mathrm{R}^+_r = \{\sim^s_{+,r}\colon s \in \mathcal{F}\}.$$

For every $s \in \mathcal{F}$ we define an equivalence relation $\sim^s$ on $\mathbf{F}$ as follows: for two elements f and g of $\mathbf{F}$ we set $f \sim^s g$ if and only if $D_f \sim^s_{+,d} D_g$ and $R_f \sim^s_{+,r} R_g$. It is easy to see that the family

$$\mathrm{R} \equiv \{\sim^s\colon s \in \mathcal{F}\}$$

is the required family of equivalence relations on $\mathbf{F}$. $\square$

6.1.3 Proposition. *The non-empty intersection of not more than τ many saturated classes of mappings is also a saturated class.*

Proof. Suppose that for every $\lambda \in \Lambda$, $I\!\!F_\lambda$ is a saturated class of mappings, where $|\Lambda| \leq \tau$, and set $I\!\!F = \cap\{I\!\!F_\lambda : \lambda \in \Lambda\}$. Let $\mathbf{F}$ be an indexed collection of elements of $I\!\!F$, $\mathbf{S}_d$ the $\mathbf{F}$-domain and $\mathbf{S}_r$ the $\mathbf{F}$-range.

For every $\lambda \in \Lambda$ denote by $(\mathbf{M}_\lambda^{d,+}, \mathbf{M}_\lambda^{r,+})$ an initial $\mathbf{F}$-comark corresponding to the class $I\!\!F_\lambda$. Let $\mathbf{M}^{r,+}$ be a co-mark of $\mathbf{S}_r$, which is a co-extension of all co-marks $\mathbf{M}_\lambda^{r,+}$, $\lambda \in \Lambda$. Denote by $\mathbf{M}^{d,+}$ any co-mark of $\mathbf{S}_d$, which is a co-extension of the $\mathbf{F}$-preimage of $\mathbf{M}^{r,+}$ and all co-marks $\mathbf{M}_\lambda^{d,+}$, where $\lambda \in \Lambda$. Then, $(\mathbf{M}^{d,+}, \mathbf{M}^{r,+})$ is a proper $\mathbf{F}$-co-mark.

Furthermore, let $(\mathbf{M}^d, \mathbf{M}^r)$ be an arbitrary proper $\mathbf{F}$-co-mark, which is a co-extension of $(\mathbf{M}^{d,+}, \mathbf{M}^{r,+})$. Denote by R_λ^+ an initial family of equivalence relations on $\mathbf{F}$ corresponding to the $\mathbf{F}$-co-mark $(\mathbf{M}^d, \mathbf{M}^r)$ and the class $I\!\!F_\lambda$. Let R_d^+ be an admissible family of equivalence relations on $\mathbf{S}_d$, which is a final refinement of all families $(\mathrm{R}_\lambda^+)^d$, where $\lambda \in \Lambda$ and let R_r^+ be an admissible family of equivalence relations on $\mathbf{S}_r$, which is a final refinement of all families $(\mathrm{R}_\lambda^+)^r$, $\lambda \in \Lambda$.

By Lemma 6.1.2 there exists an admissible family R^+ of equivalence relations on $\mathbf{F}$ such that $(\mathrm{R}^+)^d$ is a final refinement of R_d^+ and $(\mathrm{R}^+)^r$ is a final refinement of R_r^+. Consider an arbitrary admissible family R of equivalence relations on $\mathbf{F}$, which is a final refinement of R^+.

To prove the proposition it suffices to prove that for every $\mathbf{H} \in \mathrm{C}^\diamond(\mathrm{R})$ we have $f_\mathbf{H}^\mathrm{T} \in I\!\!F$, where $f_\mathbf{H}^\mathrm{T}$ is the containing mapping for $\mathbf{F}$ corresponding to the pair $(\mathbf{M}^d, \mathbf{M}^r)$, the family R, and the equivalence class $\mathbf{H}$. Since $I\!\!F_\lambda$ is saturated we have by construction that $f_\mathbf{H}^\mathrm{T} \in I\!\!F_\lambda$ for $\lambda \in \Lambda$. Therefore, $f_\mathbf{H}^\mathrm{T} \in I\!\!F$. □

6.1.4 Proposition. *In each saturated class of mappings there exist universal elements.*

Proof. Let $I\!\!F$ be a saturated class of mappings. We denote by $\mathbf{F}$ an indexed collection of elements of $I\!\!F$ such that for every $g \in I\!\!F$ there exists an element f of $\mathbf{F}$, a homeomorphism i of D_g onto D_f, and a homeomorphism j of R_g onto R_f such that $f \circ i = j \circ g$.

By Lemma 6.1.1 we need to prove that for a proper $\mathbf{F}$-co-mark $(\mathbf{M}^d, \mathbf{M}^r)$ and an $(\mathbf{M}^d, \mathbf{M}^r)$-admissible family R of equivalence relations on $\mathbf{F}$ the corresponding containing mapping

$$f^\mathrm{T} : \mathrm{T}(\mathbf{M}^d, \mathrm{R}^d) \to \mathrm{T}(\mathbf{M}^r, \mathrm{R}^r)$$

belongs to $\mathbb{F}$. But this results follows immediately by the fact that $\mathbb{F}$ is saturated. □

6.1.5 Proposition. *The domain-class and the range-class of any saturated class of mappings are saturated classes of spaces.*

Proof. Let $\mathbb{F}$ be a saturated class of mappings. Denote by $\mathbb{P}$ the domain-class of $\mathbb{F}$. Suppose that $\mathbf{S}$ is an indexed collection of elements of $\mathbb{P}$. For every element X of $\mathbf{S}$ we consider an element f of $\mathbb{F}$ such that $D_f = X$. Then, an indexed collection $\mathbf{F}$ of elements f of $\mathbb{F}$ is defined. Denote by $\mathbf{S}_r$ the $\mathbf{F}$-range. We note that the $\mathbf{F}$-domain coincides with the indexed set $\mathbf{S}$.

Since $\mathbb{F}$ is a saturated class there exists an $\mathbf{F}$-co-mark $(\mathbf{M}^{d,+}, \mathbf{M}^{r,+})$, which is an initial co-mark corresponding to the class $\mathbb{F}$. We prove that $\mathbf{M}^{d,+}$ is an initial co-mark of $\mathbf{S}$ corresponding to the class $\mathbb{P}$. To this end we consider an arbitrary co-mark $\mathbf{M}^d$ of $\mathbf{S}$, which is a co-extension of $\mathbf{M}^{d,+}$. Then, the proper $\mathbf{F}$-co-mark $(\mathbf{M}^d, \mathbf{M}^{r,+})$ is a co-extension of $(\mathbf{M}^{d,+}, \mathbf{M}^{r,+})$. Therefore, there exists an $(\mathbf{M}^d, \mathbf{M}^{r,+})$-admissible family R^+ of equivalence relations on $\mathbf{F}$, which is an initial family corresponding to the $\mathbf{F}$-co-mark $(\mathbf{M}^d, \mathbf{M}^{r,+})$ and the class $\mathbb{F}$.

To prove that $\mathbf{M}^{d,+}$ is initial it suffices to prove that $(\mathrm{R}^+)^d \equiv \mathrm{R}^{+,d}$ is an initial family of equivalence relations on $\mathbf{S}$ corresponding to the co-mark $\mathbf{M}^d$ and the class $\mathbb{P}$. To this end we consider an arbitrary family R_d of equivalence relations on $\mathbf{S}$, which is a final refinement of $\mathrm{R}^{+,d}$ and denote by $\mathbf{R}$ the admissible family of equivalence relations on $\mathbf{F}$ such that $\mathrm{R}^d = \mathrm{R}_d$.

Since $\mathbb{F}$ is saturated, for every $\mathbf{H} \in \mathrm{C}^{\diamondsuit}(\mathrm{R})$ the containing mapping $f^{\mathrm{T}}_{\mathbf{H}}$ for $\mathbf{F}$ that corresponds to the pair $(\mathbf{M}, \mathbf{M}^{r,+})$, the family R, and the equivalence class $\mathbf{H}$, belongs to $\mathbb{F}$. Therefore, the domain $D_{f^{\mathrm{T}}_{\mathbf{H}}}$ of $f^{\mathrm{T}}_{\mathbf{H}}$, which coincides with the space $\mathrm{T}(\mathbf{M}^d, \mathrm{R}^d, \mathbf{H}^d)$, belongs to the class $\mathbb{P}$ which proves that $\mathrm{R}^{+,d}$ is an initial family. Thus, $\mathbb{P}$ is saturated. Similarly we can prove that the range-class of $\mathbb{F}$ is saturated. □

6.1.6 Proposition. *Let $\mathbb{E}$ and $\mathbb{L}$ be saturated classes of spaces. Then, the class $\mathrm{C}(\mathbb{E}, \mathbb{L})$ of all mappings with the domain in $\mathbb{E}$ and the range in $\mathbb{L}$ is a saturated class of mappings.*

Proof. Let $\mathbf{F}$ be an indexed collection of elements of $\mathrm{C}(\mathbb{E}, \mathbb{L})$, let $\mathbf{S}_d$ be the $\mathbf{F}$-domain, and let $\mathbf{S}_r$ be the $\mathbf{F}$-range. Therefore, $\mathbf{S}_d$ is an indexed collection of elements of $\mathbb{E}$, and $\mathbf{S}_r$ is an indexed collection of elements of $\mathbb{L}$.

Since $\mathbb{L}$ is saturated there exists a co-mark $\mathbf{M}^{\mathbb{L},+}$ of $\mathbf{S}_r$, which is an initial co-mark corresponding to the class $\mathbb{L}$. Similarly, there exists a co-mark $\mathbf{M}^{\mathbb{E},+}$ of $\mathbf{S}_d$, which is an initial co-mark corresponding to the class $\mathbb{E}$. Without loss of generality we can suppose that $\mathbf{M}^{\mathbb{E},+}$ is a co-extension of the $\mathbf{F}$-preimage of $\mathbf{M}^{\mathbb{L},+}$, and, therefore, the $\mathbf{F}$-co-mark $(\mathbf{M}^{\mathbb{E},+}, \mathbf{M}^{\mathbb{L},+})$ is proper. Let $(\mathbf{M}^d, \mathbf{M}^r)$ be an arbitrary proper $\mathbf{F}$-co-mark, which is a co-extension of $(\mathbf{M}^{\mathbb{E},+}, \mathbf{M}^{\mathbb{L},+})$.

Denote by $\mathrm{R}^+_{\mathbb{L}}$ an initial family of equivalence relations on $\mathbf{S}_r$ corresponding to the co-mark $\mathbf{M}^r$ and the class $\mathbb{L}$. Also, denote by $\mathrm{R}^+_{\mathbb{E}}$ an initial family of equivalence relations on $\mathbf{S}_d$ corresponding to the co-mark $\mathbf{M}^d$ and the class $\mathbb{E}$. By Lemma 6.1.2 there exists an admissible family R^+ of equivalence relations on $\mathbf{F}$ such that $(\mathrm{R}^+)^d$ is a final refinement of $\mathrm{R}^+_{\mathbb{E}}$ and $(\mathrm{R}^+)^r$ is a final refinement of $\mathrm{R}^+_{\mathbb{L}}$. Obviously, R^+ is $(\mathbf{M}^d, \mathbf{M}^r)$-admissible.

Let R be an arbitrary admissible family of equivalence relations on $\mathbf{F}$, which is a final refinement of R^+. To prove the proposition it suffices to prove that for every $\mathbf{H} \in \mathrm{C}^{\diamond}(\mathrm{R})$ the containing mapping $f^{\mathrm{T}}_{\mathbf{H}}$ for $\mathbf{F}$ that corresponds to the pair $(\mathbf{M}^d, \mathbf{M}^r)$, the family R, and the equivalence class $\mathbf{H}$ belongs to $\mathbb{F}$, that is the domain $\mathrm{T}(\mathbf{M}^d, \mathrm{R}^d, \mathbf{H}^d)$ of this mapping belongs to the class $\mathbb{E}$ and the range $\mathrm{T}(\mathbf{M}^r, \mathrm{R}^r, \mathbf{H}^r)$ to the class $\mathbb{L}$. But this results follows immediately by the fact that $\mathbb{E}$ and $\mathbb{L}$ are saturated classes, and by the choise of the pair $(\mathbf{M}^d, \mathbf{M}^r)$ and the family R. □

Definition. Let f be a mapping of a space X into a space Y. The mapping f is said to be *open* (respectively, *closed*) if for every open (respectively, closed) subset G of X the set $f(G)$ is open (respectively, closed) in Y.

The mapping f is said to be *base-closed* (respectively, *c-base-closed*) if there exists a base B for X such that for every element U of B the set $f(\mathrm{Cl}_X(U))$ (respectively, the set $f(X \setminus U)$) is closed in Y.

Remark. The above given definitions of open and of closed mappings are not τ-formulations. However, for open mappings the following equivalent definition is a τ-formulation: a mapping $f : X \to Y$ is open if there exists a base B for X such that the set $f(U)$ is open for every element U of B.

6.1.7 Proposition. *Let $\mathbb{E}$ and $\mathbb{L}$ be saturated classes of spaces. Then,* *the (non-empty) class* $\mathrm{C}^{\mathrm{op}}(I\!\!E, I\!\!L)$ *of all open mappings with the domain in* $I\!\!E$ *and the range in* $I\!\!L$ *is a saturated class of mappings.*

Proof. Let $\mathbf{F}$ be an indexed collection of elements of $\mathrm{C}^{\mathrm{op}}(I\!\!E, I\!\!L)$, let $\mathbf{S}_d$ be the $\mathbf{F}$-domain, and let $\mathbf{S}_r$ be the $\mathbf{F}$-range. For every $f \in \mathbf{F}$ let

$$B^f \equiv \{W^f_\delta : \delta \in \tau\}$$

be a base for open subsets of the domain D_f of f. Then,

$$\mathbf{M}^{d,0} \equiv \{\{W^f_\delta : \delta \in \tau\} : f \in \mathbf{F}\}$$

is a co-mark of $\mathbf{S}_d$. We set

$$\mathbf{M}^{r,0} \equiv \{\{f(W^f_\delta) : \delta \in \tau\} : f \in \mathbf{F}\}.$$

We note that the $\mathbf{F}$-indexed collection $\mathbf{M}^{r,0}$ of τ-indexed families may not be a co-mark of $\mathbf{S}_r$.

As in Proposition 6.1.6 we construct the pairs

$$(\mathbf{M}^{I\!\!E,+}, \mathbf{M}^{I\!\!L,+}) \text{ and } (\mathbf{M}^d, \mathbf{M}^r)$$

and the families R^+ and R. We suppose that

$$\mathbf{M}^d = \{\{U^f_\delta : \delta \in \tau\} : f \in \mathbf{F}\},$$

$$\mathbf{M}^r = \{\{V^f_\delta : \delta \in \tau\} : f \in \mathbf{F}\}, \text{ and}$$

$$\mathrm{R} = \{\sim^s : s \in \mathcal{F}\}.$$

Therefore,

$$\mathrm{R}^d = \{\sim^s_d : s \in \mathcal{F}\} \text{ and } \mathrm{R}^r = \{\sim^s_r : s \in \mathcal{F}\}.$$

Without loss of generality we can suppose that the co-mark $\mathbf{M}^{I\!\!L,+}$ of $\mathbf{S}_r$ is a co-extension of $\mathbf{M}^{r,0}$ and the co-mark $\mathbf{M}^{I\!\!E,+}$ of $\mathbf{S}_d$ is a co-extension of $\mathbf{M}^{d,0}$. We denote by θ_d an indicial mapping from $\mathbf{M}^{d,0}$ to $\mathbf{M}^d$ and by θ_r an indicial mapping from $\mathbf{M}^{r,0}$ to $\mathbf{M}^r$.

To prove the proposition it suffices to prove that the containing mapping f^{T} of $\mathrm{T}(d) \equiv \mathrm{T}(\mathbf{M}^d, \mathrm{R}^d)$ into $\mathrm{T}(r) \equiv \mathrm{T}(\mathbf{M}^r, \mathrm{R}^r)$ is open. (In this case for every $\mathbf{H} \in \mathrm{C}^{\diamond}(\mathrm{R})$ the containing mapping $f^{\mathrm{T}}_{\mathbf{H}}$ will also be open.) For this purpose we consider the $\theta_d(\tau)$-standard base for $\mathrm{T}(\mathbf{M}^d, \mathrm{R}^d)$ and prove that for every element U of this base the set $f^{\mathrm{T}}(U)$ is open in $\mathrm{T}(\mathbf{M}^r, \mathrm{R}^r)$.

Let $U \equiv U_\delta^{\mathrm{T}(d)}(\mathbf{H}^d)$ be an element of the $\theta_d(\tau)$-standard base for $\mathrm{T}(d)$, where $\mathbf{H} \in \mathrm{C}(\mathrm{R})$. We set $\varepsilon = \theta_d^{-1}(\delta)$ and $\eta = \theta_r(\varepsilon)$. To prove that f^{T} is open it suffices to prove that

$$f^{\mathrm{T}}(U_\delta^{\mathrm{T}(d)}(\mathbf{H}^d)) = V_\eta^{\mathrm{T}(r)}(\mathbf{H}^r), \tag{6.1.3}$$

where $V_\eta^{\mathrm{T}(r)}(\mathbf{H}^r)$ is the corresponding element of the $\theta_r(\tau)$-standard base for the Containing Space $\mathrm{T}(r)$.

Let $\mathbf{a} \in U_\delta^{\mathrm{T}(d)}(\mathbf{H}^d)$ and $(x, D_f) \in \mathbf{a}$. Then, $x \in U_\delta^f = W_\varepsilon^f$ and $D_f \in \mathbf{H}^d$. Therefore, $f(x) \in f(W_\varepsilon^f) = V_\eta^f$ and $R_f \in \mathbf{H}^r$. Since $(f(x), R_f) \in f^{\mathrm{T}}(\mathbf{a}) \equiv \mathbf{b}$, we have that $\mathbf{b} \in V_\eta^{\mathrm{T}(r)}(\mathbf{H}^r)$. Thus, $f^{\mathrm{T}}(U) \subset V_\eta^{\mathrm{T}(r)}(\mathbf{H}^r)$.

Conversely, let $\mathbf{b} \in V_\eta^{\mathrm{T}(r)}(\mathbf{H}^r)$. Let also $(y, R_f) \in \mathbf{b}$. Then, $y \in V_\eta^f = f(W_\varepsilon^f)$ and $R_f \in \mathbf{H}^r$. Therefore, there exists a point $x \in W_\varepsilon^f = U_\delta^f$ such that $f(x) = y$. Denote by $\mathbf{a}$ the point of $\mathrm{T}(d)$ containing the pair (x, D_f). Since $D_f \in \mathbf{H}^d$ we have $\mathbf{a} \in U_\delta^{\mathrm{T}(d)}(\mathbf{H}^d)$. On the other hand, $f^{\mathrm{T}}(\mathbf{a}) = \mathbf{b}$, which means that $\mathbf{b} \in f^{\mathrm{T}}(U_\delta^{\mathrm{T}(d)}(\mathbf{H}^d))$. Thus, $V_\eta^{\mathrm{T}(r)}(\mathbf{H}^r) \subset f^{\mathrm{T}}(U)$, which proves relation (6.1.3). □

6.1.8 Corollary. *Let $\mathbb{E}$ and $\mathbb{L}$ be classes coinciding (independently) with one of the following classes of spaces:*

(1) *The class of all regular spaces.*

(2) *The class of all regular spaces of dimension* ind $\leq \alpha$, $\alpha \in \tau^+$.

(3) *The class of all regular countable-dimensional spaces.*

(4) *The class of all regular strongly countable-dimensional spaces.*

(5) *The class of all regular locally finite-dimensional spaces.*

Then, the class $\mathrm{C}(\mathbb{E}, \mathbb{L})$ (respectively, $\mathrm{C}^{\mathrm{op}}(\mathbb{E}, \mathbb{L})$) of all mappings (respectively, of all open mappings) with the domain in $\mathbb{E}$ and the range in $\mathbb{L}$, is saturated. Therefore, in this class there exist universal elements. □

In particular, for $\tau = \omega$ we have the following consequence.

6.1.9 Corollary. *Let $\mathbb{E}$ and $\mathbb{L}$ be classes coinciding (independently) with one of the following classes of spaces:*

(1) *The class of all separable metrizable spaces.*

(2) *The class of all separable metrizable spaces of dimension* ind $\leq \alpha$, $\alpha \in \tau^+$.

(3) *The class of all separable metrizable countable-dimensional spaces.*

(4) *The class of all separable metrizable strongly countable-dimensional spaces.*

(5) *The class of separable metrizable locally finite-dimensional spaces.* *Then, the class* $\mathrm{C}(\mathbb{E}, \mathbb{L})$ *(respectively,* $\mathrm{C}^{\mathrm{op}}(\mathbb{E}, \mathbb{L})$*) of all mappings (respectively, of all open mappings) with the domain in* $\mathbb{E}$ *and the range in* $\mathbb{L}$ *is saturated. Therefore, in this class there exist universal elements.* □

6.1.10 Proposition. *Let* $\mathbb{E}$ *and* $\mathbb{L}$ *be saturated classes of spaces. Then, the (non-empty) class* $\mathrm{C}^{\mathrm{bc}}(\mathbb{E}, \mathbb{L})$ *of all base-closed mappings with the domain in* $\mathbb{E}$ *and the range in* $\mathbb{L}$ *is a saturated class of mappings.*

Proof. The proof of this proposition in general outline is similar to the proof of Proposition 6.1.7. Let $\mathbf{F}$ be an indexed collection of elements of $\mathrm{C}^{\mathrm{bc}}(\mathbb{E}, \mathbb{L})$, $\mathbf{S}_d$ the $\mathbf{F}$-domain, and $\mathbf{S}_r$ the $\mathbf{F}$-range. As the base B^f for the domain D_f of an element f of $\mathbf{F}$ considered in the mentioned proposition we take here a base $\{W_\delta^f : \delta \in \tau\}$ such that for every $\delta \in \tau$ the set $f(\mathrm{Cl}_{D_f}(W_\delta^f))$ is closed in R_f.

Also, here we set

$$\mathbf{M}^{r,0} \equiv \{\{R_f \setminus f(\mathrm{Cl}_{D_f}(W_\delta^f)) : \delta \in \tau\} : f \in \mathbf{F}\}.$$

The pairs $(\mathbf{M}^{\mathbb{E},+}, \mathbf{M}^{\mathbb{L},+})$ and $(\mathbf{M}^d, \mathbf{M}^r)$, the families R^+ and R, the mapping f^{T}, the indicial mappings θ_d and θ_r, the ordinals δ, ε, and η, and the equivalence class $\mathbf{H}$ are the same as in Proposition 6.1.7.

To prove the proposition it suffices to show that

$$f^{\mathrm{T}}(\mathrm{Cl}_{\mathrm{T}(d)}(U_\delta^{\mathrm{T}(d)}(\mathbf{H}^d))) = \mathrm{T}(\mathbf{M}^r, \mathrm{R}^r, \mathbf{H}^r) \setminus V_\eta^{\mathrm{T}(r)}(\mathbf{H}^r). \quad (6.1.4)$$

(In this case for every $\mathbf{H} \in \mathrm{C}^{\diamond}(\mathrm{R})$ the containing mapping $f_{\mathbf{H}}^{\mathrm{T}}$ will be also belong to the class $\mathrm{C}^{\mathrm{bc}}(\mathbb{E}, \mathbb{L})$). Let

$$\mathbf{a} \in \mathrm{Cl}_{\mathrm{T}(d)}(U_\delta^{\mathrm{T}(d)}(\mathbf{H}^d)) \text{ and } (x, D_f) \in \mathbf{a}.$$

By Lemma 1.2.7,

$$x \in \mathrm{Cl}_{D_f}(U_\delta^f) = \mathrm{Cl}_{D_f}(W_\varepsilon^f) \text{ and } D_f \in \mathbf{H}^d.$$

Therefore,

$$f(x) \in f(\mathrm{Cl}_{D_f}(W_\varepsilon^f)) = R_f \setminus V_\eta^f \text{ and } R_f \in \mathbf{H}^r.$$

Since

$$(f(x), R_f) \in f^{\mathrm{T}}(\mathbf{a}) \equiv \mathbf{b},$$

we have that

$$\mathbf{b} \in \mathrm{T}(\mathbf{M}^r, \mathrm{R}^r, \mathbf{H}^r) \setminus V_\eta^{\mathrm{T}(r)}(\mathbf{H}^r).$$

Thus,

$$f^{\mathrm{T}}(\mathrm{Cl}_{\mathrm{T}(d)}(U_\delta^{\mathrm{T}(d)}(\mathbf{H}^d))) \subset \mathrm{T}(\mathbf{M}^r, \mathrm{R}^r, \mathbf{H}^r) \setminus V_\eta^{\mathrm{T}(r)}(\mathbf{H}^r).$$

Conversely, let

$$\mathbf{b} \in \mathrm{T}(\mathbf{H}^r) \setminus V_\eta^{\mathrm{T}(r)}(\mathbf{H}^r) \text{ and } (y, R_f) \in \mathbf{b}.$$

Then,

$$y \in R_f \setminus V_\eta^f = f(\mathrm{Cl}_{D_f}(W_\varepsilon^f)) \text{ and } R_f \in \mathbf{H}^r.$$

Therefore, there exists a point

$$x \in \mathrm{Cl}_{D_f}(W_\varepsilon^f) = \mathrm{Cl}_{D_f}(U_\delta^f)$$

such that $f(x) = y$. Denote by $\mathbf{a}$ the point of $\mathrm{T}(d)$ containing the pair (x, D_f). Since $D_f \in \mathbf{H}^d$ we have

$$\mathbf{a} \in \mathrm{Cl}_{\mathrm{T}(d)}(U_\delta^{\mathrm{T}(d)}(\mathbf{H}^d)).$$

On the other hand, $f^{\mathrm{T}}(\mathbf{a}) = \mathbf{b}$, which means that

$$\mathbf{b} \in f^{\mathrm{T}}(\mathrm{Cl}_{\mathrm{T}(d)}(U_\delta^{\mathrm{T}(d)}(\mathbf{H}^d))).$$

Thus,

$$\mathrm{T}(\mathbf{M}^r, \mathrm{R}^r, \mathbf{H}^r) \setminus V_\eta^{\mathrm{T}(r)}(\mathbf{H}^r) \subset f^{\mathrm{T}}(\mathrm{Cl}_{\mathrm{T}(d)}(U_\delta^{\mathrm{T}(d)}(\mathbf{H}^d))),$$

which proves relation (6.1.4). □

6.1.11 Proposition. *Let $\mathbb{E}$ and $\mathbb{L}$ be saturated classes of spaces. Then, the (non-empty) class $\mathrm{C}^{\mathrm{cbc}}(\mathbb{E}, \mathbb{L})$ of all c-base-closed mapping with the domain in $\mathbb{E}$ and the range in $\mathbb{L}$ is a saturated class of mappings.*

Proof. The proof of this proposition in general outline is similar to the proof of Proposition 6.1.10. Let $\mathbf{F}$ be an indexed collection of elements of $\mathrm{C}^{\mathrm{cbc}}(\mathbb{E}, \mathbb{L})$, $\mathbf{S}_d$ the $\mathbf{F}$-domain, and $\mathbf{S}_r$ the $\mathbf{F}$-range. As the base B^f of the domain D^f of an element f of $\mathbf{F}$ (of Proposition 6.1.10) we consider here a base $\{W_\delta^f : \delta \in \tau\}$ with the property that for every $\delta \in \tau$ the subset $f(D_f \setminus W_\delta^f)$ of R_f is closed. Also, as the set $\mathbf{M}^{r,0}$ we consider here the following:

$$\mathbf{M}^{r,0} \equiv \{\{R_f \setminus f(D_f \setminus W_\delta^f) : \delta \in \tau\} : f \in \mathbf{F}\}.$$

Finally, instead of the relation (6.1.4) of Proposition 6.1.10, here we consider the relation

$$f^{\mathrm{T}}(\mathrm{T}(\mathbf{M}^d, \mathrm{R}^d, \mathbf{H}^d) \setminus U_\delta^{\mathrm{T}(d)}(\mathbf{H}^d)) = \mathrm{T}(\mathbf{M}^r, \mathrm{R}^r, \mathbf{H}^r) \setminus V_\eta^{\mathrm{T}(r)}(\mathbf{H}^r))$$

which is proved similarly. □

6.1.12 Proposition. *Let $\mathbb{P}$ be a complete saturated class of subsets and $\mathbb{F}$ a saturated class of mappings. Then, the (non-empty) class*

$$\mathbb{K} \equiv \{(f^{-1}(P^f), D_f) : f \in \mathbb{F}, (P^f, R_f) \in \mathbb{P}\}$$

is a complete saturated class of subsets.

Proof. Let $\mathbf{S}$ be an indexed collection of spaces and

$$\mathbf{K} \equiv \{K^X : X \in \mathbf{S}\}$$

a $\mathbb{K}$-restriction of $\mathbf{S}$. Therefore, for every $X \in \mathbf{S}$ there exists $f \in \mathbb{F}$ such that

$$(K^X, X) = (f^{-1}(P^f), D_f).$$

Thus, the correspondence $X \to f$ defines:

(a) an indexed collection $\mathbf{F}$ of elements of $\mathbb{F}$ such that the $\mathbf{F}$-domain $\mathbf{S}_d$ coincides (as an indexed collection) with $\mathbf{S}$, and

(b) a $I\!P$-restriction

$$\mathbf{P} \equiv \{P^f : f \in \mathbf{F}\}$$

of the $\mathbf{F}$-range $\mathbf{S}_r$ such that

$$\mathbf{K} = \{f^{-1}(P^f) : f \in \mathbf{F}\}.$$

Since $I\!P$ is a saturated class there exists a co-mark $\mathbf{M}^{I\!P,+}$ of $\mathbf{S}_r$, which is an initial co-mark corresponding to the restriction $\mathbf{P}$ and the class $I\!P$. Also, denote by $\mathrm{R}^{I\!P,+}$ an initial family of equivalence relations on $\mathbf{S}_r$ corresponding to the co-mark $\mathbf{M}^{I\!P,+}$ of $\mathbf{S}_r$, the restriction $\mathbf{P}$, and the class $I\!P$. Without loss of generality, we can suppose that the restriction $\mathbf{P}$ is $(\mathbf{M}^{I\!P,+}, \mathrm{R}^{I\!P,+})$-complete.

On the other hand, since $I\!F$ is a saturated class there exists an initial $\mathbf{F}$-co-mark $(\mathbf{M}^{d,+}, \mathbf{M}^{r,+})$ corresponding to the class $I\!F$. Without loss of generality, we can suppose that $\mathbf{M}^{r,+}$ is a co-extension of $\mathbf{M}^{I\!P,+}$.

Let $\mathbf{M}^d$ be an arbitrary co-mark of $\mathbf{S}_d$, which is a co-extension of $\mathbf{M}^{d,+}$. Then, $(\mathbf{M}^d, \mathbf{M}^{r,+})$ is a proper $\mathbf{F}$-co-mark, which is a co-extension of the co-mark $(\mathbf{M}^{d,+}, \mathbf{M}^{r,+})$. Therefore, there exists an initial family R^+ of equivalence relations on $\mathbf{F}$ corresponding to the $\mathbf{F}$-co-mark $(\mathbf{M}^d, \mathbf{M}^{r,+})$ and the class $I\!F$. Without loss of generality, we can suppose that the family $(\mathrm{R}^+)^r$ is a final refinement of the family $\mathrm{R}^{I\!P,+}$.

Consider an arbitrary admissible family

$$\mathrm{R} \equiv \{\sim^s : s \in \mathcal{F}\}$$

of equivalence relations on $\mathbf{F}$, which is a final refinement of the family R^+.

Thus, for any $\mathbf{H} \in \mathrm{C}^\diamondsuit(\mathrm{R})$ we can consider the containing mapping

$$f^{\mathrm{T}}_{\mathbf{H}} : \mathrm{T}(\mathbf{M}^d, \mathrm{R}^d, \mathbf{H}^d) \to \mathrm{T}(\mathbf{M}^{r,+}, \mathrm{R}^r, \mathbf{H}^r)$$

corresponding to the pair $(\mathbf{M}^d, \mathbf{M}^{r,+})$, the family R, and the equivalence class $\mathbf{H}$. Since $I\!F$ is saturated all these mappings are elements of $I\!F$.

To prove that $I\!K$ is saturated it suffices to prove that if $\mathbf{L}, \mathbf{H} \in \mathrm{C}^\diamondsuit(\mathrm{R})$ and $\mathbf{L} \subset \mathbf{H}$, then

$$(\mathrm{T}(\mathbf{M}^d|_{\mathbf{K}}, \mathrm{R}^d|_{\mathbf{K}}, \mathbf{L}^d|_{\mathbf{K}}), \mathrm{T}(\mathbf{M}^d, \mathrm{R}^d, \mathbf{H}^d)) \in I\!K. \tag{6.1.5}$$

Let $\mathbf{L}, \mathbf{H} \in \mathrm{C}^\diamondsuit(\mathrm{R})$ and $\mathbf{L} \subset \mathbf{H}$. Since $I\!P$ is saturated,

$$(\mathrm{T}(\mathbf{M}^{r,+}|_{\mathbf{P}}, \mathrm{R}^r|_{\mathbf{P}}, \mathbf{L}^r|_{\mathbf{P}}), \mathrm{T}(\mathbf{M}^{r,+}, \mathrm{R}^r, \mathbf{H}^r)) \in I\!P.$$

Therefore, in order to prove relation (6.1.5) it suffices to prove that

$$(f^{\mathrm{T}}_{\mathbf{H}})^{-1}(\mathrm{T}(\mathbf{M}^{r,+}|_{\mathbf{P}}, \mathrm{R}^r|_{\mathbf{P}}, \mathbf{L}^r|_{\mathbf{P}})) = \mathrm{T}(\mathbf{M}^d|_{\mathbf{K}}, \mathrm{R}^d|_{\mathbf{K}}, \mathbf{L}^d|_{\mathbf{K}}). \qquad (6.1.6)$$

Let

$$\mathbf{a} \in \mathrm{T}(\mathbf{M}^d|_{\mathbf{K}}, \mathrm{R}^d|_{\mathbf{K}}, \mathbf{L}^d|_{\mathbf{K}}).$$

Then there exists an element (x, D_f) of $\mathbf{a}$ such that $x \in f^{-1}(P^f)$ and $D_f \in \mathbf{L}^d$. Therefore,

$$(f(x), R_f) \in f^{\mathrm{T}}_{\mathbf{H}}(\mathbf{a}) \equiv \mathbf{b} \text{ and } R_f \in \mathbf{L}^r.$$

This fact means that

$$\mathbf{b} \in \mathrm{T}(\mathbf{M}^{r,+}|_{\mathbf{P}}, \mathrm{R}^r|_{\mathbf{P}}, \mathbf{L}^r|_{\mathbf{P}})$$

and, therefore,

$$\mathbf{a} \in (f^{\mathrm{T}}_{(\mathbf{H},\mathbf{L})})^{-1}(\mathrm{T}(\mathbf{M}^{r,+}|_{\mathbf{P}}, \mathrm{R}^r|_{\mathbf{P}}, \mathbf{L}^r|_{\mathbf{P}})).$$

Thus, the right side of the relation (6.1.6) is contained in the left.

Conversely, let

$$\mathbf{a} \in (f^{\mathrm{T}}_{\mathbf{H}})^{-1}(\mathrm{T}(\mathbf{M}^{r,+}|_{\mathbf{P}}, \mathrm{R}^r|_{\mathbf{P}}, \mathbf{L}^r|_{\mathbf{P}})) \text{ and } (x, D_f) \in \mathbf{a}.$$

Therefore, $D_f \in \mathbf{L}^d$. Then,

$$f^{\mathrm{T}}_{\mathbf{H}}(\mathbf{a}) = \mathbf{b} \in \mathrm{T}(\mathbf{M}^{r,+}|_{\mathbf{P}}, \mathrm{R}^r|_{\mathbf{P}}, \mathbf{L}^r|_{\mathbf{P}})$$

and, therefore, $(f(x), R_f) \in \mathbf{b}$. Since $\mathbf{P}$ is $(\mathbf{M}^{I\!P,+}, \mathrm{R}^{I\!P,+})$-complete, $\mathbf{P}$ is also $(\mathbf{M}^{r,+}, \mathrm{R}^r)$-complete. (See Lemma 2.2.3). Therefore, $f(x) \in P^f$. Then, $x \in f^{-1}(P^f)$, which means that

$$\mathbf{a} \in \mathrm{T}(\mathbf{M}^d|_{\mathbf{K}}, \mathrm{R}^d|_{\mathbf{K}}, \mathbf{L}^d|_{\mathbf{K}}).$$

Thus, the left side of the relation (6.1.6) is contained in the right which proves this relation.

To prove that $\mathbb{K}$ is complete it suffices to prove that $\mathbf{K}$ is $(\mathbf{M}^d, \mathrm{R}^d)$-complete, that is if $\mathbf{a} \in \mathrm{T}(\mathbf{M}^d|_{\mathbf{K}}, \mathrm{R}^d|_{\mathbf{K}})$ and $(x, D_f) \in \mathbf{a}$, then $x \in f^{-1}(P^f)$. But we have already proved this fact considering relation (6.1.6). The proof of the proposition is completed. □

Remark. It is easy to verify that Propositions 6.1.6, 6.1.7, 6.1.10, 6.1.11, and 6.1.12 as well as Corollaries 6.1.8 and 6.1.9 hold if it is supposed that the classes $\mathrm{C}(\mathbb{E}, \mathbb{L})$, $\mathrm{C}^{\mathrm{op}}(\mathbb{E}, \mathbb{L})$, $\mathrm{C}^{\mathrm{bc}}(\mathbb{E}, \mathbb{L})$, $\mathrm{C}^{\mathrm{cbc}}(\mathbb{E}, \mathbb{L})$, and $\mathbb{F}$ consist only of the mappings which are onto.

6.2 Mappings with the same range

Agreement. In this section it is supposed that: (a) a fixed space denoted by Y is given and (b) two disjoint subsets of Y denoted by O and I are fixed. We note that the weight of Y may be less than τ.

Definition. A mapping f is said to be an Y-*mapping* if $R_f = \mathrm{Y}$.

Definition. Let f and g be two Y-mappings. An embedding i of D_g into D_f is said to be an Y-*embedding* of g into f if $f \circ i = g$.

An element f of a class $\mathbb{F}$ of Y-mappings is said to be *universal* in $\mathbb{F}$ if for every element g of $\mathbb{F}$ there exists an Y-embedding of g into f.

Definition. A class $\mathbb{F}$ of Y-mappings is said to be *topological* if for every Y-mappings f and g and for every homeomorphism i of D_g onto D_f such that $f \circ i = g$ the condition $f \in \mathbb{F}$ imply $g \in \mathbb{F}$. In the sequel all considered classes of Y-mappings are assumed to be topological.

Definition. Let $\mathbf{F}$ be an indexed collection of Y-mappings and

$$M \equiv \{U_\delta^{\mathrm{Y}} : \delta \in \tau\}$$

a τ-indexed family of subsets of Y. Then, the $\mathbf{F}$-indexed set

$$\{\{f^{-1}(U_\delta^{\mathrm{Y}}) : \delta \in \tau\} : f \in \mathbf{F}\}$$

of τ-indexed families is called the $\mathbf{F}$-*preimage* of M.

6.2.1 Lemma. *Let* $\mathbf{F}$ *be an indexed collection of* Y*-mappings,* $\mathbf{S}_d$ *the* $\mathbf{F}$*-domain, and* M *a mark of* Y*. Let also* $\mathbf{M}$ *be a co-mark of* $\mathbf{S}_d$*, which* *is a co-extension of the* **F***-preimage of* M *and* R *an* **M***-admissible family of equivalence relations on* $\mathbf{S}_d$*. Then, there exists a unique mapping* $f_{\mathrm{Y}}^{\mathrm{T}}$ *of the containing space* $\mathrm{T}(\mathbf{M}, \mathrm{R})$ *into the space* Y *such that for every* $f \in \mathbf{F}$ *the natural embedding of* D_f *into* $\mathrm{T}(\mathbf{M}, \mathrm{R})$ *is an* Y*-embedding of* f *into* $f_{\mathrm{Y}}^{\mathrm{T}}$*. Therefore, the mapping* $f_{\mathrm{Y}}^{\mathrm{T}}$ *is defined by the condition:* $f_{\mathrm{Y}}^{\mathrm{T}}(\mathbf{a}) = f(x)$*, where* $(x, D_f) \in \mathbf{a} \in \mathrm{T}(\mathbf{M}, \mathrm{R})$*. Moreover, if all mappings of* **F** *are onto, then* $f_{\mathrm{Y}}^{\mathrm{T}}$ *is also onto.*

Proof. In general outline the proof of the lemma is similar to the proof of Lemma 6.1.1. Let

$$\mathbf{M} = \{\{U_\delta^f : \delta \in \tau\} : f \in \mathbf{F}\},$$

$$\mathrm{R} = \{\sim^s : s \in \mathcal{F}\}, \text{ and}$$

$$M = \{V_\delta : \delta \in \tau\}.$$

Let also

$$\mathbf{M}^{-r} \equiv \{\{f^{-1}(V_\delta) \equiv W_\delta^f : \delta \in \tau\} : f \in \mathbf{F}\}$$

be the **F**-preimage of M.

Let $\mathbf{a}$ be an arbitrary point of the Containing Space $\mathrm{T} \equiv \mathrm{T}(\mathbf{M}, \mathrm{R})$. Suppose that (x, D_f) and (y, D_g) be two elements of $\mathbf{a}$. Then, for every $\delta \in \tau$ either $x \in U_\delta^f$ and $y \in U_\delta^g$ or $x \notin U_\delta^f$ and $y \notin U_\delta^g$. In particular, for every $\delta \in \tau$ either $x \in W_\delta^f$ and $y \in W_\delta^g$ or $x \notin W_\delta^f$ and $y \notin W_\delta^g$, which means that for every $\delta \in \tau$ either $f(x) \in V_\delta$ and $g(y) \in V_\delta$ or $f(x) \notin V_\delta$ and $g(y) \notin V_\delta$. Since Y is a T_0-space the last relations means that $f(x) = g(y)$. Setting $f_{\mathrm{Y}}^{\mathrm{T}}(\mathbf{a}) = f(x)$ we define a mapping $f_{\mathrm{Y}}^{\mathrm{T}}$ of T into the space Y.

Let θ be an indicial mapping from $\mathbf{M}^{-r}$ to $\mathbf{M}$. To prove that $f_{\mathrm{Y}}^{\mathrm{T}}$ is continuous it suffices to prove that for every $\delta \in \tau$,

$$(f_{\mathrm{Y}}^{\mathrm{T}})^{-1}(V_\delta) = U_\varepsilon^{\mathrm{T}}(\mathbf{S}_d), \tag{6.2.1}$$

where $\varepsilon = \theta(\delta)$ and $U_\varepsilon^{\mathrm{T}}(\mathbf{S}_d)$ the corresponding element of the standard base for T. We consider here $\mathbf{S}_d$ as an element of $\mathrm{C}(\mathrm{R})$.

Let $\delta \in \tau$ and $\mathbf{a} \in (f_{\mathrm{Y}}^{\mathrm{T}})^{-1}(V_\delta)$. Let also $(x, D_f) \in \mathbf{a}$. Then, $f_{\mathrm{Y}}^{\mathrm{T}}(\mathbf{a}) = f(x) \in V_\delta$. This fact means that $x \in W_\delta^f = U_\varepsilon^f$ and, therefore, $\mathbf{a} \in U_\varepsilon^{\mathrm{T}}(\mathbf{S}_d)$.

Conversely, let $\mathbf{a} \in U_\varepsilon^{\mathrm{T}}(\mathbf{S}_d)$ and $(x, D_f) \in \mathbf{a}$. Then, $x \in U_\varepsilon^f = W_\delta^f = f^{-1}(V_\delta)$, which means that $f(x) = f_{\mathrm{Y}}^{\mathrm{T}}(\mathbf{a}) \in V_\delta$, that is $\mathbf{a} \in (f_{\mathrm{Y}}^{\mathrm{T}})^{-1}(V_\delta)$. This result proves relation (6.2.1).

Now, let f be an element of $\mathbf{F}$ and i_f the natural embedding of D_f into $\mathrm{T}(\mathbf{M}, \mathrm{R})$. Then we have

$$f_{\mathrm{Y}}^{\mathrm{T}} \circ i_f = f. \tag{6.2.2}$$

Indeed, let $x \in D_f$, $(x, D_f) \in \mathbf{a} \in \mathrm{T}(\mathbf{M}, \mathrm{R})$. Then, $i_f(x) = \mathbf{a}$ and $f_{\mathrm{Y}}^{\mathrm{T}}(\mathbf{a}) = f(x)$, which proves relation (6.2.2). Thus, i_f is an Y-embedding of f into $f_{\mathrm{Y}}^{\mathrm{T}}$. Relation (6.2.2) uniquely determines the mapping $f_{\mathrm{Y}}^{\mathrm{T}}$. Finally, it is easy to verify that if all mappings f of $\mathbf{F}$ are onto, then $f_{\mathrm{Y}}^{\mathrm{T}}$ is also onto. □

Definition. The unique mapping $f_{\mathrm{Y}}^{\mathrm{T}} : \mathrm{T}(\mathbf{M}, \mathrm{R}) \to \mathrm{Y}$ of Lemma 6.2.1 is said to be the Y-*containing mapping* for $\mathbf{F}$ (*corresponding* to $\mathbf{M}$ and R).

Let $\mathbf{H} \in \mathrm{C}^\diamond(\mathrm{R})$. The restriction of $f_{\mathrm{Y}}^{\mathrm{T}}$ onto $\mathrm{T}(\mathbf{M}, \mathrm{R}, \mathbf{H})$ is denoted by $f_{\mathrm{Y},\mathbf{H}}^{\mathrm{T}}$ and is called the Y-*containing mapping* for $\mathbf{F}$ (*corresponding* to $\mathbf{M}$, R, and $\mathbf{H}$).

Definition. A non-empty class $I\!\!F$ of Y-mappings is said to be *saturated* if for every indexed collection $\mathbf{F}$ of elements of $I\!\!F$ there exists a co-mark $\mathbf{M}^+$ of the $\mathbf{F}$-domain $\mathbf{S}_d$, which is a co-extension of the $\mathbf{F}$-preimage of a mark of Y, satisfying the following condition: for every co-extension $\mathbf{M}$ of $\mathbf{M}^+$ there exists an $\mathbf{M}$-admissible family R^+ of equivalence relations on $\mathbf{S}_d$ such that for every family R of equivalence relations on $\mathbf{S}_d$, which is a final refinement of R^+ and for every $\mathbf{H} \in \mathrm{C}^\diamond(\mathrm{R})$, the Y-containing mapping $f_{\mathrm{Y},\mathbf{H}}^{\mathrm{T}}$ of $\mathrm{T}(\mathbf{M}, \mathrm{R}, \mathbf{H})$ into Y belongs to $I\!\!F$.

The co-mark $\mathbf{M}^+$ is said to be an *initial* co-mark for $\mathbf{F}$ (*corresponding* to the class $I\!\!F$), and the family R^+ an *initial* family for $\mathbf{F}$ (*corresponding* to the co-mark $\mathbf{M}$ and the class $I\!\!F$).

The proofs of the below given Propositions 6.2.2–6.2.5 are similar to the proofs of the corresponding propositions of Section 6.1.

6.2.2 Proposition. *The non-empty intersection of not more than τ many saturated classes of Y-mappings is also a saturated class.* □

6.2.3 Proposition. *In each saturated class of Y-mappings there exist universal elements.* □

6.2.4 Proposition. *The domain-class of any saturated class of* Y-*mappings is a saturated class of spaces.* □

6.2.5 Proposition. *Let* $\mathbb{E}$ *be a saturated class of spaces. Then, the class* $\mathrm{C}(\mathbb{E}, \mathrm{Y})$ *of all* Y-*mappings with the domain in* $\mathbb{E}$ *is a saturated class of mappings.* □

6.2.6 Proposition. *Let* $\mathbb{E}$ *be a saturated class of spaces. Then, the (non-empty) class* $\mathrm{C}^{\mathrm{op}}(\mathbb{E}, \mathrm{Y})$ *of all* Y-*mappings with the domain in* $\mathbb{E}$ *is a saturated class.*

Proof. Let $\mathbf{F}$ be an indexed collection of elements of $\mathrm{C}^{\mathrm{op}}(\mathbb{E}, \mathrm{Y})$ and $\mathbf{S}_d$ the $\mathbf{F}$-domain. Then, $\mathbf{S}_d$ is an indexed collection of elements of $\mathbb{E}$.

Denote by $\mathbf{M}^+$ an initial co-mark of $\mathbf{S}_d$ corresponding to the class $\mathbb{E}$ and let

$$\mathbf{M} \equiv \{\{U_\delta^f : \delta \in \tau\} : f \in \mathbf{F}\}$$

be an arbitrary extension of $\mathbf{M}^+$. Without loss of generality we can suppose that $\mathbf{M}^+$ is a co-extension of the $\mathbf{F}$-preimage of a mark of Y. Also, denote by R^+ an initial family of equivalence relations on $\mathbf{S}_d$ corresponding to the co-mark $\mathbf{M}$ and the class $\mathbb{E}$, and let R be an arbitrary admissible family of equivalence relations on $\mathbf{S}_d$, which is a final refinement of R^+.

By construction, for every $\mathbf{H} \in \mathrm{C}^{\diamond}(\mathrm{R})$ the space $\mathrm{T}(\mathbf{H}) \equiv \mathrm{T}(\mathbf{M}, \mathrm{R}, \mathbf{H})$ is an element of $\mathbb{E}$. Therefore, in order to prove the proposition it suffices to prove that the containing mapping $f_{\mathrm{Y},\mathbf{H}}^{\mathrm{T}}$ of $\mathrm{T}(\mathbf{H})$ into Y is open. For this it suffices to prove that for every element $U_\delta^{\mathrm{T}}(\mathbf{L})$, $\mathbf{L} \in \mathrm{C}(\mathrm{R})$, of the standard base B^{T} for T we have

$$f_{\mathrm{Y}}^{\mathrm{T}}(U_\delta^{\mathrm{T}}(\mathbf{L})) = \cup\{f(U_\delta^f) : D_f \in \mathbf{L}\}. \tag{6.2.3}$$

Let $\mathbf{a} \in U_\delta^{\mathrm{T}}(\mathbf{L}) \in B^{\mathrm{T}}$ and $(x, D_f) \in \mathbf{a}$. Then, $x \in U_\delta^f$ and $D_f \in \mathbf{L}$. Therefore, $f(x) \in f(U_\delta^f)$, which means that

$$f_{\mathrm{Y}}^{\mathrm{T}}(\mathbf{a}) \in \cup\{f(U_\delta^f) : D_f \in \mathbf{L}\}.$$

Conversely, let

$$y \in \cup\{f(U_\delta^f) : D_f \in \mathbf{L}\}.$$

Then, there exists $f \in \mathbf{F}$ such that $D_f \in \mathbf{L}$ and $y \in f(U_\delta^f)$. Therefore, there exists a point $x \in D_f$ such that $f(x) = y$. Denote by $\mathbf{a}$ the point of T containing the pair (x, D_f). Since $D_f \in \mathbf{L}$ we have $\mathbf{a} \in U_\delta^{\mathrm{T}}(\mathbf{L})$. On the other hand, $f_{\mathrm{Y}}^{\mathrm{T}}(\mathbf{a}) = f(x) = y$. Relation (6.2.3) is proved. □

6.2.7 Corollary. *Let $\mathbb{E}$ be one of the following classes of spaces:*
(1) The class of all regular spaces.
(2) The class of all regular spaces of dimension $\mathrm{ind} \leq \alpha$, $\alpha \in \tau^+$.
(3) The class of all regular countable-dimensional spaces.
(4) The class of all regular strongly countable-dimensional spaces.
(5) The class of all regular locally finite-dimensional spaces.
Then, the class $\mathrm{C}(\mathbb{E}, \mathrm{Y})$ *of all Y-mappings as well as the class* $\mathrm{C}^{\mathrm{op}}(\mathbb{E}, \mathrm{Y})$ *of all open Y-mappings with the domain in $\mathbb{E}$ are saturated. Therefore, in this class there exist universal elements.* □

In particular, if $\tau = \omega$ and Y is a separable metrizable space, then we have the following consequence.

6.2.8 Corollary. *Let $\mathbb{E}$ be one of the following classes of spaces:*
(1) The class of all separable metrizable spaces.
(2) The class of all separable metrizable spaces of dimension $\mathrm{ind} \leq \alpha$, $\alpha \in \tau^+$.
(3) The class of all separable metrizable countable-dimensional spaces.
(4) The class of all separable metrizable strongly countable-dimensional spaces.
(5) The class of separable metrizable locally finite-dimensional spaces.
Then, the class $\mathrm{C}(\mathbb{E}, \mathrm{Y})$ *of all Y-mappings (respectively,* $\mathrm{C}^{\mathrm{op}}(\mathbb{E}, \mathrm{Y})$ *of all open Y-mappings) with the domain in $\mathbb{E}$ is saturated. Therefore, in this class there exist universal elements.* □

6.2.9 Proposition. *Let $\mathbb{F}$ be a saturated class of Y-mappings and P a subset of Y satisfying the following condition: for every elements f and g of $\mathbb{F}$ for which $D_f \cap D_g = \emptyset$ there exists an element h of $\mathbb{F}$ such that $D_h = D_f \cup D_g$ and $h^{-1}(P) = f^{-1}(P)$. Then, the class*

$$\mathbb{K} \equiv \{(f^{-1}(P), D_f) : f \in \mathbb{F}\}$$

is a complete saturated class of subsets.

Proof. In general outline the proof of this proposition is similar to the proof of Proposition 6.1.12. Let $\mathbf{S}$ be an indexed collection of spaces and $\mathbf{K} \equiv \{K^X : X \in \mathbf{S}\}$ a $I\!\!K$-restriction of $\mathbf{S}$. Then, for every $X \in \mathbf{S}$ there exists an element f of $I\!\!F$ such that $(K^X, X) = (f^{-1}(P), D_f)$. The correspondence $X \to f$ defines an indexed collection $\mathbf{F}$ of elements of $I\!\!F$ such that the $\mathbf{F}$-domain $\mathbf{S}_d$ coincides with $\mathbf{S}$ and

$$\mathbf{K} = \{f^{-1}(P) : f \in \mathbf{F}\}.$$

Since $I\!\!F$ is saturated there exists a co-mark $\mathbf{M}^+$ of $\mathbf{S}_d$, which is an initial co-mark for $\mathbf{F}$ corresponding to the class $I\!\!F$. Without loss of generality we can suppose that $\mathbf{M}^+$ is a co-extension of the $\mathbf{F}$-preimage of a mark of Y. Let $\mathbf{M}$ be an arbitrary co-mark of $\mathbf{S}_d$, which is a co-extension of $\mathbf{M}^+$. Denote by R^+ an initial family for $\mathbf{F}$ corresponding to the co-mark $\mathbf{M}$ and the class $I\!\!F$ and let R be an arbitrary admissible family of equivalence relations on $\mathbf{S}_d$, which is a final refinement of the family R^+.

Thus, for every $\mathbf{H} \in \mathrm{C}^\diamondsuit(\mathrm{R})$ the Y-containing mapping

$$f^{\mathrm{T}}_{\mathrm{Y},\mathbf{H}} : \mathrm{T}(\mathbf{M}, \mathrm{R}, \mathbf{H}) \to \mathrm{Y}$$

corresponding to the co-mark $\mathbf{M}$, the family R, and the equivalence class $\mathbf{H}$, belongs to $I\!\!F$.

To prove that $I\!\!K$ is saturated it suffices to prove that if $\mathbf{H}$ and $\mathbf{L}$ are elements of $\mathrm{C}^\diamondsuit(\mathrm{R})$ and $\mathbf{H} \subset \mathbf{L}$, then

$$(\mathrm{T}(\mathbf{M}|_{\mathbf{K}}, \mathrm{R}|_{\mathbf{K}}, \mathbf{H}|_{\mathbf{K}}), \mathrm{T}(\mathbf{M}, \mathrm{R}, \mathbf{L})) \in I\!\!K. \tag{6.2.4}$$

For this purpose it suffices to prove that

$$(f^{\mathrm{T}}_{\mathrm{Y},\mathbf{H}})^{-1}(\mathrm{P}) = \mathrm{T}(\mathbf{M}|_{\mathbf{K}}, \mathrm{R}|_{\mathbf{K}}, \mathbf{H}|_{\mathbf{K}}). \tag{6.2.5}$$

Indeed, if $\mathbf{L}$ is an element of $\mathrm{C}^\diamondsuit(\mathrm{R})$ such that $\mathbf{H} \subset \mathbf{L}$, then $\mathbf{E} \equiv \mathbf{L} \setminus \mathbf{H}$ is also an element of $\mathrm{C}^\diamondsuit(\mathrm{R})$ and, therefore, the containing mapping $f^{\mathrm{T}}_{\mathrm{Y},\mathbf{E}}$ belongs to $I\!\!F$. By assumption of the lemma there exists an element h of $I\!\!F$ such that

$$D_h = \mathrm{T}(\mathbf{M}, \mathrm{R}, \mathbf{H}) \cup \mathrm{T}(\mathbf{M}, \mathrm{R}, \mathbf{E}) = \mathrm{T}(\mathbf{M}, \mathrm{R}, \mathbf{L})$$

and

$$h^{-1}(P) = \mathrm{T}(\mathbf{M}|_{\mathbf{K}}, \mathrm{R}|_{\mathbf{K}}, \mathbf{H}|_{\mathbf{K}}),$$

which means that relation (6.2.4) is true.

Now, we prove relation (6.2.5). Let $\mathbf{a} \in \mathrm{T}(\mathbf{M}|_{\mathbf{K}}, \mathrm{R}|_{\mathbf{K}}, \mathbf{H}|_{\mathbf{K}})$. Then, there exists an element (x, D_f) of $\mathbf{a}$ such that $x \in f^{-1}(P)$ and $D_f \in \mathbf{H}$. Therefore, $f^{\mathrm{T}}_{\mathrm{Y},\mathbf{H}}(\mathbf{a}) = f(x) \in P$. Thus, the right side of (6.2.5) is contained in the left.

Conversely, let $\mathbf{a} \in (f^{\mathrm{T}}_{\mathrm{Y},\mathbf{H}})^{-1}(P)$ and $(x, D_f) \in \mathbf{a}$. Then, $D_f \in \mathbf{H}$ and $f^{\mathrm{T}}_{\mathrm{Y},\mathbf{H}}(\mathbf{a}) = f(x) \in P$. Therefore, $x \in f^{-1}(P)$, which means that $\mathbf{a} \in \mathrm{T}(\mathbf{M}|_{\mathbf{K}}, \mathrm{R}|_{\mathbf{K}}, \mathbf{H}|_{\mathbf{K}})$. Thus, the left side of (6.2.5) is contained in the right proving this relation.

Moreover, the last considerations show that $\mathbf{K}$ is an $(\mathbf{M}, \mathrm{R})$-complete restriction. This proves that $I\!K$ is a complete class. □

Suppose that Y is the space of real numbers with the usual topology, $P = \{0\}$, and $I\!F$ is the class of all Y-mappings. Then, Proposition 6.2.9 implies the following consequence.

6.2.10 Corollary. *The class of all pairs (Q, X), where Q is a zero-set of a space X, is a complete saturated class of subsets.* □

Remark. It is easy to verify that the above given propositions and corollaries hold if everywhere by a Y-mapping we mean a mapping, which is onto.

Below, we define different notions of the separation of the points and the closed subsets of a space by Y-mappings. Using these notions we define some saturated classes of spaces.

Definitions. It is said that a family F of Y-mappings (Y,O,I)-*separates* the points and the closed subsets of a space X if:

(a) the cardinality of F is $\leq \tau$ and

(b) for every closed subset K of X and a point $x \in X$, which does not belongs to K, there exists a mapping $f \in F$ such that $f(x) \in \mathrm{O}$ and $f(K) \subset \mathrm{I}$.

A space X is said to be (Y,O,I)-*separated* if there exists a family F of Y-mappings of X, which (Y,O,I)-separates the points and the closed subsets of X.

6.2.11 Proposition. *The class of all* (Y,O,I)*-separated spaces is saturated.*

Proof. Let $\mathbf{S}$ be an indexed collection of (Y,O,I)-separated spaces. For every $X \in \mathbf{S}$ we denote by

$$F^X \equiv \{f^X_\varepsilon : \varepsilon \in \tau\}$$

an indexed family of Y-mappings of X, which (Y,O,I)-separates the points and the closed subsets of X. For every $\varepsilon \in \tau$ we set

$$\mathbf{F}_\varepsilon \equiv \{f^X_\varepsilon : X \in \mathbf{S}\}.$$

Let M be a mark of Y and

$$\mathbf{M} \equiv \{\{U^X_\delta : \delta \in \tau\} : X \in \mathbf{S}\}$$

a co-mark of $\mathbf{S}$, which is a co-extension of the $\mathbf{F}_\varepsilon$-preimage of M for all $\varepsilon \in \tau$. We define a family

$$\mathrm{R}^+ \equiv \{\sim^s_+ : s \in \mathcal{F}\}$$

of equivalence relations on $\mathbf{S}$ as follows. Let $s \in \mathcal{F}$ and $X, Z \in \mathbf{S}$. We say that $X \sim^s_+ Z$ if the following conditions are satisfied:

(a) X and Y are $\sim^s_{\mathbf{M}}$-equivalent, and

(b) if $\delta, \varepsilon \in s$, then $f^X_\varepsilon(X \setminus U^X_\delta) \subset \mathrm{I}$ if and only if $f^Z_\varepsilon(Z \setminus U^Z_\delta) \subset \mathrm{I}$.

It is easy to verify that R^+ is an $\mathbf{M}$-admissible family of equivalence relations on $\mathbf{S}$. Let

$$\mathrm{R} \equiv \{\sim^s : s \in \mathcal{F}\}$$

be an arbitrary family of equivalence relation on $\mathbf{S}$, which is a final refinement of R^+.

In order to prove the proposition it suffices to prove that for every $\mathbf{H} \in \mathrm{C}^\diamondsuit(\mathrm{R})$ the space $\mathrm{T}(\mathbf{H}) \equiv \mathrm{T}(\mathbf{M}, \mathrm{R}, \mathbf{H})$ is (Y,O,I)-separated. For this purpose it suffices to prove that the Containing Space $\mathrm{T} \equiv \mathrm{T}(\mathbf{M}, \mathrm{R})$ is (Y,O,I)-separated.

For every $\varepsilon \in \tau$ we denote by $f^{\mathrm{T}}_{\mathrm{Y},\varepsilon}$ the containing mapping of T into Y of Lemma 6.2.1 if in this lemma we replace the family $\mathbf{F}$ by the family $\mathbf{F}_\varepsilon$.

For every element $U_\delta^{\mathrm{T}}(\mathbf{L})$ of the standard base B^{T} of T and for every $\varepsilon \in \tau$ we denote by $f^{\mathrm{T}}_{\mathrm{Y},\varepsilon,\delta,\mathbf{L}}$ the mapping of T into Y such that $f^{\mathrm{T}}_{\mathrm{Y},\varepsilon,\delta,\mathbf{L}}(x) = f^{\mathrm{T}}_{\mathrm{Y},\varepsilon}(x)$ if $x \in \mathrm{T}(\mathbf{L})$ and

$$f^{\mathrm{T}}_{\mathrm{Y},\varepsilon,\delta,\mathbf{L}}(\mathrm{T} \setminus \mathrm{T}(\mathbf{L})) \subset \{b\},$$

where b is a fixed point of I. Obviously, $f^{\mathrm{T}}_{\mathrm{Y},\varepsilon,\delta,\mathbf{L}}$ is a continuous mapping. We prove that the family

$$F^{\mathrm{T}} \equiv \{f^{\mathrm{T}}_{\mathrm{Y},\varepsilon,\delta,\mathbf{L}} : \varepsilon \in \tau, \delta \in \tau, \mathbf{L} \in \mathrm{C}(\mathrm{R})\}$$

(Y,O,I)-separates the points and the closed subsets of T.

Indeed, let K be a closed subset of T and $\mathbf{a}$ a point of T which does not belong to K. Let $U_\delta^{\mathrm{T}}(\mathbf{H})$ be an element of B^{T}, which contains $\mathbf{a}$ and does not intersect K, and let (x, X) be an element of $\mathbf{a}$. There exists an element ε of τ such that $f_\varepsilon^X(x) \in \mathrm{O}$ and $f_\varepsilon^X(X \setminus U_\delta^X) \subset \mathrm{I}$.

Let s be an element of $\mathcal{F}$ such that $\mathbf{H} \in \mathrm{C}(\sim^s)$. We set $t = s \cup \{\delta, \varepsilon\}$ and denote by q an element of $\mathcal{F}$ such that $t \subset q$ and $\sim^q \subset \sim^t_+$. Let $\mathbf{L}$ be the $\sim^q$-equivalence class of X. Consider the mapping $f^{\mathrm{T}}_{\mathrm{Y},\varepsilon,\delta,\mathbf{L}}$ of F^{T}. By the definition of this mapping we have

$$f^{\mathrm{T}}_{\mathrm{Y},\varepsilon,\delta,\mathbf{L}}(\mathbf{a}) = f^{\mathrm{T}}_{\mathrm{Y},\varepsilon}(\mathbf{a}) = f_\varepsilon^X(x) \in \mathrm{O}.$$

Let $\mathbf{b} \in \mathrm{T} \setminus U_\delta^{\mathrm{T}}(\mathbf{L})$. If $\mathbf{b} \in \mathrm{T} \setminus \mathrm{T}(\mathbf{L})$, then we have $f^{\mathrm{T}}_{\mathrm{Y},\varepsilon,\delta,\mathbf{L}}(\mathbf{b}) = b$. Suppose that $\mathbf{b} \in \mathrm{T}(\mathbf{L}) \setminus U_\delta^{\mathrm{T}}(\mathbf{L})$ and let $(z, Z) \in \mathbf{b}$. Then, $z \in Z \setminus U_\delta^Z$ and $X \sim^q Z$. Since $\sim^q \subset \sim^t_+$, $X \sim^t_+ Z$. On the other hand, since $f_\varepsilon^X(X \setminus U_\delta^X) \subset \mathrm{I}$ by the definition of the equivalence relation $\sim^t_+$ we have $f_\varepsilon^Z(Z \setminus U_\delta^Z) \subset \mathrm{I}$, which means that $f_\varepsilon^Z(z) \in \mathrm{I}$ and, therefore, $f^{\mathrm{T}}_{\mathrm{Y},\varepsilon}(\mathbf{b}) = f_\varepsilon^Z(z) \in \mathrm{I}$. Thus, $f^{\mathrm{T}}_{\mathrm{Y},\varepsilon,\delta,\mathbf{L}}(\mathbf{b}) \in \mathrm{I}$ and T is a (Y,O,I)-separated space. □

Suppose that the space Y is the set of real numbers with the usual topology, the subset O of Y coincides with the singleton $\{0\}$, and the subset I of Y with the singleton $\{1\}$. Then, the class of all (Y,O,I)-separated spaces coincides with the class of all completely regular spaces. Thus, Proposition 6.2.10 implies the following consequence given another proof of Corollary 5.1.6.

6.2.12. Corollary. *The class of all completely regular spaces is saturated.* □

Definitions. We say that a family F of Y-mappings (Y,O*,I)-*separates* the points and the closed subsets of X if: (a) $|F| \leq \tau$ and (b) for every point $x \in X$ and for every closed subset K of X, which does not contain x, there exists an open subset U of x and a mapping $f \in F$ such that $x \in U$, $f(U) \subset \mathrm{O}$, and $f(K) \subset \mathrm{I}$.

A space X is said to be (Y,O*,I)-*separated* if there exists a family F of Y-mappings of X, which (Y,O*,I)-separates the points and the closed subsets of X.

The given below proposition is proved similarly to Proposition 6.2.11.

6.2.13 Proposition. *The class of all* (Y,O*,I)*-separated spaces is saturated.* □

Definition. Let $I\!F$ be a class of mappings. It is said that a family F^X of Y-mappings of a space X $I\!F$-(Y,O,I)-*separates* (respectively, $I\!F$-(Y,O*,I)-*separates*) the points and the closed subsets of a space X if: (a) the family F^X is contained in the class $I\!F$ and (b) the family F^X (Y,O,I)-separates (respectively, (Y,O*,I)-separates) the points and the closed subsets of X.

It is easy to verify the following proposition.

6.2.14 Proposition. *Let $I\!F$ be a saturated class of mappings. Then, the class of all $I\!F$-(Y,O,I)-separated spaces (respectively, the class of all $I\!F$-(Y,O*,I)-separated spaces) is a saturated class.* □

6.3 Diagrams and inverse spectra

Agreement 1. In this section it is supposed that: (a) a fixed set of cardinality $\leq \tau$, denoted by Λ, is given and (b) a fixed partial order on Λ, denoted by $\prec$, is given.

Definition. A *commutative diagram on* $\Lambda \cup \prec$ or for simplicity on Λ is a mapping D of $\Lambda \cup \prec$ into the class consisting of all spaces and all (continuous) mappings such that:

(a) $D(\lambda)$ is a space for every $\lambda \in \Lambda$,

(b) $D(\lambda_0, \lambda_1)$ is a mapping of $D(\lambda_1)$ into $D(\lambda_0)$ for every $(\lambda_0, \lambda_1) \in \prec$,

(c) $D(\lambda_0, \lambda_1) \circ D(\lambda_1, \lambda_2) = D(\lambda_0, \lambda_2)$ for every $(\lambda_0, \lambda_1), (\lambda_1, \lambda_2) \in \prec$ (therefore, $(\lambda_0, \lambda_2) \in \prec$), and

(d) $D(\lambda, \lambda)$ is the identical mapping of the space $D(\lambda)$, $\lambda \in \Lambda$.
The space $D(\lambda)$ is called the λ-*component* of D and the mapping $D(\lambda_0, \lambda_1)$ is called the (λ_0, λ_1)-*component* of D. For simplicity, in what follows a commutative diagram on $\Lambda \cup \prec$ will be called a *diagram*.

Definition. Let D_0 and D_1 be two diagrams. An indexed family

$$f_\Lambda \equiv \{f_\lambda : \lambda \in \Lambda\},$$

where f_λ is a mapping of the space $D_0(\lambda)$ into (onto) the space $D_1(\lambda)$, is said to be a *morphism of* D_0 *into (onto)* D_1 if

$$f_{\lambda_0} \circ D_0(\lambda_0, \lambda_1) = D_1(\lambda_0, \lambda_1) \circ f_{\lambda_1}$$

for every element $(\lambda_0, \lambda_1) \in \prec$.

A morphism $f_\Lambda \equiv \{f_\lambda : \lambda \in \Lambda\}$ of a diagram D_0 into (onto) a diagram D_1 is said to be an *isomorphism of* D_0 *into (onto)* D_1 if f_λ is an homeomorphism of $D_0(\lambda)$ into (onto) $D_1(\lambda)$ for every $\lambda \in \Lambda$. It is easy to see that in this case for every element $(\lambda_0, \lambda_1) \in \prec$ the pair $(f_{\lambda_1}, f_{\lambda_0})$ is an embedding of the mapping $D_0(\lambda_0, \lambda_1)$ into the mapping $D_1(\lambda_0, \lambda_1)$.

Definition. A class $\mathbb{D}$ of diagrams is said to be *topological* if for every diagrams D_0 and D_1 and an isomorphism f_Λ of D_0 onto D_1 the condition $D_0 \in \mathbb{D}$ implies $D_1 \in \mathbb{D}$. In what follows an arbitrary considered class of diagrams is assumed to be topological.

Definition. An element D of a class $\mathbb{D}$ of diagrams is said to be *universal in* $\mathbb{D}$ if for every element D' of $\mathbb{D}$ there exists an isomorphism of D' into D.

Definitions. Let D be a diagram. Any indexed set

$$M^D \equiv \{M^{D(\lambda)} : \lambda \in \Lambda\},$$

where $M^{D(\lambda)}$ is a mark of $D(\lambda)$, is called a D-*mark*.

Definitions and notation. Let $\mathbf{D}$ be an indexed collection of diagrams. Any indexed set

$$\overline{\mathbf{M}} \equiv \{M^D \equiv \{M^{D(\lambda)} : \lambda \in \Lambda\} : D \in \mathbf{D}\}, \tag{6.3.1}$$

where M^D is a D-mark, is called a **D**-*co-mark*. Suppose that (6.3.1) is a **D**-co-mark. For every $\lambda \in \Lambda$ we set

$$\mathbf{D}(\lambda) = \{D(\lambda) : D \in \mathbf{D}\} \text{ and}$$

$$\overline{\mathbf{M}}(\lambda) = \{M^{D(\lambda)} : D \in \mathbf{D}\}.$$

Also, for every $(\lambda_0, \lambda_1) \in \prec$ we set

$$\mathbf{D}(\lambda_0, \lambda_1) = \{D(\lambda_0, \lambda_1) : D \in \mathbf{D}\}.$$

It is easy to see that $\mathbf{D}(\lambda)$ is an indexed collection of spaces and $\overline{\mathbf{M}}(\lambda)$ is a co-mark of $\mathbf{D}(\lambda)$. Also, $\mathbf{D}(\lambda_0, \lambda_1)$ is an indexed collection of mappings such that the $\mathbf{D}(\lambda_0, \lambda_1)$-domain is $\mathbf{D}(\lambda_1)$ and the $\mathbf{D}(\lambda_0, \lambda_1)$-range is $\mathbf{D}(\lambda_0)$.

The **D**-co-mark $\overline{\mathbf{M}}$ is said to be *proper* if for every $(\lambda_0, \lambda_1) \in \prec$ the co-mark $\overline{\mathbf{M}}(\lambda_1)$ is a co-extension of the $\mathbf{D}(\lambda_0, \lambda_1)$-preimage of $\overline{\mathbf{M}}(\lambda_0)$.

Let $\overline{\mathbf{M}}_0$ and $\overline{\mathbf{M}}_1$ be two **D**-co-marks. It is said that $\overline{\mathbf{M}}_1$ is a *co-extension* of $\overline{\mathbf{M}}_0$ if for every $\lambda \in \Lambda$, $\overline{\mathbf{M}}_1(\lambda)$ is a co-extension of $\overline{\mathbf{M}}_0(\lambda)$.

Definition. Let **D** be an indexed collection of diagrams and

$$\mathrm{R} \equiv \{\sim^s : s \in \mathcal{F}\}$$

a family of equivalence relations on **D**. For every $\lambda \in \Lambda$ we denote by

$$\mathrm{R}(\lambda) \equiv \{\sim^s_\lambda : s \in \mathcal{F}\}$$

the family of equivalence relations on $\mathbf{D}(\lambda)$, which is defined as follows: for two elements $D_0(\lambda)$ and $D_1(\lambda)$ of $\mathbf{D}(\lambda)$ we set $D_0(\lambda) \sim^s_\lambda D_1(\lambda)$ if and only if $D_0 \sim^s D_1$, $s \in \mathcal{F}$. Obviously, if R is admissible, then $\mathrm{R}(\lambda)$ is also admissible.

For every $(\lambda_0, \lambda_1) \in \prec$ we denote by

$$\mathrm{R}(\lambda_0, \lambda_1) \equiv \{\sim^s_{(\lambda_0,\lambda_1)} : s \in \mathcal{F}\}$$

the family of equivalence relations on $\mathbf{D}(\lambda_0, \lambda_1)$, which is defined as follows: for two elements $D_0(\lambda_0, \lambda_1)$ and $D_1(\lambda_0, \lambda_1)$ of $\mathbf{D}(\lambda_0, \lambda_1)$ we set

$$D_0(\lambda_0, \lambda_1) \sim^s_{(\lambda_0,\lambda_1)} D_1(\lambda_0, \lambda_1)$$

if and only if $D_0 \sim^s D_1$, $s \in \mathcal{F}$.

Let $\overline{\mathbf{M}}$ be a $\mathbf{D}$-co-mark. The family R is said to be $\overline{\mathbf{M}}$-*admissible* if $\mathrm{R}(\lambda)$ is $\overline{\mathbf{M}}(\lambda)$-admissible for every $\lambda \in \Lambda$.

For every element $\mathbf{H}$ of $\mathrm{C}^{\diamondsuit}(\mathrm{R})$ we set

$$\mathbf{H}(\lambda) = \{D(\lambda) \in \mathbf{D}(\lambda) : D \in \mathbf{H}\}.$$

Obviously, $\mathbf{H}(\lambda)$ is an element of $\mathrm{C}^{\diamondsuit}(\mathrm{R}(\lambda))$.

6.3.1 Lemma. *Let $\mathbf{D}$ be an indexed collection of diagrams and $\overline{\mathbf{M}}^+$ a $\mathbf{D}$-co-mark. Then, there exists a proper $\mathbf{D}$-co-mark $\overline{\mathbf{M}}$, which is a co-extension of $\overline{\mathbf{M}}^+$.*

Proof. Suppose that

$$\overline{\mathbf{M}}^+ = \{M^{D,+} : D \in \mathbf{D}\},$$

where

$$M^{D,+} = \{M^{D(\lambda),+} \equiv \{U_\delta^{D(\lambda),+} : \delta \in \tau\} : \lambda \in \Lambda\}.$$

Denote by ϑ an one-to-one mapping of the set $\Lambda \times \tau$ onto the set τ. For every $\lambda \in \Lambda$ we denote by $\kappa(\lambda)$ the set of all elements δ of τ such that if $\vartheta^{-1}(\delta) = (\lambda', \delta')$, then $(\lambda', \lambda) \in \prec$. It is easy to verify that if $(\lambda_0, \lambda_1) \in \prec$, then $\kappa(\lambda_0) \subset \kappa(\lambda_1)$. Since $(\lambda, \lambda) \in \prec$ the cardinality of the set $\kappa(\lambda)$ is equal to τ and, therefore, there exists an one-to-one mapping ϑ_λ of $\kappa(\lambda)$ onto τ.

Let $D \in \mathbf{D}$. For every $\lambda \in \Lambda$ we define a mark

$$\{U_\delta^{D(\lambda)} : \delta \in \tau\}$$

of $D(\lambda)$ as follows. Let $\delta \in \tau$ and $\vartheta_\lambda^{-1}(\delta) = \varepsilon \in \kappa(\lambda)$. Therefore, if $\vartheta^{-1}(\varepsilon) = (\lambda', \delta')$, then $(\lambda', \lambda) \in \prec$. We set

$$U_\delta^{D(\lambda)} = (D(\lambda', \lambda))^{-1}(U_{\delta'}^{D(\lambda'),+}). \tag{6.3.2}$$

It is easy to verify that the defined indexed set $\{U_\delta^{D(\lambda)} : \delta \in \tau\}$ is a mark of $D(\lambda)$. We denote this mark by $M^{D(\lambda)}$. Therefore,

$$M^D \equiv \{M^{D(\lambda)} : \lambda \in \Lambda\}$$

is a D-mark. Then, setting

$$\overline{\mathbf{M}} = \{M^D : D \in \mathbf{D}\}$$

we obtain a **D**-co-mark. We prove that this **D**-co-mark is as required. First we prove that $\overline{\mathbf{M}}$ is proper.

Let $(\lambda_0, \lambda_1) \in \prec$, $\delta \in \tau$, $\vartheta_{\lambda_0}^{-1}(\delta) = \varepsilon$, $\vartheta^{-1}(\varepsilon) = (\lambda', \delta')$, and $\eta = \vartheta_{\lambda_1}(\varepsilon)$. Then, we have

$$U_\delta^{D(\lambda_0)} = (D(\lambda', \lambda_0))^{-1}(U_{\delta'}^{D(\lambda'),+}) \text{ and}$$

$$U_\eta^{D(\lambda_1)} = (D(\lambda', \lambda_1))^{-1}(U_{\delta'}^{D(\lambda'),+}) =$$

$$(D(\lambda', \lambda_0) \circ D(\lambda_0, \lambda_1))^{-1}(U_{\delta'}^{D(\lambda'),+}) =$$

$$(D(\lambda_0, \lambda_1))^{-1}((D(\lambda', \lambda_0))^{-1}(U_{\delta'}^{D(\lambda'),+})) =$$

$$(D(\lambda_0, \lambda_1))^{-1}(U_\delta^{D(\lambda_0)}).$$

This result means that $M^{D(\lambda_1)}$ is an extension of the $D(\lambda_0, \lambda_1)$-preimage of the mark $M^{D(\lambda_0)}$. Moreover, the mapping

$$\vartheta_{\lambda_1}^{\lambda_0} \equiv \vartheta_{\lambda_1}|_{\kappa(\lambda_0)} \circ (\vartheta_{\lambda_0})^{-1}$$

is an indicial mapping of this extension. Since this indicial mapping is independent of the element D of **D** we actually prove that the co-mark $\overline{\mathbf{M}}(\lambda_1)$ is a co-extension of the co-mark $\overline{\mathbf{M}}(\lambda_0)$, which means that $\overline{\mathbf{M}}$ is proper.

Now, we prove that $\overline{\mathbf{M}}$ is a co-extension of $\overline{\mathbf{M}}^+$. Let $\lambda \in \Lambda$. We need to prove that $\overline{\mathbf{M}}(\lambda)$ is a co-extension of $\overline{\mathbf{M}}^+(\lambda)$.

Let δ be an arbitrary element of τ. Setting $\vartheta(\lambda, \delta) = \varepsilon$ we have $\varepsilon \in \kappa(\lambda)$ and, therefore, we can consider the element $\delta' \equiv \vartheta_\lambda(\varepsilon)$ of τ. By relation (6.3.2) we have

$$U_{\delta'}^{D(\lambda)} = (D(\lambda, \lambda))^{-1}(U_\delta^{D(\lambda),+}) = U_\delta^{D(\lambda),+}. \tag{6.3.3}$$

Since the mapping $\delta \to \delta'$ is one-to-one and it is independent of D, relation (6.3.3) shows that $\overline{\mathbf{M}}(\lambda)$ is a co-extension of $\overline{\mathbf{M}}^+(\lambda)$. The proof of the lemma is complete. □

6.3.2 Lemma. *Let* **D** *be an indexed collection of diagrams. Suppose that for every* $\lambda \in \Lambda$, $\mathrm{R}^+(\lambda)$ *is an admissible family of equivalence relations* *on* $\mathbf{D}(\lambda)$. *Then, there exists an admissible family* R *of equivalence relations on* $\mathbf{D}$ *such that* $\mathrm{R}(\lambda)$ *is a final refinement of* $\mathrm{R}^+(\lambda)$ *for every* $\lambda \in \Lambda$.

Proof. Suppose that

$$\mathrm{R}^+(\lambda) = \{\sim^s_{+,\lambda} : s \in \mathcal{F}\}.$$

Let $\{\lambda_\delta : \delta \in \tau\}$ be an indication of the set Λ. For every $s \in \mathcal{F}$ on the indexed collection $\mathbf{D}$ of diagrams we define an equivalence relation $\sim^s$ as follows: for every $D_0, D_1 \in \mathbf{D}$ we set $D_0 \sim^s D_1$ if and only if

$$D_0(\lambda_\delta) \sim^s_{+,\lambda_\delta} D_1(\lambda_\delta)$$

for every $\delta \in s$. It is easy to see that the family

$$\mathrm{R} \equiv \{\sim^s : s \in \mathcal{F}\}$$

is an admissible family of equivalence relations on $\mathbf{D}$. We prove that the family $\mathrm{R}(\lambda) = \{\sim^s_\lambda : s \in \mathcal{F}\}$ is a final refinement of $\mathrm{R}^+(\lambda)$ for every $\lambda \in \Lambda$. Indeed, let $\lambda \in \Lambda$ and t an arbitrary element of $\mathcal{F}$. There exists $\delta \in \tau$ such that $\lambda = \lambda_\delta$. Setting $s = t \cup \{\delta\}$ we have $\sim^s_{\lambda,+} \subset \sim^s_\lambda \subset \sim^t_\lambda$, which proves our assertion. Thus, the family R is the required family of equivalence relation on $\mathbf{D}$. $\square$

6.3.3 Lemma. *Let* $\mathbf{D}$ *be an indexed collection of diagrams,* $\overline{\mathbf{M}}$ *is a proper* $\mathbf{D}$*-co-mark, and* R *is an* $\overline{\mathbf{M}}$*-admissible family of equivalence relations on* $\mathbf{D}$. *Let* D^{T} *be the mapping of the set* $\Lambda \cup \prec$ *into the class consisting of all spaces and all (continuous) mappings, which is defined as follows:*

(a) *for every* $\lambda \in \Lambda$, $D^{\mathrm{T}}(\lambda)$ *is considered to be the Containing Space* $\mathrm{T}(\mathbf{M}(\lambda), \mathrm{R}(\lambda))$, *and*

(b) *for every* $(\lambda_0, \lambda_1) \in \prec$, $D^{\mathrm{T}}(\lambda_0, \lambda_1)$ *is considered to be the containing mapping*

$$f^{\mathrm{T}}(\lambda_0, \lambda_1) : \mathrm{T}(\overline{\mathbf{M}}(\lambda_1), \mathrm{R}(\lambda_1)) \to \mathrm{T}(\overline{\mathbf{M}}(\lambda_0), \mathrm{R}(\lambda_0)).$$

Then, D^{T} *is a diagram such that for every element* $D \in \mathbf{D}$ *the indexed family*

$$i_{D(\Lambda)} \equiv \{i_{D(\lambda)} : \lambda \in \Lambda\},$$

where $i_{D(\lambda)}$ is the natural embedding of the space $D(\lambda)$ into the Containing Space $\overline{\mathrm{T}}(\mathbf{M}(\lambda), \mathrm{R}(\lambda))$, is an isomorphism of D into D^{T}.

Proof. We note that since $\mathrm{R}(\lambda)$ is $\overline{\mathbf{M}}(\lambda)$-admissible, the Containing Space $\mathrm{T}(\overline{\mathbf{M}}(\lambda), \mathrm{R}(\lambda))$, for $\lambda \in \Lambda$, is defined. Also, $(\overline{\mathbf{M}}(\lambda_1), \overline{\mathbf{M}}(\lambda_0))$ is a proper pair and $\mathrm{R}(\lambda_0, \lambda_1)$ is an $(\overline{\mathbf{M}}(\lambda_1), \overline{\mathbf{M}}(\lambda_0))$-admissible family for any $(\lambda_0, \lambda_1) \in \prec$. Therefore, the conditions of Lemma 6.1.1 are satisfied, which means that the containing mapping $f^{\mathrm{T}}(\lambda_0, \lambda_1)$ of $\mathrm{T}(\overline{\mathbf{M}}(\lambda_1), \mathrm{R}(\lambda_1))$ into $\mathrm{T}(\overline{\mathbf{M}}(\lambda_0), \mathrm{R}(\lambda_0))$ is defined. Moreover, if $\mathbf{a} \in \mathrm{T}(\overline{\mathbf{M}}(\lambda_1), \mathrm{R}(\lambda_1))$ and $(x, D(\lambda_1)) \in \mathbf{a}$, then $f^{\mathrm{T}}(\lambda_0, \lambda_1)(\mathbf{a}) = \mathbf{b}$, where $\mathbf{b}$ is the point of the space $\mathrm{T}(\overline{\mathbf{M}}(\lambda_0), \mathrm{R}(\lambda_0))$ containing the pair $(D(\lambda_0, \lambda_1)(x), D(\lambda_0))$.

To prove that D^{T} is a diagram we need to prove that

$$D^{\mathrm{T}}(\lambda_0, \lambda_1) \circ D^{\mathrm{T}}(\lambda_1, \lambda_2) = D^{\mathrm{T}}(\lambda_0, \lambda_2)$$

for every $(\lambda_0, \lambda_1), (\lambda_1, \lambda_2) \in \prec$. But this relation follows immediately by the relation

$$D(\lambda_0, \lambda_1) \circ D(\lambda_1, \lambda_2) = D(\lambda_0, \lambda_2).$$

Finally, the fact that the indexed family $i_{D(\Lambda)}$ is an isomorphism of D into $\mathbf{D}$ follows immediately by Lemma 6.1.1. $\square$

Definition. The diagram D^{T} of Lemma 6.3.3 is said to be the *containing diagram* (*corresponding* to the proper $\mathbf{D}$-co-mark $\overline{\mathbf{M}}$ and the family R).

Let $\mathbf{H} \in \mathrm{C}^{\diamondsuit}(\mathrm{R})$. We note that for every $(\lambda_0, \lambda_1) \in \prec$ the mapping $f^{\mathrm{T}}(\lambda_0, \lambda_1)$ satisfies the following condition:

$$f^{\mathrm{T}}(\lambda_0, \lambda_1)(\mathrm{T}(\overline{\mathbf{M}}(\lambda_1), \mathrm{R}(\lambda_1), \mathbf{H}(\lambda_1))) \subset \mathrm{T}(\overline{\mathbf{M}}(\lambda_0), \mathrm{R}(\lambda_0), \mathbf{H}(\lambda_0)).$$

We define the mapping

$$f^{\mathrm{T}}_{\mathbf{H}}(\lambda_0, \lambda_1) : \mathrm{T}(\overline{\mathbf{M}}(\lambda_1), \mathrm{R}(\lambda_1), \mathbf{H}(\lambda_1)) \to \mathrm{T}(\overline{\mathbf{M}}(\lambda_0), \mathrm{R}(\lambda_0), \mathbf{H}(\lambda_0))$$

setting

$$f^{\mathrm{T}}_{\mathbf{H}}(\lambda_0, \lambda_1)(\mathbf{a}) = f^{\mathrm{T}}(\lambda_0, \lambda_1)(\mathbf{a}),\ \mathbf{a} \in \mathrm{T}(\overline{\mathbf{M}}(\lambda_1), \mathrm{R}(\lambda_1), \mathbf{H}(\lambda_1)).$$

By $D^{\mathrm{T}}_{\mathbf{H}}$ we denote a diagram, which is defined as follows:

$$D_{\mathbf{H}}^{\mathrm{T}}(\lambda) = \mathrm{T}(\overline{\mathbf{M}}(\lambda), \mathrm{R}(\lambda), \mathbf{H}(\lambda)),\ \lambda \in \Lambda \text{ and}$$

$$D_{\mathbf{H}}^{\mathrm{T}}(\lambda_0, \lambda_1) = f_{\mathbf{H}}^{\mathrm{T}}(\lambda_0, \lambda_1),\ (\lambda_0, \lambda_1) \in \prec.$$

It is easy to verify that $D_{\mathbf{H}}^{\mathrm{T}}$ is really a diagram. This diagram is said to be the *containing diagram* (*corresponding* to $\mathbf{D}$-co-mark $\overline{\mathbf{M}}$, the family R, and the equivalence class $\mathbf{H}$).

6.3.4 Lemma. *Let $\mathbf{D}$ be an indexed collection of diagrams and $\overline{\mathbf{M}}$ a proper $\mathbf{D}$-co-mark. For a fixed element λ_0 of Λ suppose that $\mathbf{M}_{\lambda_0}$ is a co-mark of $\mathbf{D}(\lambda_0)$, which is a co-extension of $\overline{\mathbf{M}}(\lambda_0)$. Then, there exists a proper $\mathbf{D}$-co-mark $\overline{\mathbf{M}}_0$, which is a co-extension of $\overline{\mathbf{M}}$, such that $\overline{\mathbf{M}}_0(\lambda_0) = \mathbf{M}_{\lambda_0}$.*

Proof. Suppose that

$$\overline{\mathbf{M}} \equiv \{\{M^{D(\lambda)} : \lambda \in \Lambda\} : D \in \mathbf{D}\} \text{ and}$$

$$\mathbf{M}_{\lambda_0} \equiv \{M_{\lambda_0}^{D(\lambda_0)} : D \in \mathbf{D}\}.$$

Let θ_0 and θ_1 be two one-to-one mappings of τ into itself such that

$$\theta_0(\tau) \cup \theta_1(\tau) = \tau \text{ and } \theta_0(\tau) \cap \theta_1(\tau) = \emptyset.$$

Setting

$$\Lambda' = \{\lambda \in \Lambda : \lambda_0 \leq \lambda, \lambda_0 \neq \lambda\}$$

we define the $\mathbf{D}$-co-mark

$$\overline{\mathbf{M}}_0 \equiv \{\{M_0^{D(\lambda)} : \lambda \in \Lambda\} : D \in \mathbf{D}\}$$

as follows. Let D be an element of $\mathbf{D}$ and λ an element of Λ. If $\lambda \in \Lambda \setminus \Lambda'$, then we set $M_0^{D(\lambda)} = M^{D(\lambda)}$. If $\lambda \in \Lambda'$, then we consider that $M_0^{D(\lambda)}$ is the mark of $D(\lambda)$ such that:

(a) $M_0^{D(\lambda)}$ is an extension of $M^{D(\lambda)}$ and θ_0 is an indicial mapping of this extension and

(b) $M_0^{D(\lambda)}$ is an extension of the $D(\lambda_0, \lambda)$-preimage of $M_{\lambda_0}^{D(\lambda_0)}$ and θ_1 is an indicial mapping of this extension. It is easy to verify that $\overline{\mathbf{M}}_0$ is really the required proper $\mathbf{D}$-co-mark. □

Definition. A non-empty class $\mathbb{D}$ of diagrams is said to be *saturated* if for every indexed collection $\mathbf{D}$ of elements of $\mathbb{D}$ there exists a proper $\mathbf{D}$-co-mark $\overline{\mathbf{M}}^{+}$ satisfying the following condition:

for every proper $\mathbf{D}$-co-mark $\overline{\mathbf{M}}$, which is a co-extension of $\overline{\mathbf{M}}^{+}$, there exists an $\overline{\mathbf{M}}$-admissible family R^{+} of equivalence relations on $\mathbf{D}$ such that for every $\overline{\mathbf{M}}$-admissible family R of equivalence relations on $\mathbf{D}$, which is a final refinement of R^{+}, and for every element $\mathbf{L}$ of $\mathrm{C}^{\diamond}(\mathrm{R})$, the containing diagram $D^{\mathrm{T}}_{\mathbf{L}}$ that corresponds to the $\mathbf{D}$-co-mark $\overline{\mathbf{M}}$, the family R, and the equivalence class $\mathbf{L}$, belongs to $\mathbb{D}$.

The proper $\mathbf{D}$-co-mark $\overline{\mathbf{M}}^{+}$ is called an *initial* $\mathbf{D}$-co-mark (*corresponding* to the class $\mathbb{D}$), and the $\overline{\mathbf{M}}$-admissible family R^{+} is called an *initial* family (*corresponding* to the proper $\mathbf{D}$-co-mark $\overline{\mathbf{M}}$ and the class $\mathbb{D}$).

The above given Lemmas 6.3.1–6.3.4 are needed for the proofs of Propositions 6.3.5–6.3.8. Since these proofs in outline are similar to the proofs of Propositions 6.1.3–6.1.7, respectively, we prove only Proposition 6.3.8 and omit the others.

6.3.5 Proposition. *The non-empty intersection of not more than τ many saturated classes of diagrams is also a saturated class.* □

6.3.6 Proposition. *In each saturated class of diagrams there exist universal elements.* □

6.3.7 Proposition. *For every $\lambda \in \Lambda$, the class consisting of the λ-components of all elements of a saturated class of diagrams is a saturated class of spaces.* □

6.3.8 Proposition. *Suppose that for every $\lambda \in \Lambda$, $\mathbb{E}_\lambda$ is a saturated class of spaces and for every $(\lambda_0, \lambda_1) \in \prec$, $\mathbb{F}_{(\lambda_0,\lambda_1)}$ is a saturated class of mappings. Then, the class of all diagrams D such that $D(\lambda) \in \mathbb{E}_\lambda$, for $\lambda \in \Lambda$, and $D(\lambda_0, \lambda_1) \in \mathbb{F}_{(\lambda_0,\lambda_1)}$, for $(\lambda_0, \lambda_1) \in \prec$, is a saturated class.*

Proof. Denote by $\mathbb{D}$ the class of all diagrams satisfying the conditions of the proposition. Let $\mathbf{D}$ be an indexed collection of elements of $\mathbb{D}$. Then, for every $\lambda \in \Lambda$, $\mathbf{D}(\lambda)$ is an indexed collection of elements of $\mathbb{E}_\lambda$. Since $\mathbb{E}_\lambda$ is saturated there exists an initial co-mark $\mathbf{M}_0^{\mathbb{E},+}(\lambda)$ of $\mathbf{D}(\lambda)$ corresponding to $\mathbb{E}_\lambda$.

Also for every $(\lambda_0, \lambda_1) \in \prec$, $\mathbf{D}(\lambda_0, \lambda_1)$ is an indexed collection of elements of $\mathbb{F}_{(\lambda_0,\lambda_1)}$. Hence, there exists an initial $\mathbf{D}(\lambda_0, \lambda_1)$-co-mark

$$(\mathbf{M}^{d,+}(\lambda_0, \lambda_1), \mathbf{M}^{r,+}(\lambda_0, \lambda_1))$$

corresponding to the class $I\!F_{(\lambda_0,\lambda_1)}$.

For every $\lambda \in \Lambda$ we denote by $\mathbf{M}_0^+(\lambda)$ a co-mark of $\mathbf{D}(\lambda)$, which is a co-extension:

(a) of the co-mark $\mathbf{M}_0^{I\!E,+}(\lambda)$,

(b) of the co-marks $\mathbf{M}^{d,+}(\lambda_0, \lambda)$ for all elements λ_0 of Λ for which $(\lambda_0, \lambda) \in \prec$, and

(c) of the co-marks $\mathbf{M}^{r,+}(\lambda, \lambda_1)$ for all elements λ_1 of Λ for which $(\lambda, \lambda_1) \in \prec$.

Consider the $\mathbf{D}$-co-mark

$$\overline{\mathbf{M}}_0^+ \equiv \{\mathbf{M}_0^+(\lambda) : \lambda \in \Lambda\}.$$

By Lemma 6.3.1 there exists a proper $\mathbf{D}$-co-mark $\overline{\mathbf{M}}^+$, which is a co-extension of $\overline{\mathbf{M}}_0^+$. Denote by $\overline{\mathbf{M}}$ an arbitrary proper $\mathbf{D}$-co-mark, which is a co-extension of $\overline{\mathbf{M}}^+$.

For every $\lambda \in \Lambda$ we denote by $\mathrm{R}_0^{I\!E,+}(\lambda)$ an initial family of equivalence relations on $\mathbf{D}(\lambda)$ corresponding to $\overline{\mathbf{M}}(\lambda)$ and the class $I\!E_\lambda$.

Also, for every $(\lambda_0, \lambda_1) \in \prec$ we denote by $\mathrm{R}_0^{I\!F,+}(\lambda_0, \lambda_1)$ an initial family of equivalence relations on $\mathbf{D}(\lambda_0, \lambda_1)$ corresponding to the proper $\mathbf{D}(\lambda_0, \lambda_1)$-co-mark $(\overline{\mathbf{M}}(\lambda_0), \overline{\mathbf{M}}(\lambda_1))$ and the class $I\!F_{(\lambda_0,\lambda_1)}$.

For every $\lambda \in \Lambda$ we denote by $\mathrm{R}_0^+(\lambda)$ an $\overline{\mathbf{M}}(\lambda)$-admissible family of equivalence relations on $\mathbf{D}(\lambda)$, which is a final refinement:

(a) of the family $\mathrm{R}_0^{I\!E,+}(\lambda)$,

(b) of the families $(\mathrm{R}_0^{I\!F,+}(\lambda_0, \lambda))^d$ for all elements λ_0 of Λ for which $(\lambda_0, \lambda) \in \prec$, and

(c) of the families $(\mathrm{R}_0^{I\!F,+}(\lambda, \lambda_1))^r$ for all elements λ_1 of Λ for which $(\lambda, \lambda_1) \in \prec$.

By Lemma 6.3.2 there exists an $\overline{\mathbf{M}}$-admissible family R^+ of equivalence relations on $\mathbf{D}$ such that $\mathrm{R}^+(\lambda)$ is final refinement of $\mathrm{R}_0^+(\lambda)$ for every $\lambda \in \Lambda$. Denote by R an arbitrary $\overline{\mathbf{M}}$-admissible family of equivalence relations on $\mathbf{D}$, which is a final refinement of R^+.

To prove the proposition it suffices to prove that the containing diagram $D_{\mathbf{H}}^{\mathrm{T}}$ corresponding to the $\mathbf{D}$-co-mark $\mathbf{M}$, the family R, and an equivalence class $\mathbf{H}$ belongs to $I\!D$, that is $D_{\mathbf{H}}^{\mathrm{T}}(\lambda) \in I\!E_\lambda$ for every $\lambda \in \Lambda$ and $D_{\mathbf{H}}^{\mathrm{T}}(\lambda_0, \lambda_1) \in I\!F_{(\lambda_0,\lambda_1)}$ for every $(\lambda_0, \lambda_1) \in \prec$. But this follows immediately by the construction. □

Definition. In the case where the partial order $\prec$ on Λ is directed, the diagrams are called *inverse spectra.* For an inverse spectrum D the spaces $D(\lambda)$, where $\lambda \in \Lambda$, are called the *spaces of the inverse spectrum* D and the mappings $D(\lambda_0, \lambda_1)$, where $(\lambda_0, \lambda_1) \in \prec$, are called the *mappings* or *projections* of the inverse spectrum D.

Agreement 2. In what follows of this section it is assumed that the partial order $\prec$ is directed.

Proposition 6.3.8 implies the following consequence.

6.3.9 Corollary. *The class of all inverse spectra, whose spaces belong to a given saturated class of spaces and whose projections belong to a given saturated class of mappings, is saturated.* □

Notation. Let D be an inverse spectrum. By L_D we denote the set of all mappings

$$g : \Lambda \to \cup\{D(\lambda) : \lambda \in \Lambda\}$$

satisfying the following conditions:

(a) $g(\lambda) \in D(\lambda)$, where $\lambda \in \Lambda$, and

(b) if $(\lambda_0, \lambda_1) \in \prec$, then $g(\lambda_0) = D(\lambda_0, \lambda_1)(g(\lambda_1))$.

For any elements λ_1,...,λ_n of Λ, $n \in \omega \setminus \{0\}$, and open subsets U_{λ_1},...,U_{λ_n} of spaces $D(\lambda_1)$,...,$D(\lambda_n)$, respectively, we set

$$L_D(U_{\lambda_1}, ..., U_{\lambda_n}) = \{g \in L_D : g(\lambda_1) \in U_{\lambda_1}, ..., g(\lambda_n) \in U_{\lambda_n}\}.$$

By $b(L_D)$ we denote the set of all subsets of L_D of the form $L_D(U_{\lambda_1}, ..., U_{\lambda_n})$, $n \in \omega \setminus \{0\}$.

Finally, by $t(L_D)$ we denote the set of all subsets of L_D which are union of elements of $b(L_D)$.

6.3.10 Lemma. *For every inverse spectrum D the set $t(L_D)$ is a topology on L_D. The space L_D (that is, the set L_D equiped with the topology $t(L_D)$) is a T_0-space of weight $\leq \tau$.*

Proof. The set $b(L_D)$ have the following properties:

(a) the union of all elements of $b(L_D)$ is the set L_D,

(b) the empty set is an element of $b(L_D)$, and

(c) if $L_D(U_{\lambda_1}, ..., U_{\lambda_n})$ and $L_D(V_{\lambda'_1}, ..., V_{\lambda'_m})$ are elements of $b(L_D)$, then

$$L_D(U_{\lambda_1}, ..., U_{\lambda_n}) \cap L_D(V_{\lambda'_1}, ..., V_{\lambda'_m}) = L_D(U_{\lambda_1}, ..., U_{\lambda_n}, V_{\lambda'_1}, ..., V_{\lambda'_m}),$$

which are easy to verify. These properties show that the set $t(L_D)$ is a topology on L_D and the set $b(L_D)$ is a base for this topology.

Let f and g be two distinct elements of L_D. Then, there exists $\lambda \in \Lambda$ such that $f(\lambda) \neq g(\lambda)$. Therefore, there exists an open subset U_λ of $D(\lambda)$, which contains one of the points $f(\lambda)$ and $g(\lambda)$ of $D(\lambda)$ and does not contain the other. Then, the set $L_D(U_\lambda)$ contains one of the mappings f and g and does not contain the other, which means that the space L_D is a T_0-space.

To prove that the weight of L_D is $\leq \tau$ we consider a base $b(D(\lambda))$ of cardinality $\leq \tau$ for $D(\lambda)$, $\lambda \in \Lambda$. Then, the cardinality of the subset $b_0(L_D)$ of $b(L_D)$ consisting of all elements $L_D(U_{\lambda_1}, ..., U_{\lambda_n})$ of $b(L_D)$ such that $U_{\lambda_i} \in b(D(\lambda_i))$, $i = 1, ..., n$, is $\leq \tau$. Therefore, it suffices to prove that $b_0(L_D)$ is also a base for L_D. This fact follows by the following relations:

$$L_D(U_{\lambda_1}, ..., U_{\lambda_n}) = L_D(U_{\lambda_1}) \cap ... \cap L_D(U_{\lambda_n}) \text{ and}$$

$$L_D(\cup\{U^\delta_\lambda : \delta \in \tau\}) = \cup\{L_D(U^\delta_\lambda) : \delta \in \tau\},$$

which are easy to verify. □

Definition. The space L_D of Lemma 6.3.10 is called the *limit space* of the inverse spectrum D.

6.3.11 Lemma. *Let $i_\Lambda \equiv \{i_\lambda : \lambda \in \Lambda\}$ be an isomorphism of an inverse spectrum D_0 into (onto) an inverse spectrum D_1. Then, there exists an embedding $\lim i_\Lambda$ of the limit space L_{D_0} of D_0 into (onto) the limit space L_{D_1} of D_1.*

Proof. Denote by $\lim i_\Lambda$ the mapping of L_{D_0} into (onto) L_{D_1} defined as follows: if

$$g : \Lambda \to \cup\{D_0(\lambda) : \lambda \in \Lambda\}$$

is a point of L_{D_0}, then $\lim i_\Lambda(g)$ is the point f of L_{D_1} such that

$$f(\lambda) = i_\lambda(g(\lambda)).$$

It is easy to verify that f is really a point of L_{D_1}.

We prove that $\lim i_\Lambda$ is an embedding of L_{D_0} into (onto) L_{D_1}. Indeed, let g_0 and g_1 be two distinct points of L_{D_0}. Then, there exists $\lambda \in \Lambda$ such that $g_0(\lambda) \neq g_1(\lambda)$. Since i_λ is an embedding of $D_0(\lambda)$ into $D_1(\lambda)$ we have that $\lim i_\Lambda(g_0) \neq \lim i_\lambda(g_1)$, which means that the mapping $\lim i_\Lambda$ is one-to-one. It is easy to verify that if i_Λ is onto then the mapping $\lim i_\Lambda$ is also onto.

The continuity of the mappings $\lim i_\Lambda$ and $(\lim i_\Lambda)^{-1}$ follow by relation

$$L_{D_1}(U_\lambda) \cap \lim i_\Lambda(L_{D_0}) = \lim i_\Lambda(L_{D_0}(V_\lambda)),$$

where U_λ is an open subset of $D_1(\lambda)$ and $V_\lambda = (i_\lambda)^{-1}(U_\lambda)$, which is easy to verify. □

The proof of the above lemma actually contains the proof of the following lemma.

6.3.12 Lemma. *Let $f_\Lambda \equiv \{f_\lambda : \lambda \in \Lambda\}$ be a morphism of an inverse spectrum D_0 into an inverse spectrum D_1. Then, there exists a mapping $\lim f_\Lambda$ of the limit space L_{D_0} of D_0 into the limit space L_{D_1} of D_1.* □

Notation. Let $\mathbb{D}$ be a class of inverse spectra. We denote by $\mathbb{P}(\lim \mathbb{D})$ the (topological) class of all spaces X such that X is homeomorphic to the limit space of an inverse spectrum of $\mathbb{D}$. This class is called the *class of the limit spaces of $\mathbb{D}$*.

6.3.13 Proposition. *In the class of the limit spaces of a saturated class of inverse spectra there exist universal elements.*

Proof. Let $\mathbb{D}$ be a saturated class of inverse spectra. There exists an indexed collection $\mathbf{D}$ of elements of $\mathbb{D}$ such that for every $D \in \mathbb{D}$ there exists an isomorphism of D onto an element of $\mathbf{D}$. By Lemma 6.3.11 each element of $\mathbb{P}(\lim \mathbb{D})$ is homeomorphic to the limit space of an inverse spectrum of the collection $\mathbf{D}$.

Since the class $\mathbb{D}$ is saturated there exists an initial $\mathbf{D}$-co-mark $\overline{\mathbf{M}}$ corresponding to the class $\mathbb{D}$. Denote by R an initial family of equivalence relations on $\mathbf{D}$ corresponding to the $\mathbf{D}$-co-mark $\overline{\mathbf{M}}$ and the class $\mathbb{D}$.

Let D^{T} be the containing diagram corresponding to the $\mathbf{D}$-co-mark $\overline{\mathbf{M}}$ and the family R. Then, $D^{\mathrm{T}} \in \mathbb{D}$. Therefore, $L_{D^{\mathrm{T}}} \in \mathbb{P}(\lim \mathbb{D})$. By

Lemma 6.3.11 the limit space of each inverse spectrum of **D** is homeomorphic to a subset of the limit space $L_{D^{\mathrm{T}}}$ of D^{T}. Therefore, $L_{D^{\mathrm{T}}}$ is a universal element in $\mathbb{P}(\lim \mathbb{D})$. □

Corollary 6.3.9 and Proposition 6.3.13 imply the next consequence.

6.3.14. Corollary. *In the class of the limit spaces of all inverse spectra, whose spaces belong to a given saturated class of spaces and projections to a given saturated class of mappings, there exist universal elements.* □

6.4 Mappings of compact spaces

6.4.1 Lemma. *Let C^X and C^Y be normal bases of cardinality $\leq \tau$ for the closed subsets of spaces X and Y, respectively, and f a mapping of X into Y such that:*

(a) *the f-preimage of C^Y is contained in C^X and*

(b) *the f-image of C^X is contained in C^Y.*

Then, there exists a unique mapping $c(f)$ of the space $c(X, C^X)$ into the space $c(Y, C^Y)$ such that $c(f)(x) = f(x)$ for every $x \in X$. Moreover, if

(c) *the f-image of the base for the open subsets of X complementary to C^X is contained in the base for the open subsets of Y complementary to C^Y (and, therefore, f is open),*

then $c(f)$ is also open. (We note that condition (b) does not imply (c).)

Proof. First, we prove the following assertion: for every point φ of $c(X, C^X)$ there exists a point χ of $c(Y, C^Y)$ such that for every element F of χ, $f^{-1}(F) \in \varphi$. Indeed, suppose that there exists a point φ of $c(X, C^X)$, which does not satisfy the condition of the assertion. Then, for every point χ of $c(Y, C^Y)$ there exists an element F_χ of C^Y such that $F_\chi \in \chi$ and $f^{-1}(F_\chi) \notin \varphi$. By Lemma 5.3.1 there exists an element G_χ of φ such that $f^{-1}(F_\chi) \cap G_\chi = \emptyset$, that is $F_\chi \cap f(G_\chi) = \emptyset$, which means that $f(G_\chi) \notin \chi$. Therefore, the set

$$\{c(Y, C^Y) \setminus c(f(G_\chi), C^Y) : \chi \in c(Y, C^Y)\}$$

is an open cover of $c(Y, C^Y)$. Let

$$\{c(Y, C^Y) \setminus c(f(G_{\chi_i}), C^Y) : i \in n \in \omega\}$$

be a finite subcover of this cover. Then, we have

$$\cap\{c(f(G_{\chi_i}), C^Y) : i \in n\} = \emptyset.$$

This equality implies that

$$\cap\{f(G_{\chi_i}) : i \in n\} = \emptyset$$

and, therefore,

$$\cap\{G_{\chi_i} : i \in n\} = \emptyset,$$

which contradicts the fact that $G_{\chi_i} \in \varphi$, $i \in n$. This proves our assertion.

The point χ of $c(Y, C^Y)$ satisfying the condition of the above assertion is uniquely determined. Indeed, suppose that there exist two such points χ_0 and χ_1. Then, by Lemma 5.3.1 there exist elements F_0 and F_1 of χ_0 and χ_1, respectively, such that $F_0 \cap F_1 = \emptyset$. This implies that $f^{-1}(F_0) \cap f^{-1}(F_1) = \emptyset$, which contradicts the fact that $f^{-1}(F_0)$ and $f^{-1}(F_1)$ are elements of φ.

Thus, we can define a mapping $c(f)$ setting $c(f)(\varphi) = \chi$. We note the following properties of the mapping $c(f)$:

(d) if $F \in C^Y$, then $F \in \chi$ if and only if $f^{-1}(F) \in \varphi$, and

(e) if $G \in \varphi$, then $f(G) \in \chi$.

Indeed, let $F \in C^Y$ and $F \in \chi$, then by the proving assertion $f^{-1}(F) \in \varphi$. If $F \notin \chi$, then there exists an element $K \in \chi$ such that $F \cap K = \emptyset$ and, therefore, $f^{-1}(F) \cap f^{-1}(K) = \emptyset$. Since $f^{-1}(K) \in \varphi$ we have $f^{-1}(F) \notin \varphi$. To prove (e) suppose that $G \in \varphi$. If $f(G) \notin \chi$, then there exist $F \in \chi$ such that $f(G) \cap F = \emptyset$ and, therefore, $f^{-1}(f(G)) \cap f^{-1}(F) = \emptyset$. Since $G \subset f^{-1}(f(G))$ we have $G \cap f^{-1}(F) = \emptyset$, which contradicts the fact that G and $f^{-1}(F)$ are elements of φ.

Now, we prove that $c(f)$ is continuous. For this it suffices to prove that for every $F \in C^Y$,

$$(c(f))^{-1}(c(F, C^Y)) = c(f^{-1}(F), C^X). \tag{6.4.1}$$

Suppose that $F \in C^Y$ and $\varphi \in (c(f))^{-1}(c(F, C^Y))$. Then, $c(f)(\varphi) \in c(F, C^Y)$, which means that $F \in c(f)(\varphi)$. By the above assertion, $f^{-1}(F) \in \varphi$, that is $\varphi \in c(f^{-1}(F), C^X)$. Conversely, if $\varphi \in c(f^{-1}(F), C^X)$, then $f^{-1}(F) \in \varphi$. By property (d), $F \in c(f)(\varphi)$, that is $c(f)(\varphi) \in c(F, C^Y)$ or $\varphi \in (c(f))^{-1}(c(F, C^Y))$, which completes the proof of (6.4.1).

The relation $c(f)(x) = f(x)$, $x \in X$, is easy to verify. This relation imply also the uniqueness of the mapping $c(f)$.

Finally, suppose that the f-image of the base B^X for the open subsets of X complementary to C^X is contained in the base B^Y for the open subsets of Y complementary to C^Y. We need to prove that $c(f)$ is open.

First, we introduce the following notation, which concerns any space Z and any its normal base C^Z for the closed subsets. For every element U of the base B^Z for the open subsets of Z complementary to C^Z we denote by $c(U, C^Z)$ the set of all elements φ of $c(Z, C^Z)$ for which there exists an element F of φ such that $F \subset U$. It is easy to verify that

$$c(U, C^Z) = c(Z, C^Z) \setminus c(Z \setminus U, C^Z),$$

which means that the set $\{c(U, C^Z) : U \in B^Z\}$ is a base for the open subsets of $c(Z, C^Z)$, complementary to $c(C^Z)$.

Thus, in order to prove that $c(f)$ is open it suffices to prove that the set $c(f)(c(U, C^X))$ is open in $c(Y, C^Y)$ for every $U \in B^X$. Let $U \in B^X$ and $\varphi \in c(U, C^X)$. There exists an element $F \in \varphi$ such that $F \subset U$. Let (G, H) be a screening of the pair $(F, X \setminus U)$ by elements of C^X. Then, $G \subset U$, $V \equiv X \setminus H \in B^X$, and $F \subset V \subset G \in \varphi$. Since $f(F) \in c(f)(\varphi)$ and $f(F) \subset f(V)$ we have $c(f)(\varphi) \in c(f(V), C^Y)$. Therefore, to prove that $c(f)(c(U, C^X))$ is open, it suffices to prove that

$$c(f(V), C^Y) \subset c(f)(c(U, C^X)). \tag{6.4.2}$$

Let $\chi \in c(f(V), C^Y)$ and $K \in \chi$. Since $f(V) \subset f(G)$ we have $f(G) \in \chi$, which means that $K \cap f(G) \neq \emptyset$. Therefore, $G \cap f^{-1}(K) \neq \emptyset$. This fact implies that the set

$$\{G \cap f^{-1}(K) : K \in \chi\}$$

is contained in a maximal filter of C^X. Let φ' be such a filter. Then, $f^{-1}(K) \in \varphi'$ for every $K \in \chi$. By the proving assertion $c(f)(\varphi') = \chi$. On the other side, $G \in \varphi'$, which means that $\varphi' \in c(U, C^X)$ and, therefore, $\chi \in c(f)(c(U, C^X))$, which proves relation (6.4.2). Thus, the mapping $c(f)$ is open. The proof of the lemma is complete. □

Notation. We denote:

(a) by $\mathbb{P}$(compact) the class of all Hausdorff compact spaces,

(b) by $\mathbb{P}_c(\mathrm{Ind} \leq n)$, $n \in \omega$, the class of all Hausdorff compact spaces X with $\mathrm{Ind}(X) \leq n$, and

(c) by $\mathbb{P}_c(\dim \leq n)$, $n \in \omega$, the class of all Hausdorff compact spaces X with $\dim(X) \leq n$.

6.4.2 Proposition. *Suppose that $\mathbb{E}$ and $\mathbb{L}$ are independently one of the classes:*

$$\mathbb{P}(\text{compact}),\ \mathbb{P}_c(\text{Ind} \leq n),\ n \in \omega,\ \textit{and}\ \mathbb{P}_c(\dim \leq m),\ m \in \omega.$$

Then, in the class $\mathrm{C}(\mathbb{E}, \mathbb{L})$ of all mappings, respectively, in the (non-empty) class $\mathrm{C}^{\mathrm{op}}(\mathbb{E}, \mathbb{L})$ of all open mappings, with the domain in $\mathbb{E}$ and the range in $\mathbb{L}$ there exist universal elements.

Proof. We prove the proposition in the case, where $\mathbb{E} = \mathbb{P}_c(\text{Ind} \leq n)$ for some $n \in \omega$, and $\mathbb{L} = \mathbb{P}_c(\dim \leq m)$ for some $m \in \omega$. The other cases are proved similarly.

Consider the class $\mathrm{C}(\mathbb{E}, \mathbb{L})$. Let $\mathbf{F}$ be an indexed collection of elements of $\mathrm{C}(\mathbb{E}, \mathbb{L})$ such that each element of $\mathrm{C}(\mathbb{E}, \mathbb{L})$ is embedded in an element of $\mathbf{F}$. Let $\mathbf{S}_d$ and $\mathbf{S}_r$ be the $\mathbf{F}$-domain and the $\mathbf{F}$-range, respectively.

For every $f \in \mathbf{F}$ and for every $k \in \omega$ we choice a (b^n-Ind $\leq n$)-dimensional base $C_k^{f,d}$ for the closed subsets of D_f and a (b^n-dim $\leq m$)-dimensional base $C_k^{f,r}$ for the closed subsets of R_f satisfying the following conditions:

(a) $C_k^{f,d} \subset C_{k+1}^{f,d}$ and $C_k^{f,r} \subset C_{k+1}^{f,r}$

(b) the f-image of $C_k^{f,d}$ is contained in $C_{k+1}^{f,r}$, and

(c) the f-preimage of $C_k^{f,r}$ is contained in $C_{k+1}^{f,d}$.

Such bases can be constructed using Lemmas 5.3.10 and 5.4.12.

We put

$$C^{f,d} \equiv \cup\{C_k^{f,d} : k \in \omega\} \text{ and}$$

$$C^{f,r} \equiv \cup\{C_k^{f,r} : k \in \omega\}.$$

By Lemma 5.3.10, $C^{f,d}$ is a (b^n-Ind $\leq n$)-dimensional normal base for the closed subsets of D_f and by Lemma 5.4.12, $C^{f,r}$ is a (b^n-dim $\leq m$)-dimensional normal base for the closed subsets of R_f. Moreover, the f-image of $C^{f,d}$ is contained in $C^{f,r}$ and the f-preimage of $C^{f,r}$ is contained in $C^{f,d}$.

By Lemma 5.2.4 for some admissible branching triad $(\vartheta^d_\vee, \vartheta^d_\wedge, \Theta)$ and an n-expansion Δ of Θ we can construct a $(\vartheta^d_\vee, \vartheta^d_\wedge, \Theta, \Delta)$-normal indication $N^{f,d}$ of the base $C^{f,d}$. We note that the triad $(\vartheta^d_\vee, \vartheta^d_\wedge, \Theta)$ and the n-expansion Δ of Θ are independent of the mapping f.

Similarly, by Lemma 5.4.2 for some admissible tetrad $(\vartheta^r_\vee, \vartheta^r_\wedge, \vartheta, \vartheta_{\mathcal{F}})$ we can construct $(\vartheta^r_\vee, \vartheta^r_\wedge, \vartheta, \vartheta_{\mathcal{F}}, m)$-normal indication $N^{f,r}$ of the base $C^{f,r}$. (The tetrad $(\vartheta^r_\vee, \vartheta^r_\wedge, \vartheta, \vartheta_{\mathcal{F}})$ is also independent of the mapping f).

Without loss of generality, we can suppose that $N^{f,d}$ is an extension of the f-preimage of $N^{f,r}$, and $N^{f,r}$ is an extension of the f-image of $N^{f,d}$. Moreover, we can consider that the corresponding indicial mappings are independent of f. For this purpose, the constructions of the $(\vartheta^d_\vee, \vartheta^d_\wedge, \Theta, \Delta)$-normal indication $N^{f,d}$ of $C^{f,d}$ and the $(\vartheta^r_\vee, \vartheta^r_\wedge, \vartheta, \vartheta_{\mathcal{F}}, m)$-normal indication $N^{f,r}$ of $C^{f,r}$ must be carried out in such a manner that the below given conditions (additional to that of Lemmas 5.2.4, 5.4.2 and 5.1.3) to be satisfied.

In order to formulate these conditions we recall that for the construction of the $(\vartheta^d_\vee, \vartheta^d_\wedge, \Theta, \Delta)$-normal indication $N^{f,d}$ of $C^{f,d}$ we first constructed (see Lemma 5.2.4) some indexed sets $\{C^k_\delta : \delta \in \tau\}$, where $k \in \omega$, which here will be denoted by

$$\{C^{f,d,k}_\delta : \delta \in \tau\},\ k \in \omega.$$

Also, for the construction of the $(\vartheta^r_\vee, \vartheta^r_\wedge, \vartheta, \vartheta_{\mathcal{F}}, m)$-normal indication $N^{f,r}$ of $C^{f,r}$, we first construct (see Lemmas 5.4.2 and 5.1.3) some indexed sets $\{C^k_\delta : \delta \in \tau\}$, $k \in \omega$, which here will be denoted by

$$\{C^{f,r,k}_\delta : \delta \in \tau\},\ k \in \omega.$$

Now, we formulate the corresponding conditions:

(d) there exists a mapping $\overline{\theta}^r_k$, $k \in \omega$, of τ into itself, which is independent of f, such that

$$\{C^{f,r,k+1}_\delta : \delta \in \tau\}$$

is a $\overline{\theta}^r_k$-extension of the f-image of

$$\{C^{f,d,k}_\delta : \delta \in \tau\}, \text{ and}$$

(e) there exists a mapping $\overline{\theta}^d_k$, $k \in \omega$, of τ into itself, which is independent of f, such that

$$\{C_\delta^{f,d,k+1} : \delta \in \tau\}$$

is a $\overline{\theta}_k^d$-extension of the f-preimage of

$$\{C_\delta^{f,r,k} : \delta \in \tau\}.$$

Thus, we can suppose that $N^{f,d}$ is an extension of the f-preimage of $N^{f,r}$ having a mapping $\overline{\theta}^d$ as an indicial mapping and that $N^{f,r}$ is an extension of the f-image of $N^{f,d}$ having a mapping $\overline{\theta}^r$ as an indicial mapping. We note that the mappings $\overline{\theta}^d$ and $\overline{\theta}^r$ are independent of f.

We put

$$\mathbf{C}^d = \{C^{f,d} : f \in \mathbf{F}\},$$

$$\mathbf{C}^r = \{C^{f,r} : f \in \mathbf{F}\},$$

$$\mathbf{N}^d = \{N^{f,d} : f \in \mathbf{F}\}, \text{ and}$$

$$\mathbf{N}^r = \{N^{f,r} : f \in \mathbf{F}\}.$$

We observe that $\mathbf{N}^d$ is a co-extension of the $\mathbf{F}$-preimage of $\mathbf{N}^r$ having the mapping $\overline{\theta}^d$ as an indicial mapping and $\mathbf{N}^r$ is a co-extension of the $\mathbf{F}$-image of $\mathbf{N}^d$ having the mapping $\overline{\theta}^r$ as an indicial mapping.

Furthermore, we denote by $B^{f,d}$ and $B^{f,r}$ the complementary to $C^{f,d}$ and $C^{f,r}$ bases for the open subsets of D_f and R_f, respectively, and by $M^{f,d}$ and $M^{f,r}$ the indications of $B^{f,d}$ and $B^{f,r}$, which are complementary to $N^{f,d}$ and $N^{f,r}$, respectively. We put

$$\mathbf{B}^d = \{B^{f,d} : f \in \mathbf{F}\},$$

$$\mathbf{B}^r = \{B^{f,r} : f \in \mathbf{F}\},$$

$$\mathbf{M}^d = \{M^{f,d} : f \in \mathbf{F}\}, \text{ and}$$

$$\mathbf{M}^r = \{M^{f,r} : f \in \mathbf{F}\}.$$

Then, we have the following properties:

(f) $\mathbf{B}^d$ is a (b^n-Ind $\leq n$)-dimensional co-base for $\mathbf{S}_d$,

(g) $\mathbf{B}^r$ is a (b^n-dim $\leq m$)-dimensional co-base for $\mathbf{S}_r$,

(h) $\mathbf{M}^d$ is a co-mark of $\mathbf{S}_d$ and a $(\vartheta^d_\vee, \vartheta^d_\wedge, \Theta, \Delta)$-normal co-indication of $\mathbf{B}^d$, and

(i) $\mathbf{M}^r$ is a co-mark of $\mathbf{S}_r$ and a $(\vartheta^r_\vee, \vartheta^r_\wedge, \vartheta, \vartheta_{\mathcal{F}}, m)$-normal co-indication of $\mathbf{B}^r$.

We denote by R_r an $\mathbf{M}^r$-admissible family of equivalence relation on $\mathbf{S}_r$.

In Lemma 5.2.6 we replace the indexed collection $\mathbf{S}$ by $\mathbf{S}_d$, the co-base $\mathbf{B}$ by $\mathbf{B}^d$, and the $(\vartheta_\vee, \vartheta_\wedge, \Theta, \Delta)$-normal co-indication $\mathbf{N}$ by $(\vartheta^d_\vee, \vartheta^d_\wedge, \Theta, \Delta)$-normal co-indication $\mathbf{M}^d$ of $\mathbf{B}^d$ and consider the corresponding co-mark $\mathbf{M}^+$ of $\mathbf{S}_d$ of this lemma. Denote by θ^d an indicial mapping from $\mathbf{M}^d$ to $\mathbf{M}^+$. Setting $\mathbf{M} = \mathbf{M}^+$ we consider also the corresponding $\mathbf{M}$-admissible family R^+ of equivalence relations on $\mathbf{S}_d$ of this lemma.

Thus, the pair $(\mathbf{M}^+, \mathbf{M}^r)$ is an $\mathbf{F}$-co-mark. By Lemma 6.1.2 there exists an $(\mathbf{M}^+, \mathbf{M}^r)$-admissible family $\mathrm{R} \equiv \{\sim^s : s \in \mathcal{F}\}$ of equivalence relations on $\mathbf{F}$ such that R^r is a final refinement of R_r and R^d is a final refinement of R^+. Therefore, we can consider the Containing Spaces $\mathrm{T}^+ \equiv \mathrm{T}(\mathbf{M}^+, \mathrm{R}^d)$ and $\mathrm{T}^r \equiv \mathrm{T}(\mathbf{M}^r, \mathrm{R}^r)$ as well as the $(\theta^d(\tau), \diamondsuit)$-standard base $\mathrm{B}^{\mathrm{T}^+,\diamondsuit}_{\theta^d(\tau)}$ for the space T^+ and the $(\tau, \diamondsuit)$-standard base $\mathrm{B}^{\mathrm{T}^r,\diamondsuit}_\tau$ for the space T^r. Note: Concerning the $(\kappa, \diamondsuit)$-standard base $\mathrm{B}^{\mathrm{T},\diamondsuit}_\kappa$ of the Containing Spaces see Section 5.1.1. Denote by $\mathrm{C}^{\mathrm{T}^+,\diamondsuit}_{\theta^d(\tau)}$ the complementary to $\mathrm{B}^{\mathrm{T}^+,\diamondsuit}_{\theta^d(\tau)}$ base for the closed subsets of the space T^+ and by $\mathrm{C}^{\mathrm{T}^r,\diamondsuit}_\tau$ the complementary to $\mathrm{B}^{\mathrm{T}^r,\diamondsuit}_\tau$ base for the closed subsets of the space T^r.

By Lemma 5.2.6 the base $\mathrm{B}^{\mathrm{T}^+,\diamondsuit}_{\theta^d(\tau)}$ for T^+ is $(\mathrm{b}^n\text{-Ind} \leq n)$-dimensional, which means that the compact space $c(\mathrm{T}^+, \mathrm{C}^{\mathrm{T}^+,\diamondsuit}_{\theta^d(\tau)})$ belongs to the class $\mathbb{E}$. (See Lemma 5.3.6).

Similarly, by Lemma 5.4.4 the base $\mathrm{B}^{\mathrm{T}^r,\diamondsuit}_\tau$ for T^r is $(\mathrm{b}^n\text{-dim} \leq m)$-dimensional, which means that the compact space $c(\mathrm{T}^r, \mathrm{C}^{\mathrm{T}^r,\diamondsuit}_\tau)$ belongs to the class $\mathbb{L}$. (See Lemma 5.4.10).

Since $(\mathbf{M}^+, \mathbf{M}^r)$ is a proper $\mathbf{F}$-co-mark and the family R of equivalence relations on $\mathbf{F}$ is $(\mathbf{M}^+, \mathbf{M}^r)$-admissible by Lemma 6.1.1 we can consider the containing mapping f^{T} of the Containing Space T^+ into the Containing Space T^r. We prove that the base $\mathrm{C}^{\mathrm{T}^+,\diamondsuit}_{\theta^d(\tau)}$ contains the f^{T}-preimage of the base $\mathrm{C}^{\mathrm{T}^r,\diamondsuit}_\tau$ and the base $\mathrm{C}^{\mathrm{T}^r,\diamondsuit}_\tau$ contains the f^{T}-image of the base $\mathrm{C}^{\mathrm{T}^+,\diamondsuit}_{\theta^d(\tau)}$.

Indeed, in the above mensioned Lemma 6.1.1 for the continuity of the mapping f^{T} it is proved that the f^{T}-preimage of the τ-standard base $\mathrm{B}^{\mathrm{T}^r}_\tau \equiv \mathrm{B}^{\mathrm{T}^r}$ for T^r is contained in the $\theta^d(\tau)$-standard base $\mathrm{B}^{\mathrm{T}^+}_{\theta^d(\tau)}$ for T^+. Since

$(\mathrm{B}^{\mathrm{T}^r})^{\diamondsuit} = \mathrm{B}_{\tau}^{\mathrm{T}^r,\diamondsuit}$ and $(\mathrm{B}_{\theta^d(\tau)}^{\mathrm{T}^+})^{\diamondsuit} = \mathrm{B}_{\theta^d(\tau)}^{\mathrm{T}^+,\diamondsuit}$, the base $\mathrm{B}_{\theta^d(\tau)}^{\mathrm{T}^+,\diamondsuit}$ contains the f^{T}-preimage of the base $\mathrm{B}_{\tau}^{\mathrm{T}^r,\diamondsuit}$ and, therefore, the base $\mathrm{C}_{\theta^d(\tau)}^{\mathrm{T}^+,\diamondsuit}$ contains the f^{T}-preimage of the base $\mathrm{C}_{\tau}^{\mathrm{T}^r,\diamondsuit}$.

We prove that the base $\mathrm{C}_{\tau}^{\mathrm{T}^r,\diamondsuit}$ contains the f^{T}-image of the base $\mathrm{C}_{\theta^d(\tau)}^{\mathrm{T}^+,\diamondsuit}$. First, we note that $f^{\mathrm{T}}(\mathrm{T}^+) \in \mathrm{C}_{\tau}^{\mathrm{T}^r,\diamondsuit}$. By the proof of Lemma 5.1.4 each element U of $\mathrm{B}_{\theta^d(\tau)}^{\mathrm{T}^+,\diamondsuit}$ has the form:

$$U = U_{\delta_0}^{\mathrm{T}^+}(\mathbf{H}_0^d) \cup ... \cup U_{\delta_n}^{\mathrm{T}^+}(\mathbf{H}_n^d),$$

where $n \in \omega$, $\delta_0, ..., \delta_n \in \theta^d(\tau)$, and $\mathbf{H}_0, ..., \mathbf{H}_n$ are distinct $\sim^s$-equivalence classes for some $s \in \mathcal{F}$. Therefore, it suffices to prove that if $U_{\delta}^{\mathrm{T}^+}(\mathbf{H}^d) \in \mathrm{B}_{\theta^d(\tau)}^{\mathrm{T}^+}$, where $\mathbf{H}$ is an $\sim^s$-equivalence class for some $s \in \mathcal{F}$, then

$$f^{\mathrm{T}}(F) = (\mathrm{T}^r(\mathbf{H}^r) \setminus U_{\varepsilon}^{\mathrm{T}^r}(\mathbf{H}^r)) \cap f^{\mathrm{T}}(\mathrm{T}^+), \tag{6.4.3}$$

where $F = \mathrm{T}^+(\mathbf{H}^d) \setminus U_{\delta}^{\mathrm{T}^+}(\mathbf{H}^d)$ and $\varepsilon = \overline{\theta}^r((\theta^d)^{-1}(\delta))$.

Let $\mathbf{a} \in F$ and $(x, D_f) \in \mathbf{a}$, that is, $D_f \in \mathbf{H}^d$ and $x \in D_f \setminus U_{\delta}^{D_f}$. Then, $R_f \in \mathbf{H}^r$. Since $f(D_f \setminus U_{\delta}^{D_f}) \subset R_f \setminus U_{\varepsilon}^{R_f}$ we have $f(x) \in R_f \setminus U_{\varepsilon}^{R_f}$. If $(f(x), R_f) \in \mathbf{b} \in \mathrm{T}^r$, then by the definition of the mapping f^{T}, $\mathbf{b} = f^{\mathrm{T}}(\mathbf{a})$. Therefore,

$$f^{\mathrm{T}}(\mathbf{a}) \in (\mathrm{T}^r(\mathbf{H}^r) \setminus U_{\varepsilon}^{\mathrm{T}^r}(\mathbf{H}^r)) \cap f^{\mathrm{T}}(\mathrm{T}^+).$$

Conversely, let

$$\mathbf{b} \in (\mathrm{T}^r(\mathbf{H}^r) \setminus U_{\varepsilon}^{\mathrm{T}^r}(\mathbf{H}^r)) \cap f^{\mathrm{T}}(\mathrm{T}^+).$$

There exists a point $\mathbf{a}$ of T^+ such that $f^{\mathrm{T}}(\mathbf{a}) = \mathbf{b}$. Let $(x, D_f) \in \mathbf{a}$. Then, $(f(x), R_f) \in \mathbf{b}$. Therefore, $f(x) \in R_f \setminus U_{\varepsilon}^{R_f}$ and $R_f \in \mathbf{H}^r$. This fact means that $x \in D_f \setminus U_{\delta}^{\mathrm{T}^+}(\mathbf{H}^d)$ and, therefore, $\mathbf{b} \in f^{\mathrm{T}}(\mathbf{a}) \in f(F)$, which completes the proof of (6.4.3).

Thus, by Lemma 6.4.1 we can consider the mapping $c(f^{\mathrm{T}})$ of the compact space $c(\mathrm{T}^+, \mathrm{C}_{\theta^d(\tau)}^{\mathrm{T}^+,\diamondsuit})$ into the compact space $c(\mathrm{T}^r, \mathrm{C}_{\tau}^{\mathrm{T}^r,\diamondsuit})$. This mapping belongs to the class $\mathrm{C}(\mathbb{E}, \mathbb{L})$, which means that $c(f^{\mathrm{T}})$ is the required universal element in the class $\mathrm{C}(\mathbb{E}, \mathbb{L})$.

Now, we consider the class $\mathrm{C}^{\mathrm{op}}(\mathbb{E}, \mathbb{L})$. The proof of the proposition for this class is similar to the proof for the class $\mathrm{C}(\mathbb{E}, \mathbb{L})$. However, we must make the following changes. Condition (b) concerning the bases $C_k^{f,d}$ and $C_k^{f,r}$ must be replaced by the following:

(b′) the f-image of $C_k^{f,d}$ is contained in $C_{k+1}^{f,r}$ and the f-image of the base $B_k^{f,d}$, which is the complementary to $C_k^{f,d}$, is contained in the base $B_{k+1}^{f,r}$, which is the complementary to $C_k^{f,r}$.

This can be done because the mapping f is open.

Also, the above given (additional) condition (d) concerning the indexed sets $\{C_\delta^{f,d,k} : \delta \in \tau\}$ and $\{C_\delta^{f,r,k} : \delta \in \tau\}$ must be replaced by the following condition:

(d′) there exist two mappings $\overline{\theta}_k^r$ and $\overline{\theta}_k^{r,c}$, $k \in \omega$, of τ into itself, which are independent of f, such that the indexed set

$$\{C_\delta^{f,r,k+1} : \delta \in \tau\}$$

is a $\overline{\theta}_k^r$-extension of the indexed set

$$\{f(C_\delta^{f,d,k}) : \delta \in \tau\}$$

and the indexed set

$$\{R_f \setminus C_\delta^{f,r,k+1} : \delta \in \tau\}$$

is a $\overline{\theta}_k^{r,c}$-extension of the indexed set

$$\{f(D_f \setminus C_\delta^{f,d,k}) : \delta \in \tau\}.$$

The rest of the proof is the same as for the considered case. The new condition (d′) implies that $\mathbf{M}^r$ is a co-extension of the $\mathbf{F}$-image of $\mathbf{M}^d$ having a mapping $\theta^{r,c}$ as an indicial mapping.

We need only to prove that the corresponding containing mapping f^{T} of the containing space $\mathrm{T}(\mathbf{M}^+, \mathrm{R}^d)$ into the Containing Space $\mathrm{T}(\mathbf{M}^r, \mathrm{R}^r)$ is open. For this it suffices to prove that if $U_\delta^{\mathrm{T}^+}(\mathbf{H}^d)$ is an element of the θ^d-standard base $\mathrm{B}_{\theta^d(\tau)}^{\mathrm{T}+}$ for the open subsets of T^+, where $\mathbf{H}$ is an $\sim^s$-equivalence class for some $s \in \mathcal{F}$, then

$$f^{\mathrm{T}}(U_\delta^{\mathrm{T}^+}(\mathbf{H}^d)) = U_\varepsilon^{\mathrm{T}^r}(\mathbf{H}^r),$$

where $\varepsilon = \theta^{r,c}((\theta^d)^{-1}(\delta))$. This relation is proved similarly to relation (6.4.3). The proof of the proposition is completed. □

6.4.3. Lemma. *Let C^X be a normal base of cardinality $\leq \tau$ for the closed subsets of a space X, C^{Y} a normal base for the closed subsets of a compact space* Y, *and let f be a mapping of X into* Y *such that the f-preimage of C^{Y} is contained in C^X. Then, there exists a unique mapping $c_{\mathrm{Y}}(f)$ of the compact space $c(X, C^X)$ into* Y *such that $c_{\mathrm{Y}}(f)(x) = f(x)$ for every $x \in X$.*

Proof. We denote by B^{Y} the base for the open subsets of Y, which is complementary to C^{Y}.

We start proving the following assertion: for every point $\varphi \in c(X, C^X)$ there exists a point y of Y such that for every element F of C^{Y}, which contains a neighbourhood of y, we have $f^{-1}(F) \in \varphi$. Indeed, suppose that for some point φ of $c(X, C^X)$ there is no point y satisfying the condition of the assertion. Then, for every $y \in \mathrm{Y}$ there exists an element F_y of C^{Y} and an open neighbourhood U_y of y, such that $f^{-1}(F_y) \notin \varphi$ and $U_y \subset F_y$. Then, the set $\{U_y : y \in \mathrm{Y}\}$ is an open cover of Y. Denote by

$$\{U_{y_i} : i \in n \in \omega\}$$

a finite subcover of this cover. Then, we have

$$\mathrm{Y} = \cup\{F_{y_i} : i \in n\}$$

and, therefore,

$$X = \cup\{f^{-1}(F_{y_i}) : i \in n\}.$$

Since $X \in \varphi$, by Lemma 5.3.1 there exists an integer $j \in n$ such that $f^{-1}(F_{y_j}) \in \varphi$, which contradicts the choice of the sets F_y.

It is easy to verify that the point y of the assertion is uniquely determined. Therefore, we can define a mapping $c_{\mathrm{Y}}(f)$ setting $c_{\mathrm{Y}}(f)(\varphi) = y$. It is easy to see that $c_{\mathrm{Y}}(f)(x) = f(x)$ if $x \in X$.

The above assertion implies the following consequence: if $F \in C^{\mathrm{Y}}$ and $f^{-1}(F) \in \varphi \in c(X, C^X)$, then $c_{\mathrm{Y}}(f)(\varphi) \in F$. Indeed, in the opposite case there exist an open subset V' of Y and an element F' of C^{Y} such that $c_{\mathrm{Y}}(f)(\varphi) \in V' \subset F'$ and $F' \cap F = \emptyset$. This fact means that $f^{-1}(F') \cap f^{-1}(F) = \emptyset$, which is a contradiction because by the assertion, $f^{-1}(F') \in \varphi$.

Finally, we prove that $c_{\mathrm{Y}}(f)$ is continuous. To this end it suffices to prove that if $U \in B^{\mathrm{Y}}$, then the set $(c_{\mathrm{Y}}(f))^{-1}(U)$ is open. Let $U \in B^{\mathrm{Y}}$ and

φ a point of $(c_Y(f))^{-1}(U)$. Then, $c_Y(f)(\varphi) \in U$. Therefore, there exist elements V and V' of B^Y and elements F and F' of C^Y such that

$$c_Y(f)(\varphi) \in V' \subset F' \subset V \subset F \subset U.$$

By the above assertion the set $f^{-1}(F')$ is an element of φ and, therefore, $f^{-1}(Y \setminus V)$ does not belong to φ. This results means that $\varphi \in c(f^{-1}(V), C^X)$. To prove that $(c_Y(f))^{-1}(U)$ is open it suffices to prove that

$$c(f^{-1}(V), C^X) \subset (c_Y(f))^{-1}(U)$$

Let $\varphi' \in c(f^{-1}(V), C^X)$. Then, there exists an element $G' \in \varphi'$ such that $G' \subset f^{-1}(V)$ and, therefore, $f(G') \subset V$. In this case,

$$G' \subset f^{-1}(f(G')) \subset f^{-1}(V) \subset f^{-1}(F)$$

and, therefore, $f^{-1}(F) \in \varphi'$. It follows from the above assertion,

$$c_Y(f)(\varphi') \in F \subset U,$$

which means that $\varphi' \in (c_Y(f))^{-1}(U)$. □

The proof of the following proposition is similar to the proof of Proposition 6.4.2. In the proof of Proposition 6.4.4 we use Lemma 6.4.3 instead of Lemma 6.4.1 using in the proof of Proposition 6.4.2.

6.4.4 Proposition. *Suppose that $\mathbb{E}$ is one of the classes:*

$$\mathbb{P}(\text{compact}),\ \mathbb{P}_c(\text{Ind} \le n),\ n \in \omega, \text{ and } \mathbb{P}_c(\dim \le m),\ m \in \omega$$

and let Y be a fixed compact space. Then, in the class of all Y-mappings with the domain in $\mathbb{E}$ there exist universal elements. □

The proof of the following lemma is easy to verify. This lemma is needed for the proof of Proposition 6.4.6.

6.4.5 Lemma. *Let C^X, C^Y, and C^Z be normal bases of cardinality $\le \tau$ for the closed subsets of spaces X, Y, and Z, respectively. Let also*

f be a mapping of X into Y, g a mapping of Y into Z, and $h = g \circ f$. Suppose that:

(a) *the f-preimage of C^Y is contained in C^X,*

(b) *the g-preimage of C^Z is contained in C^Y,*

(c) *the f-image of C^X is contained in C^Y, and*

(d) *the g-image of C^Y is contained in C^Z. Note that the h-preimage of C^Z is contained in C^X and the h-image of C^X is contained in C^Z. Then, for the mappings*

$$c(f) : c(X, C^X) \to c(Y, C^Y),$$

$$c(g) : c(Y, C^Y) \to c(Z, C^Z), \text{ and}$$

$$c(h) : c(X, C^X) \to c(Z, C^Z)$$

constructed in Lemma 6.4.1 we have $c(h) = c(g) \circ c(f)$. □

Definition. Let Λ be a set of cardinality $\leq \tau$ with a partial order $\prec$. A commutative diagram (in particular, an inverse spectrum) D is said to be *open* if for every $(\lambda_0, \lambda_1) \in \prec$ the projection $D(\lambda_0, \lambda_1)$ is open.

In general outline the proof of the next proposition is similar to the proof of Proposition 6.4.2.

6.4.6 Proposition. *Let Λ be a set of cardinality $\leq \tau$ with a partial order $\prec$. Suppose that for every $\lambda \in \Lambda$, $\mathbb{E}_\lambda$ is one of the classes:*

$$\mathbb{P}(\text{compact}), \mathbb{P}_c(\text{Ind} \leq n), n \in \omega, \text{ and } \mathbb{P}_c(\dim \leq m), m \in \omega.$$

Then, in the (non-empty) class of all commutative diagrams (respectively, of all commutative open diagrams) on $\Lambda \cup \prec$, whose λ-components belong to $\mathbb{E}_\lambda$, there exist universal elements. □

6.4.7 Corollary. *Let Λ be a directed set of cardinality $\leq \tau$ with a partial order $\prec$. Suppose that for every $\lambda \in \Lambda$, $\mathbb{E}_\lambda$ is one of the classes:*

$$\mathbb{P}(\text{compact}), \mathbb{P}_c(\text{Ind} \leq n), n \in \omega, \text{ and } \mathbb{P}_c(\dim \leq m), m \in \omega.$$

Then, in the (non-empty) class of all inverse spectra (respectively, of all open inverse spectra) on $\Lambda \cup \prec$, whose λ-components belong to $\mathbb{E}_\lambda$, there exist universal elements. □

6.4.8 Corollary. *Let Λ be a directed set of cardinality $\leq \tau$ with a partial order $\prec$. Suppose that for every $\lambda \in \Lambda$, $\mathbb{E}_\lambda$ is one of the classes:*

$$\mathbb{P}(\text{compact}),\ \mathbb{P}_c(\text{Ind} \leq n),\ n \in \omega,\ \text{and}\ \mathbb{P}_c(\dim \leq m),\ m \in \omega.$$

Then, in the (non-empty) class of the limit spaces of all inverse spectra (respectively, of all open inverse spectra) on $\Lambda \cup \prec$, whose λ-components belong to $\mathbb{E}_\lambda$, there exist universal elements. □

Chapter 7

Actions of groups

In the present chapter we consider the actions of groups. Actually, an action of a group G on a space X is considered as an (open) mapping of $G \times X$ onto X satisfying the "group" conditions. In this case the pair (X, F) is called a G-space; the space X is called the space-coordinate; and the mapping F is called the action-coordinate of this G-space. For a given class $\mathbb{P}$ of G-spaces the class consisting of the space-coordinates (respectively, of action-coordinates) of elements of $\mathbb{P}$ is called the space-component (respectively, the action-component) of $\mathbb{P}$.

In Section 7.1 we define the notion of a saturated class of G-spaces and prove that a class $\mathbb{P}$ of G-spaces is saturated if and only if the action-component of $\mathbb{P}$ is saturated. This result implies the intersection and the universality properties of saturated classes of G-spaces. Furthermore, we prove that for every saturated class $\mathbb{R}$ of spaces (respectively, of mappings) the class $\mathbb{P}$ of all G-spaces generated by $\mathbb{R}$ is a saturated class of G-spaces. In particular, the following classes are saturated:

(a) the class of all regular (completely regular) G-spaces,

(b) the class of all regular (completely regular) countable-dimensional G-spaces,

(c) the class of all regular (completely regular) strongly countable-dimensional G-spaces,

(d) the class of all regular (completely regular) locally finite-dimensional G-spaces, and

(e) the class of all regular (completely regular) G-spaces of dimension ind less than or equal to an ordinal $\alpha \in \tau$ (in particular, α may be an element of ω).

We also prove that the class of all G-spaces (X, F^X) for which the action F^X is effective (respectively, is free) is a saturated class.

In Section 7.2 we consider the actions of groups on spaces with structure. As structure we consider a subset of a space and a $(\Phi, \Lambda, \prec, \mathbf{p})$-structure considered in Chapter 4.

Finally, in Section 7.3 we consider the actions of groups on diagrams, in particular on (inverse) spectra.

It is interesting to note that for the results of this chapter we do not use the fact that the group G as a topological space is completely regular.

Definitions. By a *group* we mean a set G with a mapping

$p : G \times G \to G$ satisfying the folowing properties:

(a) $p(p(a, b), c) = p(a, p(b, c))$ for every $a, b, c \in G$,

(b) there exists an element e of G such that $p(a, e) = p(e, a) = a$ for each $a \in G$, and

(c) for each $a \in G$ there exists an element of G, denoted by a^{-1}, such that $p(a, a^{-1}) = p(a^{-1}, a) = e$.

The mapping p is called the *product mapping* of G. For every $a, b \in G$ the element $p(a, b)$ is denoted by ab. The uniquely determined element e is called the *unit* of G and the uniquely determined element a^{-1} is called the *inverse* of a. The mapping $a \to a^{-1}$ is called the *inverse mapping* of G.

By a *topological group* we mean a group G equipped with a topology such that the product and the inverse mappings of G are continuous.

Agreement. In this chapter, unless we explicitly say otherwise, we denote by G a fixed topological group of weight $\leq \tau$.

Remark. In this chapter we shall use notions and notations of Section 6.1. In particular, we shall consider the notions of the $\mathbf{F}$-domain and $\mathbf{F}$-rang for an indexed collection $\mathbf{F}$ of mappings, the notion of a proper $\mathbf{F}$-co-mark $(\mathbf{M}^d, \mathbf{M}^r)$, the notion of an $(\mathbf{M}^d, \mathbf{M}^r)$-admissible family R of equivalence relations on $\mathbf{F}$, the notations R^d and R^r (related to the family R) and the notations $\mathbf{L}^d$ and $\mathbf{L}^r$ (related to an element $\mathbf{L}$ of $\mathrm{C}^{\diamondsuit}(\mathrm{R})$), the notion of the $\mathbf{F}$-preimage of a co-mark, and the notion of the containing mapping.

7.1 Saturated classes of G-spaces

Definition. By an *action* of G on a space X we mean a mapping F of $G \times X$ into X such that:

(1) $F(g, F(h, x)) = F(gh, x)$ for every $g, h \in G$ and $x \in X$.

(2) $F(e, x) = x$ for every $x \in X$, where e is the unit of G.

By a G-space we mean a pair $E \equiv (X, F)$, where X is a topological space and F is an action on X. In this case the space X is called the *space-coordinate* of E and the mapping F the *action-coordinate* of E.

7.1.1 Lemma. *Let (X, F) be a G-space. Then, for a fixed $g \in G$ the mapping $F_g : X \to X$ defined by relation $F_g(x) = F(g, x)$, $x \in X$, is an homeomorphism onto.*

Proof. By the definition of a G-space, for two elements g and h of G we have $F_g \circ F_h = F_{gh}$. Also, if e is the unit of G, then F_e is the identical mapping of X. In particular, we have

$$F_g \circ F_{g^{-1}} = F_{g^{-1}} \circ F_g = F_e.$$

This relation implies that for every $g \in G$ the mapping F_g is an homeomorphism onto. □

7.1.2 Corollary. *Let (X, F) be a G-space. Then, F is an open mapping.*

Proof. Let W be an open subset of G and U an open subset of X. It suffices to prove that the set $F(W \times U)$ is an open subset of X. But this follows immediately by relation

$$F(W \times U) = \cup\{F_g(U) : g \in W\}$$

and by the fact that $F_g(U)$ is open in X because by Lemma 7.1.1 F_g is an homeomorphism onto and, therefore, it is an open mapping. □

Remark. In what follows of this section and until the Lemma 7.1.4 the case (3) it is assumed that G is an arbitrary space, that is we do not use the group structure of G.

Notation. Let $\mathbf{S}$ be an indexed collection of spaces and G a fixed space. We denote by $G \otimes \mathbf{S}$ the indexed collection $\{G \times X : X \in \mathbf{S}\}$ of spaces.

Definition. Let $\mathbf{S}$ be an indexed collection of spaces, G a fixed space, and F^X, $X \in \mathbf{S}$, an open mapping of $G \times X$ into X. Consider the indexed collection

$$\mathbf{F} \equiv \{F^X : X \in \mathbf{S}\}$$

of mappings (the $\mathbf{F}$-domain of which coincides with the indexed collection $G \otimes \mathbf{S}$ and the $\mathbf{F}$-range with the indexed collection $\mathbf{S}$). A proper $\mathbf{F}$-co-mark $(\mathbf{M}^d, \mathbf{M}^r)$, where

$$\mathbf{M}^d \equiv \{\{V_\varepsilon^{G\times X} : \varepsilon \in \tau\} : X \in \mathbf{S}\} \quad \text{and}$$

$$\mathbf{M}^r \equiv \{\{U_\delta^X : \delta \in \tau\} : X \in \mathbf{S}\},$$

is said to be *product-open* if there exist an indexed base $M^G \equiv \{W_\eta^G : \eta \in \tau\}$ for the open subsets of G, an indexed base $M^X \equiv \{W_\eta^X : \eta \in \tau\}$ for the open subsets of $X \in \mathbf{S}$, a one-to-one mapping θ' of τ into τ, and two one-to-one mappings θ and ϑ of $\tau \times \tau$ into τ such that for every $X \in \mathbf{S}$ the following conditions are satisfied:

$$W_\eta^X = U_{\theta'(\eta)}^X, \quad \eta \in \tau, \tag{7.1.1}$$

$$F^X(W_{\eta_0}^G \times W_{\eta_1}^X) = U_{\theta(\eta_0,\eta_1)}^X, \quad \eta_0, \eta_1 \in \tau, \quad \text{and} \tag{7.1.2}$$

$$W_{\eta_0}^G \times W_{\eta_1}^X = V_{\vartheta(\eta_0,\eta_1)}^{G\times X}, \quad \eta_0, \eta_1 \in \tau. \tag{7.1.3}$$

It is easy to see that any proper $\mathbf{F}$-co-mark, which is a co-extension of a proper product-open $\mathbf{F}$-co-mark, is also product-open.

7.1.3 Lemma. *Let $\mathbf{S}$ be an indexed collection of spaces, G a fixed space, and F^X, for $X \in \mathbf{S}$, an open mapping of $G \times X$ into X. Then, for the indexed collection*

$$\mathbf{F} \equiv \{F^X : X \in \mathbf{S}\}$$

there exists a proper product-open $\mathbf{F}$-co-mark $(\mathbf{M}^d, \mathbf{M}^r)$. Moreover, we can suppose that for a given indexed base $M^G \equiv \{W_\eta^G : \eta \in \tau\}$ for the open subsets of G, an indexed base $M^X \equiv \{W_\eta^X : \eta \in \tau\}$ for the open subsets of $X \in \mathbf{S}$, an one-to-one mapping θ' of τ into τ, and two one-to-one mappings θ and ϑ of $\tau \times \tau$ into τ with the properties:

$$\theta'(\tau) \cap \theta(\tau \times \tau) = \emptyset \quad \text{and} \tag{7.1.4}$$

$$|\tau \setminus \vartheta(\tau \times \tau)| = \tau, \tag{7.1.5}$$

the proper **F***-co-mark* $(\mathbf{M}^d, \mathbf{M}^r)$ *satisfies conditions* (7.1.1)–(7.1.3) *(of the preceding definition), where*

$$\mathbf{M}^d \equiv \{\{V_\varepsilon^{G\times X} : \varepsilon \in \tau\} : X \in \mathbf{S}\} \quad \text{and}$$

$$\mathbf{M}^r \equiv \{\{U_\delta^X : \delta \in \tau\} : X \in \mathbf{S}\}.$$

Proof. Let $M^G \equiv \{W_\eta^G : \eta \in \tau\}$ be an indexed base for the open subsets of G and $M^X \equiv \{W_\eta^X : \eta \in \tau\}$ an indexed base for the open subsets of $X \in \mathbf{S}$. Let also $\theta' : \tau \to \tau$, $\theta : \tau \times \tau \to \tau$, and $\vartheta : \tau \times \tau \to \tau$ be one-to-one mappings satisfying the conditions (7.1.4) and (7.1.5) of the lemma. First, we denote by

$$\mathbf{M}^r \equiv \{\{U_\delta^X : \delta \in \tau\} : X \in \mathbf{S}\}$$

any co-mark of **S** satisfying relations (7.1.1) and (7.1.2). Condition (7.1.4) guarantee the existence of such a co-mark. Then, we denote by

$$\mathbf{M}^d \equiv \{\{V_\varepsilon^{G\times X} : \varepsilon \in \tau\} : X \in \mathbf{S}\}$$

any co-mark of $G \otimes \mathbf{S}$, which is a co-extension of the **F**-preimage of $\mathbf{M}^r$ and satisfies relation (7.1.3). Condition (7.1.5) guarantees the existence of such a co-mark. It is easy to see that the pair $(\mathbf{M}^d, \mathbf{M}^r)$ is the required proper product-open **F**-co-mark. □

7.1.4 Lemma. *Let* **S**, G, *and* **F** *be the same as in Lemma 7.1.3. Suppose that* $(\mathbf{M}^d, \mathbf{M}^r)$ *is a proper product-open* **F***-co-mark and* R *is an* $(\mathbf{M}^d, \mathbf{M}^r)$*-admissible family of equivalence relations on* **F**. *Then, the following conditions are satisfied:*

(1) *A set*

$$\mathbf{a} \equiv \{(x, X) : x \in X \in \mathbf{S}\}$$

is a point of the containing space $\mathrm{T}^r \equiv \mathrm{T}(\mathbf{M}^r, \mathrm{R}^r)$ *for the indexed collection* **S** *of spaces if and only if for every* $g \in G$ *the set*

$$[g, \mathbf{a}] \equiv \{((g, x), G \times X) : (x, X) \in \mathbf{a}\}$$

is a point of the containing space $\mathrm{T}^d \equiv \mathrm{T}(\mathbf{M}^d, \mathrm{R}^d)$ *for the indexed collection* $G \otimes \mathbf{S}$ *of spaces.*

(2) *The mapping* $[G, \mathrm{T}^r]$ *defined by relation*

$$[G, \mathrm{T}^r](g, \mathbf{a}) = [g, \mathbf{a}],$$

where $g \in G$ *and* $\mathbf{a} \in \mathrm{T}^r$ *is an homeomorphism of the space* $G \times \mathrm{T}^r$ *onto the space* T^d.

(3) *If* G *is a topological group and* F^X, $X \in \mathbf{S}$, *is an action of* G *on* X *then the mapping*

$$F^{G \times \mathrm{T}^r} \equiv F^{\mathrm{T}} \circ [G, \mathrm{T}^r],$$

where F^{T} *is the containing mapping*

$$F^{\mathrm{T}} : \mathrm{T}^d \to \mathrm{T}^r$$

corresponding to the **F***-co-mark* $(\mathbf{M}^d, \mathbf{M}^r)$ *and the family* R, *is also an action of* G *on* T^r. *Moreover, if* $\mathbf{L} \in \mathrm{C}^{\diamondsuit}(\mathrm{R})$, *then*

$$F^{G \times \mathrm{T}^r}(G \times \mathrm{T}^r(\mathbf{L}^r)) = \mathrm{T}^r(\mathbf{L}^r),$$

where $\mathrm{T}^r(\mathbf{L}^r) = \mathrm{T}(\mathbf{M}^r, \mathrm{R}^r, \mathbf{L}^r)$ *and, therefore, the mapping*

$$F_{\mathbf{L}}^{G \times \mathrm{T}^r} \equiv F^{G \times \mathrm{T}^r}|_{G \times \mathrm{T}^r(\mathbf{L}^r)}$$

is an action of G *on* $\mathrm{T}^r(\mathbf{L}^r)$.

Proof. Let

$$\mathbf{M}^d \equiv \{\{V_\varepsilon^{G \times X} : \varepsilon \in \tau\} : X \in \mathbf{S}\},$$

$$\mathbf{M}^r \equiv \{\{U_\delta^X : \delta \in \tau\} : X \in \mathbf{S}\},$$

$$\mathrm{R}^d = \{\sim_d^s : s \in \mathcal{F}\}, \text{ and}$$

$$\mathrm{R}^r = \{\sim_r^s : s \in \mathcal{F}\}.$$

Let also $B^G \equiv \{W_\eta^G : \eta \in \tau\}$ be an indexed base for G, $B^X \equiv \{W_\eta^X : \eta \in \tau\}$ an indexed base for $X \in \mathbf{S}$, and let $\theta' : \tau \to \tau$, $\theta : \tau \times \tau \to \tau$, and

$\vartheta : \tau \times \tau \to \tau$ be one-to-one mappings such that relations (7.1.1)–(7.1.3) are satisfied.

We prove condition (1). It suffices to show that

(a) if $x \in X \in \mathbf{S}$, $y \in Y \in \mathbf{S}$, and

$$(x, X) \sim^{\mathrm{R}^r}_{\mathbf{M}^r} (y, Y), \tag{7.1.6}$$

then for every $g \in G$,

$$((g, x), G \times X) \sim^{\mathrm{R}^d}_{\mathbf{M}^d} ((g, y), G \times Y); \tag{7.1.7}$$

(b) if $(g, x) \in G \times X \in G \otimes \mathbf{S}$ and $(g, y) \in G \times Y \in G \otimes \mathbf{S}$ such that relation (7.1.7) is satisfied, then relation (7.1.6) is also satisfied; and

(c) if $g, h \in G$ and $g \neq h$, then $((g, x), G \times X)$ is not $\sim^{\mathrm{R}^d}_{\mathbf{M}^d}$-equivalent to $((h, y), G \times Y)$ for every $x \in X \in \mathbf{S}$ and $y \in Y \in \mathbf{S}$. (About equivalence relations $\sim^{\mathrm{R}^r}_{\mathbf{M}^r}$ and $\sim^{\mathrm{R}^d}_{\mathbf{M}^d}$ see Section 1.2.)

We prove (a). Let $x \in X \in \mathbf{S}$, $y \in Y \in \mathbf{S}$, and $(x, X) \sim^{\mathrm{R}^r}_{\mathbf{M}^r} (y, Y)$. Then, $X \sim^s_r Y$ for every $s \in \mathcal{F}$, and for every $\delta \in \tau$ either $x \in U^X_\delta$ and $y \in U^Y_\delta$ or $x \notin U^X_\delta$ and $y \notin U^Y_\delta$, which means that for every $\eta \in \tau$ either $x \in W^X_\eta$ and $y \in W^Y_\eta$ or $x \notin W^X_\eta$ and $y \notin W^Y_\eta$.

Since the family R is $(\mathbf{M}^d, \mathbf{M}^r)$-admissible, $G \times X \sim^s_d G \times Y$ for every $s \in \mathcal{F}$. Let $\varepsilon \in \tau$ and $(g, x) \in V^{G\times X}_\varepsilon$ for some $g \in G$. We need to prove that $(g, y) \in V^{G\times Y}_\varepsilon$. Since the set

$$B^{G\times X}_{\vartheta(\tau\times\tau)} \equiv \{V^{G\times X}_\varepsilon : \varepsilon \in \vartheta(\tau \times \tau)\}$$

is a base for the open subsets of $G \times X$ there exist $\eta_0, \eta_1 \in \tau$ such that

$$(g, x) \in W^G_{\eta_0} \times W^X_{\eta_1} = V^{G\times X}_\eta \subset V^{G\times X}_\varepsilon,$$

where $\eta = \vartheta(\eta_0, \eta_1)$. Let s be an element of $\mathcal{F}$ such that $\eta, \varepsilon \in s$. Since $G \times X \sim^s_d G \times Y$ we have

$$W^G_{\eta_0} \times W^Y_{\eta_1} = V^{G\times Y}_\eta \subset V^{G\times Y}_\varepsilon.$$

Since $x \in W^X_{\eta_1}$ relations (7.1.1) and (7.1.6) imply that $y \in W^Y_{\eta_1}$ and, therefore, $(g, y) \in V^{G\times Y}_\varepsilon$. Similarly, if $(g, y) \in V^{G\times Y}_\varepsilon$, then $(g, x) \in V^{G\times X}_\varepsilon$. Thus, relation (7.1.7) is satisfied.

We prove (b). Let $(g, x) \in G \times X \in G \otimes \mathbf{S}$ and $(g, y) \in G \times Y \in G \otimes \mathbf{S}$ such that relation (7.1.7) is satisfied. Then, $G \times X \sim^s_d G \times Y$ for every

$s \in \mathcal{F}$, and for every $\varepsilon \in \tau$ either $(g,x) \in V_\varepsilon^{G\times X}$ and $(g,y) \in V_\varepsilon^{G\times Y}$ or $(g,x) \notin V_\varepsilon^{G\times X}$ and $(g,y) \notin V_\varepsilon^{G\times Y}$. Since the family R is $(\mathbf{M}^d, \mathbf{M}^r)$-admissible, $X \sim_{\mathrm{R}^r}^{\mathbf{M}^r} Y$ for every $s \in \mathcal{F}$. Let $\delta \in \tau$. Suppose that $x \in U_\delta^X$. We need to prove that $y \in U_\delta^Y$. There exist $\eta_0, \eta_1 \in \tau$ such that

$$(g,x) \in W_{\eta_0}^G \times W_{\eta_1}^X = V_\varepsilon^{G\times X} \text{ and } W_{\eta_1}^X = U_\eta^X \subset U_\delta^X,$$

where $\varepsilon = \vartheta(\eta_0, \eta_1)$ and $\eta = \theta'(\eta_1)$. Let s be an element of $\mathcal{F}$ such that $\eta, \delta, \varepsilon \in s$. Since $X \sim_r^s Y$ we have

$$W_{\eta_1}^Y = U_\eta^Y \subset U_\delta^Y,$$

and since $(g,x) \in V_\varepsilon^{G\times X}$ relation (7.1.7) implies that

$$(g,y) \in V_\varepsilon^{G\times Y} = W_{\eta_0}^G \times W_{\eta_1}^Y,$$

which means that $y \in W_{\eta_1}^Y$ and, therefore, $y \in U_\delta^Y$. Similarly, if $y \in U_\delta^Y$, then $x \in U_\delta^X$.

We prove (c). Suppose that $g, h \in G$ and $g \neq h$. Let also $x \in X \in \mathbf{S}$ and $y \in Y \in \mathbf{S}$. There exists an element η_0 of τ such that $W_{\eta_0}^G$ contains one of the points g and h and does not contain the other. For example, suppose that $g \in W_{\eta_0}^G$. Let $\eta_1 \in \tau$ such that $x \in W_{\eta_1}^X$. Then, $(g,x) \in V_\varepsilon^{G\times X}$ and $(h,y) \notin V_\varepsilon^{G\times Y}$, where $\varepsilon = \vartheta(\eta_0, \eta_1)$, that is the pairs $((g,x), G\times X)$ and $((g,y), G\times Y)$ do not belong to the same point of the set $\mathrm{T}(\mathbf{M}^d, \mathrm{R}^d)$. This results proves (c) and completes the proof of the condition (1).

We prove condition (2). First we prove that the mapping $[G, \mathrm{T}^r]$ is one-to-one. Let $(g, \mathbf{a})$ and $(h, \mathbf{b})$ be two distinct elements of $G \times \mathrm{T}^r$. Then, either $g \neq h$ or $\mathbf{a} \neq \mathbf{b}$.

Suppose that $g \neq h$. As in the condition (c) the pairs $((g,x), G\times X)$ and $((g,y), G\times Y)$ do not belong to the same point of the set $\mathrm{T}(\mathbf{M}^d, \mathrm{R}^d)$. By condition (1), $[g, \mathbf{a}] \neq [h, \mathbf{b}]$.

Suppose that $\mathbf{a} \neq \mathbf{b}$. Let $(x, X) \in \mathbf{a}$ and $(y, Y) \in \mathbf{b}$. Consider the $\theta'(\tau)$-standard base $\mathrm{B}_{\theta'(\tau)}^{\mathrm{T}^r}$ for the open subsets of the containing space T^r. There exists an element $U_\delta^{\mathrm{T}^r}(\mathbf{H})$ of this base, which contains one of the points $\mathbf{a}$ and $\mathbf{b}$ and does not contain the other. For example, suppose that $\mathbf{a} \in U_\delta^{\mathrm{T}^r}(\mathbf{H})$. This means that $x \in U_\delta^X$, $X \in \mathbf{H}$, and either $y \notin U_\delta^Y$ or $y \in U_\delta^Y$ and $Y \notin \mathbf{H}$. Suppose that $y \notin U_\delta^Y$. Then, $x \in W_{\eta_1}^X$ and $y \notin W_{\eta_1}^X$, where $\eta_1 = (\theta')^{-1}(\delta)$. Let $W_{\eta_0}^G \in B^G$ and $g \in W_{\eta_0}^G$. Then,

$$W_{\eta_0}^{G} \times W_{\eta_1}^{X} = V_{\varepsilon}^{G\times X},\ W_{\eta_0}^{G} \times W_{\eta_1}^{Y} = V_{\varepsilon}^{G\times Y},$$

$$(g, x) \in V_{\varepsilon}^{G\times X}, \text{ and } (h, y) \notin V_{\varepsilon}^{G\times Y},$$

where $\varepsilon = \vartheta(\eta_0, \eta_1)$. As the above, results this means that $[g, \mathbf{a}] \neq [h, \mathbf{b}]$. If $y \in U_{\delta}^{Y}$ and $Y \notin \mathbf{H}$, then X and Y are not $\sim_r^s$-equivalent for some $s \in \mathcal{F}$. Since the family R is proper we have that $G \times X$ and $G \times Y$ are not $\sim_d^s$-equivalent, which means that $[g, \mathbf{a}] \neq [g, \mathbf{b}]$. Thus, the mapping $[G, \mathrm{T}^r]$ is one-to-one.

Now, we prove that the mappings $[G, \mathrm{T}^r]$ and $([G, \mathrm{T}^r])^{-1}$ are continuous and onto. We note that the sets of the form $U_{\delta}^{\mathrm{T}^r}(\mathbf{H})$, where $\delta \in \theta'(\tau)$ and $\mathbf{H} \in \mathrm{C}(\mathrm{R}^r)$, compose a base for the open subsets of the containing space T^r and, therefore, the sets of the form $W \times U_{\delta}^{\mathrm{T}^r}(\mathbf{H})$, where $W \in B^G$, where $\delta \in \theta'(\tau)$, and $\mathbf{H} \in \mathrm{C}(\mathrm{R}^r)$, compose a base for the open subsets of the space $G \times \mathrm{T}^r$. Since the set $B_{\vartheta(\tau\times\tau)}^{G\times X}$ is a base for the open subsets of the space $G \times X$, where $X \in \mathbf{S}$, we can consider the $\vartheta(\tau \times \tau)$-standard base $\mathrm{B}_{\vartheta(\tau\times\tau)}^{\mathrm{T}^d}$ for the open subsets of T^d and its elements $U_{\varepsilon}^{\mathrm{T}^d}(\mathbf{L})$, where $\varepsilon \in \vartheta(\tau \times \tau)$, and $\mathbf{L} \in \mathrm{C}^{\diamondsuit}(\mathrm{R}^d)$. (We recall that the set $U_{\varepsilon}^{\mathrm{T}^d}(\mathbf{L})$ consists of all points $\mathbf{d}$ of the space T^d such that there exists an element $((g, x), G \times X)$ of $\mathbf{d}$ for which $(g, x) \in V_{\varepsilon}^{G\times X}$ and $G \times X \in \mathbf{L}$.)

Let $\eta_0, \eta_1 \in \tau$, $\delta = \theta'(\eta_1)$, and $\varepsilon = \vartheta(\eta_0, \eta_1)$. Let also $\mathbf{H}$ be an element of $\mathrm{C}(\mathrm{R}^r)$ and $\mathbf{L}$ the element of $\mathrm{C}(\mathrm{R}^d)$ consisting of all elements $G \times X$ of $G \otimes \mathbf{S}$ such that $X \in \mathbf{H}$ (that is, $\mathbf{L}$ corresponds to $\mathbf{H}$). To prove that $[G, \mathrm{T}^r]$ and $([G, \mathrm{T}^r])^{-1}$ are continuous and onto it suffices to prove that

$$[G, \mathrm{T}^r](W_{\eta_0}^{G} \times U_{\delta}^{\mathrm{T}^r}(\mathbf{H})) = U_{\varepsilon}^{\mathrm{T}^d}(\mathbf{L}). \tag{7.1.8}$$

Let $(g, \mathbf{a}) \in W_{\eta_0}^{G} \times U_{\delta}^{\mathrm{T}^r}(\mathbf{H})$ and $(x, X) \in \mathbf{a}$. Then, $[G, \mathrm{T}^r](g, \mathbf{a}) = [g, \mathbf{a}]$ and the point $[g, \mathbf{a}]$ of the space T^d contains the element $((g, x), G \times X)$. Since $g \in W_{\eta_0}^{G}$, $x \in U_{\delta}^{X} = W_{\eta_1}^{X}$, and $X \in \mathbf{H}$ we have $(g, x) \in V_{\varepsilon}^{G\times X}$ and $G \times X \in \mathbf{L}$, which means that $[g, \mathbf{a}] \in U_{\varepsilon}^{\mathrm{T}^d}(\mathbf{L})$. Thus, the left side of relation (7.1.8) is contained in the right.

Conversely, let $[g, \mathbf{a}]$ be a point of $U_{\varepsilon}^{\mathrm{T}^d}(\mathbf{L})$. By the definition of the mapping $[G, \mathrm{T}^r]$, $[G, \mathrm{T}^r](g, \mathbf{a}) = [g, \mathbf{a}]$. Suppose that $((g, x), G \times X)$ be an element of $[g, \mathbf{a}]$. Then, $(g, x) \in V_{\varepsilon}^{G\times X} = W_{\eta_0}^{G} \times W_{\eta_1}^{X}$ and $G \times X \in \mathbf{L}$. This means that $g \in W_{\eta_0}^{G}$, $x \in W_{\eta_1}^{X} = U_{\delta}^{X}$, and $X \in \mathbf{H}$. By condition (1), $(x, X) \in \mathbf{a}$. Therefore, $\mathbf{a} \in U_{\delta}^{\mathrm{T}^r}(\mathbf{H})$ and $(g, \mathbf{a}) \in W_{\eta_0}^{G} \times U_{\delta}^{\mathrm{T}^r}(\mathbf{H})$, which means that the right side of relation (7.1.8) is contained in the left. This fact completes the proof of this relation and the proof of condition (2).

We prove condition (3). Suppose that G is a topological group. Obviously, $F^{G\times \mathrm{T}^r}$ is continuous. We note that by construction of the containing mapping F^{T} (see Lemma 6.1.1) if $F^{\mathrm{T}}([g,\mathbf{a}]) = \mathbf{b}$ and

$$((g,x), G\times X) \in [g,\mathbf{a}] \in \mathrm{T}^d,$$

then $\mathbf{b}$ is the element of T^r containing the pair $(F^X(g,x), X)$.

Let $g, h \in G$, $\mathbf{a} \in \mathrm{T}^r$, and $(x, X) \in \mathbf{a}$. Then,

$$F^{G\times \mathrm{T}^r}(g,\mathbf{a}) = F^{\mathrm{T}}([G,\mathrm{T}^r](g,\mathbf{a})) = F^{\mathrm{T}}([g,\mathbf{a}]) = \mathbf{b},$$

where $\mathbf{b}$ is the element of T^r containing the pair $(F^X(g,x), X)$. Similarly,

$$F^{G\times \mathrm{T}^r}(h, F^{G\times \mathrm{T}^r}(g,\mathbf{a})) \equiv \mathbf{c}$$

is the element of T^r containing the pair $(F^X(h, F^X(g,x)), X)$. Since F^X is an action on X we have

$$F^X(h, F^X(g,x)) = F^X(hg, x).$$

Therefore, the element $\mathbf{c}$ contains the pair $(F^X(hg,x), X)$, which means that $F^{G\times \mathrm{T}^r}(hg,\mathbf{a}) = \mathbf{c}$. Thus,

$$F^{G\times \mathrm{T}^r}(h, F^{G\times \mathrm{T}^r}(g,\mathbf{a})) = F^{G\times \mathrm{T}^r}(hg,\mathbf{a}). \qquad (7.1.9)$$

Suppose that e is the unit of the group G. Then, we have $F^{G\times \mathrm{T}^r}(e,\mathbf{a}) = \mathbf{d}$, where $\mathbf{d}$ is the point of T^r containing the pair $(F^X(e,x), X) = (x, X)$, which means that $\mathbf{d} = \mathbf{a}$. Thus,

$$F^{G\times \mathrm{T}^r}(e,\mathbf{a}) = \mathbf{a}. \qquad (7.1.10)$$

Relations (7.1.9) and (7.1.10) show that $F^{G\times \mathrm{T}}$ is an action of G on the space T^r.

Now, suppose that $\mathbf{L} \in \mathrm{C}^{\diamondsuit}(\mathrm{R})$. Let $(g,\mathbf{a}) \in G \times \mathrm{T}^r(\mathbf{L}^r)$ (that is, $\mathbf{a} \in \mathrm{T}^r(\mathbf{L}^r)$ and therefore $X \in \mathbf{L}^r$) and $(x, X) \in \mathbf{a}$. Then, $F^{G\times \mathrm{T}^r}(g,\mathbf{a}) = F^{\mathrm{T}}([g,\mathbf{a}])$ and $((g,x), G\times X) \in [g,\mathbf{a}]$. By construction of F^{T} we have

$$(F^X(g,x), X) \in F^{\mathrm{T}}([g,\mathbf{a}].$$

Since $X \in \mathbf{L}^r$ we have $F^{\mathrm{T}}([g, \mathbf{a}]) \in \mathrm{T}^r(\mathbf{L}^r)$. Thus,

$$F^{G\times \mathrm{T}^r}(G \times \mathrm{T}^r(\mathbf{L}^r)) \subset \mathrm{T}^r(\mathbf{L}^r).$$

Since $F^{G\times \mathrm{T}^r}(\{e\} \times \mathrm{T}^r(\mathbf{L}^r)) = \mathrm{T}^r(\mathbf{L}^r)$, where e is the unit of G, we have $F^{G\times \mathrm{T}^r}(G \times \mathrm{T}^r(\mathbf{L}^r)) = \mathrm{T}^r(\mathbf{L}^r)$. This completes the proof of condition (3) and, therefore, the proof of the lemma. □

Definition. In the case, where G is a topological group, the mapping $F_{\mathbf{L}}^{G\times \mathrm{T}^r}$ constructed in Lemma 7.1.4 is called the *containing action-mapping* corresponding to the $\mathbf{F}$-co-mark $(\mathbf{M}^d, \mathbf{M}^r)$, the family R, and the element $\mathbf{L}$ of $\mathrm{C}^{\diamond}(\mathrm{R})$. In the case where $\mathbf{L} = \mathbf{F}$, $F_{\mathbf{L}}^{G\times \mathrm{T}^r}$ coincides with $F^{G\times \mathrm{T}^r}$.

Definition. Let (X, F^X) and (Y, F^Y) be two G-spaces and h a mapping of X into Y. The mapping h is said to be *equivariant* if the following condition is satisfied:

$$F^Y \circ (\mathrm{id}_G \times h) = h \circ F^X,$$

where id_G is the identical mapping of G and $\mathrm{id}_G \times h$ is the mapping of $G \times X$ into $G \times Y$ defined by relation

$$(\mathrm{id}_G \times h)(g, x) = (g, h(x)).$$

Definitions. Let $I\!\!P$ be a class of G-spaces. The class consisting of the space-coordinates of all elements of $I\!\!P$ is called the *space-component* of $I\!\!P$ and the class of mappings consisting of the action-coordinates of all elements of $I\!\!P$ is called the *action-component* of $I\!\!P$.

The class $I\!\!P$ is said to be *topological* if for every G-space (Y, F^Y) and element (X, F^X) of $I\!\!P$ the existence of an equivariant homeomorphism h of X onto Y implies that $(Y, F^Y) \in I\!\!P$.

Obviously, if $I\!\!P$ is a topological class of G-spaces, then the space-component of $I\!\!P$ is a topological class of spaces and the action-component of $I\!\!P$ is a topological class of mappings.

In what follows, whenever we consider an arbitrary class of G-spaces it is assumed that this class is topological.

Definition. Let $I\!\!P$ be a class of G-spaces. An element (T, F^T) of $I\!\!P$ is said to be *universal in* $I\!\!P$ if for every $(X, F^X) \in I\!\!P$ there exists an equivariant embedding of X into T.

Definition. A non-empty class $I\!P$ of G-spaces is said to be *saturated* if for every indexed collection $\mathbf{F}$ of elements of the action-component of $I\!P$ there exists a proper product-open $\mathbf{F}$-co-mark $(\mathbf{M}^{d,+}, \mathbf{M}^{r,+})$ satisfying the following condition: for every proper $\mathbf{F}$-co-mark $(\mathbf{M}^d, \mathbf{M}^r)$, which is a co-extension of $(\mathbf{M}^{d,+}, \mathbf{M}^{r,+})$ (and, therefore, $(\mathbf{M}^d, \mathbf{M}^r)$ is also a product-open $\mathbf{F}$-co-mark), there exists an $(\mathbf{M}^d, \mathbf{M}^r)$-admissible family R^+ of equivalence relations on $\mathbf{F}$ such that for every admissible family R of equivalence relations on $\mathbf{F}$, which is a final refinement of R^+, and an element $\mathbf{L}$ of $\mathrm{C}^{\diamond}(\mathrm{R})$ the pair

$$(\mathrm{T}(\mathbf{M}^r, \mathrm{R}^r, \mathbf{L}^r), F_{\mathbf{L}}^{G\times \mathrm{T}^r})$$

belongs to $I\!P$, where $F_{\mathbf{L}}^{G\times \mathrm{T}^r}$ is the containing action-mapping corresponding to the $\mathbf{F}$-co-mark $(\mathbf{M}^d, \mathbf{M}^r)$, the family R, and the element $\mathbf{L}$ of $\mathrm{C}^{\diamond}(\mathrm{R})$.

The proper product-open $\mathbf{F}$-co-mark $(\mathbf{M}^{d,+}, \mathbf{M}^{r,+})$ is said to be an *initial* $\mathbf{F}$-co-mark (*corresponding* to the class $I\!P$) and the $(\mathbf{M}^d, \mathbf{M}^r)$-admissible family R^+ an *initial* family (*corresponding* to the $\mathbf{F}$-co-mark $(\mathbf{M}^d, \mathbf{M}^r)$ and the class $I\!P$).

Lemma 7.1.4 implies the following consequence.

7.1.5 Corollary. *The class of all G-spaces is a saturated class.* □

7.1.6 Proposition. *A class $I\!P$ of G-spaces is saturated if and only if the action-component of $I\!P$ is a saturated class of mappings.*

Proof. Let $I\!P$ be a class of G-spaces and $I\!F$ the action-component of $I\!P$. Suppose that $I\!P$ is a saturated class. We need to prove that $I\!F$ is a saturated class of mappings. For this purpose, we consider an arbitrary indexed collection $\mathbf{F}$ of elements of $I\!F$. By the definition of a saturated class of G-spaces, there exists an initial product-open $\mathbf{F}$-co-mark $(\mathbf{M}^{d,+}, \mathbf{M}^{r,+})$ corresponding to the class $I\!P$. Obviously, this $\mathbf{F}$-co-mark can be considered also as an initial $\mathbf{F}$-co-mark corresponding to the class $I\!F$. (Here, we use the fact that $I\!F$ is a topological class of mappings.) Thus $I\!F$ is a saturated class of mappings.

Conversely, suppose that $I\!F$ is a saturated class of mappings. We need to prove that $I\!P$ is saturated. Let $\mathbf{F}$ be an indexed collection of elements of $I\!F$. There exists an $\mathbf{F}$-co-mark $(\mathbf{M}_0^{d,+}, \mathbf{M}_0^{r,+})$, which is an initial $\mathbf{F}$-co-mark corresponding to the class $I\!F$. By Lemma 7.1.3 for the indexed collection $\mathbf{F}$ there exists a proper product-open $\mathbf{F}$-co-mark $(\mathbf{M}_0, \mathbf{M}_1)$.

Let $(\mathbf{M}^{d,+}, \mathbf{M}^{r,+})$ be an arbitrary proper $\mathbf{F}$-co-mark, which is simultaneously a co-extension of the $\mathbf{F}$-co-marks $(\mathbf{M}_0^{d,+}, \mathbf{M}_0^{r,+})$ and $(\mathbf{M}_0, \mathbf{M}_1)$. Then, $(\mathbf{M}^{d,+}, \mathbf{M}^{r,+})$ is also a proper product-open $\mathbf{F}$-co-mark. It is easy to see that this co-mark is an initial $\mathbf{F}$-co-mark corresponding to the class $I\!\!P$. Thus, $I\!\!P$ is a saturated class. □

The intersection property of saturated classes of mappings and Proposition 7.1.6 imply the following proposition, which is the intersection property of saturated classes of G-spaces.

7.1.7 Proposition. *The intersection of not more than τ many saturated classes of G-spaces is also a saturated class.* □

7.1.8 Proposition. *In any non-empty saturated class of G-spaces there exist universal elements.*

Proof. Let $I\!\!P$ be a saturated class of G-spaces. Denote by $\mathbf{F}$ an indexed collection of elements of the action-component of $I\!\!P$ with the property that for every element (Y, F^Y) of $I\!\!P$ there exists an element (X, F^X) of $I\!\!P$ such that: (a) $F^X \in \mathbf{F}$, and (b) there exists an equivariant homeomorphism of Y onto X.

Let $(\mathbf{M}^d, \mathbf{M}^r)$ be a proper product-open $\mathbf{F}$-co-mark and R an $(\mathbf{M}^d, \mathbf{M}^r)$-admissible family of equivalence relations on $\mathbf{F}$ such that

$$(\mathrm{T}^r, F^{G\times \mathrm{T}^r}) \in I\!\!P,$$

where $\mathrm{T}^r \equiv \mathrm{T}(\mathbf{M}^r, \mathrm{R})$ and $F^{G\times \mathrm{T}^r}$ is the containing action-mapping corresponding to the $\mathbf{F}$-co-mark $(\mathbf{M}^d, \mathbf{M}^r)$, the family R, and the element $\mathbf{L} \equiv \mathbf{F}$ of $\mathrm{C}^{\diamondsuit}(\mathrm{R})$. The existence of such an $\mathbf{F}$-co-mark and a family of equivalence relation on $\mathbf{F}$ follows by the fact that $I\!\!P$ is saturated. We prove that this element of $I\!\!P$ is universal.

Indeed, let $(X, F^X) \in I\!\!P$. It suffices to prove that the natural embedding $i^X_{\mathrm{T}^r}$ of X into $\mathrm{T}^r \equiv \mathrm{T}(\mathbf{M}^r, \mathrm{R})$ is an equivariant mapping, that is the following relation is true:

$$i^X_{\mathrm{T}^r} \circ F^X = F^{G\times \mathrm{T}^r} \circ (\mathrm{id}_G \times i^X_{\mathrm{T}^r}). \tag{7.1.11}$$

Let $(g, x) \in G\times X$. Then, $i^X_{\mathrm{T}^r}(F^X(g, x))$ is the point $\mathbf{b}$ of T^r, which contains the pair $(F^X(g, x), X)$. On the other hand, we have $(\mathrm{id}_G \times i^X_{\mathrm{T}^r})(g, x) =$

$(g, \mathbf{a})$, where $\mathbf{a}$ is the point of T^r, which contains the pair (x, X). Therefore (see Lemma 7.1.4),

$$F^{G\times \mathrm{T}^r}(g, \mathbf{a}) = F^{\mathrm{T}}([G, \mathrm{T}^r](g, \mathbf{a}))) = F^{\mathrm{T}}([g, \mathbf{a}]),$$

where F^{T} is the containing mapping of T^d into T^r corresponding to the $\mathbf{F}$-co-mark $(\mathbf{M}^d, \mathbf{M}^r)$ and the family R. Since $((g, x), G \times X) \in [g, \mathbf{a}]$ (see Lemma 7.1.4), by Lemma 6.1.1 we have that $F^{\mathrm{T}}([g, \mathbf{a}])$ is the point of T^r, which contains the pair $(F^X(g, x), X)$, that is $F^{\mathrm{T}}[g, \mathbf{a}])$ is the point $\mathbf{b}$. Relation (7.1.11) is proved. □

7.1.9 Proposition. *The space-component of any saturated class of G-spaces is a saturated class of spaces.*

Proof. Let $\mathbb{P}$ be the considered class of G-spaces and $\mathbb{R}$ the space-component of $\mathbb{P}$. Consider an indexed collection $\mathbf{S}$ of elements of $\mathbb{R}$. Then, for every $X \in \mathbf{S}$ we have a uniquely determined action F^X on X. Denote by $\mathbf{F}$ the indexed collection $\{F^X : X \in \mathbf{S}\}$. Then, $\mathbf{F}$ is an indexed collection of elements of the action-component $\mathbb{F}$ of $\mathbb{P}$. Since $\mathbb{P}$ is saturated there exists an $\mathbf{F}$-co-mark $(\mathbf{M}^d, \mathbf{M}^r)$, which is an initial $\mathbf{F}$-co-mark corresponding to the class $\mathbb{P}$. We prove that $\mathbf{M}^r$ is an initial co-mark of $\mathbf{S}$ corresponding to the class $\mathbb{R}$.

Actually, it suffices to prove that if $\mathbf{M}$ is a co-mark of $\mathbf{S}$, which is a co-extension of $\mathbf{M}^r$, then there exists a co-mark $\mathbf{M}_0^d$ of the indexed collection $G \otimes \mathbf{S}$ of spaces such that the $\mathbf{F}$-co-mark $(\mathbf{M}_0^d, \mathbf{M})$ is product-open and a co-extension of $(\mathbf{M}^d, \mathbf{M}^r)$. It is easy to see that as the co-mark $\mathbf{M}_0^d$ we can take any co-mark of $G \otimes \mathbf{S}$, which is simultaneously a co-extension of $\mathbf{M}^d$ and the $\mathbf{F}$-preimage of the co-mark $\mathbf{M}$. □

Definition. Let $\mathbb{R}$ be a class of spaces (respectively, a class of mappings). The class consisting of all G-spaces (X, F^X), where X (respectively, F^X) is an element of $\mathbb{R}$ is called the *class of G-spaces generated by $\mathbb{R}$*.

7.1.10 Proposition. *Let $\mathbb{R}$ be a saturated class of spaces (respectively, a saturated class of mappings). Then, the (non-empty) class of all G-spaces generated by $\mathbb{R}$ is a saturated class of G-spaces.*

Proof. We prove the proposition in the case, where $\mathbb{R}$ is a saturated class of spaces. The case, where $\mathbb{R}$ is a saturated class of mappings is proved similarly.

Denote by $I\!P$ the class of G-spaces generated by $I\!R$ and by $I\!F$ the action-component of $I\!P$. Suppose that $\mathbf{F}$ is an arbitrary indexed collection of elements of $I\!F$ and let $\mathbf{S} \equiv \mathbf{S}^r$ be the $\mathbf{F}$-range and $\mathbf{S}^d$ the $\mathbf{F}$-domain. Obviously, $\mathbf{S}$ is an indexed collection of elements of $I\!R$ and $\mathbf{S}^d = G \otimes \mathbf{S}$.

Since $I\!R$ is a saturated class there exists a co-mark $\mathbf{M}_0^+$ of $\mathbf{S}$, which is an initial co-mark corresponding to the class $I\!R$. By Lemma 7.1.3 there exists a proper product-open $\mathbf{F}$-co-mark $(\mathbf{M}_0^d, \mathbf{M}_0^r)$. Denote by $\mathbf{M}^{r,+}$ an arbitrary co-mark of $\mathbf{S}$, which is simultaneously an extension of co-marks $\mathbf{M}_0^+$ and $\mathbf{M}_0^r$ of $\mathbf{S}$. Also, denote by $\mathbf{M}^{d,+}$ an arbitrary co-mark of $G \otimes \mathbf{S}$, which is simultaneously an extension of the co-mark $\mathbf{M}_0^d$ of $G \otimes \mathbf{S}$ and the $\mathbf{F}$-preimage of the co-mark $\mathbf{M}^{r,+}$ of $\mathbf{S}$. Then, the pair $(\mathbf{M}^{d,+}, \mathbf{M}^{r,+})$ is a proper product-open $\mathbf{F}$-co-mark. We prove that this co-mark is an initial $\mathbf{F}$-co-mark corresponding to the class $I\!P$.

Let $(\mathbf{M}^d, \mathbf{M}^r)$ be an arbitrary co-extension of $(\mathbf{M}^{d,+}, \mathbf{M}^{r,+})$. Since $I\!R$ is a saturated class there exists an initial family R_0^+ of equivalence relations on $\mathbf{S}$ corresponding to the co-mark $\mathbf{M}^r$ of $\mathbf{S}$ and the class $I\!R$. Denote by R^+ an $(\mathbf{M}^d, \mathbf{M}^r)$-admissible family of equivalence relations on $\mathbf{F}$ such that $(\mathrm{R}^+)^r$ is a final refinement of R_0^+. To prove that $(\mathbf{M}^{d,+}, \mathbf{M}^{r,+})$ is initial it suffices to prove that R^+ is an initial family corresponding to the $\mathbf{F}$-co-mark $(\mathbf{M}^d, \mathbf{M}^r)$ and the class $I\!P$. For this purpose, we consider an arbitrary family R of equivalence relations on $\mathbf{F}$, which is a final refinement of R^+. By Lemma 7.1.4 for a given element $\mathbf{L}$ of $\mathrm{C}^\diamondsuit(\mathrm{R})$ we can consider the containing action-mapping

$$F_{\mathbf{L}}^{G\times \mathrm{T}^r} : G \times \mathrm{T}^r(\mathbf{L}^r) \to \mathrm{T}^r(\mathbf{L}^r)$$

corresponding to the $\mathbf{F}$-co-mark $(\mathbf{M}^d, \mathbf{M}^r)$, the family R, and the element $\mathbf{L}$ of $\mathrm{C}^\diamondsuit(\mathrm{R})$, where $\mathrm{T}^r(\mathbf{L}^r) = \mathrm{T}(\mathbf{M}^r, \mathrm{R}^r, \mathbf{L}^r)$. We need to prove that this mapping belongs to the class $I\!F$, that is the space $\mathrm{T}^r(\mathbf{L}^r)$ belongs to the class $I\!R$. But this follows immediately by the fact that $I\!R$ is a saturated class and the construction of the co-mark $\mathbf{M}^r$ and the family R^r. Thus, the $\mathbf{F}$-co-mark $(\mathbf{M}^{d,+}, \mathbf{M}^{r,+})$ and the family R^+ are initial. The proof of the proposition is complete. □

Remark. In the following corollaries by the phrase "a countable-dimensional G-space" (respectively, "a strongly countable-dimensional G-space" and so on) we mean a G-space having a countable-dimensional space-coordinate (respectively, a strongly countable-dimensional space-coordinate and so on).

7.1.11 Corollary. *The following classes of G-spaces are saturated:*

1. *The class of all regular (completely regular) G-spaces.*

2. *The class of all regular (completely regular) countable-dimensional G-spaces.*

3. *The class of all regular (completely regular) strongly countable-dimensional G-spaces.*

4. *The class of all regular (completely regular) locally finite-dimensional G-spaces.*

5. *The class of all regular (completely regular) G-spaces of dimension* ind *less than or equal to an ordinal $\alpha \in \tau^+$ (in particular, α may be an element of ω).*

Therefore, in these classes there exist universal G-spaces. □

In particular, if $\tau = \omega$ we have the following consequence.

7.1.12 Corollary. *Suppose that G is a separable metrizable topological group. Then, the following classes of G-spaces are saturated:*

1. *The class of all separable metrizable G-spaces.*

2. *The class of all separable metrizable countable-dimensional G-spaces.*

3. *The class of all separable metrizable strongly countable-dimensional G-spaces.*

4. *The class of all separable metrizable locally finite-dimensional G-spaces.*

5. *The class of all separable metrizable G-spaces of dimension* ind *less than or equal to a countable ordinal α (in particular, α may be an element of ω).*

Therefore, in these classes there exist universal G-spaces. □

7.1.13 Corollary. *Let $I\!R$ be a family of spaces considered in one of the Corollaries 4.4.7, 4.4.8, 4.4.9, 4.4.11, 4.4.14, and 4.4.15. Then, the class of G-spaces generated by $I\!R$ is a saturated class of G-spaces and, therefore, in this class there exist universal elements.* □

Definition. Let (X, F) be a G-space. The action F of G on X is said to be *effective* if for every element g of G, distinct from the unit of G, there exists a point x of X such that $F(g, x) \neq x$.

The action F is said to be *free* if for every element g of G, distinct from the unit of G and a point x of X, we have $F(g, x) \neq x$.

7.1.14 Proposition. *The class consisting of all G-spaces (X, F^X), where F^X is an effective (respectively, is a free) action, is a saturated class.*

Proof. Suppose that $I\!\!P$ is the considered class and $I\!\!F$ the action-component of $I\!\!P$. Let $\mathbf{F}$ be an arbitrary indexed collection of elements of $I\!\!F$, $\mathbf{S} \equiv \mathbf{S}^r$ the $\mathbf{F}$-range, and $\mathbf{S}^d$ the $\mathbf{F}$-domain.

Consider an arbitrary proper product-open $\mathbf{F}$-co-mark $(\mathbf{M}^{d,+}, \mathbf{M}^{r,+})$. We prove that this $\mathbf{F}$-co-mark is initial. For this purpose we consider an arbitrary $\mathbf{F}$-co-mark $(\mathbf{M}^d, \mathbf{M}^r)$, which is a co-extension of $(\mathbf{M}^{d,+}, \mathbf{M}^{r,+})$ and an arbitrary $(\mathbf{M}^d, \mathbf{M}^r)$-admissible family R^+ of equivalence relations on $\mathbf{F}$. Let also R be an admissible family of equivalence relations on $\mathbf{F}$, which is a final refinement of R^+.

In order to prove the proposition it suffices to prove that R^+ is an initial family (and, therefore, $(\mathbf{M}^d, \mathbf{M}^r)$ will be an initial $\mathbf{F}$-co-mark), that is we need to prove that the containing action-mapping $\mathbf{F}_{\mathbf{L}}^{G\times\mathrm{T}^r}$ (see Lemma 7.1.4) corresponding to the $\mathbf{F}$-co-mark $(\mathbf{M}^d, \mathbf{M}^r)$, the family R, and an element $\mathbf{L}$ of $\mathrm{C}^{\diamondsuit}(\mathrm{R})$ belongs to $I\!\!F$.

Let $\mathbf{L} \in \mathrm{C}^{\diamondsuit}(\mathrm{R})$, $g \in G$, and $x \in X \in \mathbf{L} \subset \mathbf{S}$. To prove that $F_{\mathbf{L}}^{G\times\mathrm{T}^r}$ belongs to $I\!\!F$, that is $F_{\mathbf{L}}^{G\times\mathrm{T}^r}$ is effective (respectively, is free) it suffices to prove that if $F^X(g, x) \neq x$, then

$$\mathbf{F}_{\mathbf{L}}^{G\times\mathrm{T}^r}(g, \mathbf{a} \neq \mathbf{a},$$

where $\mathbf{a}$ is the element of the Containing Space $\mathrm{T}^r \equiv \mathrm{T}(\mathbf{M}^r, \mathrm{R}^r)$ containing the pair (x, X). But this follows immediately by the construction of $\mathbf{F}_{\mathbf{L}}^{G\times\mathrm{T}^r}$ (see Lemma 7.1.4) because the point $\mathbf{F}_{\mathbf{L}}^{G\times\mathrm{T}^r}(g, \mathbf{a}$ of T^r contains the pair $(F^X(g, x), X)$ while the point $\mathbf{a}$ contains the pair (x, X) and the pairs $F^X(g, x), X)$ and (x, X) are not $\sim_{\mathrm{R}^r}^{\mathbf{M}^r}$-equivalent. □

The intersection property of saturated classes of G-spaces and Propositions 7.1.10 and 7.1.14 imply the following two consequences.

7.1.15 Corollary. *Let $I\!\!P$ be a saturated class of G-spaces. Then, the subclass of $I\!\!P$ consisting of all elements (X, F^X) such that F^X is effective (respectively, is free) is a saturated class of G-spaces.* □

7.1.16 Corollary. *Corollaries 7.1.11–7.1.13 hold if the actions of all elements of considered classes of G-spaces are assumed to be effective (respectively, to be free).* □

7.2 Actions of groups on spaces with structure

Definition. Let (X, F) be a G-space. A subset Q of X is said to be *G-invariant* if $F(G \times Q) = Q$. In this case it is easy to see that the pair $(Q, F|_{G\times Q})$ is a G-space.

Notation. We denote by $\mathbb{P}(G\text{-subset})$ the class of all triads (Q, X, F) such that (X, F) is a G-space and Q is a G-invariant subset of X.

In what follows, whenever we consider two elements (Q^X, X, F^X) and (Q^Y, Y, F^Y) of $\mathbb{P}(G\text{-subset})$, then the G-invariance of a subset of X and the equivariance of a mapping of X into Y is related to G-spaces (X, F^X) and (Y, F^Y).

Definition. Any subclass of $\mathbb{P}(G\text{-subset})$ is said to be a *class of G-subsets.* Such a class $I\!\!P$ is said to be *topological* if for every element (Q^X, X, F^X) of $I\!\!P$ and an element (Q^Y, Y, F^Y) of $\mathbb{P}(G\text{-subset})$ the existence of an equivariant homeomorphism h of the space X onto the space Y such that $h(Q^X) = Q^Y$ implies that $(Q^Y, Y, F^Y) \in I\!\!P$.

In what follows an arbitrary considered class of G-subsets is assumed to be topological.

Definition. Let $E \equiv (Q^X, X, F^X) \in \mathbb{P}(G\text{-subset})$. The subset Q^X is said to be the *subset-coordinate* of E, the space X the *space-coordinate* of E, the mapping F^X the *action-coordinate* of E, the pair (Q^X, X) the *subset-space pair* of E, the pair (X, F^X) the *space-action pair* of E, and the pair $(Q^X, F^X|_{G\times Q^X})$ are said to be the *subset-action pair* of E.

Definition. Let $I\!\!P$ be a class of G-subsets. The class of subset-coordinates of all elements of $I\!\!P$ is called the *subset-component* of $I\!\!P$; the class of space-coordinates of all elements of $I\!\!P$ the *space-component* of $I\!\!P$; the class of action-coordinates of all elements of $I\!\!P$ the *action-component* of $I\!\!P$; the class of subset-space pairs of all elements of $I\!\!P$ the *adjacent to $I\!\!P$ class of subsets*; the class of space-action pairs of all elements of $I\!\!P$ the *adjacent to $I\!\!P$ class of G-spaces*; and the class of subset-action pairs of all elements of $I\!\!P$ are called the *subset-adjacent to $I\!\!P$ class of G-spaces.*

It is easy to see that all of the above defined classes are topological.

Notation. Let

$$\{(Q^{X_\gamma}, X_\gamma, F^{X_\gamma}) : \gamma \in \Gamma\} \tag{7.2.1}$$

be an indexed collection of elements of $\mathbb{P}(G\text{-subset})$. Then, this indexed collection defines the indexed collection

$$\mathbf{S} \equiv \{X_\gamma : \gamma \in \Gamma\}$$

of spaces; the indexed collection

$$\mathbf{Q} \equiv \{Q^{X_\gamma} : \gamma \in \Gamma\},$$

which is a restriction of $\mathbf{S}$; and the indexed collection

$$\mathbf{F} \equiv \{F^{X_\gamma} : \gamma \in \Gamma\}$$

of mappings, for which the $\mathbf{F}$-range is $\mathbf{S}$ and the $\mathbf{F}$-domain is the indexed collection $G \otimes \mathbf{S}$ of spaces.

The indexed collection (7.2.1) is denoted by $(\mathbf{Q}, \mathbf{S}, \mathbf{F})$. In this case, for every $X \in \mathbf{S}$ we denote by $\mathbf{Q}(X)$ the set $Q^X = Q^{X_\gamma}$ and by $\mathbf{F}(X)$ the action $F^X = F^{X_\gamma}$, where γ is the unique element γ of Γ such that $X = X_\gamma$.

7.2.1 Lemma. *Let*

$$(\mathbf{Q}, \mathbf{S}, \mathbf{F}) \tag{7.2.2}$$

be an indexed collection of elements of $\mathbb{P}(G\text{-subset})$; $(\mathbf{M}^d, \mathbf{M}^r)$ *a proper product open* $\mathbf{F}$*-co-mark; and* R *an* $(\mathbf{M}^d, \mathbf{M}^r)$*-admissible family of equivalence relation on* $\mathbf{F}$ *such that the family* R^r *of equivalence relations on* $\mathbf{S}$ *is* $(\mathbf{M}^r, \mathbf{Q})$*-admissible. Let* $\mathbf{H}$ *and* $\mathbf{L}$ *be arbitrary elements of* $\mathrm{C}^\diamondsuit(\mathrm{R})$ *with* $\mathbf{H} \subset \mathbf{L}$. *Suppose that* $\mathrm{T}^r(\mathbf{H}^r|_\mathbf{Q}) = \mathrm{T}(\mathbf{M}^r|_\mathbf{Q}, \mathrm{R}^r|_\mathbf{Q}, \mathbf{H}^r|_\mathbf{Q})$, $\mathrm{T}^r(\mathbf{L}^r) = \mathrm{T}(\mathbf{M}^r, \mathrm{R}^r, \mathbf{L}^r)$, *and* $F_\mathbf{L}^{G\times\mathrm{T}^r}$ *is the containing action-mapping (corresponding to the* $\mathbf{F}$*-co-mark* $(\mathbf{M}^d, \mathbf{M}^r)$, *the family* R, *and the element* $\mathbf{L}$ *of* $\mathrm{C}^\diamondsuit(\mathrm{R})$). *Then, the triad*

$$(\mathrm{T}^r(\mathbf{H}^r|_\mathbf{Q}), \mathrm{T}^r(\mathbf{L}^r), F_\mathbf{L}^{G\times\mathrm{T}^r}) \tag{7.2.3}$$

is an element of $\mathbb{P}(G\text{-subset})$.

Proof. Since $F_\mathbf{L}^{G\times\mathrm{T}^r}$ is an action on the space $\mathrm{T}^r(\mathbf{L}^r)$ (see Lemma 7.1.4) we need to prove only that

$$F_\mathbf{L}^{G\times\mathrm{T}^r}(G \times \mathrm{T}^r(\mathbf{H}^\mathbf{r}|_\mathbf{Q})) = \mathrm{T}^r(\mathbf{H}^r|_\mathbf{Q}).$$

It suffices to prove that if $(g, \mathbf{a}) \in G \times \mathrm{T}^r(\mathbf{H}^r|_{\mathbf{Q}})$, then

$$F_{\mathbf{L}}^{G\times \mathrm{T}^r}(g, \mathbf{a}) \in \mathrm{T}^r(\mathbf{H}^r|_{\mathbf{Q}}).$$

Let $(g, \mathbf{a}) \in G \times \mathrm{T}^r(\mathbf{H}^r|_{\mathbf{Q}})$. Then $\mathbf{a} \in \mathrm{T}^r(\mathbf{H}^{\mathbf{r}}|_{\mathbf{Q}})$. Therefore, there exists an element (x, X) of $\mathbf{a}$ such that $x \in Q^X \equiv \mathbf{Q}(X)$ and $X \in \mathbf{H}^r$.

By Lemma 7.1.4 we have

$$F^{G\times \mathrm{T}^r} = F^{\mathrm{T}} \circ [G, \mathrm{T}^r]$$

and $[G, \mathrm{T}^r](g, \mathbf{a}) = [g, \mathbf{a}]$, where $[g, \mathbf{a}]$ is the element of $\mathrm{T}^d \equiv \mathrm{T}(\mathbf{M}^d, \mathrm{R}^d)$ containing the pair $((g, x), G \times X)$. (We recall that the $\mathbf{F}$-domain is the indexed collection $G \otimes \mathbf{S}$.) By construction of F^{T} (see Lemma 6.1.1),

$$(F^X(g, x), X) \in F^{\mathrm{T}}([g, \mathbf{a}]),$$

where $F^X = \mathbf{F}(X)$. Since $x \in Q^X$ and $F^X(G \times Q^X) = Q^X$ we have that $F^X(g, x) \in Q^X$, that is $F^{\mathrm{T}}([g, \mathbf{a}]) \in \mathrm{T}^r(\mathbf{H}^r|_{\mathbf{Q}})$. On the other hand,

$$F^{G\times \mathrm{T}^r}|_{G\times \mathrm{T}^r(\mathbf{L}^r)} = F_{\mathbf{L}}^{G\times \mathrm{T}^r},$$

which means that $F_{\mathbf{L}}^{G\times \mathrm{T}^r}(g, \mathbf{a}) \in \mathrm{T}^r(\mathbf{H}^r|_{\mathbf{Q}})$. □

Definition. The triad (7.2.3) is called the *containing triad* for the indexed collection (7.2.2) *corresponding to the* $\mathbf{F}$*-co-mark* $(\mathbf{M}^d, \mathbf{M}^r)$, *the family* R, *and the elements* $\mathbf{H}$ *and* $\mathbf{L}$ of $\mathrm{C}^{\diamondsuit}(\mathrm{R})$.

Definition. A class $I\!P$ of G-subsets is said to be *saturated* if for every indexed collection $(\mathbf{Q}, \mathbf{S}, \mathbf{F})$ of elements of $I\!P$ there exists a proper product-open $\mathbf{F}$-co-mark $(\mathbf{M}^{d,+}\mathbf{M}^{r,+})$ satisfying the following condition: for every proper $\mathbf{F}$-co-mark $(\mathbf{M}^d, \mathbf{M}^r)$, which is a co-extension of $(\mathbf{M}^{d,+}\mathbf{M}^{r,+})$, there exists an $(\mathbf{M}^d, \mathbf{M}^r)$-admissible family R^+ of equivalence relation on $\mathbf{F}$ such that $(\mathrm{R}^+)^r$ is $(\mathbf{M}^r, \mathbf{Q})$-admissible and for every admissible family R of equivalence relations on $\mathbf{F}$, which is a final refinement of R^+, and elements $\mathbf{H}$ and $\mathbf{L}$ of $\mathrm{C}^{\diamondsuit}(\mathrm{R})$ with $\mathbf{H} \subset \mathbf{L}$, the containing triad

$$(\mathrm{T}^r(\mathbf{H}^{\mathbf{r}}|_{\mathbf{Q}}), \mathrm{T}^r(\mathbf{L}^{\mathbf{r}}), F_{\mathbf{L}}^{G\times \mathrm{T}^r})$$

corresponding to the $\mathbf{F}$-co-mark $(\mathbf{M}^d, \mathbf{M}^r)$, the family R, and elements $\mathbf{H}$ and $\mathbf{L}$ of $\mathrm{C}^{\diamondsuit}(\mathrm{R})$, belongs to $I\!P$.

The pair $(\mathbf{M}^{d,+}\mathbf{M}^{r,+})$ is said to be an *initial* **F**-co-mark (*corresponding to the class* $\mathbb{P}$) and the family R^+ an *initial* family (*corresponding to the* **F**-*co-mark* $(\mathbf{M}^d, \mathbf{M}^r)$ *and the class* $\mathbb{P}$).

7.2.2 Proposition. *Let* $\mathbb{P}$ *be a saturated class of* G*-subsets. Then:*
(a) *the subset-component of* $\mathbb{P}$,
(b) *the space-component of* $\mathbb{P}$,
(c) *the action component of* $\mathbb{P}$,
(d) *the adjacent to* $\mathbb{P}$ *class of subsets,*
(e) *the adjacent to* $\mathbb{P}$ *class of* G*-spaces, and*
(f) *the subset-adjacent to* $\mathbb{P}$ *class of* G*-spaces*
are saturated classes.

Proof. Let $(\mathbf{Q}, \mathbf{S}, \mathbf{F})$ be an indexed collection of elements of $\mathbb{P}$. Denote by $(\mathbf{M}^{d,+}, \mathbf{M}^{r,+})$ an initial **F**-co-mark corresponding to the class $\mathbb{P}$.

We prove that the adjacent to $\mathbb{P}$ class of G-spaces, denoted by $\mathbb{R}$, is saturated. Without loss of generality, we can suppose that **S** is an arbitrary indexed collection of the space-component of $\mathbb{R}$ and **F** is the corresponding indexed collection of action-coordinates of elements of **S**. Then, it is easy to verify that $(\mathbf{M}^{d,+}, \mathbf{M}^{r,+})$ is also an initial **F**-co-mark corresponding to the class $\mathbb{R}$. Thus, $\mathbb{R}$ is saturated.

Now, by Proposition 7.1.9 the space-component of $\mathbb{P}$, which coincides with the space-component of $\mathbb{R}$, is also saturated. Also, by Proposition 7.1.6 the action-component of $\mathbb{P}$, which coincides with the action-component of $\mathbb{R}$, is saturated.

Finally, it is not difficult to verify that the subset-component of $\mathbb{P}$, the adjacent to $\mathbb{P}$ class of subsets, and the subset-adjacent to $\mathbb{P}$ class of G-spaces are also saturated. □

7.2.3 Proposition. *Let* $\mathbb{P}$ *be a class of* G*-subsets. If the action-component of* $\mathbb{P}$ *and the adjacent to* $\mathbb{P}$ *class of subsets are saturated, then* $\mathbb{P}$ *is also saturated.*

Proof. Denote by $\mathbb{F}$ the action-component of $\mathbb{P}$ and by $\mathbb{R}$ the adjacent to $\mathbb{P}$ class of subsets.

Let $(\mathbf{Q}, \mathbf{S}, \mathbf{F})$ be an indexed collection of elements of $\mathbb{P}$. Then, **F** is an indexed collection of elements of $\mathbb{F}$ and **Q** is an $\mathbb{R}$-restriction of **S**. Since $\mathbb{F}$ is saturated there exists an **F**-co-mark $(\mathbf{M}_0^{d,+}, \mathbf{M}_0^{r,+})$, which is an initial **F**-co-mark corresponding to $\mathbb{F}$. Also, since $\mathbb{R}$ is saturated there exists an initial co-mark $\mathbf{M}^+$ of $\mathbf{S}$, which is an initial co-mark corresponding to the restriction $\mathbf{Q}$ and the class $I\!R$.

Denote by $\mathbf{M}^{r,+}$ a co-mark of $\mathbf{S}$, which is simultaneously a co-extension of $\mathbf{M}_0^{r,+}$ and $\mathbf{M}^+$, and by $\mathbf{M}^{d,+}$ a co-mark of $G \otimes \mathbf{S}$, which is simultaneously a co-extension of $\mathbf{M}_0^{d,+}$ and the $\mathbf{F}$-preimage of $\mathbf{M}^{r,+}$. Then, it is easy to verify that the $\mathbf{F}$-comark $(\mathbf{M}^{d,+}, \mathbf{M}^{r,+})$ is an initial $\mathbf{F}$-co-mark corresponding to the class $I\!P$, which proves the proposition. □

The proof of the following proposition follows by Proposition 7.2.3 and by intersection properties of the saturated classes of mappings and saturated classes of subsets.

7.2.4 Proposition. *The non-empty intersection of not more than τ many saturated classes of G-subsets is a saturated class.* □

Definition. Let $I\!P$ be a class of G-subsets. An element (Q^T, T, F^T) of $I\!P$ is said to be *universal* (respectively, *properly universal*) in $I\!P$ if for every element (Q^X, X, F^X) of $I\!P$ there exists an equivariant embedding i_X of X into T such that $Q^X \subset i_X^{-1}(Q^T)$ (respectively, $Q^X = i_X^{-1}(Q^T)$).

Definition. A class $I\!P$ of G-subsets is said to be *complete* if the adjacent to $I\!P$ class of subsets is complete.

7.2.5 Proposition. *In any (complete) saturated class of G-subsets there exist (properly) universal elements.*

Proof. Let $I\!P$ be a (complete) saturated class of G-subsets. Denote by $I\!F$ the action-component of $I\!P$ and by $I\!R$ the adjacent to $I\!P$ class of subsets. Let $(\mathbf{Q}, \mathbf{S}, \mathbf{F})$ be an indexed collection of elements of $I\!P$ with the property that for every element (Q^Y, Y, F^Y) of $I\!P$ there exist an element (Q^X, X, F^X) of $(\mathbf{Q}, \mathbf{S}, \mathbf{F})$ and an equivariant homeomorphism of Y onto X such that $h(Q^Y) = Q^X$.

Since $I\!P$ is saturated there exists a proper product-open $\mathbf{F}$-co-mark $(\mathbf{M}^d, \mathbf{M}^r)$, which is an initial $\mathbf{F}$-co-mark corresponding to the class $I\!P$. Without loss of generality we can suppose that $\mathbf{M}^r$ is an initial co-mark of $\mathbf{S}$ corresponding to the restriction $\mathbf{Q}$ and the class $I\!R$. Denote by R a family of equivalence relations on $\mathbf{F}$ such that R is an initial family corresponding to the $\mathbf{F}$-co-mark $(\mathbf{M}^d, \mathbf{M}^r)$ and the class $I\!P$, and R^r is an initial family corresponding to the co-mark $\mathbf{M}^r$, the restriction $\mathbf{Q}$, and the class $I\!R$.

Therefore, R is $(\mathbf{M}^d, \mathbf{M}^r)$-admissible and R^r is $(\mathbf{M}^r, \mathbf{Q})$-admissible. If the class $\mathbb{P}$ is complete and, therefore, the restriction $\mathbf{Q}$ of $\mathbf{S}$ is complete, then without loss of generality we can suppose that $\mathbf{Q}$ is $(\mathbf{M}^r, \mathrm{R}^r)$-complete.

Consider the containing triad $(\mathrm{T}^r|_{\mathbf{Q}}, \mathrm{T}^r, F^{G\times\mathrm{T}^r})$ corresponding to the $\mathbf{F}$-co-mark $(\mathbf{M}^d, \mathbf{M}^r)$, the family R, and the element $\mathbf{L} = \mathbf{F}$ of $\mathrm{C}^{\diamondsuit}(\mathrm{R})$, where $\mathrm{T}^r = \mathrm{T}(\mathbf{M}^r, \mathrm{R}^r)$ and $F^{G\times\mathrm{T}^r} = F_{\mathbf{L}}^{G\times\mathrm{T}^r}$). By the construction this triad belongs to $\mathbb{P}$.

As in the proof of Proposition 7.1.8, for every element (Q^X, X, F^X) of $\mathbb{P}$ the natural embedding $i^X_{\mathrm{T}^r}$ of X into T^r is an equivariant mapping. Moreover, it is easy to verify that $Q^X \subset (i^X_{\mathrm{T}^r})^{-1}(\mathrm{T}^r|_{\mathbf{Q}})$ and if $\mathbb{P}$ is complete, then $Q^X = (i^X_{\mathrm{T}^r})^{-1}(\mathrm{T}^r|_{\mathbf{Q}})$. This fact proves that $(\mathrm{T}^r|_{\mathbf{Q}}, \mathrm{T}^r, F^{G\times\mathrm{T}^r})$ is a (properly) universal element in $\mathbb{P}$. □

Definition. Let $\mathbb{P}$ be one of the classes:
(a) a class of spaces,
(b) a class of mappings,
(c) a class of subsets, and
(d) a class of G-spaces.
By the *class of G-subsets generated by* $\mathbb{P}$ we mean the class of all elements E of $\mathbb{P}(G\text{-subset})$ for which we have respectively:
(a) the space-coordinate of E belongs to $\mathbb{P}$,
(b) the action-coordinate of E belongs to $\mathbb{P}$,
(c) the subset-space pair of E belongs to $\mathbb{P}$, or
(d) the space-action pair of E belongs to $\mathbb{P}$.
Let $\mathbb{P}$ be one of the classes:
(a) a class of spaces or
(b) a class of G-spaces.
By the *class of G-subsets subset-generated by* $\mathbb{P}$ we mean the class of all elements E of $\mathbb{P}(G\text{-subset})$ for which we have respectively,
(a) the subset-coordinate of E belongs to $\mathbb{P}$, and
(b) the subset-action pair of E belongs to $\mathbb{P}$.

7.2.6 Proposition. *The following two statements are true:*
(1) Let $\mathbb{P}$ *be one of the classes:*
(a) a class of spaces,
(b) a class of mappings,
(c) a (complete) class of subsets, or
(d) a class of G-spaces.

Then, the class of G-subsets generated by $I\!P$ is also a (complete) saturated class.

(2) Let $I\!P$ be one of the classes:

(a) a class of spaces or

(b) a class of G-spaces.

Then, the class of G-subsets subset-generated by $I\!P$ is also a saturated class.

Proof. We prove the case (d) of statement (1). The other cases in general outline are proved similarly.

Denote by $I\!R$ the class of G-subsets generated by the class $I\!P$ of G-spaces and let $(\mathbf{Q}, \mathbf{S}, \mathbf{F})$ be an indexed collection of elements of $I\!R$. Therefore, $\mathbf{F} \equiv \{F^X : X \in \mathbf{S}\}$ is an indexed collection of elements of the action-component of $I\!P$. Since $I\!P$ is saturated there exists an $\mathbf{F}$-co-mark $(\mathbf{M}^{d,+}, \mathbf{M}^{r,+})$, which is an initial $\mathbf{F}$-co-mark corresponding to the class $I\!P$. Let $(\mathbf{M}^d, \mathbf{M}^r)$ be an arbitrary proper $\mathbf{F}$-co-mark, which is a co-extension of $(\mathbf{M}^{d,+}, \mathbf{M}^{r,+})$. There exists a family R^+ of equivalence relation on $\mathbf{F}$, which is an initial family corresponding to the $\mathbf{F}$-co-mark $(\mathbf{M}^{d,+}, \mathbf{M}^{r,+})$ and the class $I\!P$. Without loss of generality we can suppose that $(\mathrm{R}^+)^r$ is an $(\mathbf{M}^{r,+}, \mathbf{Q})$-admissible family. Denote by R an arbitrary admissible family of equivalence relations on $\mathbf{F}$, which is a final refinement of R^+, and let $\mathbf{H}, \mathbf{L} \in \mathrm{C}^\diamond(\mathrm{R})$ with $\mathbf{H} \subset \mathbf{L}$. Then, we can consider the triad

$$(\mathrm{T}^r(\mathbf{H}^r|_{\mathbf{Q}}), \mathrm{T}^r(\mathbf{L}^r), F_{\mathbf{L}}^{G\times \mathrm{T}^r}),$$

where $\mathrm{T}^r(\mathbf{L}^r) = \mathrm{T}(\mathbf{M}^r, \mathrm{R}^r, \mathbf{L}^r)$, $\mathrm{T}^r(\mathbf{H}^r|_{\mathbf{Q}}) = \mathrm{T}(\mathbf{M}^r|_{\mathbf{Q}}, \mathrm{R}^r|_{\mathbf{Q}}, \mathbf{H}^r|_{\mathbf{Q}})$, and $F_{\mathbf{L}}^{G\times \mathrm{T}^r}$ is the containing action-mapping corresponding to the $\mathbf{F}$-co-mark $(\mathbf{M}^d, \mathbf{M}^r)$, the family R, and the elements $\mathbf{H}$ and $\mathbf{L}$ of $\mathrm{C}^\diamond(\mathrm{R})$.

To prove that $I\!R$ is saturated it suffices to prove that

$$F_{\mathbf{L}}^{G\times \mathrm{T}^r}(G \times \mathrm{T}^r(\mathbf{H}^r|_{\mathbf{Q}})) = \mathrm{T}^r(\mathbf{H}^r|_{\mathbf{Q}}),$$

which means that the considered triad belongs to $I\!R$. For this purpose it suffices to prove that if $(g, \mathbf{a}) \in G \times \mathrm{T}^r(\mathbf{H}^r|_{\mathbf{Q}})$, then

$$F_{\mathbf{L}}^{G\times \mathrm{T}^r}(g, \mathbf{a}) \in \mathrm{T}^r(\mathbf{H}^r|_{\mathbf{Q}}).$$

Let $(g, \mathbf{a}) \in G \times \mathrm{T}^r(\mathbf{H}^r|_{\mathbf{Q}})$. There exists an element $(x, X) \in \mathbf{a}$ such that $X \in \mathbf{H}$ and $x \in Q^X \equiv \mathbf{Q}(X)$. By the construction of the containing action-mapping $F_{\mathbf{L}}^{G\times \mathrm{T}^r}$, the point $F_{\mathbf{L}}^{G\times \mathrm{T}^r}(g, \mathbf{a})$ is the point $\mathbf{b}$ of $\mathrm{T}^r(\mathbf{L}^r)$, which contains the pair $(F^X(g,x), X)$. Since $F^X(G \times Q^X) = Q^X$ we have $F^X(g,x) \in Q^X$. Therefore $\mathbf{b} \in \mathrm{T}^r(\mathbf{H}^r|_{\mathbf{Q}})$, proving the above relation. □

7.2.7 Corollary. *Let $\tau = \omega$ and $\alpha \in \omega^+$. In the class of G-subsets consisting of all elements (Q^X, X, F^X) of $\mathbb{P}(G\text{-subset})$ such that X is a separable metrizable space and Q^X is a Borel set of multiplicative (respectively, of additive) class α there exist properly universal elements.*

Proof. By Proposition 7.2.6 it sufficies to note that:

(a) the considered class coincides with the class of G-subsets generated by the class $I\!P$ of subsets consisting of all pairs (Q^X, X) such that X is a separable metrizable space and Q^X is a Borel set of multiplicative (respectively, of additive) class α and

(b) by Proposition 2.4.5 the class $I\!P$ is a complete saturated class. □

The proof of the following proposition is similar to the proof of the Proposition 7.1.14.

7.2.8 Proposition. *The class consisting of all elements (Q^X, X, F^X) of $\mathbb{P}(\text{subset})$, where F^X is an effective (respectively, is a free) action, is a saturated class.* □

Agreement. In what follows of this section we fix a cardinal ν such that $1 \le \nu \le \tau$, a ν-directing mapping Φ, a ν-set Λ of cardinality $\le \tau$ with a partial order $\prec$, and a mapping $\mathbf{p}$ of Λ into the class of all classes of subsets.

We recall (see Chapter 4) that by $\mathbb{P}(\Phi, \Lambda, \prec, \mathbf{p})$ we denote the class of all pairs (ψ_X, X), where ψ_X is a $(\Phi, \Lambda, \prec, \mathbf{p})$-structure on a space X.

Definition. Any triad (ψ_X, X, F^X), where (ψ_X, X) is an element of $\mathbb{P}(\Phi, \Lambda, \prec, \mathbf{p})$ and F^X is an action of G on X such that

$$F^X(G \times \psi_X(\lambda)) = \psi_X(\lambda),\ \lambda \in \Lambda$$

is said to be a *G-space with $(\Phi, \Lambda, \prec, \mathbf{p})$-structure* or briefly a *G-space with structure.*

In what follows if (ψ_X, X, F^X) and (ψ_Y, Y, F^Y) are G-spaces with structure, then the G-invariance of a subset of X and the equivariance of a mapping of X into Y is related to G-spaces (X, F^X) and (Y, F^Y).

Definition. A class $I\!P$ of G-spaces with structure is said to be *topological* if for any two G-spaces with structure (ψ_X, X, F^X) and (ψ_Y, Y, F^Y) and an equivariant homeomorphism h of X onto Y such that $h(\psi_X(\lambda)) = \psi_Y(\lambda)$, where $\lambda \in \Lambda$, the condition $(\psi_X, X, F^X) \in I\!P$ implies $(\psi_Y, Y, F^Y) \in I\!P$.

In what follows an arbitrary considered class of G-spaces with structure is assumed to be topological.

Below, we define some notions and give some results concerning classes of G-spaces with structure. These notions and results are similar to the corresponding notions and results concerning the classes of G-subsets. Because of that we omit the proofs of the given results.

Definition. Let $E \equiv (\psi_X, X, F^X)$ be a G-space with structure. The space X is called the *space-coordinate* of E, the mapping F^X the *action-coordinate* of E, the pair (ψ_X, X) the *structure-space pair* of E, the pair (X, F^X) the *space-action pair* of E, and the pair $(\psi_X(\lambda), F^X|_{\psi_X(\lambda)})$, where $\lambda \in \Lambda$, is called the *λ-subset-action pair* of E.

Definition. Let $I\!P$ be a class of G-spaces with structure. The class of space-coordinates of all elements of $I\!P$ is called the *space-component* of $I\!P$; the class of action-coordinates of all elements of $I\!P$, the *action-component* of $I\!P$; the class of structure-space pairs of all elements of $I\!P$ is called the *adjacent to $I\!P$ class of spaces with structure*; the class of space-action pairs of all elements of $I\!P$ the *adjacent to $I\!P$ class of G-spaces*; and for every $\lambda \in \Lambda$, the class of λ-subset-action pairs of all elements of $I\!P$ is called the *λ-adjacent class of G-spaces.*

It is easy to see that all the above defined classes are topological.

Notation. Let

$$\{\psi_{X_\gamma}, X_\gamma, F^{X_\gamma}) : \gamma \in \Gamma\} \tag{7.2.4}$$

be an indexed collection of G-spaces with structure. Then, this indexed collection defines the indexed collections

$$\mathbf{S} \equiv \{X_\gamma : \gamma \in \Gamma\},$$

$$\mathbf{\Psi} \equiv \{\psi_{X_\gamma} : \gamma \in \Gamma\}, \text{ and}$$

$$\mathbf{F} \equiv \{F^{X_\gamma} : \gamma \in \Gamma\}$$

of spaces, of structures, and of mappings, respectively. We note that $\mathbf{\Psi}$ is a co-structure on $\mathbf{S}$, the $\mathbf{F}$-range is $\mathbf{S}$, and the $\mathbf{F}$-domain is the indexed collection $G \otimes \mathbf{S}$.

The indexed collection (7.2.4) is denoted also by $(\mathbf{\Psi}, \mathbf{S}, \mathbf{F})$. In this case, for every $\lambda \in \Lambda$ we denote by $\mathbf{\Psi}(\lambda)$ the restriction $\{\psi_X(\lambda) : X \in \mathbf{S}\}$ of $\mathbf{S}$.

7.2.9 Lemma. *Let*

$$(\mathbf{\Psi}, \mathbf{S}, \mathbf{F}) \tag{7.2.5}$$

be an indexed collection of elements of G-spaces with structure; $(\mathbf{M}^d, \mathbf{M}^r)$ *a proper product open* $\mathbf{F}$*-co-mark; and* R *an* $(\mathbf{M}^d, \mathbf{M}^r)$*-admissible family of equivalence relations on* $\mathbf{F}$ *such that the family* R^r *of equivalence relations on* $\mathbf{S}$ *is* $(\mathbf{M}^r, \mathbf{\Psi})$*-admissible. Then, for every* $\mathbf{L} \in \mathrm{C}(\mathrm{R})$ *the triad*

$$(\psi_{\mathrm{T}^r(\mathbf{L}^r)}, \mathrm{T}^r(\mathbf{L}^r), F_{\mathbf{L}}^{G\times \mathrm{T}^r}), \tag{7.2.6}$$

where

(a) $\mathrm{T}^r(\mathbf{L}^r) = \mathrm{T}(\mathbf{M}^r, \mathrm{R}^r, \mathbf{L}^r)$;

(b) $F_{\mathbf{L}}^{G\times \mathrm{T}^r}$ *is the containing action-mapping (corresponding to the* $\mathbf{F}$*-co-mark* $(\mathbf{M}^d, \mathbf{M}^r)$*, the family* R*, and the element* $\mathbf{L}$ *of* $\mathrm{C}^{\diamondsuit}(\mathrm{R})$*); and*

(c) $\psi_{\mathrm{T}^r(\mathbf{L}^r)}$ *is the containing structure on* $\mathrm{T}^r(\mathbf{L}^r)$ *corresponding to the family* R^r*, the co-mark* $\mathbf{M}^r$*, the co-structure* $\mathbf{\Psi}$*, and the subspace* $\mathrm{T}^r(\mathbf{L}^r)$ *of* T^r
(see Section 4.1) is a G-space with structure. (Note: $\psi_{\mathrm{T}^r(\mathbf{L}^r)}$ *is the mapping of* Λ *into* $\mathcal{P}(\mathrm{T}^r(\mathbf{L}^r))$ *defined by relation:*

$$\psi_{\mathrm{T}^r(\mathbf{L}^r)}(\lambda) = \mathrm{T}(\mathbf{M}^r|_{\mathbf{\Psi}(\lambda)}, \mathrm{R}^r|_{\mathbf{\Psi}(\lambda)}, \mathbf{L}^r|_{\mathbf{\Psi}(\lambda)})). \ \square$$

Definition. The triad (7.2.6) is called the *containing triad* for the indexed collection (7.2.5), *corresponding to the* $\mathbf{F}$*-co-mark* $(\mathbf{M}^d, \mathbf{M}^r)$*, the family* R*, and the element* $\mathbf{L}$ *of* $\mathrm{C}^{\diamondsuit}(\mathrm{R})$.

Definition. A class $I\!P$ of G-spaces with structure is said to be *saturated* if for every indexed collection $(\mathbf{\Psi}, \mathbf{S}, \mathbf{F})$ of elements of $I\!P$ there exists a proper product-open $\mathbf{F}$-co-mark $(\mathbf{M}^{d,+}, \mathbf{M}^{r,+})$ satisfying the following condition: for every proper $\mathbf{F}$-co-mark $(\mathbf{M}^d, \mathbf{M}^r)$, which is a co-extension of

$(\mathbf{M}^{d,+}, \mathbf{M}^{r,+})$, there exists an $(\mathbf{M}^d, \mathbf{M}^r)$-admissible family R^+ of equivalence relations on $\mathbf{F}$ such that $(\mathrm{R}^+)^r$ is $(\mathbf{M}^r, \mathbf{\Psi})$-admissible and for every admissible family R of equivalence relations on $\mathbf{F}$, which is a final refinement of R^+, and element $\mathbf{L}$ of $\mathrm{C}^\diamond(\mathbf{R})$, the containing triad

$$(\psi_{\mathrm{T}^r(\mathbf{L}^r)}, \mathrm{T}^r(\mathbf{L}^r), F_{\mathbf{L}}^{G\times \mathrm{T}^r})$$

that corresponds to the $\mathbf{F}$-co-mark $(\mathbf{M}^d, \mathbf{M}^r)$, the family R, and the element $\mathbf{L}$ of $\mathrm{C}^\diamond(\mathrm{R})$), is an element of $I\!P$.

7.2.10 Proposition. *Let $I\!P$ be a saturated class of G-spaces with structure. Then, the space-component of $I\!P$, the action-component of $I\!P$, the adjacent to $I\!P$ class of G-spaces, and for every $\lambda \in \Lambda$ the λ-adjacent to $I\!P$ class of G-spaces are saturated classes.* □

7.2.11 Proposition. *The non-empty intersection of not more than τ many saturated classes of G-spaces with structure is also a saturated class of G-spaces with structure.* □

Definition. Let $I\!P$ be a class of G-spaces with structure. An element (ψ_T, T, F^T) of $I\!P$ is said to be *universal* (respectively, *properly universal*) in $I\!P$ if for every element (ψ_X, X, F^X) of $I\!P$ there exists an equivariant embedding h of X into T such that for every $\lambda \in \Lambda$ we have $\psi_X(\lambda) \subset h^{-1}(\psi_T(\lambda))$ (respectively, $\psi_X(\lambda) = h^{-1}(\psi_T(\lambda))$).

Definition. A class $I\!P$ of G-spaces with structure is said to be *complete* if the adjacent to $I\!P$ class of spaces with structure is complete.

7.2.12 Proposition. *In any (complete) saturated class of G-spaces with structure there exist (properly) universal elements.* □

Definition. Let $I\!P$ be one of the classes:
(a) a class of spaces,
(b) a class of mappings,
(c) a class of spaces with structure,
(d) a class of G-spaces, or
(e) a class of G-subsets.

By the *class of G-spaces with structure generated by $I\!P$* we mean the class consisting of all G-spaces E with structure such that:
(a) the space-coordinate of E belongs to $I\!P$,

(b) the action-coordinate of E belongs to $\mathbb{P}$,
(c) the structure-space pair of E belongs to $\mathbb{P}$,
(d) the space-mapping pair of E belongs to $\mathbb{P}$, and
(e) for every $\lambda \in \Lambda$ the λ-subset-action pair of E belongs to $\mathbb{P}$.

7.2.13 Proposition. *Let $\mathbb{P}$ be one of the classes:*
(a) a class of spaces,
(b) a class of mappings,
(c) a (complete) class of spaces with structure,
(d) a class of G-spaces, or
(e) a (complete) class of G-subsets.
Then, the class of G-spaces with structure generated by $\mathbb{P}$ is also (complete) saturated.

7.2.14 Corollary. *Let $w(G) = \omega$. There exist a separable metrizable G-space (T, F^T) and a fixed sequence of zero-dimensional G-invariant subsets $T_0, T_1, T_2, ...$ of T for which $T = T_0 \cup T_1 \cup T_2 \cup ...$ such that for every separable metrizable G-space (X, F^X) and a sequence of zero-dimensional G-invariant subsets $X_0, X_1, X_2, ..$ of X for which $X = X_0 \cup X_1 \cup X_2 \cup ...$, there exists an equivariant embedding h of X into T satisfying the following conditions:*

$$X_0 \subset h^{-1}(T_0),\ X_1 \subset h^{-1}(T_1),\ X_2 \subset h^{-1}(T_2),\ .... \tag{7.2.7}$$

Moreover, if all the above subsets $X_0, X_1, X_2, ...$ are considered to be Borel sets of multiplicative (respectively, of additive) class α, where α is a fixed countable ordinal, then the subsets $T_0, T_1, T_2, ...$ can be also considered to be Borel sets of multiplicative (respectively, of additive) class α and the conditions (7.2.7) can be replaced by the following:

$$X_0 = h^{-1}(T_0),\ X_1 = h^{-1}(T_1),\ X_2 = h^{-1}(T_2),\ ....$$

Proof. In Corollary 4.4.18 actually is considered the class of spaces with $(\Lambda, \prec, \mathbf{p})$-structure, where Λ is a countable set with $\Lambda = \Lambda^0$, the partial order $\prec$ is trivial, and for every $\lambda \in \Lambda$, $\mathbf{p}(\lambda)$ is the class of subsets consisting of all pair (Q^X, X) such that $\operatorname{ind}(Q^X) \leq 0$. Denote this class by $\mathbb{P}_0$. By Proposition 4.1.5, $\mathbb{P}_0$ is a saturated class.

For the case $\tau = \omega$, Proposition 4.1.4 implies that the class $\mathbb{P}$ consisting of all elements (ψ_X, X) of $\mathbb{P}_0$ such that X is separable metrizable (that is, X is regular) is also saturated.

Now, note that in the present corollary we consider the class of G-spaces with structure, which is generated by the class $I\!P$. By Proposition 7.2.13 this class is saturated and, therefore, in this class there exist universal elements. □

The following two corollaries can be proved as Corollary 7.1.14.

7.2.15 Corollary. *Let $w(G) = \omega$ and $n \in \omega$. There exist a separable metrizable G-space (T, F^T) and fixed zero-dimensional G-invariant subsets $T_0, ..., T_n$ of T for which $T = T_0 \cup ... \cup T_n$ such that for every separable metrizable G-space (X, F^X) and zero-dimensional G-invariant subsets $X_0, ..., X_n$ of X for which $X = X_0 \cup ... \cup X_n$, there exists an equivariant embedding h of X into T satisfying the following conditions:*

$$X_0 \subset h^{-1}(T_0), \; ..., \; X_n \subset h^{-1}(T_n). \tag{7.2.8}$$

Moreover, if all the above subsets $X_0, ..., X_n$ are considered to be Borel sets of multiplicative (respectively, of additive) class α, where α is a fixed countable ordinal, then the subsets $T_0, ..., T_n$ can be also considered to be Borel sets of multiplicative (respectively, of additive) class α and the conditions (7.2.8) can be replaced by the following:

$$X_0 = h^{-1}(T_0), \; ..., \; X_n = h^{-1}(T_n). \; \square$$

7.2.16 Corollary. *Let $w(G) = \omega$. There exists a separable metrizable G-space (T, F^T) and a fixed sequence of closed (respectively, open) G-invariant subsets $T_0, T_1, T_2, ...$ of T for which*

$$T = T_0 \cup T_1 \cup T_2 \cup ... \text{ and}$$

$$\text{ind}(T_0) \leq 0, \; \text{ind}(T_1) \leq 1, \; \text{ind}(T_2) \leq 2, \; ...$$

such that for every separable metrizable G-space (X, F^X) and a sequence of closed (respectively, open) G-invariant subsets $X_0, X_1, X_2, ...$ of X for which

$$X = X_0 \cup X_1 \cup X_2 \cup ... \text{ and}$$

$$\text{ind}(X_0) \leq 0, \; \text{ind}(X_1) \leq 1, \; \text{ind}(X_2) \leq 2, \; ...,$$

there exists an equivariant embedding h of X into T satisfying the following condition:

$$X_0 = h^{-1}(T_0),\ X_1 = h^{-1}(T_1),\ X_2 = h^{-1}(T_2),\dots.\ \square$$

The following proposition can be proved as Proposition 7.1.15.

7.2.17 Proposition. *The class consisting of all G-spaces with structure (ψ_X, X, F^X) such that the action F^X is effective (respectively, is free) is a saturated class.* $\square$

7.3 Actions of groups on diagrams

Agreement. In this section it is assumed that a fixed set Λ of cardinality $\leq \tau$ with a partial order $\prec$ is given. An arbitrary considered diagram is assumed to be defined on the set $\Lambda \cup \prec$.

Definitions. Let D be a diagram. An indexed set

$$F^D \equiv \{F^{D(\lambda)} : \lambda \in \Lambda\},$$

where $F^{D(\lambda)}$ is an action of G on the space $D(\lambda)$, is said to be an *action* of G on D if for every element (λ_0, λ_1) of $\prec$ the following condition is satisfied:

$$D(\lambda_0, \lambda_1) \circ F^{D(\lambda_1)} = F^{D(\lambda_0)} \circ (\mathrm{id}_G \times D(\lambda_0, \lambda_1))$$

(that is, if $D(\lambda_0, \lambda_1)$ is an equivariant mapping). In what follows, if an action of G on the diagram D is denoted by F^D, then by $F^{D(\lambda)}$ (that is, D is replaced by $D(\lambda)$) we will denote the corresponding action of G on $D(\lambda)$.

In the case, where Λ is directed by $\prec$, a G-diagram (D, F^D) is called a *G-spectrum*, the spaces $D(\lambda)$, $\lambda \in \Lambda$, are called *spaces* and the mappings $D(\lambda_0, \lambda_1)$, $(\lambda_0, \lambda_1) \in \prec$, *projections* of this G-spectrum.

Definition. Let (D_0, F^{D_0}) and (D_1, F^{D_1}) be two G-diagrams. A morphism

$$h_\Lambda \equiv \{h_\lambda : \lambda \in \Lambda\}$$

of D_0 into D_1 (see Section 6.3) is said to be *equivariant* if for every $\lambda \in \Lambda$ the mapping h_λ of $D_0(\lambda)$ into $D_1(\lambda)$ is equivariant (with respect to G-spaces $(D_0(\lambda), F^{D_0(\lambda)})$ and $(D_1(\lambda), F^{D_1(\lambda)})$).

Definition. A class $\mathbb{P}$ of G-diagrams is said to be *topological* if for any element (D_0, F^{D_0}) of $\mathbb{P}$ and a G-diagram (D_1, F^{D_1}) the existence of an equivariant isomorphism of D_0 onto D_1 implies that $(D_1, F^{D_1}) \in \mathbb{P}$.

In what follows, an arbitrary considered class of G-diagrams is assumed to be topological.

Notation. Let D be a diagram. By $G \otimes D$ we denote the mapping of $\Lambda \cup \prec$ into the class of all spaces and mappings, which is defined as follows: if $\lambda \in \Lambda$, then

$$(G \otimes D)(\lambda) = G \times D(\lambda),$$

and if $(\lambda_0, \lambda_1) \in \prec$, then

$$(G \otimes D)(\lambda_0, \lambda_1) = \mathrm{id}_G \times D(\lambda_0, \lambda_1)$$

(that is, $(G \otimes D)(\lambda_0, \lambda_1)$ is the mapping of the space $G \times D(\lambda_1)$ into the space $G \times D(\lambda_0)$ such that for every $(g, x) \in G \times D(\lambda_1)$ we have

$$(G \otimes D)(\lambda_0, \lambda_1)(g, x) = (g, D(\lambda_0, \lambda_1)(x))).$$

It is easy to verify that $G \otimes D$ is a diagram.

Notation. For every indexed collection $\mathbf{D} \equiv \{D_\gamma : \gamma \in \Gamma\}$ of diagrams the indexed collection

$$\{G \otimes D_\gamma : \gamma \in \Gamma\}$$

of diagrams is denoted by $G \otimes \mathbf{D}$.

Notation. Let

$$\{(D_\gamma, F^{D_\gamma}) : \gamma \in \Gamma\} \tag{7.3.1}$$

be an indexed collection of G-diagrams. This indexed collection defines the indexed collection

$$\mathbf{D} \equiv \{D_\gamma : \gamma \in \Gamma\}$$

of diagrams and the indexed collection

$$F^{\mathbf{D}} \equiv \{F^{D_\gamma} : \gamma \in \Gamma\}.$$

For the indexed collection (7.3.1) we shall use the notation $(\mathbf{D}, F^{\mathbf{D}})$.

For every $\lambda \in \Lambda$ we shall consider also the indexed collection

$$\mathbf{D}(\lambda) \equiv \{D_\gamma(\lambda) : \gamma \in \Gamma\}$$

of spaces and the indexed collection

$$F^{\mathbf{D}(\lambda)} \equiv \{F^{D_\gamma(\lambda)} : \gamma \in \Gamma\}$$

of mappings.

Obviously, the $F^{\mathbf{D}(\lambda)}$-range is $\mathbf{D}(\lambda)$ and the $F^{\mathbf{D}(\lambda)}$-domain is $G \otimes \mathbf{D}(\lambda)$.

Definitions. Let $(\mathbf{D}, F^{\mathbf{D}})$ be an indexed collection of G-diagrams. Any pair $(\overline{\mathbf{M}}^G, \overline{\mathbf{M}})$, where $\overline{\mathbf{M}}^G$ is a $G \otimes \mathbf{D}$-co-mark and $\overline{\mathbf{M}}$ is a $\mathbf{D}$-co-mark, is said to be G-$\mathbf{D}$-co-mark.

A G-$\mathbf{D}$-co-mark $(\overline{\mathbf{M}}^G, \overline{\mathbf{M}})$ is said to be *proper* if:

(a) the $G \otimes \mathbf{D}$-co-mark $\overline{\mathbf{M}}^G$ is proper,

(b) the $\mathbf{D}$-co-mark $\overline{\mathbf{M}}$ is proper, and

(c) the pair $(\overline{\mathbf{M}}^G(\lambda), \overline{\mathbf{M}}(\lambda))$, $\lambda \in \Lambda$, is a proper $F^{\mathbf{D}(\lambda)}$-co-mark.

If moreover $(\overline{\mathbf{M}}^G(\lambda), \overline{\mathbf{M}}(\lambda))$, where $\lambda \in \Lambda$, is product-open, then $(\overline{\mathbf{M}}^G, \overline{\mathbf{M}})$ is said to be also *product-open*.

Let $(\overline{\mathbf{M}}_0^G, \overline{\mathbf{M}}_0)$ and $(\overline{\mathbf{M}}_1^G, \overline{\mathbf{M}}_1)$ be two G-$\mathbf{D}$-co-marks. It is said that $(\overline{\mathbf{M}}_1^G, \overline{\mathbf{M}}_1)$ is a *co-extension* of $(\overline{\mathbf{M}}_0^G, \overline{\mathbf{M}}_0)$ if for every $\lambda \in \Lambda$, $\overline{\mathbf{M}}_1(\lambda)$ is a co-extension of $\overline{\mathbf{M}}_0(\lambda)$ and $\overline{\mathbf{M}}_1^G(\lambda)$ is a co-extension of $\overline{\mathbf{M}}_0^G(\lambda)$.

It is easy to see that if $(\overline{\mathbf{M}}_0^G, \overline{\mathbf{M}}_0)$ is a proper product-open G-$\mathbf{D}$-co-mark and a proper G-$\mathbf{D}$-co-mark $(\overline{\mathbf{M}}_1^G, \overline{\mathbf{M}}_1)$ is a co-extension of $(\overline{\mathbf{M}}_0^G, \overline{\mathbf{M}}_0)$, then $(\overline{\mathbf{M}}_1^G, \overline{\mathbf{M}}_1)$ is also product-open.

Notations. Let $(\mathbf{D}, F^{\mathbf{D}})$ be an indexed collection of G-diagrams and

$$\mathrm{R} \equiv \{\sim^s : s \in \mathcal{F}\}$$

an admissible family of equivalence relations on $(\mathbf{D}, F^{\mathbf{D}})$.

We denote by

$$\mathrm{R}^{\mathbf{D}} \equiv \{\sim^s_{\mathbf{D}} : s \in \mathcal{F}\}$$

the family of equivalence relations on $\mathbf{D}$ defined as follows: if (D_0, F^{D_0}) and (D_1, F^{D_1}) are elements of $(\mathbf{D}, F^{\mathbf{D}})$, then the elements D_0 and D_1 of $\mathbf{D}$ are $\sim^s_{\mathbf{D}}$-equivalent for some $s \in \mathcal{F}$ if and only if (D_0, F^{D_0}) and (D_1, F^{D_1}) are $\sim^s$-equivalent.

For every $\lambda \in \Lambda$ we denote by

$$\mathrm{R}^\lambda \equiv \{\sim^s_\lambda : s \in \mathcal{F}\}$$

the family of equivalence relations on $F^{\mathbf{D}(\lambda)}$ defined as follows: if (D_0, F^{D_0}) and (D_1, F^{D_1}) are elements of $(\mathbf{D}, F^{\mathbf{D}})$, then the elements $F^{D_0(\lambda)}$ and $F^{D_1(\lambda)}$ of $F^{\mathbf{D}(\lambda)}$ are $\sim^s_\lambda$-equivalent for some $s \in \mathcal{F}$ if and only if (D_0, F^{D_0}) and (D_1, F^{D_1}) are $\sim^s$-equivalent.

We recall (see Section 6.1) that for every $\lambda \in \Lambda$ we denote by

$$(\mathrm{R}^\lambda)^d \equiv \mathrm{R}^{\lambda,d} \text{ and } (\mathrm{R}^\lambda)^r \equiv \mathrm{R}^{\lambda,r}$$

the families of equivalence relations on $F^{\mathbf{D}(\lambda)}$-domain $G \otimes \mathbf{D}(\lambda)$ and on $F^{\mathbf{D}(\lambda)}$-range $\mathbf{D}(\lambda)$, respectively, induced by the family R^λ. We note that $\mathrm{R}^{\lambda,r} = \mathrm{R}^{\mathbf{D}}(\lambda)$.

Definition. Let $(\mathbf{D}, F^{\mathbf{D}})$ be an indexed collection of G-diagrams, $(\overline{\mathbf{M}}^G, \overline{\mathbf{M}})$ a proper G-$\mathbf{D}$-co-mark, and R an admissible family of equivalence relations on $(\mathbf{D}, F^{\mathbf{D}})$. We say that R is $(\overline{\mathbf{M}}^G, \overline{\mathbf{M}})$-*admissible* if for every $\lambda \in \Lambda$, the family $\mathrm{R}^{\lambda,d}$ is $\overline{\mathbf{M}}^G(\lambda)$-admissible and the family $\mathrm{R}^{\lambda,r}$ is $\overline{\mathbf{M}}(\lambda)$-admissible. It is easy to see that in this case the family $\mathrm{R}^{\mathbf{D}}$ is $\overline{\mathbf{M}}$-admissible and the family R^λ is $(\overline{\mathbf{M}}^G(\lambda), \mathbf{M}(\lambda))$-admissible.

7.3.1 Lemma. *Suppose that $(\mathbf{D}, F^{\mathbf{D}})$ is an indexed collection of G-diagrams, $(\overline{\mathbf{M}}^G, \overline{\mathbf{M}})$ is a proper product-open G-$\mathbf{D}$-co-mark, and R is an $(\overline{\mathbf{M}}^G, \overline{\mathbf{M}})$-admissible family of equivalence relations on $(\mathbf{D}, F^{\mathbf{D}})$. Consider the containing diagram D^{T} (see Lemma 6.3.3) corresponding to the proper $\mathbf{D}$-co-mark $\overline{\mathbf{M}}$ and the $\overline{\mathbf{M}}$-admissible family $\mathrm{R}^{\mathbf{D}}$ of equivalence relations on $\mathbf{D}$. Also, setting*

$$\mathrm{T}^r(\lambda) = \mathrm{T}(\overline{\mathbf{M}}(\lambda), \mathrm{R}^{\lambda,r}),\ \lambda \in \Lambda,$$

we consider the containing action-mapping $F^{G \times \mathrm{T}^r(\lambda)}$ (see Lemma 7.1.4) for the indexed collection

$$F^{\mathbf{D}(\lambda)} \equiv \{F^{D(\lambda)} : D \in \mathbf{D}\}$$

corresponding to the $F^{\mathbf{D}(\lambda)}$-co-mark $(\overline{\mathbf{M}}^G(\lambda), \overline{\mathbf{M}}(\lambda))$ and $(\overline{\mathbf{M}}^G(\lambda), \overline{\mathbf{M}}(\lambda))$-admissible family R^λ of equivalence relations on $F^{\mathbf{D}(\lambda)}$. Then, the pair $(D^{\mathrm{T}}, F^{D^{\mathrm{T}}})$, where

$$F^{D^{\mathrm{T}}} = \{F^{G\times \mathrm{T}^r(\lambda)} : \lambda \in \Lambda\},$$

is a G-diagram.

Proof. By construction of the diagram D^{T} (see Lemma 6.3.3) for every $\lambda \in \Lambda$ we have that $D^{\mathrm{T}}(\lambda) = \mathrm{T}^r(\lambda)$ and for every $(\lambda_0, \lambda_1) \in \prec$ the mapping $D^{\mathrm{T}}(\lambda_0, \lambda_1)$ is the containing mapping for the indexed collection $\mathbf{D}(\lambda_0, \lambda_1)$ of mappings corresponding to $\mathbf{D}(\lambda_0, \lambda_1)$-co-mark $(\overline{\mathbf{M}}(\lambda_1), \overline{\mathbf{M}}(\lambda_0))$ and the family $\mathrm{R}^{\mathbf{D}}(\lambda_0, \lambda_1)$ of equivalence relation on $\mathbf{D}(\lambda_0, \lambda_1)$. We note that the domain of $D^{\mathrm{T}}(\lambda_0, \lambda_1)$ is the space $D^{\mathrm{T}}(\lambda_1)$ and the range the space $D^{\mathrm{T}}(\lambda_0)$.

Also, by construction of the mapping

$$F^{G\times \mathrm{T}^r(\lambda)} \equiv F^{G\times D^{\mathrm{T}}(\lambda)}$$

(see Lemma 7.1.4) we have

$$F^{G\times D^{\mathrm{T}}(\lambda)} = F^{\mathrm{T}}_\lambda \circ [G, D^{\mathrm{T}}(\lambda)],$$

where F^{T}_λ is the containing mapping of $\mathrm{T}^d(\lambda) \equiv \mathrm{T}(\overline{\mathbf{M}}^G(\lambda), \mathrm{R}^{\lambda,d})$ into $\mathrm{T}^r(\lambda) = D^{\mathrm{T}}(\lambda)$ and $[G, D^{\mathrm{T}}(\lambda)]$ is the homeomorphism of $G \times D^{\mathrm{T}}(\lambda)$ onto $\mathrm{T}^d(\lambda)$ constructed in Lemma 7.1.4.

In order to prove that $(D^{\mathrm{T}}, F^{D^{\mathrm{T}}})$ is a G-diagram we need only to prove that for every $(\lambda_0, \lambda_1) \in \prec$ the mapping $D^{\mathrm{T}}(\lambda_0, \lambda_1)$ is an equivariant mapping, that is the following relation is true:

$$D^{\mathrm{T}}(\lambda_0, \lambda_1) \circ F^{G\times D^{\mathrm{T}}(\lambda_1)} = F^{G\times D^{\mathrm{T}}(\lambda_0)} \circ (\mathrm{id}_G \times D^{\mathrm{T}}(\lambda_0, \lambda_1)). \quad (7.3.2)$$

Let $(g, \mathbf{a}) \in G \times D^{\mathrm{T}}(\lambda_1)$ and $(x, D(\lambda_1)) \in \mathbf{a}$ for some $D \in \mathbf{D}$. Then,

$$F^{G\times D^{\mathrm{T}}(\lambda_1)}(g, \mathbf{a}) = F^{\mathrm{T}}_{\lambda_1}([G, D^{\mathrm{T}}(\lambda_1)](g, \mathbf{a})).$$

By Lemma 7.1.4, $[G, D^{\mathrm{T}}(\lambda_1)](g, \mathbf{a})$ is the point $\mathbf{b} = [g, \mathbf{a}]$ of $\mathrm{T}^d(\lambda_1)$ containing the pair $((g, x), G \times D(\lambda_1))$. By the construction of the mapping $F^{\mathrm{T}}_{\lambda_1}$,

$F^{\mathrm{T}}_{\lambda_1}(\mathbf{b})$ is the point $\mathbf{c}$ of $D^{\mathrm{T}}(\lambda_1)$ containing the pair $(F^{D(\lambda_1)}(g,x), D(\lambda_1))$. Then, $D^{\mathrm{T}}(\lambda_0,\lambda_1)(\mathbf{c})$ is the point $\mathbf{d}$ of $D^{\mathrm{T}}(\lambda_0)$ containing the pair

$$((D(\lambda_0,\lambda_1)\circ F^{D(\lambda_1)})(g,x), D(\lambda_0)).$$

On the other hand,

$$(\mathrm{id}_G \times D^{\mathrm{T}}(\lambda_0,\lambda_1))(g,\mathbf{a}) = (g, D^{\mathrm{T}}(\lambda_0,\lambda_1)(\mathbf{a})) \equiv (g,\mathbf{b}_0),$$

where $\mathbf{b}_0$ is the point of $D^{\mathrm{T}}(\lambda_0)$ containing the pair $(D(\lambda_0,\lambda_1)(x), D(\lambda_0))$. Then, $[G, D^{\mathrm{T}}(\lambda_0)](g,\mathbf{b}_0)$ is the point $\mathbf{c}_0 \equiv [g,\mathbf{b}_0]$ of $\mathrm{T}^d(\lambda_0)$, which contains the pair

$$((g, D(\lambda_0,\lambda_1)(x)), G\times D(\lambda_0)).$$

Therefore, $F^{\mathrm{T}}_{\lambda_0}(\mathbf{c}_0)$ is the point $\mathbf{d}_0$ of $D^{\mathrm{T}}(\lambda_0)$ containing the pair

$$(F^{D(\lambda_0)}(g, D(\lambda_0,\lambda_1)(x)), D(\lambda_0)) =$$

$$(F^{D(\lambda_0)}((\mathrm{id}_G\times D(\lambda_0,\lambda_1))(g,x)), D(\lambda_0)).$$

Since

$$D(\lambda_0,\lambda_1)\circ F^{D(\lambda_1)} = F^{D(\lambda_0)}\circ(\mathrm{id}_G\times D(\lambda_0,\lambda_1))$$

we have $\mathbf{d} = \mathbf{d}_0$, which proves relation (7.3.2). □

Definition. The G-diagram $(D^{\mathrm{T}}, F^{D^{\mathrm{T}}})$ constructed in Lemma 7.3.1 is called the *containing G-diagram* for the indexed collection $(\mathbf{D}, F^{\mathbf{D}})$ (*corresponding* to the proper product-open G-$\mathbf{D}$-co-mark $(\overline{\mathbf{M}}^G, \overline{\mathbf{M}})$ and the $(\overline{\mathbf{M}}^G, \overline{\mathbf{M}})$-admissible family R of equivalence relations on $(\mathbf{D}, F^{\mathbf{D}})$).

Definition. A non-empty class $I\!\!P$ of G-diagrams is said to be *saturated* if for every indexed collection $(\mathbf{D}, F^{\mathbf{D}})$ of elements of $I\!\!P$ there exists a proper G-$\mathbf{D}$-co-mark $(\overline{\mathbf{M}}^{G,+}, \overline{\mathbf{M}}^+)$ satisfying the following condition: for every proper G-$\mathbf{D}$-co-mark $(\overline{\mathbf{M}}^G, \overline{\mathbf{M}})$, which is a co-extension of $(\overline{\mathbf{M}}^{G,+}, \overline{\mathbf{M}}^+)$, there exists an $(\overline{\mathbf{M}}^G, \overline{\mathbf{M}})$-admissible family R^+ of equivalence relations on $(\mathbf{D}, F^{\mathbf{D}})$ such that for every admissible family R of equivalence relations on $(\mathbf{D}, F^{\mathbf{D}})$, which is a final refinement of R^+, the containing G-diagram

$(D^{\mathrm{T}}, F^{D^{\mathrm{T}}})$ for the indexed collection $(\mathbf{D}, F^{\mathbf{D}})$ corresponding to the G-$\mathbf{D}$-co-mark $(\overline{\mathbf{M}}^G, \overline{\mathbf{M}})$ and the family R belongs to the class $I\!P$.

The G-$\mathbf{D}$-co-mark $(\overline{\mathbf{M}}^{G,+}, \overline{\mathbf{M}}^+)$ is said to be an *initial* G-$\mathbf{D}$-co-mark *corresponding to the class* $I\!P$, and the family R^+ is said to be an *initial* family *corresponding to the* G-$\mathbf{D}$*-co-mark* $(\overline{\mathbf{M}}^G, \overline{\mathbf{M}})$ *and the class* $I\!P$.

By the standard manner we can prove the following proposition.

7.3.2 Proposition. *The non-empty intersection of not more than* τ *many saturated classes of* G*-diagrams is also a saturated class of* G*-diagrams.* □

Definition. An element (D^T, F^{D^T}) of a class $I\!P$ of G-diagrams is said to be *universal* if for every element (D, F^D) of $I\!P$ there exists an equivariant isomorphism of D into D^T.

7.3.3 Proposition. *In any saturated class of* G*-diagrams there exists universal elements.*

Proof. Let $I\!P$ be a saturated class of G-diagrams. Denote by $(\mathbf{D}, F^{\mathbf{D}})$ an indexed collection of elements of $I\!P$ such that for every element (D_0, F^{D_0}) of $I\!P$ there exists an element (D_1, F^{D_1}) of $(\mathbf{D}, F^{\mathbf{D}})$ with the property that there exists an equivariant isomorphism of D_0 onto D_1.

Denote by $(\overline{\mathbf{M}}^G, \overline{\mathbf{M}})$ an initial G-$\mathbf{D}$-co-mark corresponding to the class $I\!P$ and by R an initial family of equivalence relations on $(\mathbf{D}, F^{\mathbf{D}})$ corresponding to the G-$\mathbf{D}$-co-mark $(\overline{\mathbf{M}}^G, \overline{\mathbf{M}})$ and the class $I\!P$. Then, the containing G-diagram $(D^{\mathrm{T}}, F^{D^{\mathrm{T}}})$ for the indexed collection $(\mathbf{D}, F^{\mathbf{D}})$ corresponding to the G-$\mathbf{D}$-co-mark $(\overline{\mathbf{M}}^G, \overline{\mathbf{M}})$ and the family R (see Lemma 7.3.1) belongs to $I\!P$.

We prove that $(D^{\mathrm{T}}, F^{D^{\mathrm{T}}})$ is a universal element in $I\!P$. For this purpose it suffices to prove that for every $D \in \mathbf{D}$ the indexed collection

$$i_{D^{\mathrm{T}}}^{D} \equiv \{i_{D^{\mathrm{T}}(\lambda)}^{D(\lambda)} : \lambda \in \Lambda\},$$

where $i_{D^{\mathrm{T}}(\lambda)}^{D(\lambda)}$ is the natural embedding of $D(\lambda)$ into $D^{\mathrm{T}}(\lambda)$ has the following properties:

(a) the embedding $i_{D^{\mathrm{T}}(\lambda)}^{D(\lambda)}$, where $\lambda \in \Lambda$, is equivariant, and

(b) the indexed collection $i_{D^{\mathrm{T}}}^{D}$ is an isomorphism of D into D^{T}.

Property (a) can be proved similarly to the proof of the fact that the natural embedding $i^X_{\mathrm{T}r}$ of Proposition 7.1.8 is an equivariant mapping. To prove property (b) we need to prove that if $D \in \mathbf{D}$ and $(\lambda_0, \lambda_1) \in \prec$, then

$$D^{\mathrm{T}}(\lambda_0, \lambda_1) \circ i^{D(\lambda_1)}_{D^{\mathrm{T}}(\lambda_1)} = i^{D(\lambda_0)}_{D^{\mathrm{T}}(\lambda_0)} \circ D(\lambda_0, \lambda_1). \tag{7.3.3}$$

Let $x \in D(\lambda_1)$. Then, $i^{D(\lambda_1)}_{D^{\mathrm{T}}(\lambda_1)}(x)$ is the point $\mathbf{a}$ of $D^{\mathrm{T}}(\lambda_1)$, which contains the pair $(x, D(\lambda_1))$. Therefore, $D^{\mathrm{T}}(\lambda_0, \lambda_1)(\mathbf{a})$ is the point of $D^{\mathrm{T}}(\lambda_0)$, which contains the pair $(D(\lambda_0, \lambda_1)(x), D(\lambda_0))$. On the other hand,

$$(i^{D(\lambda_0)}_{D^{\mathrm{T}}(\lambda_0)} \circ D(\lambda_0, \lambda_1))(x)$$

is also the point of $D^{\mathrm{T}}(\lambda_0)$, which contains the pair $(D(\lambda_0, \lambda_1)(x), D(\lambda_0))$. Relation (7.3.3) is proved. □

Definition. Let $I\!\!P$ be a class of diagrams. By the *class of G-diagrams generated by* $I\!\!P$ we mean the class of all G-diagrams (D, F^D) such that $D \in I\!\!P$.

7.3.4 Proposition. *Let $I\!\!P$ be a saturated class of diagrams. Then, the class of G-diagrams generated by $I\!\!P$ is also saturated.*

Proof. Let $I\!\!R$ be the class of G-diagrams generated by $I\!\!P$. Denote by $(\mathbf{D}, F^{\mathbf{D}})$ an indexed collection of elements of $I\!\!R$. Then, $\mathbf{D}$ is an indexed collection of elements of $I\!\!P$. Let $(\overline{\mathbf{M}}^{G,+}, \overline{\mathbf{M}}^{+})$ be a proper product-open G-$\mathbf{D}$-co-mark such that $\overline{\mathbf{M}}^{+}$ is an initial co-mark of $\mathbf{D}$ corresponding to the class $I\!\!R$ and let $(\overline{\mathbf{M}}^{G}, \overline{\mathbf{M}})$ be a proper G-$\mathbf{D}$-co-mark, which is a co-extension of $(\overline{\mathbf{M}}^{G,+}, \overline{\mathbf{M}}^{+})$. Denote by R^+ an $(\overline{\mathbf{M}}^{G}, \overline{\mathbf{M}})$-admissible family of equivalence relations on $(\mathbf{D}, F^{\mathbf{D}})$ such that $(\mathrm{R}^+)^{\mathbf{D}}$ is an initial family of equivalence relations on $\mathbf{D}$ corresponding to the co-mark $\overline{\mathbf{M}}$ and the class $I\!\!R$ and by R an admissible family of equivalence relations on $(\mathbf{D}, F^{\mathbf{D}})$, which is a final refinement of R^+.

Now, for the indexed collection $(\mathbf{D}, F^{\mathbf{D}})$ we consider the containing G-diagram $(D^{\mathrm{T}}, F^{D^{\mathrm{T}}})$ corresponding to the G-$\mathbf{D}$-co-mark $(\overline{\mathbf{D}}^{G}, \overline{\mathbf{M}})$ and the family R constructed in Lemma 7.3.1. By construction D^{T} belongs to $I\!\!P$ and therefore $(D^{\mathrm{T}}, F^{D^{\mathrm{T}}})$ belongs to $I\!\!R$. This fact proves that $(\overline{\mathbf{M}}^{G,+}, \overline{\mathbf{M}}^{+})$ is an initial G-$\mathbf{D}$-co-mark and R^+ is an initial family. Thus, $I\!\!R$ is a saturated class. □

The above proposition implies the following two consequences concerning G-spectra.

7.3.5 Corollary. *In the class of all G-spectra there exist universal elements.* □

7.3.6 Corollary. *In the class of all G-spectra, whose spaces belongs to a given saturated class of spaces and projections to a given saturated class of mappings, there exist universal elements.* □

7.3.7 Lemma. *Let (D, F^D) be a G-spectrum. The mapping $F^{\lim(D)}$ of $G \times \lim(D)$ into $\lim(D)$ such that for every $g \in G$ and $f \in \lim(D)$, $F^{\lim(D)}(g, f)$ is the element h of $\lim(D)$ for which $h(\lambda) = g(f(\lambda))$, where $\lambda \in \Lambda$, is an action of G on $\lim(D)$.*

Proof. The proof of the lemma follows immediately by the following statements, which are not dificult to prove:

(1) Consider the spectra $G \otimes D$ and D. The mapping $(g, f) \to h$, where $(g, f) \in G \times \lim(D)$ and h is the point of $\lim(D)$ for which

$$h(\lambda) = (g, f(\lambda)) \in G \times D(\lambda), \ \lambda \in \Lambda,$$

is an homeomorphism of $G \times \lim(D)$ onto $\lim(G \otimes D)$. In what follows, we identify the point (g, f) of $G \times \lim(D)$ with the point h of $\lim(G \otimes D)$ considering that the spaces $G \times \lim(D)$ and $\lim(G \otimes D)$ coincide.

(2) The indexed set

$$F^D \equiv \{F^{D(\lambda)} : \lambda \in \Lambda\}$$

is a morphism of the spectrum $G \otimes D$ into the spectrum D. Therefore, $\lim(F^D)$ (see Lemma 6.3.13) is a mapping of $G \times \lim(D)$ onto $\lim(D)$.

(3) The mapping $F^{\lim(D)}$ coincides with the mapping $\lim(F^D)$.

(4) The mapping $\lim(F^D)$ is an action on $\lim(D)$. □

Definition. Let (D, F^D) be a G-spectrum. The mapping $F^{\lim(D)}$ defined in Lemma 7.3.7 is called the *limit action* of G on $\lim(D)$ and the pair $(\lim(D), F^{\lim(D)})$ is called the *limit G-space* of the given G-spectrum.

Let $I\!P$ be a class of G-spectra. The class consisting of all G-spaces (X, F^X) for which there exists an element (D, F^D) of $I\!P$ such that X and

$\lim(D)$ are equivariantly homeomorphic, is called the *class of the limit G-spaces* of $I\!P$. □

The above lemma implies the following proposition.

7.3.8 Proposition. *Let $I\!P$ be a saturated class of G-spectra. Then, in the class of the limit G-spaces of $I\!P$ there exist universal elements.* □

In particular, we have the following consequence.

7.3.9 Corollary. *In the class of the limit G-spaces of all G-spectra, whose spaces belongs to a given saturated class of spaces and projections to a given saturated class of mappings, there exist universal elements.* □

Chapter 8

Containing spaces and factorizing τ-spectra

In the present chapter we investigate the notion of a "saturated class" using the notion of a factorizing τ-spectrum.

In Section 8.1 we give the basic notions concerning the inverse spectra. For a given class $\mathbb{P}$ of spaces we consider spectra whose spaces belong to $\mathbb{P}$. Such spectra are called $\mathbb{P}$-spectra. We define the notion of a $\mathbb{P}$-τ-spectrum and the notion of a $\mathbb{P}$-factorizing spectrum with respect to a subset of its limit space. A subset of the limit space of a spectrum on which the limit projections are onto is called surjective. For $\mathbb{P}$-τ-spectra, which are $\mathbb{P}$-factorizing with respect to surjective subsets of its limit spaces, we prove a version of the so-called Spectral Theorem. In the case, where $\mathbb{P}$ is the class of all spaces (of weight $\leq \tau$), a $\mathbb{P}$-τ-spectrum is called τ-spectrum and a $\mathbb{P}$-factorizing spectrum is called factorizing.

In Section 8.2 for a given indexed collection $\mathbf{S}$ of spaces we consider the set $\mathrm{P}(\mathbf{S})$ consisting of all pairs $(\mathbf{M}, \mathrm{R})$, where $\mathbf{M}$ is a co-mark of $\mathbf{S}$ and R is an $\mathbf{M}$-admissible family of equivalence relations on $\mathbf{S}$. In this set by a natural way we define an equivalence relation denoted by $\sim^{\mathbf{S}}$. The set of all equivalence classes of $\sim^{\mathbf{S}}$ is denoted here by $\mathrm{C}(\mathbf{S})$. For every $c \in \mathrm{C}(\mathbf{S})$ the space $\mathrm{T}(\mathbf{M}, \mathrm{R})$ is independent of the element $(\mathbf{M}, \mathrm{R})$ of c and is denoted by $\mathrm{T}(c)$. On the set $\mathrm{C}(\mathbf{S})$ we define a directed partial order in such a manner that $\mathrm{C}(\mathbf{S})$ becomes a τ-complete set. On this τ-complete set we define an inverse spectrum, denoted by $\mathrm{D}[\mathbf{S}]$ whose spaces are Containing Spaces $\mathrm{T}(c)$ for the indexed collection $\mathbf{S}$.

The constructed inverse spectrum is a τ-spectrum and the limit space of this τ-spectrum contains a surjective subset S, which is homeomorphic to the free union of elements of $\mathbf{S}$. Moreover, this spectrum is factorizing with respect to the subset S. The Spectral Theorem proving in Section 8.1 implies the following result, which in some sense can be considered as the "spectral determination" of the Containing Spaces. Let D be an inverse spectrum on the set $\mathrm{C}(\mathbf{S})$ satisfying the properties: (a) D is a τ-spectrum, (b) the space $\lim(D)$ contains a surjective subset L, which is homeomorphic to the free union of elements of $\mathbf{S}$, and (c) D is factorizing with respect to the subset L. Then, the spectra D and $\mathrm{D}[\mathbf{S}]$ contain isomorphic cofinal and τ-closed subspectra. The result is that for all elements c of a cofinal and τ-closed subset of $\mathrm{C}(\mathbf{S})$ the spaces $D(c)$ are Containing Spaces for the indexed collection $\mathbf{S}$. Thus, the Containing Spaces for a given indexed collection $\mathbf{S}$ of spaces constructed in Chapter 1 "appear" here without any concrete construction of these spaces.

An application of the fact that $\mathrm{D}[\mathbf{S}]$ is factorizing is the following "covering property" of the Containing Spaces. Let h_X, $X \in \mathbf{S}$, be an embedding of X into a space Y (of weight $\leq \tau$). Then, there exists an element c of $\mathrm{C}(\mathbf{S})$ and a mapping f of the Containing Space $\mathrm{T}(c)$ into the space Y such that the following relation is satisfied:

$$f \circ i^X_{\mathrm{T}(c)} = h_X, \quad X \in \mathbf{S},$$

where $i^X_{\mathrm{T}(c)}$ is the natural embedding of X into $\mathrm{T}(c)$.

The notion of a saturated class $I\!\!P$ of spaces with structure is based on the fact that for a given indexed collection $\mathbf{S}$ of spaces the set of all elements $(\mathbf{M}, \mathrm{R})$ of $\mathrm{P}(\mathbf{S})$ such that the given on elements of $\mathbf{S}$ structure can be transfered to the Containing Space $\mathrm{T}(\mathbf{M}, \mathrm{R})$, is "sufficiently large". The most important property of "sufficiently large" subsets of $\mathrm{P}(\mathbf{S})$ is the following "intersection property": the intersection of τ many "sufficiently large" subsets of $\mathrm{P}(\mathbf{S})$ is also a "sufficiently large" subset. This property is used to prove the intersection property of saturated classes, to determine some saturated classes and to construct "new" saturated classes by the "old" ones.

In Section 8.3 we investigate the notion of a "saturated class" using "sufficiently large" subsets of $\mathrm{C}(\mathbf{S})$. There are several types of subsets of $\mathrm{C}(\mathbf{S})$ having the above mentioned "intersection property". We call these subsets the zero-type, the first-type, and the second-type subsets. We define the notion of a "saturated class" based on these sets. Such "saturated classes" are called the zero-type, the first-type, and the second-type saturated classes, respectively. The class of all spaces, the class of all regular spaces, and the class of all completely-regular spaces are zero-type saturated classes. The saturated classes considered in the preceding chapters correspond to the first-type saturated classes. (We note that a zero-type subset of $\mathrm{C}(\mathbf{S})$ is also a first-type and a second-type subset.)

The cofinal and τ-closed subsets of $\mathrm{C}(\mathbf{S})$ (which have the "intersection property") determine the second-type subsets and, therefore, the second-type saturated classes. These classes have of course the universal and the intersection properties. All the saturated classes we considered in the preceding chapters seem also to be second-type saturated classes. However, we give the corresponding proofs only for some of them. In particular, we prove that the following classes are second-type saturated: (a) the class of all regular (completely regular) spaces, (b) the class of all countable-dimensional regular (completely regular) spaces, (c) the class of all strongly countable-dimensional regular (completely regular) spaces, (d) the class of locally finite-dimensional regular (completely regular) spaces, and (e) the class of regular (completely regular) spaces of dimension ind less than or equal to an ordinal $\alpha \in \tau^{+}$ (in particular, α may be an integer).

Let $I\!\!P$ be a given second-type saturated class of spaces. Subsets of $\mathrm{C}(\mathbf{S})$ consisting of elements c such that $\mathrm{T}(\mathbf{M}, \mathrm{R}, \mathbf{L})$ is an element of $I\!\!P$ for every $(\mathbf{M}, \mathrm{R}) \in c$ and $\mathbf{L} \in \mathrm{C}^{\diamond}(\mathrm{R})$ are called $I\!\!P$-sets. By the definition of a second-type saturated class it follows the existence of cofinal and τ-closed $I\!\!P$-subsets of $\mathrm{C}(\mathbf{S})$. Therefore, for every such set $\mathrm{C}_{I\!\!P}(\mathbf{S})$ the subspectrum $\mathrm{D}[\mathbf{S}]|_{\mathrm{C}_{I\!\!P}(\mathbf{S})}$ of $\mathrm{D}[\mathbf{S}]$ is a $I\!\!P$-τ-spectrum, which is factorizing with respect to the surjective subset S of the limit space $\lim(\mathrm{D}[\mathbf{S}]|_{\mathrm{C}_{I\!\!P}(\mathbf{S})}) = \lim(\mathrm{D}[\mathbf{S}])$.

Applying the Spectral Theorem we obtain the following result, which can be considered as the "spectral determination" of the Containing Spaces (corresponding to an indexed collection of spaces and) belonging to a given second-type saturated class of spaces. Let $I\!\!P$ be a second-type saturated class of spaces and $\mathbf{S}$ an indexed collection of elements of $I\!\!P$. Suppose that D is an inverse spectrum on a cofinal and τ-closed $I\!\!P$-subset $\mathrm{C}_{I\!\!P}(\mathbf{S})$ of $\mathrm{C}(\mathbf{S})$ satisfying the following conditions: (a) D is a τ-spectrum, (b) the space $\lim(D)$ contains a surjective subset L, which is homeomorphic to the free union of elements of $\mathbf{S}$, and (c) D is $I\!\!P$-factorizing with respect to L. Then, D and $\mathrm{D}[\mathbf{S}]$ contain isomorphic cofinal and τ-closed subspectra. Therefore, D contains a cofinal and τ-closed $I\!\!P$-spectrum, which is factorizing with respect to L.

The "covering property" of the containing spaces belonging to a given second-type saturated class of spaces can be formulated as follows. Let $I\!P$ be a second-type saturated class of spaces and Y a universal element in $I\!P$. Then, there exist an indexed collection $\mathbf{S}$ of spaces; an element c of $\mathrm{C}(\mathbf{S})$, for which the Containing Space $\mathrm{T}(c)$ is a universal element in $I\!P$; a mapping f of the Containing Space $\mathrm{T}(c)$ into the space Y; and for every $X \in I\!P$ a mapping $i^X_{\mathrm{T}(c)}$ (which coincides with the natural embedding if $X \in \mathbf{S}$) such that the following relation is satisfied:

$$f \circ i^X_{\mathrm{T}(c)} = h_X, \quad X \in I\!P.$$

The above results are formulated also for some concrete classes of spaces, in particular, for some classes of separable metrizable spaces.

In Section 8.4 we generalize the results of the preceding sections concerning the classes of spaces to the so-called structural classes. We do not give the proofs of the presented results since they are similar to the corresponding results of the preceding sections.

Notations. The definition of an inverse spectrum D on a directed subset C with a partial order $\prec^{\mathrm{C}}$ and the definition of a morphism or an isomorphism of an inverse spectrum D_0 on C into an inverse spectrum D_1 on C were given in Chapter 6. (There we use the letter Λ instead of C). The related notations also introduced in Chapter 6 are as follows: $\lim(D)$ (the limit space of D); $D(c)$ (a space of D) where $c \in \mathrm{C}$; $D(c_0, c_1)$ (a mapping or a projection of D) where $(c_0, c_1) \in \prec^{\mathrm{C}}$; $\lim(D)[U]$ (an open subset of $\lim(D)$) where U is an open subset of the space $D(c)$; and

$$f_{\mathrm{C}} \equiv \{f_c : c \in \mathrm{C}\}$$

(a morphism of D_0 into D_1 or an isomorphism of D_0 onto (into) D_1) where f_c is a mapping of $D_0(c)$ into $D_1(c)$ or an homeomorphism of $D_0(c)$ onto (into) $D_1(c)$.

Remark. All spaces of the inverse spectra that are considered in the present chapter are spaces of weight $\leq \tau$. However, the weights of the limit spaces of these spectra as well as the weights of some of the considered subsets of limit spaces may be large than τ. The fact that the weight of a considered space may be large than τ will be clear from the context.

Notations. Recall that $\mathrm{I\!P}(\text{space})$, $\mathrm{I\!P}(\text{subset})$, $\mathrm{I\!P}(\text{base})$, $\mathrm{I\!P}(\text{p-base})$, and $\mathrm{I\!P}(\Phi, \Lambda, \prec, \mathbf{p})$ (see Chapter 4) denote respectively the class of all spaces, the class of all subsets, the class of all bases, the class of all p-bases, and the class of all spaces with a fixed $(\Phi, \Lambda, \prec, \mathbf{p})$-structure.

In the case, where the class $\mathbf{p}(\lambda)$ is the class $\mathbb{P}$(subset) for all $\lambda \in \Lambda$, instead of "$(\Phi, \Lambda, \prec, \mathbf{p})$", we write "$(\Phi, \Lambda, \prec)$". Also, where $\Lambda = \Lambda^0$ (about notation "Λ^0" see Section 2.4) instead of "$(\Phi, \Lambda, \prec, \mathbf{p})$" we write "$(\Lambda, \prec, \mathbf{p})$".

8.1 $I\!P$-factorizing $I\!P$-τ-spectra and the spectral theorem

Agreement. In this section it is assumed that a fixed directed set, denoted by C, is given. The partial order on C is denoted by $\prec^{\mathrm{C}}$. The partial order induced by $\prec^{\mathrm{C}}$ on any subset of C is denoted also by $\prec^{\mathrm{C}}$.

It is also assumed that an inverse spectrum on C, denoted by D, is given.

Definitions. Let C_0 be a subset of C. An element c_0 of C is said to be an *upper bound* of C_0 if for every $c \in \mathrm{C}_0$ we have $c \prec^{\mathrm{C}} c_0$.

An upper bound c_0 of C_0 is said to be the *lower upper bound* of C_0 if for every upper bound c_1 of C_0 we have $c_0 \prec^{\mathrm{C}} c_1$. The lower upper bound of C_0 (if such an element exists in C) is denoted by $\sup(\mathrm{C}_0)$. Obviously, if there exists the lower upper bound of a subset of C, then it is uniquely determined.

We say that the subset C_0 *majorates* a subset C_1 of C if for every $c \in \mathrm{C}_1$ there exists an element c' of C_0 such that $c \prec^{\mathrm{C}} c'$. We say that an element c of C *majorates* a subset C_1 of C if the singleton $\{c\}$ majorates C_1.

The subset C_0 is said to be *cofinal* in C if C_0 majorates C. Obviously, any cofinal subset of C is directed.

A subset C' of C is said to be a *chain* if every two elements of it are comparable.

The subset C_0 is said to be τ-*closed* in C if for every chain $\mathrm{C}' \subset \mathrm{C}_0$, with $|\mathrm{C}'| \leq \tau$, we have $\sup(\mathrm{C}') \in \mathrm{C}_0$ whenever the lower upper bound $\sup(\mathrm{C}')$ exists.

The set C is said to be τ-*complete* if for every chain C', with $|\mathrm{C}'| \leq \tau$, there exists the lower upper bound of C'.

8.1.1 Lemma. *Suppose that* C *is* τ*-complete and* $\{\mathrm{C}_\delta : \delta \in \tau\}$ *is an indexed set of cofinal and* τ*-closed subsets of* C. *Then, the set* $\cap\{\mathrm{C}_\delta : \delta \in \tau\}$ *is cofinal and* τ*-closed in* C.

Proof. Let c be an arbitrary element of C. Consider the set $\omega \times \tau$ with the lexicographic order denoted by $\prec_l$ (that is, for two elements (n_0, δ_1) and (n_1, δ_1) of $\omega \times \tau$ we have $(n_0, \delta_0) \prec_l (n_1, \delta_1)$ if either $n_0 < n_1$ or $n_0 = n_1$ and $\delta_0 \leq \delta_1$). By induction, we can define a subset

$$\mathrm{C}' \equiv \{c_{(n,\delta)} : (n, \delta) \in \omega \times \tau\}$$

of C such that the following conditions are satisfied:
(a) $c \prec^{\mathrm{C}} c_{(n,\delta)}$ for every $(n, \delta) \in \omega \times \tau$,
(b) $c_{(n_0,\delta_0)} \prec^{\mathrm{C}} c_{(n_1,\delta_1)}$ whenever $(n_0, \delta_0) \prec_l (n_1, \delta_1)$, and
(c) $c_{(n,\delta)} \in \mathrm{C}_\delta$ for every $(n, \delta) \in \omega \times \tau$.
Obviously, $|\mathrm{C}'| \leq \tau$. Condition (b) guarantees that C′ is a chain. For every $\delta \in \tau$ the subset

$$\mathrm{C}'_\delta \equiv \{c_{(n,\delta)} : n \in \omega\}$$

of C_δ majorates C′ and, therefore,

$$c' \equiv \sup(\mathrm{C}') = \sup(\mathrm{C}'_\delta).$$

By τ-closeness of C_δ we have $c' \in \mathrm{C}_\delta$, $\delta \in \tau$. Thus,

$$c' \in \cap\{\mathrm{C}_\delta : \delta \in \tau\} \quad \text{and} \quad c \prec^{\mathrm{C}} c'.$$

This fact show the cofinality of the set $\cap\{\mathrm{C}_\delta : \delta \in \tau\}$ in C. Its τ-closeness is obvious. □

8.1.2 Corollary. *Suppose that C is τ-complete and C′ is a subset of C such that $|\mathrm{C}'| \leq \tau$. Then, there exists an element of C which majorates the set C′.*

Proof. For every $c' \in \mathrm{C}'$ we consider the set

$$\mathrm{C}_{c'} \equiv \{c \in \mathrm{C} : c' \prec^{\mathrm{C}} c\}.$$

It is easy to see that $\mathrm{C}_{c'}$ is a cofinal and τ-closed subset of C. By Lemma 8.1.1 the intersection of all these sets is not empty. Any element of this intersection can be considered as the required element of C. □

8.1.3 Proposition. *Suppose that C is τ-complete and R is a relation on C (that is, $R \subset \mathrm{C} \times \mathrm{C}$) satisfying the following conditions:*

(1) *Existence: for every $c \in \mathrm{C}$ there exists an element c' of* C *such that* $(c, c') \in R$.

(2) *Majorantness: if* $(c, c') \in R$ *and* $c' \prec^{\mathrm{C}} c'' \in \mathrm{C}$, *then* $(c, c'') \in R$.

(3) *τ-Closeness: if c_1 is a fixed element of* C *and* C$'$ *is a chain with* $|\mathrm{C}'| \leq \tau$ *such that* $(c, c_1) \in R$ *for every* $c \in \mathrm{C}'$, *then* $(c', c_1) \in R$, *where* $c' = \sup(\mathrm{C}')$.

Then, the set $\mathrm{C}^R \equiv \{c \in \mathrm{C} : (c, c) \in R\}$ *(that is, the set of all R-reflexive elements of* C*) is a cofinal and τ-closed subset of* C.

Proof. First, we show that the set of all R-reflexive elements of C is cofinal in C. Let $c^0 \in \mathrm{C}$. By induction we construct a chain $\{c_n : n \in \omega\}$ satisfying the following condition:

(a) $c_0 = c^0$,

(b) $(c_n, c_k) \in \mathrm{C}^R$, whenever $n \leq k \in \omega$, and

(c) $c_n \prec^{\mathrm{C}} c_k$, whenever $n \leq k \in \omega$.

Suppose that the elements c_n have already been constructed for all integers $n < k \in \omega$. We construct the element c_k. Let $k = m + 1$. By the existence condition, there exists an element c^m of C such that $(c_m, c^m) \in \mathrm{C}^R$. Since C is directed we can find an element c_k of C such that $c_m \prec^{\mathrm{C}} c_k$ and $c^m \prec^{\mathrm{C}} c_k$. The majorantness condition implies that $(c_m, c_k) \in \mathrm{C}^R$. This fact finishes the construction of the set $\{c_n : n \in \omega\}$. Condition (c) implies that this set is a chain.

Let $c = \sup\{c_n : n \in \omega\}$. We show that c is an R-reflexive element of C. Indeed, by the majorantness condition, $(c_n, c) \in \mathrm{C}^R$ for all $n \in \omega$. Therefore, by the τ-closeness condition, $(c, c) \in \mathrm{C}^R$. This show cofinality of the set C^R of all R-reflexive elements in C.

Now, we prove that C^R is τ-closed. Let C$'$ be a chain consisting of R-reflexive elements and $c' = \sup(\mathrm{C}')$. Then, $(c, c) \in R$ for all $c \in \mathrm{C}'$. By the majorantness condition, $(c, c') \in \mathrm{C}^R$ for all $c \in \mathrm{C}'$. By the τ-closeness condition, $(c', c') \in \mathrm{C}^R$, that is C^R is τ-closed. □

Definition. For every $c \in \mathrm{C}$ we define a mapping, denoted by $D(c, \infty)$, of the limit space $\lim(D)$ into the space $D(c)$ as follows: if g is a point of $\lim(D)$, then we set

$$D(c, \infty)(g) = g(c).$$

The mapping $D(c, \infty)$ is called a *limit projection* of the spectrum D.

The proof of the following lemma is obvious.

8.1.4 Lemma. *For every open subset U of $D(c)$, $c \in \mathrm{C}$, we have*

$$(D(c,\infty))^{-1}(U) = \lim(D)[U].$$

Therefore, the mapping $D(c,\infty)$ is continuous. Moreover,

$$D(c_1, c_2) \circ D(c_2, \infty) = D(c_1, \infty),$$

whenever $c_1, c_2 \in \mathrm{C}$ and $c_1 \prec^{\mathrm{C}} c_2$. □

Definition. Let X be a space. A collection

$$f_{\mathrm{C}} \equiv \{f_c : c \in \mathrm{C}\},$$

where f_c is a mapping of X into $D(c)$, such that $D(c,c') \circ f_{c'} = f_c$ whenever $c, c' \in \mathrm{C}$ and $c \prec^{\mathrm{C}} c'$, is called a *mapping* of X into the spectrum D.

Let $f_{\mathrm{C}} \equiv \{f_c : c \in \mathrm{C}\}$ be a mapping of X into D. We define a mapping f of X into $\lim(D)$ as follows: for every $x \in X$ we set $f(x) = g \in \lim(D)$, where g is defined by relation $g(c) = f_c(x)$, $c \in \mathrm{C}$. It is easy to see that g is really a point of $\lim(D)$. The mapping f is said to be the *mapping of X into* $\lim(D)$ *induced by the collection* f_{C} and it is denoted by

$$\lim(f_{\mathrm{C}}) \equiv \lim\{f_c : c \in \mathrm{C}\}.$$

It is easy to verify that $f_c = D(c,\infty) \circ \lim(f_{\mathrm{C}})$, $c \in \mathrm{C}$.

The following lemma is easily proved.

8.1.5 Lemma. *Let X be a space and*

$$f_{\mathrm{C}} \equiv \{f_c : c \in \mathrm{C}\}$$

a mapping of the space X into the spectrum D. Then, for every open subset U of $D(c)$, $c \in \mathrm{C}$, we have

$$(\lim(f_{\mathrm{C}}))^{-1}(\lim(D)[U]) = f_c^{-1}(U).$$

Therefore, $\lim(f_{\mathrm{C}})$ *is a continuous mapping.* □

Definitions and notation. Let C' be a directed subset of C. If we restrict the inverse spectrum D (which is a mapping of $\mathrm{C} \cup \prec^{\mathrm{C}}$) to the set

$\mathrm{C}' \cup \prec^{\mathrm{C}}$ (here $\prec^{\mathrm{C}}$ denotes the order on C'), then we obtain a new inverse spectrum which is denoted by $D|_{\mathrm{C}'}$. This inverse spectrum is said to be a *subspectrum* of D.

The subspectrum $D|_{\mathrm{C}'}$ is said to be: (a) *cofinal* and (b) τ-*closed* if the subset C' of C is: (a) cofinal and (b) τ-closed, respectively.

By Lemma 8.1.4 the collection

$$\{D(c,\infty) : c \in \mathrm{C}'\}$$

is a mapping of the space $\lim(D)$ into the inverse spectrum $D|_{\mathrm{C}'}$. Therefore, by Lemma 8.1.5 we can consider the mapping

$$\lim\{D(c,\infty) : c \in \mathrm{C}'\}$$

of $\lim(D)$ into $\lim(D|_{\mathrm{C}'})$. This mapping is denoted also by $D(\mathrm{C}',\infty)$ and it is called the *natural* mapping of $\lim(D)$ into $\lim(D|_{\mathrm{C}'})$.

Let c' be an element of C which majorates the subset C'. Then, the collection

$$\{D(c,c') : c \in \mathrm{C}\}$$

is a mapping of the space $D(c')$ into the inverse spectrum $D|_{\mathrm{C}'}$. Therefore, by Lemma 8.1.5 we can consider the mapping

$$\lim\{D(c,c') : c \in \mathrm{C}'\}.$$

of $D(c')$ into $\lim(D|_{\mathrm{C}'})$. This mapping is denoted also by $D(\mathrm{C}',c')$ and it is called the *natural* mapping of $D(c')$ into $\lim(D|_{\mathrm{C}'})$.

8.1.6 Lemma. *Let C' be a cofinal directed subset of C. Then, the mapping $D(\mathrm{C}',\infty)$ is an homeomorphism of the space $\lim(D)$ onto the space $\lim(D|_{\mathrm{C}'})$.*

Proof. Note that if $g \in \lim(D)$ and $D(\mathrm{C}',\infty)(g) = g'$, then $g' = g|_{\mathrm{C}'}$. By the cofinality of C' it follows that if $g_1, g_2 \in \lim(D)$ and $g_1 \neq g_2$, then $g_1|_{\mathrm{C}'} \neq g_2|_{\mathrm{C}'}$. This fact means that the mapping $D(\mathrm{C}',\infty)$ is one-to-one.

On the other hand, if g' is an element of $\lim(D|_{\mathrm{C}'})$, then we can define an element g of $\lim(D)$ as follows: for every $c \in \mathrm{C}$ we take an element c' of C' such that $c \prec^{\mathrm{C}} c'$ and put

$$g(c) = D(c,c') \circ g'(c').$$

It is easy to verify that $g(c)$ is independent of c' and that g is really an element of $\lim(D)$. Moreover, by the definition of g it follows that

$$D(\mathrm{C}', \infty)(g) = g',$$

which means that the mapping $D(\mathrm{C}', \infty)$ is onto.

If U is an open subset of $D(c)$ for some element c of C', then it is easy to verify that

$$(D(\mathrm{C}', \infty))^{-1}(\lim(D|_{\mathrm{C}'})[U]) = \lim(D)[U]. \tag{8.1.1}$$

The sets of the form $\lim(D|_{\mathrm{C}'})[U]$ is a base for the open subsets of the space $\lim(D|_{\mathrm{C}'})$. Since C' is a cofinal subset of C, the sets of the form $\lim(D)[U]$ is also a base for the open subsets of $\lim(D)$. In this case, relation (8.1.1) shows that the mapping $D(\mathrm{C}', \infty)$ and the mapping $(D(\mathrm{C}', \infty))^{-1}$ are continuous, that is $D(\mathrm{C}', \infty)$ is a homeomorphism of the space $\lim(D)$ onto the space $\lim(D|_{\mathrm{C}'})$. □

Remark. Whenever we consider a cofinal subset C' of C we identify any point g of $\lim(D)$ with the point $D(\mathrm{C}', \infty)(g)$ of $\lim(D|_{\mathrm{C}'})$ and, therefore, we shall consider that the spaces $\lim(D)$ and $\lim(D|_{\mathrm{C}'})$ coincide.

Definitions. A subset L of $\lim(D)$ is said to be *surjective* if

$$D(c, \infty)(L) = D(c)$$

for every $c \in \mathrm{C}$.

Let $I\!P$ be a class of spaces (of weight less than or equal to τ). The inverse spectrum D is said to be *$I\!P$-factorizing with respect to a subset* L of $\lim(D)$ if for every mapping f of L into an element Y of $I\!P$ there exists an element c of C and a mapping f_c of $D(c)$ into Y such that $f_c \circ D(c, \infty)|_L = f$. In the case, where $I\!P$ is the class of all spaces (of weight $\leq \tau$), we shall say "factorizing" instead of "$I\!P$-factorizing". It is easy to see that if a spectrum D is $I\!P$-factorizing with respect to a subset of its limit space, then any cofinal subspectrum is also $I\!P$-factorizing with respect to the same subset of the limit space.

The inverse spectrum D is said to be *$I\!P$-factorizing* if D is $I\!P$-factorizing with respect to the whole limit space.

The inverse spectrum D is said to be *τ-continuous* if for every chain C' with $|\mathrm{C}'| \leq \tau$ and $c = \sup(\mathrm{C}')$ the natural mapping $D(\mathrm{C}', c)$ is an embedding of $D(c)$ into $\lim(D|_{C'})$. Whenever D is considered to be τ-continuous, then we shall consider the space $D(c)$ as a subset of the limit space $\lim(D|_{C'})$ identifying the point x of $D(c)$ with the point $D(C', c)(x)$ of the space $\lim(D|_{C'})$.

The inverse spectrum D is said to be a *τ-spectrum* if the following conditions are satisfied:

(1) The set C is τ-complete.

(2) $w(D(c)) \le \tau$, $c \in C$.

(3) The inverse spectrum D is τ-continuous.

If moreover $D(c) \in I\!P$ for every $c \in C$, we shall say that D is a *$I\!P$-τ-spectrum.*

8.1.7 Proposition. (The Spectral Theorem). *Let $I\!P$ be a class of spaces. Suppose that D_0 and D_1 are two $I\!P$-τ-spectra on C, which are $I\!P$-factorizing with respect to surjective subsets L_0 and L_1 of* $\lim(D_0)$ *and* $\lim(D_1)$, *respectively. Then, each mapping f of L_0 into L_1 is extended to a mapping $\overline{f}$ of* $\lim(D_0)$ *into* $\lim(D_1)$ *(that is, $\overline{f}(g) = f(g)$, $g \in L_0$) such that $\overline{f}$ is induced by a morphism of cofinal and τ-closed subspectra. Moreover, if f is a homeomorphism onto, then $\overline{f}$ is induced by an isomorphism onto of cofinal and τ-closed subspectra (and, therefore, $\overline{f}$ is also an homeomorphism onto).*

Proof. We define a subset R of the set $C \times C$ as follows: an element (c_1, c_0) belongs to R if there exists a mapping $f_{c_1}^{c_0}$ of $D_0(c_0)$ into $D_1(c_1)$ such that

$$f_{c_1}^{c_0} \circ D_0(c_0, \infty)|_{L_0} = D_1(c_1, \infty)|_{L_1} \circ f. \tag{8.1.2}$$

We prove that the set R satisfies the conditions of Proposition 8.1.3. First, we prove the existence condition. Let $c_1 \in C$. Consider the mapping $D_1(c_1, \infty)|_{L_1} \circ f$. Since D_0 is a $I\!P$-τ-spectrum, which is $I\!P$-factorizing with respect to the surjective subset L_0 and since $D_1(c_1) \in I\!P$ there exist an element c_0 of C and a mapping $f_{c_1}^{c_0}$ of $D_0(c_0)$ into $D_1(c_1)$ satisfying relation (8.1.2). Thus $(c_1, c_0) \in R$.

Now, we prove the majorantness condition. Suppose that $(c_1, c_0) \in R$ and let c_2 be an element of C such that $c_0 \prec^{C} c_2$. Let $f_{c_1}^{c_0}$ be the corresponding mapping of $D_0(c_0)$ into $D_1(c_1)$. Since

$$D(c_0, c_2) \circ D(c_2, \infty)|_{L_0} = D(c_0, \infty)|_{L_0},$$

the mapping

$$f_{c_1}^{c_2} \equiv f_{c_1}^{c_0} \circ D_0(c_0, c_2)$$

satisfies relation

$$f_{c_1}^{c_2} \circ D_0(c_2, \infty)|_{L_0} = D_1(c_1, \infty)|_{L_1} \circ f,$$

which means that $(c_1, c_2) \in R$.

Finally, we prove the τ-closeness condition. Suppose that C'' is a chain with $|\mathrm{C}''| \leq \tau$ and $c'' = \sup(\mathrm{C}'')$ such that for a fixed element c_0 of C, $(c, c_0) \in R$ for every $c \in \mathrm{C}''$. We need to prove that $(c'', c_0) \in R$. For every $c \in \mathrm{C}''$ we denote by $f_c^{c_0}$ the corresponding mapping of $D_0(c_0)$ into $D_1(c)$ such that relation

$$f_c^{c_0} \circ D_0(c_0, \infty)|_{L_0} = D_1(c, \infty)|_{L_1} \circ f, \tag{8.1.3}$$

is satisfied. Since the set L_0 is surjective, relation (8.1.3) imply that

$$f_c^{c_0} = D_1(c, c') \circ f_{c'}^{c_0},$$

whenever $c, c' \in \mathrm{C}''$ and $c \prec^{\mathrm{C}} c'$. Therefore, the collection

$$\{f_c^{c_0} : c \in \mathrm{C}''\}$$

is a mapping of the space $D_0(c_0)$ into the inverse spectrum $D_1|_{\mathrm{C}''}$. This mapping defines the limit mapping $\lim\{f_c^{c_0} : c \in \mathrm{C}''\}$ of the space $D_0(c_0)$ into the limit space $\lim(D_1|_{\mathrm{C}''})$. Since D_1 is τ-continuous $D_1(c'')$ is a subspace of the $\lim(D_1|_{\mathrm{C}''})$. We show that

$$\lim\{f_c^{c_0} : c \in \mathrm{C}''\}(D_0(c_0)) \subset D_1(c''). \tag{8.1.4}$$

Indeed, let

$$x \in D_0(c_0),$$

$$g = \lim\{f_c^{c_0} : c \in \mathrm{C}''\}(x),$$

$$y \in (D_0(c_0, \infty)|_{L_0})^{-1}(x) \subset L_0, \text{ and}$$

$$z = D_1(c'', \infty)|_{L_1}(f(y)) \in D_1(c'').$$

Then, for every $c \in \mathrm{C}''$ we have

$$g(c) = f_c^{c_0}(x) = D_1(c, \infty)|_{L_1}(f(y)) =$$

$$D_1(c, c'')(D_1(c'', \infty)|_{L_1}(f(y))) = D_1(c, c'')(z).$$

This equality means that $D_1(\mathrm{C}'', c'')(z) = g$, that is relation (8.1.4) is satisfied.

Now, we denote by $f_{c''}^{c_0}$ the mapping of $D_0(c_0)$ into $D_1(c'')$ for which

$$f_{c''}^{c_0}(x) = \lim\{f_c^{c_0} : c \in \mathrm{C}''\}(x), \quad x \in D_0(c_0).$$

The relation

$$f_{c''}^{c_0} \circ D_0(c_0, \infty)|_{L_0} = D_1(c'', \infty)|_{L_1} \circ f$$

is easy to verify. Therefore, $(c'', c_0) \in R$. Thus, R satisfies conditions of Proposition 8.1.3.

We denote by C' the set of all R-reflexive elements of C. By Proposition 8.1.3, C' is a cofinal and τ-closed subset of C.

For every $c \in \mathrm{C}'$ the mapping

$$f_c \equiv f_c^c : D_0(c) \to D_1(c)$$

satisfies the equality

$$f_c \circ D_0(c, \infty)|_{L_0} = D_1(c, \infty)|_{L_1} \circ f.$$

This fact implies that $f_c = D_1(c, c') \circ f_{c'}$ whenever $c, c' \in \mathrm{C}''$ and $c \prec^{\mathrm{C}} c'$, which means that the collection $\{f_c : c \in \mathrm{C}'\}$ is a morphism of the spectrum $D_0|_{\mathrm{C}'}$ into the spectrum $D_1|_{\mathrm{C}'}$. We set

$$\overline{f} = \lim\{f_c : c \in \mathrm{C}'\}.$$

It is easy to see that $\overline{f}|_{L_0} = f$.

Now we suppose that f is an homeomorphism of L_0 onto L_1. Therefore, we can consider the inverse homeomorphism f^{-1} of L_1 onto L_0. By the above there exist a cofinal and τ-closed subset C' of C and a morphism $\{f_c : c \in \mathrm{C}'\}$ of $D_0|_{\mathrm{C}'}$ into $D_1|_{\mathrm{C}'}$ such that for the mapping

$$\overline{f} \equiv \lim\{f_c : c \in \mathrm{C}'\}$$

we have $\overline{f}|_{L_0} = f$. Similarly, there exist a cofinal and τ-closed subset C″ of C and a morphism $\{h_c : c \in \mathrm{C}''\}$ of $D_1|_{\mathrm{C}''}$ into $D_0|_{\mathrm{C}''}$ such that for the mapping

$$\overline{h} \equiv \lim\{h_c : c \in \mathrm{C}''\}$$

we have $\overline{h}|_{L_1} = f^{-1}$. By Lemma 8.1.1 the set $\overline{\mathrm{C}} \equiv \mathrm{C}' \cap \mathrm{C}''$ is also a cofinal and τ-closed subset of C. Therefore, to prove that $\overline{f}$ is induced by an isomorphism onto it suffices to show that for every $c \in \overline{\mathrm{C}}$ the mapping

$$f_c : D_0(c) \to D_1(c)$$

is an homeomorphism onto. Let $c \in \overline{\mathrm{C}}$ and $x \in D_0(c)$. Since $D_0(c,\infty)|_{L_0}$ is onto there exists a point $g \in L_0$ such that $D_0(c,\infty)|_{L_0}(g) = x$. Then, using relation

$$h_c \circ D_1(c,\infty)|_{L_1} = D_0(c,\infty)|_{L_0} \circ f^{-1},$$

which is true for every $c \in \mathrm{C}''$, we have

$$(h_c \circ f_c)(x) = (h_c \circ f_c \circ D_0(c,\infty)|_{L_0})(g) =$$

$$(h_c \circ D_1(c,\infty)|_{L_1} \circ f)(g) = (D_0(c,\infty)|_{L_0} \circ f^{-1} \circ f)(g) =$$

$$D_0(c,\infty)|_{L_0}(g) = x,$$

that is the mapping $h_c \circ f_c$ is the identical mapping of $D_0(c)$. Similarly, we prove that the mapping $f_c \circ h_c$ is the identical mapping of $D_1(c)$. It is easy to see that these two facts imply that f_c is a homeomorphism onto, which completes the proof of the proposition. □

Remark. The space $\lim(D_1)$ of Proposition 8.1.7 may be not a Hausdorff space. This statement means that the extension of the mapping f (if such an extension exists) in general is not unique.

We note also that if D is $\mathbb{P}$-factorizing with respect to a surjective subset of $\lim(D)$, then in general D may be not $\mathbb{P}$-factorizing.

The next proposition gives a necessary and sufficient condition in order that a τ-spectrum, whose limit projections are onto, to be factorizing. (This proposition will not be used in the following).

8.1.8 Proposition. *Suppose that* C *is τ-complete and the limit projections of D are onto. Then, the spectrum D is factorizing if and only if for every open subset U of the space* $\lim(D)$ *there exist an element c of* C *and an open subset V of the space $D(c)$ such that $(D(c,\infty))^{-1}(V) = U$.*

Proof. First, we suppose that D satisfies the condition of the proposition and prove that D is factorizing. Let f be a mapping of $\lim(D)$ into a space Y (of weight $\leq \tau$). Denote by B^Y a base of cardinality $\leq \tau$ for the open subset of Y. By the condition of the proposition for every $V \in B^Y$ there exists an element c_V of C and an open subset U_V of $D(c_V)$ such that

$$(D(c_V,\infty))^{-1}(U_V) = f^{-1}(V).$$

Consider the subset $\{c_V : V \in B^Y\}$ of C. Since the cardinality of this subset is $\leq \tau$ by Corollary 8.1.2 there exists an element c of C, which majorates the set $\{c_V : V \in B^Y\}$.

Now, we construct a mapping f_c of $D(c)$ into Y. Let $x \in D(c)$. Since the mapping $D(c,\infty)$ is onto the set $(D(c,\infty))^{-1}(x)$ is not empty. Since

$$(D(c,\infty))^{-1}((D(c_V,c))^{-1}(U_V)) = f^{-1}(V),$$

for every $V \in B^Y$ either

$$(D(c,\infty))^{-1}(x) \subset f^{-1}(V) \text{ or}$$

$$(D(c,\infty))^{-1}(x) \cap f^{-1}(V) = \emptyset$$

and, therefore, either

$$f((D(c,\infty))^{-1}(x)) \subset V \text{ or}$$

$$f((D(c,\infty))^{-1}(x)) \cap V = \emptyset.$$

This results means that the set $f((D(c,\infty))^{-1}(x))$ is a singleton $\{y\}$ for some $y \in Y$. We set $f_c(x) = y$. The continuity of the mapping f_c follows by relation

$$(f_c)^{-1}(V) = (D(c_V,c))^{-1}(U_V),$$

which is easy to verify.

Conversely, suppose that D is factorizing. We prove that D satisfies the condition of the proposition. Indeed, let U be an open subset of $\lim(D)$. As the above space Y we consider the Sierpinski space, that is the set $\{0,1\}$ equiped with the topology $\{\{0,1\},\emptyset,\{1\}\}$. As the above mapping f we consider the characteristic mapping of U, that is, the mapping defined as follows: $f(x) = 1$ if $x \in U$ and $f(x) = 0$ if $x \in \lim(D) \setminus U$. It is easy to see that f is continuous. By assumption there exists an element $c \in \mathrm{C}$ and a mapping f_c of $D(c)$ into Y such that $f = f_c \circ D(c,\infty)$. This relation implies that

$$f^{-1}(\{1\}) = (D(\infty,c))^{-1}((f_c)^{-1}(\{1\})),$$

that is $U = (D(c,\infty))^{-1}(V)$, where $V = (f_c)^{-1}(\{1\})$. By continuity of f_c, V is an open subset of $D(c)$. Thus, the inverse spectrum D satisfies the condition of the lemma. □

8.2 The free union of spaces and factorizing τ-spectra

Agreement. In this section it is assumed that an indexed collection of non-empty spaces, denoted by $\mathbf{S}$, is given.

Definition. Any indexed collection $\{U^X : X \in \mathbf{S}\}$, where U^X is an open subset of $X \in \mathbf{S}$, is called an $\mathbf{S}$-*open* set. Therefore, two $\mathbf{S}$-open sets $\{U^X : X \in \mathbf{S}\}$ and $\{V^X : X \in \mathbf{S}\}$ coincide if $U^X = V^X$ for every $X \in \mathbf{S}$.

Let

$$\mathbf{M} \equiv \{\{U^X_\varepsilon : \varepsilon \in \tau\} : X \in \mathbf{S}\}$$

be a co-mark of $\mathbf{S}$. For every $\varepsilon \in \tau$ the $\mathbf{S}$-open set $\{U^X_\varepsilon : X \in \mathbf{S}\}$ is called a *component* of $\mathbf{M}$ or (more precisely) the ε-*component* of $\mathbf{M}$.

Definition. On the set of all co-marks of $\mathbf{S}$ we define:
(a) a partial order, denoted by $\prec^{cm}$ and
(b) an equivalence relation, denoted by $\sim^{cm}$, as follows.
For two co-marks

$$\mathbf{M}_0 \equiv \{\{U^X_{0,\delta} : \delta \in \tau\} : X \in \mathbf{S}\} \text{ and}$$

$$\mathbf{M}_1 \equiv \{\{U^X_{1,\delta} : \delta \in \tau\} : X \in \mathbf{S}\}$$

we write respectively:

(a) $\mathbf{M}_0 \prec^{cm} \mathbf{M}_1$ if for every $\delta \in \tau$ there exists an element ε of τ such that the δ-component of $\mathbf{M}_0$ coincides with the ε-component of $\mathbf{M}_1$ and

(b) $\mathbf{M}_0 \sim^{cm} \mathbf{M}_1$ if $\mathbf{M}_0 \prec^{cm} \mathbf{M}_1$ and $\mathbf{M}_1 \prec^{cm} \mathbf{M}_0$.

In the case, where $\mathbf{M}_0 \prec^{cm} \mathbf{M}_1$ we say that $\mathbf{M}_1$ is a *general co-extension* of $\mathbf{M}_0$. In this case any mapping θ of τ into itself (it is not necessary to be one-to-one) such that for every $\delta \in \tau$ the δ-component of $\mathbf{M}_0$ coincides with $\theta(\delta)$-component of $\mathbf{M}_1$ is said to be a *general indicial mapping* from $\mathbf{M}_0$ to $\mathbf{M}_1$.

We note that if $\mathbf{M}_1$ is a co-extension of $\mathbf{M}_0$ of $\mathbf{S}$, then $\mathbf{M}_0 \prec^{cm} \mathbf{M}_1$. However, in general the converse statement is not true.

Definition. On the set of all admissible families of equivalence relations on $\mathbf{S}$ we define:

(a) a partial order, denoted by $\prec_{af}$ and

(b) an equivalence relation, denoted by $\sim_{af}$, as follows.

For two admissible families

$$\mathrm{R}_0 \equiv \{\sim^s_0 : s \in \mathcal{F}\} \text{ and } \mathrm{R}_1 \equiv \{\sim^s_1 : s \in \mathcal{F}\}$$

we write respectively:

(a) $\mathrm{R}_0 \prec_{af} \mathrm{R}_1$ if R_1 is a final refinement of R_0 and

(b) $\mathrm{R}_0 \sim_{af} \mathrm{R}_1$ if $\mathrm{R}_0 \prec_{af} \mathrm{R}_1$ and $\mathrm{R}_1 \prec_{af} \mathrm{R}_0$, that is if R_1 is a final refinement of R_0 and R_0 is a final refinement of R_1.

Notation. We denote by $\mathrm{P}(\mathbf{S})$ the set of all pairs $(\mathbf{M}, \mathrm{R})$, where $\mathbf{M}$ is a co-mark of $\mathbf{S}$ and R is an $\mathbf{M}$-admissible family of equivalence relations on the indexed collectin $\mathbf{S}$.

Definition. On the set $\mathrm{P}(\mathbf{S})$ we define:

(a) a partial order, denoted by $\prec^{cm}_{af}$ and

(b) an equivalence relation on $\mathbf{S}$, denoted by $\sim^{cm}_{af}$, as follows.

For two elements $(\mathbf{M}_0, \mathrm{R}_0)$ and $(\mathbf{M}_1, \mathrm{R}_1)$ of $\mathrm{P}(\mathbf{S})$ we write respectively:

(a) $(\mathbf{M}_0, \mathrm{R}_0) \prec^{cm}_{af} (\mathbf{M}_1, \mathrm{R}_1)$ if $\mathbf{M}_0 \prec^{cm} \mathbf{M}_1$ and $\mathrm{R}_0 \prec_{af} \mathrm{R}_1$ and

(b) $(\mathbf{M}_0, \mathrm{R}_0) \sim^{cm}_{af} (\mathbf{M}_1, \mathrm{R}_1)$ if $\mathbf{M}_0 \sim^{cm} \mathbf{M}_1$ and $\mathrm{R}_0 \sim_{af} \mathrm{R}_1$.

For every $(\mathbf{M}, \mathrm{R}) \in \mathrm{P}(\mathbf{S})$ we denote by $\{(\mathbf{M}, \mathrm{R})\}$ the $\sim^{cm}_{af}$-equivalence class containing the pair $(\mathbf{M}, \mathrm{R})$.

In what follows for simplicity of notations the set $\mathrm{C}(\sim_{af}^{cm})$ will be denoted by $\mathrm{C}(\mathbf{S})$. (Recall that $\mathrm{C}(\sim_{af}^{cm})$ denotes the set of all $\sim_{af}^{cm}$-equivalence classes.)

Definition. On the set $\mathrm{C}(\mathbf{S})$ we define a partial order, denoted by $\prec^{\mathbf{S}}$, as follows: for two elements $c_0 \equiv \{(\mathbf{M}_0, \mathrm{R}_0)\}$ and $c_1 \equiv \{(\mathbf{M}_1, \mathrm{R}_1)\}$ of $\mathrm{C}(\mathbf{S})$ we write $c_0 \prec^{\mathbf{S}} c_1$ if $(\mathbf{M}_0, \mathrm{R}_0) \prec_{af}^{cm} (\mathbf{M}_1, \mathrm{R}_1)$.

It is easy to verify that $\prec^{\mathbf{S}}$ is well defined, that is the definition of $\prec^{\mathbf{S}}$ is independent of the pairs $(\mathbf{M}_0, \mathrm{R}_0)$ and $(\mathbf{M}_1, \mathrm{R}_1)$ taken from the elements c_0 and c_1 of $\mathrm{C}(\mathbf{S})$. Also, it is clear that $\prec^{\mathbf{S}}$ is actually a partial order.

8.2.1 Lemma. *The partial ordered set* $\mathrm{C}(\mathbf{S})$ *is* τ*-complete.*

Proof. First we note that the set $\mathrm{C}(\mathbf{S})$ is directed. Indeed, let

$$c_0 \equiv \{(\mathbf{M}_0, \mathrm{R}_0)\} \text{ and } c_1 \equiv \{(\mathbf{M}_1, \mathrm{R}_1)\}$$

be two elements of $\mathrm{C}(\mathbf{S})$. Then, for any element $c \equiv \{(\mathbf{M}, \mathrm{R})\}$ of $\mathrm{C}(\mathbf{S})$, where $\mathbf{M}$ is a co-extension of $\mathbf{M}_0$ and $\mathbf{M}_1$ and R is a final rerinement of R_0 and R_1, we have $c_0 \prec^{\mathbf{S}} c$ and $c_1 \prec^{\mathbf{S}} c$.

Now, we prove that $\mathrm{C}(\mathbf{S})$ is τ-complete. Let C be a directed subset of $\mathrm{C}(\mathbf{S})$ (in particular, C may be a chain) with $|C| \leq \tau$. We need to prove that there exists the lower upper bound $\sup(C)$ of C. Let

$$c_\delta \equiv \{\{(\mathbf{M}_\delta, \mathrm{R}_\delta)\} : \delta \in \tau\}$$

be an indication of the set C. Suppose that

$$\mathbf{M}_\delta = \{\{U_{\delta,\varepsilon}^X : \varepsilon \in \tau\} : X \in \mathbf{S}\} \text{ and}$$

$$\mathrm{R}_\delta = \{\sim_\delta^s : s \in \mathcal{F}\}.$$

Let ϑ be an one-to-one mapping of $\tau \times \tau$ onto τ. We define a co-mark

$$\mathbf{M} \equiv \{\{U_\eta^X : \eta \in \tau\} : X \in \mathbf{S}\}$$

of $\mathbf{S}$ setting $U_\eta^X = U_{\delta,\varepsilon}^X$, where $(\delta, \varepsilon) = \vartheta^{-1}(\eta)$.

Also, we define an admissible family

$$\mathrm{R} \equiv \{\sim^s : s \in \mathcal{F}\}$$

of equivalence relations on $\mathbf{S}$ setting

$$\sim^s = \cap\{\sim^s_\delta : \delta \in s\}.$$

We prove that the constructed pair $(\mathbf{M}, \mathrm{R})$ is an element of $\mathrm{P}(\mathbf{S})$, that is R is $\mathbf{M}$-admissible, and $\{(\mathbf{M}, \mathrm{R})\} = \sup(C)$.

First, we prove that R is $\mathbf{M}$-admissible. Let $t \in \mathcal{F}$. For every $\eta \in t$ there exists a pair $(\delta_\eta, \varepsilon_\eta)$ of elements of τ such that the ε_η-component of $\mathbf{M}_{\delta_\eta}$ coincides with the η-component of $\mathbf{M}$. Since C is directed there exists an element δ of τ such that $c_{\delta_\eta} \prec^{\mathbf{S}} c_\delta$, $\eta \in t$. Therefore, for every $\eta \in t$ there exists an element ε^η of τ such that the ε_η-component of $\mathbf{M}_{\delta_\eta}$ coincides with ε^η-component of $\mathbf{M}_\delta$. Let

$$t' = \{\varepsilon^\eta : \eta \in t\}.$$

Then, for every $X \in \mathbf{S}$ the t'-standard algebra $\mathrm{A}^X_{t'}$ of the marked space X with the mark $\mathbf{M}_\delta$ coincides with the t-standard algebra A^X_t of the marked space X with the mark $\mathbf{M}$. Therefore, $\sim^{t'}_{\mathbf{M}_\delta} = \sim^t_{\mathbf{M}}$. Since R_δ is $\mathbf{M}_\delta$-adsmissible there exists an element s' of $\mathcal{F}$ such that $\sim^{s'}_\delta \subset \sim^{t'}_{\mathbf{M}_\delta}$. Since R is a final refinement of R_δ there exists an element s of $\mathcal{F}$ such that $\sim^s \subset \sim^{t'}_\delta$ and, therefore, $\sim^s \subset \sim^{\mathrm{t}}_{\mathbf{M}}$, that is R is an $\mathbf{M}$-admissible family.

By the construction of the pair $(\mathbf{M}, \mathrm{R})$ the element $c \equiv \{(\mathbf{M}, \mathrm{R})\}$ of $\mathrm{C}(\mathbf{S})$ is an upper bound of C.

Now, we prove that c is the lower upper bound of C, that is $c = \sup(C)$. Suppose that

$$c' \equiv \{(\mathbf{M}', \mathrm{R}') \in \mathrm{C}(\mathbf{S}) \text{ and } c_\delta \prec^{\mathbf{S}} c'$$

for every $\delta \in \tau$, where

$$\mathrm{R}' = \{\sim^{\prime,s} : s \in \mathcal{F}\}.$$

Therefore, any component of $\mathbf{M}_\delta$, $\delta \in \tau$ is a component of $\mathbf{M}'$. Also, the family R' is a final refinement of all families R_δ, $\delta \in \tau$.

We need to prove that $c \prec^{\mathbf{S}} c'$. Since any component of $\mathbf{M}$, as a component of $\mathbf{M}_\delta$ for some $\delta \in \tau$ is a component of $\mathbf{M}'$, it suffices to prove that for every $t \in \mathcal{F}$ there exists an element $s \in \mathcal{F}$ such that $\sim^{\prime,s} \subset \sim^t$. Let $t \in \mathcal{F}$. For every $\delta \in t$ there exists an element s_δ of $\mathcal{F}$ such that $\sim^{\prime,s_\delta} \subset \sim^t_\delta$. Let s be an element of $\mathcal{F}$ such that $\sim^{\prime,s} \subset \sim^{\prime,s_\delta}$, $\delta \in t$. Then, we have

$$\sim^{\prime,s} \subset \cap\{\sim^{\prime,s_\delta}: \delta \in t\} \subset \cap\{\sim^{t}_{\delta}: \delta \in t\} = \sim^{t},$$

that is R$'$ is a final refinement of R, which means that $c \prec^{\mathbf{S}} c'$. Thus, $c = \sup(C)$. $\square$

8.2.2 Lemma. *Let* $(\mathbf{M}, \mathrm{R})$ *and* $(\mathbf{M}', \mathrm{R}')$ *be elements of* $\mathrm{P}(\mathbf{S})$, *which are* $\sim^{cm}_{af}$*-equivalent. Then,* $\mathrm{T}(\mathbf{M}, \mathrm{R}) = \mathrm{T}(\mathbf{M}', \mathrm{R}')$.

Proof. Let

$$\mathbf{M} \equiv \{\{U^X_\delta : \delta \in \tau\} : X \in \mathbf{S}\},$$

$$\mathbf{M}' \equiv \{\{U^{\prime,X}_\delta : \delta \in \tau\} : X \in \mathbf{S}\},$$

$$\mathrm{R} \equiv \{\sim^s : s \in \mathcal{F}\}, \text{ and}$$

$$\mathrm{R}' \equiv \{\sim^{\prime,s} : s \in \mathcal{F}\}.$$

In order to prove that the sets $\mathrm{T} \equiv \mathrm{T}(\mathbf{M}, \mathrm{R})$ and $\mathrm{T}' \equiv \mathrm{T}(\mathbf{M}', \mathrm{R}')$ coincide, it suffices to prove that the equivalence relations $\sim^{\mathrm{R}}_{\mathbf{M}}$ and $\sim^{\mathrm{R}'}_{\mathbf{M}'}$ (about notations see Section 1.2) defined on the set of all pairs (x, X), where $x \in X \in \mathbf{S}$, coincide.

Let (x, X) and (y, Y) be two pairs such that $(x, X) \sim^{\mathrm{R}}_{\mathbf{M}} (y, Y)$, that is $X \sim^s Y$ for every $s \in \mathcal{F}$ and the point x of X belongs to the set U^X_δ, $\delta \in \tau$, if and only if the point y of Y belongs to the set U^Y_δ. We prove that $(x, X) \sim^{\mathrm{R}'}_{\mathbf{M}'} (y, Y)$. Indeed, let t be an arbitrary element of $\mathcal{F}$. Since $\mathrm{R}' \prec_{af} \mathrm{R}$, R is a final refinement of R$'$. Therefore, there exists $s \in \mathcal{F}$ such that $\sim^s \subset \sim^{\prime,t}$ and hence $X \sim^{\prime,t} Y$. On the other hand, since each component of $\mathbf{M}'$ is a component of $\mathbf{M}$ we have that x belongs to $U^{\prime,X}_\delta$ if and only if y belongs to $U^{\prime,Y}_\delta$. Thus, $(x, X) \sim^{\mathrm{R}'}_{\mathbf{M}'} (y, Y)$. Similarly, we prove that if $(x, X) \sim^{\mathrm{R}'}_{\mathbf{M}'} (y, Y)$, then $(x, X) \sim^{\mathrm{R}}_{\mathbf{M}} (y, Y)$. Thus, $\sim^{\mathrm{R}}_{\mathbf{M}} = \sim^{\mathrm{R}'}_{\mathbf{M}'}$.

To prove that the topologies of the spaces T and T$'$ coincide it suffices to prove that: (a) each element of the standard base B^{T} of T is an open subset of T$'$ and (b) each element of the standard base $B^{\mathrm{T}'}$ of T$'$ is an open subset of T.

Let $U^{\mathrm{T}}_\delta(\mathbf{H})$ be an arbitrary element of the standard base B^{T} of the space T, where $\mathbf{H}$ is an element of $\mathrm{C}(\sim^s)$ for some $s \in \mathcal{F}$. There exists an element ε of τ such that $U^X_\delta = U^{\prime,X}_\varepsilon$ for every $X \in \mathbf{S}$. Also, there exist an element t of $\mathcal{F}$ such that $\sim^{\prime,t} \subset \sim^s$. Thus there exist elements $\mathbf{L}_0$,...,$\mathbf{L}_n$ of $\mathrm{C}(\sim^{\prime,t})$ such that

$$\mathbf{H} = \cup\{\mathbf{L}_i : i \in \{0, ..., n\}\}.$$

Then, as it is easy to verify,

$$U_\delta^{\mathrm{T}}(\mathbf{H}) = \cup\{U_\varepsilon^{\mathrm{T}'}(\mathbf{L}_i) : i \in \{0, ..., n\}\},$$

which means that $U_\delta^{\mathrm{T}}(\mathbf{H})$ is open in T'. Similarly, we can prove that each element of the standard base $B^{\mathrm{T}'}$ of T' is an open subset of T. □

Definition. Let $(\mathbf{M}_0, \mathrm{R}_0), (\mathbf{M}_1, \mathrm{R}_1) \in \mathrm{P}(\mathbf{S})$ such that $(\mathbf{M}_0, \mathrm{R}_0) \prec_{af}^{cm} (\mathbf{M}_1, \mathrm{R}_1)$. By the definition of the elements of $\mathrm{P}(\mathbf{S})$ we can consider the Containing Spaces $\mathrm{T}_0 \equiv \mathrm{T}(\mathbf{M}_0, \mathrm{R}_0)$ and $\mathrm{T}_1 \equiv \mathrm{T}(\mathbf{M}_1, \mathrm{R}_1)$. We define a mapping, denoted by $p_{\mathrm{T}_0}^{\mathrm{T}_1}$ and calling *natural*, of the space T_1 into the space T_0 as follows. Let $\mathbf{a}_1$ be a point of T_1 and $(x, X) \in \mathbf{a}_1$. There exists a point $\mathbf{a}_0$ of T_0 such that $(x, X) \in \mathbf{a}_0$. Then, we set $p_{\mathrm{T}_0}^{\mathrm{T}_1}(\mathbf{a}_1) = \mathbf{a}_0$.

8.2.3 Lemma. *Suppose that* $(\mathbf{M}_0, \mathrm{R}_0), (\mathbf{M}_1, \mathrm{R}_1) \in \mathrm{P}(\mathbf{S})$ *and*

$$(\mathbf{M}_0, \mathrm{R}_0) \prec_{af}^{cm} (\mathbf{M}_1, \mathrm{R}_1).$$

Let $\mathrm{T}_0 = \mathrm{T}(\mathbf{M}_0, \mathrm{R}_0)$ *and* $\mathrm{T}_1 = \mathrm{T}(\mathbf{M}_1, \mathrm{R}_1)$. *Then:*

(a) the mapping $p_{\mathrm{T}_0}^{\mathrm{T}_1}$ *is well defined (that is, it is independent of the element* (x, X) *of the above definition),*

(b) $p_{\mathrm{T}_0}^{\mathrm{T}_1}$ *is continuous and onto, and*

(c) $p_{\mathrm{T}_0}^{\mathrm{T}_1}$ *satisfies the condition*

$$i_{\mathrm{T}_0}^{\mathrm{T}_1} \circ i_{\mathrm{T}_1}^X = i_{\mathrm{T}_0}^X,$$

where $i_{\mathrm{T}_1}^X$ *and* $i_{\mathrm{T}_0}^X$ *are the natural embeddings of a space* $X \in \mathbf{S}$ *into the spaces* T_1 *and* T_0*, respectively.*

Proof. Let

$$\mathbf{M}_0 = \{\{U_{0,\delta}^X : \delta \in \tau\} : X \in \mathbf{S}\},$$

$$\mathbf{M}_1 = \{\{U_{1,\delta}^X : \delta \in \tau\} : X \in \mathbf{S}\},$$

$$\mathrm{R}_0 = \{\sim_0^s : s \in \mathcal{F}\},$$

$$\mathrm{R}_1 = \{\sim_1^s : s \in \mathcal{F}\}.$$

Let also $\mathbf{a}_1 \in \mathrm{T}_1$ and $(x, X), (y, Y) \in \mathbf{a}_1$. Denote by $\mathbf{a}_0$ the element of T_0 containing the pair (x, X). Since R_1 is a final refinement of R_0, $X \sim_1^s Y$ for every $s \in \mathcal{F}$. Also, since each component of $\mathbf{M}_0$ is a component of $\mathbf{M}_1$, $x \in U_{0,\delta}^X$ if and only if $y \in U_{0,\delta}^Y$ for every $\delta \in \tau$. Therefore, $(y, Y) \in \mathbf{a}_0$, which means that the mapping $p_{\mathrm{T}_0}^{\mathrm{T}_1}$ is well defined.

Let $\delta \in \tau$ and $\mathbf{H}$ be an element of $\mathrm{C}(\sim_0^s)$ for some $s \in \mathcal{F}$. Consider the element $U_\delta^{\mathrm{T}_0}(\mathbf{H})$ of the standard base $\mathrm{B}^{\mathrm{T}_0}$ of T_0. To prove that $p_{\mathrm{T}_0}^{\mathrm{T}_1}$ is continuous it suffices to prove that the set $(p_{\mathrm{T}_0}^{\mathrm{T}_1})^{-1}(U_\delta^{\mathrm{T}_0}(\mathbf{H}))$ is open in T_1. Since each component of $\mathbf{M}_0$ is a component of $\mathbf{M}_1$ there exists $\varepsilon \in \tau$ such that $U_\varepsilon^X = U_\delta^X$ for every $X \in \mathbf{S}$. Also, since R_1 is a final refinement of R_0 there exists an element t of $\mathcal{F}$ such that $\sim_1^t \subset \sim_0^s$. Therefore, there exist elements $\mathbf{L}_0$,...,$\mathbf{L}_n$ of $\mathrm{C}(\sim_1^t)$ such that

$$\mathbf{H} = \cup\{\mathbf{L}_i : i \in n + 1\}.$$

Then, it is easy to verify that

$$(p_{\mathrm{T}_0}^{\mathrm{T}_1})^{-1}(U_\delta^{\mathrm{T}_0}(\mathbf{H})) = \cup\{U_\varepsilon^{\mathrm{T}_1}(\mathbf{L}_i) : i \in n + 1\},$$

which means that $(p_{\mathrm{T}_0}^{\mathrm{T}_1})^{-1}(U_\delta^{\mathrm{T}_0}(\mathbf{H}))$ is open in T_1, that is, $p_{\mathrm{T}_0}^{\mathrm{T}_1}$ is continuous.

The rest conditions of the lemma can be easy to verify. □

The following lemma is easily proved.

8.2.4 Lemma. *Let* $(\mathbf{M}_0, \mathrm{R}_0)$, $(\mathbf{M}_1, \mathrm{R}_1)$, *and* $(\mathbf{M}_2, \mathrm{R}_2)$ *be elements of* $\mathrm{P}(\mathbf{S})$ *such that* $(\mathbf{M}_0, \mathrm{R}_0) \prec_{af}^{cm} (\mathbf{M}_1, \mathrm{R}_1)$ *and* $(\mathbf{M}_1, \mathrm{R}_1) \prec_{af}^{cm} (\mathbf{M}_2, \mathrm{R}_2)$. *Then,*

$$p_{\mathrm{T}_0}^{\mathrm{T}_1} \circ p_{\mathrm{T}_1}^{\mathrm{T}_2} = p_{\mathrm{T}_0}^{\mathrm{T}_2},$$

where $\mathrm{T}_0 = \mathrm{T}(\mathbf{M}_0, \mathrm{R}_0)$, $\mathrm{T}_1 = \mathrm{T}(\mathbf{M}_1, \mathrm{R}_1)$, *and* $\mathrm{T}_2 = \mathrm{T}(\mathbf{M}_2, \mathrm{R}_2)$. □

8.2.5 Lemma. *Let* $(\mathbf{M}_0, \mathrm{R}_0)$, $(\mathbf{M}'_0, \mathrm{R}'_0)$, $(\mathbf{M}_1, \mathrm{R}_1)$, *and* $(\mathbf{M}'_1, \mathrm{R}'_1)$, *be elements of* $\mathrm{P}(\mathbf{S})$ *such that* $(\mathbf{M}_0, \mathrm{R}_0) \sim_{af}^{cm} (\mathbf{M}'_0, \mathrm{R}'_0)$, $(\mathbf{M}_1, \mathrm{R}_1) \sim_{af}^{cm} (\mathbf{M}'_1, \mathrm{R}'_1)$, *and* $(\mathbf{M}_0, \mathrm{R}_0) \prec_{af}^{cm} (\mathbf{M}_1, \mathrm{R}_1)$. *Then, the natural mapping* $p_{\mathrm{T}_0}^{\mathrm{T}_1}$ *of the space* $\mathrm{T}_1 \equiv \mathrm{T}(\mathbf{M}_1, \mathrm{R}_1)$ *into the space* $\mathrm{T}_0 \equiv \mathrm{T}(\mathbf{M}_0, \mathrm{R}_0)$ *coincides with the natural mapping* $p_{\mathrm{T}'_0}^{\mathrm{T}'_1}$ *of the space* $\mathrm{T}'_1 \equiv \mathrm{T}(\mathbf{M}'_1, \mathrm{R}'_1)$ *into the space* $\mathrm{T}'_0 \equiv \mathrm{T}(\mathbf{M}'_0, \mathrm{R}'_0)$.

Proof. Let $\mathbf{a}$ be a point of the space T_1 and $(x, X) \in \mathbf{a}$. By Lemma 8.2.2, $\mathbf{a}$ is also a point of the space T'_1. By the definition of the mappings

$p^{T_1}_{T_0}$ and $p^{T'_1}_{T'_0}$ it follows that $(x, X) \in p^{T_1}_{T_0}(\mathbf{a})$ and $(x, X) \in p^{T'_1}_{T'_0}(\mathbf{a})$, which means that $p^{T_1}_{T_0}(\mathbf{a}) = p^{T'_1}_{T'_0}(\mathbf{a})$. Therefore, $p^{T_1}_{T_0} = p^{T'_1}_{T'_0}$. □

8.2.6 Lemma. *For every admissible family* R *of equivalence relations on* $\mathbf{S}$ *there exists a co-mark* $\mathbf{M}$ *of* $\mathbf{S}$ *such that the* $\mathbf{M}$*-standard family* $\mathrm{R}_{\mathbf{M}}$ *of equivalence relations on* $\mathbf{S}$ *is a final refinement of* R.

Proof. Let $\mathrm{R} = \{\sim^s : s \in \mathcal{F}\}$. Denote by θ an one-to-one mapping of $\mathcal{F}$ into itself such that:

(a) for every element s of $\mathcal{F}$, the number of $\sim^s$-equivalence classes is equal to $|\theta(s)|$,

(b) if s and t are distinct elements of $\mathcal{F}$, then $\theta(s) \cap \theta(t) = \emptyset$, and

(c) $|\tau \setminus (\cup\{\theta(s) : s \in \mathcal{F}\})| = \tau$.

Let f_s be an one-to-one mapping of the set of all $\sim^s$-equivalence classes onto the subset $\theta(s)$ of τ.

Now, we define the co-mark $\mathbf{M} \equiv \{M^X : X \in \mathbf{S}\}$ of $\mathbf{S}$. For this purpose, it suffices to define the mark M^X, $X \in \mathbf{S}$. Let $X \in \mathbf{S}$ and U be a non-empty open subset of X. By M^X we denote any mark $\{U^X_\delta : \delta \in \tau\}$ of X satisfying the following condition: if $s \in \mathcal{F}$ and $\mathbf{E}$ is the $\sim^s$-equivalence class containing X, then $U^X_\delta = \emptyset$ for every $\delta \in \theta(s) \setminus \{f_s(\mathbf{E})\}$, and $U^X_\delta = U$ for $\delta = f_s(\mathbf{E})$. It is not dificult to verify that for every $s \in \mathcal{F}$, $\sim^{\theta(s)}_{\mathbf{M}} \subset \sim^s$, which means that $\mathbf{M}$ is the required co-mark of $\mathbf{S}$. □

Notation. Let c be an element of $\mathrm{C}(\mathbf{S})$. By $\mathrm{T}(c)$ we denote the space $\mathrm{T}(\mathbf{M}, \mathrm{R})$, where $(\mathbf{M}, \mathrm{R})$ is an element of c. By Lemma 8.2.2 the space $\mathrm{T}(c)$ is independent of the element $(\mathbf{M}, \mathrm{R})$ of c.

Let $c_0 \equiv \{(\mathbf{M}_0, \mathrm{R}_0)\}$ and $c_1 \equiv \{(\mathbf{M}_1, \mathrm{R}_1)\}$ be elements of $\mathrm{C}(\mathbf{S})$ such that $c_0 \prec^{\mathbf{S}} c_1$. By $p^{c_1}_{c_0}$ we denote the natural mapping of the space $\mathrm{T}(c_1) \equiv \mathrm{T}(\mathbf{M}_1, \mathrm{R}_1)$ onto the space $\mathrm{T}(c_0) \equiv \mathrm{T}(\mathbf{M}_0, \mathrm{R}_0)$. By Lemma 8.2.5 this mapping is independent of the elements $(\mathbf{M}_0, \mathrm{R}_0)$ and $(\mathbf{M}_1, \mathrm{R}_1)$ of c_0 and c_1, respectively.

We denote by $\mathrm{D}[\mathbf{S}]$ the inverse spectrum on the set $\mathrm{C}(\mathbf{S})$, which is defined as follows: for every element c of $\mathrm{C}(\mathbf{S})$ we set $\mathrm{D}[\mathbf{S}](c) = \mathrm{T}(c)$ and for every pair (c_0, c_1) of $\prec^{\mathbf{S}}$ we set $\mathrm{D}[\mathbf{S}](c_0, c_1) = p^{c_1}_{c_0}$. Lemmas 8.2.1 and 8.2.3–8.2.5 imply that $\mathrm{D}[\mathbf{S}]$ is actually an inverse spectrum.

8.2.7 Proposition. *The inverse spectrum* $\mathrm{D}[\mathbf{S}]$ *is a* τ*-spectrum.*

Proof. Since the set $\mathrm{C}(\mathbf{S})$ is τ-complete (see Lemma 8.2.1) and all elements of the spectrum $\mathrm{D}[\mathbf{S}]$ are spaces of weight less than or equal to τ, it suffices to prove that the inverse spectrum $\mathrm{D}[\mathbf{S}]$ is τ-continuous.

Let C be a directed subset of the set $\mathrm{C}(\mathbf{S})$ with $|C| \le \tau$ and $c = \sup(C)$. Denote by $\{c_\delta : \delta \in \tau\}$ an indication of the set C and let $c_\delta = \{(\mathbf{M}_\delta, \mathrm{R}_\delta)\}$, where

$$\mathbf{M}_\delta = \{\{U^X_{\delta,\varepsilon} : \varepsilon \in \tau\} : X \in \mathbf{S}\} \text{ and}$$

$$\mathrm{R}_\delta = \{\sim^s_\delta : s \in \mathcal{F}\}.$$

Let $(\mathbf{M}, \mathrm{R})$ be an element of $\mathrm{P}(\mathbf{S})$, where

$$\mathbf{M} = \{\{U^X_\eta : \eta \in \tau\} : X \in \mathbf{S}\} \text{ and}$$

$$\mathrm{R} = \{\sim^t : t \in \mathcal{F}\},$$

such that:

(a) each component of $\mathbf{M}$ is a component of $\mathbf{M}_\delta$ for some $\delta \in \tau$ and each component of $\mathbf{M}_\delta$, $\delta \in \tau$, is a component of $\mathbf{M}$, and

(b) $\sim^s = \cap\{\sim^s_\delta : \delta \in s\}$.

Then (see the proof of Lemma 8.2.1), $c = \{(\mathbf{M}, \mathrm{R})\}$.

We prove that the mapping

$$\mathrm{D}[\mathbf{S}](C, c) : \mathrm{D}[\mathbf{S}](c) \to \lim(\mathrm{D}[\mathbf{S}]|_C)$$

is an embedding. First, we prove that $\mathrm{D}[\mathbf{S}](C, c)$ is one-to-one. Indeed, let $\mathbf{a}$ and $\mathbf{b}$ be two distinct points of $\mathrm{D}[\mathbf{S}](c)$, $g_0 = \mathrm{D}[\mathbf{S}](C, c)(\mathbf{a})$ and $g_1 = \mathrm{D}[\mathbf{S}](C, c)(\mathbf{b})$. For every $\delta \in \tau$ we also set $\mathrm{D}[\mathbf{S}](c_\delta, c)(\mathbf{a}) = \mathbf{a}_\delta$ and $\mathrm{D}[\mathbf{S}](c_\delta, c)(\mathbf{b}) = \mathbf{b}_\delta$. Since $g_0(c_\delta) = \mathbf{a}_\delta$ and $g_1(c_\delta) = \mathbf{b}_\delta$, in order to prove that the mapping $\mathrm{D}[\mathbf{S}](C, c)$ is one-to-one it suffices to prove that $\mathbf{a}_\delta \neq \mathbf{b}_\delta$ for some $\delta \in \tau$.

Let $(x, X) \in \mathbf{a}$ and $(y, Y) \in \mathbf{b}$. By the definition of the mapping $\mathrm{D}[\mathbf{S}](c_\delta, c)$, $(x, X) \in \mathbf{a}_\delta$ and $(y, Y) \in \mathbf{b}_\delta$. Since $\mathbf{a} \neq \mathbf{b}$ we have $(x, X) \notin \mathbf{b}$ and $(y, Y) \notin \mathbf{a}$. This fact means that either X and Y are not $\sim^s$-equivalent for some $s \in \mathcal{F}$ or $X \sim^s Y$ for all $s \in \mathcal{F}$ and there exists an element $\eta \in \tau$ such that either $x \in U^X_\eta$ and $y \notin U^Y_\eta$ or $x \notin U^X_\eta$ and $y \in U^Y_\eta$. Then, in the first case there exists an element $\delta \in s$ such that X and Y are not $\sim^s_\delta$-equivalent (see the definition of the equivalence relation $\sim^s$). Thus $\mathbf{a}_\delta \neq \mathbf{b}_\delta$. In the second case, there exists elements $\delta, \varepsilon \in \tau$ such that the

η-component of $\mathbf{M}$ coincides with the ε-component of $\mathbf{M}_\delta$, that is either $x \in U^X_{\delta,\varepsilon}$ and $y \notin U^Y_{\delta,\varepsilon}$ or $x \notin U^X_{\delta,\varepsilon}$ and $y \in U^Y_{\delta,\varepsilon}$. This result also means that $\mathbf{a}_\delta \neq \mathbf{b}_\delta$. Thus, the mapping $\mathrm{D}[\mathbf{S}](C,c)$ is one-to-one.

Now, we prove that the mapping $\mathrm{D}[\mathbf{S}](C,c)$ and the converse mapping

$$(\mathrm{D}[\mathbf{S}](C,c))^{-1} : \mathrm{D}[\mathbf{S}](C,c)(\mathrm{D}[\mathbf{S}](c)) \to \mathrm{D}[\mathbf{S}](c)$$

are continuous. For this purpose, it suffices to prove the following three statements, where by $\mathrm{B}^{\mathrm{T}_\delta}$, $\delta \in \tau$, we denote the standard base for the open subsets of the space $\mathrm{T}_\delta \equiv \mathrm{T}(\mathbf{M}_\delta, \mathrm{R}_\delta)$:

(1) The sets of the form

$$((\mathrm{D}[\mathbf{S}]|_C)(c_\delta, \infty))^{-1}(U),$$

where $U \in \mathrm{B}^{\mathrm{T}_\delta}$ and $\delta \in \tau$, compose a base for the open subsets of the space $\lim(\mathrm{D}[\mathbf{S}]|_C)$.

(2) The sets of the form

$$(\mathrm{D}[\mathbf{S}](c_\delta, c))^{-1}(U),$$

where $U \in \mathrm{B}^{\mathrm{T}_\delta}$ and $\delta \in \tau$, compose a base for the open subsets of the space $\mathrm{D}[\mathbf{S}](c)$.

(3) For every $\delta \in \tau$ and $U \in \mathrm{B}^{\mathrm{T}_\delta}$ the following relation is true:

$$\mathrm{D}[\mathbf{S}](C,c)((\mathrm{D}[\mathbf{S}](c_\delta, c))^{-1}(U)) =$$

$$((\mathrm{D}[\mathbf{S}])|_C)(c_\delta, \infty))^{-1}(U) \cap \mathrm{D}[\mathbf{S}](C,c)(\mathrm{D}[\mathbf{S}](c)). \tag{8.2.1}$$

The first statement follows by the definition of the topology on the space $\lim(\mathrm{D}[\mathbf{S}]|_C)$. We prove the second statement. Let $\eta \in \tau$, $t \in \mathcal{F}$, $\mathbf{L} \in \mathrm{C}(\sim^t)$, and let $\mathbf{a}$ be a point of the element $U^{\mathrm{T}}_\eta(\mathbf{L})$ of the standard base B^{T} for the open subsets of the space $\mathrm{T} \equiv \mathrm{T}(\mathbf{M}, \mathrm{R})$. It suffices to find elements $\delta, \varepsilon \in \tau$, $s \in \mathcal{F}$, and $\mathbf{H} \in \mathrm{C}(\sim^s_\delta)$ such that

$$\mathbf{a} \in (\mathrm{D}[\mathbf{S}](c_\delta, c))^{-1}(U^{\mathrm{T}_\delta}_\varepsilon(\mathbf{H})) \subset U^{\mathrm{T}}_\eta(\mathbf{L}). \tag{8.2.2}$$

Since C is a directed subset of $\mathrm{C}(\mathbf{S})$ there exist elements $\delta \in \tau$ and $s \in \mathcal{F}$ such that

$$\sim^s_\delta \subset \cap\{\sim^t_{\delta'} : \delta' \in t\} = \sim^t. \tag{8.2.3}$$

Without loss of generality we can suppose that there exists an element ε of τ such that the ε-component of $\mathbf{M}_\delta$ coincide with the η-component of $\mathbf{M}$. Let $(x, X) \in \mathbf{a}$. We denote by $\mathbf{H}$ the element of $\mathrm{C}(\sim_\delta^s)$ containing X.

We prove relation (8.2.2). Since $x \in U_\eta^X = U_\varepsilon^X$ we have

$$\mathrm{D}[\mathbf{S}](c_\delta, c)(\mathbf{a}) \in U_\varepsilon^{\mathrm{T}_\delta}(\mathbf{H}),$$

that is

$$\mathbf{a} \in (\mathrm{D}[\mathbf{S}](c_\delta, c))^{-1}(U_\varepsilon^{\mathrm{T}_\delta}(\mathbf{H})).$$

Let

$$\mathbf{b} \in (\mathrm{D}[\mathbf{S}](c_\delta, c))^{-1}(U_\varepsilon^{\mathrm{T}_\delta}(\mathbf{H})).$$

Suppose that $(y, Y) \in \mathbf{b}$ and let $\mathbf{b}_\delta$ be the element of T_δ containing the pair (y, Y). Then,

$$\mathrm{D}[\mathbf{S}](c_\delta, c)(\mathbf{b}) = \mathbf{b}_\delta$$

and, therefore, $\mathbf{b}_\delta \in U_\varepsilon^{\mathrm{T}_\delta}(\mathbf{H})$. This result means that $y \in U_\varepsilon^Y = U_\eta^Y$ and $Y \in \mathbf{H}$. Relation (8.2.3) imply that $\mathbf{H} \subset \mathbf{L}$ and, therefore, $Y \in \mathbf{L}$. Thus, $\mathbf{b} \in U_\eta^{\mathrm{T}}(\mathbf{L})$.

Finally, we prove the third statement, that is relation (8.2.1). Suppose that $\mathbf{a}_\delta \in U$,

$$\mathbf{a} \in (\mathrm{D}[\mathbf{S}](c_\delta, c))^{-1}(\mathbf{a}_\delta) \subset \mathrm{D}[\mathbf{S}](c), \quad \text{and}$$

$$\mathrm{D}[\mathbf{S}](C, c)(\mathbf{a}) = g \in \lim(\mathrm{D}[\mathbf{S}]|_C).$$

Then, we have $g(c_\delta) = \mathbf{a}_\delta$, which means that

$$g \in ((\mathrm{D}[\mathbf{S}]|_C)(c_\delta, \infty))^{-1}(U) \cap \mathrm{D}[\mathbf{S}](C, c)(\mathrm{D}[\mathbf{S}](c)).$$

Therefore, the left side of (8.2.1) is contained in the right.

Conversely, suppose that $\mathbf{a} \in U$ and

$$g \in ((\mathrm{D}[\mathbf{S}]|_C)(c_\delta, \infty))^{-1}(\mathbf{a}_\delta) \cap \mathrm{D}[\mathbf{S}](C, c)(\mathrm{D}[\mathbf{S}](c)).$$

Then, $g(c_\delta) = \mathbf{a}_\delta$. Let $\mathbf{a} \in \mathrm{D}[\mathbf{S}](c)$ such that $\mathrm{D}[\mathbf{S}](C, c)(\mathbf{a}) = g$. Then, $\mathrm{D}[\mathbf{S}](c_\delta, c)(\mathbf{a}) = \mathbf{a}_\delta$, that is, $\mathbf{a} \in (\mathrm{D}[\mathbf{S}](c_\delta, c))^{-1}(U)$ and, therefore,

$$g \in \mathrm{D}[\mathbf{S}](C,c)((\mathrm{D}[\mathbf{S}](c_\delta,c))^{-1}(U)).$$

Thus, the right side of (8.2.1) is contained in the left completing the proofs of relation (8.2.1) and the proposition. □

Notation. Let $c \equiv \{(\mathbf{M},\mathrm{R})\} \in \mathrm{C}(\mathbf{S})$. For every $X \in \mathbf{S}$ we denote by i_c^X the natural mapping i_{T}^X of X into the space $\mathrm{T} \equiv \mathrm{T}(\mathbf{M},\mathrm{R})$. Oviously, this mapping is independent of the element $(\mathbf{M},\mathrm{R})$ of c. By Lemma 8.2.3 the collection

$$\{i_c^X : c \in \mathrm{C}(\mathbf{S})\}$$

is a mapping of X into the inverse spectrum $\mathrm{D}[\mathbf{S}]$. The mapping of X into the limit space $\lim(\mathrm{D}[\mathbf{S}])$ induced by this collection is denoted by $i_{\lim(\mathrm{D}[\mathbf{S}])}^X$. For every $x \in X$ the point $i_{\lim(\mathrm{D}[\mathbf{S}])}^X(x)$ of $\lim(\mathrm{D}[\mathbf{S}])$ is denoted also by $g[x,X]$. It is easy to verify that for every $c \in \mathrm{C}(\mathbf{S})$, $g[x,X](c) = \mathbf{a}$, where $\mathbf{a}$ is the point of the space $\mathrm{D}[\mathbf{S}](c)$ containing the pair (x,X).

8.2.8 Lemma. *The following statements are true:*

(1) *For every $X \in \mathbf{S}$ the mapping $i_{\lim(\mathrm{D}[\mathbf{S}])}^X$ is an embedding of the space X into the space* $\lim(\mathrm{D}[\mathbf{S}])$.

(2) *For every $X \in \mathbf{S}$ the set $i_{\lim(\mathrm{D}[\mathbf{S}])}^X(X)$ is an open subset of* $\lim(\mathrm{D}[\mathbf{S}])$.

(3) *For distinct elements X and Y of $\mathbf{S}$ we have*

$$i_{\lim(\mathrm{D}[\mathbf{S}])}^X(X) \cap i_{\lim(\mathrm{D}[\mathbf{S}])}^Y(Y) = \emptyset.$$

Proof. We prove statement (1). Let $X \in \mathbf{S}$. Obviously, $i_{\lim(\mathrm{D}[\mathbf{S}])}^X$ is an one-to-one mapping of X onto the subset $i_{\lim(\mathrm{D}[\mathbf{S}])}^X(X)$ of $\lim(\mathrm{D}[\mathbf{S}])$. The continuity of the mapping $i_{\lim(\mathrm{D}[\mathbf{S}])}^X$ and the continuity of the converse mapping follow by the following relation:

$$i_{\lim(\mathrm{D}[\mathbf{S}])}^X(U_\eta^X) = (\mathrm{D}[\mathbf{S}](c,\infty))^{-1}(U_\eta^{\mathrm{T}}(\mathbf{H})) \cap i_{\lim(\mathrm{D}[\mathbf{S}])}^X(X),$$

where $c \equiv \{(\mathbf{M},\mathrm{R})\} \in \mathrm{C}(\mathbf{S})$, $\mathbf{M} = \{\{U_\eta^X : \eta \in \tau\} : X \in \mathbf{S}\}$, $\mathrm{T} = \mathrm{T}(\mathbf{M},\mathrm{R})$, and $\mathbf{H} \in \mathrm{C}(\mathrm{R})$. This relation is easy to verify.

We prove statement (2). Let $X \in \mathbf{S}$. First we prove that there exists an element $(\mathbf{M},\mathrm{R})$ of $\mathrm{P}(\mathbf{S})$ such that the singleton $\{X\}$ is an element of $\mathrm{C}(\mathrm{R})$.

Indeed, denote by R_X the family of equivalence relations on $\mathbf{S}$ defined as follows: if $s = \emptyset$, then we set $\sim^s = \mathbf{S} \times \mathbf{S}$ and if $s \neq \emptyset$, then two elements Y and Z of $\mathbf{S}$ are considered to be $\sim^s$-equivalent if and only if either $Y \neq X$ and $Z \neq X$ or $Y = Z = X$. It is easy to see that R_X is an admissible family. Therefore, by Lemma 8.2.6 there exists a co-mark $\mathbf{M}$ of $\mathbf{S}$ such that the $\mathbf{M}$-standard family $\mathrm{R}_{\mathbf{M}}$ of equivalence relations on $\mathbf{S}$ is a final refinement of R_X. Then, any pair $(\mathbf{M}, \mathrm{R})$, where R is an arbitrary $\mathbf{M}$-admissible family of equivalence relations on $\mathbf{S}$ satisfies the required condition.

Now let $c \equiv \{(\mathbf{M}, \mathrm{R})\}$ be an element of $\mathrm{C}(\mathbf{S})$ such that $\{X\} \in \mathrm{C}(\mathrm{R})$. Then, the subset $i^X_{\mathrm{T}}(X)$ of $\mathrm{T} \equiv \mathrm{T}(\mathbf{M}, \mathrm{R})$ is open in T. It is easy to verify that the open subset $(\mathrm{D}[\mathbf{S}](c, \infty))^{-1}(i^X_{\mathrm{T}}(X))$ of $\lim(\mathrm{D}[\mathbf{S}])$ coincides with the set $i^X_{\lim(\mathrm{D}[\mathbf{S}])}(X)$.

We prove statement (3). Let X and Y be two distinct elements of $\mathbf{S}$. Suppose that $c_0 \equiv \{(\mathbf{M}_0, \mathrm{R}_0)\}$ and $c_1 \equiv \{(\mathbf{M}_1, \mathrm{R}_1)\}$ are elements of $\mathrm{C}(\mathbf{S})$ such that $\{X\} \in \mathrm{C}(\mathrm{R}_0)$ and $\{Y\} \in \mathrm{C}(\mathrm{R}_1)$. Let $c \equiv \{(\mathbf{M}, \mathrm{R})\}$ be an element of $\mathrm{C}(\mathbf{S})$ such that $c_0 \prec^{\mathbf{S}} c$ and $c_1 \prec^{\mathbf{S}} c$. Then, $\{X\}$ and $\{Y\}$ are distinct elements of $\mathrm{C}(\mathrm{R})$. Therefore,

$$(\mathrm{D}[\mathbf{S}](c, \infty))^{-1}(i^X_{\mathrm{T}}(X)) \cap (\mathrm{D}[\mathbf{S}](c, \infty))^{-1}(i^Y_{\mathrm{T}}(Y)) =$$

$$i^X_{\lim(\mathrm{D}[\mathbf{S}])}(X) \cap i^Y_{\lim(\mathrm{D}[\mathbf{S}])}(Y) = \emptyset. \ \Box$$

8.2.9 Corollary. *The set*

$$S \equiv \cup\{i^X_{\lim(\mathrm{D}[\mathbf{S}])}(X) : X \in \mathbf{S}\}$$

is an open surjective (and, therefore, dense) subset of the space $\lim(\mathrm{D}[\mathbf{S}])$, *which is homeomorphic to the free union of elements of* $\mathbf{S}$. $\Box$

8.2.10 Proposition. *The τ-spectrum* $\mathrm{D}[\mathbf{S}]$ *is factorizing with respect to the subset S of* $\lim(\mathrm{D}[\mathbf{S}])$.

Proof. Let f be a mapping of S into a space Y. Let

$$M^Y \equiv \{W^Y_\delta : \delta \in \tau\}$$

be an indexed base for the open subsets of Y. For every $\delta \in \tau$ the collection

$$\{V^X_\delta \equiv (i^X_{\lim(\mathrm{D}[\mathbf{S}])})^{-1}(i^X_{\lim(\mathrm{D}[\mathbf{S}])}(X) \cap f^{-1}(W^Y_\delta)) : X \in \mathbf{S}\}$$

is an **S**-open set. We consider the following **S**-indexed set of τ-indexed families:

$$\mathbf{N} \equiv \{N^X \equiv \{V_\delta^X : \delta \in \tau\} : X \in \mathbf{S}\}.$$

(We note that this **S**-indexed set may be not a co-mark of **S**.) There exists an element $c \equiv \{(\mathbf{M}, \mathrm{R})\}$ of $\mathrm{C}(\mathbf{S})$ such that **M** is a co-extension of **N**. Suppose that

$$\mathbf{M} = \{\{U_\eta^X : \eta \in \tau\} : X \in \mathbf{S}\}$$

and let θ be an indicial mapping from **N** to **M**. We prove that there exists a mapping f_c of $\mathrm{D}[\mathbf{S}](c)$ into Y such that

$$f = f_c \circ \mathrm{D}[\mathbf{S}](c, \infty)|_S. \tag{8.2.4}$$

Indeed, consider a point $\mathbf{a}$ of $\mathrm{D}[\mathbf{S}](c)$. Since the set S is surjective we have

$$S \cap (\mathrm{D}[\mathbf{S}](c, \infty))^{-1}(\mathbf{a}) \neq \emptyset.$$

Let $\eta \in \theta(\tau)$. If $\mathbf{a} \in U_\eta^{\mathrm{T}}(\mathrm{T})$ (Here, we consider the space T as an element of $\mathrm{C}^\diamondsuit(\mathrm{R})$.), then

$$S \cap (\mathrm{D}[\mathbf{S}](c, \infty))^{-1}(\mathbf{a}) \subset f^{-1}(W_{\theta^{-1}(\eta)}^Y)$$

and if $\mathbf{a} \notin U_\eta^{\mathrm{T}}(\mathrm{T})$, then

$$(S \cap (\mathrm{D}[\mathbf{S}](c, \infty))^{-1}(\mathbf{a})) \cap f^{-1}(W_{\theta^{-1}(\eta)}^Y) = \emptyset.$$

Therefore, for every $\mathbf{a} \in \mathrm{D}[\mathbf{S}](c)$ and $\delta \in \tau$ we have that either

$$f(S \cap (\mathrm{D}[\mathbf{S}](c, \infty))^{-1}(\mathbf{a})) \subset W_\delta^Y \quad \text{or}$$

$$f(S \cap (\mathrm{D}[\mathbf{S}](c, \infty))^{-1}(\mathbf{a})) \cap W_\delta^Y = \emptyset.$$

This means that

$$f(S \cap (\mathrm{D}[\mathbf{S}](c, \infty))^{-1}(\mathbf{a})) = \{y\},$$

where y is a point of Y. In this case we set $f_c(\mathbf{a}) = y$. Relation (8.2.4) is easy to verify. The continuity of f_c follows by relation

$$(f_c)^{-1}(U_\delta^Y) = U_{\theta(\delta)}^{\mathrm{T}}(\mathrm{T}).$$

The proof of this relation can be easily done using the construction of the mapping f_c. □

Summarizing Propositions 8.2.7 and 8.2.10 and Corollary 8.2.9 we obtain the following proposition.

8.2.11 Proposition. *The inverse spectrum* $\mathrm{D}[\mathbf{S}]$ *has the following properties:*

(1) $\mathrm{D}[\mathbf{S}]$ *is a* τ*-spectrum.*

(2) *The limit space* $\lim(\mathrm{D}[\mathbf{S}])$ *contains an open surjective subset* S*, which is homeomorphic to the free union of elements of* $\mathbf{S}$.

(3) $\mathrm{D}[\mathbf{S}]$ *is factorizing with respect to* S. □

The Spectral Theorem implies immediately the following proposition. This proposition can be considered as the "spectral determination" of the Containing Spaces corresponding to a given indexed collection of spaces: the Containing Spaces "appear" here without any concrete construction of these spaces.

8.2.12 Proposition. *Suppose that* D *is an inverse spectrum on the set* $\mathrm{C}(\mathbf{S})$ *having the following properties:*

(1) D *is a* τ*-spectrum,*

(2) *The limit space* $\lim(D)$ *contains a surjective subset* L*, which is homeomorphic to the free union of elements of* $\mathbf{S}$*, and*

(3) D *is factorizing with respect to* L.

Then, the inverse spectra D *and* $\mathrm{D}[\mathbf{S}]$ *contain isomorphic cofinal and* τ*-closed subspectra. (In particular, this implies that the space* $\lim(D)$ *is homeomorphic to the space* $\lim(\mathrm{D}[\mathbf{S}])$ *and* L *is open in* $\lim(D)$*). Therefore,* D *contains a cofinal and* τ*-closed subspectrum, whose elements are Containing Spaces corresponding to the indexed collection* $\mathbf{S}$. □

The following proposition shows the "covering property" of the Containing Spaces.

8.2.13 Proposition. *Suppose that* h_X, $X \in \mathbf{S}$, *is an embedding of* X *into a space* Y *(of weight* $\leq \tau$*). Then, there exist an element* c *of* $\mathrm{C}(\mathbf{S})$ *and a mapping* f *of the Containing Space* $\mathrm{T}(c)$) *into the space* Y *such that the following relation is satisfied:*

$$f \circ i^X_{\mathrm{T}(c)} = h_X, \quad X \in \mathbf{S}, \tag{8.2.5}$$

where $i^X_{\mathrm{T}(c)}$ is the natural embedding of X into $\mathrm{T}(c)$.

Proof. Consider the τ-spectrum $\mathrm{D}[\mathbf{S}]$. This spectrum is factorizing with respect to the subset S of $\lim(\mathrm{D}[\mathbf{S}])$. Let h be a mapping of S into Y defined as follows: if $x \in X \in \mathbf{S}$, then $h(g[x, X]) = h_X(x)$. It is easy to verify that this mapping is continuous. Therefore, there exists an element c of $\mathrm{C}(\mathbf{S})$ and a mapping f of $\mathrm{T}(c)$ into Y such that

$$h = f \circ \mathrm{D}[\mathbf{S}](c, \infty)|_S.$$

Then, for every $X \in \mathbf{S}$ we have

$$h \circ i^X_S = f \circ \mathrm{D}[\mathbf{S}](c, \infty)|_S \circ i^X_S,$$

where i^X_S is the mapping of X into S such that $i^X_S(x) = g[x, X]$, $x \in X$. This relation and the equalities $i^X_{\mathrm{T}(c)} = \mathrm{D}[\mathbf{S}](c, \infty)|_S \circ i^X_S$ and $h \circ i^X_S = h_X$ imply relation (8.2.5). $\square$

8.3 Saturated classes and factorizing τ-spectra

Definitions. Let $\mathbf{S}$ be an indexed collection of spaces and $\mathrm{C}_0(\mathbf{S})$ a cofinal and τ-closed subset of the set $\mathrm{C}(\mathbf{S})$. (In particular, $\mathrm{C}_0(\mathbf{S})$ may coincide with $\mathrm{C}(\mathbf{S})$).

A subset C of $\mathrm{C}_0(\mathbf{S})$ is said to be a *zero-type* subset if there exists an element $c_0 \in C$ such that $\{c \in \mathrm{C}_0(\mathbf{S}) : c_0 \prec^{\mathbf{S}} c\} \subset C$.

A subset C of $\mathrm{C}_0(\mathbf{S})$ is said to be a *first-type* subset if there exists a co-mark $\mathbf{M}^+$ of $\mathbf{S}$ satisfying the following condition: for every co-mark $\mathbf{M}$ of $\mathbf{S}$ with $\mathbf{M}^+ \prec^{cm} \mathbf{M}$ there exists an $\mathbf{M}$-admissible family R^+ of equivalence relations on $\mathbf{S}$ such that for every admissible family R of equivalence relations on $\mathbf{S}$ with $\mathrm{R}^+ \prec_{af} \mathrm{R}$ we have $c \equiv \{(\mathbf{M}, \mathrm{R})\} \in \mathrm{C}_0(\mathbf{S})$. (Therefore, $(\mathbf{M}^+, \mathrm{R}^+) \in \mathrm{C}_0(\mathbf{S})$.)

A subset C of $\mathrm{C}_0(\mathbf{S})$ is said to be a *second-type* subset if C contains a cofinal and τ-closed subset of $\mathrm{C}_0(\mathbf{S})$.

Obviously, any zero-type subset of $\mathrm{C}_0(\mathbf{S})$ is also a first-type subset and a second-type subset of $\mathrm{C}_0(\mathbf{S})$.

Below, we define some subsets of $\mathrm{C}(\mathbf{S})$, which will play the role of the set $\mathrm{C}_0(\mathbf{S})$ of the preceding definition.

Notations. Let $\mathbf{S}$ be an indexed collection of spaces. For every restriction $\mathbf{Q}$ of $\mathbf{S}$ we denote by $\mathrm{C}_{\mathbf{Q}}(\mathbf{S})$ the subset of $\mathrm{C}(\mathbf{S})$ consisting of all elements c such that for some element $(\mathbf{M}, \mathrm{R})$ of c the family R is $(\mathbf{M}, \mathbf{Q})$-admissible.

Let $\mathbf{N}$ be an $\mathbf{S}$-indexed set of τ-indexed families. (In particular, $\mathbf{N}$ may be a co-mark of $\mathbf{S}$.) We denote by $\mathrm{C}_{\mathbf{N}}(\mathbf{S})$ the subset of $\mathrm{C}(\mathbf{S})$ consisting of all elements c such that for some (and, therefore, for every) element $(\mathbf{M}, \mathrm{R})$ of c, the co-mark $\mathbf{M}$ is a general co-extension of $\mathbf{N}$.

For every restriction $\mathbf{Q}$ of $\mathbf{S}$ and a co-indication $\mathbf{N}$ of a co-p-base $\mathbf{B}$ for $\mathbf{Q}$ in $\mathbf{S}$ we denote by $\mathrm{C}_{\mathbf{Q},\mathbf{N}}(\mathbf{S})$ the set $\mathrm{C}_{\mathbf{Q}}(\mathbf{S}) \cap \mathrm{C}_{\mathbf{N}}(\mathbf{S})$.

For every $(\Phi, \Lambda, \prec)$-co-structure $\mathbf{\Psi}$ on $\mathbf{S}$ we set

$$\mathrm{C}_{\mathbf{\Psi}}(\mathbf{S}) = \cap\{\mathrm{C}_{\mathbf{\Psi}(\lambda)}(\mathbf{S}) : \lambda \in \Lambda\}.$$

The next lemma shows that in the above definition of the subset $\mathrm{C}_{\mathbf{Q}}(\mathbf{S})$ of $\mathrm{C}(\mathbf{S})$ as the element $(\mathbf{M}, \mathrm{R})$ of c we can take any element of c.

8.3.1 Lemma. *Let $\mathbf{S}$ be an indexed collection of spaces and $\mathbf{Q}$ a restriction of $\mathbf{S}$. Suppose that c is an element of $\mathrm{C}(\mathbf{S})$ and $(\mathbf{M}_0, \mathrm{R}_0)$ is an element of c such that R_0 is $(\mathbf{M}_0, \mathbf{Q})$-admissible. Then, for every element $(\mathbf{M}, \mathrm{R})$ of c the family R is $(\mathbf{M}, \mathbf{Q})$-admissible.*

Proof. Suppose that

$$\mathrm{R}_0 = \{\sim_0^s : s \in \mathcal{F}\}$$

and let $(\mathbf{M}, \mathrm{R})$ be an arbitrary element of c, where

$$\mathrm{R} = \{\sim^s : s \in \mathcal{F}\}.$$

Consider the families

$$\mathrm{R}_0|_{\mathbf{Q}} = \{\sim_0^s|_{\mathbf{Q}} : s \in \mathcal{F}\} \quad \text{and} \quad \mathrm{R}|_{\mathbf{Q}} = \{\sim^s|_{\mathbf{Q}} : s \in \mathcal{F}\}$$

of equivalence relations on the indexed collection $\mathbf{Q}$ of spaces. (About notations see Section 1.3).

We need to prove that the family R is $(\mathbf{M}, \mathbf{Q})$-admissible, that is $\mathrm{R}|_{\mathbf{Q}}$ is $\mathbf{M}|_{\mathbf{Q}}$-admissible. Let $s \in \mathcal{F} \setminus \{\emptyset\}$. For every $\varepsilon \in s$ there exists $\delta_\varepsilon \in \tau$ such that the ε-component of $\mathbf{M}$ coincides with the δ_ε-component of $\mathbf{M}_0$. We set $s_0 = \{\delta_\varepsilon : \varepsilon \in s\}$. Then, for every $X \in \mathbf{S}$ the s_0-algebra of the marked space X with the mark $\mathbf{M}_0(X)$ coincides with the s-algebra of the marked space X with the mark $\mathbf{M}(X)$. Therefore, the s_0-algebra of the marked space $\mathbf{Q}(X)$ with the mark $\mathbf{M}_0|_{\mathbf{Q}}(\mathbf{Q}(X))$ coincides with the s-algebra of the marked space $\mathbf{Q}(X)$ with the mark $\mathbf{M}|_{\mathbf{Q}}(\mathbf{Q}(X))$, which means that $\sim^{s_0}_{\mathbf{M}_0|_{\mathbf{Q}}} = \sim^{s}_{\mathbf{M}|_{\mathbf{Q}}}$. Since the family R_0 is $(\mathbf{M}_0, \mathbf{Q})$-admissible there exists an element t_0 of $\mathcal{F}$ such that $\sim^{t_0}_0 |_{\mathbf{Q}} \subset \sim^{s_0}_{\mathbf{M}_0|_{\mathbf{Q}}}$. Since the family R is a final refinement of R_0 there exists an element t of $\mathcal{F}$ such that $\sim^t \subset \sim^{t_0}_0$ and, therefore, $\sim^t |_{\mathbf{Q}} \subset \sim^{t_0}_0 |_{\mathbf{Q}}$. Thus, $\sim^t |_{\mathbf{Q}} \subset \sim^{s}_{\mathbf{M}|_{\mathbf{Q}}}$, which means that R is $(\mathbf{M}, \mathbf{Q})$-admissible. □

8.3.2 Lemma. *Let* $\mathbf{S}$ *be an indexed collection of spaces and* $\mathbf{Q}$ *a restriction of* $\mathbf{S}$. *Then,* $\mathrm{C}_{\mathbf{Q}}(\mathbf{S})$ *is a cofinal and* τ*-closed subset of* $\mathrm{C}(\mathbf{S})$.

Proof. First we prove that $\mathrm{C}_{\mathbf{Q}}(\mathbf{S})$ is a cofinal subset of $\mathrm{C}(\mathbf{S})$. Let $c_0 \equiv \{(\mathbf{M}_0, \mathrm{R}_0)\}$ be an element of $\mathrm{C}(\mathbf{S})$. Denote by R an admissible family of equivalence relations on $\mathbf{S}$ such that R is a final refinement of R_0 and the family $\mathrm{R}|_{\mathbf{Q}}$ is a final refinement of $\mathrm{R}_{\mathbf{M}_0|_{\mathbf{Q}}}$. Then, by Lemma 1.3.1, the family R is $(\mathbf{M}_0, \mathbf{Q})$-admissible and, therefore, by Lemma 8.3.1 the element c of $\mathrm{C}(\mathbf{S})$ containing $(\mathbf{M}_0, \mathrm{R})$ belongs to $\mathrm{C}_{\mathbf{Q}}(\mathbf{S})$. Obviously, $c_0 \prec^{\mathbf{S}} c$. Thus, the set $\mathrm{C}_{\mathbf{Q}}(\mathbf{S})$ is a cofinal subset of $\mathrm{C}(\mathbf{S})$.

Now we prove that $\mathrm{C}_{\mathbf{Q}}(\mathbf{S})$ is τ-closed. Let C be a directed subset of $\mathrm{C}_{\mathbf{Q}}(\mathbf{S})$ (in particular, C may be a chain) with $|C| \le \tau$ and $c = \sup(C)$. We need to prove that $c \in \mathrm{C}_{\mathbf{Q}}(\mathbf{S})$. Let $\{c_\delta : \delta \in \tau\}$ be an indication of C, where

$$c_\delta = \{(\mathbf{M}_\delta, \mathrm{R}_\delta)\} \quad \text{and} \quad \mathrm{R}_\delta = \{\sim^s_\delta : s \in \mathcal{F}\}, \quad \delta \in \tau.$$

Denote by $\mathbf{M}$ a co-mark of $\mathbf{S}$ such that each component of any co-mark $\mathbf{M}_\delta$, $\delta \in \tau$, is a component of $\mathbf{M}$ and each component of $\mathbf{M}$ is a component of $\mathbf{M}_\delta$ for some $\delta \in \tau$. Also, consider the family $\mathrm{R} \equiv \{\sim^s : s \in \mathcal{F}\}$, such that $\sim^s = \cap\{\sim^s_\delta : \delta \in s\}$. Then, $c = \{(\mathbf{M}, \mathrm{R})\}$ (see the proof of Lemma 8.2.1). We need to prove that R is $(\mathbf{M}, \mathbf{Q})$-admissible, that is $\mathrm{R}|_{\mathbf{Q}}$ is $\mathbf{M}|_{\mathbf{Q}}$-admissible.

Let $s \in \mathcal{F} \setminus \{\emptyset\}$. For every $\varepsilon \in s$ there exist two elements δ_ε and $\eta_{\delta,\varepsilon}$ of τ such that the ε-component of $\mathbf{M}$ coincides with the $\eta_{\delta,\varepsilon}$-component of $\mathbf{M}_{\delta_\varepsilon}$. Let c_δ be an element of C such that $c_{\delta_\varepsilon} \prec^{\mathbf{S}} c_\delta$, $\varepsilon \in s$. For every $\varepsilon \in s$ there exists an element η_ε of τ such that the $\eta_{\delta,\varepsilon}$-component of $\mathbf{M}_{\delta_\varepsilon}$ coincides with the η_ε-component of $\mathbf{M}_\delta$ and, therefore, the ε-component of $\mathbf{M}$ coincides with the η_ε-component of $\mathbf{M}_\delta$.

We put $s_\delta = \{\eta_\varepsilon : \varepsilon \in s\}$. Then, for every $X \in \mathbf{S}$ the s_δ-algebra of the marked space X with the mark $\mathbf{M}_\delta(X)$ coincides with the s-algebra of the marked space X with the mark $\mathbf{M}(X)$. Therefore, the s_δ-algebra of the marked space $\mathbf{Q}(X)$ with the mark $\mathbf{M}_\delta|_\mathbf{Q}(\mathbf{Q}(X))$ coincides with the s-algebra of the marked space $\mathbf{Q}(X)$ with the mark $\mathbf{M}|_\mathbf{Q}(\mathbf{Q}(X))$, which means that $\sim^{s_\delta}_{\mathbf{M}_\delta|_\mathbf{Q}} = \sim^{s}_{\mathbf{M}|_\mathbf{Q}}$. Since R_δ is $(\mathbf{M}_\delta, \mathbf{Q})$-admissible there exists an element t_δ of $\mathcal{F}$ such that $\sim^{t_\delta}_\delta|_\mathbf{Q} \subset \sim^{s_\delta}_{\mathbf{M}_\delta|_\mathbf{Q}}$. Since the family R is a final refinement of R_δ there exists an element t of $\mathcal{F}$ such that $\sim^t \subset \sim^{t_\delta}_\delta$ and, therefore, $\sim^t|_\mathbf{Q} \subset \sim^{t_\delta}_\delta|_\mathbf{Q}$. Thus, $\sim^t|_\mathbf{Q} \subset \sim^s_{\mathbf{M}|_\mathbf{Q}}$, which means that R is $(\mathbf{M}, \mathbf{Q})$-admissible. □

8.3.3 Corollary. *Let $\mathbf{S}$ be an indexed collection of spaces, $\mathbf{N}$ an $\mathbf{S}$-indexed set of τ-indexed families, and $\mathbf{\Psi}$ a $(\Phi, \Lambda, \prec)$-co-structure on $\mathbf{S}$. Then, the subsets $\mathrm{C}_\mathbf{N}(\mathbf{S})$ and $\mathrm{C}_\mathbf{\Psi}(\mathbf{S})$ of $\mathrm{C}(\mathbf{S})$ are cofinal and τ-closed. Moreover, if $\mathbf{Q}$ is a restriction of $\mathbf{S}$ and $\mathbf{N}$ a co-p-base for $\mathbf{Q}$ in $\mathbf{S}$, then $\mathrm{C}_{\mathbf{Q},\mathbf{N}}(\mathbf{S})$ is also a cofinal and τ-closed subset of $\mathrm{C}(\mathbf{S})$.*

Proof. It is easy to see that $\mathrm{C}_\mathbf{N}(\mathbf{S})$ is a cofinal and τ-closed subset of $\mathrm{C}(\mathbf{S})$. Then, Lemmas 8.1.1 and 8.3.2 imply that $\mathrm{C}_{\mathbf{Q},\mathbf{N}}(\mathbf{S})$ and $\mathrm{C}_\mathbf{\Psi}(\mathbf{S})$ are also cofinal and τ-closed subsets of $\mathrm{C}(\mathbf{S})$. □

8.3.4 Lemma. *Let $\mathbf{S}$ be an indexed collection of spaces, and c_0 and c_1 two elements of $\mathrm{C}(\mathbf{S})$ such that $c_0 \prec^\mathbf{S} c_1$. Then, for every element $(\mathbf{M}_0, \mathrm{R}_0)$ of c_0 there exists an element $(\mathbf{M}_1, \mathrm{R}_1)$ of c_1 such that $\mathbf{M}_1$ is a co-extension of $\mathbf{M}_0$.*

Proof. Suppose that $(\mathbf{M}_0, \mathrm{R}_0)$ be an element of c_0. Consider an arbitrary element $(\mathbf{M}, \mathrm{R})$ of c_1 and let θ be a general indicial mapping from $\mathbf{M}_0$ to $\mathbf{M}$. Consider the set, which is the union of the set $\tau \setminus \theta(\tau)$ and the set of all pairs $(\delta, \varepsilon) \in \tau \times \tau$ such that $\theta(\delta) = \varepsilon$. Denote by ϑ an one-to-one mapping of this set onto τ. We construct a co-mark $\mathbf{M}_1$ of $\mathbf{S}$ considering that for every $\eta \in \tau$ the η-component of $\mathbf{M}_1$ coincides with the $\vartheta^{-1}(\eta)$-component of $\mathbf{M}$ if $\vartheta^{-1}(\eta) \in \tau \setminus \theta(\tau)$ and with ε-component of $\mathbf{M}$ if $\vartheta^{-1}(\eta) = (\delta, \varepsilon)$. It is easy to verify that $\mathbf{M}_1$ is a co-extension of $\mathbf{M}_0$. Also, any component of $\mathbf{M}$ is a component of $\mathbf{M}_1$ and conversely, that is $\mathbf{M}_1 \sim^{cm} \mathbf{M}$.

Now, we set $\mathrm{R}_1 = \mathrm{R}$. Since R_1 is $\mathbf{M}$-admissible and $\mathbf{M}_1 \sim^{cm} \mathbf{M}$, the family R_1 is also $\mathbf{M}_1$-admissible. Therefore, $(\mathbf{M}_1, \mathrm{R}_1) \in \mathrm{P}(\mathbf{S})$. By the construction, $(\mathbf{M}_1, \mathrm{R}_1) \sim^\mathbf{S} (\mathbf{M}, \mathrm{R})$, that is $(\mathbf{M}_1, \mathrm{R}_1) \in c_1$. □

It is easy to verify the following lemma.

8.3.5 Lemma. *Suppose that* R_0 *and* R_1 *are two admissible families of equivalence relations on an indexed collection* $\mathbf{S}$ *of spaces such that* $\mathrm{R}_0 \sim_{af} \mathrm{R}_1$. *Then,* $\mathrm{C}^{\diamondsuit}(\mathrm{R}_0) = \mathrm{C}^{\diamondsuit}(\mathrm{R}_1)$. $\square$

Notations. Let $\mathbf{S}$ be an indexed collecion of spaces and c an element of $\mathrm{C}(\mathbf{S})$. By Lemma 8.3.5 the set $\mathrm{C}^{\diamondsuit}(\mathrm{R})$ is independent of the element $(\mathbf{M}, \mathrm{R})$ of c. We denote this set by $\mathrm{C}^{\diamondsuit}(c)$.

For every $\mathbf{L} \in \mathrm{C}^{\diamondsuit}(c)$ we denote by $\mathrm{T}(c)(\mathbf{L})$ the subset of $\mathrm{T}(c)$ consisting of all points $\mathbf{a}$ such that for some (and, therefore, for every) element (x, X) of $\mathbf{a}$ we have $X \in \mathbf{L}$. It is easy to see that for every $(\mathbf{M}, \mathrm{R}) \in c$ we have $\mathrm{T}(\mathbf{M}, \mathrm{R}, \mathbf{L}) = \mathrm{T}(c)(\mathbf{L})$.

Let $\mathbf{Q}$ be a restriction of $\mathbf{S}$ and $\mathbf{L} \in \mathrm{C}^{\diamondsuit}(c)$. We denote by $\mathrm{T}(c)|_{\mathbf{Q}}$ (respectively, by $\mathrm{T}(c)(\mathbf{L}|_{\mathbf{Q}})$) the subset $\mathrm{T}(\mathbf{M}|_{\mathbf{Q}}, \mathrm{R}|_{\mathbf{Q}})$ (respectively, the subset $\mathrm{T}(\mathbf{M}|_{\mathbf{Q}}, \mathrm{R}|_{\mathbf{Q}}, \mathbf{L}|_{\mathbf{Q}})$) of $\mathrm{T}(c) = \mathrm{T}(\mathbf{M}, \mathrm{R})$, where $(\mathbf{M}, \mathrm{R})$ is an element of c.

8.3.6 Lemma. *Let* $\mathbf{S}$ *be an indexed collection of spaces,* c *an element of* $\mathrm{C}(\mathbf{S})$, $\mathbf{L}$ *an element of* $\mathrm{C}^{\diamondsuit}(c)$, *and* $\mathbf{Q}$ *a restriction of* $\mathbf{S}$. *Then, the sets* $\mathrm{T}(c)|_{\mathbf{Q}}$ *and* $\mathrm{T}(c)(\mathbf{L}|_{\mathbf{Q}})$ *are well defined, that is they are independent of the element* $(\mathbf{M}, \mathrm{R})$ *of* c *(considered in the definition of these sets).*

Proof. Let $(\mathbf{M}, \mathrm{R}), (\mathbf{M}', \mathrm{R}') \in c$. Then,

$$(\mathbf{M}|_{\mathbf{Q}}, \mathrm{R}|_{\mathbf{Q}}), (\mathbf{M}'|_{\mathbf{Q}}, \mathrm{R}'|_{\mathbf{Q}}) \in \mathrm{C}(\mathbf{Q}).$$

Moreover, it is easy to see that

$$(\mathbf{M}|_{\mathbf{Q}}, \mathrm{R}|_{\mathbf{Q}}) \sim^{\mathbf{Q}} (\mathbf{M}'|_{\mathbf{Q}}, \mathrm{R}'|_{\mathbf{Q}}).$$

By Lemma 8.2.2 this relation implies that

$$\mathrm{T}(\mathbf{M}|_{\mathbf{Q}}, \mathrm{R}|_{\mathbf{Q}}) = \mathrm{T}(\mathbf{M}'|_{\mathbf{Q}}, \mathrm{R}'|_{\mathbf{Q}}).$$

The last relation implies also that

$$\mathrm{T}(\mathbf{M}|_{\mathbf{Q}}, \mathrm{R}|_{\mathbf{Q}}, \mathbf{L}|_{\mathbf{Q}}) = \mathrm{T}(\mathbf{M}'|_{\mathbf{Q}}, \mathrm{R}'|_{\mathbf{Q}}, \mathbf{L}|_{\mathbf{Q}}). \ \square$$

Notations. Let $\mathbf{S}$ be an indexed collection of spaces,

$$\mathbf{N} \equiv \{\{U_\delta^X : \delta \in \tau\} : X \in \mathbf{S}\}$$

an $\mathbf{S}$-indexed set of τ-indexed families, and $c \in \mathrm{C}_{\mathbf{N}}(\mathbf{S})$. Consider an element $\mathbf{H}$ of $\mathrm{C}^{\diamondsuit}(c)$ and let $\delta \in \tau$. We denote by $U^{\mathrm{T}(c)}_{\delta,\mathbf{N}}(\mathbf{H})$ the set of all points $\mathbf{a}$ of $\mathrm{T}(c)$ such that for some (and, therefore, for every) element (x, X) of $\mathbf{a}$ we have $X \in \mathbf{H}$ and $x \in U^X_\delta$.

We denote by $\mathrm{B}^{\mathrm{T}(c)}_{\diamondsuit,\mathbf{N}}$ the set of all subsets of $\mathrm{T}(c)$ of the above defined form $U^{\mathrm{T}(c)}_{\delta,\mathbf{N}}(\mathbf{H})$.

For every $\mathbf{L} \in \mathrm{C}^{\diamondsuit}(c)$ we denote by $\mathrm{B}^{\mathbf{L}}_{\diamondsuit,\mathbf{N}}$ the set of all elements $U^{\mathrm{T}(c)}_{\delta,\mathbf{N}}(\mathbf{H})$ of $\mathrm{B}^{\mathrm{T}(c)}_{\diamondsuit,\mathbf{N}}$ such that $\mathbf{H} \subset \mathbf{L}$.

8.3.7 Lemma. *Let $\mathbf{S}$ be an indexed collection of spaces. The following statements are true:*

(1) *Suppose that $\mathbf{N}$ is a co-mark of $\mathbf{S}$ and c is an element of $\mathrm{C}_{\mathbf{N}}(\mathbf{S})$. Then the set $\mathrm{B}^{\mathrm{T}(c)}_{\diamondsuit,\mathbf{N}}$ is a base for the open subsets of $\mathrm{T}(c)$. Therefore, for every $\mathbf{L} \in \mathrm{C}^{\diamondsuit}(c)$ the set $\mathrm{B}^{\mathbf{L}}_{\diamondsuit,\mathbf{N}}$ is a base for the open subsets of $\mathrm{T}(c)(\mathbf{L})$.*

(2) *Suppose that $\mathbf{Q}$ is a restriction of $\mathbf{S}$, $\mathbf{N}$ is a co-indication of a co-p-base $\mathbf{B}$ for $\mathbf{Q}$ in $\mathbf{S}$, and c is an element of $\mathrm{C}_{\mathbf{Q},\mathbf{N}}(\mathbf{S})$. Then the set $\mathrm{B}^{\mathrm{T}(c)}_{\diamondsuit,\mathbf{N}}$ is p-base for $\mathrm{T}(c)|_{\mathbf{Q}}$ in $\mathrm{T}(c)$. Therefore, for every $\mathbf{H}, \mathbf{L} \in \mathrm{C}^{\diamondsuit}(c)$ with $\mathbf{H} \subset \mathbf{L}$ the set $\mathrm{B}^{\mathbf{H}}_{\diamondsuit,\mathbf{N}}$ is a p-base for $\mathrm{T}(c)(\mathbf{H}|_{\mathbf{Q}})$ in $\mathrm{T}(c)(\mathbf{L})$.*

Proof. We prove statement (1). Statement (2) is proved similarly.

Let $(\mathbf{M}, \mathrm{R}) \in c$. As in the proof of Lemma 8.2.1 we can prove that R is an $\mathbf{N}$-admissible family of equivalence relations on $\mathbf{S}$. Therefore, $c' \equiv \{(\mathbf{N}, \mathrm{R})\} \in \mathrm{C}(\mathbf{S})$. Obviously, $c' \prec^{\mathbf{S}} c$. By Lemma 8.3.4 there exists an element $(\mathbf{M}', \mathrm{R}')$ of c such that $\mathbf{M}'$ is a co-extension of $\mathbf{N}$. Let θ be an indicial mapping of this co-extension. Then, it is easy to verify that the $(\diamondsuit, \theta(\tau))$-standard base $\mathrm{B}^{\mathrm{T}'}_{\diamondsuit,\theta(\tau)}$ for the space $\mathrm{T}' \equiv \mathrm{T}(\mathbf{M}', \mathrm{R}')$ coincides with the set $\mathrm{B}^{\mathrm{T}(c)}_{\diamondsuit,\mathbf{N}}$, which means that $\mathrm{B}^{\mathrm{T}(c)}_{\diamondsuit,\mathbf{N}}$ is a base for the open subsets of $\mathrm{T}(c) = \mathrm{T}'$. □

Now, we give a new approach to the notion of a "saturated class" (for all considered types of classes), which is based (in the main) on cofinal and τ-closed subsets of $\mathrm{C}(\mathbf{S})$.

Definitions. A non-empty topological class $\mathbb{P}$ of spaces is said to be a *second-type saturated class* if for every indexed collection $\mathbf{S}$ of elements of $\mathbb{P}$ the subset of $\mathrm{C}(\mathbf{S})$ denoted by $\mathrm{C}^{\mathbb{P}}(\mathbf{S})$ and consisting of all elements c such that for every $\mathbf{L} \in C^{\diamondsuit}(c)$ the space $T(c)(\mathbf{L})$ belongs to $I\!P$, is a second-type subset of $C(\mathbf{S})$.

A non-empty topological class $I\!P$ of subsets is said to be a *second-type saturated class* if for every indexed collection $\mathbf{S}$ of spaces and for every $I\!P$-restriction $\mathbf{Q}$ of $\mathbf{S}$ the subset of $C_{\mathbf{Q}}(\mathbf{S})$ denoted by $C^{I\!P}_{\mathbf{Q}}(\mathbf{S})$ and consisting of all elements c such that for every $\mathbf{L}, \mathbf{H} \in C^{\diamondsuit}(c)$ with $\mathbf{H} \subset \mathbf{L}$ we have $(T(c)(\mathbf{H}|_{\mathbf{Q}}), T(c)(\mathbf{L})) \in I\!P$, is a second-type subset of $C_{\mathbf{Q}}(\mathbf{S})$.

A non-empty topological class $I\!P$ of bases is said to be a *second-type saturated class* if for every indexed collection $\mathbf{S}$ of spaces, a $I\!P$-co-base $\mathbf{B}$ for $\mathbf{S}$, and for every co-indication $\mathbf{N}$ of $\mathbf{B}$ the subset of $C_{\mathbf{N}}(\mathbf{S})$ denoted by $C^{I\!P}_{\mathbf{N}}(\mathbf{S})$ and consisting of all elements c such that for every $\mathbf{L} \in C^{\diamondsuit}(c)$ we have $(B^{\mathbf{L}}_{\diamondsuit,\mathbf{N}}, T(c)(\mathbf{L})) \in I\!P$, is a second-type subset of $C_{\mathbf{N}}(\mathbf{S})$.

A non-empty topological class $I\!P$ of p-bases is said to be a *second-type saturated class* if for every indexed collection $\mathbf{S}$ of spaces, a restriction $\mathbf{Q}$ of $\mathbf{S}$, a $I\!P$-co-base $\mathbf{B}$ for $\mathbf{Q}$ in $\mathbf{S}$, and for every co-indication $\mathbf{N}$ of $\mathbf{B}$ the subset of $C_{\mathbf{Q},\mathbf{N}}(\mathbf{S})$ denoted by $C^{I\!P}_{\mathbf{Q},\mathbf{N}}(\mathbf{S})$ and consisting of all elements c such that for every $\mathbf{L}, \mathbf{H}, \mathbf{E} \in C^{\diamondsuit}(c)$ with $\mathbf{E} \subset \mathbf{H} \subset \mathbf{L}$ we have

$$(T(c)(\mathbf{H}|_{\mathbf{Q}}), B^{\mathbf{H}}_{\diamondsuit,\mathbf{N}}, T(c)(\mathbf{L})) \in I\!P,$$

is a second-type subset of $C_{\mathbf{Q},\mathbf{N}}(\mathbf{S})$.

A non-empty topological class $I\!P$ of spaces with a $(\Phi, \Lambda, \prec)$-structure is said to be a *second-type saturated class* if for every indexed collection $\mathbf{S}$ of spaces and for every $I\!P$-co-structure $\mathbf{\Psi}$ on $\mathbf{S}$ the subset of $C_{\mathbf{\Psi}}(\mathbf{S})$ denoted by $C^{I\!P}_{\mathbf{\Psi}}(\mathbf{S})$ and consisting of all elements c such that $(\psi_{T(c)(\mathbf{H})}, T(c)(\mathbf{H})) \in I\!P$ for every $\mathbf{H} \in C^{\diamondsuit}(c)$, where $\psi_{T(c)(\mathbf{H})}$ is the mapping of Λ into $\mathcal{P}(T(c)(\mathbf{H}))$ for which $\psi_{T(c)(\mathbf{H})}(\lambda) = T(c)(\mathbf{H}|_{\mathbf{\Psi}(\lambda)})$, $\lambda \in \Lambda$, is a second-type subset of $C_{\mathbf{\Psi}}(\mathbf{S})$.

If in the above definitions we replace the words "second-type" by the words "zero-type" or by the words "first-type", then we obtain the notions of a *zero-type saturated class* and a *first-type saturated class*, respectively.

8.3.8 Proposition. *The class of all regular spaces and the class of all completely regular spaces are zero-type saturated classes. Therefore, these classes are also second-type saturated classes.*

Proof. The fact that the class $\mathbb{P}$(reg.space) of all regular spaces is saturated was proved in Proposition 1.1.6. In this proposition for an arbitrary indexed collection $\mathbf{S}$ of elements of $\mathbb{P}$(reg.space) we constructed a co-mark

$\mathbf{M}^+$ of $\mathbf{S}$ having the following property: for every co-extension $\mathbf{M}$ of $\mathbf{M}^+$, an $\mathbf{M}$-admissible family R^+, and an admissible family R, which is a final refinement of R^+, the space $\mathrm{T}(\mathbf{M}, \mathrm{R})$ is regular.

Now, we prove that the class $\mathbb{P}$(reg.space) is a zero-type saturated class. Let $\mathbf{S}$ be an indexed collection of elements of $\mathbb{P}$ and $c_0 = \{(\mathbf{M}^+, \mathrm{R}_0)\}$ an element of $\mathrm{C}(\mathbf{S})$, where $\mathbf{M}^+$ is the corresponding co-mark of Proposition 2.1.6 and R_0 is an arbitrary $\mathbf{M}^+$-admissible family of equivalence relation on $\mathbf{S}$. Suppose that $c \equiv \{(\mathbf{M}, \mathrm{R})\} \in \mathrm{C}(\mathbf{S})$ and $c_0 \prec^{\mathbf{S}} c$. By Lemma 8.3.4 there exists a co-mark $\mathbf{M}'$ of $\mathbf{S}$ such that $\mathbf{M}'$ is co-extension of $\mathbf{M}^+$ and $\mathbf{M} \sim^{cm} \mathbf{M}'$. Therefore, $c = \{(\mathbf{M}', \mathrm{R})\}$.

Furthermore, in Proposition 2.1.6 we replace the co-mark $\mathbf{M}$ by the co-mark $\mathbf{M}'$ and the family R^+ by the family R. In this case, the space $\mathrm{T}(\mathbf{M}', \mathrm{R}) = \mathrm{T}(\mathbf{M}, \mathrm{R})$ belongs to $\mathbb{P}$(reg.space) (and, therefore, for every $\mathbf{L} \in \mathrm{C}^{\diamond}(c)$ the space $\mathrm{T}(\mathbf{M}, \mathrm{R}, \mathbf{L}) = \mathrm{T}(c)(\mathbf{L})$ belongs also to $\mathbb{P}$(reg.space)), which means that $\mathbb{P}$(reg.space) is a zero-type saturated class.

Similarly, analyzing the proof of Proposition 5.1.5 we can verify that the class of all completely regular spaces is a zero-type saturated class. □

Agreement. In the next three propositions (that is, Propositions 8.3.9–8.3.11) it is assumed that one of the classes $\mathbb{P}$(space), $\mathbb{P}$(subset), $\mathbb{P}$(base), $\mathbb{P}$(p-base), and $\mathbb{P}(\Phi, \Lambda, \prec)$ is fixed and by a class we mean a topological subclass of this fixed class.

8.3.9 Proposition. *A class $\mathbb{P}$ is saturated if and only if $\mathbb{P}$ is a first-type saturated class.*

Proof. We prove the proposition in the case, where $\mathbb{P}$ is a class of bases. The other cases are proved similarly.

Suppose that $\mathbb{P}$ is a saturated class. Let $\mathbf{S}$ be an indexed collection of spaces, $\mathbf{B}$ a $\mathbb{P}$-co-base for $\mathbf{S}$, and $\mathbf{N}$ an arbitrary co-indication of $\mathbf{B}$. To prove that $\mathbb{P}$ is a first-type saturated class we need to prove that the subset $\mathrm{C}^{\mathbb{P}}_{\mathbf{N}}(\mathbf{S})$ (see the definition of a second-type saturated class of bases) of $\mathrm{C}_{\mathbf{N}}(\mathbf{S})$ is the first-type.

Denote by $\mathbf{M}^+$ an initial co-mark of $\mathbf{S}$ corresponding to the co-indication $\mathbf{N}$ of $\mathbf{B}$ and the class $\mathbb{P}$. Consider an arbitrary co-mark $\mathbf{M}$ of $\mathbf{S}$ such that $\mathbf{M}^+ \prec^{cm} \mathbf{M}$. Using Lemma 8.3.4 we can prove that there exists a co-mark $\mathbf{M}'$ of $\mathbf{S}$ such that $\mathbf{M}'$ is co-extension of $\mathbf{M}^+$, and $\mathbf{M} \sim^{cm} \mathbf{M}'$. Let θ be an indicial mapping from $\mathbf{N}$ to $\mathbf{M}'$. Denote by R^+ an initial family corresponding to co-mark $\mathbf{M}'$, the indication $\mathbf{N}$ of $\mathbf{B}$, and the class

$I\!P$. Let R be an admissible family of equivalence relations on $\mathbf{S}$ such that $\mathrm{R}^+ \prec_{af} \mathrm{R}$. Then, it is easy to verify that $c \equiv \{(\mathbf{M}', \mathrm{R})\} = \{(\mathbf{M}, \mathrm{R})\}$. Since $I\!P$ is saturated, $(\mathrm{B}^{\mathbf{L}}_{\diamondsuit,\theta(\tau)}, \mathrm{T}(\mathbf{M}', \mathrm{R}, \mathbf{L})) \in I\!P$ for every $\mathbf{L} \in \mathrm{C}^{\diamondsuit}(\mathrm{R})$. To prove that $\mathrm{C}^{I\!P}_{\mathbf{N}}(\mathbf{S})$ is a first-type subset it suffices to prove that

$$(\mathrm{B}^{\mathbf{L}}_{\diamondsuit,\mathbf{N}}, \mathrm{T}(c)(\mathbf{L})) \in I\!P, \ \mathbf{L} \in \mathrm{C}^{\diamondsuit}(c).$$

But this fact follows immediately by the relations

$$\mathrm{C}^{\diamondsuit}(\mathrm{R}) = \mathrm{C}^{\diamondsuit}(c), \quad \mathrm{B}^{\mathbf{L}}_{\diamondsuit,\theta(\tau)} = \mathrm{B}^{\mathbf{L}}_{\diamondsuit,\mathbf{N}} \text{ and}$$

$$\mathrm{T}(\mathbf{M}', \mathrm{R}, \mathbf{L}) = \mathrm{T}(\mathbf{M}, \mathrm{R}, \mathbf{L}) = \mathrm{T}(c, \mathbf{L}). \tag{8.3.1}$$

Conversely, suppose that $I\!P$ is a first-type saturated class. We need to prove that $I\!P$ is a saturated class. Let $\mathbf{S}$ be an indexed collection of spaces, $\mathbf{B}$ a $I\!P$-co-base for $\mathbf{S}$, and $\mathbf{N}$ an arbitrary co-indication of $\mathbf{B}$. Then, the subset $\mathrm{C}^{I\!P}_{\mathbf{N}}(\mathbf{S})$ is the first-type subset of $\mathrm{C}_{\mathbf{N}}(\mathbf{S})$. Let $\mathbf{M}^+$ be a co-mark of $\mathbf{S}$ considered in the definition of the first-type subsets. Without loss of generality we can suppose that $\mathbf{M}^+$ is a co-extension of $\mathbf{N}$. Suppose that $\mathbf{M}$ is an arbitrary co-extension of $\mathbf{M}^+$, and θ an indicial mapping from $\mathbf{N}$ to $\mathbf{M}$. Then, $\mathbf{M}^+ \prec^{cm} \mathbf{M}$. Since $\mathrm{C}^{I\!P}_{\mathbf{N}}(\mathbf{S})$ is a first-type subset of $\mathrm{C}_{\mathbf{N}}(\mathbf{S})$ there exists an $\mathbf{M}$-admissible family R^+ of equivalence relations on $\mathbf{S}$ such that for every admissible family R of equivalence relations on $\mathbf{S}$, which is a final refinement of R^+, we have $c \equiv \{(\mathbf{M}, \mathrm{R})\} \in \mathrm{C}^{I\!P}_{\mathbf{N}}(\mathbf{S})$, that is $(\mathrm{B}^{\mathbf{L}}_{\diamondsuit,\mathbf{N}}, \mathrm{T}(c)(\mathbf{L})) \in I\!P$ for every $\mathbf{L} \in \mathrm{C}^{\diamondsuit}(c)$. To prove that $I\!P$ is saturated it suffices to prove that

$$(\mathrm{B}^{\mathbf{L}}_{\diamondsuit,\theta(\tau)}, \mathrm{T}(\mathbf{M}, \mathrm{R}, \mathbf{L})) \in I\!P, \ \mathbf{L} \in \mathrm{C}^{\diamondsuit}(\mathrm{R}).$$

But this fact follows immediately by relations (8.3.1). □

The proof of the following proposition is similar to the proof of the corresponding proposition for the saturated classes.

8.3.10 Proposition. *In any non-empty second-type saturated class there exist universal elements.* □

The intersection property of cofinal and τ-closed subsets of a τ-complete set implies the following proposition.

8.3.11 Proposition. *The non-empty intersection of not more than τ many second-type saturated classes (of the same type) is also a second-type saturated class.* □

Definition. Let $I\!P$ be a second-type saturated class of spaces and $\mathbf{S}$ an indexed collection of elements of $I\!P$. Consider the subset $\mathrm{C}^{I\!P}(\mathbf{S})$ of $\mathrm{C}(\mathbf{S})$ (determined in the definition of a second-type saturated class of spaces). Any subset of $\mathrm{C}^{I\!P}(\mathbf{S})$ is called a *$I\!P$-subset* of $\mathrm{C}(\mathbf{S})$.

Since $I\!P$ is a second-type saturated class, the set $\mathrm{C}^{I\!P}(\mathbf{S})$ is a second-type subset of $\mathrm{C}(\mathbf{S})$. Therefore, this set contains cofinal and τ-closed $I\!P$-subsets of $\mathrm{C}(\mathbf{S})$.

By Proposition 8.2.11 and the definition of a second-type saturated class of spaces it follows immediately the following proposition.

8.3.12 Proposition. *Let $I\!P$ be a second-type saturated class of spaces and $\mathbf{S}$ an indexed collection of elements of $I\!P$. Then, for each cofinal and τ-closed $I\!P$-subset $\mathrm{C}_{I\!P}(\mathbf{S})$ of $\mathrm{C}(\mathbf{S})$, the inverse spectrum $\mathrm{D}[\mathbf{S}]|_{\mathrm{C}_{I\!P}(\mathbf{S})}$ is a $I\!P$-τ-spectrum, which is factorizing with respect to the subset S of $\lim(\mathrm{D}[\mathbf{S}]|_{\mathrm{C}_{I\!P}(\mathbf{S})}) = \lim(\mathrm{D}[\mathbf{S}])$.* □

The following proposition is the "spectral determination" of the Containing Spaces belonging to a given second-type saturated class.

8.3.13 Proposition. *Let $I\!P$ be a second-type saturated class of spaces, $\mathbf{S}$ an indexed collection of elements of $I\!P$, and $\mathrm{C}_{I\!P}(\mathbf{S})$ a cofinal and τ-closed $I\!P$-subset of $\mathrm{C}(\mathbf{S})$. Suppose that D is an inverse spectrum on the set $\mathrm{C}_{I\!P}(\mathbf{S})$ such that:*

(1) D is a τ-spectrum,

(2) The space $\lim(D)$ contains a surjective subset L, which is homeomorphic to the free union of elements of $\mathbf{S}$, and

(3) D is $I\!P$-factorizing with respect L.

Then, the spectra D and $\mathrm{D}[\mathbf{S}]$ contain isomorphic cofinal and τ-closed subspectra. Therefore, D contains a cofinal and τ-closed $I\!P$-subspectrum, which is factorizing with respect to L.

Proof. Consider the $I\!P$-τ-spectrum $\mathrm{D}[\mathbf{S}]|_{\mathrm{C}_{I\!P}(\mathbf{S})}$ and let f be a homeomorphism of the subset S of $\lim(\mathrm{D}[\mathbf{S}]|_{\mathrm{C}_{I\!P}(\mathbf{S})})$ onto the subset L of $\lim(D)$. By the Spectral Theorem (Proposition 8.1.7), D and $\mathrm{D}[\mathbf{S}]$ contain isomorphic cofinal and τ-closed subspectra. □

The "covering property" of the Containing Spaces, which are universal in a given second-type saturated class of spaces, can be formulated as follows.

8.3.14 Proposition. *Let $\mathbb{P}$ be a second-type saturated class of spaces, Y a universal element in $\mathbb{P}$, and for every $X \in \mathbb{P}$ let h_X be an embedding of X into Y. Then, there exist an indexed collection $\mathbf{S}$ of elements of $\mathbb{P}$, an element c of $\mathrm{C}(\mathbf{S})$, for which the Containing Space $\mathrm{T}(c)$ is a universal element in $\mathbb{P}$, an embedding $i^X_{\mathrm{T}(c)}$, $X \in \mathbb{P}$, of X into the Containing Space $\mathrm{T}(c)$ (which coincides with the natural embedding if $X \in \mathbf{S}$), and a mapping f of $\mathrm{T}(c)$ into the space Y such that the following relation is satisfied:*

$$f \circ i^X_{\mathrm{T}} = h_X, \quad X \in \mathbb{P}. \ \square \tag{8.3.2}$$

Proof. Denote by $\mathbf{S}$ an indexed collection of elements of $\mathbb{P}$ such that for every $Z \in \mathbb{P}$ there exist $X \in \mathbf{S}$ and an embedding i^Z_X of Z onto X such that $h_Z = h_X \circ i^Z_X$.

Let $\mathrm{C}_{\mathbb{P}}(\mathbf{S})$ be a cofinal and τ-closed $\mathbb{P}$-subset of $\mathrm{C}(\mathbf{S})$. Consider the $\mathbb{P}$-τ-spectrum $\mathrm{D}[\mathbf{S}]|_{\mathrm{C}_{\mathbb{P}}(\mathbf{S})}$. This spectrum is factorizing with respect to the subset S of $\lim(\mathrm{D}[\mathbf{S}]|_{\mathrm{C}_{\mathbb{P}}(\mathbf{S})})$. The required element c of $\mathrm{C}(\mathbf{S})$ and the mapping f of $\mathrm{T}(c)$ into the space Y can be constructed as in Proposition 8.2.13. For every $Z \in \mathbb{P}$ the embedding $i^Z_{\mathrm{T}(c)}$ is defined by relation $i^Z_{\mathrm{T}(c)} = i^X_{\mathrm{T}(c)} \circ i^Z_X$, where X is the above considered element of $\mathbf{S}$. Using this relation we can prove relation (8.3.2) as in Proposition 8.2.13. $\square$

Notation. Let $\mathbf{S}$ be an indexed collection of spaces and $\mathbf{Q}$ a restriction of $\mathbf{S}$. For every $c \equiv \{(\mathbf{M}, \mathrm{R})\} \in \mathrm{C}_{\mathbf{Q}}(\mathbf{S})$ we denote by $c|_{\mathbf{Q}}$ the element of $\mathrm{C}(\mathbf{Q})$ (here, $\mathbf{Q}$ is considered as an indexed collection of spaces) containing the pair $(\mathbf{M}|_{\mathbf{Q}}, \mathrm{R}|_{\mathbf{Q}})$. We note that since $c \in \mathrm{C}_{\mathbf{Q}}(\mathbf{S})$ the family $\mathrm{R}|_{\mathbf{Q}}$ of equivalence relations on $\mathbf{Q}$ is $\mathbf{M}|_{\mathbf{Q}}$-admissible.

The following lemma is easy to prove.

8.3.15 Lemma. *Let $\mathbf{S}$ be an indexed collection of spaces and $\mathbf{Q}$ a restriction of $\mathbf{S}$. The element $c|_{\mathbf{Q}}$, $c \in \mathrm{C}_{\mathbf{Q}}(\mathbf{S})$, is well defined, that is the definition of $c|_{\mathbf{Q}}$ is independent of the element $(\mathbf{M}, \mathrm{R})$ of c. Moreover, if $c_0, c_1 \in \mathrm{C}_{\mathbf{Q}}(\mathbf{S})$ and $c_0 \prec^{\mathbf{S}} c_1$, then $c_0|_{\mathbf{Q}} \prec^{\mathbf{Q}} c_1|_{\mathbf{Q}}$.* $\square$

8.3.16 Lemma. *Let $\mathbf{S}$ be an indexed collection of spaces and $\mathbf{Q}$ a restriction of $\mathbf{S}$. Suppose that C' is a cofinal and τ-closed subset of $\mathrm{C}(\mathbf{Q})$. Then, the subset*

$$C \equiv \{c \in \mathrm{C}_{\mathbf{Q}}(\mathbf{S}) : c|_{\mathbf{Q}} \in C'\}$$

is a cofinal and τ-closed subset of $\mathrm{C}_{\mathbf{Q}}(\mathbf{S})$.

Proof. First, we prove that C is a τ-closed subset of $\mathrm{C}_{\mathbf{Q}}(\mathbf{S})$. Let A be a chain in C with $|C| \leq \tau$ and $c_A = \sup(A)$. We need to prove that $c_A \in C$, that is $c_A|_{\mathbf{Q}} \in C'$. Consider the subset $B \equiv \{c|_{\mathbf{Q}} : c \in A\}$ of C'. By Lemma 8.3.15, B is a chain in C'. Since the set $\mathrm{C}(\mathbf{Q})$ is τ-complete and $|B| \leq \tau$ there exists the lower upper bound of B. Let $c'_B = \sup(B)$. Since C' is τ-closed $c'_B \in C'$. Therefore, it is suffices to prove that $c_A|_{\mathbf{Q}} = c'_B$.

Let

$$\{c_\delta \equiv \{(\mathbf{M}_\delta, \mathrm{R}_\delta)\} : \delta \in \tau\}$$

be an indication of A and $c_A = \{(\mathbf{M}, \mathrm{R})\}$. Then, $\{c_\delta|_{\mathbf{Q}} : \delta \in \tau\}$ is an indication of B. Let also $c'_B = \{(\mathbf{M}', \mathrm{R}')\}$.

Since $c_A = \sup(A)$ we have that (see the proof of Lemma 8.2.1) for every $\varepsilon \in \tau$ the ε-component of $\mathbf{M}$ coincides with the η-component of $\mathbf{M}_\delta$ for some $\eta, \delta \in \tau$. Therefore, the ε-component of $\mathbf{M}|_{\mathbf{Q}}$ coincides with η-component of $\mathbf{M}_\delta|_{\mathbf{Q}}$. Conversely, for every $\eta, \delta \in \tau$ the η-component of $\mathbf{M}_\delta$ coincides with the ε-component of $\mathbf{M}$ for some $\varepsilon \in \tau$. Therefore, the η-component of $\mathbf{M}_\delta|_{\mathbf{Q}}$ coincides with the ε-component of $\mathbf{M}|_{\mathbf{Q}}$. Since $c'_B = \sup(B)$ this result means that $\mathbf{M}' \sim^{cm} \mathbf{M}|_{\mathbf{Q}}$. (Here, the equivalence relation $\sim^{cm}$ is considered in the set of all co-marks of $\mathbf{Q}$).

Suppose that

$$\mathrm{R}_\delta = \{\sim^s_\delta : \delta \in \tau\}$$

and let

$$\mathrm{R}_A \equiv \{\sim^s_A : s \in \mathcal{F}\}$$

be the family of equivalence relations on $\mathbf{S}$ such that

$$\sim^s_A = \cap\{\sim^s_\delta : \delta \in s\}.$$

Then (see the proof of Lemma 8.2.1), $\mathrm{R}_A \sim_{af} \mathrm{R}$. (The equivalence relation $\sim_{af}$ is considered in the set of all admissible families of equivalence relations on $\mathbf{S}$). Similarly, if

$$\mathrm{R}_B = \{\sim^s_B : s \in \mathcal{F}\},$$

where

$$\sim_B^s = \cap\{\sim_\delta^s |_{\mathbf{Q}} : \delta \in s\},$$

then $\mathrm{R}_B \sim_{af} \mathrm{R}'$. (Here, the equivalence relation $\sim_{af}$ is considered in the set of all admissible families of equivalence relations on $\mathbf{Q}$.) On the other hand, it is easy to verify that $\mathrm{R}_B = \mathrm{R}_A|_{\mathbf{Q}}$, which means that $\mathrm{R}|_{\mathbf{Q}} \sim_{af} \mathrm{R}'$. Thus, $(\mathbf{M}|_{\mathbf{Q}}, \mathrm{R}|_{\mathbf{Q}}) \sim^{\mathbf{Q}} (\mathbf{M}', \mathrm{R}')$, that is $c|_{\mathbf{Q}} = c'$.

To prove that C is a cofinal subset of $\mathrm{C}_{\mathbf{Q}}(\mathbf{S})$ we first prove the following statement: for every $c_0 \in \mathrm{C}$ and $c' \in C'$ there exists $c \in \mathrm{C}_{\mathbf{Q}}(\mathbf{S})$ such that $c_0 \prec^{\mathbf{S}} c$ and $c' \prec^{\mathbf{Q}} c|_{\mathbf{Q}}$.

Let $c_0 = \{(\mathbf{M}_0, \mathrm{R}_0)\}$ and $c' = \{(\mathbf{M}', \mathrm{R}')\}$, where

$$\mathbf{M}' = \{\{U_\delta^{\mathbf{Q}(X)} : \delta \in \tau\} : \mathbf{Q}(X) \in \mathbf{Q}\} \text{ and}$$

$$\mathrm{R}' = \{\sim'^{,s} : s \in \mathcal{F}\}.$$

Let also θ be an one-to-one mapping of τ into itself such that $|\tau \setminus \theta(\tau)| = \tau$. Consider a co-mark

$$\mathbf{M} \equiv \{\{U_\varepsilon^X : \varepsilon \in \tau\} : X \in \mathbf{S}\}$$

of $\mathbf{S}$ such that $\mathbf{M}$ is a co-extension of $\mathbf{M}_0$ and satisfies the condition

$$U_{\theta(\delta)}^X \cap \mathbf{Q}(X) = U_\delta^{\mathbf{Q}(X)}, \ \delta \in \tau.$$

Let

$$\mathrm{R}_{\mathbf{S}} \equiv \{\sim_{\mathbf{S}}^s : s \in \mathcal{F}\}$$

be the admissible family of equivalence relations on $\mathbf{S}$ such that $\mathrm{R}_{\mathbf{S}}|_{\mathbf{Q}} = \mathrm{R}'$ (that is, $X \sim_{\mathbf{S}}^s Y$ if and only if $\mathbf{Q}(X) \sim'^{,s} \mathbf{Q}(Y)$ for every $X, Y \in \mathbf{S}$). Denote by R an $(\mathbf{M}, \mathbf{Q})$-admissible family of equivalence relations on $\mathbf{S}$, which is a final refinement of the families $\mathrm{R}_{\mathbf{S}}$ and R_0. Then, it is easy to verify that for the element $c \equiv \{(\mathbf{M}, \mathrm{R})\}$ of $\mathrm{C}_{\mathbf{Q}}(\mathbf{S})$ we have $c_0 \prec^{\mathbf{S}} c$ and $c' \prec^{\mathbf{Q}} c|_{\mathbf{Q}}$.

Now, we prove that C is a cofinal subset of $\mathrm{C}_{\mathbf{Q}}(\mathbf{S})$. Let $c_0 \in \mathrm{C}_{\mathbf{Q}}(\mathbf{S})$. Using the above statement we can construct by induction a sequence $c_0, c_1, ...$ of elements of $\mathrm{C}_{\mathbf{Q}}(\mathbf{S})$ and a sequence $c'_0, c'_1, ...$ of elements of C' satisfying the following properties:

(a) $c_i \prec^{\mathbf{S}} c_{i+1}$ and, therefore, $A \equiv \{c_i : i \in \omega\}$ is a chain in $\mathrm{C}_{\mathbf{Q}}(\mathbf{S})$,
(b) $c'_i \prec^{\mathbf{Q}} c'_{i+1}$ and, therefore, the set $B \equiv \{c'_i : i \in \omega\}$ is a chain in C',
(c) $c_i|_{\mathbf{Q}} \prec^{\mathbf{Q}} c'_{i+1}$, and
(d) $c'_{i+1} \prec^{\mathbf{Q}} c_{i+1}|_{\mathbf{Q}}$, where $i \in \omega$.

Properties (c) and (d) imply that B majorates the subset $\{c_i|_{\mathbf{Q}} : i \in \omega\}$ of $\mathrm{C}(\mathbf{Q})$ and conversely. By Lemma 8.3.15 the set $\{c_i|_{\mathbf{Q}} : i \in \omega\}$ is a chain in $\mathrm{C}(\mathbf{Q})$. Therefore,

$$c'_B \equiv \sup(\{c'_i : i \in \omega\}) = \sup(\{c_i|_{\mathbf{Q}} : i \in \omega\}).$$

Since C' is a τ-closed subset of $\mathrm{C}(\mathbf{Q})$ we have $c'_B \in C'$. On the other hand, setting

$$c = \sup\{c_i : i \in \omega\}$$

we have (see the proof of Lemma 8.2.1) $c|_{\mathbf{Q}} = c'_B$ and $c_0 \prec^{\mathbf{S}} c$, which proves that C is a cofinal subset of $\mathrm{C}_{\mathbf{Q}}(\mathbf{S})$. □

8.3.17 Corollary. *Let $\mathbf{S}$ be an indexed collection of spaces. Suppose that $\mathbf{Q}$ is a restriction of $\mathbf{S}$ such that for every $X \in \mathbf{S}$, $\mathbf{Q}(X)$ is an element of a second-type saturated class $\mathbb{P}$ of spaces (therefore, $\mathbf{Q}$ is an indexed collection of elements of $\mathbb{P}$). Let $\mathrm{C}_{\mathbb{P}}(\mathbf{Q})$ be a cofinal and τ-closed $\mathbb{P}$-subset of $\mathrm{C}(\mathbf{Q})$. Then, the subset of $\mathrm{C}_{\mathbf{Q}}(\mathbf{S})$ consisting of all elements c such that $c|_{\mathbf{Q}} \in \mathrm{C}_{\mathbb{P}}(\mathbf{Q})$ is a cofinal and τ-closed subset of $\mathrm{C}_{\mathbf{Q}}(\mathbf{S})$.* □

Definition. Let $\mathbf{S}$ be an indexed collection of spaces and $\mathbf{Q}$ a restriction of $\mathbf{S}$. The restriction $\mathbf{Q}$ is said to be *c-complete*, where $c \in \mathrm{C}_{\mathbf{Q}}(\mathbf{S})$, if there exists an element $(\mathbf{M}, \mathrm{R})$ of c such that $\mathbf{Q}$ is $(\mathbf{M}, \mathrm{R})$-complete. Lemma 8.2.2 implies that if $\mathbf{Q}$ is c-complete for some $c \in \mathrm{C}_{\mathbf{Q}}(\mathbf{S})$, then $\mathbf{Q}$ is $(\mathbf{M}, \mathrm{R})$-complete for every $(\mathbf{M}, \mathrm{R}) \in c$.

8.3.18 Proposition. *Let $\mathbb{B}$ be a second-type saturated class of bases and $\mathbb{E}$ a $\mathbb{B}$-rim-hereditary second-type saturated class of spaces. Then, for every $\alpha \in \{-1\} \cup \tau^+$ the class $\mathbb{P}(\mathrm{ind}^{\mathbb{B}}_{\mathbb{E}} \leq \alpha)$ is also a second-type saturated class of spaces.*

Proof. The proof of this proposition follows the proof of Proposition 3.1.5. We prove it by induction on α. For $\alpha = -1$ the proof is the same as in Proposition 3.1.5.

Let $\alpha \in \tau^+$. Suppose that for every $\beta \in \{-1\} \cup \alpha$ the class $\mathbb{P}(\mathrm{ind}_{\mathbb{E}}^{\mathbb{B}} \leq \beta)$ is a second-type saturated class. We prove that $\mathbb{P} \equiv \mathbb{P}(\mathrm{ind}_{\mathbb{E}}^{\mathbb{B}} \leq \alpha)$ is also such a class.

Indeed, let $\mathbf{S}$ be an indexed collection of elements of $\mathbb{P}$. By Corollary 3.1.4 for every $X \in \mathbf{S}$ there exists a $\mathbb{B}$-base B^X for X and an indication

$$N^X \equiv \{W_\eta^X : \eta \in \tau\}$$

of B^X such that

$$\mathrm{ind}_{\mathbb{E}}^{\mathbb{B}}(\mathrm{Bd}_X(W_\eta^X)) \leq \alpha[\eta]. \tag{8.3.3}$$

We set

$$\mathbf{B} = \{B^X : X \in \mathbf{S}\} \text{ and}$$

$$\mathbf{N} = \{\{W_\eta^X : \eta \in \tau\} : X \in \mathbf{S}\}.$$

Obviously, $\mathbf{B}$ is a $\mathbb{B}$-co-base for $\mathbf{S}$ and $\mathbf{N}$ is a co-indication of $\mathbf{B}$.

Since $\mathbb{B}$ is a second-type saturated class of bases the set $\mathrm{C}_{\mathbf{N}}^{\mathbb{B}}(\mathbf{S}) \subset \mathrm{C}_{\mathbf{N}}(\mathbf{S})$ contains a cofinal and τ-closed subset of $\mathrm{C}_{\mathbf{N}}(\mathbf{S})$. Denote this subset by $\mathrm{C}_{\mathbb{B}}$. Thus, for every $c \in \mathrm{C}_{\mathbb{B}}$ we have

$$(\mathrm{B}_{\diamondsuit,\mathbf{N}}^{\mathbf{L}}, \mathrm{T}(c)(\mathbf{L})) \in \mathbb{B}, \quad \mathbf{L} \in \mathrm{C}^{\diamondsuit}(c).$$

For every $\eta \in \tau$ we set

$$\mathbf{W}_\eta = \{W_\eta^X : X \in \mathbf{S}\}$$

and consider the restriction $\mathbf{S}_\eta \equiv \mathbf{Bd}(\mathbf{W}_\eta)$ of $\mathbf{S}$. By relation (8.3.3), $\mathbf{S}_\eta$ is an indexed collection of elements of the class $\mathbb{P}_\eta \equiv \mathbb{P}(\mathrm{ind}_{\mathbb{E}}^{\mathbb{B}} \leq \alpha[\eta])$. By inductive assumption, this class is a second-type saturated class. Therefore, the set $\mathrm{C}^{\mathbb{P}_\eta}(\mathbf{S}_\eta) \subset \mathrm{C}(\mathbf{S}_\eta)$ (about notation see the definition of a second-type saturated class of spaces) contains a cofinal and τ-closed subset of $\mathrm{C}(\mathbf{S}_\eta)$. Denote this subset by $\mathrm{C}'_{\mathbb{P}_\eta}$. By Corollary 8.3.17 there exists a cofinal and τ-closed subset of $\mathrm{C}_{\mathbf{S}_\eta}(\mathbf{S})$, denoted by $\mathrm{C}_{\mathbb{P}_\eta}$, such that for every element c of this set we have $c|_{\mathbf{S}_\eta} \in \mathrm{C}'_{\mathbb{P}_\eta}$.

For the restrictions $\mathbf{W}_\eta$, $\mathbf{Cl}(\mathbf{W}_\eta)$, $\mathbf{Bd}(\mathbf{W}_\eta)$, and $\mathbf{Co}(\mathbf{W}_\eta)$, $\eta \in \tau$, of $\mathbf{S}$ we consider the corresponding cofinal and τ-closed subsets $\mathrm{C}_{\mathbf{W}_\eta}(\mathbf{S})$, $\mathrm{C}_{\mathbf{Cl}(\mathbf{W}_\eta)}(\mathbf{S})$, $\mathrm{C}_{\mathbf{Bd}(\mathbf{W}_\eta)}(\mathbf{S})$, $\mathrm{C}_{\mathbf{Co}(\mathbf{W}_\eta)}(\mathbf{S})$ of $\mathrm{C}(\mathbf{S})$ and put

$$\mathrm{C}_{\mathbb{P}} = \mathrm{C}_{\mathbb{B}} \cap (\cap\{\mathrm{C}_{\mathbb{P}_\eta} : \eta \in \tau\}) \cap (\cap\{\mathrm{C}_{\mathbf{W}_\eta}(\mathbf{S}) : \eta \in \tau\}) \cap$$

$$(\cap\{\mathrm{C}_{\mathbf{Cl}(\mathbf{W}_\eta)}(\mathbf{S}) : \eta \in \tau\}) \cap (\cap\{\mathrm{C}_{\mathbf{Bd}(\mathbf{W}_\eta)}(\mathbf{S}) : \eta \in \tau\}) \cap$$

$$(\cap\{\mathrm{C}_{\mathbf{Co}(\mathbf{W}_\eta)}(\mathbf{S}) : \eta \in \tau\}).$$

By Lemma 8.1.1 $\mathrm{C}_{\mathbb{P}}$ is a cofinal and τ-closed subset of $\mathrm{C}(\mathbf{S})$. To prove the proposition it suffices to show that for every $c \in \mathrm{C}_{\mathbb{P}}$ the space $\mathrm{T}(c)(\mathbf{L})$, $\mathbf{L} \in \mathrm{C}^{\diamondsuit}(c)$, belongs to $\mathbb{P}$. For this purpose, we fix an element c of $\mathrm{C}_{\mathbb{P}}$ and an element $\mathbf{L} \in \mathrm{C}^{\diamondsuit}(c)$ and consider the base $\mathrm{B}^{\mathbf{L}}_{\diamondsuit,\mathbf{N}}$ of $\mathrm{T}(c,\mathbf{L})$. Since $c \in \mathrm{C}_{\mathbb{B}}$ this base is a $\mathbb{B}$-base.

Let $U^{\mathrm{T}(c)}_\eta(\mathbf{H})$, $\eta \in \tau$, be an arbitrary fixed element of $\mathrm{B}^{\mathbf{L}}_{\diamondsuit,\mathbf{N}}$. Therefore, we have $\mathbf{H} \subset \mathbf{L}$. We need to prove that

$$\mathrm{ind}^{\mathbb{B}}_{\mathbb{E}}(\mathrm{Bd}_{\mathrm{T}(c,\mathbf{L})}(U^{\mathrm{T}(c)}_\eta(\mathbf{H}))) \le \alpha[\eta] < \alpha. \tag{8.3.4}$$

Consider the restriction $\mathbf{W}_\eta$ of $\mathbf{S}$. Let $(\mathbf{M},\mathrm{R}) \in c$. Then, by the choice of c all conditions of Lemma 1.4.10 are satisfied if instead of the restriction $\mathbf{Q}$ of this lemma we consider the restriction $\mathbf{W}_\eta$. Therefore,

$$\mathrm{T}|_{\mathbf{Bd}(\mathbf{W}_\eta)} = \mathrm{Bd}_{\mathrm{T}}(\mathrm{T}|_{\mathbf{W}_\eta}),$$

where

$$\mathrm{T} = \mathrm{T}(\mathbf{M},\mathrm{R}),\ \mathrm{T}|_{\mathbf{W}_\eta} = \mathrm{T}(\mathbf{M}|_{\mathbf{W}_\eta}, \mathrm{R}|_{\mathbf{W}_\eta}), \text{ and}$$

$$\mathrm{T}|_{\mathbf{Bd}(\mathbf{W}_\eta)} = \mathrm{T}(\mathbf{M}|_{\mathbf{Bd}(\mathbf{W}_\eta)}, \mathrm{R}|_{\mathbf{Bd}(\mathbf{W}_\eta)}).$$

This relation and Lemma 8.3.6 imply the following relation

$$\mathrm{T}(c)|_{\mathbf{Bd}(\mathbf{W}_\eta)} = \mathrm{Bd}_{\mathrm{T}(c)}(\mathrm{T}(c)|_{\mathbf{W}_\eta}).$$

Then, as in Proposition 3.1.5 we have the following equalities:

$$\mathrm{Bd}_{\mathrm{T}(c)}(U^{\mathrm{T}(c)}_\eta(\mathbf{H})) = \mathrm{Bd}_{\mathrm{T}(c)}(\mathrm{T}(c)|_{\mathbf{W}_\eta} \cap \mathrm{T}(c)(\mathbf{H})) =$$

$$\mathrm{Bd}_{\mathrm{T}(c)}(\mathrm{T}(c)|_{\mathbf{W}_\eta}) \cap \mathrm{T}(c)(\mathbf{H}) = \mathrm{T}(c)|_{\mathbf{Bd}(\mathbf{W}_\eta)} \cap \mathrm{T}(c)(\mathbf{H}) =$$

$$\mathrm{T}(\mathbf{M}|_{\mathbf{S}_\eta}, \mathrm{R}|_{\mathbf{S}_\eta}, \mathbf{H}|_{\mathbf{S}_\eta}).$$

Since $\mathbb{P}_\eta$ is saturated, by the choice of c the subspace $\mathrm{T}(\mathbf{M}|_{\mathbf{S}_\eta}, \mathrm{R}|_{\mathbf{S}_\eta}, \mathbf{H}|_{\mathbf{S}_\eta})$ of $\mathrm{T}(\mathbf{M}|_{\mathbf{S}_\eta}, \mathrm{R}|_{\mathbf{S}_\eta})$, that is the subspace

$$\mathrm{Bd}_{\mathrm{T}(c)}(U_\eta^{\mathrm{T}(c)}(\mathbf{H})) = \mathrm{Bd}_{\mathrm{T}(c)(\mathbf{L})}(U_\eta^{\mathrm{T}(c)}(\mathbf{H}))$$

of $\mathrm{T}(c)(\mathbf{L})$ belongs to $\mathbb{P}_\eta$, or equivalently, the relation (8.3.4) is satisfied. This fact means that

$$\mathrm{ind}_{\mathbb{E}}^{\mathbb{B}}(\mathrm{T}(c)(\mathbf{L})) \le \alpha,$$

that is the space $\mathrm{T}(c)(\mathbf{L})$ belongs to $\mathbb{P}$. Thus, $\mathbb{P}$ is a second-type saturated class of spaces. □

Since the class of all bases and the class of spaces consisting of the empty space only are trivially second-type saturated classes we have the following consequence.

8.3.19 Corollary. *The class*

$$\mathbb{P}(\mathrm{ind} \le \alpha),\ \alpha \in \{-1\} \cup \tau^+,$$

is a second-type saturated class of spaces. □

8.3.20 Proposition. *Suppose that $\Lambda = \Lambda^0$ and for every $\lambda \in \Lambda$, $\mathbf{p}(\lambda)$ is a second-type saturated class of subsets. Then, the class $\mathbb{P}(\Lambda, \prec, \mathbf{p})$ is a second-type saturated class of spaces.*

Proof. The proof of this proposition follows the proof of Proposition 4.1.5. Let $\mathbf{S}$ be an indexed collection of spaces and

$$\mathbf{\Psi} \equiv \{\psi_X : X \in \mathbf{S}\}$$

a $\mathbb{P}$-co-structure on $\mathbf{S}$, where $\mathbb{P} = \mathbb{P}(\Lambda, \prec, \mathbf{p})$. For every $\lambda \in \Lambda$ we denote by $\mathrm{C}_{\mathbf{p}(\lambda)}(\mathbf{S})$ a cofinal and τ-closed $\mathbf{p}(\lambda)$-subset of $\mathrm{C}(\mathbf{S})$ (therefore, we have $\mathrm{C}_{\mathbf{p}(\lambda)}(\mathbf{S}) \subset \mathrm{C}_{\mathbf{\Psi}(\lambda)}^{\mathbf{p}(\lambda)}(\mathbf{S})$) and set

$$\mathrm{C}_{\mathbb{P}} = \cap\{\mathrm{C}_{\mathbf{p}(\lambda)}(\mathbf{S}) : \lambda \in \Lambda\}.$$

To prove the proposition it suffices to prove that for every $c \in \mathrm{C}_{\mathbb{P}}$ and $\mathbf{H} \in \mathrm{C}^{\diamond}(c)$ we have $(\psi_{\mathrm{T}(c)(\mathbf{H})}, \mathrm{T}(c)(\mathbf{H})) \in \mathbb{P}$, where $\psi_{\mathrm{T}(c)(\mathbf{H})}$ is the mapping of Λ into $\mathcal{P}(\mathrm{T}(c)(\mathbf{H}))$ such that

$$\psi_{\mathrm{T}(c)(\mathbf{H})}(\lambda) = \mathrm{T}(c)(\mathbf{H}|_{\mathbf{\Psi}(\lambda)}),\ \lambda \in \Lambda.$$

Let $c \in \mathrm{C}_{\mathbb{P}}$, $\mathbf{H} \in \mathrm{C}^{\diamondsuit}(\mathrm{c})$, and $\mathbf{a} \in \mathrm{T}(c)(\mathbf{H})$. Suppose that $(x, X) \in \mathbf{a}$. Then, $X \in \mathbf{H}$. Since ψ_X is a $(\Lambda, \prec, \mathbf{p})$-structure on X there exists $\lambda \in \Lambda$ such that $x \in \psi_X(\lambda)$. Therefore,

$$\mathbf{a} \in \mathrm{T}(c)|_{\mathbf{\Psi}(\lambda)} \cap \mathrm{T}(c)(\mathbf{H}) =$$

$$\mathrm{T}(c)(\mathbf{H}|_{\mathbf{\Psi}(\lambda)}) = \psi_{\mathrm{T}(c)(\mathbf{H})}(\lambda). \tag{8.3.5}$$

Furthermore, let $\lambda_0 \prec \lambda_1$. Since for every $X \in \mathbf{S}$, $\psi_X(\lambda_0) \subset \psi_X(\lambda_1)$ we have

$$\mathrm{T}(c)|_{\mathbf{\Psi}(\lambda_0)} \cap \mathrm{T}(c)(\mathbf{H}) \subset \mathrm{T}(c)|_{\mathbf{\Psi}(\lambda_1)} \cap \mathrm{T}(c)(\mathbf{H}),$$

that is

$$\psi_{\mathrm{T}(c)(\mathbf{H})}(\lambda_0) \subset \psi_{\mathrm{T}(c)(\mathbf{H})}(\lambda_1). \tag{8.3.6}$$

Finally, since $c \in \mathrm{C}_{\mathbb{P}}$ we have

$$(\psi_{\mathrm{T}(c)(\mathbf{H})}(\lambda), \mathrm{T}(c)(\mathbf{H})) \in \mathbf{p}(\lambda),\ \ \lambda \in \Lambda. \tag{8.3.7}$$

Relations (8.3.5)–(8.3.7) imply that $(\psi_{\mathrm{T}(c)(\mathbf{H})}, \mathrm{T}(c)(\mathbf{H})) \in \mathbb{P}$. □

The proof of the next proposition follows the proof of Proposition 4.1.6 and it can be done similarly to the preceding proposition.

8.3.21 Proposition. *Suppose that for every $\lambda \in \Lambda$, $\mathbf{p}(\lambda)$ is a second-type saturated class of subsets. If moreover $\mathbf{p}(\lambda)$ is complete for $\lambda \in \Lambda^0$, then the class $\mathbb{P}(\Phi, \Lambda, \prec, \mathbf{p})$ is a second-type saturated class of spaces.* □

Propositions 8.3.8, 8.3.11, 8.3.20, 8.3.21 and Corollary 8.3.19 imply the following corollary.

8.3.22 Corollary. *The following classes of spaces are second-type saturated classes:*

(1) *The class of all regular (completely regular) spaces.*

(2) *The class of all countable-dimensional regular (completely regular) spaces.*

(3) *The class of all strongly countable-dimensional regular (completely regular) spaces.*

(4) *The class of all locally finite-dimensional regular (completely regular) spaces.*

(5) *The class of all regular (completely regular) spaces of dimension* ind *less that or equal to α, where α is a fixed element of τ.* $\square$

In particular, for $\tau = \omega$ we have the following consequence.

8.3.23 Corollary. *The following classes of spaces are second-type saturated classes:*

(1) *The class of all separable metrizable spaces.*

(2) *The class of all countable-dimensional separable metrizable spaces.*

(3) *The class of all strongly countable-dimensional separable metrizable spaces.*

(4) *The class of locally finite-dimensional separable metrizable spaces.*

(5) *The class of all separable metrizable spaces of dimension* ind *less that or equal to α, where α is a fixed element of ω^+.* $\square$

For $\tau = \omega$ Proposition 8.3.12 and Corollary 8.3.23 imply the following consequence.

8.3.24 Corollary. *Let $\mathbb{P}$ be one of the classes mentioned in Corollary 3.23. Then, for every indexed collection $\mathbf{S}$ of elements of $\mathbb{P}$ and for every cofinal and ω-closed $\mathbb{P}$-subset $\mathrm{C}_{\mathbb{P}}(\mathbf{S})$ of $\mathrm{C}(\mathbf{S})$ the spectrum $\mathrm{D}[\mathbf{S}]|_{\mathrm{C}_{\mathbb{P}}(\mathbf{S})}$ is a $\mathbb{P}$-ω-spectrum, which is factorizing with respect to the surjective subset S of $\lim(\mathrm{D}[\mathbf{S}]|_{\mathrm{C}_{\mathbb{P}}(\mathbf{S})})$.* $\square$

For $\tau = \omega$ the Spectral Theorem and Corollaries 8.3.23 and 8.3.24 imply the following consequence.

8.3.25 Corollary. *Let $\mathbb{P}$ be one of the classes mentioned in the Corollary 3.23 and $\mathbf{S}$ an indexed collection of elements of $\mathbb{P}$. Suppose that D is an inverse spectrum on a cofinal and ω-closed $\mathbb{P}$-subset $\mathrm{C}_{\mathbb{P}}(\mathbf{S})$ of $\mathrm{C}(\mathbf{S})$ with the properties:*

(1) *D is a ω-spectrum.*

(2) *The limit space of D contains a surjective subset L, which is homeomorphic to the free union of elements of $\mathbf{S}$.*

(3) *D is $\mathbb{P}$-factorizing with respect to L.*

Then, D and $\mathrm{D}[\mathbf{S}]|_{C_{\mathbb{P}}(\mathbf{S})}$ contain isomorphic cofinal and ω-closed subspectra. Therefore, D contains a cofinal and ω-closed $\mathbb{P}$-subspectrum, which is factorizing with respect to L. □

8.4 Structural τ-spectra

Agreement. In this section we fix one of the classes $\mathbb{P}$(subset), $\mathbb{P}$(base), $\mathbb{P}$(p-base), and $\mathbb{P}(\Phi, \Lambda, \prec)$ and denote it by $\mathbb{P}$. By a *structural class* we mean any topological subclass of $\mathbb{P}$.

Notations. According to the general Agreement (see the Introductory Remarks) concerning the weights of spaces, in the definition of the elements of classes $\mathbb{P}$(subset), $\mathbb{P}$(base), $\mathbb{P}$(p-base), and $\mathbb{P}(\Phi, \Lambda, \prec)$ we restrict ourselves by consideration only spaces of weights less than or equal to τ. Without this restriction we obtain new classes, which will be denoted by $\mathbb{P}^*$(subset), $\mathbb{P}^*$(base), $\mathbb{P}^*$(p-base), and $\mathbb{P}^*(\Phi, \Lambda, \prec)$, respectively. The corresponding class for $\mathbb{P}$ will be denoted by $\mathbb{P}^*$.

Definitions. In the present section, by a *space X with structure* we mean the space-coordinate of an element E of $\mathbb{P}^*$. In this case, the element E is called the *supplement* of X and is denoted by E_X.

Any subset L of a space X with structure can be also considered as a space with structure. In this case, the supplement E_L of L is defined as follows:

(1) If $E_X = (Q^X, X) \in \mathbb{P}(\text{subset})$, then

$$E_L = (Q^X \cap L, L).$$

(2) If $E_X = (B^X, X) \in \mathbb{P}(\text{base})$, then

$$E_L = (\{U \cap L : U \in B^X\}, L).$$

(3) If $E_X = (Q^X, B^X, X) \in \mathbb{P}(\text{p-base})$, then

$$E_L = (Q^X \cap L, \{U \cap L : U \in B^X\}, L).$$

(4) If $E_X = (\psi_X, X) \in \mathbb{P}(\Phi, \Lambda, \prec)$, then

$$E_L = (\psi_L, L),$$

where $\psi_L(\lambda) = \psi_X(\lambda) \cap L$, $\lambda \in \Lambda$.

It is said that L is a *structural subset* of X whenever L is considered as a space with structure with the above defined supplement E_L.

Definition. Let X and Y be two spaces with structure. We say that a mapping f of X into Y is a *structural mapping* if the given below conditions (corresponding to the type of $\mathbb{P}^*$) are satisfied:

(1) If $\mathbb{P}^* = \mathbb{P}^*(\text{subset})$, $E_X = (Q^X, X)$ and $E_Y = (Q^Y, Y)$, then

$$f(Q^X) \subset Q^Y.$$

(2) If $\mathbb{P}^* = \mathbb{P}^*(\text{base })$, $E_X = (B^X, X)$, and $E_Y = (B^Y, Y)$, then

$$\{f^{-1}(U) : U \in B^Y\} \subset B^X.$$

(3) If $\mathbb{P}^* = \mathbb{P}^*(\text{p-base})$, $E_X = (Q^X, B^X, X)$, and $E_Y = (Q^Y, B^Y, Y)$, then

$$f(Q^X) \subset Q^Y \quad \text{and} \quad \{f^{-1}(U) : U \in B^Y\} \subset B^X.$$

(4) If $\mathbb{P}^* = \mathbb{P}^*(\Phi, \Lambda, \prec)$, $E_X = (\psi_X, X)$, and $E_Y = (\psi_Y, Y)$, then

$$f(\psi_X(\lambda)) \subset \psi_Y(\lambda), \quad \lambda \in \Lambda.$$

We say that a homeomorphism f of X onto (into) Y is a *structural homeomorphism onto (into)* if f and f^{-1} are structural mappings. (We note that in the case of the structural homeomorphism into the mapping f^{-1} is a structural mapping of the structural subset $f(X)$ of Y into the space X with structure.)

Definitions. Let C be a directed subset with a partial order $\prec^{\mathrm{C}}$. A spectrum D on C is said to be a *structural spectrum* if $D(c)$, $c \in \mathrm{C}$, is a space with structure and $D(c_0, c_1)$, $(c_0, c_1) \in \prec^{\mathrm{C}}$, is a structural mapping of the space $D(c_1)$ into the space $D(c_0)$.

A structural spectrum D on C is said to be a *structural $\mathbb{P}$-spectrum*, where $\mathbb{P}$ is a structural class, if for every $c \in \mathrm{C}$ the supplement of $D(c)$ belongs to $\mathbb{P}$.

Let D_0 and D_1 be two structural spectra on the set C and

$$f \equiv \{f_c : c \in \mathrm{C}\}$$

a morphism of D_0 into D_1 (respectively, an isomorphism of D_0 onto (into) D_1). We say that f is a *structural morphism* (respectively, a *structural isomorphism onto* (*into*)) if f_c is a structural mapping (respectively, f_c and f_c^{-1} are structural homeomorphisms onto (into)) for all $c \in \mathrm{C}$.

Below, for the limit space $\lim(D)$ of a structural spectrum D we define the so-called limit supplement of $\lim(D)$. This definition depends on the type of $\mathbb{P}$. We note that the limit supplement of $\lim(D)$ belongs to $\mathbb{P}^*$.

Definitions. Let D be a structural inverse spectrum on a directed set C with a partial order $\prec^{\mathrm{C}}$.

(1) Suppose that $\mathbb{P} = \mathbb{P}(\text{subset})$ and $E_{D(c)} = (Q^{D(c)}, D(c))$, $c \in \mathrm{C}$. We denote by Q the spectrum on the set C, which is defined as follows: $Q(c) = Q^{D(c)}$, $c \in \mathrm{C}$, and $Q(c_0, c_1)$, $(c_0, c_1) \in \prec^{\mathrm{C}}$, is the mapping of $Q^{D(c_1)}$ into $Q^{D(c_0)}$, which on the points of $Q^{D(c_1)}$ coincides with the mapping $D(c_0, c_1)$. Since the mappings of D are structural, Q is really a spectrum. The pair

$$E_{\lim(D)} \equiv (\lim(Q), \lim(D)) \in \mathbb{P}^*(\text{subset})$$

is called the *limit supplement* of $\lim(D)$. A structural subset L of $\lim(D)$ is called *structurally surjective* if

$$D(c, \infty)(L) = D(c) \quad \text{and}$$

$$D(c, \infty)(L \cap \lim(Q)) = Q^{D(c)}, \quad c \in \mathrm{C}.$$

(2) Suppose that $\mathbb{P} = \mathbb{P}(\text{base})$ and $E_{D(c)} = (B^{D(c)}, D(c))$, $c \in \mathrm{C}$. Then, the set

$$B^{\lim(D)} \equiv \cup\{\{(D(c, \infty))^{-1}(U) : U \in B^{D(c)}\} : c \in \mathrm{C}\}$$

is a base for the open subsets of $\lim(D)$. The pair

$$E_{\lim(D)} \equiv (B^{\lim(D)}, \lim(D)) \in \mathbb{P}^*(\text{base})$$

is called the *limit supplement* of $\lim(D)$. A structural subset L of $\lim(D)$ is called *structurally surjective* if

$$D(c, \infty)(L) = D(c), \; c \in \mathrm{C}$$

(that is, if the subset L of $\lim(D)$ is surjective).

(3) Suppose that $\mathbb{P} = \mathbb{P}(\text{p-base})$ and

$$E_{D(c)} = (Q^{D(c)}, B^{D(c)}, D(c)),\ c \in \mathrm{C}.$$

Then, it is easy to see that the set

$$B^{\lim(D)} = \cup\{\{(D(c,\infty))^{-1}(U) : U \in B^{D(c)}\} : c \in \mathrm{C}\}$$

is a p-base for the subset $\lim(Q)$ in $\lim(D)$, where Q is the spectrum which is defined as the above (see the case $\mathbb{P} = \mathbb{P}(\text{subset})$). The triad

$$E_{\lim(D)} \equiv (\lim(Q), B^{\lim(D)}, \lim(D)) \in \mathbb{P}^*(\text{p-base})$$

is called the *limit supplement* of $\lim(D)$. A structural subset L of $\lim(D)$ is called *structurally surjective* if

$$D(c,\infty)(L) = D(c) \text{ and}$$

$$D(c,\infty)(L \cap \lim(Q)) = Q^{D(c)},\ c \in \mathrm{C}.$$

(4) Suppose that $\mathbb{P} = \mathbb{P}(\Phi, \Lambda, \prec)$ and

$$E_{D(c)} = (\psi_{D(c)}, D(c)),\ c \in \mathrm{C}.$$

We define the mapping $\psi_{\lim(D)}$ of Λ into $\lim(D)$ as follows:

$$\psi_{\lim(D)}(\lambda) = \lim(Q(\lambda)),$$

where $Q(\lambda)$ is the spectrum on C such that

$$Q(\lambda)(c) = \psi_{D(c)}(\lambda),\ c \in C,$$

and $Q(\lambda)(c_0, c_1)$, $(c_0, c_1) \in \prec^C$, is the mapping of $\psi_{D(c_1)}(\lambda)$ into $\psi_{D(c_0)}(\lambda)$, which on the points of $\psi_{D(c_1)}(\lambda)$ coincides with the mapping $D(c_0, c_1)$. The pair

$$E_{\lim(D)} \equiv (\psi_{\lim(D)}, \lim(D)) \in \mathbb{P}^*(\Phi, \Lambda, \prec)$$

is called the *limit supplement* of $\lim(D)$. A structural subset L of $\lim(D)$ is called *structurally surjective* if

$$D(c,\infty)(L) = D(c) \text{ and } D(c,\infty)(L \cap \lim(Q(\lambda))) = \psi_{D(c)}(\lambda),$$

for every $c \in \mathrm{C}$ and $\lambda \in \Lambda$.

In particular, the above definition can be applied to define the limit supplement of the limit space of any subspectrum $D|_{C'}$ of D, where C' is a directed subset of C.

Definitions. Let D be a structural spectrum on a directed set C. The spectrum D is said to be *structurally τ-continuous* if for every chain C' of C with $|C'| \leq \tau$ and $c = \sup(C')$ the natural mapping $D(C', c)$ is a structural homeomorphism of $D(c)$ into $\lim(D|_{C'})$.

The spectrum D is said to be *structural τ-spectrum* if:

(a) D is structurally τ-continuous and

(b) D is a τ-spectrum.

The spectrum D is said to be *structurally $I\!\!P$-factorizing with respect to a structural subset* L of $\lim(D)$, where $I\!\!P$ is a structural class, if for every structural mapping f of L into a space Y with structure such that $E_{\mathrm{Y}} \in I\!\!P$ (therefore, $w(\mathrm{Y}) \leq \tau$) there exist an element $c \in \mathrm{C}$ and a structural mapping f_c of $D(c)$ into Y such that

$$f = f_c \circ D(c,\infty)|_L.$$

If $I\!\!P = \mathbb{P}$, then instead of the word "$I\!\!P$-factorizing" we use the word "factorizing".

8.4.1 Proposition. (The Structural Spectral Theorem). *Let $I\!\!P$ be a structural class. Suppose that D_0 and D_1 are two structural $I\!\!P$-τ-spectra, which are structurally $I\!\!P$-factorizing with respect to structural surjective subsets L_0 and L_1 of $\lim(D_0)$ and $\lim(D_1)$, respectively. Then, each structural mapping f of L_0 into L_1 is extended to a structural mapping $\overline{f}$ of $\lim(D_0)$ into $\lim(D_1)$ such that $\overline{f}$ is induced by a structural morphism of cofinal and τ-closed subspectra. Moreover, if f is a structural homeomorphism onto, then $\overline{f}$ is induced by a structural isomorphism onto of cofinal and τ-closed subspectra (and, therefore, $\overline{f}$ is also a structural homeomorphism onto).* □

Definitions. Let $I\!\!P$ be a structural class (therefore, $I\!\!P \subset \mathbb{P}$) and **S** an indexed collection of spaces with structure, whose supplements belong to $I\!\!P$. (About notations $\mathrm{C}^{I\!\!P}_{\mathbf{Q}}(\mathbf{S})$, $\mathrm{C}^{I\!\!P}_{\mathbf{N}}(\mathbf{S})$, $\mathrm{C}^{I\!\!P}_{\mathbf{Q},\mathbf{N}}(\mathbf{S})$, and $\mathrm{C}^{I\!\!P}_{\mathbf{\Psi}}(\mathbf{S})$ using below see the definition of a second-type saturated class in Section 8.3).

(1) Suppose that $\mathbb{P} = \mathbb{P}(\text{subset})$ and for every $X \in \mathbf{S}$, $E_X = (Q^X, X)$. Then, the indexed collection

$$\mathbf{Q} \equiv \{Q^X : X \in \mathbf{S}\}$$

is a restriction of $\mathbf{S}$ and, therefore, we can consider the subset $\mathrm{C}^{\mathbb{P}}_{\mathbf{Q}}(\mathbf{S})$ of $\mathrm{C}(\mathbf{S})$. Any subset of $\mathrm{C}^{\mathbb{P}}_{\mathbf{Q}}(\mathbf{S})$ is called a $\mathbb{P}$*-subset* of $\mathrm{C}(\mathbf{S})$. For every element c of $\mathrm{C}^{\mathbb{P}}_{\mathbf{Q}}(\mathbf{S})$ the pair $(\mathrm{T}(c)|_{\mathbf{Q}}, \mathrm{T}(c)) \in \mathbb{P}$ is called the *containing supplement* of the space $\mathrm{T}(c)$.

(2) Suppose that $\mathbb{P} = \mathbb{P}(\text{base})$ and for every $X \in \mathbf{S}$, $E_X = (B^X, X)$. Then, the indexed collection

$$\mathbf{B} \equiv \{B^X : X \in \mathbf{S}\}$$

is a co-base for $\mathbf{S}$. Suppose also that $\mathbf{N}$ is a co-indication of $\mathbf{B}$. Then, we can consider the subset $\mathrm{C}^{\mathbb{P}}_{\mathbf{N}}(\mathbf{S})$ of $\mathrm{C}(\mathbf{S})$. Any subset of $\mathrm{C}^{\mathbb{P}}_{\mathbf{N}}(\mathbf{S})$ is called a $\mathbb{P}$*-subset* of $\mathrm{C}(\mathbf{S})$. For every $c \in \mathrm{C}^{\mathbb{P}}_{\mathbf{N}}(\mathbf{S})$ the pair $(\mathrm{B}^{\mathrm{T}(c)}_{\diamondsuit,\mathbf{N}}, \mathrm{T}(c)) \in \mathbb{P}$ is called the *containing supplement* of $\mathrm{T}(c)$.

(3) Suppose that $\mathbb{P} = \mathbb{P}(\text{p-base})$ and $E_X = (Q^X, B^X, X)$, $X \in \mathbf{S}$. Then, the indexed collection

$$\mathbf{Q} \equiv \{Q^X : X \in \mathbf{S}\}$$

is a restriction of $\mathbf{S}$ and the indexed collection

$$\mathbf{B} \equiv \{B^X : X \in \mathbf{S}\}$$

is a p-co-base for $\mathbf{Q}$ in $\mathbf{S}$. Suppose also that $\mathbf{N}$ is a co-indication of $\mathbf{B}$. Then, we can consider the subset $\mathrm{C}^{\mathbb{P}}_{\mathbf{Q},\mathbf{N}}(\mathbf{S})$ of $\mathrm{C}(\mathbf{S})$. Any subset of $\mathrm{C}^{\mathbb{P}}_{\mathbf{Q},\mathbf{N}}(\mathbf{S})$ is called a $\mathbb{P}$*-subset* of $\mathrm{C}(\mathbf{S})$. For every $c \in \mathrm{C}^{\mathbb{P}}_{\mathbf{Q},\mathbf{N}}(\mathbf{S})$ the triad

$$(\mathrm{T}(c)|_{\mathbf{Q}}, \mathrm{B}^{\mathrm{T}(c)}_{\diamondsuit,\mathbf{N}}, \mathrm{T}(c)) \in \mathbb{P}$$

is called the *containing supplement* of $\mathrm{T}(c)$.

(4) Suppose that $\mathbb{P} = \mathbb{P}(\Phi, \Lambda, \prec)$ and for every $X \in \mathbf{S}$, $E_X = (\psi_X, X)$. Then, for every $\lambda \in \Lambda$ the indexed collection

$$\mathbf{\Psi}(\lambda) \equiv \{\psi_X(\lambda) : X \in \mathbf{S}\}$$

is a restriction of $\mathbf{S}$. Therefore, we can consider the subset $\mathrm{C}^{I\!P}_{\mathbf{\Psi}}(\mathbf{S})$ of $\mathrm{C}_{\mathbf{\Psi}}(\mathbf{S})$. Any subset of $\mathrm{C}^{I\!P}_{\mathbf{\Psi}}(\mathbf{S})$ is called a *$I\!P$-subset* of $\mathrm{C}(\mathbf{S})$. For every $c \in \mathrm{C}^{I\!P}_{\mathbf{\Psi}}(\mathbf{S})$ the pair $(\psi_{\mathrm{T}(c)}, \mathrm{T}(c)) \in I\!P$, where $\psi_{\mathrm{T}(c)}$ is the mapping of Λ into $\mathcal{P}(\mathrm{T}(c))$ such that

$$\psi_{\mathrm{T}(c)}(\lambda) = \mathrm{T}(c)|_{\mathbf{\Psi}(\lambda)},$$

is called the *containing supplement* of $\mathrm{T}(c)$.

Remarks. 1. In the cases, where either $I\!P \subset \mathrm{I\!P} \subset \mathrm{I\!P}(\text{base})$ or $I\!P \subset \mathrm{I\!P} \subset \mathrm{I\!P}(\text{p-base})$ the subsets $\mathrm{C}^{I\!P}_{\mathbf{N}}(\mathbf{S})$ and $\mathrm{C}^{I\!P}_{\mathbf{Q},\mathbf{N}}(\mathbf{S})$ of $\mathrm{C}(\mathbf{S})$ depend on a co-indication $\mathbf{N}$ of the corresponding co-base or co-p-base $\mathbf{B}$, respectively. Therefore, in these cases the notion of a $I\!P$-subset of $\mathrm{C}(\mathbf{S})$ is also depended upon this co-indication.

Below, whenever for a given structural class $I\!P$ we consider a fixed $I\!P$-subset or we consider the containing supplement of the space $\mathrm{T}(c)$ for some $c \in \mathrm{C}(\mathbf{S})$, where either $I\!P \subset \mathrm{I\!P}(\text{base})$ or $I\!P \subset \mathrm{I\!P}(\text{p-base})$, then we assume that a co-indication of the corresponding co-base or co-p-base has been chosen.

2. We note that if $I\!P$ is a structural second-type saturated class, then $\mathrm{C}(\mathbf{S})$ contains cofinal and τ-closed $I\!P$-subsets.

8.4.2 Lemma. *Let $I\!P$ be a structural class, $\mathbf{S}$ an indexed collection of spaces with structure whose supplements belongs to $I\!P$ and $\mathrm{C}_{I\!P}(\mathbf{S})$ a $I\!P$-subset of $\mathrm{C}(\mathbf{S})$. If $c_0, c_1 \in \mathrm{C}_{I\!P}(\mathbf{S})$ and $c_0 \prec^{\mathbf{S}} c_1$, then the natural mapping $p^{c_1}_{c_0} = \mathrm{D}[\mathbf{S}](c_0, c_1)$ of $\mathrm{T}(c_1)$ into $\mathrm{T}(c_0)$ is a structural mapping. (The Containing Spaces $\mathrm{T}(c_1)$ and $\mathrm{T}(c_0)$ are considered as spaces with structure, whose supplements are the corresponding containing supplements.)* $\square$

8.4.3 Lemma. *Let $I\!P$ be a structural class; $\mathbf{S}$ an indexed collection of spaces with structure whose supplements belongs to $I\!P$; and $\mathrm{C}_{I\!P}(\mathbf{S})$ a $I\!P$-subset of $\mathrm{C}(\mathbf{S})$. Suppose that C is a chain in $\mathrm{C}_{I\!P}(\mathbf{S})$ with $|C| \leq \tau$ and $c = \sup(C) \in \mathrm{C}_{I\!P}(\mathbf{S})$. Then, the natural mapping $\mathrm{D}[\mathbf{S}](C, c)$ of $\mathrm{D}[\mathbf{S}](c)$ into $\lim(\mathrm{D}[\mathbf{S}]|_C)$ is a structural mapping. (The limit space $\lim(\mathrm{D}[\mathbf{S}]|_C)$ is considered as a space with structure, the supplement of which is the corresponding limit supplement.)* $\square$

Definition. Let $\mathbf{S}$ be an indexed collection of spaces with structure. By the *structural free union* of elements of $\mathbf{S}$ we mean the free union F of elements of $\mathbf{S}$ considered as a space with structure, the supplement E_F of which is defined as follows:

(1) Suppose that $\mathbb{P} = \mathbb{P}(\text{subset})$ and $E_X = (Q^X, X)$, $X \in \mathbf{S}$. Denote by Q the free union of spaces Q^X, $X \in \mathbf{S}$. It is easy to see that Q is a subset of F. Then, $E_F = (Q, F)$.

(2) Suppose that $\mathbb{P} = \mathbb{P}(\text{base})$ and $E_X = (B^X, X)$, $X \in \mathbf{S}$. Suppose also that

$$N^X \equiv \{U_\delta^X : \delta \in \tau\}$$

is an indication of B^X. For every $\mathbf{S}' \subset \mathbf{S}$ and $\delta \in \tau$ we denote by $U_\delta^{\mathbf{S}'}$ the free union of spaces U_δ^X, $X \in \mathbf{S}'$, and by B the set of all such free unions. It is easy to see that B is a base for F. Then, $E_F = (B, F)$.

(3) Suppose that $\mathbb{P} = \mathbb{P}(\text{p-base})$ and $E_X = (Q^X, B^X, X)$, $X \in \mathbf{S}$. Suppose also that

$$N^X \equiv \{U_\delta^X : \delta \in \tau\}$$

is an indication of B^X. Denote by Q the subset of F defined as in the case (1) and by B the set of subsets of F defined as in the case (2). It is easy to see the set B is a p-base for Q in F. Then, $E_F = (Q, B, F)$.

(4) Suppose that $\mathbb{P} = \mathbb{P}(\Phi, \Lambda, \prec)$ and $E_X = (\psi_X, X)$, $X \in \mathbf{S}$. Then, $E_F = (\psi_F, F)$, where ψ_F is the mapping of Λ into $\mathcal{P}(F)$ such that $\psi_F(\lambda)$ is the free union of spaces $\psi_X(\lambda)$, $X \in \mathbf{S}$.

Remark. We note that in the above definitions the structural free union of elements of $\mathbf{S}$ in general is not an element of $\mathbb{P}$ because the weight of the space F may be large that τ. This union is an element of $\mathbb{P}^*$.

8.4.4 Lemma. *Let $\mathbb{P}$ be a structural second-type saturated class; $\mathbf{S}$ an indexed collection of spaces with structure whose supplements belongs to $\mathbb{P}$; and $\mathrm{C}_{\mathbb{P}}(\mathbf{S})$ a cofinal and τ-closed $\mathbb{P}$-subset of $\mathrm{C}(\mathbf{S})$. Then, the structural subset S of $\lim(\mathrm{D}[\mathbf{S}]|_{\mathrm{C}_{\mathbb{P}}(\mathbf{S})}) = \lim(\mathrm{D}[\mathbf{S}])$ is structurally homeomorphic to the structural free union of elements of $\mathbf{S}$.* $\square$

The above lemma is proved using the following lemma.

8.4.5 Lemma. *Let $\mathbf{S}$ be an indexed collection of spaces with structure and $\mathbf{S}'$ a subcollection of $\mathbf{S}$. Then, there exist a co-mark $\mathbf{M}$ of $\mathbf{S}$ and* *an* **M***-admissible family* R *of equivalence elations on* **S** *such that* **S**$'$ *is an element of* $\mathrm{C}^{\diamondsuit}(\mathrm{R})$. □

8.4.6 Proposition. *Let* $I\!\!P$ *be a structural second-type saturated class and* **S** *an indexed collection of spaces with structure whose supplements belong to* $I\!\!P$*. Then, for each cofinal and* τ*-closed* $I\!\!P$*-subset* $\mathrm{C}_{I\!\!P}(\mathbf{S})$ *of* $\mathrm{C}(\mathbf{S})$ *we have:*

(1) $\mathrm{D}[\mathbf{S}]|_{\mathrm{C}_{I\!\!P}(\mathbf{S})}$ *is a* $I\!\!P$*-*τ*-spectrum.*

(2) *The structural subset* S *of* $\lim(\mathrm{D}[\mathbf{S}]|_{\mathrm{C}_{I\!\!P}(\mathbf{S})}) = \lim(\mathrm{D}[\mathbf{S}])$ *is structurally homeomorphic to the structural free union of elements of* $I\!\!P$*.*

(3) $\mathrm{D}[\mathbf{S}]|_{\mathrm{C}_{I\!\!P}(\mathbf{S})}$ *is structurally factorizing with respect to the structural subset* S *of* $\lim(\mathrm{D}[\mathbf{S}]|_{\mathrm{C}_{I\!\!P}(\mathbf{S})})$. □

Applying the Structural Spectral Theorem we have the following proposition, which can be considered as the "spectral determination" of the containing supplements of the Containing Spaces (corresponding to an indexed collection of spaces with structure) belonging to a given structural second-type saturated class.

8.4.7 Proposition. *Let* $I\!\!P$ *be a structural second-type saturated class;* **S** *an indexed collection of spaces whose supplements belong to* $I\!\!P$*; and* $\mathrm{C}_{I\!\!P}(\mathbf{S})$ *a cofinal and* τ*-closed* $I\!\!P$*-subset of* $\mathrm{C}(\mathbf{S})$*. Suppose that* D *is a spectrum on* $\mathrm{C}_{I\!\!P}(\mathbf{S})$ *such that:*

(1) D *is a structural* τ*-spectrum on* $\mathrm{C}_{I\!\!P}(\mathbf{S})$*.*

(2) *The limit space* $\lim(D)$ *contains a structural surjective subset* L *of* $\lim(D)$ *which is structurally homeomorphic to the structural free union of elements of* **S***.*

(3) D *is structurally* $I\!\!P$*-factorizing with respect to the structural surjective subset* L*.*

Then, the spectra D *and* $\mathrm{D}[\mathbf{S}]|_{\mathrm{C}_{I\!\!P}(\mathbf{S})}$ *contain structurally isomorphic cofinal and* τ*-closed subspectra. Therefore,* D *contains a cofinal and* τ*-closed* $I\!\!P$*-subspectrum which is structurally factorizing with respect to* L*.* □

The "covering property" of the containing supplements of the Containing Spaces belonging to a given structural saturated class can be formulated as follows.

8.4.8 Proposition. *Let* $I\!\!P$ *be a structural second-type saturated class,* Y *a universal element in* $I\!\!P$*, and for every* $X \in I\!\!P$ *let* h_X *be a structural* *embedding of X into Y. Then, there exist an indexed collection $\mathbf{S}$ of spaces with structure whose supplements belong to $I\!P$; an element c of $\mathrm{C}(\mathbf{S})$, for which the containing supplement of $\mathrm{T}(c)$ is a universal element in $I\!P$; a structural mapping f of the space $\mathrm{T}(c)$ into the space Y; and for every $X \in I\!P$ a structural embedding $i^X_{\mathrm{T}(c)}$ of X into $\mathrm{T}(c)$ such that the following relation is satisfied:*

$$f \circ i^X_{\mathrm{T}} = h_X, \quad X \in I\!P. \ \square$$

Below, in the case $\tau = \omega$ we consider some concrete structural classes and apply the above given results.

8.4.9 Proposition. *The following structural classes are second-type saturated (about the notion of an ω-sequence and its relation to the notion of a $(\Phi, \Lambda, \prec, \mathbf{p})$-structure see Section 4.4):*

(1) The class of all ω-sequences $(X, X_0, ..., X_i, ...)$, where X is a separable metrizable space and X_i, $i \in \omega$, is a zero-dimensional subsets of X, such that

$$X = \cup\{X_i : i \in \omega\}.$$

(2) The class of all ω-sequences $(X, X_0, ..., X_i, ...)$, where X is a separable metrizable space and X_i, $i \in \omega$, is a closed subset of X with $\mathrm{ind}(X_i) \le i$, such that

$$X = \cup\{X_i : i \in \omega\}.$$

(3) The class of all ω-sequences $(X, X_0, ..., X_i, ...)$, where X is a separable metrizable space and X_i, $i \in \omega$, is an open subset of X with $\mathrm{ind}(X_i) \le i$, such that

$$X = \cup\{X_i : i \in \omega\}. \ \square$$

8.4.10 Corollary. *Let $I\!P$ be one of the classes mentioned in Proposition 8.4.9. Then, for every indexed collection $\mathbf{S}$ of spaces with structure whose supplements belong to $I\!P$ and for every cofinal and ω-closed $I\!P$-subset $\mathrm{C}_{I\!P}(\mathbf{S})$ of $\mathrm{C}(\mathbf{S})$ we have:*

(1) The spectrum $\mathrm{D}[\mathbf{S}]|_{\mathrm{C}_{I\!P}(\mathbf{S})}$ is a structural $I\!P$-ω-spectrum.

(2) The subset S of $\lim(\mathrm{D}[\mathbf{S}]|_{\mathrm{C}_{I\!P}(\mathbf{S})})$ is a structural surjective subset, which is structurally homeomorphic to the structurally free union of elements of $\mathbf{S}$.

(3) $\mathrm{D}[\mathbf{S}]|_{\mathrm{C}_{I\!P}(\mathbf{S})}$ *is structurally factorizing with respect to the structural surjective subset* S *of* $\lim(\mathrm{D}[\mathbf{S}]|_{\mathrm{C}_{I\!P}(\mathbf{S})})$. □

8.4.11 Proposition. *Let* $I\!P$ *be one of the classes mentioned in Proposition 8.4.9;* $\mathbf{S}$ *an indexed collection of spaces with structure whose supplement belongs to* $I\!P$*; and* $\mathrm{C}_{I\!P}(\mathbf{S})$ *a cofinal and* ω*-closed* $I\!P$*-subset of* $\mathrm{C}(\mathbf{S})$*. Suppose that* D *is an inverse spectrum on* $\mathrm{C}_{I\!P}(\mathbf{S})$ *such that:*

(1) D *is a structural* ω*-spectrum.*

(2) *The limit space of* D *contains a structural surjective subset* L *which is structurally homeomorphic to the structural free union of elements of* $\mathbf{S}$.

(3) D *is structurally* $I\!P$*-factorizing with respect to* L.

Then, the spectra D *and* $\mathrm{D}[\mathbf{S}]|_{\mathrm{C}_{I\!P}(\mathbf{S})}$ *contain structurally isomorphic cofinal and* ω*-closed subspectra. Therefore,* D *contains a cofinal a* ω*-closed* $I\!P$*-subspectrum which is structurally factorizing with respect to* L. □

Chapter 9

Isometries and universality

Agreement. In the present chapter it is assumed that $\tau = \omega$.

Notation. The elements of ω are denoted by i, j, m, n, k, l and so on instead of ε, δ, η as found in the preceding chapters. According to our assumption (see the Introductory Remarks) each element m of ω is a finite subset of ω and, therefore, it can be considered as an element of $\mathcal{F}$.

We denote by $\Re^+$ the set of all positive real numbers. In this chapter by ε and δ we denote elements of $\Re^+$.

Notation. According to the above agreement whenever we consider an arbitrary metric space it is assumed that this space is separable. For a given metric space X we denote by ϱ_X the metric of X. For every subset Q of X we denote by $\mathrm{Diam}(Q)$ the diameter of Q. It is assumed that $\mathrm{Diam}(Q) = 0$ if $Q = \emptyset$ and $\mathrm{Diam}(Q) = \infty$ if for every $m \in \omega$ there exist points x and y of Q such that $m \leq \varrho_X(x, y)$. It is also assumed that $\varrho_X(x, \emptyset) = \infty$ for every $x \in X$.

Definition. A mapping f of a metric space X into a metric space Y is said to be *isometric* or an *isometry* if

$$\varrho_X(x, y) = \varrho_Y(f(x), f(y)), \quad x, y \in X.$$

We note that every isometry is a topological embedding.

Let $I\!\!P$ be a class of metric spaces. A metric space T is said to be *isometrically containing* for $I\!\!P$ if for every $X \in I\!\!P$ there exists an isometry of X into T. If moreover T is an element of $I\!\!P$, then T is said to be *isometrically universal* in $I\!\!P$.

In the present chapter and in parallel to the Containing Spaces constructed in Section 1.2, we shall use the generalization of the Containing Spaces given in Section 1.5. The infinite cardinals ν and μ used in Section 1.5 satisfy the condition $\nu \leq \tau$ and $\mu \leq \tau^+$. Since $\tau = \omega$ we have that $\nu = \omega$ and either $\mu = \omega$ or $\mu = \omega^+$. We shall use the generalization of the Containing Spaces corresponding to the case $\nu = \omega$ and $\mu = \omega^+$. In this case the set $\mathcal{P}_\nu(\tau) = \mathcal{P}_\omega(\omega)$ coincides with the set $\mathcal{F}$ of all finite subsets of ω. (We recall that the case $\nu = \mu = \omega$ corresponds to the construction given in Section 1.2.)

Let $\mathbf{M}$ be a co-mark of an indexed collection $\mathbf{S}$ of spaces. In Section 1.5 the notion of the $(\mathbf{M}, \nu)$-standard family $\mathrm{R}^\nu_{\mathbf{M}}$ of equivalence relations on $\mathbf{S}$ and the notions of an μ-admissible and an $(\mathbf{M}, \nu, \mu)$-admissible families of equivalence relations on $\mathbf{S}$ were defined. We note that the $(\mathbf{M}, \omega)$-standard family $\mathrm{R}^\omega_{\mathbf{M}}$ of equivalence relations on $\mathbf{S}$ coincides with the $\mathbf{M}$-standard family $\mathrm{R}_{\mathbf{M}}$. Also, the only difference between an admissible family of equivalence relations on $\mathbf{S}$ and an ω^+-admissible family and, therefore, between an $\mathbf{M}$-admissible family and an $(\mathbf{M}, \omega, \omega^+)$-admissible family consists in the fact that the number of equivalence classes of any element of the admissible or $\mathbf{M}$-admissible family is finite while this number for an element of the ω^+-admissible or $(\mathbf{M}, \omega, \omega^+)$-admissible family may be countable.

In Section 9.1 for a given indexed collection $\mathbf{S}$ of metric spaces, for a co-mark $\mathbf{M}$ of $\mathbf{S}$, and for an $\mathbf{S}$-indexed collection $\mathbf{P} \equiv \{P^X : X \in \mathbf{S}\}$ of ω-indexed dense subsets, such that P^X is an ω-indexed dense subset of X if $X \neq \emptyset$ and $P^X = \emptyset$ if $X = \emptyset$, we shall define the notion of an $(\mathbf{M}, \mathbf{P})$-admissible family of equivalence relations on $\mathbf{S}$. Such a family is an $(\mathbf{M}, \omega, \omega^+)$-admissible family satisfying some additional condition. Furthermore, for a given $(\mathbf{M}, \mathbf{P})$-admissible family R, we construct a metric ϱ_{T} on the Containing Space $\mathrm{T} \equiv \mathrm{T}(\mathbf{M}, \mathrm{R})$ compatible with its topology. With respect to this metric, the natural embedding i^X_{T} of any element X of $\mathbf{S}$ into T is an isometry, that is T is an isometrically Containing Space for $\mathbf{S}$.

In Section 9.2 we define the notion of a (complete) isometrically ω-saturated class (of spaces, of subsets, of bases, of p-bases, and of spaces with structure) in the universe of discourse of metric spaces ($\equiv$ universe of metric spaces). We prove that these classes have the intersection and the universality properties. Each class $I\!\!M$ (of any considered type) in the universe of metrizable spaces defines naturally a class $I\!\!P$ (of the same type) in the universe of metric spaces. We prove that if $I\!\!M$ is a (complete)

ω-saturated class, then $\mathbb{P}$ is a (complete) isometrically ω-saturated class. We give a series of corollaries of this result concerning the existence of isometrically (and properly) universal elements. These corollaries are reformulations for separable metric spaces and isometries of some corollaries given in the preceding chapters.

In Section 9.3 we define the notion of a uniform class of metric spaces. A class $\mathbb{P}$ of metric spaces is uniform if: (a) there exists an integer $\mathbf{d} \in \omega$ such that $\mathrm{Diam}(X) \leq \mathbf{d}$ for every $X \in \mathbb{P}$ and (b) for every $\varepsilon \in \Re^+$ there exists an integer $\mathbf{n}(\varepsilon) \in \omega$ such that every element X of $\mathbb{P}$ has a finite ε-net the number of elements of which is less than or equal to $\mathbf{n}(\varepsilon)$. We give some results concerning this notion. In particular, we prove that for a class of compact metric spaces the uniformity is equivalent to the existence of an isometrically containing compact metric space for this class.

Finally, in Section 9.4 we prove that the class of all separable metric spaces with the metric dimension less than or equal to an integer $n \in \omega$ is an isometrically ω-saturated class and, therefore, in this class there exists isometrically universal spaces.

9.1 Isometrically Containing Spaces

Agreement 1. In this section it is assumed that an indexed collection of metric spaces, denoted by $\mathbf{S}$, and a co-mark of $\mathbf{S}$, denoted by

$$\mathbf{M} \equiv \{\{U_i^X : i \in \omega\} : X \in \mathbf{S}\},$$

are given.

It is also assumed that for every non-empty element X of $\mathbf{S}$ an ω-indexed dense subset of X, denoted by

$$P^X \equiv \{a_i^X : i \in \omega\},$$

is given. The $\mathbf{S}$-indexed collection

$$\{P^X : X \in \mathbf{S}\},$$

where $P^X = \emptyset$ if $X = \emptyset$, will be denoted by $\mathbf{P}$. Such a collection will be called an $\mathbf{S}$-*indexed collection of* ω-*indexed dense subsets.*

Notation. Let $n \in \omega$ and $i, j \in n+2$. For every non-empty element X of $\mathbf{S}$ we denote by $k_0(X, n, i, j)$, $k_1(X, n, i, j)$, and $k(X, n, i)$ the elements of the set $\omega \cup \{\infty\}$ (which we call also integers or numbers), defined as follows:

(a) $k_0(X, n, i, j) = \infty$ if $U_j^X = X$ (that is, if $\varrho_X(a_i^X, X \setminus U_j^X) = \infty$) and $k_0(X, n, i, j)$ is the uniquely determined element k of ω such that

$$\frac{k}{n+2} \leq \varrho_X(a_i^X, X \setminus U_j^X) < \frac{k+1}{n+2}$$

if $U_j^X \neq X$ (and, therefore, $\varrho_X(a_i^X, X \setminus U_j^X) < \infty$);

(b) $k_1(X, n, i, j)$ is the uniquely determined element k of ω such that

$$\frac{k}{n+2} \leq \varrho_X(a_i^X, a_j^X) < \frac{k+1}{n+2}; \text{ and}$$

(c) $k(X, n, i) = \infty$ if $\mathrm{Diam}(U_i^X) = \infty$ and $k(X, n, i)$ is the uniquely determined element k of ω such that

$$\frac{k}{n+2} \leq \mathrm{Diam}(U_i^X) < \frac{k+1}{n+2}$$

if $\mathrm{Diam}(U_i^X) < \infty$.

For an element X of $\mathbf{S}$, which is the empty set, we suppose that

$$k_0(X, n, i, j) = k_1(X, n, i, j) = k(X, n, i) = 0.$$

Definition. An ω^+-admissible family

$$\mathrm{R} \equiv \{\sim^s : s \in \mathcal{F}\}$$

of equivalence relations on $\mathbf{S}$ is said to be $(\mathbf{M}, \mathbf{P})$-*admissible* if for every $n \in \omega$ there exists an element m of ω, $n \leq m$, such that for every $X, Y \in \mathbf{S}$ the condition $X \sim^m Y$ implies that:

(a) $\sim^m \subset \sim^n_{\mathrm{M}}$ (and, therefore, R is $(\mathbf{M}, \omega, \omega^+)$-admissible), and

(b) for evey $i, j \in n+2$ we have

$$k_0(X, n, i, j) = k_0(Y, n, i, j),$$

$$k_1(X, n, i, j) = k_1(Y, n, i, j), \text{ and}$$

$$k(X, n, i) = k(Y, n, i).$$

The least integer m satisfying the above conditions (in particular, the condition $m \geq n$) will be denoted by $m(n)$.

We note that any ω^+-admissible family of equivalence relations on $\mathbf{S}$, which is a final refinement of an $(\mathbf{M}, \mathbf{P})$-admissible family, is also $(\mathbf{M}, \mathbf{P})$-admissible.

Agreement 2. In what follows of this section it is assumed that an $(\mathbf{M}, \mathbf{P})$-admissible family of equivalence relations on $\mathbf{S}$ is given. This family will be denoted by

$$\mathrm{R} \equiv \{\sim^s : s \in \mathcal{F}\}.$$

The proof of the next lemma follows immediately by the definition of an $(\mathbf{M}, \mathbf{P})$-admissible family of equivalence relations and the definition of the numbers $k_0(X, n, i, j)$, $k_1(X, n, i, j)$, and $k(X, n, i)$.

9.1.1 Lemma. *Let $n \in \omega$ and $i, j \in n+2$. Suppose that for two non-empty elements X and Y of $\mathbf{S}$ we have $X \sim^{m(n)} Y$. Then, the following relations are satisfied:*

$$|\varrho_X(a_i^X, X \setminus U_j^X) - \varrho_Y(a_i^Y, Y \setminus U_j^Y)| < \frac{1}{n+2} \tag{9.1.1}$$

if $U_j^X \neq X$ (and, therefore, $U_j^Y \neq Y$),

$$|\varrho_X(a_i^X, a_j^X) - \varrho_Y(a_i^Y, a_j^Y)| < \frac{1}{n+2}, \quad \text{and} \tag{9.1.2}$$

$$|\mathrm{Diam}(U_i^X) - \mathrm{Diam}(U_i^Y)| < \frac{1}{n+2} \tag{9.1.3}$$

if $\mathrm{Diam}(U_i^X) < \infty$ (and, therefore, $\mathrm{Diam}(U_i^Y) < \infty$).

Therefore, if $X \sim^{m(n)} Y$ for every $n \in \omega$, then for every $i, j \in \omega$ we have

$$\varrho_X(a_i^X, X \setminus U_j^X) = \varrho_Y(a_i^Y, Y \setminus U_j^Y), \quad \text{and} \tag{9.1.4}$$

$$\varrho_X(a_i^X, a_j^X) = \varrho_Y(a_i^Y, a_j^Y). \tag{9.1.5}$$

In particular, relation (9.1.4) implies that:

$$a_i^X \in U_j^X \text{ if and only if } a_i^Y \in U_j^Y. \;\square \tag{9.1.6}$$

Definition. On the set $\mathbf{S}$ we define an equivalence relation, denoted by $\sim^\infty$, as follows: two elements X and Y of $\mathbf{S}$ are $\sim^\infty$-equivalent if and only if they are $\sim^s$-equivalent for every $s \in \mathbf{S}$. (The set of all $\sim^\infty$-equivalent classes is denoted by $\mathrm{C}(\sim^\infty)$.)

We note that $X \sim^s Y$ for every $s \in \mathcal{F}$ if and only if $X \sim^{m(n)} Y$ for every $n \in \omega$.

Notation. Let $\mathbf{E}$ be an arbitrary $\sim^\infty$-equivalent class. We denote by $\mathrm{T}(\mathbf{E})$ the subspace of the Containing Space $\mathrm{T} \equiv \mathrm{T}(\mathbf{M}, \mathrm{R})$ consisting of all points $\mathbf{a}$ of T such that for some $(x, X) \in \mathbf{a}$, $X \in \mathbf{E}$. (Therefore, for every $(x, X) \in \mathbf{a}$ we have $X \in \mathbf{E}$.)

For every $l \in \omega$ we denote by $U_l^{\mathrm{T}(\mathbf{E})}$ the set of all points $\mathbf{a}$ of $\mathrm{T}(\mathbf{E})$ such that there exists a pair $(x, X) \in \mathbf{a}$ for which $x \in U_l^X$ (therefore, for every $(x, X) \in \mathbf{a}$, $X \in \mathbf{E}$ and $x \in U_l^X$). It is easy to see that

$$\{U_l^{\mathrm{T}(\mathbf{E})} : l \in \omega\} = \{\mathrm{T}(\mathbf{E}) \cap U : U \in \mathrm{B}^{\mathrm{T}}\},$$

which means that the set $\{U_l^{\mathrm{T}(\mathbf{E})} : l \in \omega\}$ is a base for the open subsets of $\mathrm{T}(\mathbf{E})$.

For every $X \in \mathbf{E}$ and $i \in \omega$ we denote by $\mathbf{a}_i^{\mathrm{T}(\mathbf{E})}$ the point of $\mathrm{T}(\mathbf{E})$ containing the pair (a_i^X, X). Relation (9.1.6) of Lemma 9.1.1 implies that this point is independent of the element $X \in \mathbf{E}$.

Below, for an arbitrary element $\mathbf{E}$ of the set $\mathrm{C}(\sim^\infty)$ we define a function $\varrho_{\mathrm{T}(\mathbf{E})}$ from the set $\mathrm{T}(\mathbf{E}) \times \mathrm{T}(\mathbf{E})$ to the set $\Re^+ \cup \{0\}$. We shall prove that this function is a metric on the set $\mathrm{T}(\mathbf{E})$ inducing the topology of the space $\mathrm{T}(\mathbf{E})$.

Definition. Let $\mathbf{E}$ be an arbitrary $\sim^\infty$-equivalent class. We define a function

$$\varrho_{\mathrm{T}(\mathbf{E})} : \mathrm{T}(\mathbf{E}) \times \mathrm{T}(\mathbf{E}) \to \Re^+ \cup \{0\}$$

as follows. Let $\mathbf{a}$ and $\mathbf{b}$ be two points of the set $\mathrm{T}(\mathbf{E})$. Suppose that $(x, X) \in \mathbf{a}$ and $(y, Y) \in \mathbf{b}$. Suppose also that $\{a_{i_k}^X : k \in \omega\}$ is a sequence of points of the set P^X converging to x, and $\{a_{j_k}^Y : k \in \omega\}$ is a sequence of points of P^Y converging to y. Then, we set

$$\varrho_{\mathrm{T}(\mathbf{E})}(\mathbf{a}, \mathbf{b}) = \lim_{k\to\infty}(\varrho_X(a_{i_k}^X, a_{j_k}^X)).$$

9.1.2 Lemma. *Let* $\mathbf{E}$ *be an element of* $\mathrm{C}(\sim^\infty)$. *The function* $\varrho_{\mathrm{T}(\mathbf{E})}$ *is well defined, that is, for every* $\mathbf{a}, \mathbf{b} \in \mathrm{T}(\mathbf{E})$ *the real number* $\varrho_{\mathrm{T}(\mathbf{E})}(\mathbf{a}, \mathbf{b})$

exists and is independent of the choice of the pairs (x, X) *and* (y, Y) *and the sequences* $\{a^X_{i_k} : k \in \omega\}$ *and* $\{a^Y_{j_k} : k \in \omega\}$ *(considered in the definition of* $\varrho_{\mathrm{T}(\mathbf{E})}$*).*

Proof. First, we prove that the sequence

$$\{\varrho_X(a^X_{i_k}, a^X_{j_k}) : k \in \omega\} \tag{9.1.7}$$

is a Cauchy sequence.

Indeed, let $\varepsilon \in \Re^+$. There exists an integer $k(\varepsilon) \in \omega$ such that

$$\varrho_X(x, a^X_{i_k}) < \varepsilon/4 \quad \text{and} \quad \varrho_Y(y, a^Y_{j_k}) < \varepsilon/4$$

for every integer $k \geq k(\varepsilon)$. Let $p, q \in \omega$ and $p, q \geq k(\varepsilon)$. Then, using relation (9.1.5) of Lemma 9.1.1 and supposing that $\varrho_X(a^X_{i_p}, a^X_{j_p}) \geq \varrho_X(a^X_{i_q}, a^X_{j_q})$ we have

$$|\varrho_X(a^X_{i_p}, a^X_{j_p}) - \varrho_X(a^X_{i_q}, a^X_{j_q})| = \varrho_X(a^X_{i_p}, a^X_{j_p}) - \varrho_X(a^X_{i_q}, a^X_{j_q}) \leq$$

$$\varrho_X(a^X_{i_p}, x) + \varrho_X(x, a^X_{i_q}) + \varrho_X(a^X_{i_q}, a^X_{j_q}) + \varrho_X(a^X_{j_q}, a^X_{j_p}) - \varrho_X(a^X_{i_q}, a^X_{j_q}) \leq$$

$$\varrho_X(a^X_{i_p}, x) + \varrho_X(x, a^X_{i_q}) + \varrho_Y(a^Y_{j_q}, a^Y_{j_p}) \leq$$

$$\varrho_X(a^X_{i_p}, x) + \varrho_X(x, a^X_{i_q}) + \varrho_Y(d^Y_{j_q}, y) + \varrho_Y(y, a^Y_{j_p}) <$$

$$\varepsilon/4 + \varepsilon/4 + \varepsilon/4 + \varepsilon/4 = \varepsilon.$$

Thus, the sequence (9.1.7) is a Cauchy sequence and, therefore, there exists the limit of this sequence, which is the real number $\varrho_{\mathrm{T}(\mathbf{E})}(\mathbf{a}, \mathbf{b})$.

Now, suppose that $(x', X') \in \mathbf{a}$, $(y', Y') \in \mathbf{b}$, $\{a^{X'}_{i'_k} : k \in \omega\}$ is a sequence of points of the set $P^{X'}$ converging to x', and $\{a^{Y'}_{j'_k} : k \in \omega\}$ is a sequence of points of $P^{Y'}$ converging to y'. We prove that

$$\lim_{k\to\infty}(\varrho_X(a^X_{i_k}, a^X_{j_k})) = \lim_{k\to\infty}(\varrho_{X'}(a^{X'}_{i'_k}, a^{X'}_{j'_k})). \tag{9.1.8}$$

First, we prove that the sequence $\{a^X_{i'_k} : k \in \omega\}$ converges to x. Indeed, let U^X_l, $l \in \omega$, be an open neighbourhood of x in X. Since $(x, X), (x', X') \in \mathbf{a}$, the set $U^{X'}_l$ is an open neighbourhood of x' in X'. Therefore, $U^{X'}_l$ contains almost all elements (that is, contains all elements except maybe a finite number) of the sequence $\{a^{X'}_{i'_k} : k \in \omega\}$. By property (9.1.6) of Lemma 9.1.1, U^X_l contains almost all element of the sequence $\{a^X_{i'_k} : k \in \omega\}$. Therefore, this sequence converges to x. Similarly, $\{a^Y_{j'_k} : k \in \omega\}$ converges to y.

Now, we prove relation (9.1.8). Using relation (9.1.5) of Lemma 9.1.1 we have

$$\lim_{k\to\infty}(\varrho_X(a^X_{i_k}, a^X_{j_k})) \leq$$

$$\lim_{k\to\infty}(\varrho_X(a^X_{i_k}, x) + \varrho_X(x, a^X_{i'_k}) + \varrho_X(a^X_{i'_k}, a^X_{j'_k}) + \varrho_X(a^X_{j'_k}, a^X_{j_k})) =$$

$$\lim_{k\to\infty}(\varrho_X(a^X_{i_k}, x) + \varrho_X(x, a^X_{i'_k}) + \varrho_{X'}(a^{X'}_{i'_k}, a^{X'}_{j'_k}) + \varrho_Y(a^Y_{j'_k}, a^Y_{j_k})) \leq$$

$$\lim_{k\to\infty}(\varrho_X(a^X_{i_k}, x) + \varrho_X(x, a^X_{i'_k}) + \varrho_{X'}(a^{X'}_{i'_k}, a^{X'}_{j'_k}) + \varrho_Y(a^Y_{j'_k}, y) + \varrho_Y(y, a^Y_{j_k})) \leq$$

$$\lim_{k\to\infty}(\varrho_X(a^X_{i_k}, x)) + \lim_{k\to\infty}(\varrho_X(x, a^X_{i'_k})) + \lim_{k\to\infty}(\varrho_{X'}(a^{X'}_{i'_k}, a^{X'}_{j'_k}))+$$

$$\lim_{k\to\infty}(\varrho_Y(a^Y_{j'_k}, y)) + \lim_{k\to\infty}(\varrho_Y(y, a^Y_{j_k})) = \lim_{k\to\infty}(\varrho_{X'}(a^{X'}_{i'_k}, a^{X'}_{j'_k})).$$

Similarly, we prove that

$$\lim_{k\to\infty}(\varrho_{X'}(a^{X'}_{i'_k}, a^{X'}_{j'_k})) \leq \lim_{k\to\infty}(\varrho_X(a^X_{i_k}, a^X_{j_k})),$$

which completes the proof of the relation (9.1.8) and, therefore, the proof of the lemma. □

9.1.3 Lemma. *Let* $\mathbf{E}$ *be an element of* $\mathrm{C}(\sim^\infty)$. *The function* $\varrho_{\mathrm{T}(\mathbf{E})}$ *is a metric on the set* $\mathrm{T}(\mathbf{E})$.

Proof. Let $\mathbf{a}$ and $\mathbf{b}$ be points of $\mathrm{T}(\mathbf{E})$. Obviously, $\varrho_{\mathrm{T}(\mathbf{E})}(\mathbf{a}, \mathbf{b}) \geq 0$ and

$$\varrho_{\mathrm{T}(\mathbf{E})}(\mathbf{a}, \mathbf{b}) = \varrho_{\mathrm{T}(\mathbf{E})}(\mathbf{b}, \mathbf{a}).$$

We prove that $\varrho_{\mathrm{T}(\mathbf{E})}(\mathbf{a}, \mathbf{b}) = 0$ if and only if $\mathbf{a} = \mathbf{b}$. Indeed, let $\mathbf{a} = \mathbf{b}$ and $(x, X) \in \mathbf{a}$. Let also $\{a^X_{i_k} : k \in \omega\}$ be a sequence converging to x. Then, we have

$$\varrho_{\mathrm{T}(\mathbf{E})}(\mathbf{a}, \mathbf{a}) = \lim_{k\to\infty}(\varrho_X(a^X_{i_k}, a^X_{i_k})) = 0.$$

Conversely, suppose that $\varrho_{\mathrm{T}(\mathbf{E})}(\mathbf{a},\mathbf{b}) = 0$. We prove that $\mathbf{a} = \mathbf{b}$. Let $(x, X) \in \mathbf{a}$ and $(y, Y) \in \mathbf{b}$. Since $X \sim^s Y$ for every $s \in \mathcal{F}$ it suffices to prove that for every $l \in \omega$, $x \in U_l^X$ if and only if $y \in U_l^Y$. Suppose that the sequences $\{a_{i_k}^X : k \in \omega\}$ and $\{a_{j_k}^Y : k \in \omega\}$ converge to x and y, respectively. Then, we have

$$\varrho_{\mathrm{T}(\mathbf{E})}(\mathbf{a},\mathbf{b}) = \lim_{k\to\infty}(\varrho_X(a_{i_k}^X, a_{j_k}^X)) = 0,$$

from which it follows that $\lim_{k\to\infty}(a_{j_k}^X) = x$. Let U_l^X, $l \in \omega$, be an open neighbourhood of x in X. Then, there exists an element ε of $\Re^+$ such that $\varrho_X(x, X \setminus U_l^X) > 2\varepsilon$. Without loss of generality we can suppose that for every $k \in \omega$, $\varrho_X(a_{j_k}^X, X \setminus U_l^X) > \varepsilon$ and, therefore (see relation (9.1.4) of Lemma 9.1.1), $\varrho_Y(a_{j_k}^Y, Y \setminus U_l^Y) > \varepsilon$. By the last relation it follows that $\varrho_Y(y, Y \setminus U_l^Y) \geq \varepsilon$, which means that $y \in U_l^Y$.

Similarly, we prove that if $y \in U_l^Y$, then $x \in U_l^X$, $l \in \omega$. Thus, for every $l \in \omega$, $x \in U_l^X$ if and only if $y \in U_l^Y$, which means that $\mathbf{a} = \mathbf{b}$.

Now, let $\mathbf{c}$ be another point of $\mathrm{T}(\mathbf{E})$. We prove that

$$\varrho_{\mathrm{T}(\mathbf{E})}(\mathbf{a},\mathbf{c}) \leq \varrho_{\mathrm{T}(\mathbf{E})}(\mathbf{a},\mathbf{b}) + \varrho_{\mathrm{T}(\mathbf{E})}(\mathbf{b},\mathbf{c}). \tag{9.1.9}$$

Indeed, let $(x, X) \in \mathbf{a}$, $(y, Y) \in \mathbf{b}$, and $(z, Z) \in \mathbf{c}$. Let also $\{a_{i_k}^X : k \in \omega\}$, $\{a_{j_k}^Y : k \in \omega\}$, and $\{a_{l_k}^Z : k \in \omega\}$ be sequences converging to the points x, y, and z, respectively. Then, for every $k \in \omega$ we have

$$\varrho_X(a_{i_k}^X, a_{l_k}^X) \leq \varrho_X(a_{i_k}^X, a_{j_k}^X) + \varrho_X(a_{j_k}^X, a_{l_k}^X) \leq$$

$$\varrho_X(a_{i_k}^X, a_{j_k}^X) + \varrho_Y(a_{j_k}^Y, a_{l_k}^Y)$$

and, therefore,

$$\lim_{k\to\infty}(\varrho_X(a_{i_k}^X, a_{l_k}^X)) \leq$$

$$\lim_{k\to\infty}(\varrho_X(a_{i_k}^X, a_{j_k}^X)) + \lim_{k\to\infty}(\varrho_Y(a_{j_k}^Y, a_{l_k}^Y)).$$

This implies relation (9.1.9). □

9.1.4 Lemma. *Let* $\mathbf{E}$ *be an element of* $\mathrm{C}(\sim^\infty)$. *The topology on the set* $\mathrm{T}(\mathbf{E})$ *induced by the metric* $\varrho_{\mathrm{T}(\mathbf{E})}$ *coincides with the topology of the space* $\mathrm{T}(\mathbf{E})$.

Proof. Let $\mathbf{a}$ be a point of $\mathrm{T}(\mathbf{E})$. It suffices to prove that:

(a) for every $\varepsilon \in \Re^+$ there exist an element l of ω such that $\mathbf{a} \in U_l^{\mathrm{T}(\mathbf{E})}$ and $\mathrm{Diam}(U_l^{\mathrm{T}(\mathbf{E})}) \leq \varepsilon$ and

(b) if $\mathbf{a} \in U_l^{\mathrm{T}(\mathbf{E})}$, $l \in \omega$, then there exists $\varepsilon \in \Re^+$ such that

$$\{\mathbf{b} \in \mathrm{T}(\mathbf{E}) : \varrho_{\mathrm{T}(\mathbf{E})}(\mathbf{a}, \mathbf{b}) < \varepsilon\} \subset U. \tag{9.1.10}$$

We prove (a). Consider an $\varepsilon \in \Re^+$ and let $(x, X) \in \mathbf{a}$. There exists $l \in \omega$ such that $x \in U_l^X$ and $\mathrm{Diam}(U_l^X) < \varepsilon$. Clearly, $\mathbf{a} \in U_l^{\mathrm{T}(\mathbf{E})}$. We need to prove that $\mathrm{Diam}(U^{\mathrm{T}(\mathbf{E})}) \leq \varepsilon$. Suppose that $\mathbf{b}$ and $\mathbf{c}$ are two points of $U_l^{\mathrm{T}(\mathbf{E})}$, $(y, Y) \in \mathbf{b}$, and $(z, Z) \in \mathbf{c}$. Then, $y \in U_l^Y$ and $z \in U_l^Z$. Consider two sequences $\{a_{i_k}^Y : k \in \omega\}$ and $\{a_{j_k}^Z : k \in \omega\}$ converging to y and z, respectively. Without loss of generality we can suppose that for all $k \in \omega$, $a_{i_k}^Y \in U_l^Y$ and $a_{j_k}^Z \in U_l^Z$. By property (9.1.6) of Lemma 9.1.1 we have $a_{i_k}^X \in U_l^X$ and $a_{j_k}^X \in U_l^X$, which means that $\varrho_X(a_{i_k}^X, a_{j_k}^X) < \varepsilon$. Therefore,

$$\varrho_{\mathrm{T}(\mathbf{E})}(\mathbf{a}, \mathbf{b}) = \lim_{k\to\infty}(\varrho_Y(a_{i_k}^Y, a_{j_k}^Y)) =$$

$$\lim_{k\to\infty}(\varrho_X(a_{i_k}^X, a_{j_k}^X)) \leq \varepsilon,$$

which means that $\mathrm{Diam}(U_l^{\mathrm{T}(\mathbf{E})}) \leq \varepsilon$.

We prove (b). Let $\mathbf{a} \in U_l^{\mathrm{T}(\mathbf{E})}$, $l \in \omega$, and $(x, X) \in \mathbf{a}$. Then, $x \in U_l^X$ and there exists $\varepsilon \in \Re^+$ such that $\varrho_X(x, X \setminus U_l^X) > 2\varepsilon$. We need to prove relation (9.1.10). Let $\mathbf{b}$ be a point of $\mathrm{T}(\mathbf{E})$ such that $\varrho_{\mathrm{T}(\mathbf{E})}(\mathbf{a}, \mathbf{b}) < \varepsilon$ and let $(y, Y) \in \mathbf{b}$. Consider also two sequences $\{a_{i_k}^X : k \in \omega\}$ and $\{a_{j_k}^Y : k \in \omega\}$ converging to x and y, respectively. Then, we have

$$\varrho_{\mathrm{T}(\mathbf{E})}(\mathbf{a}, \mathbf{b}) = \lim_{k\to\infty}(a_{i_k}^X, a_{j_k}^X) < \varepsilon.$$

Thus $\varrho_X(x, a_{j_k}^X) \leq \varepsilon$ and, therefore, $\varrho_X(a_{j_k}^X, X \setminus U_l^X) \geq \varepsilon$ for almost all $k \in \omega$. By relation (9.1.4) of Lemma 9.1.1, $\varrho_Y(a_{j_k}^Y, Y \setminus U_l^Y) \geq \varepsilon$ for almost all $k \in \omega$ and, therefore, $\varrho_Y(y, Y \setminus U_l^Y) \geq \varepsilon$, which implies that $y \in U_l^Y$ and, therefore, $\mathbf{b} \in U_l^{\mathrm{T}(\mathbf{E})}$. □

Notation. Let $\mathbf{E} \in \mathrm{C}(\sim^\infty)$ and $X \in \mathbf{E}$. The natural embedding i_{T}^X of X into T has the property that $i_{\mathrm{T}}^X(X) \subset \mathrm{T}(\mathbf{E})$. Therefore, we can consider the *natural embedding* of X into $\mathrm{T}(\mathbf{E})$, denoted by $i_{\mathrm{T}(\mathbf{E})}^X$, setting $i_{\mathrm{T}(\mathbf{E})}^X(x) = i_{\mathrm{T}}^X(x)$, $x \in X$.

9.1.5 Lemma. *Let* $\mathbf{E} \in \mathrm{C}(\sim^{\infty})$ *and* $X \in \mathbf{E}$. *Then, the natural embedding* $i^X_{\mathrm{T}(\mathbf{E})}$ *of* X *into* $\mathrm{T}(\mathbf{E})$ *is an isometry.*

Proof. Let x and y be two points of X, $i^X_{\mathrm{T}(\mathbf{E})}(x) = \mathbf{a}$, and $i^X_{\mathrm{T}(\mathbf{E})}(y) = \mathbf{b}$. Therefore, $(x, X) \in \mathbf{a} \in \mathrm{T}(\mathbf{E})$ and $(y, Y) \in \mathbf{b} \in \mathrm{T}(\mathbf{E})$. Suppose that $\{a^X_{i_k} : k \in \omega\}$ is a sequence converging to x and $\{a^X_{j_k} : k \in \omega\}$ is a sequence converging to y. Then, we have

$$\varrho_{\mathrm{T}(\mathbf{E})}(\mathbf{a}, \mathbf{b}) = \lim_{k\to\infty}(\varrho_X(a^X_{i_k}, a^X_{j_k})) = \varrho_X(x, y),$$

which proves the lemma. □

Notation. Let X and Y be two elements of $\mathbf{S}$, which are not $\sim^s$-equivalent for some $s \in \mathcal{F}$ (and, therefore, they are not $\sim^m$-equivalent for some element $m \in \omega$). If there exists an element n of ω such that X and Y are $\sim^{m(n)}$-equivalent, then we denote by $n(X, Y)$ the maximal such element. If there is no element n of ω such that X and Y are $\sim^{m(n)}$-equivalent, then we set $n(X, Y) = -1$.

Let $\mathbf{a}$ and $\mathbf{b}$ be two elements of T, $(x, X) \in \mathbf{a}$, and $(y, Y) \in \mathbf{b}$. If X and Y are not $\sim^s$-equivalent for some $s \in \mathcal{F}$, then we denote by $n(\mathbf{a}, \mathbf{b})$ the integer $n(X, Y)$. Obviously, $n(\mathbf{a}, \mathbf{b})$ is independent of the choice of elements (x, X) and (y, Y) of $\mathbf{a}$ and $\mathbf{b}$, respectively. Also, we have $n(\mathbf{a}, \mathbf{b}) = n(\mathbf{b}, \mathbf{a})$.

Below, we define a function ϱ_{T} from the set $\mathrm{T} \times \mathrm{T}$ to the set $\Re^+ \cup \{0\}$. We shall prove that this function is a metric on the set T inducing the topology of the space T.

Definition. We define a function

$$\varrho_{\mathrm{T}} : \mathrm{T} \times \mathrm{T} \to \Re^+ \cup \{0\}$$

as follows. Let $\mathbf{a}, \mathbf{b} \in \mathrm{T}$, $(x, X) \in \mathbf{a}$, and $(y, Y) \in \mathbf{b}$. If there exists an element $\mathbf{E}$ of $\mathrm{C}(\sim^{\infty})$ such that $\mathbf{a}, \mathbf{b} \in \mathrm{T}(\mathbf{E})$, then we set

$$\varrho_{\mathrm{T}}(\mathbf{a}, \mathbf{b}) = \varrho_{\mathrm{T}(\mathbf{E})}(\mathbf{a}, \mathbf{b}).$$

If there is no element $\mathbf{E}$ of $\mathrm{C}(\sim^{\infty})$ such that $\mathbf{a}, \mathbf{b} \in \mathrm{T}(\mathbf{E})$, then we set

$$\varrho_{\mathrm{T}}(\mathbf{a}, \mathbf{b}) = \min\{\varrho_X(x, a^X_i) + \varrho_Y(a^Y_i, y) : i \in n(\mathbf{a}, \mathbf{b}) + 2\} + \frac{1}{n(\mathbf{a}, \mathbf{b}) + 2}.$$

(We assume that if $n(\mathbf{a},\mathbf{b}) = -1$, then $n(\mathbf{a},\mathbf{b})+2 = 1$. We note that in this case $i = 0$.)

9.1.6 Lemma. *The function ϱ_{T} is well defined, that is, ϱ_{T} is independent of the choice of the pairs (x, X) and (y, Y) of $\mathbf{a}$ and $\mathbf{b}$, respectively (which are considered in the definition of ϱ_{T}).*

Proof. Let $\mathbf{E}_0$ and $\mathbf{E}_1$ be the elements of $\mathrm{C}(\sim^\infty)$ such that $\mathbf{a} \in \mathrm{T}(\mathbf{E}_0)$ and $\mathbf{b} \in \mathrm{T}(\mathbf{E}_1)$. Suppose that $\mathbf{E}_0 \neq \mathbf{E}_1$ and let $(x, X), (x', X') \in \mathbf{a}$ and $(y, Y), (y', Y') \in \mathbf{b}$. Then, $i^X_{\mathrm{T}}(x) = i^{X'}_{\mathrm{T}}(x') = \mathbf{a}$ and $i^Y_{\mathrm{T}}(y) = i^{Y'}_{\mathrm{T}}(y') = \mathbf{b}$. Therefore, by Lemma 9.1.5 for every $i \in \omega$ we have

$$\varrho_X(x, a^X_i) = \varrho_{\mathrm{T}(\mathbf{E})}(\mathbf{a}, \mathbf{a}^{\mathrm{T}(\mathbf{E})}_i) = \varrho_{X'}(x', a^{X'}_i) \text{ and}$$

$$\varrho_Y(a^Y_i, y) = \varrho_{\mathrm{T}(\mathbf{E})}(\mathbf{a}^{\mathrm{T}(\mathbf{E})}_i, \mathbf{b}) = \varrho_{Y'}(a^{Y'}_i, y').$$

These two relations imply the lemma. □

9.1.7 Proposition. *The function ϱ_{T} is a metric on the set* T.

Proof. Let $\mathbf{a}$, $\mathbf{b}$, and $\mathbf{c}$ be three points of T. Let also $(x, X) \in \mathbf{a}$, $(y, Y) \in \mathbf{b}$, and $(z, Z) \in \mathbf{c}$. By the definition of ϱ_{T} we have $\varrho_{\mathrm{T}}(\mathbf{a},\mathbf{b}) = \varrho_{\mathrm{T}}(\mathbf{b},\mathbf{a}) \geq 0$.

We prove that $\varrho_{\mathrm{T}}(\mathbf{a},\mathbf{b}) = 0$ if and only if $\mathbf{a} = \mathbf{b}$. Indeed, suppose that $\mathbf{a} = \mathbf{b}$ and let $\mathbf{E}$ be the element of $\mathrm{C}(\sim^\infty)$ such that $\mathbf{a} \in \mathrm{T}(\mathbf{E})$. Then $\varrho_{\mathrm{T}}(\mathbf{a},\mathbf{b}) = \varrho_{\mathrm{T}(\mathbf{E})}(\mathbf{a},\mathbf{b}) = 0$. Since $\varrho_{\mathrm{T}(\mathbf{E})}$ is a metric on $\mathrm{T}(\mathbf{E})$,

$$\varrho_{\mathrm{T}(\mathbf{E})}(\mathbf{a},\mathbf{b}) = \varrho_{\mathrm{T}(\mathbf{E})}(\mathbf{b},\mathbf{a}) = \varrho_{\mathrm{T}}(\mathbf{b},\mathbf{a}).$$

Conversely, suppose that $\varrho_{\mathrm{T}}(\mathbf{a},\mathbf{b}) = 0$. By the definition of ϱ_{T}, there exists $\mathbf{E} \in \mathrm{C}(\sim^\infty)$ such that $\mathbf{a},\mathbf{b} \in \mathrm{T}(\mathbf{E})$ and, therefore, $\varrho_{\mathrm{T}}(\mathbf{a},\mathbf{b}) = \varrho_{\mathrm{T}(\mathbf{E})}(\mathbf{a},\mathbf{b})$. By Lemma 9.1.3, $\mathbf{a} = \mathbf{b}$.

Thus, it suffices to prove that

$$\varrho_{\mathrm{T}}(\mathbf{a},\mathbf{c}) \leq \varrho_{\mathrm{T}}(\mathbf{a},\mathbf{b}) + \varrho_{\mathrm{T}}(\mathbf{b},\mathbf{c}). \tag{9.1.11}$$

If there exists an element $\mathbf{E} \in \mathrm{C}(\sim^\infty)$ such that $\mathbf{a}$, $\mathbf{b}$, and $\mathbf{c}$ are elements of the set $\mathrm{T}(\mathbf{E})$, then relation (9.1.11) follows by Lemma 9.1.3.

Suppose that there exists an element $\mathbf{E} \in \mathrm{C}(\sim^\infty)$ such that $\mathbf{a},\mathbf{c} \in \mathrm{T}(\mathbf{E})$ and $\mathbf{b} \notin \mathrm{T}(\mathbf{E})$. Then, $n(X, Y) = n(Y, Z)$ and, therefore, $n(\mathbf{a},\mathbf{b}) = n(\mathbf{b},\mathbf{c})$.

By the definition of ϱ_{T} there exist integers $i, j \in n(\mathbf{a}, \mathbf{b}) + 2 = n(\mathbf{b}, \mathbf{c}) + 2$ such that

$$\varrho_{\mathrm{T}}(\mathbf{a}, \mathbf{b}) = \varrho_X(x, a_i^X) + \varrho_Y(a_i^Y, y) + \frac{1}{n(\mathbf{a}, \mathbf{b}) + 2}, \tag{9.1.12}$$

$$\varrho_{\mathrm{T}}(\mathbf{b}, \mathbf{c}) = \varrho_Y(y, a_j^Y) + \varrho_Z(a_j^Z, z) + \frac{1}{n(\mathbf{b}, \mathbf{c}) + 2}. \tag{9.1.13}$$

On the other hand, by Lemma 9.1.5 we have

$$\varrho_{\mathrm{T}}(\mathbf{a}, \mathbf{c}) = \varrho_{\mathrm{T}(\mathbf{E})}(\mathbf{a}, \mathbf{c}) \leq$$

$$\varrho_{\mathrm{T}(\mathbf{E})}(\mathbf{a}, \mathbf{a}_i^{\mathrm{T}(\mathbf{E})}) + \varrho_{\mathrm{T}(\mathbf{E})}(\mathbf{a}_i^{\mathrm{T}(\mathbf{E})}, \mathbf{a}_j^{\mathrm{T}(\mathbf{E})}) + \varrho_{\mathrm{T}(\mathbf{E})}(\mathbf{a}_j^{\mathrm{T}(\mathbf{E})}, \mathbf{c}) =$$

$$\varrho_X(x, a_i^X) + \varrho_X(a_i^X, a_j^X) + \varrho_Z(a_j^Z, z). \tag{9.1.14}$$

If $n(\mathbf{a}, \mathbf{b}) \in \omega$, then by relation (9.1.2) of Lemma 9.1.1 (considering for $n = n(X, Y)$) we have

$$\varrho_X(a_i^X, a_j^X) < \varrho_Y(a_i^Y, a_j^Y) + \frac{1}{n(X, Y) + 2} \leq$$

$$\varrho_Y(a_i^Y, y) + \varrho_Y(y, a_j^Y) + \frac{1}{n(X, Y) + 2}. \tag{9.1.15}$$

If $n(\mathbf{a}, \mathbf{b}) = -1$, then relation (9.1.15) is also true because in this case $i = j = 0$. Since $n(X, Y) = n(\mathbf{a}, \mathbf{b})$, relations (9.1.12)–(9.1.15) imply (9.1.11).

Now, suppose that there exists an element $\mathbf{E}$ of $\mathrm{C}(\sim^\infty)$ such that $\mathbf{a}, \mathbf{b} \in \mathrm{T}(\mathbf{E})$ and $\mathbf{c} \notin \mathrm{T}(\mathbf{E})$. Then, there exist integers $i, j \in n(\mathbf{a}, \mathbf{c}) + 2 = n(\mathbf{b}, \mathbf{c}) + 2$ such that

$$\varrho_{\mathrm{T}}(\mathbf{a}, \mathbf{c}) = \varrho_X(x, a_i^X) + \varrho_Z(a_i^Z, z) + \frac{1}{n(\mathbf{a}, \mathbf{c}) + 2}, \tag{9.1.16}$$

$$\varrho_{\mathrm{T}}(\mathbf{b}, \mathbf{c}) = \varrho_Y(y, a_j^Y) + \varrho_Z(a_j^Z, z) + \frac{1}{n(\mathbf{b}, \mathbf{c}) + 2}. \tag{9.1.17}$$

Since $\mathbf{a}, \mathbf{b} \in \mathrm{T}(\mathbf{E})$,

$$\varrho_{\mathrm{T}}(\mathbf{a}, \mathbf{b}) = \varrho_{\mathrm{T}(\mathbf{E})}(\mathbf{a}, \mathbf{b}). \tag{9.1.18}$$

On the other hand, by the choice of the integers i and j we have

$$\varrho_{\mathrm{T}}(\mathbf{a},\mathbf{c}) \le \varrho_X(x,a_j^X) + \varrho_Z(a_j^Z,z) + \frac{1}{n(\mathbf{a},\mathbf{c})+2}. \tag{9.1.19}$$

(We note that if $n(\mathbf{a},\mathbf{b}) = -1$, then $i = j = 0$ and, therefore, in this case the last relation is also true.) Also, by Lemma 9.1.5 we have

$$\varrho_X(x,a_j^X) = \varrho_{\mathrm{T}(\mathbf{E})}(\mathbf{a},\mathbf{a}_j^{\mathrm{T}(\mathbf{E})}) \le$$

$$\varrho_{\mathrm{T}(\mathbf{E})}(\mathbf{a},\mathbf{b}) + \varrho_{\mathrm{T}(\mathbf{E})}(\mathbf{b},\mathbf{a}_j^{\mathrm{T}(\mathbf{E})}) =$$

$$\varrho_{\mathrm{T}(\mathbf{E})}(\mathbf{a},\mathbf{b})) + \varrho_Y(y,a_j^Y). \tag{9.1.20}$$

Since $n(\mathbf{a},\mathbf{c}) = n(\mathbf{b},\mathbf{c})$, relations (9.1.16)–(9.1.20) imply (9.1.11). The case, where there exists an element $\mathbf{E}$ of C(R) such that $\mathbf{b},\mathbf{c} \in \mathrm{T}(\mathbf{E})$ and $\mathbf{a} \notin \mathbf{T}(\mathbf{E})$, is proved similarly.

Finally, suppose that there is no element $\mathbf{E}$ of $\mathrm{C}(\sim^\infty)$ such that the set $\mathrm{T}(\mathbf{E})$ contains two of the points $\mathbf{a}$, $\mathbf{b}$, and $\mathbf{c}$. Without loss of generality , we can suppose that $n(\mathbf{a},\mathbf{b}) \le n(\mathbf{b},\mathbf{c})$. In this case it is easy to see that $n(\mathbf{a},\mathbf{b}) \le n(\mathbf{a},\mathbf{c})$. There exist elements $i \in n(\mathbf{a},\mathbf{b})+2$, $l \in n(\mathbf{b},\mathbf{c})+2$, and $j \in n(\mathbf{a},\mathbf{c})+2$ such that

$$\varrho_{\mathrm{T}}(\mathbf{a},\mathbf{b}) = \varrho_X(x,a_i^X) + \varrho_Y(a_i^Y,y) + \frac{1}{n(\mathbf{a},\mathbf{b})+2}, \tag{9.1.21}$$

$$\varrho_{\mathrm{T}}(\mathbf{b},\mathbf{c}) = \varrho_Y(y,a_l^Y) + \varrho_Z(a_l^Z,z) + \frac{1}{n(\mathbf{b},\mathbf{c})+2}, \tag{9.1.22}$$

$$\varrho_{\mathrm{T}}(\mathbf{a},\mathbf{c}) = \varrho_X(x,a_j^X) + \varrho_Z(a_j^Z,y) + \frac{1}{n(\mathbf{a},\mathbf{c})+2}. \tag{9.1.23}$$

Since $i \in n(\mathbf{a},\mathbf{c})+2$,

$$\varrho_{\mathrm{T}}(\mathbf{a},\mathbf{c}) \le \varrho_X(x,a_i^X) + \varrho_Z(a_i^Z,z) + \frac{1}{n(\mathbf{a},\mathbf{c})+2} \le$$

$$\varrho_X(x,a_i^X) + \varrho_Z(a_i^Z,a_l^Z) + \varrho_Z(a_l^Z,z) + \frac{1}{n(\mathbf{a},\mathbf{c})+2}. \tag{9.1.24}$$

If $n(\mathbf{b},\mathbf{c}) \in \omega$, then since $i,l \in n(\mathbf{b},\mathbf{c})+2$ relation (9.1.2) of Lemma 9.1.1 (considering for $n = n(\mathbf{b},\mathbf{c})$) implies that

$$\varrho_Z(a_i^Z, a_l^Z) < \varrho_Y(a_i^Y, a_l^Y) + \frac{1}{n(\mathbf{b}, \mathbf{c}) + 2}. \tag{9.1.25}$$

If $n(\mathbf{b}, \mathbf{c}) = -1$, then relation (9.1.25) is also true because $i = l = 0$. We have also the following relation:

$$\varrho_Y(a_i^Y, a_l^Y) \leq \varrho_Y(a_i^Y, y) + \varrho_Y(y, a_l^Y). \tag{9.1.26}$$

Since $n(\mathbf{a}, \mathbf{b}) \leq n(\mathbf{a}, \mathbf{c})$ relations (9.1.21)–(9.1.26) imply (9.1.11). Thus, ϱ_{T} is a metric on T. □

9.1.8 Proposition. *The topology on the set* T *induced by the metric* ϱ_{T} *coincides with the topology of the space* T.

Proof. We need to prove the following two properties:

(1) Let $\mathbf{a} \in U_l^{\mathrm{T}}(\mathbf{H}) \in \mathrm{B}^{\mathrm{T}}$, where $l \in \omega$ and $\mathbf{H} \in \mathrm{C}(\sim^{m(n)})$ for some $n \in \omega$. Then, there exists a number $\varepsilon \in \Re^+$ such that

$$\{\mathbf{b} \in \mathrm{T} : \varrho_{\mathrm{T}}(\mathbf{a}, \mathbf{b}) < \varepsilon\} \subset U_l^{\mathrm{T}}(\mathbf{H}).$$

(We note that the subset

$$\{U_l^{\mathrm{T}}(\mathbf{H}) \in \mathrm{B}^{\mathrm{T}} : l, n \in \omega, \mathbf{H} \in \mathrm{C}(\sim^{m(n)})\}$$

of B^{T} is a base for the space T.)

(2) Let $\mathbf{a} \in \mathrm{T}$ and $\varepsilon \in \Re^+$. Then, there exist an integer $l \in \omega$ and an element $\mathbf{H}$ of $\mathrm{C}(\sim^{m(n)})$ for some $n \in \omega$ such that

$$\mathbf{a} \in U_l^{\mathrm{T}}(\mathbf{H}) \in \mathrm{B}^{\mathrm{T}} \text{ and } \varrho_{\mathrm{T}}(\mathbf{a}, \mathbf{b}) < \varepsilon$$

for every $\mathbf{b} \in U_l^{\mathrm{T}}(\mathbf{H})$.

We prove property (1). Let $(x, X) \in \mathbf{a}$ and, therefore, $x \in U_l^X$ and $X \in \mathbf{H}$. Then, there exists a number $\delta \in \Re^+$ such that $\delta < 1$ and

$$\varrho_X(x, X \setminus U_l^X) \geq 6\delta.$$

Also, there exists an integer $i \in \omega$ such that

$$\varrho_X(a_i^X, x) < \delta \tag{9.1.27}$$

and, therefore,

$$\varrho_X(a_i^X, X \setminus U_l^X) \geq 5\delta. \tag{9.1.28}$$

We put

$$\varepsilon = \min\{\delta, \frac{1}{n+1}, \frac{1}{i}, \frac{1}{l}\}$$

and consider an element $\mathbf{b}$ of T such that $\varrho_{\mathrm{T}}(\mathbf{a}, \mathbf{b}) < \varepsilon$. Let $(y, Y) \in \mathbf{b}$. We need to prove that $\mathbf{b} \in U_l^{\mathrm{T}}(\mathbf{H})$, that is $Y \in \mathbf{H}$ and $y \in U_l^Y$.

First, we suppose that there exists an element $\mathbf{E} \in \mathrm{C}(\sim^\infty)$ such that $\mathbf{a}, \mathbf{b} \in \mathrm{T}(\mathbf{E})$. Then $\varrho_{\mathrm{T}}(\mathbf{a}, \mathbf{b}) = \varrho_{\mathrm{T}(\mathbf{E})}(\mathbf{a}, \mathbf{b})$. Let

$$\{a_{i_k}^X : k \in \omega\} \text{ and } \{a_{j_k}^Y : k \in \omega\}$$

be two sequences converging to x and y, respectively. By the definition of the metric ϱ_{T} we have

$$\lim_{k \to \infty}(\varrho_X(a_{i_k}^X, a_{j_k}^X)) = \varrho_{\mathrm{T}(\mathbf{E})}(\mathbf{a}, \mathbf{b}).$$

There exists an integer $k \in \omega$ such that $\varrho_X(x, a_{i_k}^X) < \varepsilon$, $\varrho_Y(y, a_{j_k}^Y) < \varepsilon$, and $\varrho_X(a_{i_k}^X, a_{j_k}^X) = \varrho_Y(a_{i_k}^Y, a_{j_k}^Y) < \varepsilon$. Then,

$$\varrho_X(a_{i_k}^X, X \setminus U_l^X) \geq 6\delta - \varepsilon \geq 5\delta$$

and, therefore (see Lemma 9.1.1),

$$\varrho_Y(a_{i_k}^Y, Y \setminus U_l^Y) \geq 5\delta.$$

On the other hand,

$$\varrho_Y(a_{i_k}^Y, y) \leq \varrho_Y(a_{i_k}^Y, a_{j_k}^Y) + \varrho_Y(a_{j_k}^Y, y) < 2\varepsilon \leq 2\delta,$$

which means that $y \in U_l^Y$. Obviously, in the considered case, $Y \in \mathbf{H}$. Thus, $\mathbf{b} \in U_l^{\mathrm{T}}(\mathbf{H})$.

Now, we suppose that there is no element $\mathbf{E} \in \mathrm{C}(\sim^\infty)$ such that $\mathbf{a}, \mathbf{b} \in \mathrm{T}(\mathbf{E})$. In this case, there exists an integer $j \in n(\mathbf{a}, \mathbf{b}) + 2$ such that

$$\varrho_{\mathrm{T}}(\mathbf{a}, \mathbf{b}) = \varrho_X(x, a_j^X) + \varrho_Y(a_j^Y, y) + \frac{1}{n(\mathbf{a}, \mathbf{b}) + 2}.$$

Then, we have

$$\varrho_X(x, a_j^X) < \varepsilon, \tag{9.1.29}$$

$$\varrho_Y(a_j^Y, y) < \varepsilon, \quad \text{and} \tag{9.1.30}$$

$$\frac{1}{n(\mathbf{a}, \mathbf{b}) + 2} < \varepsilon. \tag{9.1.31}$$

By the last relation and the definition of the number ε it follows that $n+1, i, l \in n(\mathbf{a}, \mathbf{b})+2$. In particular, $n(\mathbf{a}, \mathbf{b}) \neq -1$. Since $n+1 \in n(\mathbf{a}, \mathbf{b})+2$ we have $Y \in \mathbf{H}$.

Furthermore, since $i, j \in n(\mathbf{a}, \mathbf{b}) + 2$, by Lemma 9.1.1 and relations (9.1.27), (9.1.29), and (9.1.31) we have

$$\varrho_Y(a_i^Y, a_j^Y) < \varrho_X(a_i^X, a_j^X) + \frac{1}{n(\mathbf{a}, \mathbf{b}) + 2} \leq$$

$$\varrho_X(a_i^X, x) + \varrho_X(x, a_j^X) + \frac{1}{n(\mathbf{a}, \mathbf{b}) + 2} < \delta + 2\varepsilon. \tag{9.1.32}$$

Relations (9.1.30) and (9.1.32) imply that

$$\varrho_Y(a_i^Y, y) \leq \varrho_Y(a_i^Y, a_j^Y) + \varrho_Y(a_j^Y, y) < \delta + 3\varepsilon \leq 4\delta. \tag{9.1.33}$$

On the other hand, since $l \in n(\mathbf{a}, \mathbf{b}) + 2$ by Lemma 9.1.1 we have that either

$$\varrho_Y(a_i^Y, Y \setminus U_l^Y) = \infty \text{ (if } X \setminus U_l^X = \emptyset) \quad \text{or}$$

$$\varrho_X(a_i^X, X \setminus U_l^X) < \varrho_Y(a_i^Y, Y \setminus U_l^Y) + \frac{1}{n(\mathbf{a}, \mathbf{b}) + 2}$$

(if $X \setminus U_l^X \neq \emptyset$). In both cases, relations (9.1.28) and (9.1.31) imply that

$$\varrho_Y(a_i^Y, Y \setminus U_l^Y) > 5\delta - \varepsilon \geq 4\delta.$$

(We assume that $\infty < \eta$ for every $\eta \in \Re^+$.) By relation (9.1.33), $y \in U_l^Y$. Thus $\mathbf{b} \in U_l^{\mathrm{T}}(\mathbf{H})$.

We prove property (2). Let $\delta \in \Re^+$ and $\delta < \varepsilon/4$. Suppose that $(x, X) \in \mathbf{a}$. There exist integers $i, l \in \omega$ such that $x \in U_l^X$, $\mathrm{Diam}(U_l^X) < \delta$, $a_i^X \in U_l^X$, and therefore

$$\varrho_X(x, X \setminus U_l^X) > 0 \quad \text{and} \quad \varrho_X(x, a_i^X) < \delta. \tag{9.1.34}$$

Also, there exists an integer $n \in \omega$ such that $l, i \in n+2$,

$$\frac{1}{n+2} < \delta, \quad \text{and} \quad \frac{1}{n+2} < \varrho_X(x, X \setminus U_l^X). \tag{9.1.35}$$

The last relation means that $k_0(X, n, i, l) > 0$.

Denote by $\mathbf{H}$ the $\sim^{m(n)}$-equivalence class containing X and consider the element $U_l^{\mathrm{T}}(\mathbf{H})$ of B^{T}. Obviously $\mathbf{a} \in U_l^{\mathrm{T}}(\mathbf{H})$. We need to prove that $\varrho_{\mathrm{T}}(\mathbf{a}, \mathbf{b}) < \varepsilon$ for every $\mathbf{b} \in U_l^{\mathrm{T}}(\mathbf{H})$. Suppose that $\mathbf{b} \in U_l^{\mathrm{T}}(\mathbf{H})$ and $(y, Y) \in \mathbf{b}$. Then, $Y \in \mathbf{H}$ and $y \in U_l^Y$. Since $X \sim^{m(n)} Y$ and $l, i \in n+2$ we have $k_0(X, n, i, l) = k_0(Y, n, i, l)$ and therefore by relation (9.1.35), $\varrho_Y(a_i^Y, Y \setminus U_l^Y) > 0$, that is $a_i^Y \in U_l^Y$.

First, we consider the case where there exists an element $\mathbf{E}$ of $\mathrm{C}(\sim^\infty)$, such that $\mathbf{a}, \mathbf{b} \in \mathrm{T}(\mathbf{E})$. In this case, $X, Y \in \mathbf{E}$ and $\varrho_{\mathrm{T}}(\mathbf{a}, \mathbf{b}) = \varrho_{\mathrm{T}(\mathbf{E})}(\mathbf{a}, \mathbf{b})$. By Lemma 9.1.1, $\mathrm{Diam}(U_l^Y) < \delta$ and, therefore,

$$\varrho_Y(a_i^Y, y) < \delta. \tag{9.1.36}$$

Then, by Lemma 9.1.5 we have

$$\begin{aligned} \varrho_{\mathrm{T}(\mathbf{E})}(\mathbf{a}, \mathbf{b}) &\le \varrho_{\mathrm{T}(\mathbf{E})}(\mathbf{a}, \mathbf{a}_i^{\mathrm{T}(\mathbf{E})}) + \varrho_{\mathrm{T}(\mathbf{E})}(\mathbf{a}_i^{\mathrm{T}(\mathbf{E})}, \mathbf{b}) = \\ &\varrho_X(x, a_i^X) + \varrho_Y(a_i^Y, y) < 2\delta < \varepsilon. \end{aligned} \tag{9.1.37}$$

Relations (9.1.34), (9.1.36), and (9.1.37) imply that $\varrho_{\mathrm{T}}(\mathbf{a}, \mathbf{b}) < \varepsilon$.

Now, we suppose that there is no element $\mathbf{E} \in \mathrm{C}(\sim^\infty)$ such that $\mathbf{a}, \mathbf{b} \in \mathrm{T}(\mathbf{E})$. Since $X, Y \in \mathbf{H}$ we have $n \le n(X, Y) = n(\mathbf{a}, \mathbf{b})$. By the choice of the integer n we have

$$\frac{1}{n(\mathbf{a}, \mathbf{b}) + 2} < \delta. \tag{9.1.38}$$

Since $i \in n + 2 \le n(\mathbf{a}, \mathbf{b}) + 2$ we have

$$\varrho_{\mathrm{T}}(\mathbf{a}, \mathbf{b}) \le \varrho_X(x, a_i^X) + \varrho_Y(a_i^Y, y) + \frac{1}{n(\mathbf{a}, \mathbf{b}) + 2}. \tag{9.1.39}$$

By Lemma 9.1.1,

$$\mathrm{Diam}(U_l^Y) < \mathrm{Diam}(U_l^X) + \frac{1}{n(X, Y) + 2}.$$

This fact means that $\mathrm{Diam}(U_l^Y) < 2\delta$ and, therefore,

$$\varrho_Y(a_i^Y, y) < 2\delta. \tag{9.1.40}$$

Relations (9.1.34) and (9.1.38)–(9.1.40) imply that $\varrho_{\mathrm{T}}(\mathbf{a}, \mathbf{b}) < \varepsilon$, which completes the proof of the lemma. □

9.1.9 Proposition. *For every $X \in \mathbf{S}$ the natural embedding i_{T}^X of X into* T *is an isometry.*

Proof. Let **E** be the element of C($\sim^\infty$) containing X. Then,

$$i_{\mathrm{T}}^X(x) = i_{\mathrm{T}(\mathbf{E})}^X(x),\ x \in X,$$

and therefore the proof of the proposition follows immediately by Lemma 9.1.5. □

In the next section we shall use the following notation.

Notation. Let **H** be an element of $\mathrm{C}^\diamond(\mathrm{R})$. We denote by $\varrho_{\mathrm{T}(\mathbf{H})}$ the metric on the subspace T(**H**) of T induced by the metric ϱ_{T}.

9.2 Isometrically universal spaces

In the present section we study (topological) classes of the following types: classes of spaces, classes of subsets, classes of bases, classes of p-bases, and classes of spaces with structure in the universe of discourse of metric spaces. Below we give the corresponding definitions.

Definition. Let $I\!\!P$ be a class of any considered type. The class $I\!\!P$ is said to be a *class in the universe of metrizable spaces* if the space-coordinate of any its element is a metrizable space. (We assume that the space-coordinate of an element X of a class of spaces is the space X.)

The class $I\!\!P$ is said to be a *class in the universe of metric spaces* if the space-coordinate of any its element is a metric space.

In the next definition we specify the notion of a (properly) universal element for classes in the universe of metric spaces.

Definition. Assuming that the space-coordinates of the elements which we consider in the definition of a (properly) universal element (given in the preceding chapters for any considered type) are separable metric spaces and the embeddings are isometries, we obtain the notion of an *isometrically* (*and properly*) *universal element* in a given class in the universe of metric spaces.

Now, using the metrics on Containing Spaces constructed in Section 9.1 we specify the notion of an (ω, ω^+)-saturated class (see Section 2.7) for classes in the universe of metric spaces defining the notion of an isometrically ω-saturated class.

Notation. For simplicity of notation, an (ω, ω^+)-saturated class (of any considered type) will be called *ω-saturated.*

Definition. The definition of an isometrically ω-saturated class $I\!\!P$ (of any considered type) in the universe of metric spaces can be obtained by the corresponding definition of an ω-saturated class, which in turn was obtained by the corresponding definition of a saturated class (given in the preceding chapters for any considered type). For this purpose, we make the following changes on the corresponding definition of an ω-saturated class:

(1) Parallel to the considered indexed collection $\mathbf{S}$ of metric spaces, we consider also an arbitrary $\mathbf{S}$-indexed collection $\mathbf{P}$ of ω-indexed dense subsets.

(2) From the family R^+ of equivalence relations on $\mathbf{S}$ considered in the definition of an ω-saturated class, we additionally require to be $(\mathbf{M}, \mathbf{P})$-admissible. Therefore, for every $\mathbf{L} \in \mathrm{C}^{\diamondsuit}(\mathrm{R})$ the space-coordinate $\mathrm{T}(\mathbf{L})$ of the constructed element can be considered with the metric $\varrho_{\mathrm{T}(\mathbf{L})}$ (defined in Section 9.1). (We recall that the constructed element in the definition of an ω-saturated class is:

(a) the space $\mathrm{T}(\mathbf{L})$ for a class of spaces,

(b) the pair $(\mathrm{T}(\mathbf{H}|_{\mathbf{Q}}), \mathrm{T}(\mathbf{L}))$ for a class of subsets,

(c) the pair $(\mathrm{B}^{\mathbf{L}}_{\diamondsuit,\theta(\tau)}, \mathrm{T}(\mathbf{L}))$ for a class of bases,

(d) the triad $(\mathrm{T}(\mathbf{H}|_{\mathbf{Q}}), \mathrm{B}^{\mathbf{E}}_{\diamondsuit,\theta(\tau)}, \mathrm{T}(\mathbf{L}))$ for a class of p-bases,

(e) and the pair $(\psi_{\mathrm{T}(\mathbf{L})}, \mathrm{T}(\mathbf{L}))$ for a class of spaces with structure.)

Then, the given class $I\!\!P$ in the universe of metric spaces is said to be *isometrically ω-saturated* if for every $\mathbf{L} \in \mathrm{C}^{\diamondsuit}(\mathrm{R})$ the constructed element belongs to this class.

For example, we give below the definition of an isometrically ω-saturated class (in the universe) of metric spaces.

Definition. A class $\mathbb{P}$ of metric spaces is said to be *isometrically ω-saturated* if for every indexed collection $\mathbf{S}$ of elements of $\mathbb{P}$ and for every $\mathbf{S}$-indexed collection $\mathbf{P}$ of ω-indexed dense subsets there exists a co-mark $\mathbf{M}^+$ of $\mathbf{S}$ satisfying the following condition: for every co-extension $\mathbf{M}$ of $\mathbf{M}^+$ there exists an $(\mathbf{M}, \mathbf{P})$-admissible family R^+ of equivalence relations on $\mathbf{S}$ such that for every admissible family R of equivalence relations on $\mathbf{S}$, which is a final refinement of R^+, the metric space $\mathrm{T}(\mathbf{L})$ with the metric $\varrho_{\mathrm{T}(\mathbf{L})}$ (see Section 9.1) belongs to $\mathbb{P}$ for every $\mathbf{L} \in \mathrm{C}^{\diamond}(\mathrm{R})$.

The co-mark $\mathbf{M}^+$ is said to be an *initial* co-mark *corresponding* to the isometrically ω-saturated class $\mathbb{P}$. Also the family R^+ is said to be an *initial* family *corresponding* to the co-mark $\mathbf{M}$, the $\mathbf{S}$-indexed collection $\mathbf{P}$, and the isometrically ω-saturated class $\mathbb{P}$.

The isometrically ω-saturated classes (of any considered type) have the intersection and the universality properties.

9.2.1 Proposition. *The non-empty intersection of countable many isometrically ω-saturated classes (of the same type) is also isometrically ω-saturated.*

Proof. We prove the proposition for the isometrically ω-saturated classes of metric spaces. For the other considered types of classes the proposition is proved similarly.

Suppose that $\mathbb{P}_k$, $k \in \omega$, is an isometrically ω-saturated class of metric spaces and let

$$\mathbb{P} = \cap\{\mathbb{P}_k : k \in \omega\}.$$

Let $\mathbf{S}$ be a non-empty indexed collection of elements of $\mathbb{P}$. For every non-empty element X of $\mathbf{S}$ we denote by P^X an ω-indexed dense subset of X and consider the $\mathbf{S}$-indexed collection

$$\mathbf{P} \equiv \{P^X : X \in \mathbf{S}\}$$

of ω-indexed dense subsets ($P^X = \emptyset$ if $X = \emptyset$).

Denote by $\mathbf{M}_k^+$, $k \in \omega$, a co-mark of $\mathbf{S}$, which is an initial co-mark corresponding to the isometrically ω-saturated class $\mathbb{P}_k$, and let $\mathbf{M}^+$ be a co-mark of $\mathbf{S}$, which is a co-extension of all co-marks $\mathbf{M}_k^+$. Let $\mathbf{M}$ be an arbitrary co-mark of $\mathbf{S}$, which is a co-extension of $\mathbf{M}^+$.

Also, denote by R_k^+, $k \in \omega$, an initial family of equivalence relation on $\mathbf{S}$ corresponding to the co-mark $\mathbf{M}$, the $\mathbf{S}$-indexed collection $\mathbf{P}$, and the isometrically ω-saturated class $I\!P_k$, and let R^+ be an ω^+-admissible family of equivalence relations on $\mathbf{S}$, which is a final refinement of all families R_k^+, where $k \in \omega$. Let R be an arbitrary ω^+-admissible family of equivalence relations on $\mathbf{S}$, which is a final refinement of R^+.

Then, for every $\mathbf{L} \in \mathrm{C}^\diamondsuit(\mathrm{R})$ we can consider the space $\mathrm{T}(\mathbf{M}, \mathrm{R}, \mathbf{L}) \equiv \mathrm{T}(\mathbf{L})$ with the metric $\varrho_{\mathrm{T}(\mathbf{L})}$ constructed in Section 9.1. Since $I\!P_k$, $k \in \omega$ is isometrically ω-saturated, the metric space $\mathrm{T}(\mathbf{L})$ belongs to $I\!P_k$. Therefore, this space belongs also to $I\!P$, which proves the proposition. □

Definition. Let $I\!M$ be a class in the universe of metric spaces. We define the *corresponding class* in the universe of metrizable spaces as a class $I\!P$ consisting of the elements of $I\!M$ whose space-coordinates are considered as metrizable spaces (without metrics).

The class $I\!M$ is said to be *complete* if the corresponding class $I\!P$ in the univese of metrizable spaces is complete.

9.2.2 Proposition. *In any isometrically (complete) ω-saturated class there exists an isometrically (and properly) universal element.*

Proof. We prove the proposition for an isometrically ω-saturated class of metric spaces. The other cases are proved similarly.

Let $I\!P$ be an isometrically ω-saturated class of metric spaces. Denote by $\mathbf{S}$ an indexed collection of elements of $I\!P$ with the property that for every $Y \in I\!P$ there exists an isometry i_X^Y of Y onto an element X of $\mathbf{S}$, and let $\mathbf{P}$ be an $\mathbf{S}$-indexed collection of ω-indexed dense subsets.

Also, denote:

(a) by $\mathbf{M}^+$ a co-mark of $\mathbf{S}$, which is an initial co-mark corresponding to the isometrically ω-saturated class $I\!P$;

(b) by $\mathbf{M}$ an arbitrary co-mark of $\mathbf{S}$, which is a co-extension of $\mathbf{M}^+$;

(c) by R^+ an initial family of equivalence relation on $\mathbf{S}$ corresponding to the co-mark $\mathbf{M}$, the $\mathbf{S}$-indexed collection $\mathbf{P}$, and the isometrically ω-saturated class $I\!P$ (therefore, R^+ is $(\mathbf{M}, \mathbf{P})$-admissible); and

(d) by R an arbitrary admissible family of equivalence relations on $\mathbf{S}$, which is a final refinement of R^+.

Then, we can consider the Containing Space $\mathrm{T} \equiv \mathrm{T}(\mathbf{M}, \mathrm{R})$ with the metric ϱ_{T} constructed in Section 9.1. We note that for every $X \in \mathbf{S}$ the natural embedding i_{T}^{X} of X into T is an isometry (see Proposition 9.1.9). Since $I\!\!P$ is isometrically saturated, $\mathrm{T} \in I\!\!P$.

Let $Y \in I\!\!P$ and X an element of $\mathbf{S}$ such that there exists an isometry i_X^Y of Y onto X. Then, $i_{\mathrm{T}}^{X} \circ i_X^Y$ is an isometry of Y into T, which means that T is an isometrically universal element in $I\!\!P$. □

According to our agreement (that is, $\tau = \omega$) any considered space has a countable base. However, such a space may not be regular (that is, it may not be metrizable). This observation justifies the following definition.

Definition. Let $I\!\!R$ be a class (of any considered type). The subclass $I\!\!M$ of $I\!\!R$ consisting of all elements, whose space-coordinates are metrizable spaces, is called the *restriction of $I\!\!R$ on the universe of metrizable spaces.* Obviously, if $I\!\!R$ is a topological class, then $I\!\!M$ is also topological.

9.2.3 Proposition. *Let $I\!\!R$ be a (complete) ω-saturated class (of any considered type). Then, the restriction $I\!\!M$ of $I\!\!R$ on the universe of metrizable spaces is also (complete) ω-saturated.*

Proof. We prove the proposition in the case where $I\!\!R$ is a (complete) ω-saturated class of subsets. The other cases are proved similarly.

Consider the class $\mathbb{P}$(subset) of all subsets. By the definition of an ω-saturated class of subsets this class is ω-saturated. Let $\mathbb{P}$(subset+r) be the subclass of $\mathbb{P}$(subset) consisting of all elements, whose space-coordinates are regular spaces. Since the class of all regular spaces is ω-saturated (see Proposition 2.1.6) the class $\mathbb{P}$(subset+r) is also ω-saturated (see Proposition 2.2.8).

It is easy to see that $I\!\!M = I\!\!R \cap \mathbb{P}$(subset+r). By Proposition 2.2.2, $I\!\!M$ is a (complete) ω-saturated class. □

Definition. Let $I\!\!M$ be a class (of any consideed type) in the universe of metrizable spaces. We define a class $I\!\!P$ in the universe of metric spaces as follows: any element of $I\!\!M$ with an arbitrary fixed metric on its space-component compatible with the topology of this space-component is considered to be an element of $I\!\!P$. The class $I\!\!P$ is said to be the *class in the universe of metric spaces generated by the class $I\!\!M$.*

9.2.4 Proposition. *The class $I\!\!P$ in the universe of metric spaces generated by a (complete) ω-saturated class $I\!\!M$ in the universe of metrizable* *spaces is an isometrically (complete) ω-saturated class. Therefore, in this class there exist isometrically (and properly) universal elements.*

Proof. We prove the proposition for the case where $\mathbb{M}$ is an ω-saturated class of spaces with structure. The other cases are proved similarly. We shall use the notions and notation of Chapter 4.

Let $\mathbb{M}$ be an ω-saturated class of metrizable spaces with structure. We prove that $\mathbb{P}$ is an isometrically ω-saturated class of metric spaces with structure. For this purpose we consider an indexed collection $\mathbf{S}$ of metric spaces, a $\mathbb{P}$-co-structure $\mathbf{\Psi}$ on $\mathbf{S}$, and an arbitrary $\mathbf{S}$-indexed collection $\mathbf{P}$ of ω-indexed dense subsets.

Then, $\mathbf{S}$ can be considered as an indexed collection of regular spaces and $\mathbf{\Psi}$ as an $\mathbb{M}$-co-structure on $\mathbf{S}$. Since $\mathbb{M}$ is ω-saturated there exists an initial co-mark $\mathbf{M}^+$ corresponding to the $\mathbb{M}$-co-structure $\mathbf{\Psi}$ of $\mathbf{S}$.

Let $\mathbf{M}$ be an arbitrary co-extension of $\mathbf{M}^+$ and R^+ an initial family of equivalence relations on $\mathbf{S}$ corresponding to the co-mark $\mathbf{M}$ and the $\mathbb{M}$-co-structure $\mathbf{\Psi}$. Without loss of generality, we can also suppose that R^+ is $(\mathbf{M}, \mathbf{P})$-admissible.

Consider an element $\mathbf{L}$ of $\mathrm{C}^{\diamondsuit}(\mathrm{R})$. By construction, $(\psi_{\mathrm{T}(\mathbf{L})}, \mathrm{T}(\mathbf{L})) \in \mathbb{M}$. On the other hand, the space $\mathrm{T}(\mathbf{L})$ can be considered with the metric $\varrho_{\mathrm{T}(\mathbf{L})}$ constructed in Section 9.1, which means that $(\psi_{\mathrm{T}(\mathbf{L})}, \mathrm{T}(\mathbf{L})) \in \mathbb{P}$. This observation proves that $\mathbb{P}$ is an isometrically ω-saturated class. □

Below, we give a series of corollaries of the above Proposition 9.2.3 and 9.2.4. These corollaries are reformulations for the case of separable metric spaces of some corollaries given in the preceding chapters.

9.2.5 Corollary. *The following classes of separable metric spaces are isometrically ω-saturated:*

(1) *The class of all separable metric spaces.*

(2) *The class of all countably-dimensional separable metric spaces.*

(3) *The class of all strongly countably-dimensional separable metric spaces.*

(4) *The class of all locally finite-dimensional separable metric spaces.*

(5) *The class of all separable metric spaces of dimension* ind *less than or equal to n, $n \in \omega$.*

(6) *The class of all separable metric spaces of dimension* ind *less than or equal to α, $\alpha \in \omega^+$.*

Therefore, in these classes there exist isometrically universal elements. □

9.2.6 Corollary. *Let n be a natural number and β a countable ordinal. In the class of all separable metric spaces, which can be represented as the union of n (closed) subspaces of dimension* ind *less than or equal to β, there exists an isometrically universal space.* □

9.2.7 Corollary. *Let β be a countable ordinal. In the class of all separable metric spaces which can be represented as a countable union of closed subspaces of dimension* ind *less than or equal to β there exists an isometrically universal space.* □

9.2.8 Corollary. *Let β be a countable limit ordinal. In the class of all separable metric spaces which can be represented as a countable union of closed (open) subspaces of dimension* ind *less than β there exists an isometrically universal space.* □

9.2.9 Corollary. *Let $n \in \omega$. There exists a separable metric space T and fixed zero-dimensional subsets $T_0,...,T_n$, whose union is T, such that for every separable metric space X and fixed zero-dimensional subsets $X_0,...,X_n$, whose union is X, there exists an isometric embedding i_T^X of X into T with the property:*

$$(i_T^X)(X_i) \subset T_i,\ i \in \{0, ..., n\}.\ \square$$

9.2.10 Corollary. *There exists a separable metric space T and a fixed sequence $\{T_i : i \in \omega\}$ of zero-dimensional subsets of T, whose union is T, with the property that for every separable metric space X and a fixed sequence $\{X_i : i \in \omega\}$ of zero-dimensional subsets of X, whose union is X, there exists an isometric embedding i_T^X of X into T such that*

$$(i_T^X)(X_i) \subset T_i,\ i \in \omega.\ \square$$

9.2.11 Corollary. *Let $n_i \in \omega$, $i \in \omega$. There exists a separable metric space T and a fixed sequence $\{T_i : i \in \omega\}$ of closed (open) subsets of T for which*

$$\text{ind}(T_i) \leq n_i \quad \text{and} \quad \cup\{T_i : i \in \omega\} = T,$$

having the property that for every separable metric space X and a fixed sequence $\{X_i : i \in \omega\}$ of closed (open) subsets of X for which

$$\text{ind}(X_i) \leq n_i \quad \text{and} \quad \cup\{X_i : i \in \omega\} = X,$$

there exists an isometric embedding i_T^X *of* X *into* T *such that*

$$(i_T^X)^{-1}(T_i) = X_i,\ i \in \omega. \ \square$$

9.2.12 Corollary. *There exists a separable metric space* T *and a fixed increasing sequence* $\{T_i : i \in \omega\}$ *(that is,* $T_i \subset T_{i+1}$*) of zero-dimensional subspaces of* T *for which* $T = \cup\{T_i : i \in \omega\}$ *such that for every separable metric space* X *and for every increasing seguence* $\{X_i : i \in \omega\}$ *of zero-dimensional subspaces of* X *for which* $X = \cup\{X_i : i \in \omega\}$ *there exists an isometry* i_T^X *of* X *into* T *for which*

$$X_i \subset (i_T^X)^{-1}(T_i),\ i \in \omega. \ \square$$

9.2.13 Corollary. *Let* $\{n_i : i \in \omega\}$ *be a non-decreasing sequence of integers. There exists a separable metric space* T *and a fixed increasing sequence* $\{T_i : i \in \omega\}$ *of closed (open) subspaces of* T *for which*

$$T = \cup\{T_i : i \in \omega\} \quad \text{and} \quad \text{ind}(T_i) \leq n_i$$

such that for every separable metric space X *and for every increasing sequence* $\{X_i : i \in \omega\}$ *of closed (open) subspaces of* X *for which*

$$X = \cup\{X_i : i \in \omega\} \quad \text{and} \quad \text{ind}(X_i) \leq n_i$$

there exists an isometry i_T^X *of* X *into* T *for which*

$$(i_T^X)^{-1}(T_i) = X_i. \ \square$$

9.2.14 Corollary. *There exists a pair* (Q^T, T)*, where* Q^T *is a closed (open) subset of a separable metric space* T*, such that for every pair* (Q^X, X)*, where* Q^X *is a closed (open) subset of a separable metric space* X*, there exists an isometric embedding* i_T^X *of* X *into* T *with the property:*

$$(i_T^X)^{-1}(Q^T) = Q^X. \ \square$$

9.2.15 Corollary. *There exists a separable metric space* T *and a fixed countable base* B^T *for the open subsets of* T *such that for every separable metric space* X *with a fixed countable base* B^X *for the open subsets of* X *there exists an isometric embedding* i_T^X *of* X *into* T *with the property:*

$$B^X = \{(i_T^X)^{-1}(U) : U \in B^T\}. \ \square$$

9.2.16 Corollary. *There exists a triad (Q^T, B^{Q^T}, T), where Q^T is a closed (open) subset of a separable metric space T and B^{Q^T} is a p-base for Q^T in T, such that for every triad (Q^X, B^{Q^X}, X), where Q^X is a closed (open) subset of a separable metric space X and B^{Q^X} is a p-base for Q^X in X, there exists an isometric embedding i_T^X of X into T with the properties:*

$$(i_T^X)^{-1}(Q^T) = Q^X \text{ and}$$

$$B^{Q^X} = \{(i_T^X)^{-1}(U) : U \in B^{Q^T}\}. \ \square$$

9.2.17 Corollary. *For every countable limit ordinal β and for every integer $n \geq -1$ in the family of all separable metric spaces of D-dimension less than or equal to $\beta + n$ there exists an isometrically universal space.* $\square$

9.2.18 Corollary. *Let α be a countable ordinal. The Corollaries 9.2.11, 9.2.13, 9.2.14, and 9.2.16 hold if the considered closed subspaces everywhere will be replaced by Borel sets of multiplicative class α or everywhere will be replaced by Borel sets of additive class α.* $\square$

9.3 Completely bounded and compact spaces

Definition. A subset Q of a metric space X is said to be an *ε-net* of X, where $\varepsilon \in \Re^+$, if for every $x \in X$ there exists an element y of Q such that $\varrho_X(x, y) \leq \varepsilon$.

A metric space X is said to be *completely bounded* if X has a finite ε-net for every $\varepsilon \in \Re^+$. It is easy to see that any compact metric space is completely bounded.

Definition. A class $\mathbb{P}$ of metric spaces is said to be *uniform* if the following conditions are satisfied:

(a) there exists an integer $\mathbf{d} \in \omega \setminus \{0\}$ such that $\text{Diam}(X) \leq \mathbf{d}$ for every $X \in \mathbb{P}$ and

(b) for every $\varepsilon \in \Re^+$ there exists an integer $\mathbf{n}(\varepsilon) \in \omega$ such that every element X of $\mathbb{P}$ has a finite ε-net the number of elements of which is less than or equal to $\mathbf{n}(\varepsilon)$. (We note that $\mathbf{d}$ and $\mathbf{n}(\varepsilon)$ are independent of the elements X of $\mathbb{P}$.)

9.3.1 Proposition. *Let $\mathbb{M}$ be the class of metric spaces generated by a saturated class $\mathbb{R}$ of metrizable spaces. A subclass $\mathbb{P}$ of $\mathbb{M}$ is uniform if and only if there exists a completely bounded element of $\mathbb{M}$, which is an isometrically Containing Space for $\mathbb{P}$.* (We note that this element of $\mathbb{M}$ may be not an element of $\mathbb{P}$.)

Proof. Obviously, if for a subclass $\mathbb{P}$ of $\mathbb{M}$ there exists an isometrically containing element of $\mathbb{M}$, which is completely bounded, then $\mathbb{P}$ is uniform.

Now, let $\mathbb{P}$ be a uniform subclass of $\mathbb{M}$. Without loss of generality, we can suppose that all elements of $\mathbb{P}$ are not empty. There exists an integer $\mathbf{d} \in \omega \setminus \{0\}$ such that $\mathrm{Diam}(X) \leq \mathbf{d}$ for every $X \in \mathbb{P}$. Also, for every $\varepsilon \in \Re^+$ there exists an integer $\mathbf{n}(\varepsilon) \in \omega$ such that every element of $\mathbb{P}$ has a finite ε-net the number of elements of which is less than or equal to $\mathbf{n}(\varepsilon)$.

We denote by $\mathbf{S}$ an indexed collection of elements of $\mathbb{P}$ such that for every $Y \in \mathbb{P}$ there exists an isometry i_X^Y of Y onto an element X of $\mathbf{S}$. We note that $\mathbf{S}$ can also be considered as an indexed collection of elements of $\mathbb{R}$. For this purpose the elements of $\mathbf{S}$ are considered as metrizable spaces without metrics.

Also, we denote:

(a) by $\mathbf{M}^+$ an initial co-mark of $\mathbf{S}$ corresponding to the class $\mathbb{R}$;

(b) by $\mathbf{M}$ an arbitrary co-mark of $\mathbf{S}$, which is an extension of $\mathbf{M}^+$; and

(c) by $\mathrm{R}^+ \equiv \{\sim_+^s : s \in \mathcal{F}\}$ an $\mathbf{M}$-admissible family of equivalence relations on $\mathbf{S}$, which is an initial family corresponding to the co-mark $\mathbf{M}$ and the class $\mathbb{R}$.

Let X be an element of $\mathbf{S}$. We construct an ω-indexed dense subset P^X of X as follows. First, for every $n \in \omega \setminus \{0\}$ we consider an indexed subset $\{a_{(n,i)}^X : i \in \mathbf{n}(1/n)\}$ of X, which is an $1/n$-net of X, and put

$$\mathbf{s}(n) = \mathbf{n}(1) + ... + \mathbf{n}(1/n). \qquad (9.3.1)$$

Obviously, $n \leq \mathbf{s}(n)$. (We note that $\mathbf{s}(n)$ is independent of X.) Then, we set

$$P^X \equiv \{a_i^X : i \in \omega\},$$

where $a_i^X = a_{(1,i)}^X$ if $i \in \mathbf{n}(1)$, and $a_{\mathbf{s}(n)+i}^X = a_{(n,i)}^X$ if $n > 1$ and $i \in \mathbf{n}(1/n)$.

Now, we consider the $\mathbf{S}$-indexed collection of ω-indexed dense subsets

$$\mathbf{P} = \{P^X : X \in \mathbf{S}\}$$

and define a family of equivalence relations on $\mathbf{S}$, denoted by

$$\mathrm{R}_{\mathbf{P}} \equiv \{\sim^s_{\mathbf{P}} : s \in \mathcal{F}\}$$

as follows. Let s be an element of $\mathcal{F}$ and $k(s)$ the minimal element of ω such that $s \subset k(s)$. Then, two elements X and Y of $\mathbf{S}$ are considered to be $\sim^s$-equivalent if either $k(s) = 0$ or $k(s) > 0$ and $k_0(X, n, i, j) = k_0(Y, n, i, j)$, $k_1(X, n, i, j) = k_1(Y, n, i, j)$, and $k(X, n, i) = k(Y, n, i)$ for every $n \in k(s)+1$ and $i, j \in n + 2$. It is easy to see that $\mathrm{R}_{\mathbf{P}}$ is an ω^+-admissible family. Moreover, we prove that $\mathrm{R}_{\mathbf{P}}$ is actually admissible, that is for every $s \in \mathcal{F}$ the set of all $\sim^s_{\mathbf{P}}$-equivalence classes is finite.

Indeed, let s be an element of $\mathcal{F} \setminus \{\emptyset\}$, $k(s)$ the minimal element of ω such that $s \subset k(s)$, and $n \in k(s) + 1$. (Therefore, $k(s) \neq 0$.) Since $i \in n+2$, $j \in n+2$, and $\varrho_X(a^X_i, X \setminus U^X_j) \leq \mathrm{Diam}(X) \leq \mathbf{d}$ for every $X \in \mathbf{S}$, the number of distinct values of the integer $k_0(X, n, i, j)$ (n is fixed) is not more than $(n + 2)^2((n + 2)\mathbf{d} + 2)$. Similarly, the number of distinct values of the integers $k_1(X, n, i, j)$ and $k(X, n, i)$ (n is fixed) is not more than $(n+2)^2((n+2)\mathbf{d}+1)$ and $(n+2)((n+2)\mathbf{d}+2)$, respectively. Therefore, the number of equivalence classes of the equivalence relation $\sim^s_{\mathbf{P}}$ is not more than

$$\mathbf{p}(\mathbf{d}, k(s)) \equiv \textstyle\prod_{n=0}^{k(s)} (n + 2)^5((n + 2)\mathbf{d} + 2)^3. \tag{9.3.2}$$

Thus, the family $\mathrm{R}_{\mathbf{P}}$ is admissible.

Now, we denote by

$$\mathrm{R} \equiv \{\sim^s : s \in \mathcal{F}\}$$

the family of equivalence relation on $\mathbf{S}$ such that $\sim^s = \sim^s_+ \cap \sim^s_{\mathbf{P}}$. Since the families R^+ and $\mathrm{R}_{\mathbf{P}}$ are admissible, the family R is also admissible. Since R^+ is $\mathbf{M}$-admissible, R is also $\mathbf{M}$-admissible. Therefore, R can be considered as an $(\mathbf{M}, \omega, \omega^+)$-admissible family. Moreover, R is actually an $(\mathbf{M}, \mathbf{P})$-admissible family. Indeed, if k is an element of ω, then as the integer $m(k)$ (about the integer $m(k)$ see the definition of an $(\mathbf{M}, \mathbf{P})$-admissible family in Section 9.1) we can consider the integer 1 if $k = 0$ and the integer k if $k > 1$.

Thus, we can consider the Containing Space $\mathrm{T} \equiv \mathrm{T}(\mathbf{M}, \mathrm{R})$. By the choise of the co-mark $\mathbf{M}$ and the family R, the space T belongs to $I\!\!R$. Since R is also an $(\mathbf{M}, \omega, \omega^+)$-admissible family we can consider the metric ϱ_{T} on T defined in Section 9.1, which is compatible with the topology of

T. This fact means that $\mathrm{T} \in \mathbb{M}$. By the choice of the indexed collection $\mathbf{S}$ and Proposition 9.1.9, T is an isometrically Containing Space for the class $\mathbb{P}$.

To complete the proof of the proposition it suffices to prove that the metric space T is completely bounded. Let $\varepsilon \in \Re^+$. Without loss of generality, we can suppose that $\varepsilon = 1/p$, where $p \in \omega \setminus \{0\}$. Let $k = \mathbf{s}(4p)$ (see relation (9.3.1)). For every $\mathbf{H} \in \mathrm{C}(\sim^{m(k)})$ we take an element $X_{\mathbf{H}}$ of $\mathbf{H}$ and consider the set P_k^{T} consisting of all elements $\mathbf{a}$ of T such that $(x, X) \in \mathbf{a}$, where $X = X_{\mathbf{H}}$ for some $\mathbf{H} \in \mathrm{C}(\sim^{m(k)})$ and $x = a_i^X$ for some $i \in k$. Obviously,

$$|P_k^{\mathrm{T}}| \leq k|\mathrm{C}(\sim^{m(k)})|,$$

that is P_k^{T} is finite. We prove that P_k^{T} is an ε-net in T.

Let $\mathbf{b}$ be a point of T and $(y, Y) \in \mathbf{b}$. Denote by $\mathbf{H}$ the element of $\mathrm{C}(\sim^{m(k)})$ containing Y. By construction of the set P^Y and the definition of the number $\mathbf{s}(4p)$ there exists an element $i \in k$ such that $\varrho_Y(a_i^Y, y) \leq 1/(4p)$. Consider the element $\mathbf{a}$ of T containing the pair (x, X), where $X = X_{\mathbf{H}}$ and $x = a_i^X$. Then $\mathbf{a} \in P_k^{\mathrm{T}}$.

Suppose that $X \sim^s Y$ for every $s \in \mathcal{F}$ and let $\mathbf{E}$ be the element of $\mathrm{C}(\sim^\infty)$, which contains X and Y. Then, we have

$$\varrho_{\mathrm{T}}(\mathbf{a}, \mathbf{b}) = \varrho_{\mathrm{T}(\mathbf{E})}(\mathbf{a}, \mathbf{b}) \leq \varrho_Y(a_i^Y, y) \leq 1/(4p) < 1/p.$$

Now, suppose that for some $s \in \mathcal{F}$, X and Y are not $\sim^s$-equivalent. Since $X, Y \in \mathbf{H} \in \mathrm{C}(\sim^{m(k)})$, we have $k \leq n(X, Y) = n(\mathbf{a}, \mathbf{b})$. Since $i \in k$ and $4p \leq \mathbf{s}(4p) = k$, we have $i \in n(\mathbf{a}, \mathbf{b}) + 2$ and $1/(n(\mathbf{a}, \mathbf{b}) + 2) < 1/(4p)$. Therefore, we have

$$\varrho_{\mathrm{T}}(\mathbf{a}, \mathbf{b}) \leq \varrho_X(x, a_i^X) + \varrho_Y(a_i^Y, y) + \frac{1}{n(\mathbf{a}, \mathbf{b}) + 2}, \tag{9.3.3}$$

$$\varrho_X(x, a_i^X) < \varrho_Y(a_i^Y, y) + \frac{1}{n(\mathbf{a}, \mathbf{b}) + 2}. \tag{9.3.4}$$

Relations (9.3.3) and (9.3.4) imply that

$$\varrho_{\mathrm{T}}(\mathbf{a}, \mathbf{b}) < 1/(4p) + 1/(4p) + 1/(4p) + 1/(4p) = 1/p.$$

Thus, P_k^{T} is an ε-net for T, which proves that T is completely bounded. □

9.3.2 Corollary. *Let $\mathbb{P}$ be an arbitrary uniform class of metric spaces. Then, there exists an isometrically Containing Space for $\mathbb{P}$, which is completely bounded.* □

9.3.3 Corollary. *Let $\mathbb{P}$ be a uniform class of metric spaces which are:*
(a) *countable-dimensional,*
(b) *strongly countable-dimensional,*
(c) *locally finite-dimensional, and*
(d) *of dimension* ind *less than or equal to α, $\alpha \in \tau^+$.*
Then, there exists an isometrically Containing Space for $\mathbb{P}$, which is completely bounded and respectively:
(a) *countable-dimensional,*
(b) *strongly countable-dimensional,*
(c) *locally finite-dimensional, and*
(d) *of dimension* ind *less than or equal to α.* □

9.3.4 Proposition. *Let $\mathbb{P}$ be a class of compact metric spaces. The uniformity of the class $\mathbb{P}$ is a necessary and sufficient condition for the existence of an isometrically containing compact space for $\mathbb{P}$.*

Proof. It is easy to see that if there exists an isometrically containing compact space for $\mathbb{P}$, then $\mathbb{P}$ is uniform.

Conversely, suppose that $\mathbb{P}$ is uniform. By Corollary 9.3.2 there exists a completely bounded space T, which is an isometrically Containing Space for $\mathbb{P}$. Then, the metric completion of T is the required isometrically containing compact space. □

Definition. Any non-decreasing function $\mathbf{n} : \Re^+ \to \omega \setminus \{0\}$ is called a *net-function.*

We say that a net-function $\mathbf{n}_1$ is a *majorant* of a net-function $\mathbf{n}_0$ if for every $\varepsilon \in \Re^+$, $\mathbf{n}_0(\varepsilon) \leq \mathbf{n}_1(\varepsilon)$.

Definition. Let X be a completely bounded metric space. The *net-function* of X is a net-function, denoted by $\mathbf{n}^X$, such that for every $\varepsilon \in \Re^+$ the integer $\mathbf{n}^X(\varepsilon)$ coincides with the minimal of integers n for which the space X has an ε-net with n elements.

Notation. Let $\mathbb{M}$ be a class of metric spaces, $\mathbf{d} \in \omega \setminus \{0\}$, and $\mathbf{n}$ a net-function. We denote by $\mathbb{M}(\mathbf{d}, \mathbf{n})$ the subclass of $\mathbb{M}$ consisting of all completely bounded elements X such that $\mathrm{Diam}(X) \leq \mathbf{d}$ and $\mathbf{n}$ is a majorant of $\mathbf{n}^X$. Obviously, the class $\mathbb{M}(\mathbf{d}, \mathbf{n})$ is uniform.

9.3.5 Proposition. *Let $\mathbb{M}$ be the class of metric spaces generated by a saturated class $\mathbb{R}$ of metrizable spaces, $\mathbf{d}$ an integer of $\omega \setminus \{0\}$, and $\mathbf{n}$ a net-function. Then, there exists a completely bounded element T of $\mathbb{M}$, which is an isometrically Containing Space for $\mathbb{M}(\mathbf{d}, \mathbf{n})$, such that for every $p \in \omega \setminus \{0\}$ the net-function $\mathbf{n}^T$ of T satisfies condition*

$$\mathbf{n}^T(1/p) \leq 2^{2^{\mathbf{s}(4p)}} \mathbf{s}(4p)\mathbf{p}(\mathbf{d}, \mathbf{s}(4p)), \tag{9.3.5}$$

where

$$\mathbf{s}(4p) = \textstyle\sum_{n=1}^{4p} \mathbf{n}(1/n),$$

and

$$\mathbf{p}(\mathbf{d}, \mathbf{s}(4p)) = \textstyle\prod_{n=0}^{\mathbf{s}(4p)} (n+2)^5((n+2)\mathbf{d}+2)^3.$$

Proof. Let $\mathbf{S}$ be an indexed collection of non-empty elements of $\mathbb{M}$ such that for every $Y \in \mathbb{M}$ there exist an element X of $\mathbf{S}$ and an isometry of Y onto X. Let $\mathbf{M}$ be an initial co-mark of $\mathbf{S}$ corresponding to the class $\mathbb{R}$ and R_0^+ an initial family of equivalence relations on $\mathbf{S}$ corresponding to the co-mark $\mathbf{M}$ and the class $\mathbb{R}$. By Lemma 8.6.2 there exists a co-mark $\mathbf{M}_0$ such that the $\mathbf{M}_0$-standard family $\mathrm{R}_{\mathbf{M}_0}$ of equivalence relations on $\mathbf{S}$ is a final refinement of R_0^+. This fact means that the family $\mathrm{R}_{\mathbf{M}_0}$ is also an initial family corresponding to the co-mark $\mathbf{M}$ and the class $\mathbb{R}$.

Now, we denote by

$$\mathrm{R}^+ \equiv \{\sim_+^s : s \in \mathcal{F}\}$$

the family $\mathrm{R}_{\mathbf{M}_0}$ and continue the proof of the proposition similarly to the proof of Proposition 9.3.1. That is, we construct the $\mathbf{S}$-indexed collection $\mathbf{P}$ of ω-indexed dense subset and define the family

$$\mathrm{R}_{\mathbf{P}} \equiv \{\sim_{\mathbf{P}}^s : s \in \mathcal{F}\}$$

of equivalence relations on $\mathbf{S}$. We note that if $k \in \omega \setminus \{0\}$, then the number of $\sim_{\mathbf{P}}^k$-equivalence classes is given by relation (9.3.2). In particular, the family $\mathrm{R}_{\mathbf{P}}$ is admissible.

Furthermore, we define the family R of equivalence relations on $\mathbf{S}$ setting

$$\sim^s = \sim^s_{\mathbf{P}} \cap \sim^s_+.$$

This family is $\mathbf{M}$-admissible, which means that we can consider the Containing Space $\mathrm{T} \equiv \mathrm{T}(\mathbf{M}, \mathrm{R})$. Since R is also $(\mathbf{M}, \omega, \omega^+)$-admissible and $(\mathbf{M}, \mathbf{P})$-admissible, on the space T we can consider the metric ϱ_{T} constructed in Section 9.1.

Let $p \in \omega \setminus \{0\}$. We set $k = \mathbf{s}(4p)$, where

$$\mathbf{s}(4p) = \textstyle\sum_{n=0}^{4p} \mathbf{n}(1/n), \tag{9.3.6}$$

and construct the subset P_k^{T} of T, which is an $1/p$-net of T. For the number of elements of this set and since $k = \mathbf{s}(4p)$, we have

$$|P_k^{\mathrm{T}}| = k|\mathrm{C}(\sim^{m(k)})| = k|\mathrm{C}(\sim^k)|. \tag{9.3.7}$$

On the other hand, by construction of the family R, we have

$$|\mathrm{C}(\sim^k)| \leq |\mathrm{C}(\sim^k_+)||\mathrm{C}(\sim^k_{\mathbf{P}})|.$$

Since the family R^+ is $\mathbf{M}_0$-admissible by Lemma 1.1.4 we have

$$|\mathrm{C}(\sim^k_+)| \leq 2^{2^k}. \tag{9.3.8}$$

Since $k = \mathbf{s}(4p)$ relations (9.3.2), (9.3.6), (9.3.7), and (9.3.8) imply relation (9.3.5), which completes the proof of the proposition. □

9.4 The metric dimension of metric spaces

Definition. Let $\mathbb{M}$ be the class of all metric spaces. The *metric dimension* is the dimension-like function, denoted by $\mu\mathrm{dim}$, with the domain the class $\mathbb{M}$ of all metric spaces and range the set $\omega \cup \{-1, \infty\}$, which is defined as follows. For every $X \in \mathbb{M}$, $\mu\mathrm{dim}(X)$ is the least integer r of $\{-1\} \cup \omega$ (if there exists such an integer) with the property that for every $\varepsilon \in \Re^+$ there exists a locally finite open cover c of X such that the order of c is less than or equal to r (that is, the intersection of any $r + 2$ distinct elements of c is empty) and $\mathrm{Diam}(V) \leq \varepsilon$ for every $V \in c$. If there is no such integer, then we set $\mu\mathrm{dim}(X) = \infty$. (It is easy to see that the number of elements of c is not more than ω.)

Notation. We recall that for every $r \in \{-1\} \cup \omega$ by $\mathbb{P}(\mu\dim \leq r)$ is denoted the class of all metric spaces X with $\mu\dim(X) \leq r$.

9.4.1 Proposition. *For every $r \in \omega$ the class $\mathbb{P}(\mu\dim \leq r)$ is an isometrically ω-saturated class of spaces.*

Proof. Let $r \in \omega$ and $\mathbf{S}$ be an indexed collection of elements of $\mathbb{P}(\mu\dim \leq r)$ such that for every $Y \in \mathbb{M}$ there exists an isometry of Y onto an element X of $\mathbf{S}$.

Let $X \in \mathbf{S}$. For every $k \in \omega$ we denote by v_k^X a locally finite cover of X of the order $\leq r$ such that $\mathrm{Diam}(V) \leq 1/(k+1)$ for every $V \in v_k^X$. It is convenient to suppose that for every $k \in \omega$ the cover v_k^X contains the empty set. We observe that the set $\cup\{v_k^X : k \in \omega\}$ is a base of cardinality $\leq \tau$ for the open subsets of X.

Let $\{V_{k,i}^X : i \in \omega\}$ be an indication of v_k^X. Without loss of generality, we can suppose that if i_0 and i_1 are distinct elements of ω, then V_{k,i_0}^X and V_{k,i_1}^X are distinct elements of v_k^X (that is, they are distinct subsets of X) provided that one of them is not empty.

Since the cover v_k^X, where $k \in \omega$, is locally finite there exists an open cover w_k^X of X of cardinality less than or equal to ω such that any element of w_k^X intersect at most a finite number of elements of v_k^X. It is convenient to consider that the empty set is an element of w_k^X.

Let $\{W_{k,i}^X : i \in \omega\}$ be an indication of w_k^X. Without loss of generality, we can suppose that if $j, l \in \omega$ and $l < j$, then $V_{k,j}^X \cap W_{k,l}^X = \emptyset$. Indeed, let $\{W_i^{X,k} : i \in \omega\}$ be an arbitrary indication of w_k^X. First, we define a mapping ϑ of ω into itself. For this purpose, for every $i \in \omega$ we define by induction the integer $\vartheta(i)$. If $i = 0$, then $\vartheta(i)$ is an element of ω such that the condition $V_{k,j}^X \cap W_0^{X,k} \neq \emptyset$ implies that $j \in \vartheta(0)$. The existence of such an integer follows by the fact that the number of distinct elements of v_k^X, which intersect the set $W_0^{X,k}$, is finite. Furthermore, using induction we define the element $\vartheta(i)$, $i \in \omega \setminus \{0\}$, as an element of ω satisfying the properties:

(a) $\vartheta(0), ..., \vartheta(i-1) \in \vartheta(i)$ and

(b) if $V_{k,j}^X \cap W_i^{X,k} \neq \emptyset$, then $j \in \vartheta(i)$.

Now, we construct the indication $\{W_{k,i}^X : i \in \omega\}$ setting $W_{k,i}^X = W_{(\vartheta)^{-1}(i)}^{X,k}$ if $i \in \vartheta(\omega)$ and $W_{k,i}^X = \emptyset$ if $i \notin \vartheta(\omega)$. It is easy to verify that the constructed indication of w_k^X satisfies the above mentioned condition.

For every $k \in \omega$ we denote by ϑ_k^v and ϑ_k^w two one-to-one mappings of ω into itself such that

(a) $\vartheta_{k_0}^v(\omega) \cap \vartheta_{k_1}^v(\omega) = \emptyset$ if $k_0, k_1 \in \omega$ and $k_0 \neq k_1$,
(b) $\vartheta_{k_0}^w(\omega) \cap \vartheta_{k_1}^w(\omega) = \emptyset$ if $k_0, k_1 \in \omega$ and $k_0 \neq k_1$,
(c) $\vartheta_{k_0}^v(\omega) \cap \vartheta_{k_1}^w(\omega) = \emptyset$ if $k_0, k_1 \in \omega$, and
(d) $\cup\{\vartheta_k^v(\omega) \cup \vartheta_k^w(\omega) : k \in \omega\} = \omega$.

We define an indexed set $\{H_i^X : i \in \omega\}$ as follows. If $i \in \vartheta_k^v(\omega)$ for some $k \in \omega$, then we set $H_i^X = V_{k,i}^X$; and if $i \in \vartheta_k^w(\omega)$ for some $k \in \omega$, then we set $H_i^X = W_{k,i}^X$. It is easy to see that this indexed set is a mark of X.

Now, we consider the following co-mark of $\mathbf{S}$:

$$\mathbf{M}^+ \equiv \{\{H_i^X : i \in \omega\} : X \in \mathbf{S}\}$$

and denote by

$$\mathbf{M} \equiv \{\{U_i^X : i \in \omega\} : X \in \mathbf{S}\}$$

an arbitrary co-extension of $\mathbf{M}^+$ and by θ an indicial mapping of this co-extension. Also, we denote by

$$\mathrm{R}_{\mathbf{M}} = \{\sim_{\mathbf{M}}^s : s \in \mathcal{F}\}$$

the $\mathbf{M}$-standard family of equivalence relations on $\mathbf{S}$.

Let

$$\mathbf{P} = \{P^X \equiv \{a_i^X : i \in \omega\} : X \in \mathbf{S}\}$$

be an arbitrary $\mathbf{S}$-indexed set of ω-indexed dense subsets (therefore, $P^X = \emptyset$ if $X = \emptyset$).

Denote by

$$\mathrm{R}^+ \equiv \{\sim_+^s : s \in \mathcal{F}\}$$

the family of equivalence relations on $\mathbf{S}$, which is defined as follows. Let $s \in \mathcal{F} \setminus \{\emptyset\}$ and $k(s)$ be the minimal integer of $\omega \setminus \{0\}$ such that $s \subset k(s)$. Then, two elements X and Y of $\mathbf{S}$ are considered to be $\sim_+^s$-equivalent if

(a) $X \sim_{\mathbf{M}}^s Y$ and
(b) for every $n \in k(s)+1$ and $i, j \in n+2$, we have

$$k_0(X, n, i, j) = k_0(Y, n, i, j),$$

$$k_1(X,n,i,j) = k_1(Y,n,i,j), \text{ and}$$

$$k(X,n,i) = k(Y,n,i).$$

In addition, $X \sim^s_+ Y$ if $s = \emptyset$. By the above definition it follows immediately that R^+ is an $(\mathbf{M},\omega,\omega^+)$-admissible and $(\mathbf{M},\mathbf{P})$-admissible family. Let

$$\mathrm{R} \equiv \{\sim^s : s \in \mathcal{F}\}$$

be an arbitrary ω^+-admissible family of equivalence relations on $\mathbf{S}$, which is a final refinement of R^+. Then, R is also $(\mathbf{M},\omega,\omega^+)$-admissible and $(\mathbf{M},\mathbf{P})$-admissible.

Thus, we can consider the Containing Space $\mathrm{T} \equiv \mathrm{T}(\mathbf{M},\mathrm{R})$ and the metric ϱ_{T} on T constructed in Section 9.1. By Proposition 9.1.9 and the choice of $\mathbf{S}$ this metric space is an isometrically Containing Space for the class $\mathbb{M}(\mu\dim \leq r)$. Therefore, to prove the proposition, that is to prove that $\mathbf{M}^+$ is an initial co-mark corresponding to the class $\mathbb{P}(\mu\dim \leq r)$ and R^+ is an initial family corresponding to the co-mark $\mathbf{M}$ and the class $\mathbb{P}(\mu\dim \leq r)$, it suffices to prove that $\mathrm{T} \in \mathbb{M}(\mu\dim \leq r)$, that is, for every $\varepsilon \in \Re^+$ there exists a locally finite open cover of T of the Order less than or equal to r such that $\mathrm{Diam}(U) \leq \varepsilon$ for every element U of this cover.

Let $\varepsilon \in \Re^+$ and $p \in \omega \setminus \{0\}$ such that $1/p < \varepsilon/5$. We shall construct a locally finite cover u_p of T of the order $\leq r$ such that $\mathrm{Diam}(U) \leq 1/p$ for every $U \in u_p$. For this purpose, we first construct a set $u_{p,i}$, $i \in \omega$, of open subsets of T and then we put $u_p = \cup\{u_{p,i} : i \in \omega\}$.

For the construction of the above set $u_{p,i}$ we consider an arbitrary fixed element i of ω and set $j = (\vartheta^v_p)^{-1}(\theta^{-1}(i))$. Let X be a non-empty element of $\mathbf{S}$. (If all elements of $\mathbf{S}$ are empty, then $r = -1$ and, therefore, the proposition is trivial.) Then, we have $V^X_{p,j} = U^X_i$, which means that $\mathrm{Diam}(U^X_i) \leq 1/(p+1)$.

There exists an integer $l(X)$ of ω such that $a^X_{l(X)} \in U^X_i$. Denote by $n(X)$ an integer of ω such that $p, i, l(X) \in n(X)$ and

$$\varrho_X(a^X_{l(X)}, X \setminus U^X_i) > \frac{1}{n(X)+2}. \tag{9.4.1}$$

Now, we denote by $\mathbf{H}(X)$ the $\sim^{m(n(X))}$-equivalence class containing X (about the integer $m(n(X))$ see the definition of an $(\mathbf{M},\mathbf{P})$-admissible family R of Section 9.1) and consider the element $U^{\mathrm{T}}_i(\mathbf{H}(X))$ of the $\theta(\omega)$-standard base $\mathrm{B}^{\mathrm{T}}_{\theta(\omega)}$ for the open subsets of T. We prove that

$$\mathrm{Diam}(U_i^{\mathrm{T}}(\mathbf{H}(X))) \leq \varepsilon. \tag{9.4.2}$$

Indeed, let $\mathbf{a}$ and $\mathbf{b}$ be two points of $U_i^{\mathrm{T}}(\mathbf{H}(X))$ and let $(y, Y) \in \mathbf{a}$ and $(z, Z) \in \mathbf{b}$. Then,

$$y \in U_i^Y,\ z \in U_i^Z,\ n(X) \leq n(Y, Z) \leq n(\mathbf{a}, \mathbf{b}),$$

$$n(X) \leq n(X, Y), \text{ and } n(X) \leq n(X, Z).$$

The above relations and Lemma 9.1.1 imply that

$$\mathrm{Diam}(U_i^Y) < \mathrm{Diam}(U_i^X) + \frac{1}{n(X,Y)+2} < \frac{1}{p+1} + \frac{1}{p+1} < \frac{2}{p}$$

$$\mathrm{Diam}(U_i^Z) < \mathrm{Diam}(U_i^X) + \frac{1}{n(X,Z))+2} < \frac{1}{p+1} + \frac{1}{p+1} < \frac{2}{p}.$$

Relation (9.4.1) and Lemma 9.1.1 imply that

$$\varrho_Y(a_{l(X)}^Y, Y \setminus U_i^Y) > 0 \quad \text{and} \quad \varrho_Z(a_{l(X)}^Z, Z \setminus U_i^Z) > 0,$$

which means that $a_{l(X)}^Y \in U_i^Y$ and $a_{l(X)}^Z \in U_i^Z$.

First, we suppose that there exists an element $\mathbf{E}$ of $\mathrm{C}(\sim_{\mathrm{R}}^{\infty})$ such that $\mathbf{a}, \mathbf{b} \in \mathrm{T}(\mathbf{E})$. In this case,

$$\varrho_{\mathrm{T}}(\mathbf{a}, \mathbf{b}) \leq \varrho_{\mathrm{T}(\mathbf{E})}(\mathbf{a}, \mathbf{b}) \leq$$

$$\varrho_{\mathrm{T}(\mathbf{E})}(\mathbf{a}, \mathbf{a}_{l(X)}^{\mathrm{T}(\mathbf{E})}) + \varrho_{\mathrm{T}(\mathbf{E})}(\mathbf{a}_{l(X)}^{\mathrm{T}(\mathbf{E})}, \mathbf{b}) =$$

$$\varrho_Y(y, a_{l(X)}^Y) + \varrho_Z(a_{l(X)}^Z, z) < 2/p + 2/p = 4/p < \varepsilon.$$

Now, we suppose that there exists an element s of $\mathcal{F}$ such that Y and Z are not $\sim^s$-equivalent. Since $l(X) \in n(X) \leq n(\mathbf{a}, \mathbf{b})$ by the definition of the metric ϱ_{T}, we have

$$\varrho_{\mathrm{T}}(\mathbf{a}, \mathbf{b}) \leq \varrho_Y(y, a_{l(X)}^Y) + \varrho_Z(a_{l(X)}^Z, z) + \frac{1}{n(\mathbf{a}, \mathbf{b})+2} <$$

$$2/p + 2/p + 1/p = 5/p < \varepsilon.$$

Thus, relation (9.4.2) is proved.

Consider the set

$$c_{p,i} \equiv \{U_i^{\mathrm{T}}(\mathbf{H}(X)) : X \in \mathbf{S}\}.$$

Since the elements of $c_{p,i}$ belong to the countable set $\mathrm{B}^{\mathrm{T}}_{\theta(\omega)}$, the set $c_{p,i}$ is also countable. Let

$$\{U_i^{\mathrm{T}}(\mathbf{H}(X_q)) : q \in q(p,i) \in \omega + 1\}$$

be an indication of this set. We set $\mathbf{H}_0 = \mathbf{H}(X_0)$ and

$$\mathbf{H}_q = \mathbf{H}(X_q) \setminus (\cup\{\mathbf{H}_{q'} : q' \in q\}).$$

Then, we have $U_i^{\mathrm{T}}(\mathbf{H}_0) = U_i^{\mathrm{T}}(\mathbf{H}(X_0))$ and

$$U_i^{\mathrm{T}}(\mathbf{H}_q) = U_i^{\mathrm{T}}(\mathbf{H}(X_q)) \setminus (\cup\{U_i^{\mathrm{T}}(\mathbf{H}_{q'}) : q' \in q\}),\ q \in q(p,i) \setminus \{0\}.$$

Therefore, setting

$$u_{p,i} = \{U_i^{\mathrm{T}}(\mathbf{H}_q) : q \in q(p,i) \in \omega + 1\},$$

we have that the elements of $u_{p,i}$ are mutually disjoint, which means that $\mathbf{H}_q \cap \mathbf{H}_{q'} = \emptyset$ if q and q' are distinct elements of $q(p,i)$. Moreover, we have

$$\cup\{U : U \in c_{p,i}\} = \cup\{U : U \in u_{p,i}\}.$$

Now, we set

$$u_p = \cup\{u_{p,i} : i \in \omega\}$$

and prove that u_p is a locally finite open cover of T of the order $\leq r$ such that $\mathrm{Diam}(U) \leq \varepsilon$, $U \in u_p$. Indeed, we note that the elements of u_p are elements of the $(\theta(\omega), \diamondsuit)$-standard base $\mathrm{B}^{\mathrm{T}}_{\theta(\omega),\diamondsuit}$ for the open subsets of T and, therefore, they are open.

We prove that u_p is a cover. Let $\mathbf{a} \in \mathrm{T}$ and $(x, X) \in \mathbf{a}$. There exists an element j of ω such that $x \in V^X_{p,j}$. Then, $\mathbf{a} \in U_i^{\mathrm{T}}(\mathbf{H}(X))$, where $i = \theta(\vartheta^v_p(j))$, that is $\mathbf{a}$ belongs to an element of $c_{p,i}$ and, therefore, $\mathbf{a}$ belongs also to an element of $u_{p,i} \subset u_p$.

By the construction of elements of $u_{p,i}$ it follows immediately that $\mathrm{Diam}(U) \leq \varepsilon$ for every $U \in u_{p,i}$.

Now, we prove that the order of u_p is $\leq r$. Indeed, in the opposite case there exists $r+2$ distinct elements of u_p: $U_{i_0}^{\mathrm{T}}(\mathbf{H}_{q_0})$,..., $U_{i_{r+1}}^{\mathrm{T}}(\mathbf{H}_{q_{r+1}})$ and a point $\mathbf{a}$ of T such that

$$\mathbf{a} \in U_{i_0}^{\mathrm{T}}(\mathbf{H}_{q_0}) \cap ... \cap U_{i_{r+1}}^{\mathrm{T}}(\mathbf{H}_{q_{r+1}}). \tag{9.4.3}$$

Since for every $i \in \omega$ the elements of the set $u_{p,i}$ are mutually disjoint, we have that the indices i_0,...,i_{r+1} are mutually disjoint.

If $(x, X) \in \mathbf{a}$, then relation (9.4.) implies that

$$x \in U_{i_0}^X \cap ... \cap U_{i_{r+1}}^X$$

and setting

$$j_0 = (\vartheta_p^v)^{-1}(\theta^{-1}(i_0)),..., j_{r+1} = (\vartheta_p^v)^{-1}(\theta^{-1}(i_{r+1}))$$

we have

$$x \in V_{p,j_0}^X \cap ... \cap V_{p,j_{r+1}}^X,$$

which contradicts the fact that the order of v_p^X is $\leq r$ and the choice of the indication $\{V_{p,i}^X : i \in \omega\}$ of v_p^X. Thus, the order of u_p is $\leq r$.

Finally, we prove that the cover u_p is locally finite. Let $\mathbf{a} \in$ T. We must prove that there exists an open neighbourhood of $\mathbf{a}$, which intersects at most a finite number of elements of u_p.

Suppose that $(x, X) \in \mathbf{a}$. Consider the cover w_p^X and let $W_{p,j}^X$ be an element of this cover containing x. We set $i = \theta(\vartheta_p^w(j))$ and

$$s = \theta(\vartheta_p^v(\{0, ..., j\})).$$

Therefore, $x \in U_i^X = W_{p,j}^X$.

For every $i' \in s$ we consider the unique element $U_{i'}^{\mathrm{T}}(\mathbf{H}_{q(i')})$ of $u_{p,i'}$ containing x and set

$$\mathbf{H} = \cap\{\mathbf{H}_{q(i')} : i' \in s\}.$$

Then, $U_i^{\mathrm{T}}(\mathbf{H})$ is an open neighbourhood of $\mathbf{a}$. We prove that $U_i^{\mathrm{T}}(\mathbf{H})$ intersects at most a finite number of elements of u_p. Let $U_{i'}^{\mathrm{T}}(\mathbf{H}')$ be an element of u_p such that

$$U_i^{\mathrm{T}}(\mathbf{H}) \cap U_{i'}^{\mathrm{T}}(\mathbf{H}') \neq \emptyset.$$

First, we suppose that $i' \in s$. By the definition of $\mathbf{H}$ and since $\mathbf{H} \cap \mathbf{H}' \neq \emptyset$ it follows that $\mathbf{H} \subset \mathbf{H}'$. Since the elements of $u_{p,i'}$ are mutually disjoint, the set $U_{i'}^{\mathrm{T}}(\mathbf{H}')$ is the only element of $u_{p,i'}$, which intersect the set $U_i^{\mathrm{T}}(\mathbf{H})$.

Now, we suppose that $i' \notin s$. Let

$$\mathbf{b} \in U_i^{\mathrm{T}}(\mathbf{H}) \cap U_{i'}^{\mathrm{T}}(\mathbf{H}')$$

and $(y, Y) \in \mathbf{b}$. Then, $y \in U_i^Y \cap U_{i'}^Y$, that is $y \in W_{p,j}^Y \cap V_{p,j'}^Y$, where $j' = (\vartheta_p^v)^{-1}(\theta^{-1}(i'))$. Since $i' \notin s$ we have $j < j'$, which contradicts the choice of the indication $\{W_{p,i}^Y : i \in \omega\}$ of the cover w_p^Y of Y.

Thus, $U_i^{\mathrm{T}}(\mathbf{H})$ is an open neighbourhood of $\mathbf{a}$, which intersects at most a finite number of elements of the cover u_p. The proof of the proposition is complete. □

The above proposition, Proposition 9.2.1, and Corollary 9.2.5 imply the following consequences.

9.4.2 Corollary. *In the class of all separable metric spaces of metric dimension $\leq r$, $r \in \omega$, there exist isometrically universal elements.* □

9.4.3 Corollary. *In the class of all separable metric spaces of metric dimension $\leq r$, $r \in \omega$, and of dimension* ind $\leq n$, $n \in \omega$, *there exist isometrically universal elements.* □

Chapter 10

Concluding remarks and open problems

Questions concerning universality appear in Topology in all stages of its development. We recall the classical results of the following papers: [131], [374], [375], [376], [423], [424], [273], [426], [173], [274], [413], [312], [350], [239], [175], [275], [176], [314], [8] [229], [9].

With the consideration of new classes of spaces, the extension of "old" dimensions to such classes, and the investigation of infinite dimensions we have the following well known papers: [428], [104], [159], [211], [289], [291], [224], [302], [303], [385], [304], [305], [391], [392], [327], [46], [306], [328], [329], [331], [332], [447], [448], [354], [333], [334], [335], [22], [337], [338].

Some methods of construction of universal spaces appear. The most important are the methods using factorization theorems (see for example [258], [330], [22], [336], [337], [338]).

With the development of dimension theory, the further investigation of finite, infinite, and cohomological dimensions, the consideration of extension and some other new dimensions, with dimension theory's applications to more general classes of spaces, and with the further consideration of new classes of spaces and mappings, many results concerning universality appear. We refer the following papers: [228], [372], [248], [237], [439], [344], [244], [245], [250], [39], [347], [70], [348], [294], [295], [253], [403], [74], [404], [268], [444], [38], [110], [349], [433], [445], [148], [63], [446], [106], [80], [316], [317], [318], [363], [293], [405], [408], [187], [189], [406], [109], [319], [320], [321], [143], [144], [216], [407], [343], [81].

The well known concrete universal spaces now are the object of special investigations. Additional properties of such spaces are proved. We refer the reader to the following papers: [440], [12], [13], [57], [43], [394], [341], [107], [33], [342], [76], [77], [161], [78], [322], [286], [246], [281], [449], [339], [450].

In the books [177], [230], [231], [10], [114], [115] and survey papers [113], [117] one can find more details for many of the above mentioned papers.

The concept of universality plays an important role in infinite-dimensional topology especially in the investigation of questions concerning topological characterizations of different infinite-dimensional spaces. For this topic we indicatively refer the reader to the papers [399], [400], [34], [35], [96], [97], [95], [59], [58], [26] and to the books [32], [62], [93], [278], [125], [145], [27], [79], [279] for more information.

In the following sections we refer to many other papers related to universality problems for different classes of spaces. Considering applications of the method of Containing Spaces we also refer some other papers which are not directly connected with such problems. However, it seems probable that this method or some of its modifications can be applied to (classes of) spaces investigated in these papers.

The property of a given class of spaces to be saturated is useful in the construction of new classes of spaces with universal elements. In this connection, the following general problem is interesting: what classes of spaces considered in the above mentioned papers (in which there are universal elements) are saturated?

As it was mentioned in the preface there exists an (informal) connection between saturated classes and τ-properties concerning topological spaces: many classes defined by τ-properties are saturated. (A property of a space is a τ-property if in its formulation it is used not more than τ many subsets of the space.) The formal explanation of this connection is another interesting problem.

There are some "exceptions" in the above connection (which underline the "dificulties" of the posed problem). For example, the class of all metrizable spaces of weight $\leq \tau$, where $\tau > \omega$, is defined by a τ-property: a regular space belongs to this class if and only if it has a σ-discrete base (of cardinality $\leq \tau$). On the other hand, we can not prove that this class is saturated. (The situation is the same with the class of all (regular) non-metrzable Moore spaces.) As an explanation of this fact we note that for metrizable (and non-metrzable Moore) spaces ω is a specific cardinal while in the construction of Containing Spaces ω has no specific role. (In general the cardinality of a Containing Space is equal to 2^τ while the cardinality of a metrizable (or non-metrizable Moore) space is less than or equal to the cardinality of the set of all countable subsets of τ.) However, it seems probable that there exists a "modification" of the construction of Containing Spaces, which can be applied to metrizable (or to non-metrizable Moore) spaces. Thus, the existence of such a "modification" is also an interesting problem.

New classes of spaces with universal elements can be constructed using τ-properties in the following way. Suppose that we have a class of spaces (with universal or without universal elements), which is defined by a property P concerning topological spaces. If P is not a τ-property, then we can approximate it by a τ-property considering that the subsets of a space included in the formulation of P are taken from an *a priori* standard family of cardinality $\leq \tau$ of subsets of the space. Then, the new property is a τ-property and we can expect that in the new class of spaces defined by this property there exist universal elements. This method of construction of classes of spaces with universal elements was used in Chapter 3 considering dimensions Ind and dim. It is interesting to apply also this method to classes of spaces (which are not defined by τ-properties) considered in the above mentioned papers.

In this chapter we give some remarks and pose some problems concerning the construction of Containing Spaces and its applications. Most of the problems are open ended. Such problems are called general. The aim of these remarks and problems is to suggest new aspects of our considerations for a further investigation.

10.1 Remarks and problems to Chapter 1

Remark 1. Let X be a marked space and M the mark of X. We denote by c_M^X the *cardinal-algebra function* of the marked space X, which is the mapping of $\mathcal{F} \setminus \{\emptyset\}$ into ω such that $c_M^X(s) = |\mathrm{A}_s^X| \in \omega$, where A_s^X is the s-algebra of the marked space X. Obviously, the mapping c_M^X satisfies the following conditions:

(a) $c_M^X(s) \leq |2^{2^s}|$, $s \in \mathcal{F} \setminus \{\emptyset\}$, and

(b) $c_M^X(s) \leq c_M^X(t)$ if $\emptyset \neq s \subset t \in \mathcal{F}$.

Also, it is easy to see that the weight of X is infinite if and only if c_M^X satisfies the following "limit" condition:

(c) for every $m \in \omega$ there exists $s \in \mathcal{F}$ such that $m \leq c_M^X(s)$.
In this connection we pose the following problems.

10.1.1 Problems. 1. Find topological properties of marked spaces which can be characterized by the cardinal-algebra functions.

2. Find limit properties of the cardinal-algebra functions which are independent of spaces.

3. What can we say about two marked spaces which have the same cardinal-algebra function?

4. Let $\tau = \omega$. Suppose that M is a mark of the set $\mathcal{R}$ of real numbers with the usual topology. What can we say about the limit points of the sequence $\{c_M^{\mathcal{R}}(n)/2^{2^n} : n \in \omega\}$?

Remark 2. Among admissible families of equivalence relations considered on a given indexed collection $\mathbf{S}$ of spaces the *standard* families, that is, families R for which there exists a co-mark $\mathbf{M}$ of $\mathbf{S}$ such that $\mathrm{R} = \mathrm{R}_{\mathbf{M}}$ play a specific role. For such a family $\mathrm{R} \equiv \{\sim^s : s \in \mathcal{F}\}$ we have the following properties:

(a) the number of $\sim^s$-equivalence classes is $\leq |2^{2^s}|$, $s \in \mathcal{F}$, and

(b) the number of $\sim^s$-equivalence classes is less than or equal to the number of t-equivalence classes, where $s \subset t \in \mathcal{F}$.

The natural question arises if such conditions on an admissible family are sufficient in order to be standard. In this connection the following general problem (formulated for very concrete indexed collections of spaces) is interesting.

10.1.2 Problems. Let $\tau = \omega$.

1. Suppose that $\mathbf{S}$ is an indication of the singleton $\{\mathcal{R}\}$. Characterize the standard families of equivalence relations on $\mathbf{S}$ (using properties like to the above considered properties (a) and (b)).

2. Let $\mathbf{S}$ be an indication of $\{\mathcal{R}\}$. Characterize that of the admissible families R of equivalence relations on $\mathbf{S}$ for which $\mathrm{R} \sim_{af} \mathrm{R}_{\mathbf{M}}$ for some co-mark $\mathbf{M}$ of $\mathbf{S}$. (About notation $\sim_{af}$ see Section 8.2).

3. Find "non-trivial" concrete indexed collections $\mathbf{S}$ of spaces for which it is possible to characterize the standard families of equivalence relations on $\mathbf{S}$.

Concerning the above problem 10.1.2(2) we recall that in Section 8.2 it is proved (see Lemma 8.2.6) that for every admissible family R of equivalence relations on an indexed collection $\mathbf{S}$ of non-empty spaces there exists a co-mark $\mathbf{M}$ of $\mathbf{S}$ such that the family $\mathrm{R}_{\mathbf{M}}$ of equivalence relations on $\mathbf{S}$ is a final refinement of R.

Remark 3. By definition, the elements of a mark $\{U_\delta^X : \delta \in \tau\}$ of a space X are open subsets of X. However, all results of Section 1.1 and the results of Section 1.2 until Lemma 1.2.6 do not use this fact. Moreover, the construction of the Containing Spaces can be carried out supposing that the sets U_δ^X are open only for the elements δ of a fixed subset κ of τ, which is independent of X.

Definition. Let X be a space. An indexed set $N \equiv \{V_\delta^X : \delta \in \tau\}$ of subsets of X is called a *generalized mark* of X or more precisely, κ-*generalized mark* of X, where κ is a subset of τ, if the set $\{V_\delta^X : \delta \in \kappa\}$ is a base for the open subsets of X. (For $\delta \in \tau \setminus \kappa$ the subset V_δ^X may be not open.)

As for the marks, for a given generalized mark we can define the s-algebra of X, denoted also by A_s^X, $s \in \mathcal{F} \setminus \{\emptyset$.

Definition. Let $\mathbf{S}$ be an indexed collection of spaces and for every $X \in \mathbf{S}$ let $N^X \equiv \{V_\delta^X : \delta \in \tau\}$ be a κ-generalized mark of X for some $\kappa \subset \tau$. Then, the indexed collection

$$\mathbf{N} \equiv \{N^X \equiv \{V_\delta^X : \delta \in \tau\} : X \in \mathbf{S}\}$$

is called a κ-*generalized co-mark* of $\mathbf{S}$ or simply *generalized co-mark* of $\mathbf{S}$.

As for the co-marks of $\mathbf{S}$, any generalized co-mark $\mathbf{N}$ of $\mathbf{S}$ generates an admissible family of equivalence relations on $\mathbf{S}$ denoted also by $\mathrm{R}_{\mathbf{N}}$.

10.1.3 Problems. Let $\mathbf{N}$ be a generalized co-mark of an indexed collection $\mathbf{S}$ of spaces.

1. Is there a co-mark $\mathbf{M}$ of $\mathbf{S}$ such that $\mathrm{R}_{\mathbf{N}} = \mathrm{R}_{\mathbf{M}}$? Under what conditions (on $\mathbf{S}$ and $\mathbf{N}$) is this situation true?

2. Is there a co-mark $\mathbf{M}$ of $\mathbf{S}$ such that $\mathrm{R}_{\mathbf{M}}$ is a final refinement of $\mathrm{R}_{\mathbf{N}}$ and $\mathrm{R}_{\mathbf{N}}$ is a final refinement of $\mathrm{R}_{\mathbf{M}}$ (that is, $\mathrm{R}_{\mathbf{N}} \sim_{af} \mathrm{R}_{\mathbf{M}}$))? Under what conditions (on $\mathbf{S}$ and $\mathbf{N}$) is this situation true?

3 Consider Problems 1 and 2 for the case where $\tau = \omega$ and $\mathbf{S}$ is an indication of the singleton $\{\mathcal{R}\}$.

Remark 4. The generalized co-marks can be used in order to construct $(\mathbf{M}, \mathbf{Q})$-admissible families. This construction can be done as in the following proposition.

10.1.4 Proposition. *Let* $\mathbf{S}$ *be an indexed collection of spaces,* $\mathbf{Q}(\varepsilon)$, $\varepsilon \in \tau$, *a restriction of* $\mathbf{S}$, *and*

$$\mathbf{M} \equiv \{\{U_\delta^X : \delta \in \tau\} : X \in \mathbf{S}\}$$

a co-mark of $\mathbf{S}$. *Suppose that*

$$\mathbf{N} \equiv \{\{V_\eta^X : \eta \in \tau\} : X \in \mathbf{S}\}$$

is a generalized co-mark of $\mathbf{S}$ *and* θ_1 *and* θ_2 *are two one-to-one mappings of* τ *into itself satisfying conditions*

$$V_{\theta_1(\varepsilon)}^X = \mathbf{Q}(\varepsilon)(X) \text{ and } V_{\theta_2(\delta)}^X = U_\delta^X,\ X \in \mathbf{S}.$$

Then, for every $\varepsilon \in \tau$ *the family*

$$\mathrm{R}_{\mathbf{N}} \equiv \{\sim_{\mathbf{N}}^{s} : s \in \mathcal{F}\}$$

is an $(\mathbf{M}, \mathbf{Q}(\varepsilon))$*-admissible family of equivalence relations on* $\mathbf{S}$.

Proof. Let ε be a fixed element of τ. We need to prove that the family $\mathrm{R}_{\mathbf{N}}$ is $(\mathbf{M}, \mathbf{Q}(\varepsilon))$-admissible. Since $\mathbf{N}$ is a co-extension of $\mathbf{M}$ it suffices to prove that the family

$$\mathrm{R}_{\mathbf{N}}|_{\mathbf{Q}(\varepsilon)} = \{\sim_{\mathbf{N}}^{s}|_{\mathbf{Q}(\varepsilon)} : s \in \mathcal{F}\}$$

is $\mathbf{M}|_{\mathbf{Q}(\varepsilon)}$-admissible, where

$$\mathbf{M}|_{\mathbf{Q}(\varepsilon)} = \{\{\mathbf{Q}(\varepsilon)(X) \cap U_\delta^X : \delta \in \tau\} : X \in \mathbf{S}\}.$$

Let $s \in \mathcal{F}$. For every $X \in \mathbf{S}$ we consider the s-algebra $\mathrm{A}_s^{\mathbf{Q}(\varepsilon)(X)}$ of $\mathbf{Q}(\varepsilon)(X)$ corresponding to the mark

$$\mathbf{M}|_{\mathbf{Q}(\varepsilon)}(X) = \{\mathbf{Q}(\varepsilon)(X) \cap U_\delta^X : \delta \in \tau\}.$$

We recall that two elements $\mathbf{Q}(\varepsilon)(Y)$ and $\mathbf{Q}(\varepsilon)(Z)$ of $\mathbf{Q}(\varepsilon)$ are $\sim_{\mathbf{M}|_{\mathbf{Q}(\varepsilon)}}^{s}$-equivalent if and only if $\mathrm{A}_s^{\mathbf{Q}(\varepsilon)(Y)}$ and $\mathrm{A}_s^{\mathbf{Q}(\varepsilon)(Z)}$ are isomorphic by the natural way.

On the other hand setting $t = \{\theta_1(\varepsilon)\} \cup \theta_2(s)$ we have that two elements $\mathbf{Q}(\varepsilon)(Y)$ and $\mathbf{Q}(\varepsilon)(Z)$ of $\mathbf{Q}(\varepsilon)$ are $\sim_{\mathbf{N}}^{t}|_{\mathbf{Q}(\varepsilon)}$-equivalent if and only if the elements Y and Z of $\mathbf{S}$ are $\sim^t_{\mathbf{N}}$-equivalent, that is if and only if the t-algebra A^Y_t of Y corresponding to the generalized mark $\{V^Y_\eta : \eta \in \tau\}$ is isomorphic to the t-algebra A^Z_t of Z corresponding to the generalized mark $\{V^Z_\eta : \eta \in \tau\}$.

It is easy to see that for every $X \in \mathbf{S}$ the mapping i of $\mathrm{A}^{\mathbf{Q}(\varepsilon)(X)}_s$ into A^X_t for which

$$i(\mathbf{Q}(\varepsilon)(X) \cap U^X_\delta) = \mathbf{Q}(\varepsilon)(X) \cap U^X_\delta,\ \delta \in s,$$

is an isomorphism. Moreover, for two elements Y and Z of $\mathbf{S}$ the natural isomorphism of A^Y_t onto A^Z_t generates an isomorphism of algebra $\mathrm{A}^{\mathbf{Q}(\varepsilon)(Y)}_s$ onto $\mathrm{A}^{\mathbf{Q}(\varepsilon)(Z)}_s$. This fact means that

$$\sim^t_{\mathbf{N}}|_{\mathbf{Q}(\varepsilon)} \subset \sim^s_{\mathbf{M}|_{\mathbf{Q}(\varepsilon)}},$$

that is, the family $\mathrm{R}_{\mathbf{N}}|_{\mathbf{Q}(\varepsilon)}$ is $\mathbf{M}|_{\mathbf{Q}(\varepsilon)}$-admissible. □

Remark 5. For the construction of a Containing Space $\mathrm{T}(\mathbf{M}, \mathrm{R})$ for an indexed collection $\mathbf{S}$ of spaces it is assumed that the family $\mathrm{R} \equiv \{\sim^s : s \in \mathcal{F}\}$ of equivalence relations on $\mathbf{S}$ is $\mathbf{M}$-admissible. However, for the definition of the set $\mathrm{T}(\mathbf{M}, \mathrm{R})$ and the definition of the elements $U^{\mathrm{T}}_\delta(\mathbf{H})$ of the set B^{T} we do not use this fact. Moreover, for these definitions we do not use also the fact that the indexing set of R is the concrete set $\mathcal{F}$. We use only that $\mathcal{F}$ is directed.

Definition. Let $\mathbf{S}$ be an indexed collection of spaces. An indexed family

$$\mathrm{R} \equiv \{\sim^\lambda : \lambda \in \Lambda\},$$

of equivalence relations on $\mathbf{S}$ is said to be *generalized* if Λ is a directed set of cardinality $\leq \tau$ containing the empty set as an element.

For the generalized families (with the same indexing set) the definitions of the notion of an admissible family and the notion of a final refinement are the same as for the families of equivalence relations on $\mathbf{S}$.

Thus, the sets $\mathrm{T} \equiv \mathrm{T}(\mathbf{M}, \mathrm{R})$ and B^{T} for an indexed collection $\mathbf{S}$ of spaces can be defined for any co-mark $\mathbf{M}$ of $\mathbf{S}$ and any admissible generalized family R of equivalence relations on $\mathbf{S}$.

In a similar way as in Section 1.2 we can prove the following proposition.

10.1.5 Proposition. *Let* $\mathbf{S}$ *be an indexed collection of spaces,* $\mathbf{M}$ *a co-mark of* $\mathbf{S}$*, and* R *an admissible generalized family of equivalence relations on* $\mathbf{S}$*. The following statements are true:*

(1) *The set* $\mathrm{T} \equiv \mathrm{T}(\mathbf{M}, \mathrm{R})$ *equipped with the topology, for which the set* B^{T} *is a subbase (that is, the finite intersections of elements of* B^{T} *compose a base), is a* T_0*-space (of weight* $\leq \tau$*).*

(2) *For every* $X \in \mathbf{S}$ *the mapping* i_{T}^{X} *defined by the relation*

$$i_{\mathrm{T}}^{X}(x, X) = \mathbf{a},$$

where $x \in X$ *and* $\mathbf{a}$ *is the element of* T *containing the pair* (x, X)*, is an embedding of* X *into the space* T*.* □

We note that the space $\mathrm{T}(\mathbf{M}, \mathrm{R})$ of Proposition 10.1.5 may be "bad" with respect to separation axioms.

10.1.6 Problems. Let $\mathbf{S}$ be an indexed collection of regular (respectively, of completely regular) spaces and

$$\mathrm{R} \equiv \{\sim^{\lambda} : \lambda \in \Lambda\}$$

a generalized family of equivalence relations on $\mathbf{S}$ such that $\sim^{\lambda} = \mathbf{S} \times \mathbf{S}$ for every $\lambda \in \Lambda$.

1. Is there a co-mark $\mathbf{M}$ of $\mathbf{S}$ such that the space $\mathrm{T}(\mathbf{M}, \mathrm{R})$ is regular (respectively, completely regular)?
2. Under what conditions on a co-mark $\mathbf{M}$ of $\mathbf{S}$ the space $\mathrm{T}(\mathbf{M}, \mathrm{R})$ is regular (respectively, completely regular)?

Remark 6. The generalization of construction of Containing Spaces given in Section 1.5 actually concerns the definitions of the standard and admissible families of equivalence relations. The replacement of the directed set $\mathcal{F}$ considered in Sections 1.1 and 1.2 by the directed set $\mathcal{P}_{\nu}(\tau)$ considered in Section 1.5 is the main difference of the corresponding definitions. It is natural to suppose that in a feasible generalization of the construction of Containing Spaces the corresponding indexing set may be a suitable directed subset of $\mathcal{P}(\tau)$ or some other directed set. In the connection with the above we pose the following problem.

10.1.7 Problem. For a given indexed collection $\mathbf{S}$ of spaces give "the most strong" generalization of the construction of Containing Spaces based on the generalized co-marks and generalized families of equivalence relations on **S**.

Remark 7. In [23] two general problems are posed as follows:

"**Problem 7.** Let $\mathcal{P}$ be a given class of topological spaces. Determine (if possible) "standard" topological spaces Y into which it is possible to imbed each space in $\mathcal{P}$. Find the simplest such space Y.

A variant of Problem 7 is the following *universal object problem*:

Problem 8. Which classes $\mathcal{P}$ of topological spaces contain a space X into which each space in $\mathcal{P}$ imbeds?"

In our opinion the Containing Spaces as presented in Chapter 1 provide for a satisfactory answer to the above Problem 7 and the notion of a saturated class of spaces introduced in Chapter 2 gives such an answer to Problem 8.

Remark 8. During the writing of this book, the results of sections 1.1 and 1.2 were published in [192]. The results of Sections 1.3 and 1.4 are published in [193] and the results of Section 1.5 in [191]. The construction of Containing Spaces improves and generalizes the corresponding constructions of [143] (concerning the separable metrizable spaces) and [187] (concerning the regular spaces). Elements of the method can be found in the papers [179], [180], [181], [182], [184], and [185].

10.2 Remarks and problems to Chapter 2

Remark 1. By the first view, the definition of a saturated class of spaces seems to be complicated. However, with the consideration of any simplification of this definition we lose "something". Below, we give such simplifications.

Definition. If in the definition of a saturated class it is assumed that all considered elements of $\mathrm{C}^{\diamond}(\mathrm{R})$ coincide with the indexed collection **S** (which is an element of $\mathrm{C}^{\diamond}(\mathrm{R})$), then we obtain the definition of a *partially saturated class*.

It is easy to see that for the classes $I\!P$ of spaces with the property that each simultaneously open and closed subspace of an element of $I\!P$ is also an element of $\mathbb{P}$, the notions of a saturated class and a partially saturated class coincide. The same is true for the classes $\mathbb{P}$ of subsets with the property that for every element (Q, X) of $\mathbb{P}$ and a simultaneously open and closed subspace X' of X we have $(Q \cap X', X') \in \mathbb{P}$.

All results of Sections 2.1 and 2.2 remain true if by a saturated class of spaces and a saturated class of subsets we shall mean partially saturated classes. However, for some propositions the partially saturated classes are not sufficient. Such a proposition is for example Proposition 2.5.7.

Also, partially saturated classs are not convenient to use in the consideration of dimension-like functions introduced in Chapter 3. Consider for example dimension $\mathrm{ind}_{\mathbb{E}}^{\mathbb{B}}$. If the classes $\mathbb{B}$ and $\mathbb{E}$ are partially saturated, then we can not carry out the induction considered in the proof of Proposition 3.1.5. On the other hand, if $\mathbb{B}$ and $\mathbb{E}$ satisfy hereditary properties like the above properties, then we can not prove that the elements of the class $\mathbb{P}(\mathrm{ind}_{\mathbb{E}}^{\mathbb{B}} \leq \alpha)$ satisfy the same properties. Thus, the given (in Chapter 2) definitions of saturated classes, in which the "hereditary properties" are satisfied only for some open subsets (for the primary subsets $\mathrm{T}(\mathbf{H})$), in some sense are optimal.

A further simplification of the notion of a saturated class is given by the next definition. However, in this case we can not prove the intersection property of saturated classes, which is an important property of these classes.

Definitions. A non-empty class $\mathbb{P}$ of spaces is said to be *simply saturated* if for every indexed collection $\mathbf{S}$ of elements of $\mathbb{P}$ there exist a co-mark $\mathbf{M}$ of $\mathbf{S}$ and an $\mathbf{M}$-admissible family R^+ of equivalence relations on $\mathbf{S}$ such that for every admissible family R of equivalence relations on $\mathbf{S}$ the Containing Space $\mathrm{T}(\mathbf{M}, \mathrm{R})$ belongs to $\mathbb{P}$.

A non-empty class $\mathbb{P}$ of subsets is said to be *simply saturated* if for every indexed collection $\mathbf{S}$ of spaces and a $\mathbb{P}$-restriction $\mathbf{Q}$ of $\mathbf{S}$ there exist a co-mark $\mathbf{M}$ of $\mathbf{S}$ and an $(\mathbf{M}, \mathbf{Q})$-admissible family R^+ of equivalence relations on $\mathbf{S}$ such that $(\mathrm{T}|_{\mathbf{Q}}, \mathrm{T}) \in \mathbb{P}$ for every admissible family R of equivalence relations on $\mathbf{S}$, where $\mathrm{T} \equiv \mathrm{T}(\mathbf{M}, \mathrm{R})$ and $\mathrm{T}|_{\mathbf{Q}} \equiv \mathrm{T}(\mathbf{M}|_{\mathbf{Q}}, \mathrm{R}|_{\mathbf{Q}})$.

10.2.1 Problems. 1. Find ("natural") partially saturated classes and simply saturated classes, which are not saturated.

2. Find ("natural") simply saturated classes, which are not partially saturated.

Remark 2 Below, we give two new kind of "saturated" classes the so-called 2^τ-saturated and ∞-saturated classes. Any saturated class of spaces is 2^τ-saturated and ∞-saturated. However, some of non-saturated classes may be 2^τ-saturated or ∞-saturated. So, such classes in some sense can be graduated.

Notation. Let $\mathbf{S}$ be an indexed collection of spaces, $\mathbf{M}$ a co-mark of $\mathbf{S}$, and $\mathrm{R} \equiv \{\sim^s : s \in \mathcal{F}\}$ an $\mathbf{M}$-admissible family of equivalence relations on $\mathbf{S}$. We denote by $\sim^\infty_{\mathrm{R}}$ the equivalence relation on $\mathbf{S}$ defined as follows: two elements X and Y of $\mathbf{S}$ are $\sim^\infty_{\mathrm{R}}$-equivalent if and only if X and Y are s-equivalent for every $s \in \mathcal{F}$. The set of all equivalence classes of $\sim^\infty_{\mathrm{R}}$ is denoted by $\mathrm{C}(\sim^\infty_{\mathrm{R}})$.

For every element $\mathbf{E}$ of $\mathrm{C}(\sim^\infty_{\mathrm{R}})$ we denote by $\mathrm{T}(\mathbf{E})$ the subset of $\mathrm{T} \equiv \mathrm{T}(\mathbf{M}, \mathrm{R})$ consisting of all points $\mathbf{a}$ such that if $(x, X) \in \mathbf{a}$, then $X \in \mathbf{E}$.

10.2.2 Lemma. *Let $I\!\!P$ be a saturated class of spaces. Suppose that* $\mathbf{S}$ *is an indexed collection of elements of $I\!\!P$;* $\mathbf{M}^+$ *an initial co-mark of* $\mathbf{S}$ *corresponding to the class $I\!\!P$;* $\mathbf{M}$ *a co-mark of* $\mathbf{S}$*, which is a co-extension of* $\mathbf{M}^+$*;* R^+ *an initial family of equivalence relations on* $\mathbf{S}$ *corresponding to the co-mark* $\mathbf{M}$ *and the class $I\!\!P$; and* R *an admissible family of equivalence relations on* $\mathbf{S}$*, which is a final refinement of* R^+*. Then, for every element* $\mathbf{E}$ *of* $\mathrm{C}(\sim^\infty_{\mathrm{R}})$ *the space* $\mathrm{T}(\mathbf{E})$ *belongs to $I\!\!P$.*

Proof. Let $\mathbf{E}$ be an element of $\mathrm{C}(\sim^\infty_{\mathrm{R}})$. We denote by $\mathrm{R}_0 \equiv \{\sim^s_0 : s \in \mathcal{F}\}$ a family of equivalence relations on $\mathbf{S}$ defined as follows: two elements X ans Y of $\mathbf{S}$ are $\sim^s_0$-equivalent for some $s \in \mathcal{F} \setminus \{\emptyset\}$ if and only if they are $\sim^s$-equivalent and either X and Y are elements of $\mathbf{E}$ or X and Y are not elements of $\mathbf{E}$. It is easy to see that R_0 is an admissible family, which is a final refinement of R, and $\mathbf{E}$ is an element of $\mathrm{C}^\diamond(\mathrm{R}_0)$. Since $I\!\!P$ is a saturated class, the space $\mathrm{T}_0 \equiv \mathrm{T}(\mathbf{M}, \mathrm{R}_0)$ and its primary subspaces $\mathrm{T}_0(\mathbf{H}) \equiv \mathrm{T}(\mathbf{M}, \mathrm{R}_0, \mathbf{H})$, $\mathbf{H} \in \mathrm{C}^\diamond(\mathrm{R}_0)$ belong to $I\!\!P$. Therefore, the subspace $\mathrm{T}_0(\mathbf{E})$ of T_0 belongs to $I\!\!P$. On the other hand, it is easy to verify that $\mathrm{T}(\mathbf{E}) = \mathrm{T}_0(\mathbf{E})$ and, therefore, $\mathrm{T}(\mathbf{E}) \in I\!\!P$. □

Definition. A non-empty class $I\!\!P$ of spaces is said to be ∞-*saturated* if for every indexed collection $\mathbf{S}$ of elements of $I\!\!P$ there exists a co-mark $\mathbf{M}^+$ with the following property: for every co-mark $\mathbf{M}$ of $\mathbf{S}$, which is a co-extension of $\mathbf{M}^+$, there exists an $\mathbf{M}$-admissible family R^+ of equivalence relations on $\mathbf{S}$ such that for every admissible family R of equivalence relations on $\mathbf{S}$, which is a final refinement of R^+, the space $\mathrm{T}(\mathbf{E})$ belongs to $I\!P$ for every element $\mathbf{E}$ of $\mathrm{C}(\sim_{\mathrm{R}}^{\infty})$.

Definition. If in the definition of a saturated class of spaces we restrict ourselves by consideration only indexed collections $\mathbf{S}$ of cardinality less than or equal to 2^τ, then we obtain the notion of a 2^τ-*saturated* class of spaces.

Obviously, any saturated class of spaces is 2^τ-saturated. Lemma 10.2.2 implies that a saturated class is also ∞-saturated. So, we have the following proposition.

10.2.3 Proposition. *Any saturated class of spaces is 2^τ-saturated and ∞-saturated.* □

Definition. Let $I\!P$ be a class of spaces. It is said that $I\!P$ has the 2^τ-*universality* property if for every collection $\mathbf{S}$ of elements of $I\!P$ with $|\mathbf{S}| \leq 2^\tau$ there exists an element of $I\!P$ containing topologically all elements of $\mathbf{S}$.

The proofs of the next two propositions are similar to that of the corresponding propositions concerning saturated classes.

10.2.4 Proposition. *The non-empty intersection of not more than τ many 2^τ-saturated (respectively, ∞-saturated) classes of spaces is a 2^τ-saturated (respectively, an ∞-saturated) class.* □

10.2.5 Proposition. *Any 2^τ-saturated class of spaces has the 2^τ-universality property.* □

A necessary condition for a class of spaces to be saturated is the existence of universal elements in this class. Thus, the class of all T_1-spaces and the class of all Hausdorff spaces are not saturated since in these classes there are no universal elements (see [296]). However, the next proposition shows that these classes are 2^τ-saturated.

10.2.6 Proposition. *The class $\mathbb{P}(\mathrm{T}_1)$ of all T_1-spaces (respectively, the class $\mathbb{P}(\mathrm{Hs})$ of all Hausdorff spaces) is a 2^τ-saturated class.*

Proof. Let $\mathbf{S}$ be an indexed collection of elements of $\mathbb{P}(\mathrm{T}_1)$ and

$$\mathbf{M} \equiv \{\{U_\delta^X : \delta \in \tau\} : X \in \mathbf{S}\}$$

a co-mark of **S**. Since the cardinality of **S** is $\leq 2^\tau$, there exists an **M**-admissible family

$$\mathrm{R} \equiv \{\sim^s : s \in \mathcal{F}\}$$

of equivalence relations on **S** with the property: for every distinct elements X and Y of **S** there exists an element s of $\mathcal{F}$ such that X and Y are not $\sim^s$-equivalent.

To prove the proposition for $\mathbb{P}(\mathrm{T}_1)$ it suffices to prove that $\mathrm{T} \equiv \mathrm{T}(\mathbf{M}, \mathrm{R})$ is a T_1-space. Let **a** and **b** be two distict elements of T, $(x, X) \in \mathbf{a}$, and $(y, Y) \in \mathbf{b}$.

Suppose that $X = Y$ and, therefore, $x \neq y$. Then, there exists $\delta \in \tau$ such that $x \in U_\delta^X$ and $y \notin U_\delta^X$, which means that $\mathbf{a} \in U_\delta^{\mathrm{T}}(\mathbf{H})$ and $\mathbf{b} \notin U_\delta^{\mathrm{T}}(\mathbf{H})$, $\mathbf{H} \in \mathrm{C}^\diamond(\mathrm{R})$.

Now, suppose that X and Y are distinct elements of **S**. By the choice of R, there exists $s \in \mathcal{F}$ such that X and Y are not $\sim^s$-equivalent. Let **H** be the $\sim^s$-equivalence class of X and $\delta \in \tau$ such that $x \in U_\delta^X$. Then, $\mathbf{a} \in U_\delta^{\mathrm{T}}(\mathbf{H})$ and $\mathbf{b} \notin U_\delta^{\mathrm{T}}(\mathbf{H})$. Therefore, T is a T_1-space.

In the case, where **S** is an indexed collection of elements of $\mathbb{P}(\mathrm{Hs})$ the proposition is proved similarly. □

10.2.7 Proposition. *The class consisting of all Hausdorff compact spaces is ∞-saturated but not 2^τ-saturated.*

Proof. The fact that the considered class is not 2^τ-saturated is proved similarly to the proof of Proposition 10.7.1. On the other hand, Lemma 10.4.2 implies that this class is ∞-saturated. □

10.2.8 Proposition. *The class of all connected spaces is ∞-saturated but not 2^τ-saturated.*

Proof. It is easy to see that the considered class is not 2^τ-saturated. However, Lemma 10.4.2 implies that this class is ∞-saturated. □

10.2.9 Problems. 1. Let $\mathbb{P}$ be a ∞-saturated and 2^τ-saturated class of spaces. Under what conditions on $\mathbb{P}$ is saturated this class?

2. Is it true that the class of all T_1-spaces and the class of all Hausdorff spaces are not ∞-saturated?

3. Let **S** be an indexed collection of T_1-spaces (respectively, of Hausdorff spaces), **M** a co-mark of **S**, and R an **M**-admissible family of equivalence

relations on **S**. Under what conditions on **S**, **M**, and R the Containing Space $\mathrm{T}(\mathbf{M}, \mathrm{R})$ is a T_1-space (respectively, a Hausdorff space)?

4. Is it true that the class of all completely metrizable spaces (of weight $\leq \tau$) is ∞-saturated?

5. Let $\tau = \omega$. Is it true that the class of all separable completely metrizable spaces is ∞-saturated?

Remark 3. In this remark we shall consider totally disconnected spaces. Such spaces are considered by many authors. We refer the papers [375], [267], [118], [344], [361], [346], [315], [325], [215], [94] and propose the readers to the books [115] and [279] for more details.

Definition. A space X is said to be *totally disconnected* if for every distinct points x and y of X there exists a simultaneously open and closed subset U of X containing one of the points x and y and does not contain the other.

10.2.10 Lemma. *A space X is totally disconnected if there exists a collection of cardinality $\leq \tau$ of simultaneously open and closed subsets of X such that for every distinct elements x and y of X there exists an element of this collection containing one of the points x and y and does not contain the other.* (We note that the mentioned collection in general is not a base for the open subsets of X).

Proof. Let X be a totally disconnected space and B a base of cardinality $\leq \tau$ for the open subsets of X. During the proof of the lemma we shall say that a pair (U, V) of elements of B is separated if there exists a simultaneously open and closed subset W of X such that $U \subset W$ and $W \cap V = \emptyset$. In this case the set W will be called separator of the pair (U, V). For every separated pair $p \equiv (U, V)$ of elements of B we choose a fixed separator and denote it by W_p.

Let P be the set of all separated pairs of elements of B. Obviously, the cardinality of the collection $\{W_p : p \in P\}$ is $\leq \tau$. We prove that this collection satisfies the condition of the lemma. Indeed, let x and y two distinct points of X. There exists a simultaneously open and closed subset W of X containing x and not containing y. We choose elements U and V of B such that $x \in U$, $y \in V$, $U \subset W$, and $W \cap V = \emptyset$. Therefore, W is a separator of the pair (U, V). Then, the separator W_p of this pair contains x and does not contain y. This completes the proof of the lemma. □

In [344] it is proved that in the class of all separable metrizable totally disconnected spaces there are no universal elements and, therefore, this class is not saturated. However, we have the following proposition.

10.2.11 Proposition. *The class of all totally disconnected spaces is 2^τ-saturated and, therefore, this class has the 2^τ-universality property.*

Proof. Let $\mathbf{S}$ be an indexed collection of totally disconnected spaces such that $|\mathbf{S}| \leq 2^\tau$. By Lemma 10.2.10 for every $X \in \mathbf{S}$ we can consider an indexed collection

$$M^X \equiv \{V_\delta^X : \delta \in \tau\}$$

of simultaneously open and closed subsets of X satisfying the property that for every two distinct points x and y of X there exists an element of this collection containing x and does not containing y. We put

$$\mathbf{M}_\tau \equiv \{M^X : X \in \mathbf{S}\}.$$

Let $\mathbf{M}^+$ be a co-mark of $\mathbf{S}$, which is a co-extension of $\mathbf{M}_\tau$ and

$$\mathbf{M} \equiv \{\{U_\delta^X : \delta \in \tau\} : X \in \mathbf{S}\}$$

an arbitrary co-mark of $\mathbf{S}$, which is a co-extension of $\mathbf{M}^+$. Suppose that

$$\mathrm{R}^+ \equiv \{\sim_+^s : s \in \mathcal{F}\}$$

is an $\mathbf{M}$-admissible family of equivalence relations on $\mathbf{S}$ satisfying the following condition: for every distinct element X and Y of $\mathbf{S}$ there exists an element s of $\mathcal{F}$ such that X and Y are not $\sim_+^s$-equivalent. To prove the proposition it suffices to show that the space $\mathrm{T} \equiv \mathrm{T}(\mathbf{M}, \mathrm{R})$ is totally disconnected for every admissible family R of equivalence relations on $\mathbf{S}$, which is a final refinement of R^+. Let R be such a family.

We denote by θ an indicial mapping from $\mathbf{M}_\tau$ to $\mathbf{M}$ and by M^T the collection consisting of the following sets: (a) all primary subsets $\mathrm{T}(\mathbf{H}) \equiv \mathrm{T}(\mathbf{M}, \mathrm{R}, \mathbf{H})$ of T, where $\mathbf{H} \in \mathrm{C}^\diamondsuit(\mathrm{R})$, and (b) all elements $U_\delta^\mathrm{T}(\mathbf{H})$ of the standard base for the open subsets of T, where $\delta \in \theta(\tau)$ and $\mathbf{H} \in \mathrm{C}^\diamondsuit(\mathrm{R})$. Obviously, the cardinality of M^T is $\leq \tau$. It is not difficult to verify that all elements of M^T are simultaneously open and closed and that for two distinct point $\mathbf{a}$ and $\mathbf{b}$ of T there exists an element U of M^T containing one of the points **a** and **b** and not containing the other. This observation means that T is a totally disconnected space. □

10.2.12 Problems 1. Let $\tau \neq \omega$. Is it true that the class of all totally disconnected spaces (of weight $\leq \tau$) is not saturated?

2. Is it true that in the class of all totally disconnected T_0-spaces (respectively, regular or completely regular spaces) there are no universal elements?

3. Is it true that the class of all totally disconnected spaces is not ∞-saturated?

Remark 4. The existence of universal elements is not a sufficient condition in order that a class of spaces to be saturated. For example, the class of all Hausdorff compact spaces is not saturated (see Section 10.7) while in this class there exists universal elements. Moreover, for some "usual" classes of spaces (in which there exist universal elements), it is difficult to prove if they are saturated or not. Such classes are for example the class of all normal spaces and the class of all metric spaces. For these classes we put the following problems.

10.2.13 Problems. Suppose that $\tau \neq \omega$.

1. Is it true that the class of all normal spaces is not saturated?

2. Is it true that the class of all normal spaces is not 2^τ-saturated nor ∞-saturated?

3. Let **S** be an indexed collection of normal spaces, **M** a co-mark of **S**, and R an **M**-admissible family of equivalence relations on **S**. Under what conditions on **S**, **M**, and R the Containing Space $\mathrm{T}(\mathbf{M}, \mathrm{R})$ is a normal space?

10.2.14 Problems. Suppose that $\tau \neq \omega$.

1. Is it true that the class of all metrizable spaces is not saturated?

2. Is it true that the class of all metrizable spaces is not 2^τ-saturated nor ∞-saturated?

3. Let **S** be an indexed collection of metrizable spaces, **M** a co-mark of **S**, and R an **M**-admissible family of equivalence relations on **S**. Under what conditions on **S**, **M**, and R the Containing Space $\mathrm{T}(\mathbf{M}, \mathrm{R})$ is a metrizable space?

In view of Proposition 10.6.2 and the fact that the product of a normal space X and a zero-dimensional normal space Y in general is not normal (see for example [114]), the answers to Problems 10.3.13(1,2) seem to be negative. We note also that for $\tau = \omega$ the answers to these problems are affirmative.

Remark 5. In [373] (see also [116]) for a given countable ordinal α very simple examples of Borel sets M_α and A_α (lying in the Hilbert cube H) which are exactly of the multiplicative and additive class α respectively are constructed. These sets satisfy the following property: if X is a metric space and $B \subset X$ is a Borel set of the multiplicative (additive) class α in X, then there exists a continuous mapping φ of X into H such that $\varphi^{-1}(M_\alpha) = B$ (such that $\varphi^{-1}(A_\alpha) = B$).

Actually, the authors in [373] consider the class $\mathbb{P}$(mt-α-mult) (respectively, $\mathbb{P}$(mt-α-add)) of subsets consisting of all pairs (B, X), where B is a Borel set of multiplicative (respectively, of additive) class α in a metrizable space X. In the connection to the above and additionally to Problems 10.2.14 we pose the following problems.

10.2.15 Problems. Suppose that $\tau \neq \omega$ and $\alpha \in \omega^+$.

1. Is it true that the classes $\mathbb{P}$(mt-α-mult) and $\mathbb{P}$(mt-α-add) are not (complete) saturated?

2. Is it true that in the classes $\mathbb{P}$(mt-α-mult) and $\mathbb{P}$(mt-α-add) there are no (properly) universal elements?

In the case where $\tau = \omega$ the classes $\mathbb{P}$(mt-α-mult) and $\mathbb{P}$(mt-α-add) are complete saturated. (And, therefore, in these classes there are properly universal elements.) This statement follows by Propositions 2.1.6, 2.2.8, and 2.4.6.

10.2.16 Problem. Suppose that $\tau = \omega$. Are the pairs (M_α, H) and (A_α, H) properly universal in the above defined classes $\mathbb{P}$(mt-α-mult) and $\mathbb{P}$(mt-α-add), respectively?

Remark 6. Although we consider complete classes of subsets we do not yet prove the existence of non-complete classes of subsets. This proof is done by the next proposition.

10.2.17 Proposition. *The class of all subsets (that is, the class of subsets consisting of all pairs (Q, X), where Q is a subset of a space X) is not complete.*

Proof. Let X be a space such that there exists a family $\mathcal{Q}$ of cardinality 2^{2^τ} consisting of subspaces of X with the property that any two distinct elements of $\mathcal{G}$ are not homeomorphic. For every element $Q \in \mathcal{Q}$ we denote by X_Q a copy of a space X and consider the indexed collection

$$\mathbf{S} \equiv \{X_Q : Q \in \mathcal{Q}\}.$$

Denote by $\mathbf{Q}$ the restriction of $\mathbf{S}$ such that $\mathbf{Q}(X_Q) = Q$ for every $X_Q \in \mathbf{S}$. To prove the proposition it suffices to prove that there are no co-mark $\mathbf{M}$ of $\mathbf{S}$ and an $(\mathbf{M}, \mathbf{Q})$-admissible family R of equivalence relations on $\mathbf{S}$ such that $\mathbf{Q}$ is an $(\mathbf{M}, \mathrm{R})$-complete restriction of $\mathbf{S}$.

Suppose that there exist a co-mark

$$\mathbf{M} \equiv \{\{U_\delta^{X_Q} : \delta \in \tau\} : X_Q \in \mathbf{S}\}$$

of $\mathbf{S}$ and an $(\mathbf{M}, \mathbf{Q})$-admissible family R of equivalence relations on $\mathbf{S}$ such that $\mathbf{Q}$ is an $(\mathbf{M}, \mathrm{R})$-complete restriction. Let

$$\mathbf{M}_p \equiv \{\{V_\delta^{X_Q} : \delta \in \tau\} : X_Q \in \mathbf{S}\}$$

be a co-mark of $\mathbf{S}$ such that $V_\delta^{X_{Q_0}} = V_\delta^{X_{Q_1}}$ for every elements Q_0 and Q_1 of $\mathcal{Q}$ and $\delta \in \tau$. Without loss of generality, we can suppose that $\mathbf{M}$ is co-extension of $\mathbf{M}_p$ (see Lemma 2.2.4) and θ is an indicial mapping of this co-extension.

Consider the set $\mathrm{C}(\sim_{\mathrm{R}}^{\infty})$. The cardinality of this set is $\leq 2^\tau$. Let $\mathbf{E}$ be an element of $\mathrm{C}(\sim_{\mathrm{R}}^{\infty})$ containing two distinct elements X_{Q_0} and X_{Q_1} of $\mathbf{S}$ (that is, Q_0 and Q_1 are not homeomorphic subsets of X). We denote by $\mathbf{Q}(\mathbf{E})$ the subset of the Containing Space $\mathrm{T} \equiv \mathrm{T}(\mathbf{M}, \mathrm{R})$ consisting of all points $\mathbf{a}$ such that there exists a pair (x, X_Q) of $\mathbf{a}$ with $X_Q \in \mathbf{E}$ and $x \in Q$, that is, $\mathbf{Q}(\mathbf{E}) = \mathrm{T}|_{\mathbf{Q}} \cap \mathrm{T}(\mathbf{E})$, where $\mathrm{T}|_{\mathbf{Q}} = \mathrm{T}(\mathbf{M}|_{\mathbf{Q}}, \mathrm{R}|_{\mathbf{Q}})$.

Consider the natural embedding $e_{\mathrm{T}}^{X_{Q_0}}$ of X_{Q_0} into T. Obviously, the set $e_{\mathrm{T}}^{X_{Q_0}}(X_{Q_0})$ is a subset of $\mathrm{T}(\mathbf{E})$. We prove that $e_{\mathrm{T}}^{X_{Q_0}}(Q_0) = \mathbf{Q}(\mathbf{E})$. By Corollary 1.3.6 we have $e_{\mathrm{T}}^{X_{Q_0}}(Q_0) \subset \mathbf{Q}(\mathbf{E})$. Therefore, we need to prove the converse relation. Let $\mathbf{a} \in \mathbf{Q}(\mathbf{E})$. Consider an element (x, X_Q) of $\mathbf{a}$. Then, the pair (x, X_{Q_0}) is also an element of $\mathbf{a}$. Indeed, in the opposite case since X_{Q_0} and X_Q are elements of $\mathbf{E}$ there exists $\delta \in \tau$ such that either $x \in U_\delta^{X_Q}$ and $x \notin U_\delta^{X_{Q_0}}$ or $x \notin U_\delta^{X_Q}$ and $x \in U_\delta^{X_{Q_0}}$. In the first case, there exists $\varepsilon \in \theta(\tau)$ such that $x \in U_\varepsilon^{X_Q} \subset U_\delta^{X_Q}$. Since $\varepsilon \in \theta(\tau)$

we have $x \in U_\varepsilon^{X_{Q_0}}$, which means that the set $U_\varepsilon^{X_{Q_0}}$ is not a subset of $U_\delta^{X_{Q_0}}$. This conclusion contradicts the fact that X_Q and X_{Q_0} are elements of $\mathbf{E}$. Therefore, $(x, X_{Q_0}) \in \mathbf{a}$. Similarly we prove this relation in the second case. Since $\mathbf{Q}$ is an $(\mathbf{M}, \mathrm{R})$-complete restriction we have $x \in \mathrm{Q}_0$ and, therefore, $e_{\mathrm{T}}^{X_{Q_0}}(x) = \mathbf{a}$. This means that $\mathbf{Q}(\mathbf{E}) \subset e_{\mathrm{T}}^{X_{Q_0}}(Q_0)$. Thus, $e_{\mathrm{T}}^{X_{Q_0}}(Q_0) = \mathbf{Q}(\mathbf{E})$. This relation implies that the subspace Q_0 of X is homeomorphic to the subspace $\mathbf{Q}(\mathbf{E})$ of T. Similarly we prove that the subspace Q_1 of X is homeomorphic to the subspace $\mathbf{Q}(\mathbf{E})$ of T. Therefore, Q_0 and Q_1 are homeomorphic, which contradicts the choice of the family $\mathcal{Q}$. □

By the definition of a complete class of subsets it follows immediately that any subclass of a complete class is also complete. The next proposition shows that the notion of the completeness of classes of subsets is preserved also under the union of not more than τ many classes of subsets.

10.2.18 Proposition. *Suppose that $\mathbb{P}_\delta$, $\delta \in \tau$, is a complete class of subsets. Then, the class $\mathbb{P} = \cup\{\mathbb{P}_\delta : \delta \in \tau\}$ is also complete.* (The classes $\mathbb{P}_\delta$ are not necessary to be saturated).

Proof. Let $\mathbf{S}$ be an indexed collection of spaces and $\mathbf{Q}$ a $\mathbb{P}$-restriction of $\mathbf{S}$. We need to find a co-mark $\mathbf{M}$ of $\mathbf{S}$ and an $(\mathbf{M}, \mathbf{Q})$-admissible family R of equivalence relations on $\mathbf{S}$ such that $\mathbf{Q}$ is $(\mathbf{M}, \mathrm{R})$-complete.

For every $\delta \in \tau$ we denote by $\mathbf{S}_\delta$ the indexed subcollection of $\mathbf{S}$ consisting of all elements X such that $(\mathbf{Q}(X), X) \in \mathbb{P}_\delta$. Without loss of generality we can suppose that $\mathbf{S}_{\delta_0} \cap \mathbf{S}_{\delta_1} = \emptyset$ if $\delta_0 \neq \delta_1$. By $\mathbf{Q}_\delta$ we denote the restriction $\{\mathbf{Q}(X) : X \in \mathbf{S}_\delta\}$ of $\mathbf{S}_\delta$. Obviously, $\mathbf{Q}_\delta$ is a complete restriction of $\mathbf{S}_\delta$.

We denote by $\mathbf{M}_\delta$ a co-mark of $\mathbf{S}_\delta$ and by R_δ an $(\mathbf{M}_\delta, \mathbf{Q}_\delta)$-admissible family of equivalence relations on $\mathbf{S}_\delta$ such that the restriction $\mathbf{Q}_\delta$ of $\mathbf{S}_\delta$ is $(\mathbf{M}_\delta, \mathrm{R}_\delta)$-complete.

Now, we denote by $\mathbf{M}$ the co-mark of $\mathbf{S}$ such that for every $X \in \mathbf{S}_\delta$, $\delta \in \tau$, we have $\mathbf{M}(X) = \mathbf{M}_\delta(X)$. Also, we denote by $\mathrm{R} \equiv \{\sim^s : s \in \mathcal{F}\}$ an $(\mathbf{M}, \mathbf{Q})$-admissible family of equivalence relations on $\mathbf{S}$ such that: (a) for every $\delta \in \tau$ the family $\mathrm{R}^\delta \equiv \{\sim^{\delta,s} : s \in \mathcal{F}\}$, where $\sim^{\delta,s}$ is the restriction of $\sim^s$ on $\mathbf{S}_\delta$, is a final refinement of R_δ, and (b) if $X \in \mathbf{S}_{\delta_0}$, $Y \in \mathbf{S}_{\delta_1}$, and $\delta_0 \neq \delta_1$, then X and Y are not $\sim^s$-equivalent for some $s \in \mathcal{F}$. Such a family is easy to construct. Then, it is easy to verify that the restriction $\mathbf{Q}$ is $(\mathbf{M}, \mathrm{R})$-complete. □

In addition to the above results it is easy to prove the following proposition.

10.2.19 Proposition. *If the cardinality of an indexed collection* **S** *of spaces is less than or equal to* 2^τ, *then any restriction of* **S** *is complete.* □

10.2.20 Problem. Suppose that $\tau = \omega$. Is it true that the class of all pairs (Q, X), where Q is a Borel set of a space X, is not complete?

The more general problem is the following.

10.2.21 Problem. Let Φ be a τ-directing mapping. Is it true that the classes of subsets

$$\cup\{\mathbb{P}(\gamma, \tau, \Phi, \mathrm{Cl}) : \gamma \in \tau^+\} \text{ and } \cup\{\mathbb{P}(\gamma, \tau, \Phi, \mathrm{Op}) : \gamma \in \tau^+\}$$

are not complete?

Remark 7. It is easy to prove that a finite union of saturated classes is saturated. However, an infinite union of saturated classes may not be saturated. For example, the class $\mathbb{P}(\mathrm{ind} < \omega)$ of all finite dimensional spaces (in the sense of dimension ind) is not saturated since in this class there exist no universal elements. However, this class is the union of the classes $\mathbb{P}(\mathrm{ind} \leq n)$, $n \in \omega$, which are saturated.

Below, we pose some problems concerning the product of saturated classes.

Notation. Let $\mathbb{P}_\delta$, $\delta \in \tau$, be a class of non-empty spaces. By

$$\otimes\{\mathbb{P}_\delta : \delta \in \tau\}$$

we denote the topological class consisting of all spaces which are homeomorphic to the product $\times\{X_\delta : \delta \in \tau\}$ for some elements X_δ of $\mathbb{P}_\delta$.

For every space X we denote by $X \otimes \mathbb{P}_0$ the topological class consisting of all spaces which are homeomorphic to the product $X \times Y$ for some element Y of $\mathbb{P}_0$.

10.2.22 Problems. Let $\mathbb{P}_\delta$, $\delta \in \tau$, is a saturated class of non-empty spaces.

1. Is the class $\mathbb{P}_0 \otimes \mathbb{P}_1$ saturated?

2. Is the class $\otimes\{\mathbb{P}_\delta : \delta \in \tau\}$ saturated?
3 Let X be a space. Is the class $X \otimes \mathbb{P}_0$ saturated?

Definition. Let $\mathbb{P}$ be a class of spaces. A saturated class $\mathbb{P}_0$ of spaces is said to be a *minimal saturated class of spaces containing* $\mathbb{P}$ if: (a) $\mathbb{P} \subset \mathbb{P}_0$ and (b) the condition $\mathbb{P} \subset \mathbb{P}'$, where $\mathbb{P}'$ is a saturated class of spaces, implies that $\mathbb{P}_0 \subset \mathbb{P}'$.

By the intersection property of saturated classes it follows that if there exists a minimal saturated class of spaces containing a given class, then it is uniquely determined.

Similarly, we can define the minimal class for other types of saturated classes.

10.2.23 Problems. 1. Let $\mathbb{P}$ be a class of spaces. Is there the minimal saturated class of spaces containing $\mathbb{P}$? (We note that the class of all spaces is a saturated class containing $\mathbb{P}$.)

2. How to construct the minimal saturated class of spaces (if such a class exists) containing the given class of spaces?

3. Consider questions 1 and 2 for other type of classes.

Remark 8. The commutative co-operators considered in Section 2.3 are connected with the co-operators **Cl**, **Bd**, **Int**, and **Co**. So, it is interesting to find some others.

Definition. Let $\mathbf{O}^1 \equiv \{\mathrm{O}_X^1 : X \in \mathbb{R}\}$ and $\mathbf{O}^2 \equiv \{\mathrm{O}_X^2 : X \in \mathbb{R}\}$ be two co-operators (on a (saturated) class $\mathbb{R}$ of spaces). The co-operator

$$\mathbf{O} \equiv \{\mathrm{O}_X \equiv \mathrm{O}_X^1 \circ \mathrm{O}_X^2 : X \in \mathbb{R}\}$$

is called the *composition* of $\mathbf{O}^1$ and $\mathbf{O}^2$.

It is easy to verify the following proposition.

10.2.24 Proposition. *The composition of commutative co-operators on a saturated class of spaces is also commutative.* □

10.2.25 Problem. Construct (completely) commutative co-operators (on a given saturated class of spaces), which are distinct from co-operators considered in Section 2.3 and its compositions.

Remark 9. Concerning the generalization of the notion of a saturated class given in Section 2.7 we pose the following problems.

10.2.26 Problems. Let ν and μ be two cardinals satisfying conditions (1)–(3) of Section 1.5 such that that $(\omega, \omega) \neq (\nu, \mu)$.

1. Is there a saturated class (that is, (ω, ω)-saturated class) which is not (ν, μ)-saturated?
2. Is there a (ν, μ)-saturated class which is not saturated?

The more general problem is the following.

10.2.27 Problem. Suppose that (ν, μ) and (ν', μ') are distinct pairs of cardinals satisfying conditions (1)–(3) of Section 1.5. Is there a class of spaces which is (ν, μ)-saturated but not (ν', μ')-saturated?

10.2.28 Problem. Suppose that $\omega \neq \nu$. Is the class of all (regular) P_ν-spaces saturated? (About P_ν-spaces mentioned in Section 2.7 see for example [85].)

Remark 10. During the writing of the book, the results of Section 2.1 were published in [192], the results of Sections 2.2 and 2.3 in [193], the results of Sections 2.5 and 2.6 in [194], and the results of Section 2.7 in [191]. We note that in [192] as a saturated class is considered a partially saturated class.

10.3 Remarks and problems to Chapter 3

Remark 1. The standard theorem concerning dimensions ind, Ind, and dim (in the realm of regular or normal spaces) one can find, for example, in [115].

As it is mentioned in Chapter 3 we did not investigate the relations between the defferent dimension-like functions that were introduced in this chapter (which are called below *new dimension-like functions* or *new dimensions*). We also did not investigate the relations between the well known dimensions and the dimension-like functions that we introduced. For the new dimension-like functions we did not investigate also what standard theorems of dimension theory hold. So, the following general problems have arisen.

10.3.1 Problems. 1. Find relations between new dimension-like functions.

2. Find relations between new dimensions and well known dimensions.

3 Formulate the standard theorems of dimension theory for the new dimensions and investigate which of them hold for these dimensions. This problem is interesting especially for base dimension-like functions.

Remark 2. In [4] (see also [151], [241], [386], [152], [311], [307], [1], [2], [3]) many space dimension-like functions, which are called proper or generalized dimension functions, are considered. It is interesting to see these dimension-like functions from the point of view of the universality problem. Since many of them are not defined by τ-property it is interesting to give different variations of these dimensions (defined by τ-properties) to which we can apply the method of Containing Spaces. Thus, we can pose the following problem.

10.3.2 Problem. Apply the method of Containing Spaces to proper and generalized dimension functions given in [4] or to its variations.

Remark 3. In this remark we apply the method of Containing Spaces to the *order dimension modulo* $\mathcal{P}$ considered in [4] and denoted there by $\mathcal{P}$-Odim. Here, we shall denote this dimension by $\mathrm{Odim}^{\mathbb{B}}_{\mathbb{E}}$, where $\mathbb{B}$ is a class of bases and $\mathbb{E}$ is a class of spaces. The class $\mathbb{E}$ corresponds to $\mathcal{P}$ and the class $\mathbb{B}$ corresponds to σ-locally finite bases considered in the definition of $\mathcal{P}$-Odim given in [4]. (The metrizability of the space is a part of the definition given in [4].) Since we do not use σ-locally finite bases, the definition of $\mathrm{Odim}^{\mathbb{B}}_{\mathbb{E}}$ given below can be applied to any space. However, we restrict ourself by considering only the universality problem for this dimension-like function.

Definition. Let $\mathbb{B}$ be a class of bases and $\mathbb{E}$ a $\mathbb{B}$-rim-hereditary class of spaces (see Section 3.1) containing the empty space. We denote by $\mathrm{Odim}^{\mathbb{B}}_{\mathbb{E}}$ the function with the class of all spaces as domain and as range the set $\{-1, \infty\} \cup \omega$ satisfying the following conditions: (a) $\mathrm{Odim}^{\mathbb{B}}_{\mathbb{E}}(X) = -1$ if and only if $X \in \mathbb{E}$ and (b) $\mathrm{Odim}^{\mathbb{B}}_{\mathbb{E}}(X) \leq n$, where $n \in \omega$, if and only if there exists a base $B \in \mathbb{B}$ for the open subsets of the space X such that if $U_0, ..., U_n$ are distinct elements of B, then

$$\mathrm{Bd}_X(U_0) \cap ... \cap \mathrm{Bd}_X(U_n) \in \mathbb{E}.$$

Definition. Let ν be a cardinal with $\nu \leq \tau$. A base B for the open subsets of a space X is said to be *ν-dividing* if for every $\delta \in \nu$ there exists a base B_δ for the open subsets of X such that: (a) $B = \cup\{B_\delta : \delta \in \nu\}$ and (b) if $\delta_0, \delta_1 \in \nu$, $\delta_0 \neq \delta_1$, $\emptyset \neq U_0 \in B_{\delta_0}$, and $\emptyset \neq U_1 \in B_{\delta_1}$, then $U_0 \neq U_1$.

A space X is said to be *base ν-diviting* if each base of cardinality $\leq \tau$ for the open subsets of X is ν-dividing.

10.3.3 Proposition. *Let $I\!\!R$ be a saturated class consisting of base τ-dividing spaces. Let also $I\!\!B$ be a saturated class of bases and $I\!\!E$ a $I\!\!B$-rim-hereditary saturated class of spaces containing the empty space. Then, in the (non-empty) class $I\!\!P_{I\!\!R}(\mathrm{Odim}^{I\!\!B}_{I\!\!E} \leq n)$, where $n \in \{-1\} \cup \omega$, consisting of all elements X of $I\!\!R$ such that $\mathrm{Odim}^{I\!\!B}_{I\!\!E}(X) \leq n$ there exists universal elements.*

Proof. Suppose that $\mathbf{S}$ is an indexed collection of elements of $I\!\!P \equiv I\!\!P_{I\!\!R}(\mathrm{Odim}^{I\!\!B}_{I\!\!P} \leq n)$ such that each element of $I\!\!P$ is homeomorphic to an element of $\mathbf{S}$.

For every $\delta \in \tau$ we denote by θ_δ an one-to-one mapping of τ into itself such that: (a) $\tau = \cup\{\theta_\delta(\tau) : \delta \in \tau\}$ and (b) $\theta_{\delta_0}(\tau) \cap \theta_{\delta_1}(\tau) = \emptyset$ if $\delta_0 \neq \delta_1$.

Consider an element X of $\mathbf{S}$ and a base B^X for the open subsets of X with the property that

$$\mathrm{Bd}_X(V_0) \cap ... \cap \mathrm{Bd}_X(V_n) \in I\!\!E$$

if V_0,...,V_n are distinct elements of B^X. Since X is base τ-dividing, for every $\delta \in \tau$ there exists a base B^X_δ such that: (a) $B^X = \cup\{B^X_\delta : \delta \in \tau\}$ and (b) if $\delta_0, \delta_1 \in \tau$, $\delta_0 \neq \delta_1$, $\emptyset \neq V_0 \in B^X_{\delta_0}$, and $\emptyset \neq V_1 \in B^X_{\delta_1}$, then $V_0 \neq V_1$.

Let $M^X_\delta \equiv \{V^X_{\delta,\varepsilon} : \varepsilon \in \tau\}$ be an indication of B^X_δ such that $V^X_{\delta,\varepsilon_0} \neq V^X_{\delta,\varepsilon_1}$ for every $\delta, \varepsilon_0, \varepsilon_1 \in \tau$ with $\varepsilon_0 \neq \varepsilon_1$. We set $M^X = \{V^X_\eta : \eta \in \tau\}$, where $V^X_\eta = V^X_{\delta,\varepsilon}$ if $\eta = \theta_\delta(\varepsilon)$.

Let $\mathbf{M}_0$ be a co-mark of $\mathbf{S}$ such that $\mathbf{M}_0(X) = M^X$ for every $X \in \mathbf{S}$.

Denote by τ^{n+1}_* the subset of τ^{n+1} consisting of all elements $(\eta_0, ..., \eta_n)$ of τ^{n+1} such that $\eta_i \neq \eta_j$ if $i, j \in \{0, ..., n\}$ and $i \neq j$.

Let $\overline{\eta} \equiv (\eta_0, ..., \eta_n) \in \tau^{n+1}_*$. For every $X \in \mathbf{S}$ we set

$$Q^X_{\overline{\eta}} = \mathrm{Bd}_X(U^X_{\eta_0}) \cap ... \cap \mathrm{Bd}_X(U^X_{\eta_n}).$$

By the conditions of the proposition, the indexed collection

$$\mathbf{Q}_{\overline{\eta}} \equiv \{Q^X_{\overline{\eta}} : X \in \mathbf{S}\}$$

of spaces is an $I\!\!E$-restriction of $\mathbf{S}$.

We consider the co-mark $\mathbf{M}_0|_{\mathbf{Q}_{\overline{\eta}}}$ of $\mathbf{Q}_{\overline{\eta}}$, which is the trace on $\mathbf{Q}_{\overline{\eta}}$ of the co-mark $\mathbf{M}_0$ of $\mathbf{S}$. We denote by $\mathbf{M}_{0,\mathbf{Q}_{\overline{\eta}}}$ an initial co-mark of $\mathbf{Q}_{\overline{\eta}}$ corresponding to the class $I\!\!E$, which is a co-extension of $\mathbf{M}_0|_{\mathbf{Q}_{\overline{\eta}}}$, and by $\vartheta_{\overline{\eta}}$ an indicial mapping of this co-extension.

Also, we denote by $\mathrm{R}_{0,\mathbf{Q}_{\overline{\eta}}}$ an initial family of equivalence relations on $\mathbf{Q}_{\overline{\eta}}$ corresponding to the co-mark $\mathbf{M}_{0,\mathbf{Q}_{\overline{\eta}}}$ and the class $I\!\!E$.

It is easy to construct an $\mathbf{S}$-indexed set $\mathbf{M}_{0,\overline{\eta}}$ of τ-indexed families of open subsets such that: (a) $\mathbf{M}_{0,\overline{\eta}}$ is a co-extension of $\mathbf{M}_0$ and the mapping $\vartheta_{\overline{\eta}}$ is an indicial mapping of this co-extension and (b) the trace of $\mathbf{M}_{0,\overline{\eta}}$ on $\mathbf{Q}_{\overline{\eta}}$ is the co-mark $\mathbf{M}_{0,\mathbf{Q}_{\overline{\eta}}}$. We note that $\mathbf{M}_{0,\overline{\eta}}$ in general is not uniquely determined.

Also, we denote by $\mathrm{R}_{0,\overline{\eta}}$ an admissible family of equivalence relations on $\mathbf{S}$ such that the trace of $\mathrm{R}_{0,\overline{\eta}}$ on $\mathbf{Q}_{\overline{\eta}}$ is $\mathrm{R}_{0,\mathbf{Q}_{\overline{\eta}}}$.

Now, let

$$\mathbf{M} \equiv \{\{U_\delta^X : \delta \in \tau\} : X \in \mathbf{S}\}$$

be an initial co-mark of $\mathbf{S}$ corresponding to the class $I\!\!R$, which is simultaneously a co-extension of $\mathbf{M}_{0,\overline{\eta}}$ for all $\overline{\eta} \in \tau_*^{n+1}$. We denote by θ an indicial mapping from $\mathbf{M}_0$ to $\mathbf{M}$.

Also, let R be an initial family of equivalence relations on $\mathbf{S}$ corresponding to the co-mark $\mathbf{M}$ and the class $I\!\!R$, which is a final refinement of $\mathrm{R}_{0,\overline{\eta}}$ for all $\overline{\eta} \in \tau_*^{n+1}$.

Then, we can consider the Containing Space $\mathrm{T} \equiv \mathrm{T}(\mathbf{M}, \mathrm{R})$. We prove that T is a universal element in $I\!\!P$. By the construction, T belongs to $I\!\!R$. Therefore, it suffices to prove that $\mathrm{Odim}_{I\!\!P}^{I\!\!B}(\mathrm{T}) \leq n$. For this purpose we need to find a base B for the open subsets of T such that

$$\mathrm{Bd}_{\mathrm{T}}(U_0) \cap ... \cap \mathrm{Bd}_{\mathrm{T}}(U_n) \in I\!\!E$$

if $U_0, ..., U_n$ are distinct elements of B.

We construct the required base B as a subset of the standard base B^{T} of T. Let ϑ be an one-to-one mapping of τ onto the set $\mathcal{F}$. We denote by B the subset of B^{T} consisting of all elements $U_\varepsilon^{\mathrm{T}}(\mathbf{H})$ such that $\varepsilon \in \theta(\theta_\delta(\tau))$ for some $\delta \in \tau$ and $\mathbf{H}$ is an $\sim^{\vartheta(\delta)}$-equivalence class.

First we prove that B is a base for the open subsets of T. Let $\mathbf{a}$ be a point of T, $(x, X) \in \mathbf{a}$, and U an open neighbourhood of $\mathbf{a}$ in T. Without loss of generality we can suppose that U belongs to B^{T}, that is U has the form $U_\eta^{\mathrm{T}}(\mathbf{L})$, where $\eta \in \tau$ and $\mathbf{L}$ is an $\sim^s$-equivalence class for some $s \in \mathcal{F}$. Let $\delta = \vartheta^{-1}(s)$. Since B_δ^X is a base for the open subsets of X there exists an element $\varepsilon_0 \in \tau$ such that $x \in V_{\delta,\varepsilon_0}^X \subset U_\eta^X$. Since $X \in \mathbf{L}$ we have $\mathbf{a} \in U_\varepsilon^{\mathrm{T}}(\mathbf{L}) \subset U_\eta^{\mathrm{T}}(\mathbf{L})$, where $\varepsilon = \theta(\theta_\delta(\varepsilon_0))$. It is easy to verify that $U_\varepsilon^{\mathrm{T}}(\mathbf{L}) \in B$, which means that B is a base for the open subsets of T.

Now, we prove that $\mathrm{Odim}_{\mathbb{P}}^{\mathbb{B}}(\mathrm{T}) \leq n$. Let $U_{\varepsilon_0}^{\mathrm{T}}(\mathbf{H}_0)$,...,$U_{\varepsilon_n}^{\mathrm{T}}(\mathbf{H}_n)$ be distinct elements of B. We need to prove that

$$\mathrm{Bd}_{\mathrm{T}}(U_{\varepsilon_0}^{\mathrm{T}}(\mathbf{H}_0)) \cap ... \cap \mathrm{Bd}_{\mathrm{T}}(U_{\varepsilon_n}^{\mathrm{T}}(\mathbf{H}_n)) \in \mathbb{E}. \tag{10.3.1}$$

We can suppose that ε_0,...,ε_n are mutually distinct elements of τ. Indeed, in the opposite case for two distinct elements i and j of the set $\{0, ..., n\}$ we have $\varepsilon_i = \varepsilon_j$. Since $U_{\varepsilon_i}^{\mathrm{T}}(\mathbf{H}_i) \neq U_{\varepsilon_j}^{\mathrm{T}}(\mathbf{H}_j)$ the subsets $\mathbf{H}_i$ and $\mathbf{H}_j$ of $\mathbf{S}$ are distinct. If $\mathbf{H}_i \cap \mathbf{H}_j = \emptyset$, then $U_{\varepsilon_i}^{\mathrm{T}}(\mathbf{H}_i) \cap U_{\varepsilon_j}^{\mathrm{T}}(\mathbf{H}_j) = \emptyset$ and therefore $\mathrm{Bd}_{\mathrm{T}}(U_{\varepsilon_0}^{\mathrm{T}}(\mathbf{H}_0)) \cap ... \cap \mathrm{Bd}_{\mathrm{T}}(U_{\varepsilon_n}^{\mathrm{T}}(\mathbf{H}_n)) = \emptyset$. Since $\emptyset \in \mathbb{E}$, relation (10.3.1) are satisfied. Hence, we can suppose that $\mathbf{H}_i \cap \mathbf{H}_j \neq \emptyset$. In this case, if $\mathbf{H}_i$ is an $\sim^{s_i}$-equivalence class and $\mathbf{H}_j$ is an $\sim^{s_j}$-equivalence class, then $s_i \neq s_j$. By the choice of the mapping ϑ it follows that if $\varepsilon_i \in \theta(\theta_{\delta_i}(\tau))$ and $\varepsilon_j \in \theta(\theta_{\delta_j}(\tau))$, then $\delta_i \neq \delta_j$. By the choice of the mappings θ_δ, $\delta \in \tau$, we have $\varepsilon_i \neq \varepsilon_j$, which is a contradiction.

Thus, $\bar{\varepsilon} = (\varepsilon_0, ..., \varepsilon_n) \in \tau_*^{n+1}$ and, therefore, we can consider the $\mathbb{E}$-restriction $\mathbf{Q}_{\bar{\varepsilon}}$ of $\mathbf{S}$. This restriction can be considered as an indexed collection of elements of $\mathbb{E}$. By the construction, $\mathbf{M}|_{\mathbf{Q}_{\bar{\varepsilon}}}$ is an initial co-mark of $\mathbf{Q}_{\bar{\varepsilon}}$ corresponding to the class $\mathbb{E}$ and $\mathrm{R}|_{\mathbf{Q}_{\bar{\varepsilon}}}$ is an initial family of equivalence relations on $\mathbf{Q}_{\bar{\varepsilon}}$ corresponding to the co-mark $\mathbf{M}|_{\mathbf{Q}_{\bar{\varepsilon}}}$ and the class $\mathbb{E}$. Since $\mathbb{E}$ is a saturated class, the subset

$$\mathrm{T}|_{\mathbf{Q}_{\bar{\varepsilon}}}(\mathbf{L}) \equiv \mathrm{T}(\mathbf{M}|_{\mathbf{Q}_{\bar{\varepsilon}}}, \mathrm{R}|_{\mathbf{Q}_{\bar{\varepsilon}}}, \mathbf{L})$$

of the Containing Space

$$\mathrm{T}|_{\mathbf{Q}_{\bar{\varepsilon}}} \equiv \mathrm{T}(\mathbf{M}|_{\mathbf{Q}_{\bar{\varepsilon}}}, \mathrm{R}|_{\mathbf{Q}_{\bar{\varepsilon}}}) \subset \mathrm{T}$$

belongs to $\mathbb{E}$ for every element $\mathbf{L}$ of the set $\mathrm{C}(\mathrm{R}|_{\mathbf{Q}_{\bar{\varepsilon}}})$.

Let $\mathbf{H} = \mathbf{H}_0 \cap ... \cap \mathbf{H}_n$ and $\mathbf{L} = \mathbf{H}|_{\mathbf{Q}_{\bar{\varepsilon}}}$. In order to complete the proof of the proposition it suffices to prove that

$$\mathrm{T}|_{\mathbf{Q}_{\bar{\varepsilon}}}(\mathbf{L}) = \mathrm{Bd}_{\mathrm{T}}(U_{\varepsilon_0}^{\mathrm{T}}(\mathbf{H}_0)) \cap ... \cap \mathrm{Bd}_{\mathrm{T}}(U_{\varepsilon_n}^{\mathrm{T}}(\mathbf{H}_n)) \tag{10.3.2}$$

Let $\mathbf{a} \in \mathrm{T}|_{\mathbf{Q}_{\bar{\varepsilon}}}(\mathbf{L})$. Consider an element $(x, Q^X_{\bar{\varepsilon}})$ of $\mathbf{a}$. Then, $x \in Q^X_{\bar{\varepsilon}}$, that is

$$x \in \mathrm{Bd}_X(U^X_{\varepsilon_0}) \cap ... \cap \mathrm{Bd}_X(U^X_{\varepsilon_n}). \tag{10.3.3}$$

Since $Q^X_{\bar{\varepsilon}} \in \mathbf{L}$, $X \in \mathbf{H} = \mathbf{H}_0 \cap ... \cap \mathbf{H}_n$. By Lemma 1.2.7, relation (10.3.3) implies that

$$\mathbf{a} \in \mathrm{Bd}_{\mathrm{T}}(U^{\mathrm{T}}_{\varepsilon_0}(\mathbf{H}_0)) \cap ... \cap \mathrm{Bd}_{\mathrm{T}}(U^{\mathrm{T}}_{\varepsilon_n}(\mathbf{H}_n)),$$

that is the left side of (10.3.2) is contained in the right.

Conversely, let

$$\mathbf{a} \in \mathrm{Bd}_{\mathrm{T}}(U^{\mathrm{T}}_{\varepsilon_0}(\mathbf{H}_0)) \cap ... \cap \mathrm{Bd}_{\mathrm{T}}(U^{\mathrm{T}}_{\varepsilon_n}(\mathbf{H}_n)). \tag{10.3.4}$$

Consider an element (x, X) of $\mathbf{a}$. Then, we have $X \in \mathbf{H}_0 \cap ... \cap \mathbf{H}_n = \mathbf{H}$, that is $Q^X_{\bar{\varepsilon}} \in \mathbf{L}$. By Lemma 1.2.7, relation (10.3.4) implies relation (10.3.3). This fact means that $x \in Q^X_{\bar{\varepsilon}}$ which implies that $\mathbf{a} \in \mathrm{T}|_{\mathbf{Q}_{\bar{\varepsilon}}}(\mathbf{L})$. Thus, the right side of (10.3.2) is contained in the left completing the proof of this relation and the proof of the proposition. □

10.3.4 Proposition. *Let $\tau = \omega$. A separable metrizable space X is base ω-dividing if and only if X contains no isolated points.*

Proof. It is easy to see that if a separable metrizable space X is base ω-dividing, then X is a space without isolated points.

Conversely, suppose that a separable metrizable space X is a space without isolated points. To prove the proposition it suffices to show that any base B of cardinality $\leq \omega$ for the open subsets of X contains two bases B_0 and B_1 for the open subsets of X with the property: if $\emptyset \neq U \in B_0$ and $\emptyset \neq V \in B_1$, then $U \neq V$.

Denote by ϱ_X a metric on X and let B be a base of cardinality $\leq \omega$ for the open subsets of X. Suppose that $c \equiv \{U_n : n \in \omega\}$ is an indexed cover of X satisfying the following conditions:

(a) $c \subset B$, and

(b) U_m is not contained in U_n if $n < m$.

We prove that $B \setminus c$ is a base for the open subsets of X. Indeed, let $x \in X$ and W be an open neighbourhood of x. It suffices to find an element U of $B \setminus c$ such that $x \in U \subset W$. Let V be an element of B such that $x \in V \subset W$. If $V \notin c$, then we set $U = V$. If $V \in c$, then there exists an element $m \in \omega$ such that $V = U_m$. Since x is a non-isolated point of X there exists an element U of B such that $x \in U \subset U_m$ and $U \neq U_n$ for every $n \in \omega$ with $n \leq m$. Then, $U \notin c$ and, therefore, $U \in B \setminus c$.

Now, using the above property of B, for every $i \in \omega$ we can construct by induction an indexed cover $c^i \equiv \{U^i_n : n \in \omega\}$ of X such that:

(a) $c^0 \subset B$ and $c^i \subset B \setminus (\cup\{c^j : j \in i\})$, $i \in \omega \setminus \{0\}$,

(b) $\mathrm{Diam}(U^i_n) \leq 1/(i+1)$, $i, n \in \omega$, and

(c) for every $i \in \omega$, U^i_m is not contained in U^i_n if $n < m$.

Then, it is easy to verify that the sets

$$B_0 \equiv \{\emptyset\} \cup (\cup\{c^i : i = 2k, k \in \omega\}) \text{ and}$$

$$B_1 \equiv \{\emptyset\} \cup (\{c^i : i = 2k+1, k \in \omega\})$$

are the required bases for open subsets of X. □

The following corollary follows by Propositions 10.3.3 and 10.3.4 and by the fact that the class of all separable metrizable spaces without isolated points is a saturated class.

10.3.5 Corollary. *Let $\tau = \omega$. Let also $\mathbb{B}$ be a saturated class of bases and $\mathbb{E}$ a $\mathbb{B}$-rim-hereditary saturated class of spaces containing the empty space. Then, the class of all separable metrizable spaces X without isolated points such that $\mathrm{Odim}^{\mathbb{B}}_{\mathbb{E}}(X) \leq n \in \omega$ is a saturated class of spaces.* □

10.3.6 Problems. 1. Is Proposition 10.3.3 true for classes $\mathbb{R}$ consisting of base ν-dividing spaces, where ν is a cardinal with $\omega \leq \nu < \tau$?

2. Is Proposition 10.3.4 true for non-metrizable spaces with a countable base?

3. Let $\tau = \omega$. Let also $\mathbb{B}$, $\mathbb{E}$, and n are the same as in Corollary 10.3.5. Is it true that any separable metrizable space Y is contained in a separable metrizable space X without isolated points such that $\mathrm{Odim}^{\mathbb{B}}_{\mathbb{E}}(X) \leq n \in \omega$?

4. Let ν be a cardinal such that $\omega \leq \nu \leq \tau$. Characterize the base ν-dividing spaces.

5. We recall (see [114]) that the minimum of cardinals μ for which a space X has a base of cardinality $\leq \mu$ at a point x of X is called the *character* of X at x and is denoted by $\chi(x, X)$. Let $\tau \neq \omega$ and ν be a cardinal such that $\omega \leq \nu \leq \tau$. Is it true that a space X is base ν-dividing if and only if $\chi(x, X) \geq \nu$ for every $x \in X$? (We note that if a space X

has a point x such that $\chi(x, X) < \nu$, then it is easy to prove that X is not base ν-dividing.)

Remark 4. In this remark we apply the method of Containing Spaces to weakly n-dimensional separable metrizable spaces and its generalizations.

Definition. Let $\alpha \in \tau^+ \cup \{\infty\}$. We recall that (see [114]) the *dimension* ind *at a point x of a space X is α* if α is the minimum of all ordinals β such that for every open neighbourhood U of x there exists an open neighbourhood V of x satisfying conditions: $V \subset U$ and $\mathrm{ind}(\mathrm{Bd}_X(V)) < \alpha$. The dimension ind of a space X at a point $x \in X$ is denoted by $\mathrm{ind}_x(X)$.

Let X be a space for which $\mathrm{ind}(X) = \alpha$. The set of all points x of a space X such that $\mathrm{ind}_x(X) = \alpha$ is said to be the *dimensional kernel* of X and is denoted by $K(X)$.

It is easy to see that for every non-empty space X we have

$$\mathrm{ind}(X) = \sup\{\mathrm{ind}_a(X) : a \in X\}.$$

This equality implies that if α is an isolated ordinal and $\mathrm{ind}(X) = \alpha$, then $K(X) \neq \emptyset$. On the other hand, if α is a non-zero limit ordinal, then there exists a space X such that $\mathrm{ind}(X) = \alpha$ and $K(X) = \emptyset$.

Definition. A space X is said to be *weakly n-dimensional*, where $n \in \omega \setminus \{0\}$, if $\mathrm{ind}(X) = n$ and $\mathrm{ind}(K(X)) \leq n - 1$.

The first examples of weakly n-dimensional separable metrizable spaces are constructed in [375] and [266]. In [273] it is proved that if X is a weakly n-dimensional separable metrizable space, then $\mathrm{ind}(K(X)) = n - 1$ and in [272] and [425] that $K(X)$ is an F_σ set. Weakly n-dimensional separable metrizable spaces are considered also in [398] and [280]. (For more details see [115])

In [282] it is proved that in the class of all weakly n-dimensional separable metrizable spaces there exist universal elements.

10.3.7 Proposition. *Let $n \in \omega \setminus \{0\}$. The class of subsets consisting of all pairs $(K(X), X)$, where X is a weakly n-dimensional space (and $K(X)$ is the dimensional kernel of X), is a saturated class of subsets. Therefore, in this class there exist universal elements.*

Proof. Denote by $\mathbb{P}_n$ the considered class of subsets and by $\mathbb{P}(\text{ind} \leq k)$ the saturated class of all spaces of dimension $\text{ind} \leq k \in \tau$.

Let $\mathbf{S}$ be an indexed collection of spaces such that the restriction

$$\mathbf{K} \equiv \{K(X) : X \in \mathbf{S}\}$$

is a $\mathbb{P}_n$-restriction of $\mathbf{S}$. This restriction can also be considered as an indexed collection $\mathbf{S_K}$ of elements of $\mathbb{P}(\text{ind} \leq n-1)$. Denote by

$$\mathbf{M}^+_{\mathbf{K}} \equiv \{\{W^{K(X)}_\delta : \delta \in \tau\} : X \in \mathbf{S}\}$$

an initial co-mark of $\mathbf{S_K}$ corresponding to the class $\mathbb{P}(\text{ind} \leq n-1)$. Let

$$\mathbf{M_S} \equiv \{\{W^X_\delta : \delta \in \tau\} : X \in \mathbf{S}\}$$

be an $\mathbf{S}$-indexed set of τ-indexed families such that for every $X \in \mathbf{S}$ and $\delta \in \tau$ we have:

(a) W^X_δ is an open subset of X, and

(b) $W^X_\delta \cap K(X) = W^{K(X)}_\delta$.

Now, we consider the restriction

$$\mathbf{Q} \equiv \{X \setminus K(X) : X \in \mathbf{S}\}$$

of $\mathbf{S}$. We denote by

$$\mathbf{M_Q} \equiv \{\{V^X_\delta : \delta \in \tau\} : X \in \mathbf{S}\}$$

an $\mathbf{S}$-indexed set of τ-indexed families such that for every $X \in \mathbf{S}$ and $\delta \in \tau$ we have:

(c) $\{V^X_\delta : \delta \in \tau\}$ is a p-base for $X \setminus K(X)$ in X, and

(d) $\text{ind}(\text{Bd}_X(V^X_\delta)) \leq n-2$.

For every $\delta \in \tau$ we consider the indexed collection

$$\mathbf{S}_\delta \equiv \{\text{Bd}_X(V^X_\delta) : X \in \mathbf{S}\}$$

of spaces. This collection can be considered as an indexed collection of elements of the saturated class $\mathbb{P}(\text{ind} \leq n-2)$. We denote by

$$\mathbf{M}^+_\delta \equiv \{\{W^{\delta,X}_\varepsilon : \varepsilon \in \tau\} : X \in \mathbf{S}\}$$

an initial co-mark of $\mathbf{S}_\delta$ corresponding to the class $\mathbb{P}(\text{ind} \leq n-2)$. Let

$$\mathbf{M}_{\delta,\mathbf{S}} \equiv \{\{V_\varepsilon^{\delta,X} : \varepsilon \in \tau\} : X \in \mathbf{S}\}$$

be an $\mathbf{S}$-indexed set of τ-indexed families such that for every $X \in \mathbf{S}$ and $\delta, \varepsilon \in \tau$ we have:

(e) $V_\varepsilon^{\delta,X}$ is an open subset of X, and

(f) $V_\varepsilon^{\delta,X} \cap \mathrm{Bd}_X(V_\delta^X) = W_\varepsilon^{\delta,X}$.

Let $\mathbf{M}^+$ be a co-mark of $\mathbf{S}$ such that

(g) $\mathbf{M}^+$ is an initial co-mark corresponding to the class $\mathbb{P}(\mathrm{ind} \leq n)$,

(h) $\mathbf{M}^+$ is a co-extension of $\mathbf{M}_\mathbf{S}$,

(i) $\mathbf{M}^+$ is a co-extension of $\mathbf{M}_\mathbf{Q}$, and

(j) $\mathbf{M}^+$ is a co-extension of $\mathbf{M}_{\delta,\mathbf{S}}$ for every $\delta \in \tau$.

We prove that $\mathbf{M}^+$ is an initial co-mark corresponding to the restriction $\mathbf{K}$ of $\mathbf{S}$ and the class $\mathbb{P}_n$. For this purpose we consider an arbitrary co-mark

$$\mathbf{M} \equiv \{\{U_\delta^X : \delta \in \tau\} : X \in \mathbf{S}\}$$

of $\mathbf{S}$, which is a co-extension of $\mathbf{M}^+$. Then,

$$\mathbf{M}_{\mathbf{K},0} \equiv \{\{K(X) \cap U_\delta^X : \delta \in \tau\} : X \in \mathbf{S}\}$$

is a co-mark of $\mathbf{S}_\mathbf{K}$, which is a co-extension of $\mathbf{M}_\mathbf{K}^+$.

We denote by

$$\mathrm{R}_{\mathbf{K},0}^+ \equiv \{\sim_{\mathbf{K},0,+}^s : s \in \mathcal{F}\}$$

an initial family of equivalence relations on $\mathbf{S}_\mathbf{K}$ corresponding to the co-mark $\mathbf{M}_{\mathbf{K},0}$ and the class $\mathbb{P}(\mathrm{ind} \leq n-1)$ and by

$$\mathrm{R}_{\mathbf{K},0,\mathbf{S}}^+ \equiv \{\sim_{\mathbf{K},0,\mathbf{S},+}^s : s \in \mathcal{F}\}$$

the family of equivalence relations on $\mathbf{S}$ defined as follows: two elements X and Y of $\mathbf{S}$ are $\sim_{\mathbf{K},0,\mathbf{S},+}^s$-equivalent for some $s \in \mathcal{F}$ if and only if the elements $K(X)$ and $K(Y)$ of $\mathbf{S}_\mathbf{K}$ are $\sim_{\mathbf{K},0,+}^s$-equivalent.

For every $\delta \in \tau$ the indexed collection

$$\mathbf{M}_{\delta,0} \equiv \{\{\mathrm{Bd}_X(V_\delta^X) \cap U_\varepsilon^X : \varepsilon \in \tau\} : X \in \mathbf{S}\}$$

is a co-mark of $\mathbf{S}_\delta$, which is a co-extension of $\mathbf{M}_\delta^+$. Denote by

$$\mathrm{R}_{\delta,0}^+ \equiv \{\sim_{\delta,0,+}^s : s \in \mathcal{F}\}$$

an initial family of equivalence relations on $\mathbf{S}_\delta$ corresponding to the co-mark $\mathbf{M}_{\delta,0}$ and the class $\mathbb{P}(\text{ind} \leq n-2)$ and by

$$\mathrm{R}^+_{\delta,0,\mathbf{S}} \equiv \{\sim^s_{\delta,0,\mathbf{S},+} : s \in \mathcal{F}\}$$

the family of equivalence relations on $\mathbf{S}$ defined as follows: two elements X and Y of $\mathbf{S}$ are $\sim^s_{\delta,0,\mathbf{S},+}$-equivalent for some $s \in \mathcal{F}$ if and only if $\mathrm{Bd}_X(V^X_\delta)$ and $\mathrm{Bd}_Y(V^Y_\delta)$ are $\sim^s_{\delta,0,+}$-equivalent.

Now, we consider an $(\mathbf{M},\mathbf{K})$-admissible family

$$\mathrm{R}^+ \equiv \{\sim^s_+ : s \in \mathcal{F}\}$$

of equivalence relations on $\mathbf{S}$ such that:

(k) R^+ is an initial family corresponding to the co-mark $\mathbf{M}$ and the class $\mathbb{P}(\text{ind} \leq n)$, and

(l) R^+ is a final refinement of all families $\mathrm{R}^+_{\mathbf{K},0,\mathbf{S}}$ and $\mathrm{R}^+_{\delta,0,\mathbf{S}}$, $\delta \in \tau$.

Let also

$$\mathrm{R} \equiv \{\sim^s : s \in \mathcal{F}\}$$

be an arbitrary admissible family of equivalence relations on $\mathbf{S}$, which is a final refinement of R^+.

To prove that $\mathbf{M}^+$ is an initial co-mark it suffices to prove that the family R^+ is an initial family corresponding to the co-mark $\mathbf{M}$, the restriction $\mathbf{K}$, and the class $\mathbb{P}_n$. For this purpose it suffices to prove that the pair $(\mathrm{T}|_\mathbf{K}, \mathrm{T})$, where $\mathrm{T} = \mathrm{T}(\mathbf{M},\mathrm{R})$ and $\mathrm{T}|_\mathbf{K} = \mathrm{T}(\mathbf{M}|_\mathbf{K}, \mathrm{R}|_\mathbf{K})$, belongs to $\mathbb{P}_n$. That is, we need to prove the following conditions:

(m) $\text{ind}(\mathrm{T}) = n$,

(n) $\text{ind}(\mathrm{T}|_\mathbf{K}) \leq n-1$,

(o) if $\mathbf{a} \in \mathrm{T}|_\mathbf{K}$, then $\text{ind}_\mathbf{a}(\mathrm{T}) = n$, and

(p) if $\mathbf{a} \in \mathrm{T} \setminus \mathrm{T}|_\mathbf{K}$, then $\text{ind}_\mathbf{a}(\mathrm{T}) \leq n-1$.

Since T contains n–dimensional subspaces we have $\text{ind}(\mathrm{T}) \geq n$. (Here we use the fact that the dimension ind is hereditary on arbitrary subsets, which can easily be proved by induction). Therefore, to prove condition (m) it suffices to prove that $\text{ind}(\mathrm{T}) \leq n$. This follows by the fact that $\mathbf{M}^+$ is an initial co-mark corresponding to the class $\mathbb{P}(\text{ind} \leq n)$; $\mathbf{M}$ is a co-extension of $\mathbf{M}^+$; R^+ is an initial family corresponding to the co-mark $\mathbf{M}$ and the class $\mathbb{P}(\text{ind} \leq n)$; and R is a final refinement of R^+.

Condition (n) follows by the fact that $\mathbf{M}^+_\mathbf{K}$ is an initial co-mark corresponding to the class $\mathbb{P}(\text{ind} \leq n-1)$; $\mathbf{M}|_\mathbf{K} \equiv \mathbf{M}_{\mathbf{K},0}$ is a co-extension of

$\mathbf{M}^+_{\mathbf{K}}$; $\mathrm{R}^+_{\mathbf{K},0}$ is an initial family corresponding to the co-mark $\mathbf{M}_{\mathbf{K},0}$ and the class $\mathbb{P}(\mathrm{ind} \leq n-1)$; and $\mathrm{R}|_{\mathbf{K}}$ is a final refinement of $\mathrm{R}^+_{\mathbf{K},0}$.

Let $\mathbf{a} \in \mathrm{T}|_{\mathbf{K}}$. There exists a pair $(x, X) \in \mathbf{a}$ such that $x \in K(X)$. Let e^X_{T} be the natural embedding of X into T. Then, $e^X_{\mathrm{T}}(x) = \mathbf{a}$. Since $\mathrm{ind}_x(X) = n$ we have $\mathrm{ind}_{\mathbf{a}}(e^X_{\mathrm{T}}(X)) = n$. Therefore, $\mathrm{ind}_{\mathbf{a}}(\mathrm{T}) \geq n$. By condition (m), $\mathrm{ind}_{\mathbf{a}}(\mathrm{T}) = n$. This equality proves condition (o).

Finally, let $\mathbf{a} \in \mathrm{T} \setminus \mathrm{T}|_{\mathbf{K}}$ and U an arbitrary open neighbourhood of $\mathbf{a}$ in T. Without loss of generality we can suppose that $U = U^{\mathrm{T}}_{\varepsilon}(\mathbf{H})$ for some $\varepsilon \in \tau$, $t \in \mathcal{F}$, and $\mathbf{H} \in \mathrm{C}(\sim^t)$. Consider an element (x, X) of $\mathbf{a}$. Then, $e^X_{\mathrm{T}}(x) = \mathbf{a}$, which means that $x \in X \setminus K(X)$. Also, we have $x \in U^X_{\varepsilon}$. Since $\{V^X_{\delta} : \delta \in \tau\}$ is a p-base for $X \setminus K(X)$ in X there exists an element δ of τ such that $x \in V^X_{\delta} \subset U^X_{\varepsilon}$. Let θ be an indicial mapping from $\mathbf{M}_{\mathbf{Q}}$ to $\mathbf{M}$ and $\eta = \theta(\delta)$. Then, $x \in U^X_{\eta} \subset U^X_{\varepsilon}$. Denote by $\mathbf{E}$ the element of $\mathrm{C}(\sim^s)$, where $s = \{\eta, \varepsilon\} \cup t$, containng the space X. Then, $\mathbf{a} \in U^{\mathrm{T}}_{\eta}(\mathbf{E}) \subset U^{\mathrm{T}}_{\varepsilon}(\mathbf{H})$.

Let $\mathrm{R}_{\delta,0} \equiv \{\sim^s_{\delta,0} : s \in \mathcal{F}\}$ be the family of equivalence relations on $\mathbf{S}_{\delta}$ defined as follows: two elements $\mathrm{Bd}_X(V^X_{\delta})$ and $\mathrm{Bd}_Y(V^Y_{\delta})$ are $\sim^s_{\delta,0}$-equivalent for some $s \in \mathcal{F}$ if and only if X and Y are $\sim^s$-equivalent. Then, we can consider the Containing Space $\mathrm{T}(\mathbf{M}_{\delta,0}, \mathrm{R}_{\delta,0})$ constructed for the indexed collection $\mathbf{S}_{\delta}$ of spaces. By the construction, the subspace $\mathrm{T}(\mathbf{E}_{\delta,0}) \equiv \mathrm{T}(\mathbf{M}_{\delta,0}, \mathrm{R}_{\delta,0}, \mathbf{E}_{\delta,0})$ of this space belongs to $\mathbb{P}(\mathrm{ind} \leq n-2)$, where $\mathbf{E}_{\delta,0}$ is the equivalence class of the relation $\sim^s_{\delta,0}$ corresponding to the class $\mathbf{E}$ of the relation $\sim^s$. This fact means that $\mathrm{ind}(\mathrm{T}(\mathbf{E}_{\delta,0})) \leq n-2$. On the other hand $\mathrm{Bd}(U^{\mathrm{T}}_{\eta}(\mathbf{E})) = \mathrm{T}(\mathbf{E}_{\delta,0})$. Therefore, $\mathrm{ind}(\mathrm{Bd}(U^{\mathrm{T}}_{\eta}(\mathbf{E}))) \leq n-2$. This relation means that $\mathrm{ind}_{\mathbf{a}}(\mathrm{T}) \leq n-1$ proving condition (p) and completing the proof of the proposition. □

10.3.8 Corollary. *Let $\tau = \omega$ and $n \in \omega \setminus \{0\}$. The class of subsets consisting of all pairs $(K(X), X)$, where X is a weakly n-dimensional separable metrizable space (and $K(X)$ is the dimensional kernel of X), is a complete saturated class of subsets. Therefore, in this class there exist universal elements.* □

We note that the completeness of the class of subsets considered in Corollary 10.3.8 follows by the fact that for every weakly n-dimensional separable metrizable space X its dimensional kernel $K(X)$ is an F_{σ}-set.

10.3.9 Corollary. *Let $n \in \omega \setminus \{0\}$. The class of all weakly n-dimensional spaces is saturated. Therefore, in this class there exists universal elements.* □

10.3.10 Corollary. *Let $\tau = \omega$ and $n \in \omega \setminus \{0\}$. The class of all weakly n-dimensional separable metrizable spaces is saturated. Therefore, in this class there exists universal elements.* □

Below, we give a generalization of the notion of a weakly n-dimensional space.

Definition. (In this definition the dimensional kernel of a weakly m-dimensional space X is denoted by $K_m(X)$). Let $k, n \in \omega \setminus \{0\}$ and $k \leq n$. A space X is said to be *k-weakly n-dimensional* if:

(0) $\mathrm{ind}(X) = n$,
(1) $\mathrm{ind}(K_n(X)) = n - 1$,
(2) $\mathrm{ind}(K_{n-1}(K_n(X))) = n - 2$,
⋮
(k) $\mathrm{ind}(K_{n-k+1}(...(K_n(X))...)) = n - k$.

Therefore, a weakly n-dimensional space is 1-weakly n-dimensional.

The following two propositions in general outline are proved similarly to Proposition 10.3.7.

10.3.11 Proposition. *Let $k, n \in \omega \setminus \{0\}$, where $k \leq n$. The (non-empty) class of all k-weakly n-dimensional spaces is a saturated class of spaces. Therefore, in this class there exist universal elements.* □

10.3.12 Proposition. *Let $\alpha \in \tau^+ \setminus \omega$ be an isolated ordinal and $\beta \in \alpha$. Then, the (non-empty) class of subsets consisting of all pairs $(K(X), X)$, where X is an α-dimensional space and $K(X)$ is the dimensional kernel of X such that $\mathrm{ind}(K(X)) \leq \beta$, is saturated. Therefore, in this class there exist universal elements.* □

10.3.13 Problems. 1. Is it possible to prove directly the completeness of the class of subsets considered in Corollary 10.3.8 without using the fact that for a weakly n-dimensional separable metrizable space X its dimensional kernel $K(X)$ is an F_σ-set?

2. Is it true that the class $\mathbb{P}_n$ of Proposition 10.3.7 for the case $\tau \neq \omega$ is a complete class?

3. Let $k, n \in \omega \setminus \{0\}$ and $1 < k \leq n$. Is there a k-weakly n-dimensional (separable metrizable) space?

4. Let $\tau \neq \omega$ and X be a weakly n-dimensional space, where $n \in \omega\setminus\{0\}$. Is it true that:

(a) $\mathrm{ind}(K(X)) = n - 1$?

(b) the subset $K(X)$ of X is an F_σ-set?

Remark 5. Some other problems concerning new dimension-like functions are posed below.

10.3.14 Problem. Prove that Propositions 3.3.2(2), 3.3.4(2), 3.3.6, 3.3.8(2), 3.3.10(2), and 3.3.14 are not true if instead of the integer $n \in \omega$ we consider an ordinal $\alpha \in \tau^+ \setminus \omega$.

10.3.15 Problems. 1. Apply the method of Containing Spaces to known relative dimensions (see for example [71], [73]) and to classes of the pairs considered in [421] and [128].

2. Is it possible to apply the method of Containing Spaces to the notion of the dimension given in a general form in [309] (see also [310])?

3. Is it possible to characterize axiomatically (using saturation classes) some of the new dimensions?

10.3.16 Problems. 1. Find standard methods to determine disjoint pairs of closed subsets of a space X such that:

(a) The number of the determined pairs is not more than τ.

(b) The determined pairs in some sense approximate any given pair of disjoint closed subsets of the space X.

(c) If X is a compact space, then for every pair (K, Q) of disjoint closed subsets of X there exists a determined pair (G, F) for which $K \subset G$ and $Q \subset F$.

2. Find standard methods to determine finite open covers of a space X such that:

(a) The number of determined covers is not more than τ.

(b) The determined covers in some sense approximate any given finite open cover of the space X.

(c) If X is a compact space, then for every finite open cover c of X there exists a determined cover, which is a refinement of c.

10.4 Completeness and completion of marked spaces.

The notions of the completeness and completion of metric spaces (see [130],[162]) and its generalizations to uniform spaces, that is to completely regular spaces with uniform structure (see [437],[47]) are well known. The modifications and generalizations of these notions are investigated by many authors. We refer to some of the corresponding papers: [390], [65], [340], [388], [389], [51], [171], [56], [355], [166], [55], [136], [132], [29], [66], [133], [67], [68], [247], [219], [98], [99], [90], [100], [101], [102], [92], [356], [103]. For many other papers and for more details see the following books and survey papers: [292], [362], [129], [91], [232], [30], [69], [52], [233], [234].

In this section we apply a special type of Containing Spaces to define a completion of a given marked space. A pecularity of this completion is the fact that it depends on the class of marked spaces containing as element the given marked space, that is we actually define the notion of a completion of elements of a given class of marked spaces. Another pecularity of the defined completion is the fact that for its construction we do not use any notion of a convergent sequence or filter. The completeness and completion of marked spaces are used to define some kind of completeness and completion of (regular) topological spaces.

In what follows, a marked space X with a mark M^X will be denoted by (X, M^X).

Definition. Two marked spaces (X, M^X) and (Y, M^Y) are called *similar* if they are $\sim^s_m$-equivalent for every $s \in \mathcal{F}$. (About the equivalence relation $\sim^s_m$ see Chapter 1).

Two marked spaces (X, M^X) and (Y, M^Y), where

$$M^X = \{U^X_\delta : \delta \in \tau\} \text{ and } M^Y = \{U^Y_\delta : \delta \in \tau\},$$

are called *similarly homeomorphic* if there exists a homeomorphism h of X onto Y such that $U^X_\delta = h^{-1}(U^Y_\delta)$ for every $\delta \in \tau$. In this case h is called a *similar homeomorphism* of (X, M^X) onto (Y, M^Y). Obviously, similarly homeomorphic marked spaces are similar.

10.4.1 Lemma. *The similar homeomorphism between two similarly homeomorphic marked spaces is unique determined.*

Proof. Let (X, M^X) and (Y, M^Y) be two similarly homeomorphic marked spaces, where $M^X = \{U_\delta^X : \delta \in \tau\}$ and $M^Y = \{U_\delta^Y : \delta \in \tau\}$. The proof of the lemma follows immediately by the fact that for every point x of X the image y of this point under a similar homeomorphism satisfies the following condition: y belongs to U_δ^Y for some $\delta \in \tau$ if and only if x belongs to U_δ^X for the same δ. □

Definition. Let $\mathbb{M}$ be a class of marked spaces. The class $\mathbb{M}$ is said to be *topological* if for every similarly homeomorphic marked spaces (X, M^X) and (Y, M^Y) the condition $(X, M^X) \in \mathbb{M}$ implies $(Y, M^Y) \in \mathbb{M}$. In the sequel, an arbitrary class of marked spaces is assumed to be topological.

An indexed collection of elements of $\mathbb{M}$ will be denoted by $(\mathbf{S}, \mathbf{M})$, where $\mathbf{S}$ is an indexed collection of spaces; $\mathbf{M}$ is a co-mark of $\mathbf{S}$; and for every $X \in \mathbf{S}$ the pair $(X, \mathbf{M}(X))$ is an element of $\mathbb{M}$. Below, we shall consider indexed collection $(\mathbf{S}, \mathbf{M})$ of similar elements of $\mathbb{M}$ which means that each element of $(\mathbf{S}, \mathbf{M})$ is similar to each of others, that is, $\mathbf{S}$ is the only element of the set $\mathrm{C}(\mathrm{R}_\mathbf{M})$.

An indexed collection $(\mathbf{S}, \mathbf{M})$ of similar elements of $\mathbb{M}$ is said to be *full* if for every element (Y, M^Y) of $\mathbb{M}$, which is similar to elements of $(\mathbf{S}, \mathbf{M})$, there exists an element (X, M^X) of $(\mathbf{S}, \mathbf{M})$, which is similarly homeomorphic to (Y, M^Y).

10.4.2 Lemma. *Let $\mathbb{M}$ be a class of marked spaces, $(\mathbf{S}, \mathbf{M})$ an indexed collection of similar elements of $\mathbb{M}$, and (X, M^X) an arbitrary element of $(\mathbf{S}, \mathbf{M})$ with $M^X = \{U_\delta^X : \delta \in \tau\}$. The following statements are true:*

(1) *The standard base B^T for the open subsets of the Containing Space $\mathrm{T} \equiv \mathrm{T}(\mathbf{M}, \mathrm{R}_\mathbf{M})$ coincides with the set $\{U_\delta^\mathrm{T}(\mathbf{S}) : \delta \in \tau\}$.*

(2) *The set $i_\mathrm{T}^X(X)$ is a dense subset of* T.

(3) *The marked spaces (X, M^X) and $(\mathrm{T}, \mathrm{B}^\mathrm{T})$ are similar.* (Here, B^T is considered to be the indication $\{U_\delta^\mathrm{T}(\mathbf{S}) : \delta \in \tau\}$ of the standard base for open subsets of T called the *natural* indication).

Proof. The proof of statement (1) follows immediately by the fact that the only element of the set $\mathrm{C}(\mathrm{R}_\mathbf{M})$ is the set $\mathbf{S}$. By the same fact the proof of statement (3) follows immediately by Corollary 1.2.5.

We prove statement (2). Let $U_\delta^\mathrm{T}(\mathbf{S})$ be a non-empty element of B^T. We need to prove that $U_\delta^\mathrm{T}(\mathbf{S}) \cap i_\mathrm{T}^X(X) \neq \emptyset$. Consider the set U_δ^X. If this set is empty, then $U_\delta^Y = \emptyset$ for every $Y \in \mathbf{S}$ because (X, M^X) is similar to any element of $(\mathbf{S}, \mathbf{M})$. In this case $U_\delta^\mathrm{T}(\mathbf{S})$ is also empty, which is a contradiction. Therefore, $U_\delta^X \neq \emptyset$. Let $x \in U_\delta^X$. Then, the point $\mathbf{a}$ of T containing the pair (x, X) belongs to $U_\delta^{\mathrm{T}}(\mathbf{S})$. On the other hand, $\mathbf{a} \in i_{\mathrm{T}}^X(X)$, that is, $U_\delta^{\mathrm{T}}(\mathbf{S}) \cap i_{\mathrm{T}}^X(X) \neq \emptyset$. □

Definition. Let $\mathbb{M}$ be a class of marked spaces, $(\mathbf{S}, \mathbf{M})$ an indexed collection of similar elements of $\mathbb{M}$, and (X, M^X) an arbitrary element of $(\mathbf{S}, \mathbf{M})$. The pair $(\mathrm{T}, \mathrm{B}^{\mathrm{T}})$ of Lemma 10.4.2 is called the $(\mathbf{S}, \mathbf{M})$-*completion* of (X, M^X) and the mapping i_{T}^X the *natural* embedding of (X, M^X) in this $(\mathbf{S}, \mathbf{M})$-completion.

In the case where $(\mathbf{S}, \mathbf{M})$ is a full indexed collection of similar elements of $\mathbb{M}$ (containing (X, M^X)), the pair $(\mathrm{T}, \mathrm{B}^{\mathrm{T}})$ is called an $\mathbb{M}$-*completion* of (X, M^X).

We note that for some indexed collection $(\mathbf{S}, \mathbf{M})$ of similar elements of $\mathbb{M}$ the $(\mathbf{S}, \mathbf{M})$-completion of an element of this collection may not be an element of $\mathbb{M}$. Therefore, an $\mathbb{M}$-completion of an element of $\mathbb{M}$ in general is not an element of $\mathbb{M}$.

10.4.3 Lemma. *Let $\mathbb{M}$ be an indexed collection of marked spaces and (X, M^X) an element of $\mathbb{M}$. Any two $\mathbb{M}$-completions $(\mathrm{T}, \mathrm{B}^{\mathrm{T}})$ and $(\mathrm{T}', \mathrm{B}^{\mathrm{T}'})$ of (X, M^X) are similarly homeomorphic. Moreover, if h is the similar homeomorphism of $(\mathrm{T}, \mathrm{B}^{\mathrm{T}})$ onto $(\mathrm{T}', \mathrm{B}^{\mathrm{T}'})$ and i_{T}^X and $i_{\mathrm{T}'}^X$ are the natural embeddings of (X, M^X) into $(\mathrm{T}, \mathrm{B}^{\mathrm{T}})$ and $(\mathrm{T}', \mathrm{B}^{\mathrm{T}'})$, respectively, then $i_{\mathrm{T}'}^X = h \circ i_{\mathrm{T}}^X$.*

Proof. Suppose that $(\mathbf{S}, \mathbf{M})$ and $(\mathbf{S}', \mathbf{M}')$ are two full indexed collections of similar elements of $\mathbb{M}$ containing the element (X, M^X). We need to prove that the marked spaces $(\mathrm{T}, \mathrm{B}^{\mathrm{T}})$ and $(\mathrm{T}', \mathrm{B}^{\mathrm{T}'})$ are similarly homeomorphic, where $\mathrm{T} = \mathrm{T}(\mathbf{M}, \mathrm{R}_{\mathbf{M}})$, $\mathrm{T}' = \mathrm{T}(\mathbf{M}', \mathrm{R}_{\mathbf{M}'})$, B^{T} is the natural indication of the standard base for T, and $\mathrm{B}^{\mathrm{T}'}$ is the natural indication of the standard base for T′.

Let $\mathbf{a}$ be an element of T and $y \in Y \in \mathbf{S}$ such that $(y, Y) \in \mathbf{a}$. There exists an element (Z, B^Z) of $(\mathbf{S}', \mathbf{M}')$, which is similarly homeomorphic to (Y, M^Y). Let g be the similar homeomorphism of (Y, M^Y) onto (Z, M^Z). Denote by $\mathbf{b}$ the point of T′ containing the pair $(g(y), Z)$. We define a mapping h of T into T′ setting $h(\mathbf{a}) = \mathbf{b}$ and prove that h is the similar homeomorphism of $(\mathrm{T}, \mathrm{B}^{\mathrm{T}})$ onto $(\mathrm{T}', \mathrm{B}^{\mathrm{T}'})$.

First we prove that the point $\mathbf{b}$ is independent of the considered element (Z, M^Z) of $(\mathbf{S}', \mathbf{M}')$ and the pair (y, Y) of $\mathbf{a}$. Indeed, let $(Z', M^{Z'})$ be another element of $(\mathbf{S}', \mathbf{M}')$, which is similarly homeomorphic to (Y, M^Y), and g' the corresponding similar homeomorphism. Then, $g'(y)$ belongs to an element $U_\delta^{Z'}$ for some $\delta \in \tau$ if and only if $g(y)$ belongs to U_δ^Z for the same δ. This fact means that the pair $(g'(y), Z')$ belongs to $\mathbf{b}$, that is, $\mathbf{b}$ is independent of (Z, M^Z). Similarly, we can prove that the point $\mathbf{b}$ is independent of the element (y, Y) of $\mathbf{a}$.

Now, we prove that the mapping $h : \mathrm{T} \to \mathrm{T}'$ is one-to-one. Let $\mathbf{a}_0$ and $\mathbf{a}_1$ be distinct points of T, $(y_0, Y_0) \in \mathbf{a}_0$ and $(y_1, Y_1) \in \mathbf{a}_1$. Since (Y_0, M^{Y_0}) and (Y_1, M^{Y_1}) are similar and $\mathbf{a}_0 \neq \mathbf{a}_1$ there exists $\delta \in \tau$ such that either $y_0 \in U_\delta^{Y_0}$ and $y_1 \notin U_\delta^{Y_1}$ or $y_0 \notin U_\delta^{Y_0}$ and $y_1 \in U_\delta^{Y_1}$. Consider elements (Z_0, M^{Z_0}) and (Z_1, M^{Z_1}) of $\mathbf{S}'$, which are similarly homeomrphic to (Y_0, M^{Y_0}) and (Y_1, M^{Y_1}), respectively. Denote by g_0 and g_1 the similar homeomorphisms of (Y_0, M^{Y_0}) onto (Z_0, M^{Z_0}) and of (Y_1, M^{Y_1}) onto (Z_1, M^{Z_1}), respectively, and set $g_0(y_0) = z_0$ and $g_1(y_1) = z_1$. Then, z_0 belongs to $U_\delta^{Z_0}$ if and only if y_0 belongs to $U_\delta^{Y_0}$. Similarly, z_1 belongs to $U_\delta^{Z_1}$ if and only if y_1 belongs to $U_\delta^{Y_1}$. Therefore, either $z_0 \in U_\delta^{Z_0}$ and $z_1 \notin U_\delta^{Z_1}$ or $z_0 \notin U_\delta^{Z_0}$ and $z_1 \in U_\delta^{Z_1}$. Obviously, $(z_0, Z_0) \in h(\mathbf{a}_0)$ and $(z_1, Z_1) \in h(\mathbf{a}_1)$. Therefore, $h(\mathbf{a}_0) \neq h(\mathbf{a}_1)$, which means that h is one-to-one.

We prove that $h : \mathrm{T} \to \mathrm{T}'$ is onto. Let $\mathbf{b}_0$ be a point of T' and $(z_0, Z_0) \in \mathbf{b}_0$. There exists an element (Y_0, M^{Y_0}) of $(\mathbf{S}, \mathbf{M})$, which is similarly homeomorphic to the element (Z_0, M^{Z_0}). Let g be the similar homeomorphism of (Y_0, M^{Y_0}) onto (Z_0, M^{Z_0}). Denote by $\mathbf{a}_0$ the point of T containing the pair $(g^{-1}(z_0), Y_0)$. Then, it is easy to see that $h(\mathbf{a}_0) = \mathbf{b}_0$.

Furthermore, we prove that h and h^{-1} are continuous and h is the similar homeomorphism of $(\mathrm{T}, \mathrm{B}^{\mathrm{T}})$ onto $(\mathrm{T}', \mathrm{B}^{\mathrm{T}'})$. Let $(y, Y) \in \mathbf{a} \in \mathrm{T}$, $(z, Z) \in \mathbf{b} \in \mathrm{T}'$ and $h(\mathbf{a}) = \mathbf{b}$. Suppose that (Y, M^Y) and (Z, M^Z) are similarly homeomorphic and g is the similar homeomorphism of (Y, M^Y) onto (Z, M^Z) such that $g(y) = z$. It suffices to prove that $\mathbf{a}$ belongs to an element $U_\delta^{\mathrm{T}}(\mathbf{S})$ of B^{T} for some $\delta \in \tau$ if and only if $\mathbf{b}$ belongs to the element $U_\delta^{\mathrm{T}'}(\mathbf{S}')$ of $\mathrm{B}^{\mathrm{T}'}$ for the same δ. But this assertion follows immediately by the above conditions and the definition of the elements of the standard base of a Containing Space.

Finally, relation $i_{\mathrm{T}'}^X = h \circ i_{\mathrm{T}}^X$ is not difficult to verify. □

Definition. Let $I\!M$ be a class of marked spaces and $(\mathbf{S}, \mathbf{M})$ an indexed collection of similar elements of $I\!M$. An element (X, M^X) of $(\mathbf{S}, \mathbf{M})$ is said to be $(\mathbf{S}, \mathbf{M})$-*complete* if $i_{\mathrm{T}}^X(X) = \mathrm{T} \equiv \mathrm{T}(\mathbf{M}, \mathrm{R}_{\mathbf{M}})$. In this case i_{T}^X is the similar homeomorphism of (X, M^X) onto $(\mathrm{T}, \mathrm{B}^{\mathrm{T}})$.

An element (X, M^X) of $I\!M$ is said to be $I\!M$-*complete* if (X, M^X) is $(\mathbf{S}, \mathbf{M})$-complete for every indexed collection $(\mathbf{S}, \mathbf{M})$ of similar elements of

$\mathbb{M}$ containing (X, M^X).

10.4.4 Lemma. *Let $\mathbb{M}$ be a class of marked spaces. An element (X, M^X) of $\mathbb{M}$ is $\mathbb{M}$-complete if (X, M^X) is $(\mathbf{S}, \mathbf{M})$-complete for some full indexed collection $(\mathbf{S}, \mathbf{M})$ of similar elements of $\mathbb{M}$ containing (X, M^X).*

Proof. Suppose that an element (X, M^X) with $M^X = \{U^X_\delta ; \delta \in \tau\}$ is $(\mathbf{S}, \mathbf{M})$-complete, where $(\mathbf{S}, \mathbf{M})$ is a full indexed collection of elements of $\mathbb{M}$. We prove that (X, M^X) is $(\mathbf{S}', \mathbf{M}')$-complete for any indexed collection $(\mathbf{S}', \mathbf{M}')$ of elements of $\mathbb{M}$ containing (X, M^X). Indeed, let $(\mathbf{S}', \mathbf{M}')$ be such an indexed collection. Suppose that (X, M^X) is not $(\mathbf{S}', \mathbf{M}')$-complete. This assumption means that the set $\mathrm{T}' \setminus i^X_{\mathrm{T}'}(X)$ is not empty, where T' is the Containing Space $\mathrm{T}(\mathbf{M}', \mathrm{R}_{\mathbf{M}'})$ corresponding to the indexed collection $\mathbf{S}'$. Let $\mathbf{a}$ be a point of this set. There exist an element (Y, M^Y) of $(\mathbf{S}', \mathbf{M}')$ with $M^Y = \{U^Y_\delta : \delta \in \tau\}$ and a point y of Y such that $(y, Y) \in \mathbf{a}$.

Since $(\mathbf{S}, \mathbf{M})$ is a full indexed collection there exists an element (Z, M^Z) of $(\mathbf{S}, \mathbf{M})$ with $M^Z = \{U^Z_\delta : \delta \in \tau\}$, which is similarly homeomorphic to (Y, M^Y). Suppose that h is the similar homeomorphism of (Y, M^Y) onto (Z, M^Z). Let $\mathbf{b}$ be the point of the Containing space $\mathrm{T} \equiv \mathrm{T}(\mathbf{M}, \mathrm{R}_{\mathbf{M}})$ containing the pair (z, Z), where $z = h^{-1}(y)$. Therefore, the point z of Z belongs to U^Z_δ for some $\delta \in \tau$ if and only if the point y of Y belongs to U^Y_δ for the same δ.

Since (X, M^X) is $(\mathbf{S}, \mathbf{M})$-complete there exists a point x of X such that $(x, X) \in \mathbf{b}$. Then for every $\delta \in \tau$, x belongs to U^X_δ if and only if z belongs to U^Z_δ. Therefore, x belongs to U^X_δ if and only if y belongs to U^Y_δ. This fact implies that $(x, X) \in \mathbf{a}$, that is, $\mathbf{a} \in i^X_{\mathrm{T}'}(X)$, which is a contradiction. Thus, (X, M^X) is $(\mathbf{S}', \mathbf{M}')$-complete. $\square$

10.4.5 Lemma. *Let $\mathbb{M}$ be an indexed collection of marked spaces and $(\mathbf{S}, \mathbf{M})$ a full indexed collection of similar elements of $\mathbb{M}$ such that the pair $(\mathrm{T}, \mathrm{B}^{\mathrm{T}})$ belongs to $\mathbb{M}$, where $\mathrm{T} = \mathrm{T}(\mathbf{M}, \mathrm{R}_{\mathbf{M}})$ and B^{T} is the natural indication of the standard base for the open subsets of T. Then, $(\mathrm{T}, \mathrm{B}^{\mathrm{T}})$ is $\mathbb{M}$-complete. (Thus, if an $\mathbb{M}$-completion of an element of $\mathbb{M}$ belongs to $\mathbb{M}$, then it is $\mathbb{M}$-complete).*

Proof. We need to prove that if $(\mathbf{S}', \mathbf{M}')$ is an indexed collection of similar elements of $\mathbb{M}$ containing the element $(\mathrm{T}, \mathrm{B}^{\mathrm{T}})$ of $\mathbb{M}$, then the natural embedding $i^{\mathrm{T}}_{\mathrm{T}'}$ of T into $\mathrm{T}' \equiv \mathrm{T}(\mathbf{M}', \mathrm{R}_{\mathbf{M}'})$ is onto. Let $\mathbf{b}$ be an element of T' and $(y, Y) \in \mathbf{b}$. There exists an element (X, M^X) of $(\mathbf{S}, \mathbf{M})$, which is similarly homeomorphic to (Y, M^Y). Let h be the corresponding similar homeomorphism of (X, M^X) onto (Y, M^Y) and $x = h^{-1}(y)$. Denote by $\mathbf{a}$ the point of T containing the pair (x, X). Then, it is easy to verify that $i^{\mathrm{T}}_{\mathrm{T}'}(\mathbf{a}) = \mathbf{b}$. □

Below we consider a concrete class of marked spaces.

Notation. Let θ_0 and θ_1 be two fixed one-to-one mappings of τ into itself such that $\theta_0(\tau) \cap \theta_1(\tau) = \emptyset$ and $\theta_0(\tau) \cup \theta_1(\tau) = \tau$ and let $\kappa = \theta_0(\tau)$ and $\kappa_1 = \theta_1(\tau)$.

We denote by $\mathbb{M}(\mathrm{rs})$ the class of marked spaces consisting of all pairs (X, M^X) such that X is a regular space and $M^X \equiv \{U^X_\delta : \delta \in \tau\}$ satisfies the following conditions:

(a) the set $\{U^X_\delta : \delta \in \kappa\}$ is a base for the open subsets of X and

(b) for every $\delta \in \kappa$, $U^X_{\theta_1(\theta_0^{-1}(\delta))} = X \setminus \mathrm{Cl}(U^X_\delta)$.

For $\tau = \omega$ the class $\mathbb{M}(\mathrm{rs})$ is denoted by $\mathbb{M}(\mathrm{sms})$. We note that for any element (X, M^X) of $\mathbb{M}(\mathrm{sms})$ the space X is separable metrizable.

10.4.6 Proposition. *Any* $\mathbb{M}(\mathrm{rs})$*-completion of each element of* $\mathbb{M}(\mathrm{rs})$ *belongs to* $\mathbb{M}(\mathrm{rs})$.

Proof. First we note the following fact. Let $(\mathbf{S}, \mathbf{M})$ be any indexed collection of elements of $\mathbb{M}(\mathrm{rs})$. If we spell out the proof of the Proposition 2.1.5 we can see that actually it is proved that the co-mark $\mathbf{M}$ is an initial co-mark of $\mathbf{S}$ corresponding to the class of regular spaces and the family $\mathrm{R}_{\mathbf{M}}$ is an initial family corresponding to the co-mark $\mathbf{M}$ and the class of regular spaces. This observation means that the space $\mathrm{T}(\mathbf{M}, \mathrm{R}_{\mathbf{M}})$ is a regular space.

Now, suppose that $(\mathbf{S}, \mathbf{M})$ is a full indexed collection of similar elements of $\mathbb{M}(\mathrm{rs})$. Then, by the above the space $\mathrm{R} \equiv \mathrm{T}(\mathbf{M}, \mathrm{R}_{\mathbf{M}})$ is regular. Thus, to prove the proposition we need to prove that the natural indication of the standard base for the open subsets of T satisfies the conditions (a) and (b) of the definition of the class $\mathbb{M}(\mathrm{rs})$.

For every $(X, M^X) \in (\mathbf{S}, \mathbf{M})$ we have that the set $\{U^X_\delta : \delta \in \kappa\}$ is a base for the open subsets of X. By Corollary 1.2.8 the set $\{U^{\mathrm{T}}_\delta(\mathbf{S}) : \delta \in \kappa\}$ is a base for the open subsets of T, which means that condition (a) is satisfied.

We prove condition (b). Let $\delta \in \kappa$. We need to prove that

$$U^{\mathrm{T}}_{\theta_1(\delta)}(\mathbf{S}) = \mathrm{T} \setminus \mathrm{Cl}_{\mathrm{T}}(U^{\mathrm{T}}_\delta(\mathbf{S})).$$

Let $\mathbf{a} \in \mathrm{Cl}(U_\delta^{\mathrm{T}}(\mathbf{S}))$ and $(x, X) \in \mathbf{a}$. By Lemma 1.2.7, $x \in \mathrm{Cl}_X(U_\delta^X)$. Therefore, $x \notin U_{\theta_1(\delta)}^X$, which means that $\mathbf{a} \notin U_{\theta_1(\delta)}^{\mathrm{T}}(\mathbf{S})$. Conversely, let $\mathbf{a} \notin \mathrm{Cl}_{\mathrm{T}}(U_\delta^{\mathrm{T}}(\mathbf{S}))$ and $(x, X) \in \mathbf{a}$. By Lemma 1.2.7, $x \notin \mathrm{Cl}(U_\delta^X)$ and, therefore, $x \in U_{\theta_1(\delta)}^X$, which means that $\mathbf{a} \in U_{\theta(\delta)}^{\mathrm{T}}(\mathbf{S})$. This relation completes the proof of the proposition. □

10.4.7 Proposition. *Let $\tau = \omega$. If X is a separable completely metrizable space, then there exists a co-mark M^X of X such that the pair (X, M^X) is an* $\mathbb{M}(\mathrm{sms})$*-complete element of the class* $\mathbb{M}(\mathrm{sms})$.

Proof. We denote by ϑ an one-to-one mapping of $\omega \times \omega$ onto $\omega \setminus \{0\}$. Let X be a separable completely metrizable space and ϱ a complete metric on X. For every $n \in \omega$ we consider the open cover $C_n \equiv \{S(x, n) : x \in X\}$ of X, where $S(x, n) = \{y \in X : \varrho(x, y) < 1/n\}$. Let $c_n \equiv \{V_{(n,i)}^X : i \in \omega\}$ be a star-finite refinement of C_n (see [209]). We put $M_0^X \equiv \{V_\delta^X : \delta \in \omega\}$, where $V_0^X = \emptyset$ and $V_\delta^X = V_{\vartheta^{-1}(\delta)}^X$ if $\delta \in \omega \setminus \{0\}$. It is easy to see that M_0^X is a mark of X.

Now, we define a mark $M^X \equiv \{ U_\delta^X : \delta \in \omega\}$ of X setting $U_{\theta_0(\delta)}^X = V_\delta^X$ and $U_{\theta_1(\delta)}^X = X \setminus \mathrm{Cl}_X(V_\delta^X)$, where θ_0 and θ_1 are the mapping of ω into itself, which are used in the definition of the class $\mathbb{M}(\mathrm{sms})$. We shall prove that the pair (X, M^X) is an $\mathbb{M}(\mathrm{sms})$-complete element of $\mathbb{M}(\mathrm{sms})$.

Indeed, let $(\mathbf{S}, \mathbf{M})$ be a full indexed collection of similar elements of $\mathbb{M}(\mathrm{sms})$ containing the pair (X, M^X). Consider the Containing Space $\mathrm{T} \equiv \mathrm{T}(\mathbf{M}, \mathrm{R}_{\mathbf{M}})$. We need to prove that the natural embedding i_{T}^X of X into T is onto. Let $\mathbf{a}$ be a point of T and $\kappa(\mathbf{a})$ be a subset of $\kappa \equiv \theta_0(\omega)$ consisting of all elements δ such that $\mathbf{a}$ belongs to the element $U_\delta^{\mathrm{T}}(\mathbf{S})$ of the standard base B^{T} of T. Let κ' be a subset of $\kappa(\mathbf{a})$ such that for every $\delta \in \kappa(\mathbf{a})$ there exists a unique element δ' of κ' for which $U_\delta^{\mathrm{T}}(\mathbf{S}) = U_{\delta'}^{\mathrm{T}}(\mathbf{S})$. Thus, the intersection of any finite number of elements of the set $\{U_\delta^{\mathrm{T}}(\mathbf{S}) : \delta \in \kappa'\}$ is non-empty and for distinct elements δ and δ' of κ' we have $U_\delta^{\mathrm{T}}(\mathbf{S}) \neq U_{\delta'}^{\mathrm{T}}(\mathbf{S})$. If the set κ' is finite, then $\mathbf{a}$ is an isolated point of T. In this case, since $i_{\mathrm{T}}^X(X)$ is a dense subset of T (see Lemma 10.4.2) there exists a point x of X such that $i_{\mathrm{T}}^X(x) = \mathbf{a}$. Therefore, we can suppose that κ' is an infinite set. Since the pairs (X, M^X) and $(\mathrm{T}, \mathrm{B}^{\mathrm{T}})$ are similar the intersection of any finite number of elements of the set $\{U_\delta^X : \delta \in \kappa'\}$ is non-empty and for distinct elements δ and δ' of κ' we have $U_\delta^X \neq U_{\delta'}^X$. Let ϑ' be an one-to-one mapping of κ' onto ω. Consider the indexed set $\{U_j : j \in \omega\}$, where $U_j = U_{(\vartheta')^{-1}(j)}^X$. For every $j \in \omega$ we consider a point x_j of $U_1 \cap \ldots \cap U_j$.

By the construction of M_0^X it follows that $\{x_j : j \in \omega\}$ is a fundamental sequence. Since ϱ is a complete metric on X there exists a point x of X, which is a limit point of this sequence. We note that $x \in \mathrm{Cl}_X(U_j)$ for every $j \in \omega$.

We prove that $i_{\mathrm{T}}^X(x) = \mathbf{a}$. Indeed, if $i_{\mathrm{T}}^X(x) = \mathbf{b} \neq \mathbf{a}$, then there exists $\delta \in \theta_0(\omega)$ such that $\mathbf{a} \in \mathrm{Cl}_{\mathrm{T}}(U_\delta^{\mathrm{T}}(\mathbf{S}))$ and $\mathbf{b} \notin \mathrm{Cl}_{\mathrm{T}}(U_\delta^{\mathrm{T}}(\mathbf{S}))$. In this case $x \notin \mathrm{Cl}_X(U_\delta^X)$. On the other hand, there exists $j \in \omega$ such that $U_j = U_\delta^X$ and, therefore, $x \notin \mathrm{Cl}_X(U_j)$, which contradicts the choice of x. Thus, $i_{\mathrm{T}}^X(x) = \mathbf{a}$. □

Definition. Let X be a regular space. A space X is said to be $\mathbb{M}(\mathrm{rs})$-*complete* if there exists a mark M^X of X such that the pair (X, M^X) is an $\mathbb{M}(\mathrm{rs})$-complete element of $\mathbb{M}(\mathrm{rs})$.

A regular space Y is said to be an $\mathbb{M}(\mathrm{rs})$-*completion* of the space X if there exists a mark M^X of X and a mark M^Y of Y such that:

(a) the pairs (X, M^X) and (Y, M^Y) are similar elements of $\mathbb{M}(\mathrm{rs})$ and
(b) the pair (Y, M^Y) is $\mathbb{M}(\mathrm{rs})$-complete.

10.4.8 Problems. 1. Give an internal characterization of $\mathbb{M}(\mathrm{rs})$-complete (marked) spaces.

2. Is it true that a space X is completely metrizable if and only if X is $\mathbb{M}(\mathrm{rs})$-complete?

3. Is there a connection between metric completions of a metrizable space and its $\mathbb{M}(\mathrm{rs})$-completions?

4. Is there a connection between uniform completions of a completely metrizable space and its $\mathbb{M}(\mathrm{rs})$-completions?

5. Is there a connection between quasi-uniform completions of a regular space and its $\mathbb{M}(\mathrm{rs})$-completions?

10.4.9 Problems. Let $\tau = \omega$.

1. Give an internal characterization of $\mathbb{M}(\mathrm{sms})$-complete spaces.

2. Is it true that if a separable metrizable space X is $\mathbb{M}(\mathrm{sms})$-complete, then X is a separable completely metrizable space?

3. Let (X, M^X) be an element of $\mathbb{M}(\mathrm{sms})$ and $(\mathrm{T}, \mathrm{B}^{\mathrm{T}})$ its $\mathbb{M}(\mathrm{sms})$-completion. Is there a metric d on X such that the metric completion of X is homeomorphic to T?

4. Let X be a separable completely metrizable space. Characterize the marks M^X of X such that the element (X, M^X) of $\mathbb{M}(\mathrm{sms})$ is $\mathbb{M}(\mathrm{sms})$-complete.

10.4.10 Problem. 1. Find classes $\mathbb{M}$ of marked spaces with "interesting" $\mathbb{M}$-complete elements.

2. Is there any connection between the notions of completness and completion of marked spaces and that of the papers mentioned in the beginning of this section?

10.5 Containing Spaces and connectedness

The connectedness is a property of spaces, which is not transferable to the Containing Spaces. We have the following proposition, which is easy to prove.

10.5.1 Proposition. *Any class consisting of non-empty connected spaces is not saturated.* □

In spite of the above result, the universality problem for some classes of connected spaces can be investigated by Containing Spaces. Below, we give a method for such an investigation. In general outline, this method consists of the identification of some points of the Containing Space $\mathrm{T}(\mathbf{M}, \mathrm{R})$ constructed for an indexed collection $\mathbf{S}$ of spaces, a suitable co-mark $\mathbf{M}$ of $\mathbf{S}$, and a family R of equivalence relations on $\mathbf{S}$. With respect to the points, which are identified, we consider the simplest case. In some sense more complicated cases are considered in the papers [184] and [185] for $\tau = \omega$ and special classes of spaces. (In these papers the notion of a co-mark and other related notions are not defined and the corresponding Containing Spaces, in the context of this book, are not explicitly determined).

Definition. Let $\mathbf{S}$ be an indexed collection of spaces. A restriction $\mathbf{Q}$ of $\mathbf{S}$ is said to be a *co-singleton* of $\mathbf{S}$ if for every $X \in \mathbf{S}$ the subset $\mathbf{Q}(X)$ of X is a singleton. A co-singleton $\mathbf{Q}$ of $\mathbf{S}$ is said to be *closed* if for every $X \in \mathbf{S}$, $\mathbf{Q}(X)$ is a closed subset of X. Obviously, if all elements of $\mathbf{S}$ are T_1-spaces, then any co-singleton of $\mathbf{S}$ is closed.

Definition. Let $\mathbf{S}$ be an indexed collection of spaces and

$$\mathbf{Q} \equiv \{\{q(X)\} : X \in \mathbf{S}\}$$

a co-singleton of $\mathbf{S}$. A co-mark

$$\mathbf{M} \equiv \{\{U_\delta^X : \delta \in \tau\} : X \in \mathbf{S}\}$$

of $\mathbf{S}$ is said to be $\mathbf{Q}$-*proper* if there exist two disjoint infinite subsets κ_0 and κ_1 of the set τ, whose union is τ, and a mapping θ of $\kappa_0 \times \kappa_0$ into κ_0 such that for every $X \in \mathbf{S}$ the following conditions are satisfied:

(a) $q(X) \in U^X_\delta$ if $\delta \in \kappa_0$,

(b) $q(X) \notin U^X_\delta$ if $\delta \in \kappa_1$, and

(c) for every $(\delta_0, \delta_1) \in \kappa_0 \times \kappa_0$ we have $U^X_{\delta_0} \cap U^X_{\delta_1} = U^X_{\theta(\delta_0,\delta_1)}$.

In this case the set $\{\kappa_0, \kappa_1, \theta\}$ is called the *defining set* of $\mathbf{M}$.

Notation. Let $\mathbf{S}$ be an indexed collection of spaces, $\mathbf{Q}$ a co-singleton of $\mathbf{S}$, $\mathbf{M}$ a $\mathbf{Q}$-proper co-mark of $\mathbf{S}$, and R an $(\mathbf{M}, \mathbf{Q})$-admissible family of equivalence relations on $\mathbf{S}$. We put

$$\mathrm{T}/\mathrm{T}|_{\mathbf{Q}} = (\mathrm{T} \setminus \mathrm{T}|_{\mathbf{Q}}) \cup \{\mathrm{T}|_{\mathbf{Q}}\},$$

where $\mathrm{T} \equiv \mathrm{T}(\mathbf{M}, \mathrm{R})$ and $\mathrm{T}|_{\mathbf{Q}} \equiv \mathrm{T}(\mathbf{M}|_{\mathbf{Q}}, \mathrm{R}|_{\mathbf{Q}})$.

The mapping p of T onto $\mathrm{T}/\mathrm{T}|_{\mathbf{Q}}$ for which $p(\mathbf{a}) = \mathbf{a}$ if $\mathbf{a} \notin \mathrm{T}|_{\mathbf{Q}}$ and $p(\mathbf{a}) = \mathrm{T}|_{\mathbf{Q}}$ if $\mathbf{a} \in \mathrm{T}|_{\mathbf{Q}}$ is called the *natural projection* of T onto $\mathrm{T}/\mathrm{T}|_{\mathbf{Q}}$.

10.5.2 Lemma. *Let $\mathbf{S}$ be an indexed collection of spaces, $\mathbf{Q}$ a co-singleton of $\mathbf{S}$,*

$$\mathbf{M} \equiv \{\{U^X_\delta : \delta \in \tau\} : X \in \mathbf{S}\}$$

a $\mathbf{Q}$-proper co-mark of $\mathbf{S}$ with the defining set $\{\kappa_0, \kappa_1, \theta\}$, and

$$\mathrm{R} \equiv \{\sim^s : s \in \mathcal{F}\}$$

an $(\mathbf{M}, \mathbf{Q})$-admissible family of equivalence relations on $\mathbf{S}$. Then, the set

$$\mathrm{B}^{\mathrm{T}/\mathrm{T}|_{\mathbf{Q}}} \equiv \{p(U) : U = U^{\mathrm{T}}_\delta(\mathbf{S}) \in \mathrm{B}^{\mathrm{T}}_{\kappa_0}\} \cup \mathrm{B}^{\mathrm{T}}_{\kappa_1}$$

is a base for a topology on the set $\mathrm{T}/\mathrm{T}|_{\mathbf{Q}}$. (We note that in the above relation the set $\mathbf{S}$ is considered as an element of the set $\mathrm{C}(\mathrm{R})$.)

Proof. Since the union of all elements of $\mathrm{B}^{\mathrm{T}/\mathrm{T}|_{\mathbf{Q}}}$ is the set $\mathrm{T}/\mathrm{T}|_{\mathbf{Q}}$ it suffices to prove that if $\mathbf{a} \in \mathrm{T}/\mathrm{T}|_{\mathbf{Q}}$, $V_0, V_1 \in \mathrm{B}^{\mathrm{T}/\mathrm{T}|_{\mathbf{Q}}}$, and $\mathbf{a} \in V_0 \cap V_1$, then there exists $W \in \mathrm{B}^{\mathrm{T}/\mathrm{T}|_{\mathbf{Q}}}$ such that $\mathbf{a} \in W \subset V_0 \cap V_1$. Let $\mathbf{a} \in \mathrm{T}/\mathrm{T}|_{\mathbf{Q}}$, $V_0, V_1 \in \mathrm{B}^{\mathrm{T}/\mathrm{T}|_{\mathbf{Q}}}$, and $\mathbf{a} \in V_0 \cap V_1$.

First we suppose that $V_0, V_1 \in \mathrm{B}^{\mathrm{T}}_{\kappa_1}$. Then, $\mathbf{a} \neq \mathrm{T}|_{\mathbf{Q}} \in \mathrm{T}/\mathrm{T}|_{\mathbf{Q}}$. There exist $\delta_0, \delta_1 \in \kappa_1$, $s_0, s_1 \in \mathcal{F}$, $\mathbf{H} \in \mathrm{C}(\sim^{s_0})$, and $\mathbf{L} \in \mathrm{C}(\sim^{s_1})$ such that

$V_0 = U^{\mathrm{T}}_{\delta_0}(\mathbf{H})$ and $V_1 = U^{\mathrm{T}}_{\delta_1}(\mathbf{L})$. There exist $\delta \in \tau$ and $\mathbf{E} \in \mathrm{C(R)}$ such that $\mathbf{a} \in U^{\mathrm{T}}_{\delta}(\mathbf{E}) \subset U^{\mathrm{T}}_{\delta_0}(\mathbf{H}) \cap U^{\mathrm{T}}_{\delta_1}(\mathbf{L})$. Obviously $\delta \in \kappa_1$ and, therefore, we can put $W = U^{\mathrm{T}}_{\delta}(\mathbf{E})$.

Now, we suppose that $V_0 \in \{p(U) : U \in \mathrm{B}^{\mathrm{T}}_{\kappa_0}\}$ and $V_1 \in \mathrm{B}^{\mathrm{T}}_{\kappa_1}$. In this case, $\mathbf{a} \neq \mathrm{T}|_{\mathbf{Q}} \in \mathrm{T}/\mathrm{T}|_{\mathbf{Q}}$. There exist $\delta_0 \in \kappa_0$, $\delta_1 \in \kappa_1$, and $\mathbf{H} \in \mathrm{C}(\sim^s)$ for some $s \in \mathcal{F}$ such that $V_0 = p(U^{\mathrm{T}}_{\delta_0}(\mathbf{S}))$ and $V_1 = U^{\mathrm{T}}_{\delta_1}(\mathbf{H})$. Since $\mathbf{a} \neq \mathrm{T}|_{\mathbf{Q}}$, $\mathbf{a} \in U^{\mathrm{T}}_{\delta_0}(\mathbf{S}) \cap U^{\mathrm{T}}_{\delta_1}(\mathbf{H})$. Then, there exist $\delta \in \tau$ and $\mathbf{L} \in \mathrm{C(R)}$ such that $\mathbf{a} \in U^{\mathrm{T}}_{\delta}(\mathbf{L}) \subset U^{\mathrm{T}}_{\delta_0}(\mathbf{S}) \cap U^{\mathrm{T}}_{\delta_1}(\mathbf{H})$. Obviously, $\delta \in \kappa_1$. Therefore, setting $W = U^{\mathrm{T}}_{\delta}(\mathbf{L})$ we have $\mathbf{a} \in W \subset V_0 \cap V_1$.

Finally, we suppose that $V_0, V_1 \in \{p(U) : U \in \mathrm{B}^{\mathrm{T}}_{\kappa_0}\}$. In this case, it suffices to prove that $V_0 \cap V_1 \equiv W \in \mathrm{B}^{\mathrm{T}/\mathrm{T}|_{\mathbf{Q}}}$. There exist $\delta_0, \delta_1 \in \kappa_0$ such that $V_0 = p(U^{\mathrm{T}}_{\delta_0}(\mathbf{S}))$ and $V_1 = p(U^{\mathrm{T}}_{\delta_1}(\mathbf{S}))$. Since the mapping θ is independent of the elements of $\mathbf{S}$, for every $Y \in \mathbf{S}$ we have $U^{Y}_{\delta} = U^{Y}_{\delta_0} \cap U^{Y}_{\delta_1}$, where $\delta = \theta(\delta_0, \delta_1) \in \kappa_0$. This fact implies that $U^{\mathrm{T}}_{\delta}(\mathbf{S}) = U^{\mathrm{T}}_{\delta_0}(\mathbf{S}) \cap U^{\mathrm{T}}_{\delta_1}(\mathbf{S})$ and, therefore, $p(U^{\mathrm{T}}_{\delta}(\mathbf{S})) = p(U^{\mathrm{T}}_{\delta_0}(\mathbf{S})) \cap p(U^{\mathrm{T}}_{\delta_1}(\mathbf{S})) = W$, which means that the set $W \in \mathrm{B}^{\mathrm{T}/\mathrm{T}|_{\mathbf{Q}}}$. □

Definition. The set $\mathrm{T}/\mathrm{T}|_{\mathbf{Q}}$ with the topology defined in Lemma 10.5.2 is called a *quotient Containing Space* or more precisely the $\mathbf{Q}$-*quotient Containing Space* (*corresponding to the indexed collection* $\mathbf{S}$ *of spaces*) and is denoted also by $\mathrm{T}/\mathrm{T}|_{\mathbf{Q}}$.

We note that in general the space $\mathrm{T}/\mathrm{T}|_{\mathbf{Q}}$ is not a T_0-space and its topology is not the quotient topology. (About the quotient topology see for example [114]).

10.5.3 Proposition. *Suppose that* $\mathbf{S}$ *is an indexed collection of (connected) spaces;* $\mathbf{Q}$ *is a closed co-singleton of* $\mathbf{S}$*;* $\mathbf{M}$ *is a* $\mathbf{Q}$*-proper co-mark of* $\mathbf{S}$*; and* R *is an* $(\mathbf{M}, \mathbf{Q})$*-admissible family of equivalence relations on* $\mathbf{S}$*. Then, the* $\mathbf{Q}$*-quotient Containing Space* $\mathrm{T}/\mathrm{T}|_{\mathbf{Q}}$ *is a (connected)* T_0*-space of weight* $\leq \tau$ *such that for every* $X \in \mathbf{S}$ *the mapping* $p \circ e^{X}_{\mathrm{T}}$ *is an embedding of* X *into* $\mathrm{T}/\mathrm{T}|_{\mathbf{Q}}$*, where* p *is the natural projection of* T *onto* $\mathrm{T}|_{\mathbf{Q}}$ *and* e^{X}_{T} *is the natural embedding of* X *into* T.

Proof. Let $\{\kappa_0, \kappa_1, \theta\}$ be the defining set of $\mathbf{M}$. Suppose that $\mathbf{Q}(X) = \{q(X)\}$, $X \in \mathbf{S}$. By Lemma 10.5.2 the weight of the space $\mathrm{T}/\mathrm{T}|_{\mathbf{Q}}$ is $\leq \tau$. We prove that $\mathrm{T}/\mathrm{T}|_{\mathbf{Q}}$ is a T_0-space. Let $\mathbf{a}$ and $\mathbf{b}$ be two distinct points of $\mathrm{T}/\mathrm{T}|_{\mathbf{Q}}$. One of these points is distinct from $\mathrm{T}|_{\mathbf{Q}} \in \mathrm{T}/\mathrm{T}|_{\mathbf{Q}}$. Let $\mathbf{a} \neq \mathrm{T}|_{\mathbf{Q}}$ and $(x, X) \in \mathbf{a}$. Then, $x \neq q(X)$.

First we suppose that $\mathbf{b} = \mathrm{T}|_{\mathbf{Q}}$. Since the set $\{q(X)\}$ is closed in X there exists $\delta \in \tau$ such that $x \in U_\delta^X$ and $q(X) \notin U_\delta^X$, which means that $\delta \in \kappa_1$. Then, the open set $U_\delta^{\mathrm{T}}(\mathbf{S})$ of $\mathrm{T}/\mathrm{T}|_{\mathbf{Q}}$ contains $\mathbf{a}$ and does not contain the point $\mathbf{b}$.

Now, we suppose that $\mathbf{b} \neq \mathrm{T}|_{\mathbf{Q}}$. Since the space T is a T_0-space there exist $\delta \in \tau$ and $\mathbf{H} \in \mathrm{C}(\mathrm{R})$ such that the set $U_\delta^{\mathrm{T}}(\mathbf{H})$ contains one of the points $\mathbf{a}$ and $\mathbf{b}$ and does not contain the other. Without loss of generality we can suppose that $\mathbf{a} \in U_\delta^{\mathrm{T}}(\mathbf{H})$ and $q(X) \notin U_\delta^X$, that is, $\delta \in \kappa_1$. Therefore, the set $U_\delta^{\mathrm{T}}(\mathbf{H})$ is an open subset of $\mathrm{T}/\mathrm{T}|_{\mathbf{Q}}$ containing $\mathbf{a}$ and not containing $\mathbf{b}$. Thus, the space $\mathrm{T}/\mathrm{T}|_{\mathbf{Q}}$ is a T_0-space.

Furthermore, let $X \in \mathbf{S}$. It is easy to see that $p \circ e_{\mathrm{T}}^X$ is an one-to-one mapping of X into $\mathrm{T}/\mathrm{T}|_{\mathbf{Q}}$. The fact that this mapping is an embedding follows by the following relations:

$$(p \circ e_{\mathrm{T}}^X)(U_\delta^X) = p(U_\delta^{\mathrm{T}}(\mathbf{S})) \cap (p \circ e_{\mathrm{T}}^X)(X) \text{ and}$$

$$(p \circ e_{\mathrm{T}}^X)^{-1}(p(U_\delta^{\mathrm{T}}(\mathbf{S}))) = U_\delta^X \text{ for } \delta \in \kappa_0,$$

and

$$(p \circ e_{\mathrm{T}}^X)(U_\delta^X) = U_\delta^{\mathrm{T}}(\mathbf{H}) \cap (p \circ e_{\mathrm{T}}^X)(X) \text{ and}$$

$$(p \circ e_{\mathrm{T}}^X)^{-1}(U_\delta^{\mathrm{T}}(\mathbf{H})) = U_\delta^X \text{ for } \delta \in \kappa_1 \text{ and } \mathbf{H} \in \mathrm{C}(\mathrm{R}).$$

These relations are easy to prove.

Finally, if all elements of $\mathbf{S}$ are connected, then the space $\mathrm{T}/\mathrm{T}|_{\mathbf{Q}}$ is also connected since $\mathrm{T}/\mathrm{T}|_{\mathbf{Q}}$ is the union of the connected subsets $(p \circ e_{\mathrm{T}}^X)(X)$, $X \in \mathbf{S}$, and $\mathrm{T}|_{\mathbf{Q}}$ is the common point of these subsets. □

Let $I\!P$ be a class of connected spaces. Considering the universality problem for this class we try to find for every indexed collection $\mathbf{S}$ of elements of $I\!P$ a co-singleton $\mathbf{Q}$ of $\mathbf{S}$, a $\mathbf{Q}$-proper co-mark $\mathbf{M}$ of $\mathbf{S}$ and an $\mathbf{M}$-admissible family R of equivalence relations on $\mathbf{S}$ such that the $\mathbf{Q}$-quotient Containing Space $\mathrm{T}/\mathrm{T}|_{\mathbf{Q}}$, where $\mathrm{T} = \mathrm{T}(\mathbf{M}, \mathrm{R})$ and $\mathrm{T}|_{\mathbf{Q}} = \mathrm{T}(\mathbf{M}|_{\mathbf{Q}}, \mathrm{R}|_{\mathbf{Q}})$, is an element of $I\!P$. In this direction we can develop an approach, which is similar to that of saturated classes. However, we restrict ourselves by considering only the case where $I\!P$ is the class of all regular connected spaces and prove that in this case the space $\mathrm{T}/\mathrm{T}|_{\mathbf{Q}}$ can be assummed to be regular.

Definition. Let $\mathbf{S}$ be an indexed collection of regular spaces, $\mathbf{Q}$ a co-singleton of $\mathbf{S}$, and

$$\mathbf{M} = \{\{U_\delta^X : \delta \in \tau\} : X \in \mathbf{S}\}$$

a $\mathbf{Q}$-proper co-mark of $\mathbf{S}$ with defining set $\{\kappa_0, \kappa_1, \theta\}$. The co-mark $\mathbf{M}$ is said to be $\mathbf{Q}$-*regular* if there exists a mapping ϑ of κ_0 into itself such that for every $X \in \mathbf{S}$ we have

$$\mathrm{Cl}_X(U_{\vartheta(\delta)}^X) \subset U_\delta^X,\ \delta \in \kappa_0.$$

In this case the set $\{\kappa_0, \kappa_1, \theta, \vartheta\}$ is called the *defining set* of $\mathbf{M}$.

The proof of the following lemma can be carried out using the methods of Lemmas 5.1.2 and 5.1.3.

10.5.4 Lemma. *Let* $\mathbf{S}$ *be an indexed collection of regular spaces and* $\mathbf{Q} \equiv \{\{q(X)\} : X \in \mathbf{S}\}$ *a co-singleton of* $\mathbf{S}$*. Then, there exists a* $\mathbf{Q}$*-regular co-mark, which is an initial co-mark of* $\mathbf{S}$ *corresponding to the class of all regular spaces.* □

10.5.5 Proposition. *Let* $\mathbf{S}$ *be an indexed collection of regular (connected) spaces and* $\mathbf{Q}$ *a co-singleton of* $\mathbf{S}$*. Then, for every* $\mathbf{Q}$*-regular co-mark* $\mathbf{M}$ *of* $\mathbf{S}$*, which is an initial co-mark corresponding to the class of all regular spaces, and an* $(\mathbf{M}, \mathbf{Q})$*-admissible family* R *of equivalence relations on* $\mathbf{S}$*, which is an initial family corresponding to the co-mark* $\mathbf{M}$ *and the class of all regular spaces, the* $\mathbf{Q}$*-quotient Containing Space* $\mathrm{T}/\mathrm{T}|_{\mathbf{Q}}$*, where* $\mathrm{T} \equiv \mathrm{T}(\mathbf{M}, \mathrm{R})$ *and* $\mathrm{T}|_{\mathbf{Q}} = \mathrm{T}(\mathbf{M}|_{\mathbf{Q}}, \mathrm{R}|_{\mathbf{Q}})$*, is a regular (connected) space. (Therefore, in the class of all regular connected spaces there exist universal elements.)*

Proof. Let $\mathbf{M}$ be a $\mathbf{Q}$-regular co-mark of $\mathbf{S}$, which is an initial co-mark corresponding to the class of all regular spaces, and R an $(\mathbf{M}, \mathbf{Q})$-admissible family, which is an initial family corresponding to the co-mark $\mathbf{M}$ and the class of all regular spaces. Then, the space T is a regular space. Since $\mathbf{Q}$ is a closed co-singleton by Proposition 10.5.3, the $\mathbf{Q}$-quotient space $\mathrm{T}/\mathrm{T}|_{\mathbf{Q}}$ is a T_0-space of weight $\leq \tau$. By construction of $\mathbf{Q}$-quotient spaces, this space is regular at all points of its subset $\mathrm{T} \setminus \mathrm{T}|_{\mathbf{Q}}$. Therefore, it suffices to prove that $\mathrm{T}/\mathrm{T}|_{\mathbf{Q}}$ is regular at the point $\mathrm{T}|_{\mathbf{Q}}$, that is, for every $\delta \in \kappa_0$ there exists $\varepsilon \in \kappa_0$ such that

$$\mathrm{Cl}_{\mathrm{T}/\mathrm{T}|_{\mathbf{Q}}}(p(U_\varepsilon^{\mathrm{T}}(\mathbf{S}))) \subset p(U_\delta^{\mathrm{T}}(\mathbf{S})).$$

This relation can be proved by a similar manner as the corresponding relation in the proof of Proposition 2.1.6. □

10.5.6 Problems. 1. Generalize the construction of **Q**-quotient Containing Spaces for an indexed collection **S** of spaces for more complicated restrictions **Q** of **S**.

2. Generalize the construction of a quotient Containing Space for an indexed collection **S** of spaces considering simultaneously τ many co-singletons of **S**.

3. Generalize the construction of a quotient Containing Space for an indexed collection **S** consisting of countable spaces considering simultaneously countable many co-singletons of **S**.

In the theory of separable metrizable continua (that is, compact connected Hausdorff spaces) and related spaces, the existence (or not existence) of universal elements for different classes of (one-dimensional) spaces is proved. Such classes are considered for example in the following papers: [374], [436], [357], [313], [275], [240], [368], [276], [37], [414], [227], [255], [149], [178], [257], [61], [256], [287], [226], [263], [182], [183], [288], [197], [210], [264], [265], [297], [198], [261], [94], [141], [186], [142], [54], [126], [127], [140], [215], [60] (a survey paper), [89], [443].

In the bibliography there are also many papers related to connectedness in more general classes of spaces than completely regular or even regular spaces. Below we shall consider some of them.

The papers [167], [120], [165], [87], [105], [442], [134], [49], [196], and [418] are related to spaces on which all continuous functions (or mappings into a fixed space) are constant. Such spaces are necessarily connected and not completely regular while they may be regular. In some of the above papers it is proved that every regular (respectively, Hausdorff) space is contained in a regular (respectively, Hausdorff) space on which all continuous functions are constant. Since in the class of all regular spaces there exist universal elements, this implies that in the class of all spaces on which all continuous functions are constant, there exist universal elements. Similarly, in the class of all Hausdorff spaces on which all continuous functions are constant, there are no universal element because in the class of all Hausdorff spaces there are no universal elements.

In the papers [6], [112], [352], [419], [434], and [435] the property of spaces to be connectifiable is considered. (A Hausdorff space X is said to be *connectifiable* if X can be embedded topologically as a dense subset into a connected Hausdorff space.)

In the papers [222], [221], [441], [108], [260], [284], [204], [205], [429], [82], [214], and [155] connected spaces with a dispersion point are considered. (A point a of a connected space X is said to be a *dispersion* point if the set $X \setminus \{a\}$ does not contain connected subsets.) If such a point exists, then it is unique (see [221]). A space with a dispersion point may be separable and metrizable. (See for example the well known space constructed in [222].)

In the papers [86], [206], [207], [208], and [417] strongly rigid spaces are considered. (A space X is said to be *strongly rigid* if each continuous mapping of X into itself is either identical or constant.) Such a space, which may be separable metrizable (see [86]), is necessarily connected.

Some of the above mentioned papers are related also to countable connected spaces. There are many other papers related to such spaces. For example: [423], [154], [36], [50], [146], [259], [147], [360], [393], [220], [277], [285], [202], [205], [238], [24], [409], [358], [359], [410], [48], [323], [415], [416]. We note that a countable connected space is not necessarily regular and has the property that all continuous functions on it are constant.

Usually the connectedness of a space is considered with some additional properties of the space. Such properties, except the above mentioned, are for example the following:

(a) The property of a space to be Urysohn. (A space X is said to be *Urysohn* if for every distinct point x and y of X there exist open neighbourhoods U and V of x and y, respectively, such that $\mathrm{Cl}_X(U) \cap \mathrm{Cl}_X(V) = \emptyset$.)

(b) The property of a space to be almost regular. (A space X is said to be *almost regular* if there exists a dense open subset U of X such that the space X is regular at all points of U.)

(c) The property of a space to be locally connected, and so on.

Other aspects of connectedness, distinct from the above, are considered in the papers: [395], [396], [326], [283], [119], [11], [353], [397], [160], [156], [157], [24], [379], [158], [111], and [153].

The method of quotient Containing Spaces can be applied to some classes consisting of connected spaces, which are considered in the above mentioned papers. Below, using this method we give some results concerning the universality problems for such classes. It is probable that some possible generalizations of the quotient Containing Spaces can be used for further investigation (from the universality point of view) of connectedness in general classes of spaces.

Notation. Let Y be a space. We denote by $\mathbb{P}$(rs-Y) the class of all regular spaces X with the property that each continuous mapping of X into Y is constant.

If instead of regular spaces we consider Hausdorff spaces, then the corresponding class will be denoted by $\mathbb{P}$(Hs-Y).

Below, using the quotient Containing Spaces we give another proof (distinct from the above mentioned) that in the class $\mathbb{P}$(rs-Y) there exist universal elements.

10.5.7 Proposition. *For every space* Y *in the class* $\mathbb{P}$(rs-Y) *there exist universal elements.*

Proof. Let Y be a space. Denote by **S** an indexed collection of elements of $\mathbb{P}$(rs-Y) such that for every element Z of $\mathbb{P}$(rs-Y) there exists an element X of **S**, which is homeomorphic to Z. Let $\mathbf{Q} \equiv \{q(X) : X \in \mathbf{S}\}$ be a co-singleton of **S**. By Lemma 10.5.4 there exists a **Q**-regular co-mark **M** of **S**, which is an initial co-mark corresponding to the class of all regular spaces. Denote by R an $(\mathbf{M}, \mathbf{Q})$-admissible family of equivalence relations on **S**. Then, by Propositions 10.5.3 and 10.5.5 the **Q**-quotient Containing Space $\mathrm{T}/\mathrm{T}|_{\mathbf{Q}}$, where $\mathrm{T} = \mathrm{T}(\mathbf{M}, \mathrm{R})$ and $\mathrm{T}|_{\mathbf{Q}} = \mathrm{T}(\mathbf{M}|_{\mathbf{Q}}, \mathrm{R}|_{\mathbf{Q}})$, is a regular space containing topologically all elements of $\mathbb{P}$(rs-Y). Therefore, to prove the proposition it suffices to prove that each continuous mapping of $\mathrm{T}/\mathrm{T}|_{\mathbf{Q}}$ into Y is constant.

Let f be a continuous mapping of $\mathrm{T}/\mathrm{T}|_{\mathbf{Q}}$ into Y. Consider an arbitrary point **a** of $\mathrm{T}/\mathrm{T}|_{\mathbf{Q}}$ distinct from $\mathrm{T}|_{\mathbf{Q}}$. Let (x, X) be an element of **a**. Denote by f^X the restriction of f at the subset $(p \circ i^X_{\mathrm{T}})(X)$ of $\mathrm{T}/\mathrm{T}|_{\mathbf{Q}}$, where p is the natural projection of T onto $\mathrm{T}/\mathrm{T}|_{\mathbf{Q}}$. Since this subset is homeomorphic to X we have that the mapping f^X is constant, that is $f^X(x) = f^X(q(X))$. This equality implies that $f(\mathbf{a}) = f(\mathrm{T}|_{\mathbf{Q}})$, which means that the mapping f is constant. □

Definition. Let $\mathbb{K}$ be a class of subsets. A space X is said to be $\mathbb{K}$-*connectifiable* if there exists a connected space $\widetilde{X}$ such that $(X, \widetilde{X}) \in \mathbb{K}$. (Therefore, a space X is connectifiable if and only if X is $\mathbb{K}$-connectifiable, where $\mathbb{K}$ is the class of all pairs $(Y, \widetilde{Y})$ such that $\widetilde{Y}$ is a connected Hausdorff space and Y is a dense subset of $\widetilde{Y}$.)

Notations. Let $\mathbb{K}$ be a class of subsets. We denote by $\mathbb{P}(\mathbb{K}$-cnfl) the class of all spaces, which are $\mathbb{K}$-connectifiable.

If $I\!\!K$ is the class of subsets consisting of all pairs $(Y, \widetilde{Y})$ such that $\widetilde{Y}$ is a regular space and Y is: (a) a dense subset of $\widetilde{Y}$, (b) an open subset of $\widetilde{Y}$, and (c) a closed subset of $\widetilde{Y}$, then the class $I\!\!P(I\!\!K\text{-cnfl})$ is denoted respectively by:
(a) $I\!\!P$(rs-cnfl),
(b) $I\!\!P$(rs-open-cnfl), and
(c) $I\!\!P$(rs-closed-cnfl).

If in the above notations instead of regular spaces we consider completely regular spaces, then we define the classes:
(a) $I\!\!P$(crs-cnfl),
(b) $I\!\!P$(crs-open-cnfl), and
(c) $I\!\!P$(crs-closed-cnfl), respectively.

Similarly, if we consider Hausdorff spaces, then we obtain the classes:
(a) $I\!\!P$(Hs-cnfl),
(b) $I\!\!P$(Hs-open-cnfl), and
(c) $I\!\!P$(Hs-closed-cnfl), respectively.

10.5.8 Proposition. *In the class* $I\!\!P$(rs-cnfl) *there exist universal elements.*

Proof. Let $\mathbf{S}$ be an indexed collection of elements of $I\!\!P$(rs-cnfl) such that for every element Y of $I\!\!P$(rs-cnfl) there exists an element X of $\mathbf{S}$, which is homeomorphic to Y. For every element X of $\mathbf{S}$ we consider a regular connected space, denoted by $\widetilde{X}$, which contains topologically the space X as a dense subset. Without loss of generality we can suppose that X is a dense subset of $\widetilde{X}$. We put

$$\widetilde{\mathbf{S}} \equiv \{\widetilde{X} : X \in \mathbf{S}\}.$$

Let $\mathbf{Q}$ be co-singleton of $\mathbf{S}$. Then, $\mathbf{Q}$ can be also considered as a closed co-singleton of $\widetilde{\mathbf{S}}$. By Lemma 10.5.4 there exists a $\mathbf{Q}$-regular co-mark

$$\widetilde{\mathbf{M}} \equiv \{\{U_\delta^{\widetilde{X}} : \delta \in \tau\} : \widetilde{X} \in \widetilde{\mathbf{S}}\}$$

of $\widetilde{\mathbf{S}}$ with a defining set $\{\kappa_0, \kappa_1, \theta, \vartheta\}$, which is an initial co-mark corresponding to the class of all regular spaces. Then, it is to verify that

$$\mathbf{M} \equiv \{\{U_\delta^X \equiv X \cap U_\delta^{\widetilde{X}} : \delta \in \tau\} : X \in \mathbf{S}\}$$

is a $\mathbf{Q}$-regular co-mark of $\mathbf{S}$ with the same defining set.

Now, we denote by

$$\widetilde{\mathrm{R}} \equiv \{\sim_*^s : s \in \mathcal{F}\}$$

an $(\widetilde{\mathbf{M}}, \mathbf{Q})$-admissible family of equivalence relations on $\widetilde{\mathbf{S}}$ satisfying the following conditions:

(a) $\widetilde{\mathrm{R}}$ is an initial family of equivalence relations on $\widetilde{\mathbf{S}}$ corresponding to the co-mark $\widetilde{\mathbf{M}}$ and the class of all regular spaces and

(b) the family

$$\mathrm{R} \equiv \{\sim^s : s \in \mathbf{F}\}$$

of equivalence relations on $\mathbf{S}$ such that for every $s \in \mathcal{F}$ two elements X and Y of $\mathbf{S}$ are $\sim^s$-equivalent if and only if $\widetilde{X}$ and $\widetilde{Y}$ are $\sim_*^s$-equivalent, is an $(\mathbf{M}, \mathbf{Q})$-admissible family of equivalence relations on $\mathbf{S}$.

Consider the Containing Spaces $\widetilde{\mathrm{T}} = \mathrm{T}(\widetilde{\mathbf{M}}, \widetilde{\mathrm{R}})$ and $\mathrm{T} = \mathrm{T}(\mathbf{M}, \mathrm{R})$. We prove that T can be considered as a dense subset of $\widetilde{\mathrm{T}}$. For this purpose, to every point $\mathbf{a}$ of T we correspond the point $\widetilde{\mathbf{a}}$ of $\widetilde{\mathrm{T}}$ defined as follows: if $(x, X) \in \mathbf{a}$, then $\widetilde{\mathbf{a}}$ is the point of $\widetilde{\mathrm{T}}$ containing the pair $(x, \widetilde{X})$. It is easy to see that:

(a) the point $\widetilde{\mathbf{a}}$ is independent of the element (x, X) of $\mathbf{a}$ and

(b) if $\mathbf{a}_0$ and $\mathbf{a}_1$ are distinct points of T, then $\widetilde{\mathbf{a}}_0$ and $\widetilde{\mathbf{a}}_1$ are distinct points of $\widetilde{\mathrm{T}}$.

Thus, the correspondence $\mathbf{a} \to \widetilde{\mathbf{a}}$ is an one-to-one mapping $i_{\widetilde{\mathrm{T}}}^{\mathrm{T}}$ of T into $\widetilde{\mathrm{T}}$. To prove that $i_{\widetilde{\mathrm{T}}}^{\mathrm{T}}$ is an homeomorphism we consider an element $\widetilde{\mathbf{H}}$ of $\mathrm{C}(\widetilde{\mathrm{R}})$ and denote by $\mathbf{H}$ the element of $\mathrm{C}(\mathrm{R})$ consisting of all elements X of $\mathbf{S}$ such that $\widetilde{X} \in \widetilde{\mathbf{S}}$. Then, it is easy to verify that for every $\delta \in \tau$ we have

$$U_\delta^{\widetilde{\mathrm{T}}}(\widetilde{\mathbf{H}}) \cap i_{\widetilde{\mathrm{T}}}^{\mathrm{T}}(\mathrm{T}) = i_{\widetilde{\mathrm{T}}}^{\mathrm{T}}(U_\delta^{\mathrm{T}}(\mathbf{H})), \tag{10.5.1}$$

where $U_\delta^{\widetilde{\mathrm{T}}}(\widetilde{\mathbf{H}})$ is the corresponding element of the standard base for the open subsets of $\widetilde{\mathrm{T}}$ and $U_\delta^{\mathrm{T}}(\mathbf{H})$ is the corresponding element of the standard base for the open subsets of T. This relation implies that $i_{\widetilde{\mathrm{T}}}^{\mathrm{T}}$ is an homeomorphism such that $i_{\widetilde{\mathrm{T}}}^{\mathrm{T}}(\mathrm{T})$ is a dense subset of $\widetilde{\mathrm{T}}$. Thus, T can be considered as a dense subset of $\widetilde{\mathrm{T}}$ identifying each point $\mathbf{a}$ of T with the point $\widetilde{\mathbf{a}}$ of $\widetilde{\mathrm{T}}$.

On the other hand, we can consider the $\mathbf{Q}$-quotient Containing Spaces $\widetilde{\mathrm{T}}/\widetilde{\mathrm{T}}|_{\mathbf{Q}}$ and $\mathrm{T}/\mathrm{T}|_{\mathbf{Q}}$, where $\mathrm{T}|_{\mathbf{Q}} = \mathrm{T}(\mathbf{M}|_{\mathbf{Q}}, \mathrm{R}|_{\mathbf{Q}})$ and $\widetilde{\mathrm{T}}|_{\mathbf{Q}} = \mathrm{T}(\widetilde{\mathbf{M}}|_{\mathbf{Q}}, \widetilde{\mathrm{R}}|_{\mathbf{Q}})$.

By Proposition 10.5.5, $\widetilde{\mathrm{T}}/\widetilde{\mathrm{T}}|_{\mathbf{Q}}$ is a regular connected space and by Proposition 10.5.3 the space $\mathrm{T}/\mathrm{T}|_{\mathbf{Q}}$ contains topologically all elements of $\mathbf{S}$ and, therefore, all elements of $\mathbb{P}$(rs-cnfl). Thus, to prove the proposition it suffices to prove that the space $\mathrm{T}/\mathrm{T}|_{\mathbf{Q}}$ can be considered as a dense subset of $\widetilde{\mathrm{T}}/\widetilde{\mathrm{T}}|_{\mathbf{Q}}$. However, this follows immediately by the fact that T is a dense subset of $\widetilde{\mathrm{T}}$ and by relation (10.5.1). □

Using some modifications of the notion of a $\mathbf{Q}$-regular co-mark we can prove the following two propositions.

10.5.9 Proposition. *In the classes*

$$\mathbb{P}(\text{rs-open-cnfl}) \text{ and } \mathbb{P}(\text{rs-closed-cnfl})$$

there exist universal elements. □

10.5.10 Proposition. *In the classes*

$$\mathbb{P}(\text{crs-cnfl}), \mathbb{P}(\text{crs-open-cnfl}) \text{ and } \mathbb{P}(\text{crs-closed-cnfl})$$

there exist universal elements. □

In [434] it is shown that any T_1-space without isolated points (respectively, any T_0-space) is contained topologically as a dense subset in a connected T_1-space (respectively, T_0-space). Therefore, in the class of all spaces, which can be embedded in a connected T_1-space as a dense subset, there are no universal elements because this class coincides with the class of all T_1-spaces in which there are no universal elements.

On the other hand, there exists a completely regular space which can not be embedded into a regular connected space as a dense subset while it can be embedded into a Hausdorff such space (see [112]).

The following two proposition are proved similarly to Proposition 10.5.7 and Proposition 10.5.8, respectively.

10.5.11 Proposition. *The classes* $\mathbb{P}$(Hs-Y) *and* $\mathbb{P}(\mathrm{T}_1\text{-Y})$ *have the* 2^{τ}*-universality property for every space* Y. □

10.5.12 Proposition. *The classes*

$$\mathbb{P}(\text{Hs-cnfl}) \ \mathbb{P}(\text{Hs-open-cnfl})\mathbb{P}(\text{Hs-closed-cnfl})$$

have the 2^τ-universality property. □

Notation. We denote by $\mathbb{P}$(rs-disp) the class of all regular spaces with a dispersion point. If instead of regular spaces we consider T_1-spaces or Hausdorff spaces or completely regular spaces, then we obtain the classes $\mathbb{P}(\mathrm{T}_1\text{-disp})$, and $\mathbb{P}$(Hs-disp), and $\mathbb{P}$(crs-disp), respectively.

10.5.13 Proposition. *The classes*

$$\mathbb{P}(\mathrm{T}_1\text{-disp}),\ \mathbb{P}(\text{Hs-disp}),\ \mathbb{P}(\text{rs-disp}),\ \textit{and}\ \mathbb{P}(\text{crs-disp})$$

have the 2^τ-universality property.

Proof. We prove the proposition for the class $\mathbb{P}$(rs-disp). For the other classes the proof is similar.

Let $\mathbf{S}$ be an indexed collection of elements of $\mathbb{P}$(rs-disp) such that $|\mathbf{S}| \leq 2^\tau$. For every $X \in \mathbf{S}$ denote by $q(X)$ the dispersion point of X and consider the co-singleton $\mathbf{Q} \equiv \{q(X) : X \in \mathbf{S}\}$ of $\mathbf{S}$. Let $\mathbf{M}$ be a $\mathbf{Q}$-regular initial co-mark of $\mathbf{S}$ corresponding to the class of all regular spaces. Since the cardinality of $\mathbf{S}$ is less than or equal to 2^τ there exists an $(\mathbf{M}, \mathbf{Q})$-admissible family R of equivalence relations on $\mathbf{S}$ such that each class of the equivalence relation $\sim^\infty_{\mathrm{R}}$ contains only one element of $\mathbf{S}$. Then, the $\mathbf{Q}$-quotient Containing Space $\mathrm{T}/\mathrm{T}|_{\mathbf{Q}}$, where $\mathrm{T} = \mathrm{T}(\mathbf{M}, \mathrm{R})$ and $\mathrm{T}|_{\mathbf{Q}} = \mathrm{T}(\mathbf{M}|_{\mathbf{Q}}, \mathrm{R}|_{\mathbf{Q}})$, is a regular space containing topologically all elements of $\mathbf{S}$.

To prove the proposition it suffices to prove that the point $\mathrm{T}|_{\mathbf{Q}}$ is a dispersion point of $\mathrm{T}/\mathrm{T}|_{\mathbf{Q}}$, that is the subspace $(\mathrm{T}/\mathrm{T}|_{\mathbf{Q}}) \setminus \{\mathrm{T}|_{\mathbf{Q}}\}$ of $\mathrm{T}/\mathrm{T}|_{\mathbf{Q}}$ does not contain non-trivial connected sets. By the construction of the $\mathbf{Q}$-quotient Containing Space $(\mathrm{T}/\mathrm{T}|_{\mathbf{Q}}) \setminus \{\mathrm{T}|_{\mathbf{Q}}\}$ is also a subspace of T. If X and Y are two distinct elements of $\mathbf{S}$, then there exists an element $\mathbf{H}$ of $\mathrm{C}(\mathrm{R})$ such that $X \in \mathbf{H}$ and $Y \notin \mathbf{H}$. The set $\mathrm{T}(\mathbf{H}) = \mathrm{T}(\mathbf{M}, \mathrm{R}, \mathbf{H})$ is an open and closed subset of T. Therefore, the set $U \equiv \mathrm{T}(\mathbf{H}) \setminus \mathrm{T}|_{\mathbf{Q}}$ is an open and closed subset $\mathrm{T} \setminus \mathrm{T}|_{\mathbf{Q}}$. Then, U is also an open and closed subset of $(\mathrm{T}/\mathrm{T}|_{\mathbf{Q}}) \setminus \{\mathrm{T}|_{\mathbf{Q}}\}$. It is easy to see that U contains the set $(p \circ i^X_{\mathrm{T}})(X \setminus \{q(X)\})$ and does not contain the set $(p \circ i^Y_{\mathrm{T}})(Y \setminus \{q(Y)\})$. This observation means that if there exists a non-trivial connected set of $(\mathrm{T}/\mathrm{T}|_{\mathbf{Q}}) \setminus \{\mathrm{T}|_{\mathbf{Q}}\}$, then this set must be contained in the set $(p \circ i^X_{\mathrm{T}})(X \setminus \{q(X)\})$ for some $X \in \mathbf{S}$. However, the set $(p \circ i^X_{\mathrm{T}})(X \setminus \{q(X)\})$ is homeomorphic to $X \setminus \{q(X)\}$, which does not contain non-trivial connected subsets. Thus, $(\mathrm{T}/\mathrm{T}|_{\mathbf{Q}}) \setminus \{\mathrm{T}|_{\mathbf{Q}}\}$ does not contain non-trivial connected subsets. □

10.5.14 Problems. 1. Is it true that there are no universal elements in the class of all completely regular spaces (respectively, of all regular or of all Hausdorff spaces) with the dispersion point?

2. Is it true that there are no universal elements in the class of all strongly rigid completely regular spaces (respectively, regular or Hausdorff spaces)?

3. Is it true that the class of all strongly rigid completely regular spaces (respectively, regular or Hausdorff spaces) have the property of 2^τ-universality?

4. Is it true that there are no universal elements in the class of all countable connected Hausdorff spaces (respectively, almost regular or Urysohn spaces)?

5. Is it true that the class of all countable connected Hausdorff spaces (respectively, of all countable connected almost regular spaces or of all countable connected Urysohn spaces or of all countable connected T_1-spaces) has the property of 2^τ-universality?

We note that in [237] it is proved that in the class of all Hausdorff countable spaces with the first axiom of separability, there exists a universal element which is a connected space.

Definition. Let $I\!P$ be a class of spaces (respectively, of regular spaces). Let also κ_0 and κ_1 be subsets of τ such that $|\kappa_0| = |\kappa_1| = \tau$ and $\kappa_0 \cup \kappa_1 = \tau$, θ a mapping of $\kappa_0 \times \kappa_0$ into κ_0, and ϑ a mapping of κ_0 into κ_0. A space X is called $(\kappa_0, \kappa_1, \theta)$-*quotient space of* $I\!P$ (respectively, $(\kappa_0, \kappa_1, \theta, \vartheta)$-*quotient space of* $I\!P$) if there exist an indexed collection $\mathbf{S}$ of elements of $I\!P$, a closed co-singleton $\mathbf{Q}$ of $\mathbf{S}$, a $\mathbf{Q}$-proper (respectively, a $\mathbf{Q}$-regular) co-mark $\mathbf{M}$ of $\mathbf{S}$, and an $(\mathbf{M}, \mathbf{Q})$-admissible family R of equivalence relations on $\mathbf{S}$ such that X is homeomorphic to the $\mathbf{Q}$-quotient Containing Space $\mathrm{T}/\mathrm{T}|_{\mathbf{Q}}$.

10.5.15 Problems. Let $I\!P$ be a saturated class of spaces with closed points (respectively, a saturated class of regular spaces) and κ_0, κ_1, θ, and ϑ as in the above definition.

1. Is there a universal element in the class of all $(\kappa_0, \kappa_1, \theta)$-quotient (respectively, $(\kappa_0, \kappa_1, \theta, \vartheta)$-quotient) spaces of $I\!P$?

2. Is it true that the class of all $(\kappa_0, \kappa_1, \theta)$-quotient (respectively, of all $(\kappa_0, \kappa_1, \theta, \vartheta)$-quotient) spaces of $I\!P$ have the property of 2^τ-universality?

3. Under what conditions on $I\!P$ do the above questions 1 and 2 have affirmative answers?

10.6 Universal sets parametrised by regular zero-dimensional spaces

In this section we consider the connection between (complete) saturated classes of subsets and the so-called universal sets parametrised by topological spaces. Such sets are considered in the papers: [236], [324], [218], [83], [84], [137], [139], [138]. For its relation to Descriptive Set Theory see for example the books [217] and [290].

Notation. Suppose that $\mathrm{R} \equiv \{\sim^s : s \in \mathcal{F}\}$ is an admissible family of equivalence relations on an indexed collection $\mathbf{S}$ of spaces and $\mathbf{H}$ is an $\sim^s$-equivalence class for some $s \in \mathcal{F}$. We denote by $\mathbf{H}(\sim^\infty_\mathrm{R})$ the set of all elements of $\mathrm{C}(\sim^\infty_\mathrm{R})$, which are contained in $\mathbf{H}$. The set of all sets of the form $\mathbf{H}(\sim^\infty_\mathrm{R})$ is denoted by $\mathrm{B}(\sim^\infty_\mathrm{R})$.

Since the union of all elements of $\mathrm{B}(\sim^\infty_\mathrm{R})$ is the set $\mathrm{C}(\sim^\infty_\mathrm{R})$, Lemma 1.2.7(1) implies the following lemma.

10.6.1 Lemma. *Let* R *be an admissible family of equivalence relations on an indexed collection* $\mathbf{S}$ *of spaces. Then, the set* $\mathrm{B}(\sim^\infty_\mathrm{R})$ *is a base for a topology on the set* $\mathrm{C}(\sim^\infty_\mathrm{R})$ *and the corresponding topological space is a regular zero-dimensional (in the sence of dimension ind) space.* □

Definition. By the *space* $\mathrm{C}(\sim^\infty_\mathrm{R})$ we mean the set $\mathrm{C}(\sim^\infty_\mathrm{R})$ equipped with the topology for which the set $\mathrm{B}(\sim^\infty_\mathrm{R})$ is a base for the open subsets.

Now, we consider Containing Spaces $\mathrm{T}(\mathbf{M}, \mathrm{R})$ for special indexed collections $\mathbf{S}$ of spaces and co-marks $\mathbf{M}$ of $\mathbf{S}$.

Definition. A set $\{X\}$ consisting of a space X only is called a *space-singleton.*

Let $\mathbf{S}$ be an indication of a space-singleton $\{X\}$ (that is, each element of $\mathbf{S}$ coincides with the space X). A co-mark $\mathbf{M}$ of $\mathbf{S}$ is said to be *straight* if there exists a mark $M^X \equiv \{V^X_\delta : \delta \in \tau\}$ of X such that for every $Y \in \mathbf{S}$ we have $\mathbf{M}(Y) = M^X$ (that is, if $\mathbf{M}(Y) = \{V^Y_\delta : \delta \in \tau\}$, then $V^X_\delta = V^Y_\delta$ for every $\delta \in \tau$). We say also that the straight co-mark $\mathbf{M}$ is *generated* by the mark M^X of X.

A co-mark $\mathbf{M}$ of $\mathbf{S}$ is said to be *semi-straight* if it is a co-extension of a straight co-mark of $\mathbf{S}$.

10.6.2 Proposition. *Let* $\mathbf{S}$ *be an indication of a space-singleton* $\{X\}$ *and* $\mathbf{M}$ *a semi-straight co-mark of* $\mathbf{S}$. *Then, for any* $\mathbf{M}$*-admissible family* R *of equivalence relations on* $\mathbf{S}$ *the Containing Space* $\mathrm{T}(\mathbf{M}, \mathrm{R})$ *is homeomorphic to the product of the spaces* X *and* $\mathrm{C}(\sim_{\mathrm{R}}^{\infty})$.

Proof. Suppose that $\mathbf{M}$ is a co-extension of a straight co-mark $\mathbf{M}_0$ generated by a mark $M^X \equiv \{V_\delta^X : \delta \in \tau\}$ of X and let

$$\mathbf{M} = \{\{U_\delta^Y : \delta \in \tau\} : Y \in \mathbf{S}\} \text{ and}$$

$$\mathbf{M}_0 = \{\{V_\delta^Y : \delta \in \tau\} : Y \in \mathbf{S}\}.$$

We note that $V_\delta^Y = V_\delta^Z$ for every $Y, Z \in \mathbf{S}$ and $\delta \in \tau$. Denote by θ an indicial mapping from $\mathbf{M}_0$ to $\mathbf{M}$.

Let $\mathrm{R} \equiv \{\sim^s : s \in \mathcal{F}\}$ be an $\mathbf{M}$-admissible family of equivalence relations on $\mathbf{S}$. We need to prove that the Containing Space $\mathrm{T} \equiv \mathrm{T}(\mathbf{M}, \mathrm{R})$ is homeomorphic to $X \times \mathrm{C}(\sim_{\mathrm{R}}^{\infty})$. For this purpose we define a mapping h of the product space $X \times \mathrm{C}(\sim_{\mathrm{R}}^{\infty})$ onto the space T as follows: if $(x, \mathbf{E}) \in X \times \mathrm{C}(\sim_{\mathrm{R}}^{\infty})$, then we set $h(x, \mathbf{E}) = \mathbf{a}$, where $\mathbf{a}$ is the point of T containing the pair (x, Y) for some $Y \in \mathbf{E}$. The point $\mathbf{a}$ is independent of the element Y of $\mathbf{E}$. Indeed, let Z be another element of $\mathbf{E}$. We must prove that the pairs (x, Y) and (x, Z) are $\sim_{\mathbf{M}}^{\mathrm{R}}$-equivalent, that is $(x, Y), (x, Z) \in \mathbf{a}$. Since $X, Y \in \mathbf{E}$ it suffices to prove that if x belongs to U_δ^Y for some $\delta \in \tau$, then x belongs also to U_δ^Z. Let $x \in U_\delta^Y$, $\delta \in \tau$. There exists $\varepsilon \in \theta(\tau)$ such that $x \in U_\varepsilon^Y \subset U_\delta^Y$. Since $U_\varepsilon^Y = V_\eta^Y = V_\eta^Z = U_\varepsilon^Z$, where $\eta = \theta^{-1}(\varepsilon)$, we have $x \in U_\varepsilon^Z$. On the other hand, if $x \notin U_\delta^Z$, then the set U_ε^Z is not contained in U_δ^Z, which contradicts the fact that X and Y are elements of $\mathbf{E}$.

It is easy to verify that h is an one-to-one mapping of $X \times \mathrm{C}(\sim_{\mathrm{R}}^{\infty})$ onto T. Moreover, for every $Z \in \mathbf{S}$, $\delta \in \tau$, $s \in \mathcal{F}$, and $\mathbf{H} \in \mathrm{C}(\sim^s)$ the following relation is true:

$$h(V_\delta^X \times \mathbf{H}(\sim_{\mathrm{R}}^{\infty})) = U_{\theta(\delta)}^{\mathrm{T}}(\mathbf{H}).$$

Since h is one-to-one and onto, this relation implies that h and h^{-1} are continuous, that is, h is an homeomorphism of the space $X \times \mathrm{C}(\sim_{\mathrm{R}}^{\infty})$ onto the space T. □

Definition. Let $I\!P$ be a class of subsets and X and Y two spaces. A subset $\widetilde{Q}$ of $X \times Y$ is said to be a *strongly* $I\!P$*-universal set for* X *parametrised by the space* Y if the following conditions are satisfied:

(a) The pair $(\widetilde{Q}, X \times Y)$ belongs to $I\!P$.
(b) For every $y \in Y$ we have $(Q_y, X) \in I\!P$, where

$$Q_y = \{x \in X : (x, y) \in \widetilde{Q}\}.$$

(c) For every subset Q of X for which $(Q, X) \in I\!P$ there exists a unique point y of Y such that $Q = Q_y$.

If in the above condition (c) we omit the word unique, then we obtain the definition of a *$I\!P$-universal set for X parametrised by the space Y*

Notation. Let $I\!P$ be a class of subsets and X an arbitrary space. We denote by $I\!P(X)$ the family of all pairs (Q, X) such that $(Q, X) \in I\!P$. Therefore, $I\!P(X) \neq \emptyset$ if and only if X belongs to the space-component of $I\!P$.

We denote by $\mathbf{S}(X, I\!P)$ the indexed collection

$$\{X_Q : (Q, X) \in I\!P(X)\},$$

where X_Q is a copy of the space X. We note that $|\mathbf{S}(X, I\!P)| = |I\!P(X)|$.

The restriction $\mathbf{Q}$ of $\mathbf{S}(X, I\!P)$ for which $\mathbf{Q}(X_Q) = Q$ will be denoted by $\mathbf{Q}(X, I\!P)$.

10.6.3 Proposition. *Let $I\!P$ be a saturated class of subsets and X an element of the space-component of $I\!P$. Then, the relation $|I\!P(X)| \leq 2^\tau$ is a necessary and sufficient condition for the existence of a strongly $I\!P$-universal set for X parametrized by a regular zero-dimensional (in the sense of dimension* ind*) space of weight $\leq \tau$.*

Proof. Suppose that there exists a regular zero-dimensional space Y of weight $\leq \tau$ and a subset $\widetilde{Q}$ of $X \times Y$, which is strongly $I\!P$-universal for X. By the definition of strongly $I\!P$-universal sets it follows that there exists a one-to-one correspondence between the points of the space Y and the elements of $I\!P(X)$, that is $|Y| = |I\!P(X)|$. On the other hand, $|Y| \leq 2^\tau$. Thus, $|I\!P(X)| \leq 2^\tau$.

Conversely, suppose that $|I\!P(X)| \leq 2^\tau$. Let $\mathbf{M}_0$ be a straight co-mark of $\mathbf{S}(X, I\!P)$; $\mathbf{M}$ an initial co-mark of $\mathbf{S}(X, I\!P)$ corresponding to the class $I\!P$, which is a co-extension of $\mathbf{M}_0$ (and, therefore, $\mathbf{M}$ is a semi-stright co-mark); and R_0 an initial family of equivalence relation on $\mathbf{S}(X, I\!P)$ corresponding to the co-mark $\mathbf{M}$, the restriction $\mathbf{Q}(X, I\!P)$ of $\mathbf{S}(X, I\!P)$, and the class $I\!P$. Since the cardinality of $\mathbf{S}(X, I\!P)$ is less than or equal to 2^τ there exists an admissible family $\mathrm{R}_1 \equiv \{\sim_1^s : s \in \mathcal{F}\}$ of equivalence relations on $\mathbf{S}(X, I\!P)$ such that if Y and Z are distinct elements of $\mathbf{S}(X, I\!P)$, then Y and Z are not $\sim_1^s$-equivalent for some $s \in \mathcal{F}$. (We note that this family may not be $\mathbf{M}$-admissible.) Denote by R an admissible family of equivalence relations on $\mathbf{S}(X, I\!P)$, which is simultaneously a final refinement of R_0 and R_1. Then, we can consider the space $\mathrm{T} \equiv \mathrm{T}(\mathbf{M}, \mathrm{R})$. Since the co-mark $\mathbf{M}$ is semi-straight we can consider also the homeomorphism h of the space $X \times \mathrm{C}(\sim_{\mathrm{R}}^{\infty})$ onto the space T constructed in Proposition 10.6.2.

We put $Y = \mathrm{C}(\sim_{\mathrm{R}}^{\infty})$ and $\widetilde{Q} = h^{-1}(\mathrm{T}|_{\mathbf{Q}})$, where $\mathbf{Q} = \mathbf{Q}(X, I\!P)$ and $\mathrm{T}|_{\mathbf{Q}} = \mathrm{T}(\mathbf{M}|_{\mathbf{Q}}, \mathrm{R}|_{\mathbf{Q}})$. By Lemma 10.6.1, Y is a zero-dimensional space of weight $\leq \tau$. By the construction of the family R_1 the correspondence

$$Q \to \{X_Q\},\ Q \in I\!P(X),$$

is a one-to-one correspondence between the elements of $\mathbf{S}(X, I\!P)$ and the elements of the set $\mathrm{C}(\sim_{\mathrm{R}_1}^{\infty})$. Since the family R is a final refinement of R_1 the same correspondence is also a one-to-one correspondence between the elements of $\mathbf{S}(X, I\!P)$ and the elements of $\mathrm{C}(\sim_{\mathrm{R}}^{\infty})$.

We prove that $\widetilde{Q}$ is a strongly $I\!P$-universal set for X parametrized by Y. Since $I\!P$ is a saturated class by the construction of $\mathbf{M}$ and R it follows that $(\mathrm{T}|_{\mathbf{Q}}, \mathrm{T}) \in I\!P$, that is $(\widetilde{Q}, X \times Y) \in I\!P$. Thus, condition (a) of the definition of strongly $I\!P$-universal sets is satisfied.

Let $y \in Y$. Then, there exists a subset Q of X such that $(Q, X) \in I\!P$ and $y = \{X_Q\}$. Consider the natural embedding $e_{\mathrm{T}}^{X_Q}$ of the space X_Q into the space T. It is easy to verify that if $x \in X$, then we have

$$(h^{-1} \circ e_{\mathrm{T}}^{X_Q})(x) = (x, y) \in X \times Y.$$

Moreover, a point x of X belongs to Q if and only if $e_{\mathrm{T}}^{X_Q}(x) \in \mathrm{T}|_{\mathbf{Q}}$, that is if and only if (x, y) belongs to $\widetilde{Q}$. By the above considerations it follows that conditions (b) and (c) of the definition of strongly $I\!P$-universal sets are also satisfied. This completes the proof of the proposition. □

10.6.4 Proposition. *Let $I\!P$ be a complete saturated class of subsets and X an element of the space-component of $I\!P$. Then, there exists a strongly $I\!P$-universal set for X parametrized by a regular zero-dimensional (in the sense of dimensiion* ind*) space of weight $\leq \tau$.*

Proof. Let $\mathbf{M}_0$ be a straight co-mark of $\mathbf{S}(X, I\!P)$. Let also

$$\mathbf{M} \equiv \{\{U_\delta^{X_Q} : \delta \in \tau\} : X_Q \in \mathbf{S}(X, I\!P)\}$$

be an initial co-mark of $\mathbf{S}(X, I\!P)$ corresponding to the class $I\!P$, which is a co-extension of $\mathbf{M}_0$ (and, therefore, $\mathbf{M}$ is a semi-stright co-mark), and $\mathrm{R} \equiv \{\sim^s : s \in \mathcal{F}\}$ an initial family of equivalence relation on $\mathbf{S}(X, I\!P)$ corresponding to the co-mark $\mathbf{M}$, the restriction $\mathbf{Q}(X, I\!P)$, and the class $I\!P$. Without loss of generality, we can suppose that the restriction $\mathbf{Q}(X, I\!P)$ is $(\mathbf{M}, \mathrm{R})$-complete. Thus, we can consider the space $\mathrm{T} \equiv \mathrm{T}(\mathbf{M}, \mathrm{R})$. Since the co-mark $\mathbf{M}$ is semi-straight we can consider also the homeomorphism h of the space $X \times \mathrm{C}(\sim_{\mathrm{R}}^\infty)$ onto the space T constructed in Proposition 10.6.2.

As in Theorem 10.6.3 we put $Y = \mathrm{C}(\sim_{\mathrm{R}}^\infty)$ and $\widetilde{Q} = h^{-1}(\mathrm{T}|_{\mathbf{Q}})$, where $\mathbf{Q} = \mathbf{Q}(X, I\!P)$ and $\mathrm{T}|_{\mathbf{Q}} = \mathrm{T}(\mathbf{M}|_{\mathbf{Q}}, \mathrm{R}|_{\mathbf{Q}})$. By Lemma 10.6.1, Y is a zero-dimensional space of weight $\leq \tau$.

Now, we prove that each element of $\mathrm{C}(\sim_{\mathrm{R}}^\infty)$ consists of an element of $\mathbf{S}(X, I\!P)$. Indeed, suppose that X_{Q_0} and X_{Q_1} are distinct elements of $\mathbf{S}(X, I\!P)$ (that is, Q_0 and Q_1 are distinct subsets of X) belonging to an element of $\mathrm{C}(\sim_{\mathrm{R}}^\infty)$, which means that X_{Q_0} and X_{Q_1} are $\sim^s$-equivalent for all $s \in \mathcal{F}$. Let x be a point of one of the subsets Q_0 and Q_1 which does not belong to the other. Suppose, for example, that $x \in Q_0$ and $x \notin Q_1$. Consider the point $\mathbf{a}$ of $\mathrm{T}|_{\mathbf{Q}}$ containing the pair (x, X_{Q_0}). We show that $(x, X_{Q_1}) \in \mathbf{a}$. Since X_{Q_0} and X_{Q_1} are $\sim^s$-equivalent for every $s \in \mathcal{F}$ we need only to show that $x \in U_\delta^{X_{Q_0}}$ for some $\delta \in \tau$ if and only if $x \in U_\delta^{X_{Q_1}}$. This fact can be prove as the corresponding result in the proof of Proposition 10.6.2. Thus, $(x, X_{Q_1}) \in \mathbf{a}$. Since the restriction $\mathbf{Q}$ is $(\mathbf{M}, \mathrm{R})$-complete we have $x \in Q_1$, which contradicts the choice of x. Therefore, each element of $\mathrm{C}(\sim_{\mathrm{R}}^\infty)$ consists of an element of $\mathrm{S}(X, I\!P)$.

The rest of the proof is similar to that of Proposition 10.6.3. □

Propositions 10.6.3 and 10.6.4 imply the following consequence.

10.6.5 Corollary. *Let $I\!P$ be a complete saturated class of subsets and X an element of the space-component of $I\!P$. Then, $|I\!P(X)| \leq 2^\tau$.* □

In the connections with the Natural Problem 2.3 and Natural Problem 3.2 of [83] we give the following two corollaries.

10.6.6 Corollary. *Let ν be an infinite cardinal with $\nu \leq \tau$, $\gamma \in \nu^+$, Φ a ν-directing mapping, and $I\!F$ a complete saturated class of subsets.*

Then, for each space X there exists a strongly $\mathbb{P}(\gamma, \nu, \Phi, \mathbb{F})$-universal set parametrized by a regular zero-dimensional (in the sense of dimension ind*) space (of weight $\leq \tau$).* (About the definition of the complete class $\mathbb{P}(\gamma, \nu, \Phi, \mathbb{F})$ of subsets see Section 2.4.) □

10.6.7 Corollary. *Suppose that $\tau = \omega$ and $\alpha \in \omega^+$. Then, for every space X with a countable base (metrizable or not) there exists a strongly $\Sigma(\alpha)$-universal (respectively, a strongly $\Pi(\alpha)$-universal) set for X parametrized by a separable metrizable zero-dimensional space, where $\Sigma(\alpha)$ (respectively, $\Pi(\alpha)$) is the class of subsets consisting of all pairs (Q, X) such that Q is a subset of the space X of additive (respectively, of multiplicative) class α.* □

10.6.8 Problems. 1. Is it possible to characterize the regular zero-dimensional spaces considered in Propositions 10.6.3 and 10.6.4?

2. Under what conditions on the (complete) saturated class $\mathbb{P}$ of subsets and the space X of Propositions 10.6.3 and 10.6.4 can we consider the space 2^τ or the space $(\mathrm{D}(\tau))^\omega$ (where $\mathrm{D}(\tau)$ is the discrete space of weight τ) instead of the corresponding regular zero-dimensional space of weight $\leq \tau$ (replacing if it necessary the strong $\mathbb{P}$-universality by $\mathbb{P}$-universality)?

3. Is it possible to prove Corollary 10.6.5 directly by the definition of a complete saturated class of subsets?

10.7 Compactifications and extensions of Containing Spaces

Remark 1. As we already have mentioned, the class of all Hausdorff compact spaces is not saturated. This fact follows by the next more general result.

10.7.1 Proposition. *Any class which consists of non-empty compact spaces is not saturated.*

Proof. Let $\mathbb{P}$ be a class consisting of non-empty compact spaces. By Proposition 7.2.3 it follows that if:

(a) $\mathbf{M}_0$ and $\mathbf{M}_1$ are two co-marks of an indexed collection $\mathbf{S}$ of spaces such that $\mathbf{M}_1$ is a co-extension of $\mathbf{M}_0$ and

(b) R_0 and R_1 are respectively $\mathbf{M}_0$-admissible and $\mathbf{M}_1$-admissible families of equivalence relations on $\mathbf{S}$ such that R_1 is a final refinement of R_0, then there exists a continuous mapping of the space $\mathrm{T}(\mathbf{M}_1, \mathrm{R}_1)$ onto the space $\mathrm{T}(\mathbf{M}_0, \mathrm{R}_0)$. This fact implies that if $\mathrm{T}(\mathbf{M}_0, \mathrm{R}_0)$ is not compact, then $\mathrm{T}(\mathbf{M}_1, \mathrm{R}_1)$ also is not compact. Therefore, in order to prove that the class $I\!\!P$ is not saturated it suffices to find an indexed collection $\mathbf{S}$ of elements of $I\!\!P$, a co-mark $\mathbf{M}$ of $\mathbf{S}$, and an $\mathbf{M}$-admissible family R of equivalence relations on $\mathbf{S}$ such that the Containing Space $\mathrm{T}(\mathbf{M}, \mathrm{R})$ is not compact.

Let X be an element of $I\!\!P$ and $\mathbf{S} \equiv \{X_i : i \in \omega\}$ an indication of the singleton $\{X\}$. Consider an arbitrary co-mark

$$\mathbf{M} \equiv \{\{U_\delta^{X_i} : \delta \in \tau\} : X_i \in \mathbf{S}\},$$

of $\mathbf{S}$ such that $U_\delta^{X_i} = X$ if $\delta = i \in \omega \subset \tau$ and $U_\delta^{X_i} = \emptyset$ if $\delta \in \omega \setminus \{i\} \subset \tau$. Then, it is easy to verify that the set $\mathrm{C}(\sim_{\mathrm{R}_\mathbf{M}}^\infty)$ is infinite.

Let $\mathrm{R} \equiv \{\sim^s : s \in \mathcal{F}\}$ be an $\mathbf{M}$-admissible family of equivalence relations on $\mathbf{S}$. Then, the set $\mathrm{C}(\sim_\mathrm{R}^\infty)$ is also infinite. Suppose that for every element $\mathbf{E}$ of $\mathrm{C}(\sim_\mathrm{R}^\infty)$ there exists an element $\mathbf{H}_\mathbf{E}$ of $\mathrm{C}(\mathrm{R})$ such that the set $\mathbf{H}_\mathbf{E}(\sim_\mathrm{R}^\infty)$ coincides with the set $\mathbf{E}$. (About notation $\mathbf{H}_\mathbf{E}(\sim_\mathrm{R}^\infty)$ see Section 10.1.) In this case, it is easy to verify that the set

$$\{\mathrm{T}(\mathbf{M}, \mathrm{R}, \mathbf{H}_\mathbf{E}(\mathrm{C} \sim_\mathrm{R}^\infty)) : \mathbf{E} \in \mathrm{C}(\sim_\mathrm{R}^\infty)\}$$

is an infinite cover of the space $\mathrm{T} \equiv \mathrm{T}(\mathbf{M}, \mathrm{R})$ by disjoint non-empty open subsets, which means that the space T is not compact.

Therefore, without loss of generality, we can suppose that there exists an element $\mathbf{E}$ of $\mathrm{C}(\sim_\mathrm{R}^\infty)$ such that any element $\mathbf{H}$ of $\mathrm{C}(\mathrm{R})$ containing $\mathbf{E}$ contains also another element of $\mathrm{C}(\sim_\mathrm{R}^\infty)$ distinct from $\mathbf{E}$. We denote by $\mathbf{H}_{s,\mathbf{E}}$, $s \in \mathcal{F}$, the $\sim^s$-equivalence class containing the set $\mathbf{E}$.

Now, we define a family $\mathrm{R}' \equiv \{\sim'^{,s} : s \in \mathcal{F}\}$ of equivalence relations on $\mathbf{S}$ as follows. Let $s \in \mathcal{F} \setminus \{\emptyset\}$ and $X, Y \in \mathbf{S}$. Then, X and Y are considered to be $\sim'^{,s}$-equivalent if either $X, Y \notin \mathbf{E}$ and $X \sim^s Y$ or $X, Y \in \mathbf{E}$. It is easy to verify that R' is an admissible family, which is a final refinement of R such that: (a) the set $\mathbf{E}$ is an $\sim'^{,s}$-equivalence class for every $s \in \mathcal{F} \setminus \{\emptyset\}$ and (b) any $\sim^s$-equivalence class not containing $\mathbf{E}$ is a $\sim'^{,s}$-equivalence class.

Thus, we can consider the space $\mathrm{T}' \equiv \mathrm{T}(\mathbf{M}, \mathrm{R}')$. It is not difficult to prove that the set

$$\{\mathrm{T}(\mathbf{M}, \mathrm{R}', \mathbf{E})\} \cup \{\mathrm{T}(\mathbf{M}, \mathrm{R}') \setminus \mathrm{T}(\mathbf{M}, \mathrm{R}', \mathbf{H}_{s,\mathbf{E}}) : s \in \mathcal{F} \setminus \{\emptyset\}\},$$

is an open cover of T', which does not have any finite subcover. Thus, T' is not compact, which completes the proof of the proposition. □

10.7.2 Problem. Let $\mathbf{S}$ be an indexed collection of compact Hausdorff spaces and $\mathbf{M}$ a co-mark of $\mathbf{S}$. Under what conditions on $\mathbf{S}$ and $\mathbf{M}$ the Containing Space $\mathrm{T}(\mathbf{M}, \mathrm{R}_{\mathbf{M}})$ is a compact Hausdorff space?

The following proposition gives an answer to the above problem.

10.7.3 Proposition. *Let* $\mathbf{S}$ *be an indexed collection of compact Hausdorff spaces,* $\mathbf{M}$ *a co-mark of* $\mathbf{S}$*, which is an initial co-mark corresponding to the class of all regular spaces, and* $\mathrm{R} \equiv \{\sim^s : s \in \mathcal{F}\}$ *an* $\mathbf{M}$*-admissible family of equivalence relations on* $\mathbf{S}$*, which is an initial family corresponding to the co-mark* $\mathbf{M}$ *and the class of all regular spaces. (Therefore, the Containing Space* $\mathrm{T} \equiv \mathrm{T}(\mathbf{M}, \mathrm{R})$ *is a regular space.) Then,* T *is compact if and only if the space* $\mathrm{C}(\sim^{\infty}_{\mathrm{R}})$ *is compact.*

Proof. Let T be a compact space. We define a mapping p_{R} of T onto $\mathrm{C}(\sim^{\infty}_{\mathrm{R}})$ as follows. Let $\mathbf{a}$ be a point of T and $(x, X) \in \mathbf{a}$. Then, we set $p_{\mathrm{R}}(\mathbf{a}) = \mathbf{E}$, where $\mathbf{E}$ is the element of $\mathrm{C}(\sim^{\infty}_{\mathrm{R}})$ containing the space X. Obviously, p_{R} is well defined, that is $\mathbf{E}$ is independent of (x, X). It is easy to verify that p_{R} is a continuous mapping of T onto $\mathrm{C}(\sim^{\infty}_{\mathrm{R}})$. Therefore, $\mathrm{C}(\sim^{\infty}_{\mathrm{R}})$ is a compact Hausdorff space.

Conversely, let $\mathrm{C}(\sim^{\infty}_{\mathrm{R}})$ be a compact Hausdorff space. Suppose that c is an open cover of T. We need to prove that c contains a finite subcover. Without loss of generality we can suppose that the elements of c belong to the standard base for the open subsets of T.

Since all elements of $\mathbf{S}$ are compact, for every element $\mathbf{E}$ of $\mathrm{C}(\sim^{\infty}_{\mathrm{R}})$ the subspace $\mathrm{T}(\mathbf{E})$ of T is also compact. Therefore, there exists a finite subset

$$c_{\mathbf{E}} \equiv \{U^{\mathrm{T}}_{\delta_0}(\mathbf{H}_0), ..., U^{\mathrm{T}}_{\delta_k}(\mathbf{H}_k)\}$$

of c such that $\mathrm{T}(\mathbf{E})$ is contained in the union of elements of $c_{\mathbf{E}}$. Suppose that $\mathbf{H}_i$ is an $\sim^{s_i}$-equivalence class and let $s = \cup\{s_i : i \in \{0, ..., k\}\}$. Denote by $\mathbf{H}_{\mathbf{E}}$ the $\sim^s$-equivalence class containing $\mathbf{E}$. Then,

$$\{\mathbf{H}_{\mathbf{E}} : \mathbf{E} \in \mathrm{C}(\sim^{\infty}_{\mathrm{R}})\}$$

is an open cover of the space $\mathrm{C}(\sim^{\infty}_{\mathrm{R}})$. Let $\mathbf{E}_0, ..., \mathbf{E}_m$ be elements of $\mathrm{C}(\sim^{\infty}_{\mathrm{R}})$ such that

$$\mathrm{C}(\sim^{\infty}_{\mathrm{R}}) = \cup\{\mathbf{H}_{\mathbf{E}_i} : i \in \{0, ..., m\}\}.$$

Then, it is easy to verify that

$$\mathrm{T} = \cup\{c_{\mathbf{E}_i} : i \in \{0, ..., m\}\},$$

which means that c contains a finite subcover. Thus, T is compact. □

10.7.4 Problems. 1. Let **S** be an indexed collection of spaces, **M** a co-mark of **S**, and R an **M**-admissible family of equivalence relations on **S** such that the space $\mathrm{T} \equiv \mathrm{T}(\mathbf{M}, \mathrm{R})$ is regular. Suppose that: (a) for every $\mathbf{E} \in \mathrm{C}(\sim^{\infty}_{\mathrm{R}})$ the subspace $\mathrm{T}(\mathbf{E})$ of T is completely metrizable and (b) the space $\mathrm{C}(\sim^{\infty}_{\mathrm{R}})$ is completely metrizable. Is the space T completely metrizable?

In particular,

2. Let $\tau = \omega$. Let also **S** be an indexed collection of spaces, **M** a co-mark of **S**, and R an **M**-admissible family of equivalence relations on **S** such that the space $\mathrm{T} \equiv \mathrm{T}(\mathbf{M}, \mathrm{R})$ is regular. Suppose that: (a) for every $\mathbf{E} \in \mathrm{C}(\sim^{\infty}_{\mathrm{R}})$ the subspace $\mathrm{T}(\mathbf{E})$ of T is separable completely metrizable and (b) the space $\mathrm{C}(\sim^{\infty}_{\mathrm{R}})$ is separable completely metrizable. Is the space T separable completely metrizable?

Definition. An admissible family R of equivalence relations on an indexed collection of spaces is said to be *compact* if the space $\mathrm{C}(\sim^{\infty}_{\mathrm{R}})$ is compact.

It is said that a class $I\!P$ of spaces has a *compact resolution* if for some *full* indexed collection **S** of elements of $I\!P$ (that is, each element of $I\!P$ is homeomorphic to an element of **S**) and some co-mark **M** of **S** there exists a compact **M**-admissible family R of equivalence relations on **S**.

In this case, the Containing Space $\mathrm{T}(\mathbf{M}, \mathrm{R})$ is called a *compact resolution* of $I\!P$.

Such a resolution is called *universal* if the Containing Space $\mathrm{T}(\mathbf{M}, \mathrm{R})$ belongs to $I\!P$.

In the above definitions instead of the word compact we shall use other words characterizing a space. For example: normal, separable completely metrizable and so on.

Notation. Let $m \in \omega$. We denote by:

(a) $\mathbb{P}(\text{s-b}^{\text{n}}\text{-Ind} \leq m)$ (see Section 5.2) the class of all (completely regular) spaces X such that $\text{s-b}^{\text{n}}\text{-Ind}(X) \leq m$,

(b) $\mathbb{P}_n(\text{s-b}^{\text{n}}\text{-Ind} \leq m)$ the class of all normal spaces, which are elements of $\mathbb{P}(\text{s-b}^{\text{n}}\text{-Ind} \leq m)$, and

(c) $\mathbb{P}_c(\text{Ind} \leq m)$ (see Section 6.4) the class of all compact spaces, which are elements of $\mathbb{P}(\text{s-b}^{\text{n}}\text{-Ind} \leq m)$. (Obviously, this class coincides with the class of all compact spaces X such that $\text{Ind}(X) \leq m$.)

If in the above definitions we replace dimension Ind by dimension dim, then we obtain the classes:

(a) $\mathbb{P}(\text{s-b}^{\text{n}}\text{-dim} \leq m)$,

(b) $\mathbb{P}_n(\text{s-b}^{\text{n}}\text{-dim} \leq m)$, and

(c) $\mathbb{P}_c(\dim \leq m)$, respectively.

10.7.5 Problems. 1. Have the class of all completely regular spaces universal compact resolution?

2. Have the class of all compact Hausdorff spaces universal compact resolution?

3. Let $m \in \omega$. Which of the classes

$$\mathbb{P}(\text{s-b}^{\text{n}}\text{-Ind} \leq m),\ \mathbb{P}_n(\text{s-b}^{\text{n}}\text{-Ind} \leq m),\ \mathbb{P}_c(\text{Ind} \leq m) \text{ and}$$

$$\mathbb{P}(\text{s-b}^{\text{n}}\text{-dim} \leq m),\ \mathbb{P}_n(\text{s-b}^{\text{n}}\text{-dim} \leq m),\ \mathbb{P}_c(\dim \leq m)$$

have universal compact resolutions?

10.7.6 Problems. 1. Has the class of all completely metrizable spaces a completely metrizable resolution?

2. Let $\tau = \omega$. Has the class of all separable completely metrizable spaces a universal separable completely metrizable resolution?

Definition. Let:

(a) $\mathbf{S} \equiv \{X_\lambda : \lambda \in \Lambda\}$ and $\mathbf{S}' \equiv \{X'_\lambda : \lambda \in \Lambda\}$ be two indexed collections of spaces,

(b) $\mathbf{M} \equiv \{\{U_\delta^{X_\lambda} : \delta \in \tau\} : X_\lambda \in \mathbf{S}\}$ and $\mathbf{M}' \equiv \{\{U_\delta^{X'_\lambda} : \delta \in \tau\} : X'_\lambda \in \mathbf{S}\}$ be co-marks of $\mathbf{S}$ and $\mathbf{S}'$, respectively, and

(c) $\mathrm{R} \equiv \{\sim^s : s \in \mathcal{F}\}$ an $\mathbf{M}$-admissible family of equivalence relations on $\mathbf{S}$ and $\mathrm{R}' \equiv \{\sim'^{,s} : s \in \mathcal{F}\}$ an $\mathbf{M}'$-admissible family of equivalence relations on $\mathbf{S}'$.

It is said that the Containing space $\mathrm{T}(\mathbf{M}', \mathrm{R}')$ is a *proper extension* of the Containing Space $\mathrm{T}(\mathbf{M}, \mathrm{R})$ if there exist a one-to-one mapping ϑ of Λ into Λ' and an embedding $i(\lambda, \vartheta(\lambda))$ of X_λ into $X'_{\vartheta(\lambda}$ (In this case, we shall identify the point x of X_λ with the point $i(\lambda, \vartheta(\lambda))(x)$ and consider the space X_λ as a subset of the space $X'_{\vartheta(\lambda)}$.) such that the following conditions are satisfied:

(1) $U_\delta^{X_\lambda} = U_\delta^{X'_{\vartheta(\lambda)}} \cap X_\lambda$ for every $\lambda \in \Lambda$ and $\delta \in \tau$.

(2) For every element $\mathbf{H}'$ of $\mathrm{C}(\sim'^{,s})$, $s \in \mathcal{F}$, the set of all elements X_λ of $\mathbf{S}$ such that $X'_{\vartheta(\lambda)} \in \mathbf{H}'$, is an element of $\mathrm{C}(\sim^s)$.

10.7.7 Problems. 1. Let $\mathbf{S}$ be an indexed collection of completely regular spaces, $\mathbf{M}$ an initial co-mark of $\mathbf{S}$ corresponding to the class of all completely regular spaces, and R an initial family of equivalence relations on $\mathbf{S}$ corresponding to $\mathbf{M}$ and the class of all completely regular spaces. Under what conditions on $\mathbf{S}$, $\mathbf{M}$, and R does the Containing Space $\mathrm{T}(\mathbf{M}, \mathrm{R})$ have a proper compactification (that is, a proper extension, which is a compact space)?

2. Consider the question 1 in the case where all elements of $\mathbf{S}$ are compact spaces.

10.8 Remarks and problems to Chapters 4-9

Remark 1. For separable metrizable spaces the notions of a countable-dimensional space and of a strongly countable-dimensional space were defined in [174]: a separable metrizable space is countable-dimensional (respectively, strongly countable-dimensional) if it can be represented as a countable union of its finite-dimensional (respectively, closed finite-dimensional) subspaces. Locally finite-dimensional spaces were defined and studied in [438] and [439]. A separable metrizable countable-dimensional space can be characterized also as follows:

(1) A space X is countable-dimensional if and only if X is a countable union of its zero-dimensional subspaces.

This characterization follows by the fact that an n-dimensional space, $n \in \omega$, can be represented as a union of $n+1$ zero-dimensional subsets (see [425] - for compact spaces, [411], [172]).

(2) A space X is countable-dimensional if X can be represented as a union of zero-dimensional subsets X_i, $i \in \omega$, such that $X_i \subset X_{i+1}$ (see [412], [298]).

(3) A space X is countable-dimensional if and only if X can be represented as a union of its subspaces X_i, $i \in \omega$, such that $X_i \subset X_{i+1}$ and $\mathrm{ind}(X_i) \leq i$.

This characterization follows by the fact that the dimension of the union of $i+1$ zero-dimensional spaces is $\leq i$ (see [425] - for compact spaces, [411], [172]).

The countable-dimensionality and stongly countable-dimensionality are considered by many authors for more general classes of spaces. However, in this case as a base dimension it is considered dimension dim (that is, a space X is considered to be zero-dimensional or finite-dimensional if $\dim(X) = 0$ or $\dim(X) = n \in \omega$, respectively). The most natural such class is the class of metrizable spaces. (For this class the dimension dim coincides with the dimension Ind (see [211] and [289]).)

For the class of metrizable spaces the above given characterizations (1)–(3) also coincide (see for example [117]). However, for the class of completely regular or normal spaces the characterization (1) is not valid. In [121] (see also [122], [364]) (under the Continuum Hypothesis) a perfectly normal compact space X is constructed such that: (a) $\dim(X) = 2$, that is X is countable-dimensional and (b) X can not be represented as a countable union of its zero-dimensional subspaces.

Thus, considering characterizations (1)–(3) as definitions in more general classes than the class of metrizable spaces we define distinct types of countable-dimensionality. All of them can be considered as a generalizations of the countable-dimensionality in separable metrizable spaces.

In this book for the definition of the notions of countable-dimensionality, strongly countable-dimensionality, and locally finite-dimensionality it is used the dimension ind. This dimension is defined for any space and is convenient to use in the universality problems. However, instead of dimension ind we can consider some other dimensions (mensioned in Corollary 4.4.26).

The classes of spaces, which are (explicitly or implicitly) considered in corollaries of section 4.4, can be considered as generalizations (in different directions) of countable-dimensionality, strongly countable-dimensionality, and locally finite-dimensionality.

Chapter 4 contains all results of the paper [189]. (In this paper the notion of a saturated class of spaces is not defined.)

We note that the class of spaces considered in Corollary 4.4.7 for the case of closed subspaces and $\nu = n \in \omega$ (that is ν is a finite integer) does not coincides with the class of spaces of dimension $\leq \beta$. This follows by the fact that there exists (see [401] and [242]) a separable metrizable space X and two closed subspaces A and B of this space such that $\operatorname{ind}(X) = \omega + 1$ and $\operatorname{ind}(A) = \operatorname{ind}(B) = \omega$.

Also we note that the class of spaces considered in Corollary 4.4.8 for the case of closed subspaces and $\nu = \omega^+$ (and, therefore, $\omega^+ \leq \tau$) contains a separable metrizable space X with $\dim(X) = 1$ (see [345]). The same class contains all metrizable spaces if ν is the cardinal of continuum (see [21]). The class of spaces considered in Corollary 4.4.23 contains all metrizable spaces (see [384], [21], [298]).

The D-dimension (based on Ind) were introduced and studied in [163] and [164]. This dimension is considered also in the papers [225], [249], [251], [252], [317], [88], [64]. The D-dimension coincides (for metrizable and some more general spaces) with the dimension trker considered in particular in [115], where one can find more details.

The D-dimension considered in Chapter 4 is based on the dimension ind. Using Lemma 1 of [163] one can prove that as considered here, D-dimension coincides for separable metrizable spaces with that of [163].

10.8.1 Problems. 1. Find relations between the notions of countable-dimensionality, of strongly countable-dimensionality, and of locally finite-dimensionality which are based on distinct new dimension-like functions.

2. Is it true that Corollaries 4.4.16, 4.4.18, 4.4.19, 4.4.21, and 4.4.23 are not valid if in the formulations of these corollaries we replace the symbol "$\subset$" ("is a subset of") by "$=$" ("is equal to")?

3. Consider the above question 2 in the case where all spaces of the mentioned corollaries are separable metrizable.

4. Let $n \in \omega$. Is it true that there is no separable metrizable space T and its zero-dimensional subsets T_0,....,T_n with $T = T_0 \cup ... \cup T_n$ having the property that for every separable metrizable space X and its zero-dimensional subsets X_0,...,X_n with $X = X_0 \cup ... \cup X_n$ there exists an embedding i of X into T satisfying conditions:

$$X_0 = i^{-1}(T_0), ..., \; X_n = i^{-1}(T_n)?$$

(Compare this question with Corollary 4.4.25.) (We note that such a space T and zero-dimensional subsets exist if it is additionally assumed that all considered zero-dimensional subsets of T and X are of multiplicative class (or of additive class) α, where α is a fixed countable ordinal.)

5. What standard theorems concerning transfinite dimensions (see [115]) are valid for D-dimension considered in Chapter 4?

6. Is it possible to apply the method of Containing Spaces to uniform spaces?

7. Is it possible to apply the method of Containing Spaces to quasi-uniform spaces?

Remark 2. In Chapter 5 for our investigation we use the notion of a normal base for the closed (open) subsets of a completely regular space and construct the compactification corresponding to such a base. This compactification is known as the Wallman compactification (associated with the given normal base) and its construction now is a standard material of General Topology and can be found for example in the books [4] and [279]. The term "normal base" we take from [4]. In [135], where the Wallman compactifications were defined, it was used he term "semi-normal" and in [279] the term "Wallman base".

The well known Wallman extension (compactification) of a space (associated with the set of all closed subsets) was the first extension (compactification) constructed (see [432]) via closed subsets of a space (see also [114]). Compactifications constructed via bases of closed subsets of a space were discussed in [369], [370], [371]. The notion of a Wallman compactification of a space was defined in the paper [135]. In this paper the author also poses the problem if any compactification of a space is also a Wallmal compactification. Many papers were devoted to this problem. In [420] this problem is answered in the negative. More information about the Wallman compactifications can be found for example in [4] and [279].

Below we pose some problems concerning the defined in Chapter 5 dimension-like functions.

10.8.2 Problems. 1. What standard theorem of dimension theory are true for dimension-like functions s-b^n-Ind and s-b^n-dim defined in the class of all completely regular spaces?

2. Formulate the standard theorems of dimension theory (see [115]) for base dimension-like functions and consider which of them are true for the functions b^n-Ind and b^n-dim.

10.8.3 Problems. Let $\alpha \in \tau^+ \setminus \omega$.

1. Is a countable union of (s-b^n-Ind $\leq \alpha$)-uniform classes of spaces an (s-b^n-Ind $\leq \alpha$)-uniform class?

2. Is the union of τ many (s-b^n-Ind $\leq \alpha$)-uniform classes of spaces an (s-b^n-Ind $\leq \alpha$)-uniform class?

3. Consider the class $\mathbb{P}$(s-b^n-Ind $\leq \alpha$) of all completely regular spaces X such that s-b^n-Ind$(X) \leq \alpha$. What is the minimum of cardinal ν such that the class $\mathbb{P}$(s-b^n-Ind $\leq \alpha$) is a union of ν many (s-b^n-Ind $\leq \alpha$)-uniform classes?

Remark 3. During the writing of the book some results of Chapter 6 were published in [195].

For the construction of universal elements in classes of mappings we use the construction of universal elements in classes of spaces (separately in the domain-class and in the range-class of spaces). As for the spaces, we can use τ-properties to define classes of mappings in which there are universal elements. We note that properties of mappings defined by some conditions on the preimages of points in general is not τ-properties because the points of a space usually is large than τ. So, we can pose the following general problem.

10.8.4 Problem. Find new saturated classes of mappings using τ-properties.

10.8.5 Problems. Let Λ be a directed set with a partial order $\prec$.

1. Suppose that $I\!D$ is a saturated class of inverse spectra on $\Lambda \cup \prec$. Is the class of the limit spaces of all elements of $I\!D$ saturated?

In particular,

2. Let $I\!P_\lambda$, $\lambda \in \Lambda$, be a fixed saturated class of spaces. Consider the (saturated) class $I\!D$ of all inverse spectra on $\Lambda \cup \prec$ such that $D(\lambda) \in I\!P_\lambda$. Is the class of the limit spaces of all elements of $I\!D$ saturated?

3. Let $I\!P$ be a fixed saturated class of spaces. Consider the (saturated) class $I\!D$ of all inverse spectra D such that $D(\lambda) \in I\!P$. Is the limit spaces of all elements of $I\!D$ saturated?

4. Under what conditions are the above questions true?

Definition. Let X and Y be two spaces. A mapping $f : X \to Y$ (not necessarily continuous) is said to be *Borel* if for every Borel set U of Y the set $f^{-1}(U)$ is a Borel set of X. (About Borel functions (mappings) see for example [217].)

Lt $\alpha \in \tau^+$. A Borel mapping $f : X \to Y$ is said to be α*(mult)-Borel* (respectively, α*(add)-Borel*) if there exists a base B (of cardinality $\leq \tau$) for the open subsets of Y such that the subset $f^{-1}(U)$ of Y is of multiplicative (respectively, of additive) class α for every $U \in B$.

We note that any Borel mapping is α(mult)-Borel or α(add)-Borel for some $\alpha \in \tau^+$.

10.8.6 Problems. 1. Find saturated classes of Borel mappings.

2. Let $I\!P$ and $I\!R$ be two saturated classes of spaces. Consider the class $I\!F$ of all Borel mappings f from an element X of $I\!P$ to an element Y of $I\!R$.

Is it true that in general in $\mathbb{F}$ there are no universal elements? (We note that in the definition of an embedding of a mapping into another mapping and, therefore, in the definition of a universal mapping we actually do not use the continuity of mappings.)

The above question is related with the following question.

3. Let $\mathbb{P}$ and $\mathbb{R}$ be two saturated classes of spaces and let $\alpha \in \tau^+$. Do there exist spaces $X \in \mathbb{P}$ and $Y \in \mathbb{R}$ and an α(mult)-Borel or α(add)-Borel mapping $f : X \to Y$, which is neither β(mult)-Borel nor β(add)-Borel for all $\beta \in \alpha$?

4. Let $\mathbb{P}$ and $\mathbb{R}$ be two saturated classes of spaces and let $\alpha \in \tau^+$. Consider the class $\mathbb{F}$ of all α(mult)-Borel (respectively, of all α(add)-Borel) mappings from elements of $\mathbb{P}$ to elements of $\mathbb{R}$. Is it true that the class $\mathbb{F}$ is a saturated class of mappings?

Remark 4. Basic information on G-spaces can be found for example in [431]. Universality problems for classes of G-spaces and Borel G-spaces (see below the definition) are considered in the following works: [254], [427], [430], [5], [14], [19], [15], [20], [269], [16], [28], [17], [270], [168], [365], [271], [18]. In these works usually it is assumed that G satisfies some additional conditions.

10.8.7 Problems. Let G be an arbitrary topological group.

1. Is there a universal element in the class of all G-spaces (X, F), where X is a metrizable space?

2. Is there a universal element in the class of all G-spaces (X, F), where X is a metrizable space and $\mathrm{ind}(X) \leq n \in \omega$? (We note that for locally compact σ-compact group G and the dimension dim the answer is positive (see [269]).)

3. Is there a universal element in the class of all G-spaces (X, F), where X is: (a) a compact Hausdorff space, (b) a compact Hausdorff space of dimension ind (respectively, of Ind or dim) $\leq n \in \omega$?

4. Is there a universal element in the class of all G-spaces (X, F), where X is a normal space?

5. Under what conditions on G the above questions are true?

Definition. Let G be a topological group, X a topological space, and $\alpha \in \tau^+$. A Borel mapping (respectively, α(mult)-Borel mapping or α(add)-Borel mapping) F of $G \times X$ into X is said to be a *Borel action* (respectively, an *α(mult)-Borel action* or an *α(add)-Borel action*) of G on X if:

(a) $F(e, x) = x$ for every $x \in X$, where e is the unit of G, and

(b) $F(gh, x) = F(g, F(h, x))$ for every $g, h \in G$ and $x \in X$.

Any pair (X, F), where X is a topological space and F is a Borel action (respectively, an α(mult)-Borel action or an α(add)-Borel action) on X, is said to be a *BG-space* (respectively, an *α(mult)-BG-space* or an *α(add)-BG-space*.

(In the case, where G (as a topological space) and X are complete separable metrizable space, a BG-space (X, F) is said to be *Borel G-space*. About Borel G-spaces see for example [28].)

10.8.8 Problems. 1. Aply the method of Containing Spaces to Borel actions of groups.

2. Let $I\!P$ be a saturated class of spaces and $I\!R$ the class of all BG-spaces (X, F) such that $X \in I\!P$. Is it true that in general in the class $I\!R$ there are no universal elements?

3. Let $I\!P$ be a saturated class of spaces, $\alpha \in \tau^+$, and $I\!R$ the class of all α(mult)-BG-spaces (respectively, of all α(add)-BG-spaces) (X, F) such that $X \in I\!P$. Is it true that in the class $I\!R$ there exist universal elements?

Remark 5. The notion of a factorizing τ-spectrum and the Spectral Theorem for such spectra were considered in [367] (see also [366]). Versions of Spectral Theorems are considered in [72], [75], [123], and [124]. More informations and some historical comments about factorizing τ-spectra and Spectral Theorems one can found in [125] and [79].

10.8.9 Problem. Let $\mathbf{S}$ be an indexed collection of spaces and $\mathrm{C}(\mathbf{S})$ the corresponding τ-complete set. Is there a version of the Spectral Theorem based on the first-type subsets of $\mathrm{C}(\mathbf{S})$ (instead of the cofinal τ-closed subsets, that is instead of the second-type subsets of $\mathrm{C}(\mathbf{S})$)?

10.8.10 Problems. 1. Is it true that all saturated classes of spaces considered in the book (that is, the first type saturated classes) are also second-type saturated?

2. Find zero-type saturated classes of spaces (except that mensioned in Chapter 8).

Remark 6. The first separable metric space, which is isometrically universal in the class of all separable metric spaces were constructed in [422] and [426]. In [25] a normed linear space is indicated as an isometrically universal space for all separable metric spaces. Isometrically universal spaces of weight $\tau > \omega$ having some additional properties (like to that of the Urysohn space) are constructed (under a condition on τ) in [213].

The following works are related to isometrically universal spaces: [377], [378], [170], [31], [199], [243], [235], [201], [53], [32], [200], [203], [44], [45], [150], [387], [169], [188], [190], [7], [40], [41], [402], [42].

The papers [383], [380], [381], [382], [212], [302], [300], and [301] (see also the books [299] and [308]) are related to the metric dimension or to the metric-depended dimension functions.

The net-function of a completely bounded metric space defined in Chapter 9 is a slight variation of the corresponding function considered in the papers [351] and [223]. (For a given completely bounded metric space X and $\varepsilon > 0$ it is considered the smallest number $\mathcal{N}_\varepsilon(X)$ of sets of diameter $\leq \varepsilon$ which cover the space X.)

Chapter 9 contains all results of the papers [188] and [190].

In Chapter 9 we consider (ω, ω^+)-saturated classes (that is, ω-saturated classes) in parallel to saturated classes. However, we do not know any (ω, ω^+)-saturated class, which is not a saturated class. So, we can formulate the following problem which is the simplest particular case of Problem 10.2.26(2).

10.8.11 Problem. Is there an (ω, ω^+)-saturated class which is not a saturated class?

10.8.12 Problem. Let $\mathbf{S}$ be an indexed collection of separable metric spaces. Are there a co-mark $\mathbf{M}$ of $\mathbf{S}$, an $\mathbf{M}$-admissible family R of equivalence relations on $\mathbf{S}$, and a metric on the Containing Space $\mathrm{T}(\mathbf{M}, \mathrm{R})$ such that for every $X \in \mathbf{S}$ the natural embedding e^X_{T} of X into T is an isometry?

10.8.13 Problems. 1. Is it true that Corollaries 9.2.10 and 9.2.12 are not valid if in the formulations of these corollaries we replace the symbol "$\subset$" ("is a subset of") by "$=$" ("is equal to")?

2. Let $n \in \omega$. Is it true that there is no separable metric space T and its zero-dimensional subsets T_0,....,T_n with $T = T_0 \cup ... \cup T_n$ having the property that for every separable metric space X and its zero-dimensional subsets X_0,...,X_n with $X = X_0 \cup ... \cup X_n$ there exists an isometry i of X into T satisfying conditions:

$$X_0 = i^{-1}(T_0), \ldots, X_n = i^{-1}(T_n)?$$

(Compare this question with Corollary 9.2.9.) (We note that such a space T and zero-dimensional subsets exist if it is additionally assumed that all considered zero-dimensional subsets of T and X are of multiplicative class (or of additive class) α, where α is a fixed countable ordinal.)

10.8.14 Problems. 1. Is it possible to improve (in some sense) the result of Proposition 9.3.5?

In particular,

2. Is there an isometrically containing completely bounded metric space K for the class $\mathbb{M}(\mathbf{d}, \mathbf{n})$ of metric spaces such that

$$\lim_{p \to \infty} \frac{\mathbf{n}^K(1/p)}{\mathbf{n}^T(1/p)} = 0?$$

3. Find an isometrically containing completely bounded metric space K for $\mathbb{M}(\mathbf{d}, \mathbf{n})$ with the "best" net-function.

10.8.15 Problems. 1. What is the minimum of cardinals ν such that the class of all compact metric spaces can be represented as a union of ν many uniform classes? (In other words, what is the minimum of cardinals ν such that there is a set of cardinality ν consisting of compact metric spaces, with the property that every compact metric space is isometrically embedded in an element of this set?)

3. Let $\mathbb{M}(\mathbf{d}, \mathbf{n})$ be a fixed uniform class of compact metric spaces and K a fixed compact metrizable space. Under what conditions does there exist a metric on K such that the compact metric space K is an element of $\mathbb{M}(\mathbf{d}, \mathbf{n})$?

10.8.16 Problems. 1. Is it possible to characterize topologically some saturated classes $\mathbb{R}$ of metrizable spaces using:

(a) the notion of a uniform subfamily of the class $\mathbb{M}$ of metric spaces generated by $\mathbb{R}$ and

(b) the notion of a net-function of a space?

In particular,

2. Consider the question 1 for the classes:

(a) of all countable-dimensional spaces,

(b) of all strongly countable-dimensional spaces,

(c) of all locally finite-dimensional spaces, and

(d) of all spaces of dimension ind $\leq \alpha$, $\alpha \in \omega^+$.

10.8.17 Problems. 1. Is it possible to extend (some of) the results of Chapter 9 to metric spaces of weight larger that ω?

2. Apply the method of Containing Spaces to other metric-depended dimension functions or to its versions.

Bibliography

[1] J.M. Aarts, *Completeness degree. A generalization of dimension*, Fund. Math. 63 (1968), 27-41.

[2] J.M. Aarts, *A characterization of strong inductive dimension*, Fund. Math. 70 (1971), 147-155.

[3] J.M. Aarts and T. Nishiura, *Kernels in dimension theory*, Trans. of Amer. Math. Soc. 178 (1973), 227-240.

[4] J.M. Aarts and T. Nishiura, *Dimension and extensions*, North-Holland 1993.

[5] H. Abels, *A universal proper G-space*, Math. Z. 159 (1978), 143-158.

[6] O.T. Alas, M.G. Tkačenko, V.V. Tkachuk, R.G. Wilson, *Connectifying some spaces*, Topology Appl. 71 (1996), 203-215.

[7] F. Alessi and P. Baldan, *A characterization of distance between 1-bounded compact ultrametric spaces through a universal space*, Theoretical Computer Science 193 (1998), 113-127.

[8] P.S. Alexandroff, *On the theory of topological spaces*, Dokl. Acad. Nauk SSSR 2 (1936), 51-54 (Russian).

[9] P.S. Alexandroff, *On the notion of a space in topology*, Uspekhi Mat. Nauk 2 (1947), no. 1, 5-57 (Russian).

[10] P. Alexandroff and B. Pasynkov, *Introduction to Dimension Theory*, Moscow 1973 (Russian).

[11] D.R. Anderson, *On connected irresolvable Hausdorff spaces*, Proc. Amer. Math. Soc. 16 (1965), 463-466.

[12] R.D. Anderson, *A characterization of the universal curve and a proof of its homogeneity*, Ann. of Math. 20 (1958), 313-324.

[13] R.D. Anderson, *One-dimensional continuous curves and a homogeneity theorem*, Ann. of Math. (2) 68 (1958), 1-16.

[14] S. Antonyan, *Tychonov's theorem in the category of topological transformation groups*, Dokl. Acad. Nauk Arm. SSR. 71 (1980), no. 4, 212-216 (Russian).

[15] S. Antonyan, *Equivariant embeddings into G-AR's*, Glasnik Matem. 22 (1987), no. 2, 503-533.

[16] S. Antonyan, *Equivariant embeddings and $\aleph_0$-bounded groups*, Vestnik Mosk. Univ. Ser I, Matem. Mekh. 1 (1994), 16-22 (Russian); English translation in: Moscow Univ. Math. Bull. 49 (1994), no. 1, 13-16.

[17] S. Antonyan, *Equivariant embeddings of metrizable G-spaces*, Uspekhi Matem. Nauk 46 (1996), no. 4, 145-147 (Russian); English translation in: Russian Math. Surv. 51 (1996), no. 5, 999-1001.

[18] S. Antonyan, *Universal proper G-spaces*, Topology Appl. 117 (1) (2002), 23-43.

[19] S. Antonyan and Yu. Smirnov, *Universal objects and bicompact extensions for topological transformation groups*, Doklady Akad. Nauk SSSR 257 (1981), no. 3, 521-526 (Russian); English translation in: Soviet Math. Doklady 23 (1981), no. 2, 279-284.

[20] S. Antonyan and J. de Vries, *Tychonov theorem for G-spaces*, Acta Math. Hung. 50 (1987), no. 3-4, 253-257.

[21] A. V. Arhangel'skiĭ, *On ranks of families of sets and dimension of spaces*, Fund. Math. 52 (1963), 257-275.

[22] A. Arhangel'skii, *Factorization of mappings according to weight and dimension*, Dokl. Akad. Nauk SSSR 174 (1967), 1243-1246 (Russian); English translation in: Soviet Math. Dokl. 8 (1967), 731-734.

[23] A.V. Arkhangel'skii and V.V. Fedorchuk, *The basic concepts and constructions of General Topology*, Itogi nauki i tekhniki, Sovremennye problemy matematiki, Fundamental'nye napravleniya, Vol 17, Obshchaya topologiya 1, Publisher VINITI, Moscow 1988 (Russian); English translation in: Encyclopedia of Mathematical Science Vol. 17 (General Topology I), Berlin 1990, 1-90.

[24] I. Baggs, *A connected Hausdorff space which is not contained in a maximal connected space*, Pacif. J. Math. 51 (1974), no. 1, 11-18.

[25] S. Banach, *Thèorie des opérations linéaires*, PWN, Warszawa 1932.

[26] T. Banakh and R. Cauty, *Universalité forte pour les sous-ensembles totalement bornés. Applications aux espaces $C_p(X)$*, Colloq. Math. 73 (1997), 25-33.

[27] T. Banakh, T. Radul, and M. Zarichnyi, *Absorbing Sets in Infinite-Dimensional Manifolds*, Mathematical Studies, Monograph Series, Vol. 1, VNTL Publishers, Lviv, 1996.

[28] H. Becker and A. Kechris, *The Descriptive Set Theory of Polish Group Actions*, Cambridge Univ. Press, Campridge, 1996.

[29] H.L. Bentley and H. Herrrlich, *Completeness of nearness spaces*, Topological Structure II, Proc. Sympos. Amsterdam 1978, Mathematisch Centrum, 29-40.

[30] H.L. Bentley, H. Herrrlich and M. Hušek, *The historical development of uniform, proximal, and nearness consepts in topology*, in: Handbook of the History of General Topology, Vo. 2, C.E. Aull and R. Lowen (eds), Dordrecht, Kluwer, 577-629.

[31] C. Bessaga, *A note on universal Banach spaces of finite dimension*, Bull Acad. Polon. Sci., Sér. Sci. Math. Astr. Phys. 6 (1958), 97-101.

[32] C. Bessaga and A. Pełczyński, *Selected Topics in Infinite-dimensional Topology*, Monograf. Mat. 58, Polish Scientific Publishers, Warsawa 1975.

[33] M. Bestvina, *Characterizing k-dimensional universal Menger compacta*, Memoirs AMS 71 (1988), no. 380, 1-110.

[34] M. Bestvina, P. Bowers, J. Mogilski, and J.J. Walsh, *Characterizing Hilbert space manifolds revisited*, Top. Appl. 24 (1986), 53-69.

[35] M. Bestvina and J. Mogilski, *Characterizing certain incomplete infinite-dimensional absolute retracts*, Michigan Math. J. 33 (1986), 291-313.

[36] R.H. Bing, *A connected countable Hausdorff space*, Proc. Amer. Math. Soc. 4 (1953), 474.

[37] L.M. Blumenthal and K. Menger, *Studies in geometry*, W.H. Freeman 1970.

[38] L.Yu. Bobkov, B.A. Pasynkov, and D.A. Stavrova, *On spaces near to locally finite-dimensional*, V Tiraspol Simposium on General Topology and its Applications, Kishinew 1985, 25-26 (Russian).

[39] L.Yu. Bobkov, *On universal locally finite-dimensional spaces*, Uspekhi Mat. Nauk 37 (1982), no. 2, 191-192 (Russian); English translation in: Russian Math. Surveys 37 (1982), no. 2, 205-206.

[40] S.A. Bogatyi, *Compact homogeneity of Urysohn's universal metric space*, Uspekhi Mat. Nauk 55 (2000), no. 2, 131-132 (Russian); English translation in: Russ. Math. Surv. 55 (2000), no. 2, 332-333.

[41] S.A. Bogatyi, *A universal homogeneous rational ultrametric on the space of irrationals*, Vestnik Mosk. Univ. Ser. I Mat. Mekh. (2000), no. 6, 20-24 (Russian); English translation in: Moscow Univ. Math. Bull. 55 (2000), no. 6, 20-24.

[42] S.A. Bogatyi, *Metrically homogeneous spaces*, Uspekhi Mat. Nauk 57 (2002), no. 2, 3-22 (Russian); English translation in: Russian Math. Servey 57 (2002), no. 2, 221-240.

[43] K. Borsuk, *Remark on the Cartesian product of two 1-dimensional spaces*, Bull. Acad. Pol. Sci. Sér. Math. 23 (1975), 971-973.

[44] Á. P. Bosznay, *A counterexample to a general stability result in universal spaces*, Ann. Univ. Sci. Budapest. Eötvös Sect. Math. 22/23 (1979/80), 221-223.

[45] Á. P. Bosznay, *A remark on universal spaces*, Publ. Math. Debrecen 27 (1980), no. 3-4, 251-253.

[46] H.G. Bothe, *Eine Einbettung m-dimensionaler Mengen in einem $(m+1)$-dimensionalen absoluten Retrakt*, Fund. Math. 52 (1963), 209-224.

[47] N. Bourbaki, *Topologie générale ch. I et II*, Paris 1940.

[48] S.M. Boyles and G.X. Ritter, *A connected locally connected countable space which is almost regular*, Colloq. Math. 46 (1982), 189-195.

[49] H. Brandenburg and A. Mysior, *For every Hausdorff space Y there exists a non-trivial Moore space on which all continuous functions are constant*, Pasific J. III, 1 (1984), 1-8.

[50] M. Brown, *A countable connected Hausdorff space*, Bull. Amer. Math. Soc. 59 (1953), 367.

[51] G.C.L. Brümmer, *Initial quasi-uniformities*, Nederl. Akad. Wetensch. Proc. Ser. A 72 = Indag Math. 31 (1969), 403-409.

[52] G.C.L. Brümmer, *Categorical aspects of the theory of quasi-uniform spaces*, Rend.Ist. Mat. Univ. Trieste 30 Suppl. (1999), 45-74.

[53] P. Buneman, *A note on the metric properties of trees*, J. Comb. Th. (B) 17 (1974), 48-50.

[54] R.D. Buskirk, *A universal totally regular continuum*, Houston J. Math. 20 (1994), no. 4, 745-755.

[55] J.W. Carlson, *B-completeness in nearness spaces*, Gen Top. Appl. 5 (1975), 263-268.

[56] J.W. Carlson and T.L. Hicks, *On completeness in quasi-uniform spaces*, J. Math. Anal. Appl. 34 (1971), 618-627.

[57] J.W. Cannon, *A positional characterization of the (n-1)-dimensional Sierpiński curve in $S^n, (n \neq 4)$*, Fund. Math. 79 (1973), 107-112.

[58] R. Cauty, *The Property of Strong Universality and its Applications*, Proccedings of Steklov Mathematical Institut of RAN 212 (1996), 95-122 (Russian).

[59] R. Cauty, T. Dobrowolski, and W. Marciszewski, *A contribution to the topological classification of the spaces* $C_p(X)$, Fund. Math. 142 (1993), 269-301.

[60] J.J. Charatonik, *History of continuum theory*, C.E. Aull and R. Lowen (eds), Handbook of the History of General Topology, Vol. 2 (1998), 703-786, Kluwer Academic Publishers.

[61] W.J. Charatonik, *Smooth dendroids as inverse limits of dendrites*, Fund. Math. 124 (1984), 163-168.

[62] T.A. Chapman, *Lectures on Hilbert cube manifolds*, American Mathematical Society, Providence, R.I., 1976, Expository lectures from the CBMS Regional Conference held at Guilford College, October 11-15, 1975, Regional Conference Series in Mathematics, No. 28.

[63] M.G. Charalambous, *Universal spaces for locally finite-dimensional Tychonoff spaces*, Proc. Amer. Math. Soc. 106 (1989), no. 2, 507-514.

[64] M.M. Criado and J. Tarres, *On strong small transfinite and D-dimension*, Math. Japonica 34 (1989), 909-917.

[65] Á. Császár, *Complétion et compactification D'espaces syntopogènes*, General Topology and its Relations to Modern Analysis and Algebra, Academic Press 1962, 133-137. (Proceedings of the Prague Symposium 1961).

[66] Á. Császár, *Regular extensions of quasi-uniformities*, Studia Sci. Math. Hung. 14 (1979), 15-26.

[67] Á. Császár, *Extensions of quasi-uniformities*, Acta Math. Hung. 37 (1981), 121-145.

[68] Á. Császár, *Complete extensions of quasi-uniform spaces*, ed. J. Novak, General Topology and its Relations to Modern Analysis and Algebra V, Heldermann, Berlin, 1982, 104-113. (Proceedings Fifth Prague Topology Symposium, 1981.)

[69] Á. Császár, *Old and new results on quasi-uniform extensions*, Rend. Ist. Mat. Univ. Trieste 30 Suppl. (1999), 75-85.

[70] V.A. Chatyrko, *On inductive dimensions*, Vestnik Mosk. Univ. 38 (1983), no. 2, 7-10 (Russian).

[71] A. Chigogidze, *On the relative dimensions of completely regular spaces*, Bull. Georgian Acad. Sci. 85 (1977), 45-48 (Russian).

[72] A. Chigogidze, *On the spectral theorem of E.V. Shchepin*, Bull. Georgian Acad. Sci. 194 (1981), 25-28.

[73] A. Chigogidze, *On relative dimensions*, General Topology. Spaces of functions and dimension, Moscow, Moscow Univ. Press (1985), 67-117 (Russian).

[74] A. Chigogidze, *Characteristic of the dimension* dim *of R-compacta, and universal spaces of a given weight and a given dimension*, Uspekhi Mat. Nauk 40 (1985), no. 5(245), 265-266 (Russian).

[75] A. Chigogidze, *On Baire isomorphisms of non-metrizable compacta*, Comm. Math. Univ. Carolinae 26 (1985), 811-820.

[76] A. Chigogidze, *Compacta lying in the n-dimensional universal Menger compactum and having homeomorphic complememts in it*, Math. USSR-Sb. 61 (1988), 471-484.

[77] A. Chigogidze, *n-soft mappings of n-dimensional spaces*, Mat. Notes 46 (1989), 88-95.

[78] A. Chigogidze, *Autohomeomorphisms of the universal Menger compacta are stable*, Bull. Georgian Acad. Sci. 142 (1991), 477-479.

[79] A. Chigogidze, *Inverse Spectra*, North-Holland Mathematical Library, Vol. 53, North-Holland Publishing Co., Amsterdam, 1996.

[80] A. Chigogidze and V. Valov, *Universal maps and surjective characterizations of completely metrizable LC^n-spaces*, Proc. Amer. Math. Soc. 109 (1990), 1125-1133.

[81] A. Chigogidze and V. Valov, *Universal metric spaces and extension dimension*, Topology Appl. 113 (2001), 23-27.

[82] J. Cobb and W. Voxman, *Dispersion points and fixed points*, Amer. Math. Monthly, 87 (1980), no. 4, 278-281.

[83] P.J. Collins, *Universal in our time*, Topology Proceedings 25 (Summer 2000), 487-494.

[84] P.J. Collins, *Problems on universal*, Topology Appl. 140 (2004), 33-36.

[85] W.W. Comfort and S. Negrepontis, *The Theory of Ultrafilters*, Springer-Verlag 1974.

[86] H. Cook, *Continua which admit only the identity mapping onto nondegenerate subcontinua*, Fund. Math. 60 (1967), 241-249.

[87] C.A. Coppin, *Continuous functions from a connected locally connected space into a connected space with a dispersion point*, Proc. Amer. Math. Soc. 32(2) (1972), 625-626.

[88] E. Cuchillo-Ibáñez, *A relation between the strong small transfinite and the D-dimension*, Q&A in General Topology 9 (1991), 81-88.

[89] U.B. Darji, *Complexity of hereditarily decomposable continua*, Topology and Applications 103 (2000), 243-248.

[90] J. Deák, *Extensions of quasi-uniformities for prescribed bitopologies I, II*, Studia Sci. Math. Hung. 25 (1990), 45-67 and 69-91.

[91] J. Deák, *A survey of compatible extensions (presenting 77 unsolved problems)*, Colloq. Math. Soc. János Bolyai, vol. 55, Top., Pécs (Hungary) 1989, 1993, 127-175.

[92] J. Deák, *A bitopological view of quasi-uniform completeness, I, II, III*, Studia Sci. Math. Hung. 30 (1995), 389-409; 411-431 and 31 (1996), 385-404.

[93] J.J. Dijkstra, *Fake topological Hilbert spaces and characterizations of dimension in terms of negligibility*, CWI Tract, vol 2, Centre for Mathematics and Computer Science, Amsterdam, 1984.

[94] J. Dijkstra and J. van Mill, *Erdős space and homeomorphism groups of manifolds*, preprint.

[95] J.J. Dijkstra, J. van Mill, and J. Mogilski, *The space of infinite-dimensional compact spaces and other topological copies of* $(l_f^2)^\omega$, Pac. J. Math. 152 (1992), 255-273.

[96] T. Dobrowolski and J. Mogilski, *Problems on Topological Classifications of Incomplete Metric Spaces*, Open Problems in Topology, J. van Mill and G.M. Reed (Editors), Elsevier Science Publishers B.V. (North-Holland), 1990.

[97] T. Dobrowolski, S.P. Gul'ko, and J. Mogilski, *Function spaces homeomorphic to the countable product of* l_2^f, Top. Appl. 34 (1990), 153-160.

[98] D. Doichinov, *On completeness of quasi-uniform spaces*, C.R. Acad. Bulgare Sci. 41 (1988), no. 7, 5-8.

[99] D. Doichinov, *On completeness in quasi-metric spaces*, Topology Appl. 30 (1988), no. 2, 127-148.

[100] D. Doichinov, *Another class of completable quasi-uniform spaces*, C. R. Acad. Bulgare Sci. 44 (1991), no. 3, 5-6.

[101] D. Doichinov, *A concept of completeness of quasi-uniform spaces*, Topology Appl. 38 (1991), no. 3, 205-217.

[102] D. Doitchinov, *Completeness and completion of quasi-uniform spaces*, Trudy Mat. Inst. Steklov 193 (1992), 103-107 (Russian); English translation: Proc. Steklov Inst. Math. 1993, no. 3, 111-115.

[103] D. Doichinov, *E-completions of quasi-uniform spaces*, Proc. Symposium on Categorical Topology, University of Cape Town 1994, B. Banaschewki, C.R.A. Gilmour and H. Herrlich (eds), Dept. Math. University of Cape Town 1999, 89-102.

[104] C.H. Dowker, *An imbedding theorem for paracompact metric spaces*, Duke Math. J. 14 (1947), 639-645.

[105] E.K. van Douwen, *A regular space on which every continuous real-valued function is constant*, Nieuw Archief voor Wiskunde 20 (1972), 143-145.

[106] E. van Douwen, *Closed images of σ-discrete metrizable spaces*, unpublished manuscript of [vD u 98] in the list of J. van Mill, In memoriam: Eric Karel van Douwen (1946-1987), Topology Appl. 31 (1989), 1-18.

[107] A.N. Dranishnikov, *Universal Menger compacta and universal mappings*, Mat. Sb. (N.S.) 129(171) (1986), no. 1, 121-139 (Russian); English translation in: Math. USSR-Sb. 57 (1987), no. 1, 131-149.

[108] R. Duda, *On biconnected sets with dispersion points*, Rozprawy Mat. 37 (1964), 1-58.

[109] J. Dydak and J. Mogilski, *Universal cell-like maps*, Proc. Amer. Math. Soc. 122 (1994), 943-948.

[110] R.D. Edwards, *Approximating codimension ≥ 3 σ-compacta with locally homotopically unknotted embeddings*, Topology and Appl. 24 (1986), 95-122.

[111] A.G. El'kin, *Maximal connected Hausdorff spaces*, Matem. Zametki 26 (6) (1979), 939-948 (Russian); English Translation in: Math. Notes 26 (5-6) (1979), 974-978.

[112] A. Emeryk and W. Kulpa, *The Sorgenfrey line has no connected compactification*, Comm. Math. Univ. Carolin. 18 (1977), 483-487.

[113] R. Engelking, *Transfinite dimension*, Surveys in General Topology, ed. G.M. Reed, New York 1980, 131-161.

[114] R. Engelking, *General Topology*, Heldermann Verlag, Berlin, second ed., 1989.

[115] R. Engelking, *Theory of Dimensions, Finite and Infinite*, Sigma Series in Pure Mathematics Vol. 10, Heldermann Verlag 1995.

[116] R. Engelking, W. Holsztyński and R. Sikorski, *Some examples of Borel sets*, Coll. Math. 15 (1966), 271-274.

[117] R. Engelking and E. Pol, *Countable-dimensional spaces: a survey*, Dissertationes Mathematicae CCXVI (1983).

[118] P. Erdős, *The dimension of the rational points in Hilbert space*, Annals of Math. 41 (1940), 734-736.

[119] P. Erdős, *Some remarks on connected sets*, Bull. Amer. Math. Soc. 59 (1944), 442-446.

[120] W.T. van Est und U.H. Freudenthal, *Trenung durch stetige Funktionen in Topologischen Raumen*, Indagationes Math. 13 (1951), 359-368.

[121] V.V. Fedorčuk, *Infinite-dimensional bicompacta*, Izv. Akad. Nauk SSSR, Ser. Mat. 42 (1978), 1162-1178 (Russian); English translation in: Math. USSR-Izv. 13 (1979), 445-460.

[122] V.V. Fedorčuk, *On the dimension of hereditarily normal spaces*, Proc. London Math. Soc. 36 (1978), 163-175.

[123] V.V. Fedorčuk, *Soft mappings, multivalued retractions and functors*, Uspekhi Mat. Nauk 41 (1986), no. 6, 121-159 (Russian).

[124] V.V. Fedorčuk, *On characterization of n-soft mappings*, Dokl. Akad. Nauk SSSR 228 (1986), no. 2, 313-316 (Russian)

[125] V.V. Fedorchuk and A. Chigogidze, *Absolyutnye retrakty i beskonechnomernye mnogoobraziya*, "Nauka", Moscow, 1992 (Russian).

[126] L.E. Feggos, S.D. Iliadis, and S.S. Zafiridou, *Some families of planar rim-scattered spaces and universality*, Houston J. Math. 20 (1994), 351-364.

[127] L.E. Feggos, S.D. Iliadis, and S.S. Zafiridou, *Planar rational compacta*, Colloq. Math., LXVIII (1995), 49-54.

[128] V.V. Filippov and L. Úry, *A theorem on relative universal spaces with given weight and dimension*, Acta Math. Acad. Sci. Hungar. 32 (1978), no. 3-4, 371-375.

[129] P. Fletcher and W.F. Lindgren, *Quasi-uniform spaces*, Marcel Dekker, New York, 1982.

[130] M. Fréchet, *Sur quelquespoints du Calcul fonctionnel*, Thése, Rend. del Circolo Matem. di Palermo, 22 (1906).

[131] M. Frechet, *Les dimension d'un ensemble abstrait*, Math. Ann. 68 (1910), 145-168.

[132] R. Frič and D. Kent, *Completion functors for Cauchy spaces*, Intern. J. Math. Sci. 2 (1979), 589-604.

[133] R. Frič and D. Kent, *A nice completion need to be strict*, Proc. Fifth Prague Topol. Sympos. 1981, Heldermann Verlag, 189-192.

[134] L. Friedler, *Open connected functions*, Can. Math. Bull. 16(1) (1973), 57-60.

[135] O. Frink, *Compactifications and semi-normal spaces*, Amer. J. Math. 86 (1964), 602-607.

[136] S. Gähler, W. Gähler and G. Kneis, *Completion of pseudo-topological vector spaces*, Math. Nachrichten 75 (1976), 185-206.

[137] P.M. Gartside, R.W. Knight and J.T.H. Lo, *Parametrising open universal*, Topology Appl. 119 (2002), no 2, 131-145.

[138] P.M. Gartside and J.T.H. Lo, *The hierarchy of Borel universal sets*, Topology Appl. 119 (2002), no. 2, 117-129.

[139] P.M. Gartside and J.T.H. Lo, *Open universal sets*, Topology Appl. 129 (2003), 89-101.

[140] D.N. Georgiou, *A note on a question of Y. Hattori*, Questions & Answers in General Topology, 13 (1995), no. 2, 217-219.

[141] D.N. Georgiou and S.D. Iliadis, *Containing spaces and the property of finite intersections*, Rendiconti del Circolo Matematico di Palermo, Serie II, (1992), no. 29, 415-424. (V Convergno Internazionale di Topologia in Italia, Lecce-Otranto, 17-21 Settembre 1992.)

[142] D.N. Georgiou and S.D. Iliadis, *Containing spaces and α-uniformities*, Topology. Theory and Applications, II (Pećs, 1989) North-Holland, Amsterdam (1993), 245-282. (Colloquia Mathematica Societatis Janos Bolyai 55, Topology, Pećs, Hungary, 1989.)

[143] D.N. Georgiou and S.D. Iliadis, *Universal elements in some families of regular spaces with countable bases*, Zapiski Nauchnykh Seminarov S.-Peterburg Otdel. Mat. Inst. Steclov (POMI) 231 (1995), 88-134 (Russian); English translation in: Journal Mathematical Sciences 91 (1998), no. 6, 3387-3415.

[144] D.N. Georgiou and S.D. Iliadis, *Rational n-dimensional spaces and the property of universality*, Fundamentalnaya i prikladnaya matematika 4 (1998), no. 4, 941-955 (Russian).

[145] H. Gladdines, *Absorbing systems in infinite-dimensional manifolds and applications*, Ph.D. thesis, Vrije Universiteit, Amsterdam, the Netherlands, 1994.

[146] S.W. Golomb, *A connected topology for the integers*, Amer. Math. Monthly 66 (1959), 663-665.

[147] S.W. Golomb, *Arithmetica Topologica*, General Topology and its Relations to Modern Analysis and Algebra, Academic Press 1962, 179-186 (Proccedings of the Prague Symposium 1961).

[148] T. Goto, *Universal spaces for countable dimensional metric spaces*, Proc. Amer. Math. Soc. 103 (1988), no. 4, 1290-1292.

[149] J. Grispolakis and E.D. Tymchatyn, *A universal smooth dendroid*, Bull. Acad. Polon. Sci. Ser. Sci. Math. Astronom. Phys. 26 (1978), 991-998.

[150] M. Gromov, *Groups of polynomial growth and expanding maps*, Publ. I.H.E.S. 53 (1981), 53-78.

[151] J. de Groot, *Topologische Studiën. Compactificatie, Voortzetting van Afbeeldingen en Samenhang*, Thesis, University of Groningen, 1942, 102pp.

[152] J. de Groot and T. Nishiura, *Inductive compactness as a generalization of semi-compactness*, Fund. Math. 58 (1966), 201-218.

[153] C. Guido, *Connected spaces which are not strongly connected*, Rend. Circ. Mat. Palermo (2) 1984, Suppl. No 6, 149-153 (Proceedings of the 12th winter school on Abstract Analysis, Srni 1984).

[154] W. Gustin, *Countable connected spaces*, Bull. Amer. Soc. 52 (1946), 101-106.

[155] A. Gutek, *A connected space with a dispersion point without fixed-point proprety*, Topology Appl. 30 (1988), 101-105.

[156] J.A. Guthrie, D.F. Reynolds and H.E. Stone, *Connected expansions of topologies*, Bull Austral. Math. Soc. 9 (1973), 259-265.

[157] J.A. Guthrie and H.E. Stone, *Spaces whose connected expansions preserve connected subsets*, Fund. Math. 80 (1973), 91-100.

[158] J.A. Guthrie, H.E. Stone and M.L. Wage, *Maximal connected expansions of the reals*, Proc. Amer. Math. Soc. 69 (1978), 159-165.

[159] S.W. Hahn, *Universal Spaces Under Strong Homeomorphisms*, Trans. Amer. Math. Soc. 70 (1951), no. 2, 301-311.

[160] P.C. Hammer and W.E. Singletary, *Connectedness-equivalent spaces on the line*, Pendiconti Rendiconti del Circolo Mathematico di Palermo 17 (1968), 343-355.

[161] Y. Hattori, *A note on universal spaces for finite dimensional complete metric spaces*, Glasnik Matematički 24 (1989), 139-147.

[162] F. Hausdorff, *Grundzüge der Mengenlehre*, Leipzig (Vien), 1914.

[163] D.W. Henderson, *D-dimension I. A new transfinite dimension*, Pasific J. Mat. 26 (1968), no. 1, 91-107.

[164] D.W. Henderson, *D-dimension II. Separable spaces and compactifications*, Pacific J. Math. 26 (1968), no. 1, 109-113.

[165] H. Herrlich, *Wann sind alle stetigen Abbildungen in Y konstant?*, Math. zeitscr. 90 (1965), 152-154.

[166] H. Herrlich, *Topological structures*, Topological Structures I, Symposium in honour of J. de Groot (1914-1972), Amsterdam 1973, Mathematisch Centrum, 59-122.

[167] E. Hewitt, *On two problems of Urysohn*, Annals of Mathematics 47 (1946).

[168] G. Hjorth, *A universal Polish G-space*, Topology Appl. 91 (1999), 141-150.

[169] M.R. Holmes, *The universal separable metric space of Urysohn and isometric embeddings thereof in Banach spaces*, Fund. Math. 140 (1992), 199-223.

[170] G.E. Huhunaishvili, *A property of the universal metric space of Urysohn*, Dokl. Akad. Nauk SSSR 101 (1955), 607-610 (Russian).

[171] W.N. Hunsaker and W.F. Lindgren, *Construction of quasi-uniformities*, Math. Ann. 188 (1970), 39-42.

[172] W. Hurewicz, *Normalbereiche und Dimensionstheorie*, Math. Ann. 96 (1927), 736-764.

[173] W. Hurewicz, *Über das Verhältniss separabler Räume zu kompakten Räumen*, Proc. Koninkl Akad. Amsterdam, Ser. A, 30 (1927), no. 1, 425-430.

[174] W. Hurewicz, *Über unendlich-dimensionale Punktmengem*, Proc. Acad. Amsterdam 31 (1928), 916-922.

[175] W. Hurewicz, *Dimensionstheorie und Cartesische Räume*, Proc. Akad. Amsterdam 34 (1931), 399-400.

[176] W. Hurewicz, *Über Abbildungen von endlich dimensionalen Räumen auf Teilmengen Cartesischer Räume*, Sitzungsberichte Preuss. Akad. Wiss. Phys.-Math. Klasse (1933), 754-768.

[177] W. Hurewicz and H. Wallman, *Dimension Theory*, Princeton 1941.

[178] S.D. Iliadis, *Universal continuum for the class of completely regular continua*, Bull. de L'Academie Polonaise Des Science, Vol. XXVIII (1980), no. 11-12, 603-607.

[179] S.D. Iliadis, *On rim-type of spaces*, Lecture Notes in Mathematics, 1960, 45-54. (Proceedings of the International Topological Conference, Leningrad, August, 1982.)

[180] S.D. Iliadis, *Rational spaces of a given rim-type and the property of universality*, Topology Proceedings 11 (1986), 65-113.

[181] S.D. Iliadis, *Rim-finite spaces and the property of universality*, Houston J. of Math. 12 (1986), no. 1, 55-78.

[182] S.D. Iliadis, *Continua that are almost locally a bundle of arcs and the property of universality*, General Topology and its Relations to Modern Analysis and Algebra VI, 281-290. (Proceedings of the Sixth Prague Topological Symposium 1986.)

[183] S.D. Iliadis, *The rim-type of spaces and the property of universality*, Houston J. of Math. 13 (1987), no. 3, 373-388.

[184] S.D. Iliadis, *Rational spaces and the property of universality*, Fund. Math. 131 (1988), 167-184.

[185] S.D. Iliadis, *Universal spaces for some families of rim-scattered spaces*, Tsukuba J. of Math. 16 (1992), no. 1, 123-159.

[186] S.D. Iliadis, *Some problems concerning universality in families of n-dimensional rational spaces*, Zapiski Nauchnykh Seminarov S.-Peterburg Otdel. Mat. Inst. Steclov (POMI) 208 (1992), 98-102 (Russian).

[187] S.D. Iliadis, *On universal objects in some classes of topological and metric spaces*, Dissertation for Dr. Sci. degree, Moscow State University, 1993 (Russian).

[188] S.D. Iliadis, *Isometric embeddings and universality*, Uspekhi Mat. Nauk RAN 48 (1993), no. 5, 173-174 (Russian).

[189] S.D. Iliadis, *Universal regular spaces of weight τ*, Dokl. Ross. Akad. Nauk 334 (1) (1994) 12-15 (Russian); English translation in: Russ. Acad. Sci. Dokl. Math. 49 (1) (1994), 10-14.

[190] S.D. Iliadis, *Compact and completely bounded metric spaces and isometric embeddings*, Uspekhi Mat. Nauk RAN 50 (1995), no. 6, 179-180 (Russian).

[191] S.D. Iliadis, *A generalization of the construction of containing spaces*, Topology Proceedings Volume 25, Summer 2000, 539-551. (Proceedings of the 1st Turkish International Conference on Topology and Its Applications, Istambul University, Turkey.)

[192] S.D. Iliadis, *A construction of containing spaces*, Topology and its Applications 107 (2000), 97-116.

[193] S.D. Iliadis, *Some properties of the containing spaces and saturated classes of spaces*, Applied General Topology, Vol 4 (2003), no. 2, 487-507. (Proceedings of the International Conference "Function spaces, proximities and quasi-uniformities", Caserta, Italy, 2002.)

[194] S.D. Iliadis, *Saturated classes of bases*, Note di Matematica 22 (2003), no. 2.

[195] S.D. Iliadis, *Mappings and universality*, Topology and its Applications 137 (2004), 175-186.

[196] S. Iliadis and V. Tzannes, *Spaces on which every continuous map into a given space is constant*, Can. J. Math. 38 (1986), 1281-1298.

[197] S.D. Iliadis and S.S. Zafiridou, *Rim-scattered spaces and the property of universality*, Colloquium Mathematica Societatis János Bolyai 55, Topology, 321-347. (Proceedings of the Colloquium on Topology, Pécs, Hungary, 1989.)

[198] S.D. Iliadis and S.S. Zafiridou, *Planar rational compacta and universality*, Fund. Math. 141 (1992), 109-118.

[199] R. Jajte, *Remarks on an universal space for spaces of the type* B_0, Bull. Soc. Sci. Letters Lódź 12 (1961), no. 4, 4pp.

[200] W.B. Johnson and A Szankowski, *Complementably universal Banach spaces*, Studia Math. 58 (1976), 91-97.

[201] C. Joiner, *On Urysohn's universal separable metric space*, Fund. Math. LXXIII (1971), 51-58.

[202] F.B. Jones and A.H. Stone, *Countable locally connected Urysohn spaces*, Colloq. Math. 22 (1971), 239-244.

[203] N.J. Kalton, *Universal spaces and universal bases in linear spaces*, Studia Math. 61 (1977), 161-191.

[204] V. Kannan, *A countable connected Urysohn space with a dispersion point*, Proc. Amer. Soc. 35(1) (1972), 289-290.

[205] V. Kannan and M. Rajagopalan, *Regularity and dispersion in countable spaces*, Duke Math. J. 39 (1972), 729-734.

[206] V. Kannan and M. Rajagopalan, *Construction and Applications of Rigid Spaces I*, Advances in Math. 29 (1978), 89-130.

[207] V. Kannan and M. Rajagopalan, *Construction and Applications of Rigid Spaces II*, Amer. J. Math. 100 (1978), 1139-1172.

[208] V. Kannan and M. Rajagopalan, *Construction and Applications of Rigid Spaces III*, Can. J. Math. 30(5) (1978), 926-932.

[209] S. Kaplan, *Homology properties of arbitrary subsets of Euclidean spaces*, Trans. Amer. Math. Soc. 62 (1947), 248-271.

[210] Z.Karno and J. Krasinkiewicz, *On some famous examples in dimension theory*, Fund. Math. 134 (1990), 213-220.

[211] M. Katětov, *On the dimension of non-separable spaces. I*, Csech. Math. Journ. 2 (1952), 333-368.

[212] M. Katětov, *On the relation between the metric and topological dimension*, Czech. Math. J. 8 (1958), no. 2, 163-166 (Russian).

[213] M. Katětov, *On universal metric spaces*, General Topology and its Relations to Modern Analysis and Algebra VI, Z. Frolík (ed.), Heldermann Verlag Berlin 1988, 323-330. (Proc. Sixth Prague Topological Symposium 1986.)

[214] H. Katsuura, *Dispersion points and continuous functions*, Topology Appl. 28 (1988), 233-240.

[215] K. Kawamura, L.G. Oversteegen, and E.D. Tymchatyn, *On homogeneous totally disconnected 1-dimensional spaces*, Fund. Math. 150 (1996), 97-112.

[216] K. Kawamura and K. Tsuda, *Universal spaces for finite-dimensional closed images of locally compact metric spaces*, Topology Appl. 85 (1998), 175-193.

[217] A.S. Kechris, *Classical Descriptive Set Theory*, Springer-Verlag 1995.

[218] A.S. Kechris and D.A. Martin, *A note on universal sets for classes of countable* G'_δ*s*, Mathematika 22 (1975), 43-45.

[219] D. Kent and G. Richardson, *Cauchy spaces with regular completions*, Pasific J. Math. 111 (1984), 105-116.

[220] A.M. Kirch, *A countable connected locally connected Hausdorff space*, Amer. Math. Monthly 76 (1969), 169-171.

[221] J.R. Kline, *A theorem concerning connected point sets*, Fund. Math. 2 (1922), 238-239.

[222] B. Knaster and K. Kuratowski, *Sur les ensembles connexes*, Fund. Math. 2 (1921), 206-255.

[223] A.N. Kolmogorov, *On some asymptotic characteristics of totally bounded metric spaces*, Dokl. Akad. Nauk SSSR 108 (1956), 385-388 (Russian).

[224] H.J. Kowalsky, *Einbettung metrischer Räume*, Arch. Math. 8 (1957), 336-339.

[225] I.M. Kozlovskiĭ, *Two theorems on metric spaces*, Dokl. Akad. Nauk SSSR 204 (1972), 784-787 (Russian); English translation in: Soviet Math Dokl. 13 (1972), 743-747.

[226] J. Krasinkiewicz, *On two theorems of Dyer*, Coll. Math. 50 (1986), no. 2, 201-208.

[227] J. Krasinkiewicz and P. Minc, *Nonexistence of universal continua for certain classes of curves*, Bull. Acad. Pol. Sci. Sér. Math. 9 (1976), 733-741.

[228] W. Kulpa, *On uniform universal spaces*, Fund. Math. LXIX (1970), 243-251.

[229] K. Kuratowski, *Sur les theéorèmes du "plongement" dans la theorie de la dimension*, Fund. Math. 28 (1937), 336-342.

[230] K. Kuratowski, *Topology I*, New York 1966.

[231] K. Kuratowski, *Topology II*, New York 1968.

[232] H.P.A. Künzi, *Quasi-uniform spaces – eleven years later*, Topology Proc. 18 (1993), 143-171.

[233] H.P.A. Künzi, *Nonsymmetric distances and their associated topologies: about the origins of basic ideas in the area of asymmetric topology*, C.E. Aull and R. Lowen (eds), Handbook of the History of General Topology, Volume 3, 853-968, Kluwer Academic Publishers 2001.

[234] H.P.A. Künzi, *Quasi-Uniform Spaces in the Year 2001*, M. Hušek and J. van Mill (eds), Recent Progress in General Topology II, 315-344. Elsevier Science B.V. 2002.

[235] H.E. Lacey and P.D. Morris, *On universal Spaces of the Type $C(X)$*, Proc. Amer. Math. Soc. 19 (1968), no. 2, 350-353.

[236] D.G. Larman and C.A. Rogers, *The descriptive character of certain universal sets*, Proc. London Math. Soc. (3) 27 (1973), 385-401.

[237] L.L. Larmore, *A universal countable first-countable Hausdorff space*, Duke Math. J. 39 (1972), 179-181.

[238] L.L. Larmore, *A connected countable Hausdorff space, S_α for every countable ordinal α*, Bol. Soc. Mat. Mexicana 2(17) (1972), 14-17.

[239] S. Lefschetz, *On compact spaces*, Ann. of Math. (2) 32 (1931), 521-538.

[240] A. Lelek, *On weakly chainable continua*, Fund. Math. 51 (1962), 271-282.

[241] A. Lelek, *Dimension and mappings of spaces with finite deficiency*, Colloq. Math. 12 (1964), 221-227.

[242] B.T. Levšenko, *On infinite-dimensional spaces*, Dokl. Acad. Nauk SSSR 139 (1961), 286-289 (Russian); English translation in: Soviet Math. Dokl. 2 (1961), 915-918.

[243] J. Lindenstrauss, *On the extension of operators with a finite-dimensional range*, Illinois J. Math. 8 (1964), 488-499.

[244] S.L. Lipscomb, *A universal one-dimensional metric space*, TOPO 72 - General Topology and its Applications, Second Pittsburgh Internat. Conf., Vol. 378 of Lecture Notes in Math., Springer-Verlag, New York, 1974.

[245] S.L. Libscomb, *On imbedding finite-dimensional metric spaces*, Trans. Amer. Math. Soc. 211 (1975), 143-160.

[246] S.L. Lipscomb and J.C. Perry, *Lipscomb's $L(A)$ space fractalized in Hilbert's $l^2(A)$ space*, Proc. Am. Math. Soc. 115 (1992), 1157-1165.

[247] E. Lowen-Colebunders, *On the regularity of the Kowalsky completion*, Canad. J. Math. 36 (1984), 58-70.

[248] L.A. Luxemburg, *On infinite-dimensional spaces with transfinite dimension*, Dokl. Akad. Nauk SSSR 199 (1971), no. 6, 1243-1246 (Russian).

[249] L.A. Luxemburg, *On transfinite D-dimension*, Abstracts of VI All-Union Topological Conference, Tbilisi 1972, 80-81 (Russian).

[250] L.A. Luxemburg, *On universal metric locally finite dimensional spaces*, General Topology and Appl.10 (1979), 283-290.

[251] L.A. Luxemburg, *On compact metric spaces with non-coinciding transfinite dimensions*, Pasific J. Math. 93 (1981), 339-386.

[252] L.A. Luxemburg, *On compactifications of metric spaces with transfinite dimensions*, Pasific J. Math. 101 (1982), 399-450.

[253] L.A. Luxemburg, *On universal infinite-dimensional spaces*, Fund. Math. 122 (1984), 129-147.

[254] G.W. Mackey, *Point realizations of transformations groups*, Illinois J. Math. 6 (1962), 327-335.

[255] T. Maćkowiak, *Planable and smooth dendroids*, General Topology and its Relations to Modern Analysis and Algebra IV, Proc. Fourth Prague Topological Symposium, 1976, Part B Contributed Papers, Society of Czechosklovak Mathematicians and Physicits, Prague 1977, 260-267.

[256] T. Maćkowiak, *A universal hereditarily indecomposable continuum*, Proc. Amer. Math. Soc. 94 (1985), 167-172.

[257] T. Maćkowiak and E.D. Tymchatyn, *Continuous mappings on continua II*, Dissertationes Math. (Rozprawy Mat.) 225 (1984), 1-57.

[258] S. Mardešić, *On covering dimension and inverse limits of compact spaces*, Ill. J. Math. 4 (1960), no. 2, 278-291.

[259] J.M. Martin, *Homogeneous countable connected Hausdorff spaces*, Proc. Amer. Math. Soc. 12 (1961), 308-314.

[260] J. Martin, *A countable Hausdorff space with a dispersion point*, Duke Math. J. 33 (1966), 165-167.

[261] J.C. Mayer, J.Nikiel and L.G. Oversteegen, *Universal spaces for R-trees*, Trans. Amer. Math. Soc. 334 (1992), no.1, 411-432.

[262] J.C. Mayer and L.G. Oversteegen, *Continuum theory*, Recent Progress in General Topology, M. Hušek and J. van Mill, editors, Elsevier Science Publishers B.V. (North-Holland) (1992), 453-492.

[263] J.C. Mayer, L.G. Oversteegen and E.D. Tymchatyn, *The Menger curve: Characterization and extension of homeomorphisms of non-locally separating closed subsets*, Dissertationes Math. CCLII (1986), 1-50.

[264] J.C. Mayer and E.D. Tymchatyn, *Containing spaces for planar rational compacta*, Dissertattiones Mathematicae, CCC, Warszawa, 1990.

[265] J.C. Mayer and E.D. Tymchatyn, *Universal rational spaces*, Dissertattiones Mathematicae, CCXCIII, Warszawa, 1990.

[266] S. Mazurkiewicz, *Sur les ensembles de dimension faible*, Fund. Math. 13 (1929), 210-217.

[267] S. Mazurkiewicz, *Sur les problème κ et λ de Urysohn*, Fund. Math. 10 (1927), 311-319.

[268] S.V. Medvedev, *Zero-dimensional homogeneous Borel sets*, Soviet Math. Dokl. 32 (1985), 144-147.

[269] M. Megrelishvili, *Compactification and factorization in the category of G-spaces*, Categorical Topology, ed. J. Adamek and S. MacLane, Word Scientific, Singapore, 1989, 220-237.

[270] M. Megrelishvili, *Free topological G-groups*, New Zealand J. Math. 25 (1996), 59-72.

[271] M. Megrelishvili and T. Scarr, *The equivariant universality and couniversality of the Cantor cube*, Fund. Math. 167 (2001), 269-275.

[272] K. Menger, *Überdie Dimension von Punktmengen II*, Monatsh. für Math. und Phys. 34 (1926), 137-161.

[273] K. Menger, *Allgemeine Räume und Cartesische Räume*, Erste Mitteilung: Proceedings of the Section of Sciences. Koninklijke Akademie van Wetenschappen te Amsterdam 29 (1926), 476-482; Zweite Mitteilung: *Über um fassendste n-dimensionale Mengen*, Proc. Akad. Wetensch. Amsterdam 29 (1926), 1125-1128.

[274] K. Menger, *Dimensionstheorie*, Teubner, Leipzig-Berlin 1928.

[275] K. Menger, *Kurventheorie*, Teubner, Leipzig-Berlin 1932 (reprinted by Chelsea Publ. Co., Bronx, N.Y. 1967).

[276] M.C. McCord, *Universal $\mathcal{P}$-like conpacta*, Michigan Math. J. 13 (1966), 71-85.

[277] G.J. Michaelides, *On countable connected Hausdorff spaces*, Studies and Essays, Mathematics Research Center, National Taiwan University Taipei, Taiwan, China, (October 1970), 183-189.

[278] J. van Mill, *Infinite-dimensional topology: prerequisites and introduction*, North-Holland Publishing Co., Amsterdam, 1989.

[279] J. van Mill, *The Infinite-Dimensional Topology of Function Spaces*, North-Holland Mathematical Library, vol 64, North-Holland Publishing Co., Amsterdam, 2001.

[280] J. van Mill and R. Pol, *On the existence of weakly n-dimensional spaces*, Proc. Amer. Math. Soc. 113 (1991), 581-585.

[281] J. van Mill and R. Pol, *Remark on products of 1-dimensional compacta*, Q&A in General Topology 13 (1995), no. 1, 97-98.

[282] J. van Mill and R. Pol, *Note on weakly n-dimensional spaces*, Monatsh. Math. 132 (2001), no. 1, 25-33.

[283] E.W. Miller, *Concerning biconnected sets*, Fund. Math. 29 (1937), 123-133.

[284] G.G. Miller, *A countable Urysohn space with an explosion point*, Notices Amer. Math. Soc. 13 (1966), 589.

[285] G.G. Miller, *Countable connected spaces*, Proc. Amer. Soc. 26 (1970), 355-360.

[286] Ur. Milutinović, *Completeness of the Lipscomb universal space*, Glasnik Matematički 27(47) (1992), 343-364.

[287] L. Mohler, *The depth of tranches in lambda-dendroids*, Proc. Amer. Math. Soc. 96 (1986), 715-720.

[288] L. Mohler and J. Nikiel, *A universal smooth dendroid answering a question of J. Krasinkiewicz*, Houston J Math. 14 (1988), 535-541.

[289] K. Morita, *Normal families and dimension theory for metric spaces*, Math. Ann. 128 (1954), 350-362.

[290] Y.N. Moschovakis, *Descriptive Set Theory*, North-Holland Publishing Company, Amsterdam, 1980.

[291] S. Mrówka, *On Universal Spaces*, Bull. de L'Acagémie Pol. des Sciences Cl. III- Vol. IV (1956), no. 8, 479-481.

[292] M.G. Murdeshwar and S.A. Naimpally, *Quasi-uniform topological spaces*, Noordhoff, Groningen, 1966.

[293] A. Mysior, *Universal compact T_1-spaces*, Canad. Math. Bull. 35 (1992), 261-266.

[294] A. Mysior, *A universal space for metrizable σ-discrete spaces*, Handwritten manuscript.

[295] A. Mysior, *A universal regular space*, Handwritten manuscript.

[296] A. Mysior, *There exist no universal spaces for* T_0- *and* T_1-*spaces.* Handwritten manuscript.

[297] S.B. Nadler, Jr, *Continuum theory*, Marcel Dekker 1992.

[298] K. Nagami, *Monotone sequences of 0-dimensional subsets of metric spaces*, Proc. Japan Acad. 41 (1965), 771-772.

[299] K. Nagami, *Dimension theory*, New York 1970.

[300] K. Nagami and J.H. Roberts, *Metric-dependent dimension functions*, Proc. Amer. Math. Soc. 16 (1965), 601-604.

[301] K. Nagami and J.H. Roberts, *Study of metric-dependent dimension functions*, Trans. Amer. Math. Soc. 129 (1967), 414-435.

[302] J. Nagata, *Note on dimension theory for metric spaces*, Fund. Math. 45 (1958), 143-181.

[303] J. Nagata, *On countable-dimensional spaces*, Proc. Japan Acad. 34 (1958), 146-149.

[304] J. Nagata, *On the countable sum of zero-dimensional metric spaces*, Fund. Math. 48 (1960), no. 1, 1-14.

[305] J. Nagata, *On universal n-dimensional set for metric spaces*, Journal Reine und Angew. Math. 204 (1960), no. 1-4, 132-138.

[306] J. Nagata, *A remark on general imbedding theorems in dimension theory*, Proc. Japan Acad. 36 (1963), 197-199.

[307] J. Nagata, *Some aspects of extension theory in general topology*, Internt. Sympos. on Extension Theory, Berlin, 1967, 157-161.

[308] J. Nagata, *Modern dimension theory*, Heldermann Verlag, Berlin, 1983.

[309] A.G. Nemets, *On a general method of approach to factorization theorems in dimension theory*, Seminar on General Topology, Moscow, Moscow Univ. Press (1981), 59-78 (Russian).

[310] A.G. Nemets and B.A. Pasynkov, *On two general methods of aproach to factorization theorems in dimension theory*, Dokl. Akad. Nauk SSSR 233 (1977), no. 5, 788 (Russian).

[311] T. Nishiura, *Inductive invariants and dimension theory*, Fund. Math. 59 (1966), 243-262.

[312] G. Nöbeling, *Über eine n-dimensionale Universalmenge im* R^{2n+1}, Math. Ann. 104 (1931), 71-80.

[313] G. Nöbeling, *Über regular eindimensionale Räme*, Math. Ann. 104 (1931), 81-91.

[314] G. Nöbeling, *Über die rationale Dimension*, Math. Ann. 109 (1934), 353-375.

[315] A. O'Connor and A. Rogers, *A topological dimension*, Journ. London Math. Soc. 46 (1992), 180-192.

[316] W. Olszewski, *Universal spaces for locally finite-dimensional and strongly countable-dimensional metrizable spaces*, Fund. Math. 135 (1990), 97-109.

[317] W. Olszewski, *On D-dimension of metrizable spaces*, Fund. Math. 140 (1991), 35-48.

[318] W. Olszewski, *Embeddings of finite-dimensional spaces into finite products of* 1*-dimensional spaces*, Topology and Appl. 40 (1991), 93-99.

[319] W. Olszewski, *Universal spaces in the theory of transfinite dimension, I*, Fund Math. 144 (1994), 243-258.

[320] W. Olszewski, *Universal spaces in the theory of transfinite dimension, II*, Fund. Math. 145 (1994), 121-139.

[321] W. Olszewski, *Universal separable metrizable spaces of given cohomological dimension*, Topology Appl. 61 (1995), 293-299.

[322] W. Olszewski and L. Piatkiewicz, *Closed embeddings of completely metrizable spaces into universal spaces*, Glasnik Matematički 27(47) (1992), 175-181.

[323] R.G. Ori and M. Rajagopalan, *On countable connected locally connected almost regular Urysohn space*, Topol. Appl. 17 (1984), 157-171.

[324] A.J. Ostaszewski, *Families of compact sets and their universals*, Mathematika 21 (1974), 116-127.

[325] L.G. Oversteegen and E.d. Tymchatyn, *On the dimension of certain totally disconnected spaces*, Proc. Amer. Math. Soc. 122 (1994), 885-891.

[326] K. Padmavally, *An example of a connected irresolvable Hausdorff space*, Duke Math. J. 20 (1953). 513-520.

[327] B. Pasynkov, *Relations between zero-dimensional mappings, universal spaces, dimensionality, open zero-dimensional mappings and inverse spectra*, Dokl. Akad. Nauk SSSR 144 (1962), 1217-1220 (Russian).

[328] P.A. Pasynkov, *On a generalization of topological products*, Dokl. Akad. Nauk SSSR 150 (1963), no. 1, 40-43 (Russian).

[329] B.A. Pasynkov, *On universal spaces for certain classes of spaces*, Dokl. Akad. Nauk SSSR 153 (1963), no. 5, 1009-1012 (Russian).

[330] B.A. Pasynkov, *On ω-mappings and inverse spectra*, Dokl. Akad. Nauk SSSR 150 (1963), 488-491 (Russian); English translation in: Soviet Math. Dokl. 4 (1963), 706-709.

[331] B.A. Pasynkov, *On universal bicompacta of given weight and dimension*, Dokl. Akad. Nauk SSSR 154 (1964), no. 5, 1042-1043 (Russian); English translation in: Soviet Math. Dokl. 5 (1964), 245-246.

[332] B.A. Pasynkov, *Partial topological products*, Dokl. Akad. Nauk SSSR 154 (1964), no. 4, 767-770 (Russian).

[333] B.A. Pasynkov, *Partial topological products*, Tr. Mosk. Mat. Ob. 13 (1965), 136-245.

[334] B.A. Pasynkov, *On universal compacta*, Uspehi Mat. Nauk 21 (1966), no. 4, 91-100 (Russian); English translation in: Russian Math Surveys 21 (1966), no. 4, 77-86.

[335] B.A. Pasynkov, *On universal bicompacta and metric spaces of given dimension*, Fund. Math. 60 (1967), 285-308.

[336] B.A. Pasynkov, *Factorization of mappings onto metric spaces*, Dokl. Akad. Nauk SSSR 182 (1968), 268-271 (Russian); English translation in: Soviet Math. Dokl. 9 (1968), 1140-1143.

[337] B.A. Pasynkov, *On the dimension of normal spaces*, Dokl. Akad. Nauk SSSR 201 (1971), 1049-1052 (Russian); English translation in: Soviet Math. Dokl. 12 (1971), 1784-1787.

[338] B.A. Pasynkov, *A factorization theorem for nonclosed sets*, Dokl. Akad. Nauk SSSR 202(1972), 1274-1276 (Russian); English translation in: Soviet Math. Dokl. 13 (1972), 292-295.

[339] J.C. Perry, *Lipscomb's universal space is the attractor of an infinite iterated function system*, Proc. Amer. Math. Soc. 124 (1996), 2479-2489.

[340] W.J. Pervin and J.L. Sieber, *Completeness in quasi-uniform spaces*, Math. Ann. 158 (1965), 79-81.

[341] E. Pol, *On J. Nagata's universal spaces*, Topology Appl. 26 (1987), 33-41.

[342] E. Pol, *Residuality of the set of embeddings into Nagata's n-dimensional universal spaces*, Fund. Math. 129 (1988), 59-67.

[343] E. Pol, R. Pol and M Reńska, *On countable-dimensional spaces with the Menger property, rational dimension and a question of S.D. Iliadis*, Monatsh. Math. 128 (1999), no. 4, 331-348.

[344] R. Pol, *There is no universal totally disconnected space*, Fund. Math. 79 (1973), 265-267.

[345] R. Pol, *On category-raising and dimension-raising open mappings with discrete fibers*, Coll. Math. 44 (1981), 65-76.

[346] R. Pol, *A weakly infinite-dimensional compactum which is not countable-dimensional*, Proc. Amer. Math. soc. 82 (1981), 634-636.

[347] R. Pol, *A remark on A-weakly infinite-dimensional spaces*, Topology and Appl. 13 (1982), 97-101.

[348] R. Pol, *On classification of weakly infinite-dimensional compacta*, Fund. Math. 116 (1983), 169-188.

[349] R. Pol, *Countable dimensional universal sets*, Trans. Amer. Math. Soc. 297 (1986), no. 1, 255-268.

[350] L.S. Pontrjagin and G.V. Tolstowa, *Beweis des Mengerschen Einbettungssatzes*, Math. Ann. 105 (1931), 734-747.

[351] L.S. Pontrjagin and L.G. Schnirelmann, *Sur une propriété métrique de la dimension*, Ann. of Math. 33 (1932), 156-162.

[352] J.R. Porter and R.G. Woods, *Subspaces of connected spaces*, Topology Appl. 68 (1996), 113-131.

[353] M. Rajagopalan and A. Wilansky, *Resersible Topological Spaces*, J. of Austral. Math. Soc. 6 (1966), 129-138.

[354] A. Ramer, *Some Problems on Universal Spaces*, Bull. Acad. Pol. Sci. Sér. Math. XIII (1965), no. 4, 291-294.

[355] E. Reed, *Completion of uniform convergence spaces*, Math. Ann. 194 (1971), 83-108.

[356] H. Render, *Nonstandard methods of completing quasi-uniform spaces*, Topology Appl. 62 (1995), 101-125.

[357] H. Reschowsky, *Über rationale Kurven*, Fund. Math. 15 (1930), 18-37.

[358] G.X. Ritter, *A connected, locally connected countable Hausdorff space*, Amer. Math. Monthly 83 (1976), 185-186.

[359] G.X. Ritter, *A connected, locally connected countable Urysohn space*, Gen. Top. Appl. 7 (1977), 65-70.

[360] P. Roy, *A countable connected Urysohn space with a dispersion point*, Duke Math. J. 33 (1966), 331-334.

[361] L. Rubin, R.M. Schori, and J.J. Walsh, *New dimension-theory techniques for constructing infinite-dimensional examples*, Gen. Top. Appl. 10 (1979), 93-102.

[362] S. Salbany, *Bitopological Spaces, Compactifications and Completions*, Math. Monographs, No. 1, Dept. Math. Univ. Cape Town 1974.

[363] J.B. Sancho and M.T. Sancho, *Dimension of distributive lattices and universal spaces*, Topology Appl. 42 (1991), 25-36.

[364] N.V. Savinov, *An example of a perfectly normal bicompactum without intermediate dimensions*, Vest. Mosk. Univ. Ser. Mat. (1976), no. 3, 52-56 (Russian); English translation in: Moscow Univ. Math. Bull. 31 (1976), no. 3-4, 40-43.

[365] T. Scarr, *Topological and Borel compactifications of Polish G-spaces*, Topology and its Applications 105 (2000), 113-119.

[366] E.V. Ščepin, *Topology of limit spaces of uncountable inverse spectra*, Uspekhi Mat. Nauk 31 (1976), 191-226 (Russian).

[367] E.V. Ščepin, *Functors and uncountable powers of compacta*, Uspekhi Mat. Nauk 36 (1981), 3-62 (Russian).

[368] R.M. Schori, *A universal snake-like continuum*, Proc. Amer. Math. Soc. 16 (1965), 1313-1316.

[369] N.A. Shanin, *On special extensions of topological spaces*, Dokl. Akad. Nauk SSSR 38 (1943), 6-9 (Russian).

[370] N.A. Shanin, *On special extensions of topological spaces*, Dokl. Akad. Nauk SSSR 38 (1943), 110-113 (Russian).

[371] N.A. Shanin, *On the theory of bicompact extensions of topological spaces*, Dokl. Akad. Nauk SSSR 38 (1943), 154-156 (Russian).

[372] M.A. Shtan'ko, *Solution of Menger's problem in the class of compacta*, Dokl. Akad. Nauk SSSR 201 (1971), 1299-1302 (Russian); English translation in: Soviet Math. Dokl. 12 (1971), 1846-1849.

[373] R. Sikorski, *Some examples of Borel sets*, Coll. Math. 5 (1958), 170-171.

[374] W. Sierpiński, *Sur une courbe cantorienne qui contient une image biunivoque et continue de toute courbe donnée*, Comptes Rendus de l'Acadé mie de paris 162 (1916), 629-632; Mat. Sb. 30 (1916),

267-287 (Russian); French translation in: Wacław Sierpiński, Oeuvres choisies, vol. II, PWN Warszawa 1974, 99-106.

[375] W. Sierpiński, *Sur les ensembles connexes et non connexes*, Fund. Math. 2 (1921), 81-95.

[376] W. Sierpiński, *Sur une propriété des ensembles frontières*, Fund. Math. 3 (1922), 7-13.

[377] W. Sierpiński, *Sur un espace métrique séparable universel*, Fund. Math. 33 (1945), 115-122.

[378] W. Sierpiński, *General Topology*, Univ. of Toronto Press, Toronto 1952.

[379] P. Simon, *An example of maximal connected Hausdorff space*, Fund. Math. 100 (2) (1978), 157-163.

[380] K.A. Sitnikov, *An example of two-dimensional set in three-dimensional Euclidean space allowing arbitrarily small deformations into a one-dimensional polyedron*, Dokl. Akad. Nauk SSSR 67 (1949), 229-232 (Russian).

[381] K.A. Sitnikov, *An example of two-dimensional set in three-dimensional Euclidean space allowing arbitrarily small deformations into a one-dimensional polyedron and some new character of the dimension of the sets in Euclidean spaces*, Dokl. Akad. Nauk SSSR 88 (1953), 21-24 (Russian).

[382] K.A. Sitnikov, *Combinatorial topology of nonclosed sets. I*, Mat. Sb., Nov. Ser. 34 (1954), no. 1, 3-54 (Russian).

[383] Yu.M. Smirnov, *On the metric dimension in the sence of P.S. Alexandrov*, Izv. Akad. Nauk SSSR, Ser. Mat. 20 (1956), 679-684 (Russian).

[384] Yu.M. Smirnov, *An example of a zero-dimensional space which has infinite covering dimension*, Dokl. Akad. Nauk SSSR 123 (1958), 40-42 (Russian).

[385] Yu.M. Smirnov, *On universal spaces for certain classes of infinite-dimensional spaces*, Izv. Akad. Nauk SSSR Ser. Mat. 23 (1959),

185-196 (Russian); English translation in: Amer. Math. Soc. Transl. Ser. 2, 21 (1962), 21-33.

[386] Yu.M. Smirnov, *Über die Dimension der Adjunkten bei Kompaktifizierungen*, Monatsb. Deutsch. Akd Wiss. Berlin 7 (1965), 230-232.

[387] B. Sołtys, *Complementably universal spaces for p-Banach spaces of finite dimension*, Annales Societatis Mathematicae Polonae, Series I: Commentationes Mathematicae XXIII (1983), 133-141.

[388] R.A. Stoltenberg, *Some properties of quasi-uniform spaces*, Proc. London Math. Soc. 17 (1967), 226-240.

[389] R.A. Stoltenberg, *Completion for quasi-uniform spaces*, Proc. Amer. Math. Soc. 18 (1967), 864-867.

[390] A.H. Stone, *Paracompactness and product spaces*, Proc. Amer. Math. Soc., 54 (1948), 977-982.

[391] A.H. Stone, *Universal spaces for some metrizable uniformities*, Quart. J. Math. Oxford (2), 11 (1960), 105-115.

[392] A.H. Stone, *Non-separable Borel sets*, Rozprawy Matematyczne 28 (1962), 1-41.

[393] A.H. Stone, *A countable connected locally connected Hausdorff space*, Notices Amer. Math. Soc. 16 (1966), 422.

[394] M.A. Swardson, *A short proof of Kowalsky's hedgehog theorem*, Proc. Amer. Math. Soc. 75 (1979), 188.

[395] P.M. Swingle, *Generalizations of biconnected sets*, Amer. Journal of Math. 53 (1931), 385-400.

[396] P.M. Swingle, *Two types of connected sets*, Bull. Amer. Math. Soc. 37 (1931), 254-258.

[397] J.P. Thomas, *Maximal connected topologies*, J. Austral. Math. Soc. 8 (1968), 700-705.

[398] B. Tomaszewski, *On weakly n-dimensional spaces*, Fund. Math. 103 (1979), 1-8.

[399] H. Toruńczyk, *On CE-images of the Hilbert cube and characterizations of Q-manifolds*, Fund. Math. 106 (1980), 31-40.

[400] H. Toruńczyk, *Characterizing Hilbert space topology*, Fund. Math. 111 (1981), 247-262.

[401] C.H. Toulmin, *Shuffling ordinals and transfinite dimension*, Proc. London Math. Soc. 4, (1954), no. 4, 177-196.

[402] V.I. Trofimov, *On a property of Urysohn's universal metric space*, Mat. Zam. 69 (2001), no. 2, 319-320.

[403] K. Tsuda, *Non-existence of universal spaces for some stratifiable spaces*, Topology Proceedings 9 (1984), 165-171.

[404] K. Tsuda, *Dimension theory of general spaces*, Doctoral dissertation, University of Tsukuba (1985).

[405] K. Tsuda, *A black box left by van Douwen inside out*, Questions Answers Gen. Topology 11 (1993), 15-18.

[406] K. Tsuda, *Universal spaces for closed images of σ-discrete metric spaces*, Canad. Math. Bull. 37 (1994), 419-427.

[407] K. Tsuda, *Universal spaces for zero-dimensional closed images of metric spaces*, Yokohama Math. J. 46 (1998), 29-69.

[408] K. Tsuda and S. Watson, *A universal space for closed images of rationals*, Top. Appl. 54 (1993), 65-76.

[409] S.F. Tsvid (Čvid), *Countable connected spaces*, Current questions in Mathematical Logic and Set Theory (1975), 276-284. Moskov. Gos. Ped. Inst. im. Lenina, Moscow (Russian).

[410] S.F. Tsvid (Čvid), *A countable strongly unicoherent space*, Mat. Zametki 24 (1978), no. 2, 289-294, 303 (Russian); English translation in: Math. Notes 24 (1978), no. 1-2, 655-657).

[411] L.A. Tumarkin, *Beitrag zur allgemeinen Dimensionstheorie*, Mat. Sb. 33 (1926), 57-86.

[412] L.A. Tumarkin, *Sur la decomposition des espaces en une somme denombrable d'ensembles de dimension null*, Bull. Moscow Univ. 1 (1960), 25-32.

[413] A.N. Tychonoff, *Über die topologische Erweiterung von Räumen*, Math. Ann. 102 (1930), 544-561.

[414] E.D. Tymchatyn, *On the rim-types of hereditarily locally connected continua*, Fund. Math. LXXXIX (1975), 93-97.

[415] V. Tzannes, *Three countable connected spaces*, Colloq. Math. 2 (1988), 267-279.

[416] V. Tzannes, *A note on countable connected locally connected Urysohn almost regular spaces*, Colloq. Math. LXII (2) (1991), 321-323.

[417] V. Tzannes, *A Moore strongly rigid space*, Canad. Math. Bull. 34(4), (1991), 547-552.

[418] V.Tzannes, *Two Moore spaces on which every continuous real-valued function is constant*, Tsukuba J. Math. 16 (1992), no. 1, 203-210.

[419] V. Tzannes, *Every countable regular space without isolated points is connectifiable*, Can. Math. Bull. 37 (4) (1994), 556-559.

[420] V.M. Ul'janov, *Solution of a basic problem on compactifications of Wallman type*, Soviet Math. Doklady 18 (1977), 567-571 (Russian).

[421] L. Úry, *On relative universal embedding spaces*, Acta Math. Acad. Sci. Hungar. 28 (1976).

[422] P. Urysohn, *Sur un espace métrique universel*, C.R. Acad. Sci. Paris 180 (1925), 803-806.

[423] P. Urysohn, *Über die Machitigeit der zusammenhangenden Mengen*, Math. Ann. 94 (1925), 262-295.

[424] P.S. Urysohn, *Zum Metrisationsproblem*, Math. Ann. 94 (1925), 309-315.

[425] P. Urysohn, *Mémoire sur les multiplicités Cantoriennes (suite)*, Fund. Math. 8 (1926), 225-359.

[426] P. Urysohn, *Sur un espace métrique universel*, Bull. Sci. Math. 51 (1927), 1-38.

[427] V.S. Varadarajan, *Groups of automorphisms of Borel spaces*, Trans. Amer. Math. Soc. 109 (1963), 191-220.

[428] N.B. Vedenisov, *Remarks on dimension of topological spaces*, Uchen. Zap. Mosk. Univ. 30 (1939), 131-140 (Russian).

[429] E.J. Vought, *A countable connected Urysohn space with a dispersion point that is regular almost everywhere*, Colloq. Math. 28 (1973), 205-209.

[430] J. de Vries, *Universal topological transformation groups*, General Topology Appl. 5 (1975), 107-122.

[431] J. de Vries, *Elements of Topological Dynamics*, Kluwer Academic Publishers, Dordrecht, 1993.

[432] H. Wallman, *Lattices and topological spaces*, Annals of Math. 39 (1938), 112-126.

[433] A. Waśko, *Spaces universal under closed embeddings for finite dimensional complete metric spaces*, Bull. London Math. Soc. 18 (1986), 293-298.

[434] S. Watson, *Embedding a Space in a Connected Space*, York University Report no. 91-03.

[435] S. Watson and R.G. Wilson, *Embeddings in Connected Spaces*, Houston J. Mah. 19 (3) (1993), 469-481.

[436] T. Wazewski, *Sur les courbes de Jordan ne refermant aucune courbe fermée de Jordan*, Annales de la Société Polanaise de Mathématique 2 (1923), 49-170.

[437] A. Weil, *Sur les espaces à structure uniforme et sur la topologie générale*, Paris, 1938.

[438] B.R. Wenner, *Finite-dimensional properties of infinite-dimensional spaces*, Pasific J. Math. 42 (1972), 267-276.

[439] B.R. Wenner, *A universal separable metric locally finite-dimensional spaces*, Fund. Math. 80 (1973), 283-286.

[440] G.T. Whyburn, *Topological characterization of the Sierpiński curve*, Fund. Math. 45 (1958), 320-324.

[441] R. Wilder, *A point set which has no true quasicomponents and which becomes connected on the addition of a single point*, Bull. Amer. Math. Soc., 33 (1927), 423-427.

[442] J. Williams, *Nested sequences of local uniform spaces*, Trans. Amer. Math. Soc. 168 (1972), 471-481.

[443] S. Zafiridou, *A note about planar completely regular continua with structure*, Topology Appl. 123 (2002), 199-203.

[444] I.P. Zaharov, *On universal infinite-dimensional spaces*, V Tiraspol Simposium on General Topology and its Applications, Kishinew 1985, 99-100 (Russian).

[445] I.P. Zaharov, *On universal metrizable spaces of given weight and depth*, Vestnik Moscow University, Ser. Mat., 6 (1987), 6-9 (Russian).

[446] I.P. Zaharov, *On universal locally n-dimensional and locally finite-dimensional spaces*, General Topology, Spaces and Mappings, Moscow 1989, 152-155 (Russian).

[447] A.V. Zarelua, *A universal bicompact of the given weight and dimension*, Dokl. Akad. Nauk SSSR 154 (1964), no. 5, 1015-1018 (Russian); English translation in: Soviet Math. Dokl. 5 (1964), 214-218.

[448] A. Zarelua, *On extending mappings over extensions which have some special properties*, Siberian Math. J. 5 (1964), 532-548 (Russian).

[449] A. Zarelua, *Algebras of continuous functions of some universal compacta*, Fourth Russian-Japanese Colloquium on General Topology, May 1995, Moscow State University, 1995, 35.

[450] A.V. Zarelua, *The algebraic structure of the ring of functions of some universal spaces*, Fundam. i Prikl. Mat. 4 (1998), no. 1, 81-100 (Russian).

Special Symbols

A_s^X, 11, 38, 435
$\mathrm{A}_s^{\mathbf{H}}$, 16, 42
$\mathrm{A}(2^s_X)$, 39
$\mathrm{A}(2^s_{\mathbf{H}})$, 42
(A^X, X), 164
$\mathbf{a}_i^{\mathrm{T}(\mathbf{E})}$, 396
a^{-1}, 292

$\mathbf{B}$, 6, 78
$\mathbf{B}(X)$, 6
$\mathbf{Bd}$, 64
$\mathbf{Bd}(\mathbf{Q})$, 31
$\mathrm{Bor}(X)$, 65
$\mathrm{Bd}_X(Q)$, 6
$\mathrm{B}(\sim^\infty_{\mathrm{R}})$, 487
$\mathrm{B}^{\mathrm{T}}_*$, 18, 43
$\mathrm{B}^{\mathrm{T}}_{*,\kappa}$, 18, 43
$\mathrm{B}^{\mathrm{T}}_\kappa$, 21, 43
$\mathrm{B}^{\mathrm{T}}_{\diamondsuit,\kappa}$, 21, 44
$\mathrm{B}^{\mathbf{L}}_\kappa$, 22
$\mathrm{B}^{\mathbf{L}}_{\diamondsuit,\kappa}$, 22
B^{T}, 22, 43
$\mathrm{B}^{\mathrm{T}}_\diamondsuit$, 22, 43
$\mathrm{B}^{\mathbf{L}}$, 22
$\mathrm{B}^{\mathbf{L}}_\diamondsuit$, 22
$\mathrm{B}^{\mathrm{T}|\mathbf{Q}}$, 25
$\mathrm{B}^{\mathrm{T}|\mathbf{Q}}_\kappa$, 25
$\mathrm{B}^{\mathrm{T}|\mathbf{Q}}_\diamondsuit$, 25
$\mathrm{B}^{\mathrm{T}|\mathbf{Q}}_{\diamondsuit,\kappa}$, 25
$\mathrm{B}^{\mathbf{E}}$, 25
$\mathrm{B}^{\mathbf{E}}_\kappa$, 25
$\mathrm{B}^{\mathbf{E}}_\diamondsuit$, 25
$\mathrm{B}^{\mathbf{E}}_{\diamondsuit,\kappa}$, 25
$\mathrm{B}^{\mathrm{T},\diamondsuit}_k$, 186
$\mathrm{B}^{\mathbf{L},\diamondsuit}_k$, 186
$\mathrm{B}^{\mathrm{T}(c)}_{\diamondsuit,\mathbf{N}}$, 366
$\mathrm{B}^{\mathbf{L}}_{\diamondsuit,\mathbf{N}}$, 366
$\mathrm{B}^{\mathrm{T}/\mathrm{T}|\mathbf{Q}}$, 475
$B^{\lim(D)}$, 382, 383
B^Q, 5
$B^\gamma_\nu(X,\Phi,\mathcal{G})$, 66
(B, X), 74
b-dim$^{\mathbb{F}}$, 145
b-dim^{Op}, 149
b-dim^{Cl}, 149
b-dim^{rCl}, 149
b^n-dim, 231
b$^\vee$-Ind$^{\mathbb{F}}_{\mathbb{B}}$, 133
b$^\vee$-Ind, 134
b-Ind$^{\mathbb{F}}_{\mathbb{B}}$, 143
b$^\diamondsuit$-Ind$^{\mathbb{F}}_{\mathbb{B}}$, 143
b^n-Ind, 209
b$_0$-ind$^{\mathbb{B}}_{\mathbb{L}}$, 107

b_0-ind, 107
b_1-ind$^{\mathbb{B}}_{\mathbb{L}}$, 112
b_1-ind, 112
b^0-ind$^{\mathbb{B}}_{\mathbb{L}}$, 113
b^0-ind, 113
b^1-ind$^{\mathbb{B}}_{\mathbb{L}}$, 114
b^1-ind, 114
b^p-ind$^{\mathbb{D}}_{\mathbb{L}}$, 115
b^{pos}-ind$^{\mathbb{D}}_{\mathbb{L}}$, 116
b^{ps}-ind$^{\mathbb{D}}_{\mathbb{L}}$, 116
b^p-ind, 116
b^{pos}-ind, 116
b^{ps}-ind, 116
b(0)-ind$_{\mathbb{L}}$, 117
b(0)-ind, 117
b(1)-ind$^{\mathbb{D}}_{\mathbb{L}}$, 118
b(1)-ind, 118
b_0-p_i-ind$^{\mathbb{B}}_{\mathbb{N}}$, $i \in \{0,1\}$, 125
b_0-p_i-ind, $i \in \{0,1\}$, 126
b_1-p_i-ind$^{\mathbb{B}}_{\mathbb{N}}$, $i \in \{0,1\}$, 127
b_1-p_i-ind, $i \in \{0,1\}$, 127
b^0-p_i-ind$^{\mathbb{B}}_{\mathbb{N}}$, $i \in \{0,1\}$, 128
b^0-p_i-ind, $i \in \{0,1\}$, 129
b^1-p_i-ind$^{\mathbb{B}}_{\mathbb{N}}$, $i \in \{0,1\}$, 129
b^1-p_i-ind, $i \in \{0,1\}$, 130
$b(L_D)$, 275

Cl, 64
Co, 64
Cl(**Q**), 31
Co(**Q**), 31
C, 335
C^R, 337
C(∼), 4
$Cl_X(Q)$, 6
$Co_X(Q)$, 6
C(R), 7, 41
$C^{\diamondsuit}(R)$, 7
$C^{\diamondsuit}_{\nu}(R)$, 41
$C^{T,\diamondsuit}_k$, 186
$C^{\mathbf{L},\diamondsuit}_k$, 186
$C(\mathbb{E},\mathbb{L})$, 247
$C^{op}(\mathbb{E},\mathbb{L})$, 249
$C^{bs}(\mathbb{E},\mathbb{L})$, 251
$C^{cbc}(\mathbb{E},\mathbb{L})$, 253
$C(\mathbf{S})$, 348
$C_0(\mathbf{S})$, 361
$C_{\mathbf{Q}}(\mathbf{S})$, 362
$C_{\mathbf{N}}(\mathbf{S})$, 362
$C_{\mathbf{\Psi}}(\mathbf{S})$, 362
$C_{\mathbf{Q},\mathbf{N}}(\mathbf{S})$, 362
$C^{\mathbb{P}}(\mathbf{S})$, 366
$C^{\mathbb{P}}_{\mathbf{Q}}(\mathbf{S})$, 367
$C^{\mathbb{P}}_{\mathbf{N}}(\mathbf{S})$, 367
$C^{\mathbb{P}}_{\mathbf{Q},\mathbf{N}}(\mathbf{S})$, 367
$C^{\mathbb{P}}_{\mathbf{\Psi}}(\mathbf{S})$, 367
$C_{\mathbb{P}}(\mathbf{S})$, 370
$C^{\diamondsuit}(c)$, 365
C^Q, 215
$c(f)$, 278
$c_Y(f)$, 287
$c(G,C)$, 212
$c(C)$, 212
c^X_M, 433
$c(\lambda)$, 67
$c|_{\mathbf{Q}}$, 371

D, 266
$\mathbf{D}(\lambda)$, 267
$\mathbf{D}(\lambda_0,\lambda_1)$, 267
$(\mathbf{D}, F^{\mathbf{D}})$, 323
D, 166
D[**S**], 353
Diam(Q), 391
D, 265
D_f, 2

D^{T}, 271
$D^{\mathrm{T}}_{\mathbf{H}}$, 271
$D(c, \infty)$, 337
$D|_{\mathrm{C}'}$, 339
$D(\mathrm{C}', c')$, 339
$(D^{\mathrm{T}}, F^{D^{\mathrm{T}}})$, 325
$\mathbf{d}$, 417
dim, 144
d^X_s, 11, 38

E_X, 380
$E_{\lim(D)}$, 382, 383
e, 292
e^X_{T}, 23, 45
$e^{\mathrm{T}|\mathbf{Q}}_{\mathrm{T}}$, 26
$e^{\mathrm{T}|\mathbf{K}}_{\mathrm{T}|\mathbf{Q}}$, 29

$\mathcal{F}$, 1
$\mathcal{F}_k$, 182
F^D, 321
$F^{D^{\mathrm{T}}}$, 325
$F^{\lim(D)}$, 329
$F^{G\times\mathrm{T^r}}$, 296
$F^{G\times\mathrm{T^r}}_{\mathbf{L}}$, 296
f_{C}, 338
f^{T}, 244
$f^{\mathrm{T}}_{\mathbf{H}}$, 244
$f^{\mathrm{T}}_{\mathrm{Y}}$, 257
$f^{\mathrm{T}}_{\mathrm{Y},\mathbf{H}}$, 258
f_Λ, 266
$f^{\mathrm{T}}_{\mathbf{H}}(\lambda_0, \lambda_1)$, 271
$\overline{f}$, 341

$\mathcal{G}^\diamondsuit$, 5
G, 292
$G \otimes D$, 322
$G \otimes \mathbf{D}$, 322
$G \otimes \mathbf{S}$, 293
$[G, \mathrm{T^r}]$, 296
$g[x, X]$, 357
$[g, \mathbf{a}]$, 295

$\mathbf{H}|_{\mathbf{Q}}$, 8
$\mathbf{H}^d$, 242
$\mathbf{H}^r$, 242
$\mathbf{H}(\lambda)$, 268
$\mathbf{H}(\sim^\infty_{\mathrm{R}})$, 487
$h_\wedge$, 321

Int(**Q**), 31
Int, 64
Ind, 132
$\mathrm{Int}_X(Q)$, 6
I_f, 2
id_G, 301
$\mathrm{id}_G \times h$, 301
ind, 91
$\mathrm{ind}^{\mathbb{B}}_{\mathbb{E}}$, 92
$\mathrm{ind}_x(X)$, 459
i^s_X, 12
i^X_C, 215
i^X_c, 357
$i^X_{\lim(\mathrm{D}[\mathbf{S}])}$, 357
i^X_{T}, 400

$\mathbb{K}$-df, 105
$K(N, v)$, 133
$K(X)$, 459
$k_0(X, n, i, j)$, 393
$k_1(X, n, i, j)$, 393
$k(X, n, i)$, 393

L_D, 275
$L_D(U_{\lambda_1}, ..., U_{\lambda_n})$, 275
$\lim i_\Lambda$, 276
$\lim f_\Lambda$, 277

$\lim(D)$, 337
$\lim(f_C)$, 338
$(\lim(D), F^{\lim(D)})$, 329
$l(\lambda)$, 67

$\mathbb{M}(\mathbf{d}, \mathbf{n})$, 421
$\mathbb{M}(\mathrm{rs})$, 471
$\mathbb{M}$-df, 123
$\mathbb{M}^s$-df, 131
$\mathbb{M}^p$-df, 131
$\mathbf{M}$, 6
$\mathbf{M}(X)$, 6
$\mathbf{M}|_{\mathbf{Q}}$, 7
$\mathbf{M}^d$, 241
$\mathbf{M}^r$, 241
$\overline{\mathbf{M}}$, 266
$\overline{\mathbf{M}}(\lambda)$, 267
$(\mathbf{M}^d, \mathbf{M}^r)$, 242
M^D, 266
M^Q, 5

$\mathbf{n}$, 421
$\mathbf{n}(\varepsilon)$, 417
$\mathbf{n}^X$, 421
$n(\alpha)$, 162
$n(X, Y)$, 401
$n(\mathbf{a}, \mathbf{b})$, 401

$\mathcal{O}$, 3
$\mathcal{O}^*$, 162
$\mathbf{O}$, 61
$\mathbf{O}(\mathbf{Q})$, 62
$\mathbf{O}(\mathbb{P})$, 62
$\mathbf{O}^{-1}(\mathbb{P})$, 62
$\mathrm{Odim}_{\mathbb{E}}^{\mathbb{B}}$, 453
O_X, 61

$\mathcal{P}(X)$, 4
$\mathcal{P}_\nu(\tau)$, 36
$\mathbb{P}(\mathrm{Cl})$, 57
$\mathbb{P}(\mathrm{Op})$, 57
$\mathbb{P}(\mathrm{n.dense})$, 57
$\mathbb{P}(\gamma, \nu, \Phi, \mathbb{F})$, 71
$\mathbb{P}(G_\delta\text{-set})$, 71
$\mathbb{P}(F_\sigma\text{-set})$, 71
$\mathbb{P}(\mathbb{K}\text{-cnfl})$, 481
$\mathbb{P}(\mathrm{base})$, 74
$\mathbb{P}(\mathrm{p\text{-}base})$, 78
$\mathbb{P}(\mathrm{pos\text{-}base})$, 78
$\mathbb{P}(\mathrm{ps\text{-}base})$, 78
$\mathbb{P}(\mathrm{T}_3\text{-p-base})$, 82
$\mathbb{P}(\mathrm{T}_3^s\text{-p-base})$, 82
$\mathbb{P}(\mathrm{T}_3^c\text{-p-base})$, 82
$\mathbb{P}(\mathrm{T}_3^h\text{-p-base})$, 82
$\mathbb{P}(df \leq \alpha)$, 90
$\mathbb{P}(\mathrm{subset})$, 91
$\mathbb{P}(\emptyset)$, 91
$\mathbb{P}(\emptyset, \emptyset)$, 91
$\mathbb{P}(\emptyset, \mathrm{space})$, 91
$\mathbb{P}(\{\emptyset\}, \emptyset)$, 91
$\mathbb{P}(\emptyset, \{\emptyset\}, \emptyset)$, 91
$\mathbb{P}(\emptyset, \{\emptyset\}, \mathrm{space})$, 91
$\mathbb{P}(\mathrm{rCl})$, 149
$\mathbb{P}(\mathrm{crs\text{-}cnfl})$, 482
$\mathbb{P}(\mathrm{crs\text{-}closed\text{-}cnfl})$, 482
$\mathbb{P}(\mathrm{crs\text{-}open\text{-}cnfl})$, 482
$\mathbb{P}(\mathrm{crs\text{-}cnfl})$, 482
$\mathbb{P}(\mathrm{rs\text{-}disp})$, 485
$\mathbb{P}(\mathrm{rs\text{-}closed\text{-}cnfl})$, 482
$\mathbb{P}(\mathrm{rs\text{-}open\text{-}cnfl})$, 482
$\mathbb{P}(\mathrm{rs\text{-}disp})$, 485
$\mathbb{P}(\mathrm{rs\text{-}Y})$, 481
$\mathbb{P}_\lambda$, 152
$\mathbb{P}(\Phi, \Lambda, \prec, \mathbf{p})$, 153
$\mathbb{P}(\Lambda, \prec, \mathbf{p})$, 153
$\mathbb{P}((\beta + n)\text{-rep})$, 164

$\mathbb{P}(\mathrm{D} \leq \beta + n)$, 166
$\mathbb{P}(\Lambda, \prec_t, \mathbf{p}_{\mathrm{O}})$, 169
$\mathbb{P}(\Lambda, \prec_t, \mathbf{p}_{\mathrm{Cl}})$, 169
$\mathbb{P}(\Lambda, \prec_t, \mathbf{p}_{\mathrm{Op}})$, 169
$\mathbb{P}(\Lambda, \prec_o, \mathbf{p}_{\mathrm{Cl}})$, 169
$\mathbb{P}(\Lambda, \prec_o, \mathbf{p}_{\mathrm{Op}})$, 169
$\mathbb{P}(\Lambda, \prec_o, \mathbf{p}_{\mathrm{O}})$, 169
$\mathbb{P}(\subset \text{normal base})$, 191
$\mathbb{P}(\mathrm{b}^n\text{-Ind} \leq m)$, 210
$\mathbb{P}(\text{s-b}^n\text{-Ind} \leq m)$, 210
$\mathbb{P}_n(\text{s-b}^n\text{-Ind} \leq m)$, 495
$\mathbb{P}(\mathrm{Ind}, \nu, \xi, \prec)$, 222
$\mathbb{P}(\mathrm{b}^n\text{-dim} \leq m)$, 231
$\mathbb{P}(\text{s-b}^n\text{-dim} \leq m)$, 232
$\mathbb{P}_n(\text{s-b}^n\text{-dim} \leq m)$, 495
$\mathbb{P}(\dim, \nu, \xi, \prec)$, 237
$\mathbb{P}(\lim \mathbb{D})$, 277
$\mathbb{P}(\text{compact})$, 280
$\mathbb{P}_c(\mathrm{Ind} \leq n)$, 280
$\mathbb{P}_c(\dim \leq n)$, 281
$\mathbb{P}(G\text{-subset})$, 308
$\mathbb{P}(\mathrm{Hs})$, 442
$\mathbb{P}(\text{Hs-disp})$, 445
$\mathbb{P}(\text{Hs-Y})$, 481
$\mathbb{P}(\text{Hs-cnfl})$, 482
$\mathbb{P}(\text{Hs-closed-cnfl})$, 482
$\mathbb{P}(\text{Hs-open-cnfl})$, 482
$\mathbb{P}(\text{mt-}\alpha\text{-add})$, 447
$\mathbb{P}(\text{mt-}\alpha\text{-mult})$, 447
$\mathbb{P}(\text{space})$, 334
$\mathbb{P}(\text{subset})$, 334
$\mathbb{P}(\mathrm{T}_1)$, 442
$\mathbb{P}(\mathrm{T}_1\text{-disp})$, 445
$\mathbb{P}(\Phi, \Lambda, \prec)$, 335
$\mathbb{P}(\Lambda, \prec, \mathbf{p})$, 335
$\mathbb{P}^*(\text{subset})$, 380
$\mathbb{P}^*(\text{base})$, 380
$\mathbb{P}^*(\text{p-base})$, 380
$\mathbb{P}^*(\Phi, \Lambda, \prec)$, 380
$\mathbb{P}(\mu\dim \leq r)$, 424
$\mathbb{P}(\text{subset} + \mathrm{r})$, 413
$\mathbb{P}(X)$, 489
$\mathrm{P}(\mathbf{S})$, 347
P_k, 182
P^X, 393
$\mathbf{p}$, 418
$\mathbf{p}_{\mathrm{O}}$, 169
$\mathbf{p}_{\mathrm{Cl}}$, 169
$\mathbf{p}_{\mathrm{Op}}$, 169
$\mathbf{p}(\mathbf{d}, k(s))$, 419
$\text{p-b}_i\text{-p}_j\text{-ind}$, $i, j \in \{0,1\}$, 131
$\text{p-b}^i\text{-p}_j\text{-ind}$, $i, j \in \{0,1\}$, 131
$\mathrm{p}_i\text{-ind}_{\mathbb{F}}^{\mathbb{D}}$, $i \in \{0,1\}$, 98
$\mathrm{p}_i\text{-ind}$, $i \in \{0,1\}$, 98
$\mathrm{pos}_i\text{-ind}_{\mathbb{F}}^{\mathbb{D}}$, $i \in \{0,1\}$, 98
$\mathrm{ps}_i\text{-ind}_{\mathbb{F}}^{\mathbb{D}}$, $i \in \{0,1\}$, 98
$\mathrm{pos}_i\text{-ind}$, $i \in \{0,1\}$, 98
$\mathrm{ps}_i\text{-ind}$, $i \in \{0,1\}$, 98
$\mathrm{p}_i\text{-ind}_{\mathbb{F}}^{\mathbb{B}}$, 104
p, 292

$\mathbf{Q}$, 7
$\mathbf{Q}(X)$, 7
$\mathbf{Q}(X, \mathbb{P})$, 489
$(\mathbf{Q}, \mathbf{S}, \mathbf{F})$, 309
(Q, X), 56
(Q, B, X), 78
(Q, X, F), 308

$\mathcal{R}^+$, 391
R, 7, 41
$\mathrm{R}|_{\mathbf{Q}}$,
$\mathrm{R}_{\mathbf{M}}$, 14
$\mathrm{R}_{\mathbf{N}}$, 435
R^d, 242
R^r, 242

$\mathrm{R}(\lambda)$, 267
$\mathrm{R}(\lambda_0, \lambda_1)$, 267
$\mathrm{R}^{\mathbf{D}}$, 323
R^{λ}, 324
$\mathrm{R}_{\mathbf{p}}$, 419
$\mathrm{R}^{\nu}_{\mathbf{M}}$, 40
R_f, 2

$\mathbf{S}$, 6
$\mathbf{S}(X, I\!P)$, 489
$\mathbf{S}_d$, 241
$\mathbf{S}_r$, 241
$\mathrm{S}(\nu, \xi)$, 170
$\mathrm{S}(\nu, \xi, \mathrm{Cl})$, 170
$\mathrm{S}(\nu, \xi, \mathrm{Op})$, 170
$\mathrm{S}(\nu, \leq, \xi)$, 170
$\mathrm{S}(\nu, \leq, \xi, \mathrm{Cl})$, 170
$\mathrm{S}(\nu, \leq, \xi, \mathrm{Op})$, 170
$\overline{\mathrm{S}}(\nu, \xi)$, 168
$\overline{\mathrm{S}}(\nu, \xi, \mathrm{Cl})$, 168
$\overline{\mathrm{S}}(\nu, \xi, \mathrm{Op})$, 168
$\overline{\mathrm{S}}(\nu, \leq, \xi)$, 168
$\overline{\mathrm{S}}(\nu, \leq, \xi, \mathrm{Cl})$, 168
$\overline{\mathrm{S}}(\nu, \leq, \xi, \mathrm{Op})$, 168
$\overline{\mathrm{S}}$, 170
S, 358
s-df, 123
s-$\mathrm{b_i}$-ind, $i \in \{0, 1\}$, 123
s-$\mathrm{b^i}$-ind, $i \in \{0, 1\}$, 123
s-$\mathrm{b^p}$-ind, 123
s-$\mathrm{b^{pos}}$-ind, 123
s-$\mathrm{b^{ps}}$-ind, 123
s-b(0)-ind, 123
s-b(1)-ind, 123
s-b_i-p_j-ind, $i, j \in \{0, 1\}$, 131
s-b^i-p_j-ind, $i, j \in \{0, 1\}$, 131
s-b-Ind, 143
s-$\mathrm{b}^{\vee}$-Ind, 143
s-$\mathrm{b}^{\diamondsuit}$-Ind, 143
s-b-$\mathrm{dim}^{\mathrm{Op}}$, 150
s-b-$\mathrm{dim}^{\mathrm{Cl}}$, 150
s-b-$\mathrm{dim}^{\mathrm{rCl}}$, 150
s-b^n-Ind, 210
s-b^n-dim, 232
s-p_i-ind, $i \in \{0, 1\}$, 105
s-pos_i-ind, $i \in \{0, 1\}$, 105
s-ps_i-ind, $i \in \{0, 1\}$, 105
str-$\mathrm{ind}^{I\!B}_{I\!R}$, 157

T, 15, 42
$\mathrm{T}(\mathbf{M}, \mathrm{R})$, 15, 42
$\mathrm{T}(\mathbf{M}, \mathrm{R}, \mathbf{H})$, 15 ,42
$\mathrm{T}(\mathbf{H})$, 15, 42
$\mathrm{T}(\mathbf{M}|_{\mathbf{Q}}, \mathrm{R}|_{\mathbf{Q}})$, 25
$\mathrm{T}(\mathbf{M}|_{\mathbf{Q}}, \mathrm{R}|_{\mathbf{Q}}, \mathbf{L}|_{\mathbf{Q}})$, 25
$\mathrm{T}|_{\mathbf{Q}}$, 25
$\mathrm{T}(\mathbf{E})$, 25
$\mathrm{T}(\mathbf{L}|_{\mathbf{Q}})$, 25
$\mathrm{T}(c)$, 353
$\mathrm{T}(c)(\mathbf{L})$, 365
$\mathrm{T}(c)|_{\mathbf{Q}}$, 365
$\mathrm{T}(c)(\mathbf{L}|_{\mathbf{Q}})$, 365
$\mathrm{T}(\mathbf{E})$, 396
$\mathrm{T^r}(\lambda)$, 324
$\mathrm{T}_{(s,u)}(\mathbf{H})$, 16, 42
$\mathrm{T}_{(\delta,0)}(\mathbf{H})$, 16, 42
$\mathrm{T}/\mathrm{T}|_{\mathbf{Q}}$, 475
$t(L_D)$, 275

$\mathbf{U}_{\delta}$, 31
$U^{\mathrm{T}(c)}_{\delta,\mathbf{N}}(\mathbf{H})$, 366
$U^{\mathrm{T}(\mathbf{E})}_{l}$, 396
$U^{\mathrm{T}}_{\delta}(\mathbf{H})$, 18, 43
$U^{\mathrm{T}|_{\mathbf{Q}}}_{\delta}(\mathbf{H})$, 25
$u(X, s, \delta)$, 11, 38
$u(\mathbf{H}, s, \delta)$, 16, 42

$V(N, v)$, 133

$w(X)$, 5

X^{n+1}, 2
$X \setminus Q$, 6
$X_{(\delta,0)}$, 11, 38
$X_{(\delta,1)}$, 11, 38
$X_{(s,f)}$, 11, 38
$X(s, u)$, 12, 38
X_ψ^N, 195
$|X|$, 4
(X, F), 293
(X, M^X), 466
$\{x_0, ..., x_n\}$, 2
(x, y), 2
$(x_0, ..., x_n)$, 2

Y^X, 2

$\alpha[*]$, 90
$\beta(\alpha)$, 162
Δ, 194
Δ^χ, 195
Θ, 194
Θ^χ, 195
$\vartheta_{\mathcal{F}}$, 225
$\vartheta_{\vee,k}$, 196
$\vartheta_{\psi,k}$, 197
$\vartheta_{\vee,k}$, 183
$\vartheta_{\wedge,k}$, 183
ϑ_k, 183
$(\vartheta_\vee, \vartheta_\wedge, \vartheta, \vartheta_{\mathcal{F}})$, 225
$(\vartheta_\vee, \vartheta_\wedge, \Theta)$, 195
$(\vartheta_\vee, \vartheta_\wedge, \vartheta)$, 181
θ, 6, 7
$\{\kappa_0, \kappa_1, \theta\}$, 475
$\{\kappa_0, \kappa_1, \theta, \vartheta\}$, 478

$\Lambda \cup \prec$, 321
Λ^γ, 67
μ, 36
μdim, 423
ν, 36
ν^+, 3
ξ_Λ, 168
π_i^k, $i \in \{1, 2\}$, 183
ϱ_X, 391
ϱ_{T}, 401
$\varrho_{\mathrm{T}(\mathbf{H})}$, 409
$\varrho_{\mathrm{T}(\mathbf{E})}$, 396
τ, 1
$\overline{\tau}$, 193
$\mathbf{\Phi}$, 66
$(\mathbf{\Phi}, \Lambda, \prec, \mathbf{p})$-$\mathrm{ind}_{\mathbb{R}}^{\mathbb{B}}$, 157
φ_x^C, 212
$\chi\psi$, 194
$\chi(x, X)$, 458
$\boldsymbol{\Psi}$, 153
$\boldsymbol{\Psi}(\lambda)$, 153
$(\boldsymbol{\Psi}, \mathbf{S}, \mathbf{F})$, 317
Ψ, 194
Ψ^χ, 195
Ψ_k, 196
$\psi_{\mathrm{Bd}_X(U)}$, 157
$\psi_k(\lambda)$, 157
$\psi_{\lim(D)}$, 383
ψ_X, 69
$\psi_{\mathrm{T}(\mathbf{L})}$, 153
(ψ_X, X), 152
(ψ_X, X, F^X), 315
ω, 3

$\vee\{\mathbf{F}(\lambda) : \lambda \in \Lambda\}$, 30
$\wedge\{\mathbf{F}(\lambda) : \lambda \in \Lambda\}$, 30
$\sim$, 4
$\sim_{\mathbf{M}}^{\mathrm{R}}$, 15, 42

$\sim|_{\mathbf{Q}}$, 8
$\sim^s_m$, 13, 39
$\sim^s_{\mathbf{M}}$, 14, 40
$\sim^s_{(\lambda_0,\lambda_1)}$, 267
$\sim^{cm}_{af}$, 347
$\sim^{cm}$, 346
$\sim_{af}$, 347
$\sim^{\infty}$, 395
$\prec$, 4
$\prec^{cm}$, 346
$\prec_{af}$, 347
$\prec^{cm}_{af}$, 347
$\prec^{\mathrm{C}}$, 335
$\prec^{\mathbf{S}}$, 348
$\prec_t$, 168
$\equiv$, 1
$\emptyset$, 1
-1, 3
$2^s_{\mathbf{H}}$, 16, 42
2^s_X, 11, 38
∞, 3

Index

The symbols and letter-symbols are not considered in the lexicographic order. The prefix "co-" is not considered as a symbol.

action of a group, 293
  Borel, 502
  α(add)-Borel, 502
  α(mult)-Borel, 502
  effective, 306
  free, 306
  limit, 329
  on a G-diagram, 321
action-component of a class
  of G-spaces, 301
  of G-spaces with structure, 316
  of G-subsets, 308
action-coordinate
  of a G-space, 293
  of a G-space with structure, 316
  of an element of a class of G-subsets, 308
adjacent class
  of G-spaces, 308, 316
  of spaces with structure, 316
  of subsets, 79, 308
λ-adjacent class of G spaces, 316
admissible
  branching triad, 195
  tetrad, 225
  triad, 181
algebra, 4
  generated, 4–5
  minimal, 4
  of a set, 4
ν-algebra of a set, 37
  minimal, 37
s-algebra of a space, 11

base
  base-normal, 180
  (b^n-Ind $\leq \alpha$)-dimensional, 193
  (b^n-dim $\leq m$)-dimensional, 225
  disjunctive, 180
  ν-dividing, 454
  τ-indexed, 5
  positional, 28
  standard, 22
  κ-standard, 22
  $\diamondsuit$-standard, 22
  $(\diamondsuit, \kappa)$-standard, 22
  $(\kappa, \diamondsuit)$-standard, 186
base for
  a space, 5
  the closed subsets, 5
    normal, 179
  the open subsets, 5
    normal, 180
$\mathbb{P}$-base, 74
p-base, 28
$\mathbb{P}$-p-base, 78
pos-base, 28
ps-base, 28
T_3-p-base, 82
T_3^c-p-base, 82
T_3^s-p-base, 82
branching family of mappings, 194

cardinal, 3

- chain, 335
- character, 458
- class, 1
  - additive, 66
  - corresponding, 412
  - $\mathbb{F}$-hereditary, 143
  - $\mathbb{F}$-$\diamondsuit$-hereditary, 143
  - $\mathbb{F}$-$\vee$-hereditary, 133
  - multiplicative, 66
  - $\mathbb{B}$-rim-hereditary, 92
  - $\mathbb{B}$-b_0-rim-hereditary, 106
  - $\mathbb{B}$-b_1-rim-hereditary, 112
  - $\mathbb{B}$-b^0-rim-hereditary, 113
  - $\mathbb{B}$-b^1-rim-hereditary, 114
  - $\mathbb{B}$-b_0-p_0-rim-hereditary, 125
  - $\mathbb{B}$-b_0-p_1-rim-hereditary, 125
  - $\mathbb{B}$-b_1-p_0-rim-hereditary, 126
  - $\mathbb{B}$-b_1-p_1-rim-hereditary, 126
  - $\mathbb{B}$-b^0-p_0-rim-hereditary, 128
  - $\mathbb{B}$-b^0-p_1-rim-hereditary, 128
  - $\mathbb{B}$-b^1-p_0-rim-hereditary, 129
  - $\mathbb{B}$-b^1-p_1-rim-hereditary, 129
  - $\mathbb{B}$-p_0-rim-hereditary, 104
  - $\mathbb{B}$-p_1-rim-hereditary, 104
  - $\mathbb{B}$-str-rim-hereditary, 157
  - b(0)-rim-hereditary, 117
  - $\mathbb{D}$-b(1)-rim-hereditary, 117
  - $\mathbb{D}$-b^p-rim-hereditary, 115
  - $\mathbb{D}$-p_0-rim-hereditary, 97
  - $\mathbb{D}$-p_1-rim-hereditary, 97
  - structural, 380
- class of
  - all bases, 74
  - all p-bases, 78
  - all pos-bases, 78
  - all ps-bases, 78
  - bases, 74
  - p-bases, 78
    - closed, 79
    - complete, 79
    - open, 79
  - G-diagrams generated, 328
  - spaces with structure, 152
    - complete, 154
  - G-spaces generated, 304
  - G-spaces with structure, 315
    - complete, 318
    - generated, 318
  - subsets, 56
    - closed, 59
    - complete, 59
    - open, 59
  - G-subsets, 308
    - complete, 312
  - G-subsets generated, 313
  - G-subsets subset-generated, 313
  - the limit spaces, 277
- class in the universe of
  - metric spaces, 409
    - complete, 412
  - metric spaces generated, 413
  - metrizable spaces, 409
- co-base
  - (b^n-Ind $\leq \alpha$)-dimensional, 203
  - (b^n-dim $\leq m$)-dimensional, 229
  - for an indexed collection, 6
    - closed, 6
    - open, 6
  - normal, 185
- $\mathbb{P}$-co-base, 74
- co-extension
  - general, 347
  - of a **D**-co-mark, 267
  - of a pair $(\overline{\mathbf{M}}_0^G, \overline{\mathbf{M}}_0)$, 323
  - of an **F**-co-mark, 242
  - of an **S**-indexed set, 7
- co-indication
  - of an **S**-indexed set, 6
  - $(\vartheta_\vee, \vartheta_\wedge, \vartheta)$-normal, 182
  - $(\vartheta_\vee, \vartheta_\wedge, \Theta, \Delta)$-normal, 204
  - $(\vartheta_\vee, \vartheta_\wedge, \vartheta, \vartheta_{\mathcal{F}}, m)$-normal, 229
- collection, 1
  - full indexed, 467
  - **S**-indexed, 393
- co-mark, 6
  - **O**-commutative, 63
  - generalized, 435

κ-generalized, 435
Q-proper, 475
Q-regular, 478
semi-straight, 487
straight, 487
F-co-mark, 242
product-open, 294
proper, 242
D-co-mark (for diagrams), 267
proper, 267
compact resolution, 495
universal, 495
complementary to
a base, 179
a co-base, 185
a co-indication, 185
a family, 179
an indication, 179
$I\!\!M$-completion, 468
IM(rs)-completion, 473
$(\mathbf{S}, \mathbf{M})$-completion, 468
completion of an element, 67
component of a co-mark, 346
ε-component of a co-mark, 346
λ-component of a diagram, 266
(λ_0, λ_1)-component of a diagram, 266
composition, 451
containing
action-mapping, 301
diagram, 271, 272
G-diagram, 326
mapping, 244–245
space, vii
structure, 153
triad, 310, 317
Containing Space, 19, 44
quotient, 476
Q-quotient, 476
Y-containing mapping, 258
co-operator on a class of spaces, 62
commutative, 62, 63
topological, 62
co-p-base, 78
$I\!\!P$-co-p-base, 78
co-pos-base, 78
co-ps-base, 78
co-singleton, 474
closed, 474
co-structure, 153
$I\!\!P$-co-structure, 153
$(\Phi, \Lambda, \prec, \mathbf{p})$-co-structure, 153
cover of a space, 144

defining classes, 89
diagram (commutative), 266
open, 289
dimension, 89
at a point, 459
dim, 144
large inductive, 132
metric, 423
new, 452
order, 453
small inductive, 91
dimensional kernel, 459
dimension-like function, 89
base, 89
base positional, 89
new, 452
positional, 89
space, 89
dispersion point, 480
domain of a mapping, 2
F-domain, 241
domain-class, 241

element
free (of a ν-set), 67
isometrically universal, 410
properly, 410
majorating, 335
minimal (of an order), 4
equivalence class, 4
induced, 242
$\sim$-equivalence class, 4
equivalence relation, 4
standard, 13
s-standard, 13

$\sim^{\mathrm{R}}_{\mathbf{M}}$-equivalence relation, 15, 42
$\sim^{s}_{\mathbf{M}}$-equivalence relation, 14
$\sim^{s}_{m}$-equivalence relation, 13,39
    standard, 13
    s-standard, 13
embedding
    (for classes $\mathbb{P}((\beta+n)$-rep.)), 164
    (for mappings), 241
Y-embedding, 256
Empty Subset Condition, 145
expansion of a branching family, 194
α-expansion, 194
extension of an indexed set, 6
θ-extension of an indexed set, 6

family, 1
    $(\mathrm{Y}, \mathrm{O}, \mathrm{I})$-separating, 262
    $(\mathrm{Y}, \mathrm{O}^*, \mathrm{I})$-separating, 265
    $\mathbb{F}$-$(\mathrm{Y}, \mathrm{O}, \mathrm{I})$-separating, 265
    $\mathbb{F}$-$(\mathrm{Y}, \mathrm{O}^*, \mathrm{I})$-separating, 265
    standard, 22
    κ-standard, 22
    $\diamondsuit$-standard, 22
    $(\diamondsuit, \kappa)$-standard, 22
family of equivalence relations
    admissible, 8
    μ-admissible, 41
    $\mathbf{M}$-admissible, 15
    $\overline{\mathbf{M}}$-admissible, 268
    $(\mathbf{M}^d, \mathbf{M}^r)$-admissible, 242
    $(\mathbf{M}, \mathbf{P})$-admissible, 394
    $(\mathbf{M}, \mathbf{Q})$-admissible, 24
    $(\mathbf{M}, \mathbf{\Psi})$-admissible, 153
    $(\overline{\mathbf{M}}^G, \overline{\mathbf{M}})$-admissible, 324
    $(\mathbf{M}, \nu, \mu)$-admissible, 41
    $\mathbf{O}$-commutative, 63
    compact, 495
    generalized, 437
    induced, 242
    $(\kappa, \diamondsuit)$-standard, 186
    $\mathbf{M}$-standard, 15
    $(\mathbf{M}, \nu)$-standard, 41
filter, 212
    maximal, 212

final refinement, 8, 41
Finite Free Union Condition, 134
Finite Union Condition, 145
τ-formulation, xii
function, 2
    cardinal-algebra, 433

group (topological), 292

homomorphism
    of a ν-algebra, 37
    of an algebra, 5
    structural, 381

$\mathbf{F}$-image, 242
f-image of an (indexed) set, 3
image-range of a mapping, 2
$\mathbf{S}$-indexed set
    of families, 6
    of τ-indexed families, 6
    of ω-indexed subsets, 393
indication, 3
    natural, 467
    $(\vartheta_\vee, \vartheta_\wedge, \Theta)$-normal, 196
    $(\vartheta_\vee, \vartheta_\wedge, \vartheta)$-normal, 182
    $(\vartheta_\vee, \vartheta_\wedge, \Theta, \Delta)$-normal, 196
    $(\vartheta_\vee, \vartheta_\wedge, \vartheta, \vartheta_{\mathcal{F}}, m)$-normal, 225
indicial mapping
    general, 347
    of a co-extension, 7
    of an extension, 6
initial co-mark
    for classes of bases, 74
    for classes of p-bases, 79
    for classes of diagrams, 273
    for classes of mappings, 258
    for classes of spaces, 51
    for classes of spaces with structure, 153
    for classes of subsets, 56
    for (ν, μ)-saturated classes, 85
    for ω-saturated classes, 411
initial $\mathbf{F}$-co-mark
    for classes of mappings, 245

for classes of Y-mappings, 258
for classes of G-spaces, 302
for classes of G-subsets, 311
initial G-**D**-co-mark, 327
initial **D**-co-mark (for diagrams), 273
initial family
for classes of bases, 74
for classes of p-bases, 79
for classes of diagrams, 273
for classes of G-diagrams, 327
for classes of mappings, 245
for classes of Y-mappings, 258
for classes of spaces, 51
for classes of G-spaces, 302
for classes of spaces with structure, 153
for classes of subsets, 56
for classes of G-subsets, 311
for (ν, μ)-saturated classes, 85
for ω-saturated classes, 411
intersection
of classes, 2
of restrictions, 30
property, x
inverse
element of a group, 292
mapping of a group, 292
of a mapping, 2
isometry, 391
isomorphic sets, 68
isomorphism
of algebras, 5
of s-algebras, 13, 39
natural, 13, 39
of ν-algebras, 37
of diagrams, 266
structural, 382

level of an element, 67

mapping, 2
base-closed, 248
c-base-closed, 248
Borel, 501
α(add)-Borel, 501
α(mult)-Borel, 501
closed, 248
ν-directing, 66
equivariant, 301
general indicial, 347
isometric, 391
one-to-one, 2
onto, 2
open, 248
Φ-operational, 69
mapping of, 2
a space into a spectrum, 338
a space into the limit space, 338
induced, 338
Containing Spaces, 351
equivariant, 321
natural, 351
limit spaces, 339
natural, 339
spectrum, 275
structural, 381
Y-mapping, 256
mark of a space, 5
generalized, 435
κ-generalized, 435
D-mark (for diagrams), 266
morphism of diagrams, 266
equivariant, 321
structural, 382

natural embedding
of a space, 23, 45, 400
of a specific subset, 27
ε-net, 417
net-function, 421
of a space, 421

order, 4
of a cover, 144
partial, 4
well ordering, 4
ordinal, 3
ν-accessible, 171

countable, 4
isolated, 3
limit, 3

pair, 2
ordered, 2
B-proper, 143
$B^{\diamondsuit}$-proper, 143
$B^{\vee}$-proper, 133
$N^{\vee}$-proper, 133
space-action, 308, 316
subset-action, 308
λ-subset-action, 316
subset-space, 308
$(\overline{\mathbf{M}}^G, \overline{\mathbf{M}})$, 323
product-open, 323
proper, 323
peak of a set, 67
peaked union of ν-sets, 68
F-preimage
of a co-mark, 241
of an indexed family, 256
f-preimage of an (indexed) set, 3
preorder, 4
primary subspace, 19
product mapping of a group, 292
product of classes, 2
projection
limit, 337
of an inverse spectrum, 275
proper extension, 496
τ-property, xii

F-range, 241
range of a mapping, 2
range-class, 241
range-restriction of a mapping, 157
refinement of a cover, 144
$(\beta + n)$-representation, 162
restriction
of a class on the universe, 413
of an indexed collection, 7
closed, 31
complete, 59
$(\mathbf{M}, \mathrm{R})$-complete, 30
open, 31
$I\!\!P$-restriction, 56
ring, 4
generated, 4–5
minimal, 4
of (subsets of) a set, 4

saturated class
first-type, 367
minimal, 451
partially, 439
second-type, 366-367
simply, 440
zero-type, 367
∞-saturated class, 441
2^{τ}-saturated class, 442
saturated class of
bases, 74
weakly, 74
p-bases, 79
weakly, 79
diagrams, 273
G-diagrams, 326
mappings, 245
Y-mappings, 258
spaces, 51
G-spaces, 302
spaces with structure, 153
G-spaces with structure, 317
subsets, 56
G-subsets, 310
ω-saturated class, 410
isometrically, 410–411
(ν, μ)-saturated class of spaces, 85
saturation property, xi
for an ordinal, 90
weak, 90
sequence, 3
ν-sequence, 167
set, 1
Borel, 65
τ-complete, 335
defining, 475, 478

directed, 4
indexed, 2
Λ-indexed, 3
indexing, 3
S-open, 346
peaked, 67
γ-peaked, 67
strongly $\mathbb{P}$-universal, 488
$\mathbb{P}$-universal, 489
well ordered, 4
F_σ-set, 65
G_δ-set, 65
ν-set, 67
free, 68
screening of a pair, 179
similar homeomorphism, 466
space
almost regular, 480
base ν-dividing, 454
$\mathbb{M}$-complete, 473
completely bounded metric, 417
connectifiable, 480
$\mathbb{K}$-connectifiable, 481
containing, vii
countable-dimensional, 170
strongly, 170
isometrically containing, 391
locally finite-dimensional, 171
marked, 5
$\mathbb{M}$-complete, 469
$(\mathbf{S}, \mathbf{M})$-complete, 469
similar, 466
similarly homeomorphic, 466
of an inverse spectrum, 275
$(\kappa_0, \kappa_1, \theta)$-quotient, 486
$(\kappa_0, \kappa_1, \theta, \vartheta)$-quotient, 486
$(\mathrm{Y}, \mathrm{O}, \mathrm{I})$-separated, 262
$(\mathrm{Y}, \mathrm{O}^*, \mathrm{I})$-separated, 265
strongly rigid, 480
totally disconnected, 444
universal, vii, 391
Urysohn, 480
weakly n-dimensional, 459
k-weakly n-dimensional, 464
with structure, 152, 380
P_σ-space, 85
G-space
Borel, 503
limit, 329
with structure, 315
with $(\Phi, \Lambda, \prec, \mathbf{p})$-structure, 315
BG-space, 503
$\alpha(\mathrm{add})$-BG-space, 503
$\alpha(\mathrm{mult})$-BG-space, 503
space-component of a class
of bases, 76
of p-bases, 79
of ν-sequences, 170
of G-spaces, 301
of spaces with structure, 154
of G-spaces with structure, 316
of subsets, 61
of G-subsets, 308
space-coordinate of an element
of a class of bases, 76
of a class of p-bases, 79
of a class of ν-sequences, 170
of a class of G-spaces, 293
of a class of spaces with structure, 154
of a class of G-spaces with structure, 316
of a class of subsets, 61
of a class of G-subsets, 308
space-singleton, 487
spectrum
τ-continuous, 340
structurally, 384
$\mathbb{P}$-factorizing, 340
structurally, 384
with respect to a subset, 340
inverse, 275
structural, 381
G-spectrum, 321
τ-spectrum, 341
structural, 384
$\mathbb{P}$-spectrum, 331
structural, 381

$I\!\!P$-τ-spectrum, 341
$(\Lambda, \prec, \mathbf{p})$-structure, 152
$I\!\!P$-structure, 152
$(\Phi, \Lambda, \prec, \mathbf{p})$-structure, 152
structure on a space, 152
structure-space pair, 316
subset
  τ-closed, 335
  cofinal, 335
  first-type, 361
  G-invariant, 308
  majorating, 335
  second-type, 361
  specific, 27
  structural, 381
  surjective, 340
    structurally, 382–383
  zero-type, 361
$I\!\!P$-subset, 370, 385–386
subset-adjacent class, 308
subset-component of a class
  of p-bases, 79
  of subsets, 61
  of G-subsets, 308
subset-coordinate of an element of a class of G-subsets, 308
subspectrum, 339
  τ-closed, 339
  cofinal, 339
supplement, 380
  containing, 385–386
  limit, 382–383

tetrad, 2
topological class
  of bases, 74
  of p-bases, 78
  of diagrams, 266
  of G-diagrams, 322
  of mappings, 241
  of Y-mappings, 256
  of spaces, 51
  of G-spaces, 301
  of spaces with structure, 152
  of G-spaces with structure, 316
  of subsets, 56
  of G-subsets, 308
trace on a restriction
  of a co-mark, 7
  of a family of equivalence relations, 8
  of an equivalence class, 8
  of an equivalence relation, 8
trace on a subset
  of a base, 5
  of an indexed set, 5
triad, 2
n-tuple, 2

(s-b^n-Ind $\leq \alpha$)-uniform class, 210
uniform class of metric spaces, 417
union
  of classes, 2
  of restrictions, 30
  of ν-sets, 68
    peaked, 68
  structural free, 386
unit of a group, 292
universal element in a class
  of bases, 75
  of p-bases, 80
    properly, 80
  of diagrams, 266
  of G-diagrams, 327
  of ν-sequences, 168
    properly, 168
  of G-spaces, 301
  of spaces with struture, 154
    properly, 154
  of G-spaces with structure, 318
    properly, 318
  of subsets, 57
    properly, 60
  of G-subsets, 312
    properly, 312
  $\mathbb{P}((\beta + n)$-rep), 164
universal space, vii
  isometrically, 391

universality problem, vii
2^{τ}-universality property, 442
upper bound, 335
    lower, 335

zero-set, 180